Water Treatment and Desalination

Water Treatment and Desalination

Membrane Separation using Renewable Energy Resources

Edited by
Mohamed Bassyouni
Professor, Chemical Engineering, Port Said University, Port Said, Egypt
and
Chemical Engineering, Islamic University of Madinah, Madinah, Saudi Arabia

WILEY

Published by John Wiley & Sons, Inc., Hoboken, New Jersey.
Published simultaneously in Canada.

Library of Congress Cataloging-in-Publication Data applied for:

Hardback ISBN: 9781394300075
ePDF ISBN: 9781394300099
epub ISBN: 9781394300082
oBook ISBN: 9781394300105

Cover Design: Wiley
Cover Image: Courtesy of the Author

Set in 9.5/12.5pt STIXTwoText by Straive, Chennai, India

Contents

List of Contributors *xxi*
Preface *xxxi*

1 Membrane Technologies for Water Desalination *1*
Ibrahim A. Said
1.1 Introduction *1*
1.2 Fundamentals of Membrane Technologies for Water Desalination *2*
1.3 Membrane Distillation (MD) *4*
1.3.1 Membrane Distillation Configurations *4*
1.3.2 Advantages and Challenges of Membrane Distillation *5*
1.4 Forward Osmosis (FO) *6*
1.4.1 Advantages and Challenges of Forward Osmosis *8*
1.5 Reverse Osmosis (RO) *9*
1.5.1 Advantages and Challenges of Reverse Osmosis *10*
1.5.2 Osmotically Assisted Reverse Osmosis (OARO) *11*
1.5.2.1 Advantages and Challenges of OARO *12*
1.6 Membrane Capacitive Deionization *13*
1.7 Microfiltration (MF), Ultrafiltration (UF), and Nanofiltration (NF) *15*
1.7.1 Microfiltration (MF) *15*
1.7.2 Ultrafiltration (UF) *15*
1.7.3 Nanofiltration (NF) *16*
1.8 Conclusion *16*
Questions *17*
References *18*

2 Materials, Fabrication Techniques, and Characterization for Membrane Desalination *23*
Zhuo Fan Gao, Ji Wu, Min Lei, Xin Cheng Jin, Zhuo Huang, and Xingdong Lv
2.1 Introduction *23*
2.2 Current State-of-the-Art Membrane Processes for Desalination *23*
2.3 Material Choice for Desalination Membranes *24*

2.3.1 Types of Membrane Materials *24*
2.4 Fabrication of Membranes for Desalination *26*
2.4.1 Thin-Film Composite (TFC) Membrane via Interface Polymerization *28*
2.4.2 Electrospun Nanofiber Membranes *29*
2.5 Characterization and Performance Evaluation *31*
2.5.1 Common Characterization Techniques for Desalination Membranes *31*
2.5.2 Performance Evaluation *32*
2.5.3 Desalination Mechanisms *33*
2.6 Concluding Remarks *34*
Abbreviations/Nomenclature *34*
Acknowledgment *35*
Questions *35*
References *36*

3 Exergy Analysis of Membrane Desalination Systems *39*
Noha M. Sayed, Mohamed Rabie, Asmaa Elrasheedy, Raya Al-Dadah, and Marwa Elkady
3.1 Introduction *39*
3.1.1 Exergy Concept *40*
3.1.2 Exergy Analysis *40*
3.1.2.1 Conventional Exergy Analysis *42*
3.1.2.2 Advanced Exergy Analysis *43*
3.2 Exergy Analysis of Membrane Desalination Systems *46*
3.2.1 Reverse Osmosis *46*
3.2.2 Forward Osmosis *48*
3.2.3 Membrane Distillation *48*
3.2.4 Electrodialysis *49*
3.2.5 Hybrid Systems *51*
3.3 Conclusion *54*
Nomenclature *54*
Questions *55*
References *56*

4 Water Desalination via Solar Energy and Photovoltaic *63*
Yunqian Dai, Xiangyu Meng, Haoxuan Li, Ziyu Tao, and Xiao Shen
4.1 Introduction and Development Outline of Solar Water Desalination *63*
4.1.1 Introduction and Definition of the Solar Water Evaporation Process *63*

4.1.1.1 Photon Absorption Process 64
4.1.1.2 Photothermal Conversion Process 65
4.1.1.3 Seawater Evaporation and Collection Process 66
4.1.2 Historical Development 67
4.1.2.1 Stage One: Direct Solar Heating 69
4.1.2.2 Stage Two: Photothermal Heating 69
4.1.2.3 Stage Three: Interfacial Photothermal Heating 70
4.1.2.4 Stage Four: Three-Dimensional Interfacial Photothermal Heating 72
4.2 Introduction to Material and Device Design Strategies for Solar Water Evaporation 73
4.2.1 Introduction to the Design Strategies of Photothermal Water Evaporation Materials 73
4.2.1.1 Metal Nanoparticles 74
4.2.1.2 Semiconductor Materials 77
4.2.1.3 Carbon-Based Materials 79
4.2.1.4 Hybridized Materials 81
4.2.2 Introduction to the Design Strategies of Water Condensation and Collection Devices 83
4.2.2.1 Single-Stage Condensation Device 83
4.2.2.2 Multistage Condensation Device 89
4.3 The Development of Integrating Solar Water Evaporation with Sustainable Energy Conversion Applications 94
4.3.1 Combining Solar Water Evaporation with Photovoltaic Power Generation 95
4.3.2 Combining Solar Water Evaporation with Thermoelectric Power Generation 102
4.3.3 Combining Solar Water Evaporation with Mechanical Energy Conversion for Electricity Generation 104
4.3.4 Solar Water Evaporation-Induced Hydrovaltic Effect on Electricity Generation 105
4.3.5 Combining Solar Water Evaporation with Salinity Gradients for Electricity Generation 107
4.4 Summary and Outlook 109
4.4.1 Enhancing the Efficiency of Solar-to-Freshwater Conversion 109
4.4.2 Reducing Costs 109
4.4.3 Developing Smart and Automated Photothermal Evaporation Materials 110
4.4.4 Interdisciplinary Research 110
4.4.5 Leveraging the Unique Physical and Chemical Behaviors of Photothermal Interface Evaporation 110
4.4.6 Eco-Friendliness and Sustainability 111

4.4.7 Multifunctional Integration and Extended Applications of Photothermal Evaporation Technology *111*
4.5 Conclusion *111*
Questions *112*
References *112*

5 Membrane-Based Desalination Utilizing Wind and Wave Energy *121*
Mohamed S. Salem, Mahmoud A. Shouman, Osama Abdelrehim, and Asmaa A. Khater
5.1 Introduction *121*
5.2 Wind-Powered Membrane Desalination *122*
5.2.1 Wind-Powered Desalination Plant Design Challenges *123*
5.2.2 Wind-Powered Desalination Plant Design Considerations *123*
5.2.3 Reverse Osmosis (RO)—Wind-Powered Desalination Systems *124*
5.2.3.1 Autonomous Wind-Powered Desalination Systems with Backup *126*
5.2.3.2 Autonomous Wind-Powered Desalination Without Backup *129*
5.2.3.3 Direct-Driven Desalination *130*
5.2.4 Electrodialysis (ED)–Wind-Powered Desalination Systems *132*
5.2.5 Membrane Distillation (MD) via Harnessing Waste Heat of Wind Turbines *134*
5.3 Wave-Powered Membrane Desalination *135*
5.3.1 Integration in Ocean Environment *136*
5.3.2 Onshore Deployment Strategies *137*
5.3.3 Energy Recovery in Wave-Powered Desalination (WPD) Systems *140*
5.4 Conclusion *141*
Questions *142*
References *143*

6 Utilization of Geothermal Energy in Water Desalination Systems *149*
Amr Mansi, Maggie Nasry Attia, Noran Ashraf, Hamad AlMohamadi, and Mohamed Bassyouni
6.1 Introduction *149*
6.2 Global Status of Geothermal Energy *150*
6.3 Geothermal Desalination *152*
6.3.1 Geothermal Energy Utilization Schemes *153*
6.3.2 Geothermal Desalination for Thermophysical Systems *154*
6.3.3 Geothermal Desalination for Membrane Processes *156*
6.4 Future Perspectives *157*
6.4.1 Geothermal Desalination Potential in the USA *158*

6.4.2 Geothermal Desalination Potential in the Middle East and Africa *159*
6.4.3 Geothermal Desalination Potential in Australia *160*
6.5 Considerations and Challenges for Geothermal Desalination *161*
6.5.1 Land Use *164*
6.5.2 Geological Hazards *164*
6.5.3 Waste Heat Releases *164*
6.5.4 Atmospheric Emissions *164*
6.5.5 Water Footprint *165*
6.5.6 Noise and Social Impacts *165*
6.6 Conclusion *165*
Questions *166*
References *166*

7 Bioenergy and Biogas Utilization for Water Desalination *171*
Naira Meky and Amor Abdelkader
7.1 Introduction *171*
7.1.1 Objectives of the Chapter *172*
7.1.2 Structure of the Chapter *173*
7.2 Overview of Bioenergy *173*
7.2.1 Bioenergy Sources *174*
7.2.2 Biogas Production *175*
7.2.3 Anaerobic Digestion Process *178*
7.2.4 Factors Affecting the Anaerobic Digestion Process *180*
7.2.5 Biogas Composition, Properties, and Applications *181*
7.2.5.1 Electricity Generation *182*
7.2.5.2 Heat Production *182*
7.2.5.3 Combined Heat and Power (CHP) Systems *183*
7.3 Water Desalination Techniques and Integration with Bioenergy *184*
7.3.1 Conventional Desalination Methods *184*
7.3.1.1 Distillation *184*
7.3.1.2 Reverse Osmosis *184*
7.3.1.3 Electrodialysis *184*
7.3.2 Challenges of Conventional Desalination *185*
7.3.2.1 Energy Intensity *185*
7.3.2.2 Environment Impacts *185*
7.3.2.3 Economic Considerations *186*
7.3.3 Biogas Utilization in Desalination *186*
7.3.3.1 Energy Generation for Desalination (Heat and Electricity) *186*
7.3.3.2 Combined and Hybrid Systems for Desalination *187*
7.4 Environmental and Economic Aspects *188*

7.4.1 Sustainability of Bioenergy in Desalination *189*
7.4.1.1 Reduction in Carbon Footprint *189*
7.4.1.2 Waste Management Benefits *189*
7.4.2 Economic Analysis *189*
7.5 Challenges, Limitations, and Future Research Directions *190*
7.5.1 Technical Challenges and Limitations *190*
7.5.2 Research Gaps and Needs *190*
7.6 Conclusion and Recommendations *192*
List of Abbreviations *193*
Questions *193*
References *194*

8 Emerging Desalination Techniques Powered by Waste Heat *199*
Jesús A. Claudio-Rizo, Denis A. Cabrera-Munguía, Tirso E. Flores-Guía, and Lucía F. Cano-Salazar
8.1 Introduction *199*
8.2 Multi-Effect Distillation (MED) and Membrane Distillation (MD) *209*
8.2.1 Multiple-Effect Distillation *210*
8.2.1.1 Principle and Operating Conditions *210*
8.2.1.2 MED Configurations *211*
8.2.1.3 Limitations and MED Economics *211*
8.2.2 Membrane Distillation *213*
8.2.2.1 Principle and Operating Conditions *213*
8.2.2.2 MD Configurations *216*
8.2.2.3 MD Module Types *217*
8.2.2.4 Membrane Classification *219*
8.2.2.5 Fouling and Cleaning of Membranes *221*
8.2.2.6 Pretreatment for the Feed *223*
8.3 Adsorption Desalination and Hybrid Desalination Systems *225*
8.3.1 Adsorption Desalination *226*
8.3.1.1 Mechanics of Adsorption Desalination *226*
8.3.1.2 Energy and Heat Management in AD *226*
8.3.1.3 System Overview *227*
8.3.1.4 Research on Adsorption-Based Water Desalination *227*
8.3.1.5 Desalinating Using Carbon Nanotubes *227*
8.3.1.6 Natural Zeolites and Innovative Adsorbent Materials in Desalination *228*
8.3.1.7 Enhanced Adsorption Capacity with Metal–Organic Networks of the ZIF Family *228*

8.3.1.8 Advances in Adsorbent Composites and Innovations in Desalination Systems *229*
8.3.1.9 Theoretical Study on Heat Transfer Enhancement *229*
8.3.1.10 Capacitive Deionization Technology *229*
8.3.1.11 Vermiculite as a Promising 2D Material *229*
8.3.1.12 Mechanisms of Water Vapor Adsorption and Capillary Adsorption *230*
8.3.1.13 Enhancements in Adsorption-Cooling and Desalination Systems *230*
8.3.2 Hybrid Desalination Systems *230*
8.3.2.1 Benefits of Hybrid Desalination Systems *231*
8.3.2.2 Challenges and Considerations *231*
8.3.2.3 Advantages of Hybrid Desalination Systems *231*
8.3.2.4 Energy Sources Used in Hybrid Desalination Systems *232*
8.3.2.5 Advances in Hybrid Desalination Systems Utilizing Renewable Energy *232*
8.3.2.6 Innovations in Hybrid Desalination Systems *233*
8.4 Low-Temperature Thermal Desalination (LTTD) *235*
8.4.1 Challenges and Innovations in Implementing LTTD Technology *236*
8.4.2 Innovations and Efficiency Enhancements in LTTD Technology *238*
8.4.2.1 Key Factors Influencing LTTD Efficiency *238*
8.4.3 Experimental Findings and Design Improvements *239*
8.4.3.1 Enhancements and Innovations in LTTD Systems *239*
8.4.3.2 Reducing Energy Consumption in LTTD Systems *240*
8.4.3.3 Environmental and Ecological Considerations *240*
8.4.3.4 Optimizing Water Quality in LTTD Systems *240*
8.4.4 Advanced Mathematical Models and Heat Transfer Optimization in LTTD *241*
8.4.4.1 Evaporation Rate Models *241*
8.4.4.2 Heat Transfer in Condensers *241*
8.4.4.3 Heat Exchanger Design and Optimization *241*
8.4.4.4 Enhancing Condenser Efficiency *241*
8.4.5 Integrating Renewable Energy and Industrial Heat Sources in LTTD *242*
8.4.5.1 Utilizing Waste Heat from Industrial Processes *242*
8.4.5.2 Solar Energy Integration *242*
8.4.5.3 Case Studies and Prototypes *244*
8.4.6 Advantages and Future Trends in LTTD *244*
8.5 Perspectives and Conclusion *244*
Questions *245*
References *246*

9 **Minerals Recovery from Seawater by Using Membrane Crystallization** *257*
Ali Bahadar and Mohamed Helmy Abdel-Aziz
9.1 Introduction *257*
9.1.1 Importance of Minerals and Their Applications *258*
9.1.2 Challenges in Traditional Mineral Extraction Methods *258*
9.1.3 Potential of Seawater for Mineral Recovery *258*
9.2 Membrane Crystallization: Principles and Mechanisms *259*
9.2.1 Crystallization *259*
9.2.2 Membrane-assisted Crystallization (MCr) *259*
9.3 Targeted Minerals and Their Significance *262*
9.4 Membrane Crystallization Process Design *265*
9.4.1 Membrane Distillation Crystallization (MDCr) *266*
9.4.2 Osmotic Membrane Crystallization (OMC) *269*
9.4.3 Concentrative Crystallization *270*
9.4.4 Antisolvent Crystallization *271*
9.5 Membrane Materials and Their Performance *272*
9.5.1 Polysulfone (PSF) *272*
9.5.2 Polyacrylonitrile (PAN) *273*
9.5.3 Polyvinylidene Fluoride (PVDF) *273*
9.5.4 Polypropylene (PP) *275*
9.5.5 Polyamide (PA) *275*
9.6 Operating Parameters and Process Optimization *276*
9.6.1 Temperature *277*
9.6.2 Solution Supersaturation *277*
9.6.3 Impact of Feed Flow, Recirculation Rates, and Air Gap *277*
9.6.4 Membrane Characteristics *278*
9.6.5 Fouling *278*
9.7 Challenges and Mitigation Strategies *280*
9.8 Case Studies and Applications *282*
9.9 Economic Viability and Sustainability *284*
9.10 Conclusion *286*
Questions *287*
References *288*

10 **Integrated Hybrid Systems for Water Desalination** *303*
Heba Abdallah and Eman Sobhy
10.1 Introduction *303*
10.2 Desalination by Membrane Technology *304*
10.2.1 Reverse Osmosis (RO) *304*
10.2.2 Electrodialysis and Electrodialysis Reversal (ED/EDR) *305*

10.2.3 Membrane Distillation (MD) *305*
10.2.4 Other Membrane Processes *308*
10.3 Hybrid Energy Sources for Water Desalination *309*
10.3.1 High-Pressure Recovery of Brine Using Energy Recovery Device *309*
10.3.2 Integrated Renewable Energy with Membrane Technology *312*
10.4 Desalination Using Nuclear Energy *314*
10.5 Hybrid Desalination Systems *315*
10.5.1 Hybrid Reverse Osmosis, Membrane Distillation, and Reverse Electrodialysis *315*
10.5.2 Integrated Membrane System Between RO and PRO *316*
10.5.3 Hybrid System of Capacitive Deionization with Pressure Retard Osmosis *318*
10.6 Hybrid Water Desalination System for ZLD *320*
10.7 Case Study: China's Mega-Desalination Plant *323*
10.8 Conclusion *325*
Questions *326*
References *326*

11 Pressure-Retarded Osmosis for Blue Energy Generation *331*
Giti Nouri, Sara Pakdaman, and Catherine N. Mulligan
11.1 Introduction to Pressure-Retarded Osmosis *331*
11.1.1 Historical Development and Evolution *332*
11.1.2 Principles of Operation *333*
11.1.3 Configurations of Pressure-Retarded Osmosis Systems *334*
11.1.4 Factors Influencing Power Density in Pressure-Retarded Osmosis Systems *335*
11.1.5 Free Energy of Mixing *336*
11.1.6 The Flux of Water and Salt Across a Pressure-Retarded Osmosis Membrane *336*
11.1.7 Actual Energy Production of Pressure-Retarded Osmosis *337*
11.1.8 Maximum Energy Production by Pressure-Retarded Osmosis *338*
11.2 Fundamentals of Pressure-Retarded Osmosis *339*
11.2.1 Osmotic Pressure and Water Flux *339*
11.2.2 Comparison with Different Osmosis-Based Technologies *340*
11.2.3 Water Transport and Theory of Rejection *342*
11.2.4 Phenomenon of Concentration Polarization *343*
11.2.5 Fouling Challenges in Osmosis Processes *344*
11.3 Membrane Technology in the Pressure-Retarded Osmosis System *346*
11.3.1 Development of Membranes for the Pressure-Retarded Osmosis Process *346*

11.3.2 Flat-Sheet Membranes *347*
11.3.3 Hollow-Fiber Membranes *347*
11.3.4 Roles of Nanomaterials in Membrane Modification for the Pressure-Retarded Osmosis Process *348*
11.4 Energy, Economic, and Environmental Considerations for Pressure-Retarded Osmosis *351*
11.4.1 Energy Assessment *351*
11.4.2 Cost Contributors *352*
11.4.3 Environmental Sustainability of the Pressure-Retarded Osmosis System *353*
11.4.3.1 Comparative Analysis of Greenhouse Gas Emissions *353*
11.4.4 Life Cycle Assessment and Life Cycle Cost Analysis of the Pressure-Retarded Osmosis System *354*
11.5 Pressure-Retarded Osmosis Hybrid Systems *356*
11.5.1 Reverse Osmosis-Pressure-Retarded Osmosis Hybrid System *356*
11.5.2 Membrane Distillation-Pressure-Retarded Osmosis Hybrid System *358*
11.5.3 Membrane Bioreactor-Pressure-Retarded Osmosis Hybrid System *359*
11.5.4 Liquid-Phase Ion-Stripping-Pressure-Retarded Osmosis Hybrid System *359*
11.5.5 Thermosiphon-Pressure-Retarded Osmosis Powered by Solar Energy *359*
11.5.6 Solar Pond-Pressure-Retarded Osmosis Hybrid System *360*
11.5.7 Geothermal Pressure-Retarded Osmosis Hybrid System *361*
11.5.8 Multigeneration Pressure-Retarded Osmosis System *361*
11.6 Pressure-Retarded Osmosis Applications *363*
11.7 Challenges and Future Directions *364*
11.7.1 Technical Challenges and Solutions *364*
11.7.2 Large-Scale Implementation *365*
11.8 Conclusion *367*
Questions *367*
References *368*

12 Modeling and Simulation of Water Desalination Systems Using TRNSYS *375*
Abdelfatah Marni Sandid, Essam Rabea Ibrahim Mahmoud, Mohamed Bassyouni, and Yasser Elhenawy
12.1 Introduction *375*
12.2 Materials and Methods *377*
12.2.1 Air-Gap Membrane Distillation *377*

12.2.2 Solar Collectors *378*
12.2.3 Thermal Desalination Systems Using Different Heat Exchanger Fluids *380*
12.2.3.1 Water as a Heat Exchanger Fluid *380*
12.2.3.2 Ethylene Glycol as a Heat Exchanger Fluid *381*
12.2.3.3 Water–Ethylene Glycol Mixture as a Heat Exchanger Fluid *381*
12.2.3.4 Comparative Analysis of Heat Exchanger Fluids *382*
12.3 Weather Characteristics of Different Locations *383*
12.4 TRNSYS Modeling of Desalination Systems *384*
12.5 Results and Discussion *385*
12.5.1 Three-Dimensional Visualization of Air-Gap Membrane Distillation Parameters *385*
12.5.2 Comparative Analysis of Meteorological Parameters *386*
12.5.3 Comparative Distillate Water, Inlet and Outlet Temperatures under Different Climatic Conditions *388*
12.5.4 Temperature Variations in Thermal Collector Systems *390*
12.5.5 Influence of Heat Capacity Fluids on Distillate Production *391*
12.5.6 Effect of Heat Exchanger Efficiency on Membrane Distillation Performance *392*
12.6 Conclusion *393*
Nomenclature *395*
Questions *395*
References *396*

13 Economic and Environmental Aspects *399*
Moataz El-Bagoury, Dina Aboelela, Mohammed Alyoubi, and Shereen M. S. Abdel-Hamid
13.1 Introduction *399*
13.2 Economic Aspects *400*
13.2.1 Capital Costs *400*
13.2.2 Operational Costs *401*
13.2.3 Economic Viability *402*
13.2.4 Life Cycle Cost Assessments *404*
13.3 Environmental Aspects *406*
13.3.1 Water Treatment *407*
13.3.2 Energy Consumption *407*
13.3.3 Waste Treatment *409*
13.3.4 Environmental Impact Assessment *411*
13.4 Renewable Energy Integration *413*
13.4.1 Renewable Energy Costs *413*
13.4.2 Technical Challenges and Solutions *414*

13.4.3 Economic Benefits *414*
13.5 Comparative Study *416*
13.6 Conclusion *418*
Questions *419*
References *419*

14 Water Treatment by Microbial Fuel Cell System *425*
Ilse Alejandra Ortiz Mora, Patricia Balderas Hernández, Gabriela Roa Morales, and Eduardo Campos Medina
14.1 Introduction *425*
14.1.1 History and Development of Microbial Fuel Cells *426*
14.1.1.1 The First Observation *426*
14.1.1.2 Mid-Twentieth Century *426*
14.1.1.3 Understanding Microbial Fuel Cells *426*
14.1.1.4 Technological Progress During the Late Twentieth Century *426*
14.1.1.5 Expanding Applications *427*
14.1.1.6 Early Twenty-First Century and the Rapid Advancements in the Applications of Microbial Fuel Cells *427*
14.1.1.7 Microbial Fuel Cell's Scalability as a New Concept *427*
14.1.1.8 More Recent Developments *427*
14.1.1.9 Optimization and Sustainability of Microbial Fuel Cells as Main Interest *428*
14.2 Wastewater *428*
14.2.1 Domestic Wastewater *428*
14.2.2 Industrial Wastewater *428*
14.2.3 Municipal Wastewater *429*
14.2.4 Agricultural Wastewater *429*
14.2.5 Storm Wastewater *429*
14.2.6 Combined Wastewater *429*
14.3 Microorganisms *431*
14.4 Electrochemical Cells *433*
14.5 Biofuel Cells *434*
14.6 Yeast Fuel Cells *434*
14.7 Microbial Fuel Cells *435*
14.7.1 Exoelectrogenic Bacteria *437*
14.7.2 Electron Transport Chain *439*
14.7.3 Cytochromes *440*
14.8 Case Studies: Exoelectrogenic Bacteria in Microbial Fuel Cells *442*
14.8.1 Microbial Communities in Microbial Fuel Cells *442*
14.8.2 Onion Residues and Microbial Fuel Cells *443*
14.8.3 Biofilm Analysis of Sewage Wastewater Microbial Fuel Cells *443*

14.8.4 Electrical Power with Isolated Electrogenic Bacteria in Microbial Fuel Cells *443*
14.9 Design of a Microbial Fuel Cell *444*
14.9.1 Anode and Cathode of a Microbial Fuel Cell *444*
14.9.2 Single-Chamber Microbial Fuel Cell *445*
14.9.3 Double-Chamber Microbial Fuel Cell *446*
14.9.4 Proton-Exchange Membrane in Microbial Fuel Cells *447*
14.9.5 Types of Substrates in Microbial Fuel Cells *447*
14.9.6 Electrodes *449*
14.9.7 Mediator Microbial Fuel Cells *452*
14.9.8 Un-Mediator or Mediator-Less Microbial Fuel Cells *453*
14.10 Microbial Fuel Cell's Areas of Improvement *453*
14.11 Case Studies: Applications of Microbial Fuel Cells for Water Treatment *454*
14.11.1 Microbial Fuel Cells for Drug Removal *454*
14.11.2 Microbial Fuel Cells for Copper Removal *455*
14.11.3 Microbial Fuel Cells for Dye Removal *455*
14.11.4 Microbial Fuel Cells for Domestic Wastewater Treatment *455*
14.12 Closing Argument and Perspectives of Microbial Fuel Cells *455*
Questions *457*
References *457*

15 Economic Feasibility of Membrane Technology *467*
Asmaa Ahmed, Mussad M. Alzahrani, Muhammad Mubashir, Ahmed Refaat, and Mohamed Elsakka
15.1 Introduction *467*
15.2 Integration of Renewable Energy Resources into Membrane Systems *471*
15.2.1 Common Renewable Energy Sources for Membrane Technologies *471*
15.2.1.1 Solar Energy *471*
15.2.1.2 Wind Energy *474*
15.2.1.3 Geothermal Energy *474*
15.2.2 Technological Advancements Enhancing Renewable Integration into Membrane Systems *475*
15.2.2.1 Energy Recovery Devices *477*
15.2.2.2 Hybrid Energy Systems *477*
15.2.2.3 Advanced Membrane Materials *478*
15.3 Cost Dynamics and Economic Feasibility of Membrane Systems *479*
15.3.1 Investment Requirements *479*
15.3.1.1 Equipment, Infrastructure, and Installation Costs *479*

15.3.1.2 Economic Considerations in Plant Design and Sizing *480*
15.3.1.3 Operational and Maintenance Costs *481*
15.3.1.4 Downtime and Lifespan of Renewable Energy Components *481*
15.3.2 Levelized Cost of Water Models *482*
15.3.2.1 Conventional Energy-Based Membrane Systems *482*
15.3.2.2 Renewable Energy-Based Membrane Systems *483*
15.3.2.3 Levelized Water Cost Model for Reverse Osmosis System *484*
15.3.3 Sensitivity Analysis of Energy Prices and Other Key Factors *486*
15.3.4 Economic and Environmental Benefits of Renewable Energy Integration *486*
15.3.4.1 Impact on Greenhouse Gas Emissions *486*
15.3.4.2 Long-Term Financial Benefits *487*
15.4 Challenges and Barriers to Economic Feasibility: An Analysis *487*
15.4.1 Technological Barriers *487*
15.4.2 Financial Barriers *488*
15.4.3 Market and Infrastructure Limitations *488*
15.5 Suggested Strategies to Overcome Challenges *488*
15.6 Conclusion *489*
Questions *489*
Acknowledgments *490*
References *490*

16 Practical Insights, Commercial Installations, and Future Projections *499*
Asmaa Ahmed, Mohamed Elsakka, Yasser Elhenawy, Asif Taher, Tapas Mallick, and Ahmed Amer
16.1 Introduction *499*
16.2 Technical Aspects of Membrane Systems *500*
16.2.1 Membrane Performance *500*
16.2.2 Fouling and Scaling Challenges *500*
16.2.3 Energy Consumption and Efficiency *501*
16.2.4 System Design and Configuration *501*
16.2.5 Operation and Maintenance *502*
16.2.6 Cleaning Protocols *502*
16.2.6.1 Chemical Cleaning *502*
16.2.6.2 Backwashing *502*
16.2.6.3 Air Scouring *503*
16.2.7 Membrane Longevity and Replacement *503*
16.2.8 System Monitoring and Optimization *505*
16.2.9 Cost Management in Operation and Maintenance *505*
16.3 Novel Membrane Materials *505*

16.3.1 Graphene-Based Membranes *505*
16.3.2 Carbon Nanotube Membranes *506*
16.3.3 Aquaporin-Based Biomimetic Membranes *507*
16.3.4 Superhydrophobic Membranes *507*
16.3.5 Functionalized Nanocomposite Membranes *508*
16.3.6 Zwitterionic Membranes *508*
16.3.7 Nanofiber-Based Membranes *508*
16.3.8 Metal–Organic Framework Membranes *509*
16.3.9 Mixed-Matrix Membranes *509*
16.4 Real-World Installations *510*
16.4.1 Al Montazah Solar-Powered Desalination Plant, Egypt *510*
16.4.2 Al Khafji Solar-Powered Desalination Plant, Saudi Arabia *511*
16.4.3 Perth Desalination Plant, Australia *511*
16.4.4 Masdar City Solar Desalination Plant, UAE *511*
16.4.5 Small-Scale Solar-Powered Desalination, Cape Verde *512*
16.4.6 Jebel Ali Desalination Plant, UAE *512*
16.4.7 Carlsbad Desalination Plant, USA *512*
16.4.8 Shuwaikh Desalination Plant, Kuwait *512*
16.4.9 Salalah Desalination Plant, Oman *513*
16.4.10 Gold Coast Desalination Plant, Queensland *513*
16.5 Future Projections for Membrane Desalination *513*
16.5.1 Technological Advancements *513*
16.5.2 Renewable Energy Integration *514*
16.5.3 Environmental and Social Impacts *515*
16.5.4 Policy and Regulatory Trends *515*
16.6 Conclusion *516*
Questions *517*
Acknowledgments *517*
References *518*

17 Advancements in Intensified Membrane Technology for Enhanced Desalination Process *527*
Febio Dalanta, Osi Arutanti, and Tutuk Djoko Kusworo
17.1 Introduction *527*
17.2 Process Intensification Applications in Membrane Technologies *528*
17.3 Applications of Intensified Membranes in Seawater Desalination *529*
17.3.1 Two-Dimensional (2D) Nanostructured Hybrid Membranes for Improvement of Desalination Processes *532*
17.3.1.1 MXene-Based Composite Membranes *532*
17.3.1.2 Carbon-Based Composite Membranes *536*
17.3.1.3 Graphene-Based Nanocomposite Membranes for Desalination *538*

17.3.2 Membrane Distillation for Desalination Process *542*
17.3.3 Photothermal and Solar-Powered Membrane Distillation for Desalination *545*
17.3.4 Electrodialysis Membranes for Desalination Process *549*
17.3.5 Future Challenges and Opportunities of Intensified Membrane Technology for Desalination *553*
17.4 Conclusion *558*
Questions *559*
Conflict of Interests *560*
Acknowledgments *560*
References *560*

18 Nuclear Power for Water Desalination *573*
Mauricio Chocrón, María Cecilia Conti, and Natalia Quici
18.1 Introduction *573*
18.2 Seawater Desalination and Related Technologies *574*
18.3 Nuclear Desalination *575*
18.3.1 Uses and Consumption of Water in a Nuclear Power Plant (NPP) *576*
18.3.2 Hybrid Systems Coupled with a Nuclear Power Plant *579*
18.3.3 Case Study: Technical and Economic Aspects of a Desalination Plant Integrated to a Small Modular Reactor for Nuclear Cogeneration *579*
18.3.3.1 Connection of an Integrated Desalination Plant to the Secondary Circuit of the NPP in a Hybrid Configuration *579*
18.3.3.2 Main Characteristics of the Small Size Reactor Plant *579*
18.3.3.3 Main Aspects of the NPP Secondary Circuit *580*
18.3.3.4 Description of the Intermediate Circuit *583*
18.3.3.5 Main Components and Dimensions *587*
18.3.3.6 Pretreatment *588*
18.3.3.7 Siting and Cost of Water Production *588*
18.3.3.8 Pilot Demonstration Project *590*
18.4 Conclusion *598*
Acknowledgment *599*
Questions *599*
References *599*

Key Answers for Questions *603*
Index *631*

List of Contributors

Heba Abdallah
Chemical Engineering Department
Engineering and Renewable Energy Research Institute
National Research Centre
Dokki
Giza
Egypt

Mohamed Helmy Abdel-Aziz
Chemical and Materials Engineering Department
King Abdulaziz University
Raabigh
Saudi Arabia

and

Chemical Engineering Department
Faculty of Engineering
Alexandria University
Alexandria
Egypt

Shereen M. S. Abdel-Hamid
Faculty of Energy and Environmental Engineering
The British University in Egypt (BUE)
El-Sherouk City
Egypt

and

Department of Basic Sciences
Innovation University
Tenth of Ramadan City
Egypt

Amor Abdelkader
Faculty of Science and Technology
Bournemouth University
Poole
UK

Osama Abdelrehim
Mechanical Power Engineering Department
Faculty of Engineering
Mansoura University
Mansoura
Egypt

Dina Aboelela
Faculty of Energy and Environmental Engineering
The British University in Egypt (BUE)
El-Sherouk City
Egypt

Asmaa Ahmed
Mechanical Power Engineering Department
Faculty of Engineering
Port Said University
Port Said
Egypt

and

Faculty of Engineering
East Port Said National University
Salam Misr City
Port Said
Egypt

Raya Al-Dadah
Department of Mechanical Engineering
University of Birmingham
Birmingham
UK

Hamad AlMohamadi
Department of Chemical Engineering
Faculty of Engineering
Islamic University of Madinah
Madinah
Saudi Arabia

and

Sustainability Research Center
Islamic University of Madinah
Madinah
Saudi Arabia

Mohammed Alyoubi
Department of Design and Engineering
Faculty of Science and Technology
Bournemouth University
Poole
UK

Mussad M. Alzahrani
Mechanical and Energy Engineering Department
Imam Abdulrahman Bin Faisal University
Dammam
Saudi Arabia

Ahmed Amer
Mechanical Power Engineering Department
Faculty of Engineering
Port Said University
Port Said
Egypt

Osi Arutanti
Research Center for Chemistry
National Research and Innovation Agency
South Tangerang
Indonesia

Noran Ashraf
Center of Excellence for Membrane Testing and Characterization (CEMTC)
Port Said University
Port Said
Egypt

and

Department of Chemical Engineering
Faculty of Engineering
Port Said University
Port Said
Egypt

Maggie Nasry Attia
Center of Excellence for Membrane Testing and Characterization (CEMTC)
Port Said University
Port Said
Egypt

and

Department of Chemical Engineering
Faculty of Engineering
Port Said University
Port Said
Egypt

Ali Bahadar
Chemical and Materials Engineering Department
King Abdulaziz University
Raabigh
Saudi Arabia

Mohamed Bassyouni
Center of Excellence for Membrane Testing and Characterization (CEMTC)
Port Said University
Port Said
Egypt

and

Department of Chemical Engineering
Faculty of Engineering
Port Said University
Port Said
Egypt

and

Department of Chemical Engineering
Faculty of Engineering
Islamic University of Madinah
Madinah
Saudi Arabia

and

Sustainability Research Center
Islamic University of Madinah
Madinah
Saudi Arabia

Denis A. Cabrera-Munguía
Department of Advanced Materials
Faculty of Chemical Sciences
Autonomous University of Coahuila
Saltillo
Coahuila
Mexico

Lucía F. Cano-Salazar
Department of Advanced Materials
Faculty of Chemical Sciences
Autonomous University of Coahuila
Saltillo
Coahuila
Mexico

Mauricio Chocrón
Departamento Química y Procesos en Instalaciones Nucleares
Gerencia de Coordinación de Proyectos Tecnológicos Nucleares
Gerencia de Energía Nuclear
Comisión Nacional de Energía Atómica
Buenos Aires
Argentina

and

Instituto Balseiro
Universidad de Cuyo
Comisión Nacional de Energía Atómica
Rio Negro
Argentina

and

Facultad de Ingeniería
Universidad de Buenos Aires
Buenos Aires
Argentina

Jesús A. Claudio-Rizo
Department of Advanced Materials
Faculty of Chemical Sciences
Autonomous University of Coahuila
Saltillo
Coahuila
Mexico

María Cecilia Conti
Instituto Balseiro
Universidad de Cuyo
Comisión Nacional de Energía Atómica
Rio Negro
Argentina

and

Gerencia de Coordinación de Proyectos Tecnológicos Nucleares
Gerencia de Energía Nuclear
Comisión Nacional de Energía Atómica
Buenos Aires
Argentina

Yunqian Dai
School of Chemistry and Chemical Engineering
Southeast University
Nanjing
China

Febio Dalanta
Research Center for Chemistry
National Research and Innovation Agency
South Tangerang
Indonesia

Moataz El-Bagoury
Faculty of Energy and Environmental Engineering
The British University in Egypt (BUE)
El-Sherouk City
Egypt

Yasser Elhenawy
Center of Excellence for Membrane Testing and Characterization (CEMTC)
Port Said University
Port Said
Egypt

and

Mechanical Power Engineering Department
Faculty of Engineering
Port Said University
Port Said
Egypt

and

School of Chemical and Metallurgical Engineering
University of the Witwatersrand
Johannesburg
South Africa

Marwa Elkady
Chemical and Petrochemical Engineering Department
Egypt-Japan University of Science and Technology
New Borg El-Arab
Egypt

Asmaa Elrasheedy
Chemical Engineering Department
Faculty of Engineering
Port Said University
Port Said
Egypt

Mohamed Elsakka
Mechanical Power Engineering Department
Faculty of Engineering
Port Said University
Port Said
Egypt

and

Faculty of Engineering
East Port Said National University
Salam Misr City
Port Said
Egypt

Tirso E. Flores-Guía
Department of Advanced Materials
Faculty of Chemical Sciences
Autonomous University of Coahuila
Saltillo
Coahuila
Mexico

Zhuo Fan Gao
Science and Technology Achievement Promotion Center
Changjiang River Scientific Research Institute
Wuhan
China

Patricia Balderas Hernández
Facultad de Química UAEMéx
CCIQS UAEM-UNAM
Toluca
Mexico

Zhuo Huang
Science and Technology Achievement Promotion Center
Changjiang River Scientific Research Institute
Wuhan
China

Xin Cheng Jin
Environmental Division
Wuhan Changjiang Kechuang Technology Development Co., Ltd.
Wuhan
China

Asmaa A. Khater
Mechanical Power Engineering Department
Faculty of Engineering
Mansoura University
Mansoura
Egypt

Tutuk Djoko Kusworo
Department of Chemical Engineering
Faculty of Engineering
Diponegoro University
Semarang
Indonesia

Min Lei
College of Computer Science
South-Central University for Nationalities
Wuhan
China

Haoxuan Li
Key Laboratory of Eco-Textiles
Jiangnan University
Wuxi
China

Xingdong Lv
Science and Technology Achievement Promotion Center
Changjiang River Scientific Research Institute
Wuhan
China

Essam Rabea Ibrahim Mahmoud
Department of Mechanical Engineering
Faculty of Engineering
Islamic University of Madinah
Madinah
Saudi Arabia

Tapas Mallick
Environment and Sustainability Institute
University of Exeter
Penryn Campus
Cornwall
UK

Amr Mansi
Center of Excellence for Membrane Testing and Characterization (CEMTC)
Port Said University
Port Said
Egypt

and

Department of Chemical Engineering
Faculty of Engineering
Port Said University
Port Said
Egypt

Eduardo Campos Medina
Facultad de Planeación Urbana y Regional UAEMéx
Colonia Universidad
Toluca
Mexico

Naira Meky
Civil Engineering Department
Faculty of Engineering
Aswan University
Aswan
Egypt

Xiangyu Meng
School of Chemistry and Chemical Engineering
Southeast University
Nanjing
China

Xiao Shen
Key Laboratory of Eco-Textiles
Jiangnan University
Wuxi
China

Ilse Alejandra Ortiz Mora
Facultad de Química UAEMéx
CCIQS UAEM-UNAM
Toluca
Mexico

Gabriela Roa Morales
Facultad de Química UAEMéx
CCIQS UAEM-UNAM
Toluca
Mexico

Muhammad Mubashir
Water Technologies Innovation Institute and Research Advancement
Jubail
Saudi Arabia

Catherine N. Mulligan
Department of Building, Civil and Environmental Engineering
Concordia University
Quebec
Montreal
Canada

Giti Nouri
Department of Building, Civil and Environmental Engineering
Concordia University
Quebec
Montreal
Canada

Sara Pakdaman
Department of Building, Civil and Environmental Engineering
Concordia University
Quebec
Montreal
Canada

Natalia Quici
División Química de la Remediación Ambiental
Gerencia Química
Centro Atómico Constituyentes
Comisión Nacional de Energía Atómica
CONICET
Buenos Aires
Argentina

and

Centro de Tecnologías Químicas
Departamento de Ingeniería Química
FRBA-UTN
Buenos Aires
Argentina

Mohamed Rabie
Mechanical Power Engineering Department
Mansoura University
Mansoura
Egypt

Ahmed Refaat
Electrical Engineering Department
Faculty of Engineering
Port Said University
Port Said
Egypt

Ibrahim A. Said
Nanosystems Engineering Research Center for Nanotechnology-Enabled Water Treatment (NEWT)
Rice University
Houston
TX
USA

and

Chemical Engineering Department
Faculty of Engineering
Alexandria University
Alexandria
Egypt

Mohamed S. Salem
Mechanical Power Engineering Department
Faculty of Engineering
Mansoura University
Mansoura
Egypt

Abdelfatah Marni Sandid
Center of Research in Mechanics (CRM)
BP N73B
Freres Ferrad
Ain El Bey
Constantine
Algeria

Noha M. Sayed
Chemical Engineering Department
Minia University
Minia
Egypt

Mahmoud A. Shouman
Mechanical Power Engineering Department
Faculty of Engineering
Mansoura University
Mansoura
Egypt

Eman Sobhy
Water Pollution Research Department
Environmental and Climate Change Research Institute
National Research Centre
Dokki
Giza
Egypt

Asif Taher
Environment and Sustainability Institute
University of Exeter
Penryn Campus
Cornwall
UK

Ziyu Tao
School of Chemistry and Chemical Engineering
Southeast University
Nanjing
China

Ji Wu
Department of Chemical and Biomolecular Engineering
National University of Singapore
Singapore

Preface

Population growth, industrialization, and climate change have made water scarcity one of the most urgent global issues. Innovative and sustainable approaches to water purification and desalination are crucial as freshwater resources become more scarce. Modern developments in water treatment and desalination technology were studied in this book, with particular attention paid to important developments in materials, fabrication methods, and the incorporation of renewable energy sources into water desalination and treatment procedures. Important aspects were thoroughly covered by experts in chemical engineering, environmental engineering, mechanical engineering, and renewable energy throughout the chapters. The most recent membrane materials and fabrication techniques, including thin-film composite and nanofiber membranes, were thoroughly explained to readers, along with their important effects on improving the sustainability and efficiency of water purification. The incorporation of diverse renewable energy sources, such as geothermal, solar, wind, nuclear, biogas, and bioenergy, into the processes of water treatment and desalination was given special attention. This novel combination presented encouraging opportunities to reduce costs, minimize environmental impacts, and decrease energy consumption. This book clarified realistic routes towards sustainable and financially feasible water treatment solutions by connecting renewable energy systems with desalination and water treatment technologies. Designed to serve both academics and industry professionals, this book provided an essential resource for researchers, specialists, policymakers, and students aiming to develop and implement sustainable water treatment and desalination practices. Gratitude is extended to the contributing authors and institutions whose invaluable research and collaboration have made this comprehensive work possible. The editor hopes readers will find this book both informative and inspiring, driving continued innovation and commitment towards addressing global water scarcity through sustainable desalination technologies.

Mohamed Bassyouni
Professor of Chemical Engineering

1

Membrane Technologies for Water Desalination

Ibrahim A. Said[1,2]

[1] *Nanosystems Engineering Research Center for Nanotechnology-Enabled Water Treatment (NEWT), Rice University, Houston, TX, USA*
[2] *Chemical Engineering Department, Faculty of Engineering, Alexandria University, Alexandria, Egypt*

1.1 Introduction

Approximately 2.1 billion people, more than one-fourth of the world, lack access to safe drinking water [1]. Population growth increases water demand, but global climate change and aging or lack of infrastructure compromise water supply. As a result, more than half of the world suffers from water scarcity for at least one month of the year [2]. Currently, at least 37 countries face "very high" levels of water stress, and at least 33 countries depend on other countries to source over 50% of their water resources. Pollution, overpopulation, inefficient agriculture irrigation methods, and insufficient funds compound the issue. To create a circular water economy for the sustainability and resiliency of our society, we must tap into nontraditional water sources like wastewater, brackish groundwater, seawater, oil and gas produced water, and nutrient-rich agricultural water using desalination technologies [3]. Some countries, such as Qatar and Kuwait in the Middle East and North Africa (MENA), already rely on desalination for 100% of their water needs [4]. There are two main ways for water desalination: thermal and membrane processes. In thermal water desalination, heat is used to evaporate water into vapor, leaving behind all the impurities and dissolved salts as concentrated brine. The vapor is then condensed into a liquid for use as fresh water. Examples of key thermal desalination technologies include multistage flash distillation (MSF), multi-effect distillation (MED), mechanical vapor compression (MVC), and mechanical vapor recompression (MVR). These different thermal technologies employ different techniques of distillation to achieve the same goal of separating pure water from salts and other impurities through the application of heat. In membrane-based desalination, a semi-permeable membrane is used to

Water Treatment and Desalination: Membrane Separation using Renewable Energy Resources, First Edition. Edited by Mohamed Bassyouni.

separate dissolved salts and impurities from water through fine filters, trapping salts and producing pure water on one side of the membrane and a waste stream on the other side of the membrane. Membrane-based desalination technologies rely on various types of membranes to separate dissolved salts and other impurities from water. These membranes can be broadly categorized into hydrophilic and hydrophobic membranes. Hydrophilic membranes are characterized by their affinity for water, making them ideal for processes where water needs to pass through the membrane while salts and other contaminants are retained, and they are exemplified by forward osmosis (FO), reverse osmosis (RO), osmotically assisted reverse osmosis (OARO), microfiltration (MF), ultrafiltration (UF), and nanofiltration (NF). Membranes that are hydrophobic do not let the flow of liquid water but allow the passage of water vapor through them, and they are employed by membrane distillation (MD) [5], electrodialysis (ED) [6], and membrane capacitive deionization (MCDI) [7]. Consequently, membrane desalination has a number of advantages over thermal desalination that make it an attractive method for producing potable water. First, on average, capital expenditures (CapEx) in membrane-based technologies, particularly RO and OARO, tend to be lower than those of thermal desalination techniques like MSF and MED. In addition to leaving a small environmental footprint, these systems have relatively simpler operations compared to their thermal counterparts, requiring fewer moving parts with less need for high temperatures or heating during a changeover from one complexity to another associated with thermal technologies. Moreover, in addition to being able to remove a wide variety of contaminants with minimal posttreatment requirements, these systems lead to the production of higher quality permeate water with low salt waste volumes and consume less energy which results in low greenhouse gas emissions. Hence, membrane-based technologies are considered a cost-effective and efficient choice for a variety of water sources, including produced water, brackish water, seawater, and wastewater.

1.2 Fundamentals of Membrane Technologies for Water Desalination

Membrane-based desalination technologies use different driving forces for separation such as pressure force, concentration gradient, electrical potential, and temperature difference to remove dissolved salts and contaminants from water through a semi-permeable porous membrane. In basic principle, these forces are employed through the membrane and hold solutes on one side while allowing the water to flow through the membrane (Figure 1.1). Reverse osmosis, a common pressure-driven technology for desalination, uses high pressure to relieve osmotic pressure, allowing water molecules to pass through hydrophilic

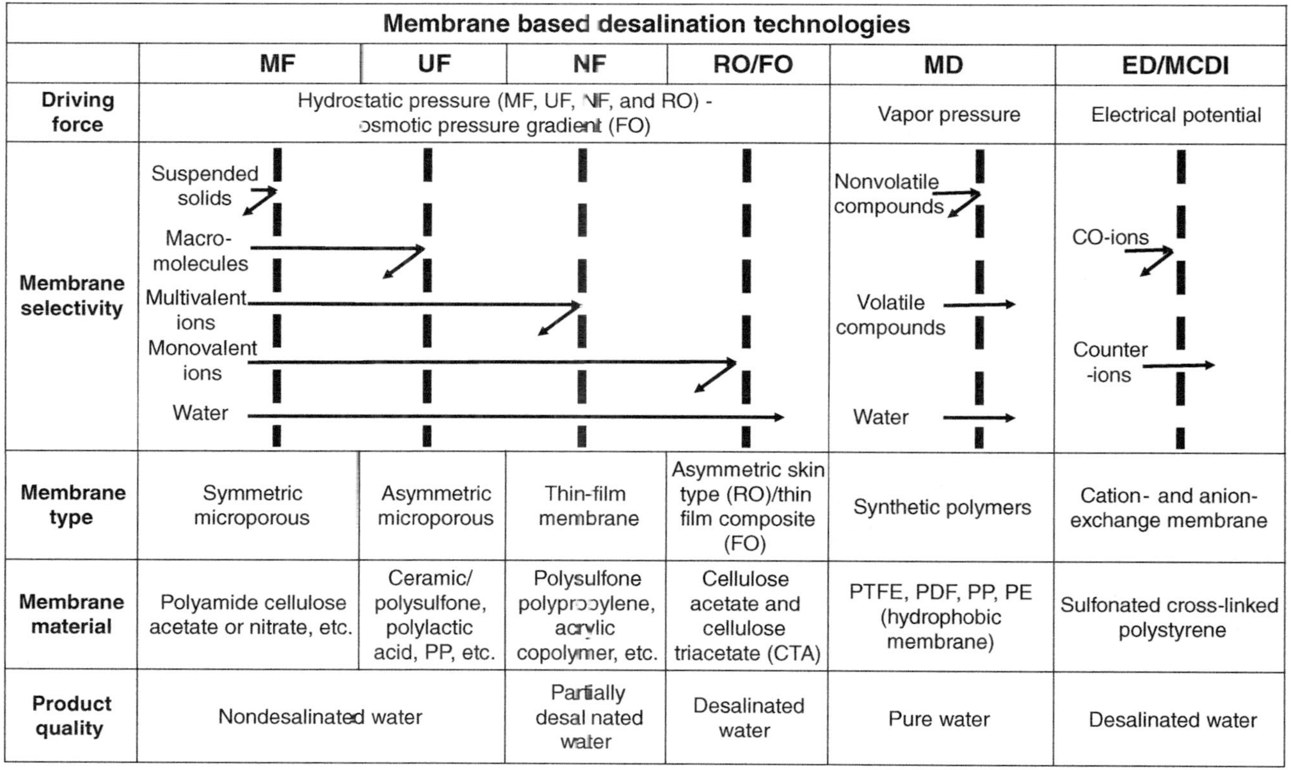

Figure 1.1 Types of membranes and characteristics.

membranes, preventing salts and other contaminants. Other methods, such as forward osmosis and nanofiltration, rely on osmotic and size-exclusion principles to separate dissolved salts from water. Membrane distillation uses hydrophobic membranes to remove water vapor from brine by achieving temperature differences across the membrane, and consequently vapor pressure differences. There are key parameters that impact the salt rejection rate, water flux, and efficiency, including membrane materials, pore size, and operating conditions.

1.3 Membrane Distillation (MD)

Membrane distillation is a hybrid desalination process that combines thermal evaporation and membrane separation. A microporous, hydrophobic membrane separates the hot feed solution from the cold product water. The feedwater evaporates on the membrane surface (i.e., feed–membrane interface). A vapor pressure difference established between the feed–membrane interface and the permeate–membrane interface drives water vapor transport through the membrane. Condensation of the vapor inside or outside the MD reactor forms purified water. As the separation process is realized through a phase change, nonvolatile species including salt ions, colloids, and nonvolatile organics can theoretically be rejected at 100% [8, 9].

1.3.1 Membrane Distillation Configurations

Four conventional MD configurations are used to create a vapor pressure difference across the MD membrane (Figure 1.2) to drive vapor transport. Hot feed steam flows in direct contact with the membrane in all four configurations, but the design on the permeate side varies. In direct contact membrane distillation (DCMD) [10], a clean, cold stream flows in direct contact with the membrane on the opposite side of the feedwater; the temperature difference between the feed–membrane and permeate–membrane interfaces creates the vapor pressure difference, and the vapor transporting through the membrane is captured by and condenses in the permeate flow (Figure 1.2a). By adding a stagnant air layer and a condensing plate between the membrane and the cold stream (i.e., coolant), air gap membrane distillation (AGMD) (Figure 1.2b) allows the separation of the condensate from the coolant and reduces heat loss by conductive heat transfer through the membrane [5, 11]. Sweeping gas membrane distillation (SGMD) employs a cold, inert gas flow to reduce the vapor pressure on the permeate side (Figure 1.2c) [12–14], while a vacuum is applied to the permeate channel in vacuum membrane distillation (VMD) to reduce the vapor pressure while minimizing conductive heat loss (Figure 1.2d) [15]. It is worth mentioning that

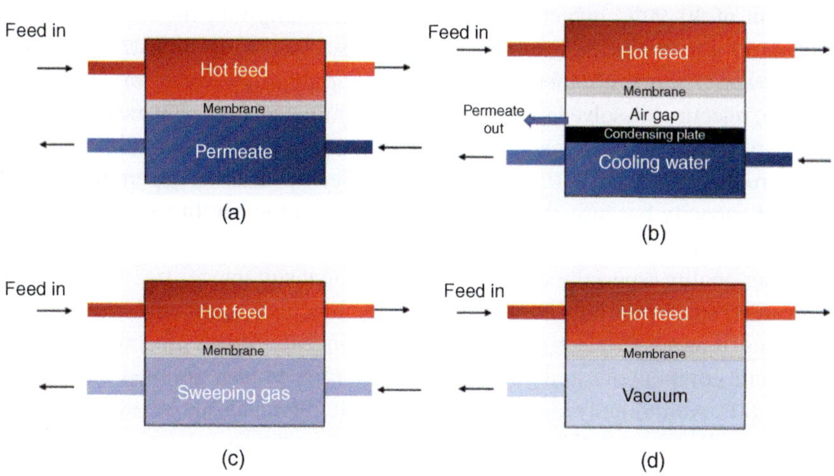

Figure 1.2 Four most common types of membrane distillation (MD) technology. (a) Direct contact membrane distillation (DCMD), (b) air gap membrane distillation (AGMD), (c) sweeping gas membrane distillation (SGMD), and (d) vacuum membrane distillation (VMD).

the VMD and SGMD systems require an external condenser for condensing the permeated vapor. While, in VMD, a waterproof vacuum pump or a venturi is needed to provide a vacuum [5, 11]. DCMD is the most commonly used configuration in laboratory tests due to its simplicity [16, 17], while AGMD is commonly used in pilot systems because of its higher energy efficiency and the separation of coolant and permeate [18]. More recent developments have resulted in variations in these conventional configurations by combining advantageous features. For example, permeate or water gap membrane distillation (PGMD/WGMD) combines features in DCMD and AGMD and fills the gap on the permeate side with distilled water instead of air to improve the condensation and eliminate the need for an external condenser for the permeated vapor [19, 20]. Moreover, SGMD combined with AGMD represents the thermostatic sweeping gas membrane distillation (TSGMD), where gas sweeps between the air gap and the condensing plate to increase the rate of mass transfer [21]. Each MD configuration has its own advantages and disadvantages depending on the desired process, feedwater quality, operating conditions [22].

1.3.2 Advantages and Challenges of Membrane Distillation

There are various benefits associated with using membrane distillation technology, which include low energy requirement, low operation pressure, high rejection, and scalability. Moreover, this process is based on a low-temperature operation of 30–90°C, which reduces energy consumption, thereby making it possible to use low-grade heat sources like waste heat from industries or even solar power. The membranes used in this case only allow water vapor to pass through, rejecting dissolved salts or other contaminants like heavy metals, which results in high-quality permeate water that can be used for desalination and clean water production. These MD systems can be scaled up or down depending on the required capacity because they are modular and can easily fit into an existing water treatment train. Unlike other membrane processes, MD is less affected by feedwater quality especially salinity levels, making it suitable for treating difficult feedwaters, such as produced water, seawater, brackish water, and wastewater. Operating at atmospheric pressure or slightly above it reduces mechanical stress on system components and minimizes membrane fouling as well as scaling risks in MD. Additionally, minimal pretreatment chemicals are required in MD, resulting in lower operational costs and reduced environmental impacts due to fewer cleaning chemical requirements before desalting takes place. Membrane distillation, however, is plagued with several issues, such as membrane scaling and fouling, energy efficiency, thermal polarization, and cost. In the case of a hydrophobic membrane being partially or completely wetted by water, a process called "wetting" the membrane ensues, in which the infiltrating water can cross through. Conversely, thermal polarization is a scenario in which a temperature gradient exists between the bulk feed solution and the membrane's interface, thereby lowering the driving force for vapor transport and affecting overall flux. Even if MD exhibits fewer scaling tendencies than other types of membranes, it could be scaled by certain scales and inorganic salts as well as get fouled with organic compounds or biological materials. Regular cleaning coupled with the right pretreatment is one way to address these problems. Extended exposure to high temperatures or aggressive chemicals causes degradation of the membrane materials, shortening their life span and affecting their performance negatively. MD uses less power than traditional distillation but is generally not more efficient than RO for desalination purposes. The initial capital outlay for MD systems may be high, especially in large-scale installations. Additionally, system costs per unit area are relatively high compared to RO membranes. Therefore, the industry's greatest challenge at this point involves designing processes that are both optimized and integrated into renewable sources aiming at enhancing energy efficiency.

1.4 Forward Osmosis (FO)

Forward osmosis is an advanced membrane technology that uses osmotic pressure differentials for water treatment as well as desalination. With respect to reverse osmosis, which employs hydraulic force to push water through a semi-permeable membrane, FO utilizes the natural osmotic pressure gradient established by a highly concentrated draw solution. Consequently, the feedwater concentrates, while the draw solution dilutes, making it an energy-efficient and probably cost-effective alternative to traditional desalting methods. Another closely related process called pressure-retarded osmosis (PRO) has been tested and demonstrated as a potential technology for power generation [23–26]. PRO uses the osmotic pressure difference between high-salinity brine and fresh water to pressurize the brine stream, thereby converting the osmotic pressure of high-salinity brine into a hydrostatic pressure that can be used to generate electricity. The main difference between FO and PRO is that the draw solution in the PRO process is hydraulically pressurized, whereas the draw solution in the FO process is not pressurized.

The FO process is governed by the osmotic pressure (π), which is the pressure required to prevent the solvent from moving through a semi-permeable membrane. This directly depends on the solute concentration in a solution. Unlike RO where an applied hydraulic pressure (ΔP) drives water flux (J_v) across the membrane, in FO, it occurs due to an osmotic pressure differential ($\Delta\pi$). These can be expressed mathematically as follows:

$$J_w = A(\sigma \cdot \Delta\pi - \Delta P) \tag{1.1}$$

where J_w is defined as the water flux, σ the reflection coefficient, A the membrane water permeability constant, and ΔP the applied pressure. For FO, ΔP equals zero; for PRO, $\Delta\pi > \Delta P$; and for RO, $\Delta P > \Delta\pi$. The water flux directions in RO, FO, and PRO are illustrated in Figure 1.3. The water flux directions and driving forces for RO, FO, and PRO were characterized by Lee and coworkers [28]. The PRO zone, FO point, and RO zone, along with the water flux reversal point, are demonstrated in Figure 1.4.

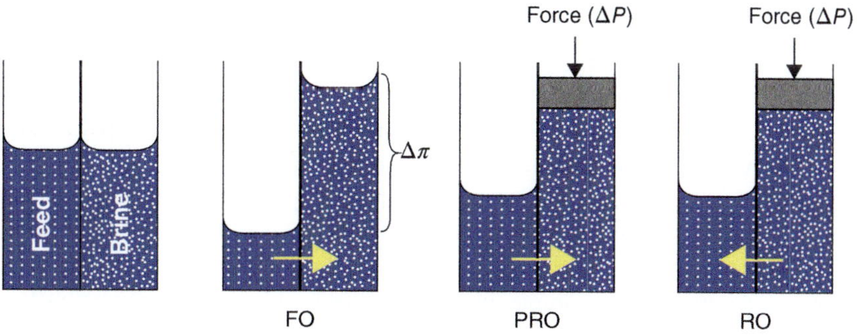

Figure 1.3 The solvent flow in FO, PRO, and RO. In FO, water diffuses through the membrane to the saline side of the solution with almost zero pressure difference ($\Delta P = 0$). In PRO, water moves from a less saline liquid under positive pressure ($\Delta\pi > \Delta P$) to that which is more saline. In RO, water flows under hydraulic pressure from a more to a less saline side ($\Delta P > \Delta\pi$). Source: Mansour et al. [30]/with permission of Elsevier.

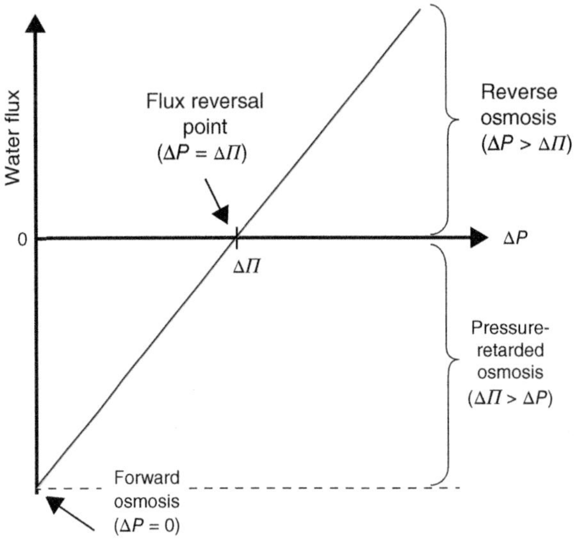

Figure 1.4 The direction and value of water flow vs. applied pressure in the three processes (FO, PRO, and RO). When the hydraulic pressure difference is zero, FO occurs; in the PRO region, the applied pressure difference ranges between zero and the flux reversal point, while in the RO area, it exceeds the osmotic pressure difference. Source: Mansour et al. [30]/with permission of Elsevier.

1.4.1 Advantages and Challenges of Forward Osmosis

FO outweighs other types of desalting methods, such as RO, due to several reasons, among them: lower energy demand because it does not require high-pressure pumps. Therefore, this results in reduced operational costs and thus decreased electricity usage. Moreover, RO membranes are more prone to fouling than FO membranes because the absence of hydraulic pressure (ΔP) minimizes the compaction of foulants and scaling ions on the membrane surface. This leads to a longer life span of the FO membrane and reduced maintenance and replacement costs. The versatile applications of FO in various disciplines have attracted attention. For instance, FO can be integrated with other processes to enhance water recovery and reduce energy consumption in seawater desalination. Similarly, in wastewater treatment applications, FO helps to concentrate waste streams and recover clean water. For example, in food industries, it is used for concentration and dehydration purposes, while in the pharmaceutical sector, it is employed for the extraction or purification of sensitive compounds. Lately, the incorporation of renewable energy sources into FO systems has increased its sustainability. Hybrid systems combine FO with solar or wind energy. However, several major issues related to FO need to be addressed before it's widely adopted. Among such problems is internal concentration polarization (ICP), which reduces the effective osmotic pressure driving force by decreasing the water flux due to a change in the concentration gradient across the support layer of the membrane. Also, "reverse solute diffusion" occurs when draw solutes diffuse back into the feed solution, thereby leading to the dilution of the draw solution, making it less efficient process-wise. The choice of proper draw solutions is a key element: they have to be nontoxic, stable, inexpensive, and instantly regenerative to sustain a viable FO system.

1.5 Reverse Osmosis (RO)

Reverse Osmosis is a well-known method of desalinating water by using a semi-permeable membrane to separate the solvent from substances with a high solute concentration, transporting from a region with a higher solute concentration to another with a lower solute concentration with external pressure [29]. This removes salts and other impurities, resulting in clean water. RO works through hydraulic pressure that surpasses the osmotic pressure of feedwater, thereby driving water molecules across the barrier and leaving dissolved salts and pollutants like heavy metals behind. Osmotic pressure occurs when water moves through a semipermeable membrane from a dilute solution to a concentrated solution so as to achieve equalization in the product concentration on both sides of the membrane. At pressures greater than osmotic pressure, RO can reverse this movement through membranes, thus allowing the separation of low-molecular-weight organic constituents and ions out of the water bulk.

It is common for mechanical energy recovery devices to be used on the pressurized brine stream so as to reduce their overall energy consumption during operation [30, 31]. As compared to seawater, salt concentration increases with the osmotic pressure of a brine stream since the latter does not have any significant influence on the osmotic pressure of a product water. In practice, however, Δp is usually <85 bars due to the adverse effects of high hydraulic pressures on membrane transport properties. Typically, Δp in most conventional RO systems is limited to 70 bars, resulting in a maximum of ~82,000 mg/L total dissolved solids (TDS) [32]. A typical RO seawater desalination plant is composed of four major stages: pretreatment, high-pressure pumping, membrane separation, and posttreatment. In terms of module types commercially used for industrial purposes today, these stages involve membraning by using hollow fine fiber filtration as well as spiral-wound technology that has found the greatest commercial success till now. The pressure vessels contain hollow fine fiber membranes made from cellulose triacetate and polyamide materials [33]. Similarly, spiral-wound membranes are fabricated to achieve high water flux and high salt rejection and also meet the stringent public health and environmental safety regulations. The feedwater is subjected to hydraulic pressure using high-pressure pumps, forcing water molecules through a semi-permeable membrane to the permeate side while retaining salts and other impurities. The desalinated water, known as permeate, can be further processed or distributed, while the waste brine is discharged or treated elsewhere.

1.5.1 Advantages and Challenges of Reverse Osmosis

As a technology for desalting water, RO offers some advantages that make it preferable. The most important thing about it is that it can remove impurities and dissolved salts, providing high-quality potable water. In this case, RO systems have been designed in a way, making it possible for them to be used at different scales according to the need. Therefore, the RO process requires less energy compared to thermal distillation methods as there is no change in the phase [34]. Also, there have been advancements in membrane technology that have improved the performance and life span of RO membranes, leading to increased water flow rates, resistance to fouling or compression, and salt rejection levels. However, some problems are also encountered by this technology. The most common among these challenges is membrane fouling which refers to a situation whereby undesired substances such as dirt, microbes, and other unwanted materials on its surface as well as within its pores reduce the water flux performance and salt rejection [35]. The operational costs could increase, maintenance may be required frequently, and potential membrane damage could arise as a result of fouling. It can be microbial fouling that irreversibly destroys the cellulose acetate membrane or affects the performance of other types of membranes, such as polyamide. For instance, fouling control calls for proper pretreatment, regular cleaning, as well as at some point chemical dosing, which may come with more operational complexities and increased operational costs. Another challenge is energy consumption and costs associated with RO systems. They consume less energy than thermal distillation methods but demand high-pressure pumps that consume significant amounts during desalination. This may lead to high operational costs, especially in large-scale plants mainly concerned with desalination processes. The initial capital cost of installing RO systems is also substantial, while ongoing membrane replacement only adds to this whole figure. Further Research and Development (R&D) efforts are concentrating on improving membrane materials, system design optimization, and finding alternative methods through which fouling can be mitigated and energy use can be reduced, ensuring RO remains a viable technology for addressing global water scarcity. Of note, however, is that conventional Reverse Osmosis systems have major technical challenges while handling high-salinity waters. such as highly saline brines exceeding 75 g/L. New generations of RO have been introduced in recent years as a result of R&D efforts into RO technology, such as Osmotically Assisted Reverse Osmosis (OARO) [36] which is sometimes known as low salt rejection reverse osmosis (LSRO). These new generations of RO technologies overcome the limitations of traditional RO systems in different ways. To begin with, the OARO technology is built to handle high salt levels, which makes it useful for seawater desalination, oil- and gas-produced water, and treatment of industrial brines. Furthermore, the two technologies improve the energy efficiency of the desalination process, a concern worth addressing, given that conventional RO systems use a lot of energy, especially for high-salinity water sources. By utilizing its pressure optimization and osmotic pressure gradient, OARO also helps reduce overall energy requirements during desalting while mitigating other restrictions associated with conventional RO, such as impediments to mass transport between feed and permeate sides [37–39].

1.5.2 Osmotically Assisted Reverse Osmosis (OARO)

The OARO technology is a new generation of RO as an advanced method of desalinating high-salinity feedwater that combines the principles of Forward Osmosis and conventional Reverse Osmosis to improve water recovery and save energy [39]. The feed solution in FO is less concentrated than the draw one, unlike in OARO, where a little more concentrated sweep solution is employed to develop an osmotic pressure difference across the hydrophilic membrane. This approach assists the RO process by lowering the hydraulic pressure required for desalting water and overcoming the osmotic pressure limitation. This means that compared to conventional RO, OARO can operate at lower hydraulic pressures, thus enabling the dewatering of highly saline feed streams while staying within the maximum permitted hydraulic pressures of the membrane module [38]. A mixed approach enables OARO to harness both processes' advantages, making it a suitable alternative for desalinating high salinity water. The usual procedure involves exerting high hydraulic pressure on feedwater during RO such that its natural osmotic pressure is negated and forcing water molecules through this semi-permeable material while leaving out salts and impurities. It helps in pulling more water through membranes with lowered hydraulic pressure via an additional osmotic driving force provided by this sweep solution, hence becoming more energy-efficient, leading to higher water recovery rates [36].

Figure 1.5 illustrates a schematic diagram of the OARO process that involves multiple stages to produce desalinated water. In this process, a high saline feed enters an OARO module at an elevated hydraulic pressure. The opposite permeate side of the module is simultaneously washed by a low-salinity sweep solution. A high-pressure feed and a low-pressure sweep are different from each other

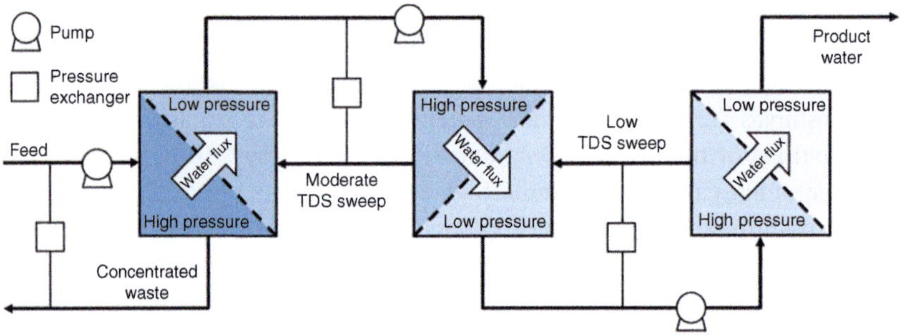

Figure 1.5 Process flow diagram of the OARO system. The first two stages in the system are OARO, while the final stage is RO. Source: Bartholomew et al. [38]/with permission of Elsevier.

in such a way that they create a hydraulic pressure difference, which exceeds the osmotic pressure difference across membranes. Water flux is driven by this pressure differential to concentrate the feed while diluting the sweep solution. The concentrated feed forms a residual stream from the OARO process, whose disposal or crystallization can be done. The diluted sweep solution remains relatively high in concentration and if the target recovery cannot be achieved by RO, it can be pressurized and sent back to a second OARO module for further processing. This re-concentration cycle involves all new solute-containing sweeping solutions as its inlet into the downstream OARO system. Consequently, it is followed by another OARO unit for receiving the diluted effluent produced from the second stage. Consequently, each next stage will have a less concentrated sweep solution until it reaches such a concentration when only reverse osmosis final treatment is possible for them. This multistage design eventually incorporates a product water outlet, a waste concentrate outlet, closed cycles for saline sweeps, and a feed inlet. The iterative manner of doing this allows optimization of water recovery and energy consumption minimization during efficient desalination as well as will involve a feed inlet, a concentrated waste outlet, closed cycles of saline sweeps, and a product water outlet [37].

Ideally, for the successful operation of this process, it requires selective membranes that are characterized by a low solute permeability coefficient and a high water permeability coefficient. Some typical components of an OARO system are a feed solution, a sweep solution, a semipermeable membrane, and a pumping section.

1.5.2.1 Advantages and Challenges of OARO

For water desalination purposes, OARO is far better than traditional RO, and as such it is an attractive option. The most important of these is that it uses less energy in desalinating high-salinity brines. OARO can obtain brine concentrations exceeding 200,000 mg/L, which are 50–80% higher than the 130,000 mg/L achievable by ultra-high-pressure reverse osmosis (UHP RO). OARO taps into the difference in osmotic pressures between the draw and feed solutions to achieve higher permeate fluxes with lower hydraulic pressures compared to conventional reverse osmosis systems. Consequently, this process remains cost-effective and environmentally friendly since it can be combined with renewable sources of power. Moreover, one more difficulty faced by scientists consists in finding suitable draw solutions. A good draw solution for efficient water extraction should be nontoxic, stable, and possess high osmolality properties. In addition, cost considerations continue to haunt the adoption of OARO [37]. Although the operating expenses are usually less than those associated with common RO membranes, there will still be significant upfront costs regarding specialized membranes and draw solution regeneration facilities. This calls for improvements in membrane technologies as well as drawing solution chemistry and system designs so that large-scale desalination through OARO can become a reality at reduced costs.

1.6 Membrane Capacitive Deionization

MCDI is an innovative electrochemical process aimed at desalination, which removes ions from water and stores them in capacitive porous electrodes [40]. The MCDI module features a housing that contains stacks of cells (Figure 1.6). Each cell comprises two current collectors connected to two porous carbon electrodes separated by a polymeric spacer. The current collectors, which can be made of nonporous conductive materials like titanium and graphite foils, connect the porous carbon electrodes to the power source, preventing electrocorrosion. The spacer acts as a passage for deionized water through the porous electrodes. The MCDI process causes ionized water to pass through a small electric field generated across the carbon electrode, usually around 1.2 V or below. Dissolved ions are attracted toward the porous electrode due to the electric field applied across it [7]. Four factors determine how easily ions can be attracted to electrodes within cells, which include valence, size, solvation enthalpy, and charge density. Polarity reversal or switching off electrical power regenerates these electrodes, eliminating chemical addition, salts that are usually used in conventional water treatment processes, such as softeners. MCDI systems upgrade the basic capacitive deionization (CDI) process by incorporating ion-exchange membranes with functional groups that allow only positive or negative ions to flow through. Porous electrodes face the anion-exchange membrane (AEM) and

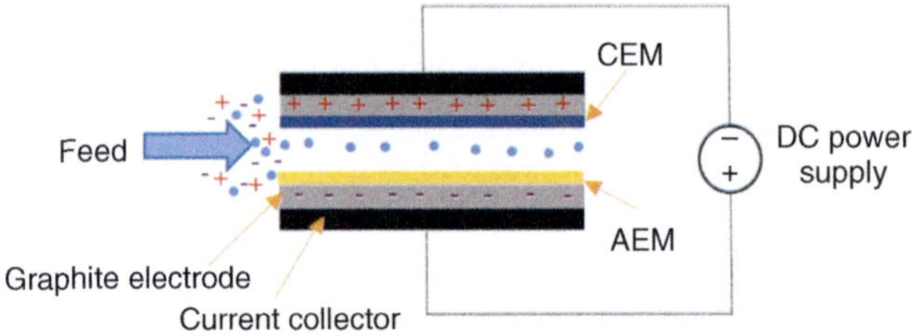

Figure 1.6 Membrane capacitive deionization process. Source: Fuentes et al. [40]/with permission of Elsevier.

cation-exchange membrane (CEM) [41]. This arrangement increases salt removal and charge efficiency by blocking co-ions at the electrode surface while minimizing undesired Faradaic reactions that could lead to deterioration. Particularly, when compared with ion removal technologies such as distillation and reverse osmosis where water is removed instead of ions, MCDI holds much promise for energy efficiency. This is especially true for streams having relatively low concentrations of ions requiring salt removal; for example, in freshwaters where the upper limit is about 10 mM, there are roughly 3,000 water molecules per ion. These comparatively few ions can be efficiently removed from a large solvent volume due to the charge that is inherent in any ion. When a solution flows through cells, ions are separated from the solvent and stored temporarily in porous electrode assemblies on each side of the cell by applying an electrical field across it. Ion-exchange membranes are directly placed ahead of electrodes to prevent the release of co-ions (ions with the same polarity as the electrode), which would decrease the salt removal efficiency. In particular, a negatively biased electrode is placed against a cation-exchange membrane, while a positively biased electrode faces an anion-exchange membrane. Consequently, there is counterion mobility into and off the electrode as opposed to co-ion transfer that occurs, enhancing the overall process efficacy. MCDI systems can operate in two main charging modes: constant current (CC) and constant voltage (CV) [40]. The choice of the charging mode influences energy consumption as well as recovery patterns. As the literature suggests, CC mode has much lower energy consumption than CV mode, due mainly to better control of charging currents in CC mode, which dissipate less power through resistive components. Furthermore, it reduces the time when Faradaic reactions could occur during operation at non-Faradaic conditions. Overall, MCDI provides notable benefits for the desalination process, such as high performance, minimal chemical use, and energy conservation. These methods are particularly attractive for sustainable water treatment, especially when there is a need to remove ions efficiently with minimal damage to the environment.

1.7 Microfiltration (MF), Ultrafiltration (UF), and Nanofiltration (NF)

MF, UF, and NF are major membrane technologies used for desalination, and they differ with respect to the size of their membrane pores. These processes are crucial in water treatment and purification as they remove various impurities while at the same time improving the efficiency of downstream desalination steps, such as RO.

1.7.1 Microfiltration (MF)

The MF membranes usually have larger pore sizes between 0.08 and 0.5 μm, which make MF highly effective in eliminating large microorganisms and suspended solids from feed solutions [42]. The technology was developed for water treatment and purification; since then, it has been mainly applied for that purpose. MF is frequently combined with other separation techniques like reverse osmosis (RO) and UF, thus enhancing the overall process of treatment. MF can separate microorganisms from suspended solids, thereby playing a vital role as a pretreatment phase in seawater desalination. In essence, there has been remarkable development regarding pretreatment membranes for seawater desalination through the adoption of MF. One key advantage is its reliability in handling difficult water sources, along with easy design and operation, as well as lower operating and capital costs compared to traditional pretreatment methods [43]. However, fouling problems can limit its capacity to produce fresh water from desalting plants.

1.7.2 Ultrafiltration (UF)

Ultrafiltration membranes have smaller pores ranging from about 0.001 to about 0.01 μm than those of MF. Also, UF can hold back bacteria, viruses, and soluble macromolecules and hence is useful for treating solutions and suspensions having intermediate-molecular-weight solutes [44]. UF has also contributed significantly to developments in water treatment technology since it began being used around 1960. UF has a high permeate quality, resulting in a stable filtrate that does not undergo colloidal fouling or lead to increased turbidity on RO membranes. Thus, the overall efficiency and yield of desalination processes can be improved due to less contamination of the colloidal nature and turbidities of RO membranes. Like other pretreatment steps, UF also suffers from fouling, which impairs its efficiency. However, through Integrated Membrane Systems (IMS) formation, UF is still an essential component of desalination.

1.7.3 Nanofiltration (NF)

The average NF pore sizes range between 0.5 and 5 nm, which enables the nanofiltration process to filter particles that are within the size ranges processed by both UF and RO [45, 46]. This pressure-driven process can withstand differential pressures ranging between 500 and 2,000 kPa, lower than those required for RO, but still give out a high permeate flux [45].

For instance, both divalent ions such as Ca^{2+} and Mg^{2+} and monovalent ions like Na^+, K^+, and Cl^- can be effectively retained by NF. As a result, applications involving water softening, removal of organic matter, and wastewater treatment/recovery are appropriate for NF. Besides having a different network structure from RO, this helps in the removal of a wide range of contaminants in the operation while maintaining high efficiency.

MF/UF/NF in integrated membrane systems have led to the development of more advanced desalting techniques. Consequently, membrane fouling issues have been greatly reduced, improving the yield and performance of desalting processes using MF/UF/NF [47]. There are many advantages, including

1) **stable quality filtration**: The consistency in the quality of an MF/UF filtrate means that there will be no reason for fouling on the RO membranes, leading to reduced fouling;
2) **lessened colloidal fouling**: It decreases the occurrence of colloidal fouling in RO/NF systems, thereby ensuring better overall functionality and life span for the membranes;
3) **disposal convenience of concentrates**: Concentrates from MF/UF are easier to dispose of compared to those generated by traditional chemically upgraded pretreatment processes.

MF, UF, and NF are therefore important aspects of modern desalination technologies. All these processes have specific benefits such as contaminant removal, operational efficiency, and integration with other desalination steps.

1.8 Conclusion

The lack of access to fresh water can be tackled using novel solutions like membrane technologies that target water desalination directly. The significance of desalination plants has increased recently, as the world has been suffering from a shortage of fresh water due to frequent droughts and environmental pollution, forcing researchers to investigate new technologies to secure fresh water. Conventional nonmembrane desalination technologies present challenges, which is why moving the needle to a membrane solution is vital. The analysis of membrane technologies, which include membrane distillation (MD), forward osmosis (FO), reverse osmosis (RO), osmotically-assisted reverse osmosis (OARO), membrane capacitive deionization (MCDI), microfiltration (MF), ultrafiltration (UF), and nanofiltration (NF), provides insights into membrane operations, characteristics, and advantages that make these approaches a viable option for desalination processes. Other performance parameters such as permeate flux, energy efficiency, capital expenditures (CapEX), operational expenditures (OpEx), and environmental effects are relevant for assessing the performance effectiveness of each membrane technology in detail. However, within the last 10 years, new membrane materials, configurations, and designs have significantly improved these parameters and have demonstrated advanced membrane performance requisites, energy performances, and cost-effective features. Nonetheless, the future for membrane desalination technologies is optimistic, despite significant barriers like fouling, high costs, and environmental challenges. These advancements will be significant in improving the efficiency of these advanced technologies.

Questions

1. Why is the importance of advanced membrane water desalination technologies increasing?
2. What are the different types of membrane-based technologies?
3. What are the fundamental principles behind reverse osmosis (RO) and forward osmosis (FO)? How do they differ?
4. How is membrane distillation (MD) different from conventional distillation methods?
5. What is osmotically assisted reverse osmosis (OARO)? How does it enhance desalination processes?
6. What are the operational mechanisms of membrane capacitive deionization (MCDI)? What advantages does it offer over traditional desalination methods?
7. What factors should be considered while selecting a membrane type for desalination purposes?
8. How can membrane technology revolutionize water desalination to make it more efficient and sustainable?

References

1 Osseiran N. (2017). 2.1 billion people lack safe drinking water at home, more than twice as many lack safe sanitation. World Health Organization.

2 Mekonnen, M.M. and Hoekstra, A.Y. (2016). Four billion people facing severe water scarcity. *Sci. Adv.* 2 (2): https://doi.org/10.1126/sciadv.1500323.

3 Mauter, M.S. and Fiske, P.S. (2020). Desalination for a circular water economy. *Energy Environ. Sci.* 13: 3180–3184.

4 Ghaffour, N. (2009). The challenge of capacity-building strategies and perspectives for desalination for sustainable water use in MENA. *Desalin. and Water Treat.* 5 (1–3): 48–53.

5 Darwish, N., Hilal, N., and Alkhudhiri, A. (2012). Membrane distillation: a comprehensive review. *Desalination* 287: 2–18.

6 Al-Amshawee, S., Yunus, M.Y.B.M., Azoddein, A.A.M. et al. (2020). Electrodialysis desalination for water and wastewater: a review. *Chem. Eng. J.* 380: 122231.

7 Elewa, M.M., El Batouti, M., and Al-Harby, N.F. (2023). A comparison of capacitive deionization and membrane capacitive deionization using novel fabricated ion exchange membranes. *Materials* 16 (13): https://doi.org/10.3390/ma16134872.

8 Lu, K.J., Cheng, Z.L., Chang, J. et al. (2019). Design of zero liquid discharge desalination (ZLDD) systems consisting of freeze desalination, membrane distillation, and crystallization powered by green energies. *Desalination* 458: 66–75. https://doi.org/10.1016/j.desal.2019.02.001.

9 Zuo, J., Chung, T.-S., and Lu, K.-J. (2016). Tri-bore PVDF hollow fibers with a super-hydrophobic coating for membrane distillation. *J. Memb. Sci.* 514: 165–175.

10 Matsuura, T., Kruczek, B., Khayet, M., and Qtaishat, M. (2008). Heat and mass transfer analysis in direct contact membrane distillation. *Desalination* 219 (1–3): 272–292.

11 Khayet, M., Matsuura, T., and Souhaimi, M.K. (2011). *Membrane Distillation: Principles and Applications*. Elsevier.

12 Godino, P., Mengual, J.I., and Khayet, M. (2000). Theory and experiments on sweeping gas membrane distillation. *J. Memb. Sci.* 165: 261–272.

13 Chomiak, T., Floyd, J., Li, Q., and Said, I.A. (2020). Sweeping gas membrane distillation (SGMD) for wastewater treatment, concentration, and desalination: a comprehensive review. *Chem. Eng. Process. Process Intensif.* 153: 107960.

14 Said, I.A., Fuentes, N., He, Z. et al. (2021). Treatment of brackish water reverse osmosis brine using only solar energy. *Water Res. Technol.* 7 (10): 1840–1851. https://doi.org/10.1039/D1EW00291K.

15 Zhang, Y., Dong, H., Zhang, L. et al. (2015). A comprehensive review of vacuum membrane distillation technique. *Desalination* 356: 1–14.

16 Xin, R., Huang, X., Zuo, K. et al. (2021). Wetting-resistant photothermal nanocomposite membranes for direct solar membrane distillation. *J. Memb. Sci.* 620: 118913.

17 Wang, W., Deshmukh, A., Jia, S. et al. (2020). Multifunctional nanocoated membranes for high-rate electrothermal desalination of hypersaline waters. *Nat. Nanotechnol.* 15: 1025–1032.

18 Ruiz-Aguirre, A., Acién, F.G., Zaragoza, G., and Andrés-Mañas, J.A. (2020). Performance increase of membrane distillation pilot scale modules operating in vacuum-enhanced air-gap configuration. *Desalination* 475: 114202.

19 Khayet, M. and Essalhi, M. (2014). Application of a porous composite hydrophobic/hydrophilic membrane in desalination by air gap and liquid gap membrane distillation: a comparative study. *Sep. Purif. Technol.* 133: 176–186.

20 Ghaffour, N., Alsaadi, A.A., Amy, G.L., and Francis, L. (2013). Material gap membrane distillation: a new design for water vapor flux enhancement. *J. Memb. Sci.* 448: 240–247.

21 Rivier, C.A., Marison, I.W., von Stockar, U., and Garcı´ a-Payo, M.C. (2002). Separation of binary mixtures by thermostatic sweeping gas membrane distillation: II. experimental results with aqueous formic acid solutions. *J. Membr. Sci.* 198 (2): 197–210.

22 Sitter, K.D., Dotremont, C., Pinoy, L. et al. (2017). Membrane synthesis for membrane distillation: a review. *Sep. Purif. Technol.* 182: 36–51.

23 Aaberg, R.J. (2003). Osmotic power—a new and powerful renewable energy source. *Refocus* 4 (6): 48–50.

24 Loeb, S. (1974). Osmotic power plants. *Science* 189: 350.

25 Loeb, S. (1976). Production of energy from concentrated brines by pressure retarded osmosis. I. preliminary technical and economic correlations. *J. Membr. Sci.* 1: 49.

26 Loeb, S. and Mehta, G.D. (1978). Internal polarization in the porous substructure of a semi-permeable membrane under pressure-retarded osmosis. *J. Membr. Sci.* 4: 261.

27 Childress, A.E., Elimelech, M., and Cath, T.Y. (2006). Forward osmosis: principles, applications, and recent developments. *J. Memb. Sci.* 281 (1–2): 70–87.

28 Baker, R.W., Lonsdale, H.K., and Lee, K.L. (1981). Membranes for power generation by pressure-retarded osmosis. *J. Memb. Sci.* 8 (2): 141–171.

29 Badrelzaman, M., Darwish, N.N., Darwish, N.A. et al. (2019). Reverse osmosis desalination: a state-of-the-art review. *Desalination* 459 (1): 59–104.

30 Mansour, T.M., Ismail, T.M., Ramzy, K., and Abd El-Salam, M. (2020). Energy recovery system in small reverse osmosis desalination plant: experimental and theoretical investigations. *Alex. Eng. J.* 59 (5): 3741–3753.

31 Schunke, A.J., Hernandez Herrera, G.A., Padhye, L., and Berry, T.-A. (2020). Energy recovery in SWRO desalination: current status and new possibilities. *Front. Sustain. Cities* 2: https://doi.org/10.3389/frsc.2020.00009.

32 Wang, Z., Deshmukh, A., Yuhao, D., and Elimelech, M. (2020). Minimal and zero liquid discharge with reverse osmosis using low-salt-rejection membranes. *Water Res.* 170: 115317.

33 Koch Membrane Systems Inc (KMS) (2005). Pressure vessels to house RO elements. *Membr. Technol.* 2005 (10): 5.

34 Jiang, L., Yue, T., Li, X., and Li, H. (2018). Application of reverse osmosis in purifying drinking water. *E3S Web of Conf—ICEMEE* 38: 01037.

35 Ahmed, M.A., Amin, S., and Mohamed, A.A. (2023). Fouling in reverse osmosis membranes: monitoring, characterization, mitigation strategies and future directions. *Heliyon* 9 (4): e14908.

36 Beigi, B.H.M., Gadkari, S., and Sadhukhan, J. (2022). Osmotically assisted reverse osmosis, simulated to achieve high solute concentrations, at low energy consumption. *Sci. Rep.* 12: 13741.

37 Bartholomew, T.V., Siefert, N.S., and Mauter, M.S. (2018). Cost optimization of osmotically assisted reverse osmosis. *Environ. Sci. Technol.* 52 (20): 11813–11821.

38 Bartholomew, T.V., Mey, L., Arena, J.T. et al. (2017). Osmotically assisted reverse osmosis for high salinity brine treatment. *Desalination* 421: 3–11.

39 Peters, C.D. and Hankins, N.P. (2019). Osmotically assisted reverse osmosis (OARO): five approaches to dewatering saline brines using pressure-driven membrane processes. *Desalination* 458: 1–13.

40 Fuentes, N., Li, Q., and Said, I.A. (2022). Energy recovery in electrified capacitive deionization systems for wastewater treatment and desalination: a comprehensive review. *Chem. Eng. Process. Process Intensif.* 178: 109030.

41 McNair, R., Szekely, G., and Dryfe, R.A.W. (2021). Ion-exchange materials for membrane capacitive deionization. *ACS EST Water* 1 (2): 217–239.

42 Hashaikeh, R., Hilal, N., and Anis, S.F. (2019). Microfiltration membrane processes: a review of research trends over the past decade. *J. Water Proc. Eng.* 32: 100941.

43 Bing, W., Suwarno, S.R., Tan, H.S. et al. (2017). Gravity-driven microfiltration pretreatment for reverse osmosis (RO) seawater desalination: microbial community characterization and RO performance. *Desalination* 418 (15): 1–8.

44 Lai, Z. and Kammakakam, I. (2023). Next-generation ultrafiltration membranes: a review of material design, properties, recent progress, and challenges. *Chemospher* 316: 137669.

45 Chollom, M.N., Rathilal, S., Pillay, V.L., and Alfa, D. (2015). The applicability of nanofiltration for the treatment and reuse of textile reactive dye effluent. *Water SA* 41 (3): https://doi.org/10.4314/wsa.v41i3.12.

46 Teow, Y.H., Ang, W.L., Chung, Y.T. et al. (2015). Nanofiltration membranes review: recent advances and future prospects. *Desalination* 356 (15): 226–254.

47 Khawaji, A.D., Kutubkhanah, I.K., and Wie, J.-M. (2008). Advances in seawater desalination technologies. *Desalination* 221 (1–3): 47–69.

2

Materials, Fabrication Techniques, and Characterization for Membrane Desalination

Zhuo Fan Gao[1], Ji Wu[2], Min Lei[3], Xin Cheng Jin[4], Zhuo Huang[1], and Xingdong Lv[1]

[1]*Science and Technology Achievement Promotion Center, Changjiang River Scientific Research Institute, Wuhan, China*
[2]*Department of Chemical and Biomolecular Engineering, National University of Singapore, Singapore*
[3]*College of Computer Science, South-Central University for Nationalities, Wuhan, China*
[4]*Environmental Division, Wuhan Changjiang Kechuang Technology Development Co., Ltd., Wuhan, China*

2.1 Introduction

Membrane desalination technology has come a long way since the introduction of cellulose acetate membranes in the 1960s. It stands today as a critical technology in the fight against global water scarcity, enabling the separation of salts, impurities, and contaminants from seawater and brackish water through semipermeable membranes. The academic and industrial sectors have recently directed increased attention toward the innovation of materials, fabrication methods, and characterization techniques for membrane desalination. This rising research trend can be seen in Figure 2.1, which illustrates the growing number of publications on membrane desalination in the Web of Science database, reflecting the expanding interest in the field. The evolution of membrane materials, progress in fabrication techniques, advancement of characterization tools, and rigorous assessment of desalination performance are the key research areas revolving around membrane desalination technology.

2.2 Current State-of-the-Art Membrane Processes for Desalination

Desalination of seawater and brackish water reveals an important route to alleviating the shortage of fresh water in many parts of the world, and membrane

Water Treatment and Desalination: Membrane Separation using Renewable Energy Resources, First Edition. Edited by Mohamed Bassyouni.

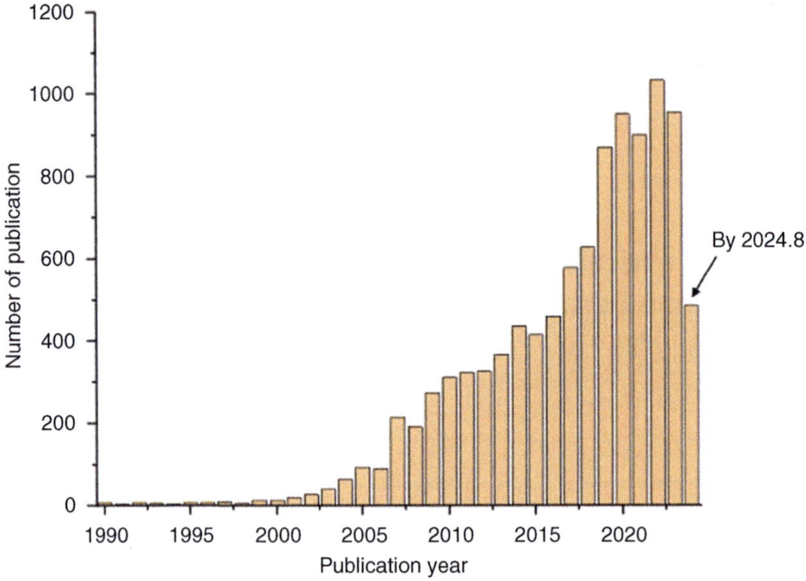

Figure 2.1 Number of publications on materials, fabrication methods, and characterization related to membrane desalination since the late 1990s.

technology plays a pivotal role in enabling energy-efficient desalination processes, especially when compared with the traditional thermal-intensive distillation methods. Along its development, a variety of membrane processes, primarily differed by their driving force, effective pore size and desalination efficiency, have been invented and applied for desalination, which are summarized in Table 2.1.

2.3 Material Choice for Desalination Membranes

The selection of membrane materials is crucial because it directly influences the separation efficiency, selectivity, and durability of the desalination process. This section explores the various membrane materials utilized in desalination processes.

2.3.1 Types of Membrane Materials

Organic polymers are the most widely used membrane materials in desalination due to their fairly balanced water permeability and water/salt selectivity. Polyamide membranes are particularly popular because their syntheses and separation properties can be tailored for specific separation tasks through

Table 2.1 Key features of different state-of-the-art membrane desalination processes.

Type of membrane process	Reverse osmosis (RO)	Forward osmosis (FO)	Electrodialysis (ED)	Nanofiltration (NF)
Desalination capacity (%)	80–99	85–95	80–90	50–80
Average pore size (nm)	0.1–1	0.25–0.30	1–10	0.2–2
Energy input	Mechanical energy	Thermal energy	Electrical energy	Mechanical energy
Advantages	High rejection, high-quality water, energy efficiency, versatility in design, environmental friendliness	Lower energy requirements, wide range of feed water options, compatibility with low-quality heat sources	Selective ion removal, energy efficiency, scalability, continuous operation, less impact on the environment	Selective salt removal, low energy requirement, reduced potential for fouling, less pretreatment required
Limitations	High operating pressure, maintenance and membrane fouling, concentrated brine disposal, high cost	Limited availability of suitable draw solutions, membrane fouling by draw solutes, lower water recovery	Limited availability of suitable draw solutions, membrane fouling by draw solutes, lower water recovery	Limited salt rejection, reduced water flux, high cost, not fully developed

Source: Adapted from Ali et al. [1]/reproduced with permission from Elsevier.

interfacial polymerization. However, their susceptibility to fouling and chemical degradation remains a challenge. In comparison, membranes made of inorganic materials, such as ceramics, metals, and their oxides, offer superior thermal and chemical stability and are also less prone to biofouling. However, inorganic membranes are generally more expensive to produce and can have lower permeability than polymeric membranes. At the intersection of polymers and inorganics, mixed matrix membranes (MMMs) that integrate inorganic or hybrid fillers into the organic polymer matrices emerge as a highly versatile and popular membrane material design. Common filler choices include zeolites, carbon nanotubes, or metal-organic frameworks (MOFs). MMMs aim to enhance the performance of polymeric membranes by utilizing the more favorable separation properties of the incorporated inorganic nanomaterials, such as the well-defined pore structure, stronger molecular-sieving and high selectivity [1].

The advancement of membrane materials is of critical importance to improving the efficiency and sustainability of desalination technologies. Ongoing research has been largely focused on developing materials that can deliver high water flux, superior selectivity, and robustness against harsh conditions. The integration of novel nanoarchitectures like covalent organic frameworks (COFs), a newer class of framework structure with better solution-processability and affinity for organic polymers than MOFs, is a promising avenue for future research.

COF-based membranes are at the forefront of advanced materials research, harnessing the unique properties of COF structures, which are crystalline porous polymers with well-defined 2D or 3D architectures. COFs are a class of functional materials capable of utilizing, converting, and storing energy. For example, 2D COFs exhibit compelling properties as photocatalysts/electrocatalysts, proton conductors, and cathode materials in batteries. However, the physiochemical properties of COF materials are highly dependent on their synthesis conditions. In order to achieve a rational design of the synthesis conditions, an in-depth understanding of the underlying mechanisms involved in the formation of COF is necessary. During the synthesis process, a catalyst plays a crucial role in kickstarting the polymerization reaction and continues to be active in its early stages and beyond. However, understanding this aspect remains an area where research is still in the dark. Therefore, it is crucial to monitor the reaction at the beginning of solvent mixing and during the initial stages of molecular interactions. However, current characterization methods face many challenges in capturing these transient stages. Christoph et al. [2] studied the COF polymerization and backbone formation operationally using an optical technique called the interference scattering microscopy (iSCAT). Liquid–liquid phase separation was observed, revealing a specially structured solvent in the form of surfactant-free emulsions present in the reaction mixture during conventional COF syntheses (Figure 2.2). The findings suggest that the solvents can do much more than just dissolving the monomers as they can also act as kinetic modulators through the separation of reactants and catalysts. With these findings uncovered, a scheme for the synthesis of COFs at room temperature rather than high temperature was developed due to its simpler and greener reaction conditions. By probing COF syntheses through the use of liquid-phase imaging techniques, this study provides a better understanding of the design of COF reaction environments. It also puts an emphasis on the application of light-scattering-based techniques as a powerful visualization tool for chemical reactions to more rational material syntheses.

2.4 Fabrication of Membranes for Desalination

The development of membrane materials for desalination is driven by the increasing performance requirements including improved water flux, higher

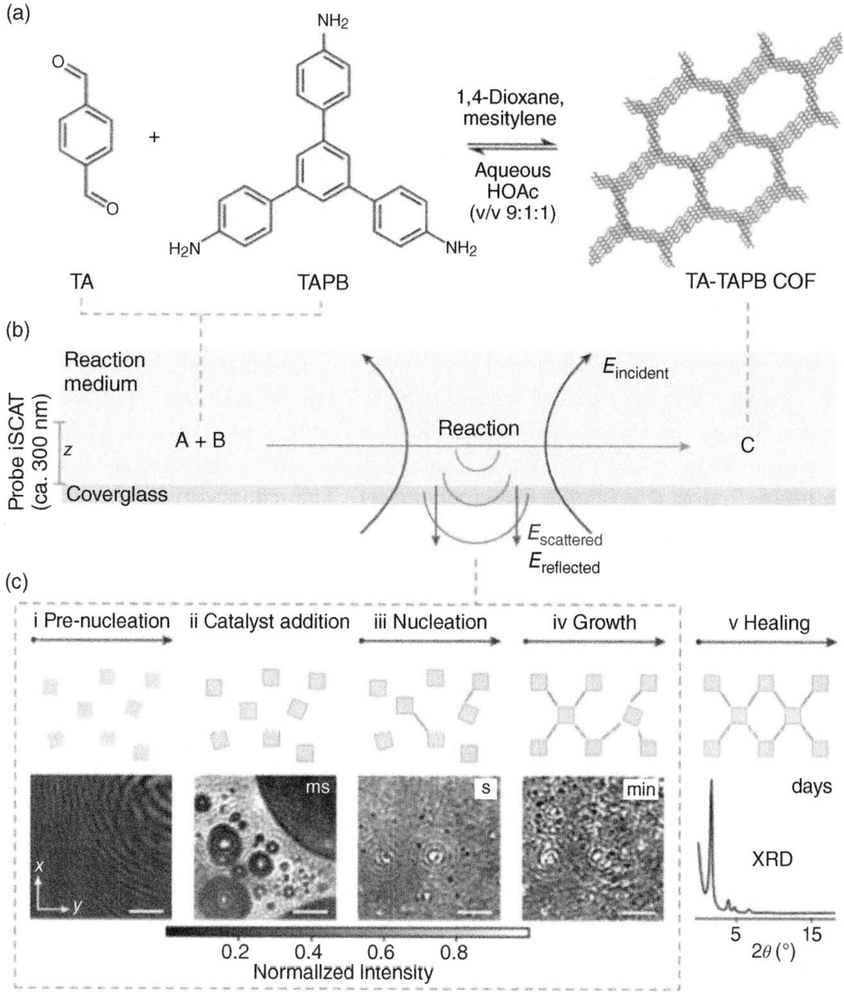

Figure 2.2 Mechanisms of COF formation. Source: [2]/with Permission of Springer Nature.

salt rejection, and better membrane longevity. The development of desalination membranes began with the cellulose acetate membranes introduced by Loeb and Sourirajan in the early 1960s. These membranes laid the foundation for pressure-driven desalination technology, especially through the reverse osmosis (RO) process. In the 1980s, thin-film composite (TFC) membranes, as another breakthrough in desalination membrane design, were invented that offered substantially improved performance in terms of water flux and salt rejection, allowing them to replace the earlier cellulose acetate membranes.

2.4.1 Thin-Film Composite (TFC) Membrane via Interface Polymerization

As shown in Figure 2.3, the TFC membranes are fabricated via interfacial polymerization, forming a thin and compact selective layer deposited on top of a porous substrate that acts as the support layer. This thin selective layer, predominantly made by polyamide (PA), is typically obtained with a thickness in the range of a few hundred nanometers. Known for their excellent salt rejection and water permeability, TFC membranes have now become the most widely adopted design for membrane desalination [3–8].

The emergence of nanomaterials, distinguished by their more attractive permeability, selectivity, chemical resilience, and minimal fouling tendencies, has sparked a new wave of research in TFC membranes. By incorporating either non-porous or porous nanoparticles into the selective layer to form a nanocomposite or MMM layer, TFC membranes can be transformed into the thin-film nanocomposite (TFN) membranes. The characteristic features of the particles can then be introduced into the membrane, including enhanced adsorption, optical/quantum effects, catalytic activity, electrical properties, and antifouling nature. A successful integration of particles and polymers requires the careful selection of the nanoparticle chemistry, morphology, and size. Notably, as compared with the non-porous ones, porous nanomaterials, such as carbon nanotubes (CNTs), nanoporous polymers, and metallic nanoparticles, present more intriguing opportunities for the TFN membrane design. These materials feature distinct nanopores that can facilitate water transport while effectively impeding the permeation of larger, undesired solutes [10, 11]. A schematic illustration in Figure 2.4 showcases the fabrication of TFN membranes via interfacial polymerization, highlighting how the functionalized nanofillers are actually being incorporated [12].

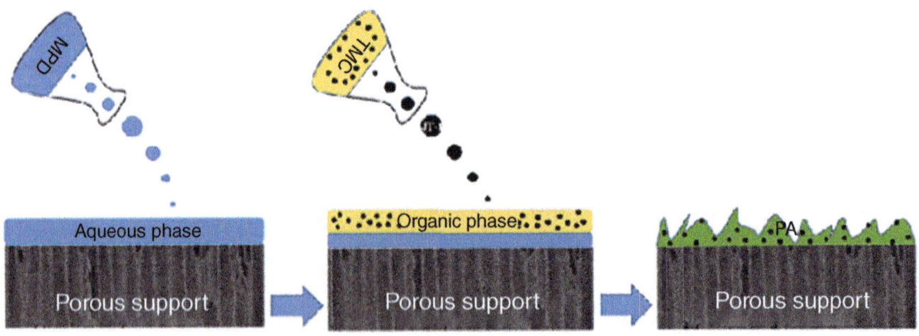

Figure 2.3 Schematic representation of the process for manufacturing TFC membranes through interfacial polymerization. Source: Reproduced from Han Zhang et al. [9], with permission of John Wiley & Sons.

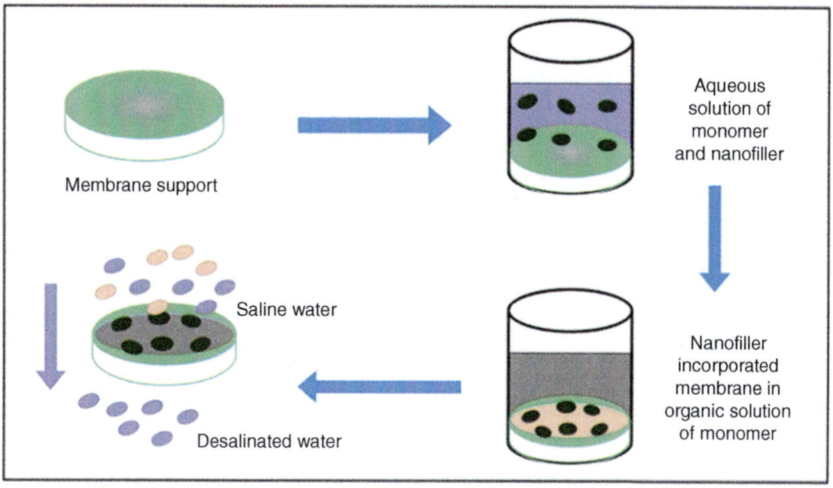

Figure 2.4 Schematic representation of the process for manufacturing TFN membranes through interfacial polymerization in the presence of functionalized nanofillers. Source: Kumar et al. [13]/with permission of American Chemical Society.

As the predominant membrane design for desalination, continuous research efforts have been dedicated to further enhancing the performance and stability of TFC membranes. As introduced previously, a TFC membrane consists of a thin PA selective layer interfacially polymerized on top of a porous support, both of which are important components that determine the membrane's desalination performance. As a result, researchers have explored methods to modify the support layer, refine the PA active layer or engineer the membrane surface in order to boost the separation efficiency of TFC membranes.

Key factors influencing the water permeation properties of a membrane include their surface hydrophilicity, roughness, and charge. To enhance the water permeability of their membranes, Li et al. [8] and Weng et al. [10] employed hydrophilic materials in the selective layer, like cellulose and chitosan. Besides, researchers have incorporated novel materials into the support layer or during the IP process, trying to enhance the water permeability. On top of that, surface modifications of the PA layer have also been explored. Shown in Figure 2.5 is a milestone diagram summarizing the advancements and modifications that have achieved a superior water/salt selectivity to the early commercial TFC membranes [8].

2.4.2 Electrospun Nanofiber Membranes

Electrospun nanofiber membranes (NFMs) have emerged as a significant advancement in the field of desalination, offering enhanced separation characteristics

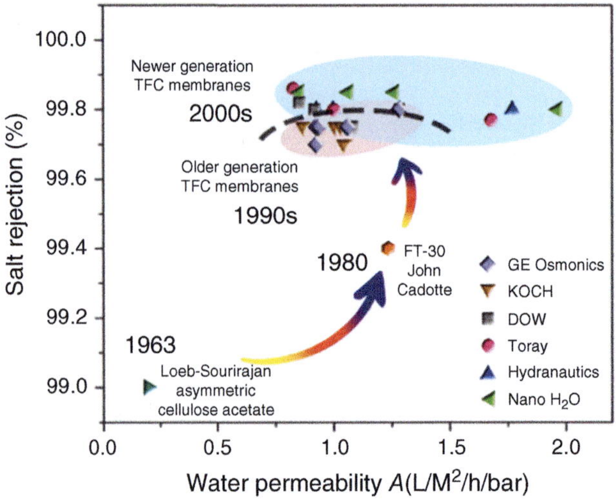

Figure 2.5 Advances of TFC membranes in their water permeability and salt rejection. Source: Lim et al. [11]/with permission of Elsevier.

due to their unique structure and tunable properties. The development of this type of membrane was driven by the demand for efficient, sustainable solutions to the global challenge of water scarcity. This innovative process harnesses a high-voltage electric field to stretch polymer solutions or melts into ultrafine nanofibers before being assembled into membranes. The electrospinning method enables the creation of thin fibers with nanoscale diameters, giving rise to more favorable membrane structures characterized by high porosity and an extensive surface area. NFMs mark a groundbreaking advancement in fibrous membrane materials as they exhibit many attractive features, including large surface area, superior porosity, tunable wettability, and high permeability. They hold a great potential to revolutionize desalination systems and enhance water and wastewater treatment processes. More specifically, the 3D interconnected structure in NFMs offers substantial advantages over traditional polymer membranes in terms of operational efficiency, cost-effectiveness, and energy consumption [10]. However, the main challenges faced by NFMs lie in their inherent fragility and susceptibility to fouling, which can deteriorate their separation performance over time. To address these issues, researchers have explored surface modification techniques that could mitigate fouling and reinforce nanofiber crosslinking, thereby strengthening the membrane's mechanical stability [14]. As an illustration, Figure 2.6 provides a schematic view of an electrospun nanofiber membrane designed for wastewater treatment [11]. It clearly captures the intricate architecture of these advanced fibrous building blocks.

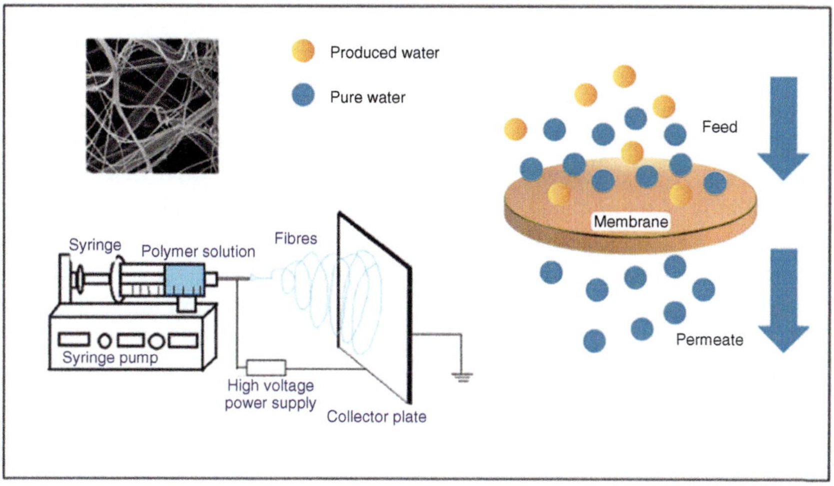

Figure 2.6 Schematic representation of electrospun nanofiber membrane for water treatment. Source: Reproduced from Lu et al. [17], with permission of American Chemical Society.

2.5 Characterization and Performance Evaluation

The characterization and performance evaluation of membranes for desalination have undergone significant advancements to keep up with the evolution of membrane materials and fabrication techniques. These analysis tools are essential to understanding the behavior of membranes under given operating conditions and optimizing their performance for desalination processes. The early development of desalination membrane characterization focused on the simple measurement of permeability and salt rejection. However, as the complexity of membrane materials increased, characterization techniques have become more sophisticated and have extended beyond the simple measurement of separation. This trend was accelerated by the introduction of TFC membranes, which required a more detailed analysis of their multilayer structure and the properties of each component.

2.5.1 Common Characterization Techniques for Desalination Membranes

Morphological, Packing, and Chemical Structures: Scanning electron microscopy (SEM) is employed to examine the surface and cross-sectional morphology as well as the pore structure of membranes. The preparation process involves slicing the dried membranes under cryogenic conditions using liquid nitrogen, followed by affixing them to a brass stub. Subsequently, a thin gold layer is deposited onto the sample surface through sputtering to enhance the conductivity and resolution of the imaging. Transmission electron microscopy (TEM) is utilized to obtain detailed information about the internal structure of the membranes, providing insights into the spatial arrangement of materials and identifying the presence of defects or irregularities within the membrane. X-ray diffraction (XRD) spectroscopy is employed to determine the crystallinity of membranes which reflects the packing structure of polymers. Pore size distribution analysis could be carried out using techniques such as mercury porosity or bubble point tests to provide detailed information about the pore structure of the membranes. Fourier transform infrared (FTIR) spectroscopy is used to analyze the chemical composition and identify the key functional groups present in the membrane materials, providing important information about the types of chemical bonds in the samples as well as their overall chemical structures.

Surface Properties: Atomic force microscopy (AFM) offers high-resolution imaging of membrane surfaces, allowing for the investigation of their topographical roughness and surface texture. Water contact angle measurements are conducted, usually using a contact angle goniometer, to assess the hydrophilicity or hydrophobicity of membrane surfaces which is crucial for understanding their interaction with liquids and how it influences the water permeability. The zeta potential (ζ) measurements are performed to determine the surface charge of the membrane as a function of varying pH values, which is an important parameter for evaluating the membrane's stability and its interaction with charged particles in a solution.

Thermal and Mechanical Properties: Thermogravimetric analysis (TGA) is employed to study the thermal stability and degradation behavior of the membrane materials. Mechanical testing, including tensile strength and elongation at break, could be performed to evaluate the mechanical properties and robustness of the membrane.

2.5.2 Performance Evaluation

To evaluate the desalination efficiency of a membrane, its water permeability and salt rejection are to be tested and obtained. Nanofiltration (NF) and RO membranes are generally pressure-driven, and their separation performance is primarily based on the water flux (J, L/m^2/h, LMH) and salt rejection ($R\%$) at different pressures. Their calculations are shown in equations (2.1) and (2.2) below [16]:

$$J = V/(A \cdot t) \tag{2.1}$$

$$R = \frac{C_f - C_p}{C_f} \times 100\% \tag{2.2}$$

where $V(L)$ represents the total volume of water permeated, $A(m^2)$ denotes the effective membrane filtration surface area, $t(h)$ represents the duration of filtration, and C_p and C_f(μS/cm) refer to the ionic conductivity of the permeate and the feed solution, respectively.

In the case of permeation-based FO membranes, the primary metrics for assessing their separation performance are the water flux (J_w, LMH) and reverse salt flux (J_s, g/m²/h, GMH). Their calculations are outlined in equations (2.3) and (2.4) below, respectively [17, 18]:

$$J_w = \Delta m/(\rho \cdot A_m \cdot \Delta t) \tag{2.3}$$

$$J_s = (V_t \times C_t - V_0 \times C_0)/(A_m \cdot A_t) \tag{2.4}$$

where Δm(g) represents the mass of water permeated, ρ(g/cm³) denotes the density of water, A_m(m²) is the effective area of the membrane, $\Delta t(h)$ indicates the measured time interval, C_0 and C_t(g/L) represent the concentration of the feed solution at the beginning and at time = t, respectively, and V_0 and $V_t(L)$ are the initial volume and the volume at time = t of the withdrawn solution, respectively.

2.5.3 Desalination Mechanisms

The transport of water through dense membranes during separation in general follows the solution-diffusion theory where the selective permeation is primarily achieved through two mechanisms: size exclusion and electrostatic repulsion [19–25]. The incorporation of nanomaterials into the membrane can lead to the coexistence of various mechanisms, depending on the pore size of the filler material. As their nanochannel or aperture size decreases to the molecular level, the primary mode of mass transfer transitions from being interface-driven to molecular-sieving and eventually solution-diffusion [26–30].

Membranes made from 2D nanomaterials follow a similar separation principle whereby the stacked nanosheets give rise to selective water transport [31–34]. In these materials, size exclusion occurs via either the nanopores on the nanosheet itself (Figure 2.7a and b) or through the interlayer spacing between the stacked nanosheets (Figure 2.7e), which enables the preferential transport of smaller water molecules while rejecting the larger hydrated ions (Figure 2.7d). The electrostatic repulsion mechanism involves the charged nanosheet surfaces hindering the permeation of hydrated ions with like charge [8, 10, 11, 35]. This electrostatic effect is key to achieving selective water transport and efficient desalination by 2D stacked membranes and monolayer nanoporous membranes as the charge residing on pore edges (Figure 2.7c) or on the nanosheet boundaries and surfaces can effectively impede ion passage via electrostatic repulsion (Figure 2.7f) [36–38].

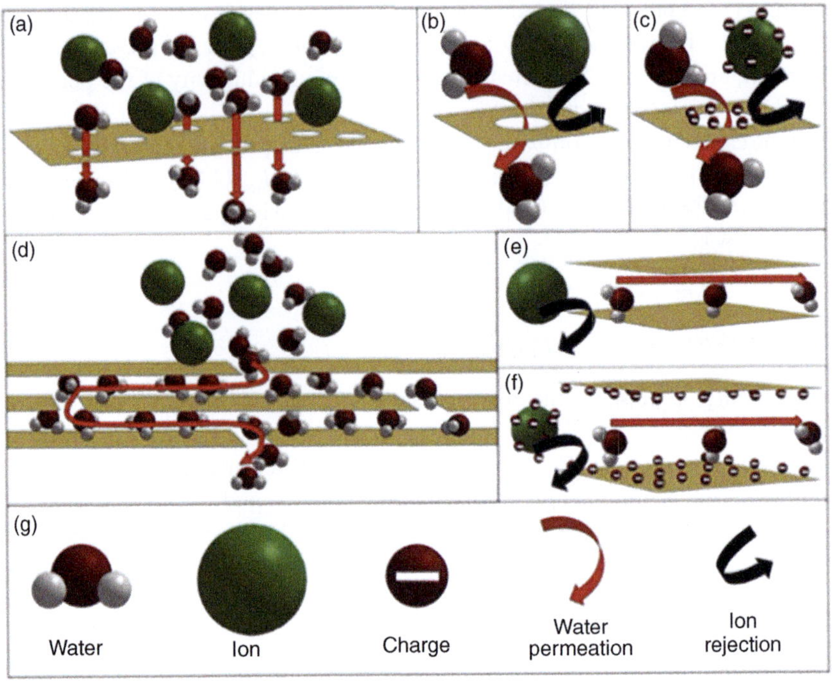

Figure 2.7 The size exclusion and electrostatic repulsion mechanisms in membranes of 2D nanosheets for desalination. Source: Safaei et al. [39]/Permission of Elsevier/CC BY 4.0.

2.6 Concluding Remarks

TFC membranes mark a significant progress in the development of membrane desalination technology. Opportunities for further advancement of the TFC membrane lies in the invention of novel, higher-performing and more robust materials which can be found at the intersection of polymers and functional nanomaterials with attractive transport properties. This integrative approach that draws from both polymers and nanofillers triggers new desalination membrane research directions toward MMMs and TFN membranes. It has subsequently sparked the extensive exploration of various classes of nanomaterials, such as MOFs, COFs, and 2D nanosheets, in the field of membrane desalination.

Abbreviations/Nomenclature

AFM	atomic force microscopy
COFs	covalent organic frameworks

iSCAT	interference scattering microscopy
MMMs	mixed matrix membranes
MOFs	metal-organic frameworks
SEM	scanning electron microscopy
TEM	transmission electron microscopy
TFC	thin-film composite
TFN	thin-film nanocomposite
XRD	x-ray diffraction

Acknowledgment

This work was supported by the Ministry of Science and Technology of the People's Republic of China (No. 2022YFE0117000), the Knowledge Innovation Program of Wuhan-Basic Research (No. 2022020801010237 and No. CKSD2022360/CL), and the Fundamental Research Funds for Central Public Welfare Research Institutes (No. CKSF20241025/TG8).

Questions

1. What are some of the state-of-the-art membrane desalination processes? Please name four of them. Among these processes, which one offers the highest desalination efficiency?

2. What is the most common material for making desalination membranes? What are its disadvantages and what other emerging materials are there as alternatives?

3. What is a thin-film composite (TFC) membrane? What advantages does it offer?

4. What are some characterization tools that can be used to analyze the surface morphology of desalination TFC membranes?

5. What are the important parameters used to measure a membrane's desalination performance? How are they calculated respectively?

References

1 Ali, I., Hasan, S.Z., Garcia, H. et al. (2024). Recent advances in graphene-based nano-membranes for desalination. *Chem. Eng. J.* 483: 149108.

2 Gruber, C.G., Frey, L., Guntermann, R. et al. (2024). Early stages of covalent organic framework formation imaged in operando. *Nature* 630: 872–877.

3 Culp, T.E., Khara, B., Brickey, K.P. et al. (2021). Nanoscale control of internal inhomogeneity enhances water transport in desalination membranes. *Forensic Sci.* 371: 72–75.

4 Lu, X. and Elimelech, M. (2021). Fabrication of desalination membranes by interfacial polymerization: history, current efforts, and future directions. *Chem. Soc. Rev.* 50: 6290–6307.

5 Ding, L., Li, L., Liu, Y. et al. (2020). Effective ion sieving with Ti_3C_2T_x_ MXene membranes for production of drinking water from seawater. *Nat. Sustainability* 3: 296–302.

6 Wang, H., Wang, M., Liang, X. et al. (2021). Organic molecular sieve membranes for chemical separations. *Chem. Soc. Rev.* 50: 5468–5516.

7 Zhao, S., Jiang, C., Fan, J. et al. (2021). Hydrophilicity gradient in covalent organic frameworks for membrane distillation. *Nat. Mater.* 20: 1551–1558.

8 Li, H., Wang, K., Sun, Y. et al. (2018). Recent advances in gas storage and separation using metal–organic frameworks. *Mater. Today* 21 (2): 108–121.

9 Zhang, H., Wang, Y., Wei, Y. et al. (2020). Fabrication of polyamide thin film nanocomposite reverse osmosis membrane incorporated with a novel graphite-based carbon material for desalination. *J. Appl. Polym. Sci.* 137 (35): 49030.

10 Weng, R., Chen, L., Lin, S. et al. (2017). Preparation and characterization of antibacterial cellulose/chitosan nanofiltration membranes. *Polymers* 9 (4).

11 Lim, Y.J., Goh, K., Kurihara, M., and Wang, R. (2021). Seawater desalination by reverse osmosis: current development and future challenges in membrane fabrication – a review. *J. Membr. Sci.* 629: 119292.

12 Yang, Z., Guo, H., and Tang, C.Y. (2019). The upper bound of thin-film composite (TFC) polyamide membranes for desalination. *J. Membr. Sci.* 590: 117297.

13 Kumar, M., Khan, M.A., and Arafat, H.A. (2020). Recent developments in the rational fabrication of thin film nanocomposite membranes for water purification and desalination. *ACS Omega* 5: 3792–3800.

14 Zhao, D.L., Japip, S., Zhang, Y. et al. (2020). Emerging thin-film nanocomposite (TFN) membranes for reverse osmosis: a review. *Water Res.* 173: 115557.

15 Casanova, S., Liu, T.Y., Chew, Y.M.J. et al. (2020). High flux thin-film nanocomposites with embedded boron nitride nanotubes for nanofiltration. *J. Membr. Sci.* 597: 117749.

16 Shirazi, M.M.A., Kargari, A., Ramakrishna, S. et al. (2017). Electrospun membranes for desalination and water/wastewater treatment: a comprehensive review. *J. Membr. Sci. Res.* 3: 209–227.

17 Lu, T., Liang, H., Cao, W. et al. (2022). Blow-spun nanofibrous composite self-cleaning membrane for enhanced purification of oily wastewater. *J. Colloid Interface Sci.* 608: 2860–2869.

18 Lu, T., Cao, W., Liang, H. et al. (2022). Blow-Spun nanofibrous membrane for simultaneous treatment of emulsified oil/ water mixtures, dyes, and bacteria. *Langmuir* 38: 15729–15739.

19 Wang, D., Yuan, H., Chen, Y. et al. (2020). A cellulose-based nanofiltration membrane with a stable three-layer structure for the treatment of drinking water. *Cellulose* 27 (14): 8237–8253.

20 Hu, W., Chen, Y., Dong, X. et al. (2021). Positively charged membranes constructed via complexation for chromium removal through micellar-enhanced forward osmosis. *Chem. Eng. J.* 420: 129837.

21 Yuan, H., Hao, R., Sun, H. et al. (2022). Engineered Janus cellulose membrane with the asymmetric-pore structure for the superhigh-water flux desalination. *Carbohydr. Polym.* 291: 119601.

22 Li, Y., Zhao, W., Weyland, M. et al. (2019). Thermally reduced nanoporous graphene oxide membrane for desalination. *Environ. Sci. Technol.* 53: 8314–8323.

23 Cohen-Tanugi, D. and Grossman, J.C. (2012). Water desalination across nanoporous graphene. *Nano Lett.* 12: 3602–3608.

24 Cohen-Tanugi, D., Lin, L.C., and Grossman, J.C. (2016). Multilayer nanoporous graphene membranes for water desalination. *Nano Lett.* 16: 1027–1033.

25 Mohammed, K. (2023). A review on the polyamide thin film composite (TFC) membrane used for desalination: Improvement methods, current alternatives, and challenges. *Chem. Eng. Res. Des.* 191: 472–492.

26 Mona, M.N., Abeer, A.M., Mahmoud, M.E., and Ahmed, A.E. (2019). Desalination using modified configuration of supported liquid membrane with enhancement of mass transfer of NaCl. *Water Sci. Technol.* 79 (1): 175–187.

27 Pham, M.X., Le, T.M., Tran, T.T. et al. (2021). Fabrication and characterization of polyamide thin-film composite membrane via interfacial polycondensation for pervaporation separation of salt and arsenic from water. *RSC Adv.* 11: 39657–39665.

28 Gu, J.E., Lee, S., Stafford, C.M. et al. (2013). Molecular layer-by-layer assembled thin-film composite membranes for water desalination. *Adv. Mater.* 25 (34): 4778–4782.

29 Goh, P.S., Ismail, A.F., and Hilal, N. (2016). Nano-enabled membranes technology: sustainable and revolutionary solutions for membrane desalination? *Desalination* 380: 100–104.

30 Kim, K.C., Lin, X., and Li, C. (2022). Structural design of the electrospun nanofibrous membrane for membrane distillation application: a review. *Environ. Sci. Pollut. Res.* 29: 82632–82659.

31 Li, X.Q., Zhu, B., and Zhu, J. (2019). Graphene oxide based materials for desalination. *Carbon* 146: 320–328.

32 Wang, M., Zhang, P., Liang, X. et al. (2022). Ultrafast seawater desalination with covalent organic framework membranes. *Nat. Sustainability* 5: 518–526.

33 Ali, A.S.M., Soliman, M.M., Kandil, S.H. et al. (2022). Tailoring nanocomposite membranes of cellulose acetate/silica nanoparticles for desalination. *J. Materiomics* 8 (6): 1122–1130.

34 Aani, S.A., Haroutounian, A., Wright, C.J., and Hilal, N. (2018). Thin film nanocomposite (TFN) membranes modified with polydopamine coated metals/carbon-nanostructures for desalination applications. *Desalination* 427: 60–74.

35 Raza, A., Hassan, J.Z., Mahmood, A. et al. (2022). Recent advances in membrane-enabled water desalination by 2D frameworks: graphene and beyond. *Desalination* 531: 115684.

36 Goh, K., Karahan, H.E., Wei, L. et al. (2016). Carbon nanomaterials for advancing separation membranes: a strategic perspective. *Carbon* 109: 694–710.

37 Li, B., Japip, S., and Chung, T.S. (2020). Molecularly tunable thin-film nanocomposite membranes with enhanced molecular sieving for organic solvent forward osmosis. *Nat. Commun.* 11 (1): 1198.

38 Zhang, M., Guan, K., Ji, Y. et al. (2019). Controllable ion transport by surface-charged graphene oxide membrane. *Nat. Commun.* 10 (1): 1253.

39 Safaei, J., Xiong, P., and Wang, G. (2020). Progress and prospects of two-dimensional materials for membrane-based water desalination. *Mater. Today Adv. Ther.* 8.

3

Exergy Analysis of Membrane Desalination Systems

Noha M. Sayed[1], Mohamed Rabie[2], Asmaa Elrasheedy[3], Raya Al-Dadah[4], and Marwa Elkady[5]

[1] Chemical Engineering Department, Minia University, Minia, Egypt
[2] Mechanical Power Engineering Department, Mansoura University, Mansoura, Egypt
[3] Chemical Engineering Department, Faculty of Engineering, Port Said University, Port Said, Egypt
[4] Department of Mechanical Engineering, University of Birmingham, Birmingham, UK
[5] Chemical and Petrochemical Engineering Department, Egypt-Japan University of Science and Technology, New Borg El-Arab, Egypt

3.1 Introduction

Desalination has proved to be the best solution for the freshwater scarcity challenge by producing freshwater from saline water. Desalination technologies are mainly divided into thermal and membrane processes [1]. Thermal desalination processes such as multi-stage flash (MSF), multi-effect distillation (MED), and thermal vapor compression (TVC) require excessive amounts of thermal and electrical energy for heating, which widens the gap to the thermodynamic limit (which is the minimum amount of energy required to desalinate 1 m^3 of seawater with a specific recovery ratio). Accordingly, several countries are now shifting to desalination technologies that are membrane-based like reverse osmosis (RO) as they have potentially lower specific energy consumption (SEC) [2]. Other developing membrane-based desalination techniques also include forward osmosis (FO), electrodialysis (ED), and pervaporation (PV). Nevertheless, membrane distillation (MD) is considered a hybrid desalination technique that combines the merits of thermal and membrane desalination processes. Research interest has risen in decreasing energy consumption by relying on FO, MD, and hybrid processes, which provide low-energy demand technologies.

The desalination process is an irreversible energy-intensive process that operates far from the least work for separation. To minimize energy consumption, it is crucial to understand the minimum separation energy of the reversible desalination processes. Thermodynamics bound the least work required for the reversible

Water Treatment and Desalination: Membrane Separation using Renewable Energy Resources, First Edition. Edited by Mohamed Bassyouni.

desalination process of seawater with a salinity of 35 g/L to 1.06 kWh/m^3 with a 50% recovery ratio. RO operates near the thermodynamic limit, by consuming about 2.5 to 4 kWh/m^3. However, other membrane-based processes consume loads of energy that are far from the thermodynamic limit [2]. Subsequently, thermodynamic analysis, such as exergy analysis, is inevitable for the irreversibility minimization of membrane-based desalination processes.

This chapter introduces the exergy concept and the difference between conventional and advanced exergy analyses. It also reviews previous exergy (conventional and advanced) analysis studies for stand-alone and integrated desalination processes.

3.1.1 Exergy Concept

The first law of thermodynamics (FLT) states that energy can be transferred from one form to another but can neither be created nor destroyed. It quantifies energy transferred in the system regardless of its quality using energy analysis. This balance does not give any details regarding the direction of energy flow and energy destruction. Accordingly, the limitations of FLT led to the second law of thermodynamics (SLT) that identifies energy flow direction from the higher energy content to the lower one, which gives a clear explanation of the energy quality or exergy [3, 4]. Exergy is a Greek word consisting of ex, which means from, and ergon, which means work. So, exergy is the maximum obtainable work for the systems that produce work and/or the minimum work for systems that consume work when the system reaches the dead state with the environment [5].

There are two common types of exergies: chemical and thermomechanical exergies. Chemical exergy is defined as the maximum achievable (theoretical) work from the system when the concentration (W) of each component in the system varies from its initial concentration (W_0) at environmental temperature and pressure. The thermomechanical (physical) exergy, on the other hand, can be defined as the maximum work that can be obtained from the system when the temperature (T) and pressure (P) of the system vary from the temperature (T_0) and pressure (P_0) of the dead state [6].

3.1.2 Exergy Analysis

Thermodynamic analysis is a known procedure for optimizing thermal systems' performance using energy or exergy methods. Energy analysis is based on the FLT, which characterizes desalination systems in terms of energetic efficiency and compares the output energy relative to the input energy. However, exergy analysis evaluates the quality of the usable energy by comparing system and environmental conditions [7].

A conventional exergy analysis provides useful details about the system's thermodynamic performance and effectively determines each of the components' exergy destruction, i.e., exergy loss or energy quality loss. In other words, each component is treated as an independent entity, but in fact, integrated system components are interlinked.

However, this conventional analysis does not distinguish between unavoidable and avoidable exergy destruction.

As an alternative, such comprehensive data are provided by the so-called advanced exergy analysis. Each system component's exergy degradation is divided into endogenous and exogenous, as well as avoidable and unavoidable components, using advanced exergy analysis. This analysis can be used to identify system components that affect the overall performance of the process. While attempting to distinguish between the avoidable components of endogenous and external exergy destruction—a distinction that a conventional exergy analysis says nothing about—this knowledge is particularly helpful [8]. Figure 3.1 shows the difference between conventional and advanced exergy analyses.

Exergy measures energy quality and degradation due to irreversibility (losses) in real systems. The SLT states that some energy entering the system is destroyed because of the generation of entropy through the desalination system. Exergy analysis yields valuable insights into usable energy and the potential of process optimization. An example of an exergy analysis of a RO system is provided to help understand the conventional and advanced exergy equations. Exergy is divided into physical, chemical, kinetic, and potential exergies.

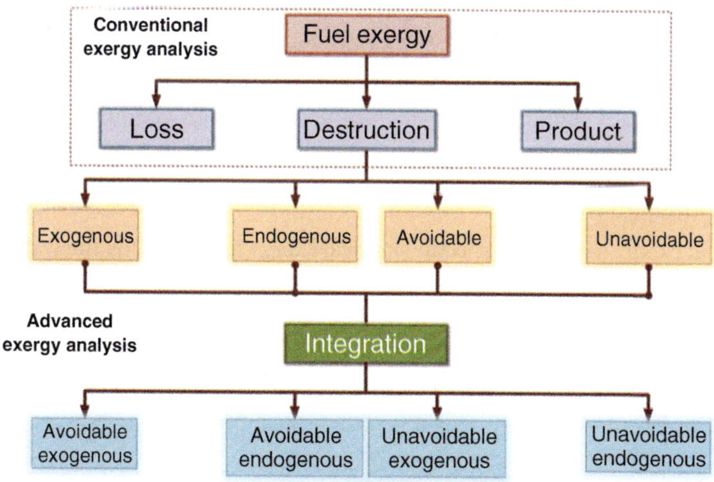

Figure 3.1 Difference between conventional and advanced exergy analyses.
Source: Esmaeilion et al. [9]/Elsevier/CC BY 4.0.

Assumptions:

- The kinetic and potential exergy flows are negligible.
- Reference state conditions are $T_0 = 298.15$ K, $P_0 = 1$ atm, and $S_{wo} = 35$ g/L, where T_0 is the environment temperature, P_0 is the environment pressure, and S_{wo} is the seawater salinity reference.
- e_1^{PH}, e_2^{PH}, and e_3^{PH} are physical exergies of seawater, permeate, and brine streams, respectively.
- e_1^{CH}, e_2^{CH}, and e_3^{CH} are chemical exergies of seawater, permeate, and brine streams, respectively.
- Steady-state conditions are considered.

3.1.2.1 Conventional Exergy Analysis

The exergy balance can be written as [11]:

$$\Sigma Ex_{in} - \Sigma\, Ex_{out} - Ex_s - Ex_d = 0$$

where Ex_{in}, Ex_{out}, Ex_s, and Ex_d are inlet flow exergy, outlet flow exergy, stored exergy, and exergy destruction, respectively.

At steady-state conditions:

$$Ex_d = \Sigma\, Ex_{in} - \Sigma\, Ex_{out}$$

In this case, flow exergy (ex_f) equals the summation of physical and chemical exergies:

$$ex_f = ex_{PH} + ex_{CH} = (h - h_0) - T_0(s - s_0) + \Sigma w_i(\mu_i - \mu_{i,0})$$

where ex_{PH}, ex_{CH}, h, h_0, s, s_0, w_i, μ_i, and $\mu_{i,0}$ are specific physical exergy, specific chemical exergy, specific enthalpy at system conditions, specific enthalpy at dead-state conditions, specific entropy at system conditions, specific entropy at dead-state conditions, species mass fraction, the specific chemical potential at system conditions, and the specific chemical potential at dead-state conditions, respectively.

The physical exergy rate equals specific physical exergy multiplied by the water mass flowrate, as follows:

$$Ex_{CH} = M_w * ex_{ph}$$

For the RO system, as shown in Figure 3.2, exergetic efficiency can be determined as follows [10]:

$$\eta_{ex} = \frac{\Sigma\, Ex_{useful}}{\Sigma Ex_{in}} = \frac{\left(Ex_{permeate}^{CH} + Ex_{brine}^{CH}\right) - Ex_{feed}^{PH}}{Ex_{feed}^{PH} - \left(Ex_{permeate}^{PH} + Ex_{brine}^{PH}\right)}$$

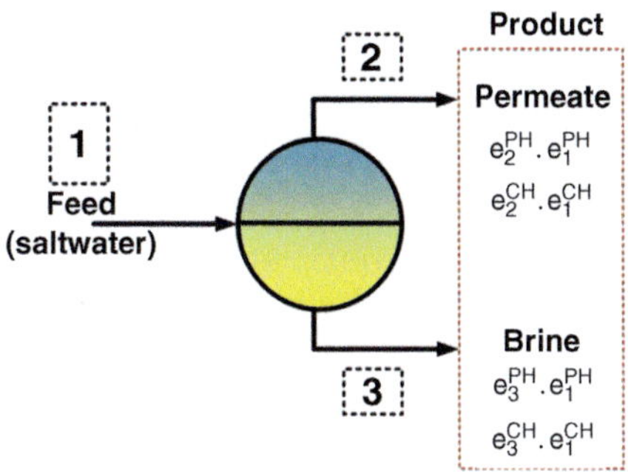

Figure 3.2 General schematic diagram of the RO system. Source: Blanco-Marigorta et al. [10]/with permission of Elsevier.

3.1.2.2 Advanced Exergy Analysis

Advanced exergy analysis provides additional details compared to conventional exergy analysis. In advanced exergy analysis, each system component's rate of destruction is split into avoidable, unavoidable, endogenous, and exogenous categories. This division of the destruction rate into avoidable and unavoidable components describes the discrepancy between theoretical operating conditions and the real ones. The potential for thermodynamic efficiency improvement can be estimated from the previous data. On the other hand, the segmentation of the exergy destruction into endogenous and exogenous categories specifies whether the rate of exergy destruction is due to the component of interest or other connected system components. This information provides insights into whether the engineering redesign should be carried out for the component before the surrounding components or vice versa. The last categorization into endogenous and exogenous exergy destruction can be further categorized into endogenous avoidable, endogenous unavoidable, exogenous avoidable, and exogenous unavoidable subcategories, hence providing information on which subcategory affects the improvement of the overall thermodynamic performance of the system the most [9] (Figure 3.3). Table 3.1 illustrates the exergy analysis terminologies.

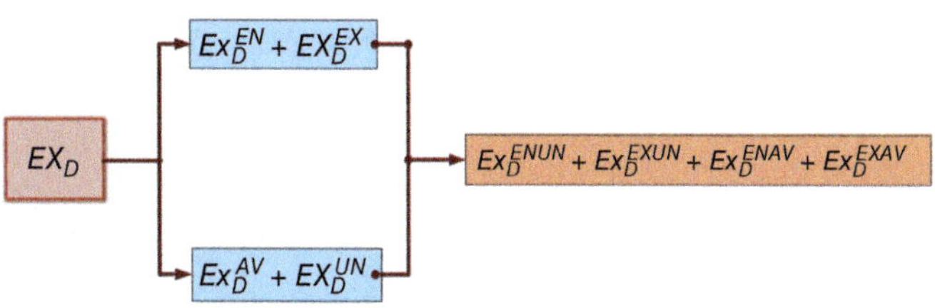

Figure 3.3 Exergy destruction splitting according to the advanced exergy analysis. Source: Xia et al. [12]/with permission of Elsevier.

Table 3.1 Exergy analysis terminologies.

Terminologies	Definition	Symbols/equations	References
Exergy	The maximum/minimum theoretical work produced/consumed in the case of the system and the surroundings to reach complete equilibrium (physical, chemical, and mechanical). It is divided into physical, chemical, kinetic, and potential exergies.	$Ex_t = Ex_{PH} + Ex_{CH} + Ex_{po} + Ex_{ke}$	[13]
Deadstate	It is the environmental reference conditions.	$P_0 = 1$ atm, $S_{wo} = 35$ g/L, $T_0 = 25°C$	[14]
Exergy destruction	It is considered as the exergy loss due to the entropy generation in the system.	$Ex_{d,k} = Ex_{d,k}^{AV} + Ex_{d,k}^{UV}$	[8]
Fuel exergy	It is the exergy that is consumed in the system to produce a desired product.	$Ex_f = Ex_{in}$	[10]
Product exergy	It is the useful output exergy of the required product.	$Ex_p = Ex_{out}$	[10]
Second law efficiency	It can be described as the least work needed for the desalination process relative to the actual work used.	$\eta_{II} = \frac{W_{min}}{W_{act}}$	[15]
Minimum work of separation	It is the least reversible work that is required for the separation process of a desalination system when the destruction of the exergy equals zero.	$W_{min} = \Sigma Ex_{out} - \Sigma Ex_{in}$	[15]
Unavoidable destruction	It is defined as the least amount of destruction of the exergy that cannot be reduced.	$Ex_{d,k}^{UV} = Ex_p \left(\frac{Ex_{d,k}}{Ex_p} \right)^{UN}$	[16]
Avoidable destruction	It is the part of the destruction of exergy that reveals the largest potential for system performance enhancement.	$Ex_{d,k}^{AV} = Ex_{d,k} - Ex_{d,k}^{UV}$	[16]

Endogenous destruction	It is the destruction of exergy that occurs in the k-th component when working under actual conditions while other components work under ideal conditions.	$Ex_{d,k}^{EN}$	[8]
Exogenous destruction	This type of destruction of the k-th component occurs due to the destruction of exergy in other components when working under actual conditions while the component operates under ideal conditions.	$Ex_{d,k}^{Ex}$	[8]
Endogenous unavoidable exergy destruction	It is the destruction of exergy which cannot be minimized by technical enhancement.	$Ex_{d,k}^{EN,UV} = Ex_{p,k}^{EN}\left(\frac{Ex_{d,k}}{Ex_{p}}\right)^{UN}$	[8]
Endogenous-avoidable exergy destruction	It is the destruction of exergy that occurs due to the irreversibilities in the k-th component. This type can be decreased by k-th component performance improvement.	$Ex_{d,k}^{EN,AV} = Ex_{d,k}^{EN} - Ex_{d,k}^{EN,UV}$	[16]
Exogenous-avoidable exergy destruction	It is the destruction of exergy that takes place due to irreversibilities in the system and other components excluding the k-th component, which can be avoided by system configuration change.	$Ex_{d,k}^{Ex,AV} = Ex_{d,k}^{EX} - Ex_{d,k}^{EX,UV}$	[16]
Exogenous unavoidable exergy destruction	It is the exergy destruction that occurs due to irreversibilities in the system and other components excluding the k-th component, which cannot be avoided.	$Ex_{d,k}^{Ex,UV} = Ex_{d,k}^{UN} - Ex_{d,k}^{EN,UV}$	[16]

3.2 Exergy Analysis of Membrane Desalination Systems

Exergy analysis of membrane-based desalination units can be categorized as follows:

- RO, FO, ED, and MD.
- Hybridization of these membrane-based technologies with each other or with thermal-based technologies, such as humidification-dehumidification (HDH) and MSF systems.

Various previous investigations, including energy, conventional and advanced exergy, and exergy/economic studies, provide a substantial understanding of actual energy consumption, minimum exergy destruction, and cost related to the exergy of each system, which contributes to system optimization.

3.2.1 Reverse Osmosis

RO is the dominant membrane desalination process in various countries owing to the less energy used compared to its thermal desalination counterpart [1]. The RO system consists of a water feed well, a feed tank, two cartridge filters for pretreatment, a high-pressure pump, two stages of RO units, and a permeate tank, as shown in Figure 3.4. One of the primary shortcomings of the RO process is the intensive energy consumption, as the separation process

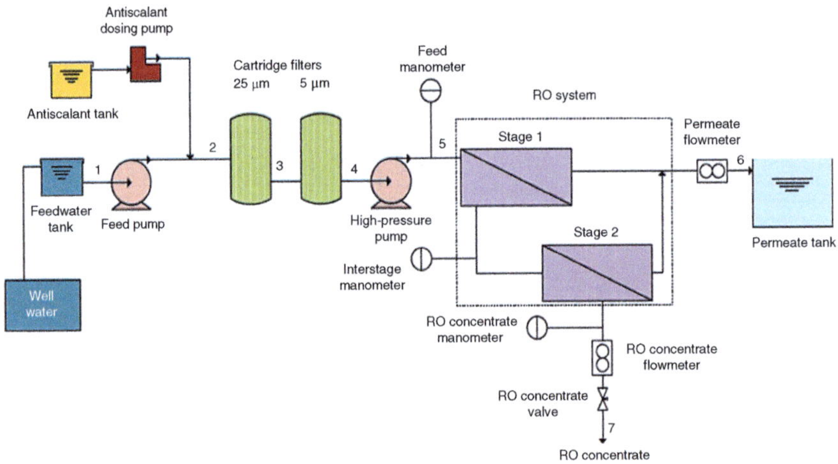

Figure 3.4 Schematic diagram of RO for brackish water desalination. Source: Fellaou et al. [14]/with permission of Elsevier.

occurs using a high-pressure pump that forces water to permeate through a semi-permeable membrane. Accordingly, various studies have been performed to optimize RO effectiveness using exergy analysis. The performance of a full-scale brackish water plant was investigated using conventional and advanced exergy analyses [14]. The conventional exergy study found that the RO system (membrane modules), the high-pressure pump, and the feed pump have the highest irreversibilities, accounting for approximately 64.28%, 40.84%, and 38.48%, respectively.

To enhance RO performance and identify the primary factors that cause irreversibilities, an exergetic analysis was performed for an RO plant with a production capacity of 7,250 m^3/d and 500 ppm potable water quality [17]. The system was designed to extract saline feedwater by the use of low-pressure pumps, which then divided it into process water and blended water. To avoid RO membrane deterioration, saline process water was first passed through a static mixer that contained HCl and antiscalant. Afterward, it was fed to cartridge filters to separate and get rid of suspended solids and then fed to the RO membrane unit using a high-pressure pump. The desalination unit containing the RO membrane separated saline water into permeate (freshwater) and brine (rejected water), which were released by means of two throttling valves. This unit exergy analysis revealed that the high-pressure pump supplied 80.8% of the exergy entering the system, with the low-pressure pump supplying the remaining exergy. Exergy destructions that occurred in the membrane, static mixer, filters, and throttling valves were caused by the pressure drops across these units. The RO membrane was found to contain the highest exergy destruction, and this result was found to comply with the study by Romero-Ternero et al. [18], which reported that 80% of exergy destruction took place in RO units, high-pressure pumps, and valves. On the other hand, Aljundi [19] reported that the highest exergy destruction took place in throttling valves (58.8%). The second law efficiency of the system was found to be about 4.3% and 4.1% due to irreversibilities occurring in the system, which provided an opportunity to propose an alternative design or modifications. The proposed design was to replace throttling valves with pressure exchangers that convey pressure energy from a high-pressure stream to a low-pressure stream; this modification decreases the losses in exergy by 94% and enhances second law efficiency by increasing it to 4.9% instead of 4.3%. Eshoul et al. [13] utilized a pressure exchanger instead of the throttling valves, which led to an enhancement in the exergy efficiency to 77%, and the exergy destruction was reduced to 53%. Sadri et al. [20] simulated the RO desalination unit via exergy analysis to investigate its performance. The study comprised the effect of changing different operating parameters, such as the temperature and pressure of feed saline water. The results showed that increasing the pressure enhanced the exergetic efficiency, while the exergy destruction was reduced. The highest exergy destruction value was found to be in the brine stream. They also reported that increasing the membrane area enhances both the exergetic efficiency and freshwater production.

3.2.2 Forward Osmosis

FO utilizes a highly selective semi-permeable membrane to recover water as a solvent from two streams with different concentrations and osmotic pressure. In contrast to the famous RO technique, in FO, the flow of water is from a lower-concentration feed solution to a higher-concentration saline solution. This technology has shown promise in treating wastewater and reusing water for industrial applications. A domestic desalination system was examined thermodynamically using energy, exergy, and economic analyses to determine minimum or zero-liquid discharge and minimum SEC [21]. The exergy analysis results showed higher destruction in the membrane from the draw solution side owing to the higher difference in concentration. On the other hand, the highest exergy destruction was found in the pumping system.

3.2.3 Membrane Distillation

MD is a versatile hybrid membrane–thermal separation process that was developed in 1960 but still attracts recent interest [22–24]. In this process, volatile compounds are separated using the partial water vapor difference between the feed and permeate sides, at which only water vapor is transported through the hydrophobic membrane [25]. MD has various developed configurations as reported in numerous studies [26–29]. Although the process has promising merits over traditional desalination technologies [30–32], the high SEC compared to the thermodynamic limit [2] and low thermal efficiency are the main limitations that hinder its commercial application [33–36]. Accordingly, assessing the thermal performance of such a system based on exergy analysis is inevitable.

Various exergy analysis investigations of MD modules have been conducted to evaluate the thermal effectiveness based on the SLT. Banat and Jwaied [37] assessed the thermal performance of large and compact solar MD systems. The exergetic efficiencies were 0.3% and 0.5% of the compact and the large systems, respectively. The MD technology was identified as the primary source for water desalination with the highest energy destruction, followed by the heat exchange between the permeate and brine discharge, and finally the MD module.

An MD pilot plant powered by geothermal energy was the subject of another study that assessed its thermal efficiency [38]. According to the study, the discharged permeate stream had the maximum exergy degradation. Consequently, series permeate recycling was carried out to reduce energy losses. Exergy analysis

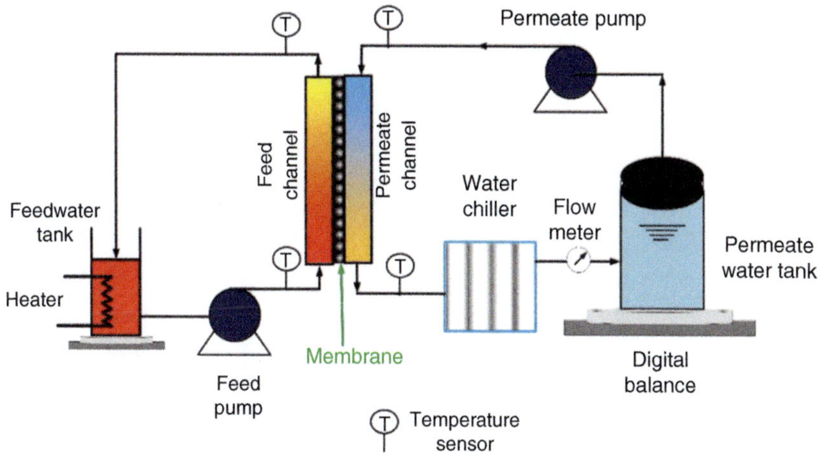

Figure 3.5 Membrane distillation schematic diagram. Source: Choi et al. [52]/MDPI/CC BY 4.0.

has been used in several attempts to identify irreversibilities in vacuum MD [39, 40], air-gap MD [41], direct-contact membrane distillation (DCMD) [42–44], MD-integrated systems [45–49], and multi-effect MD systems [50].

Likewise, a detailed review [51] reported that the most prominent MD performance evaluation criteria and exergy analysis were deduced as a performance indicator for promoting thermal energy usage. A comprehensive review [51] reported that exergy analysis was the most prominent evaluation tool for thermal systems. Additionally, a computational fluid dynamics (CFD)-based exergy study of the DCMD unit was performed to indicate the exergy destructions of each component under various operating conditions. The exergy destruction distribution through the DCMD unit has also been identified [52] (Figure 3.5).

3.2.4 Electrodialysis

ED is considered an important membrane desalination process for groundwater and salt production from brine. Figure 3.6 presents a schematic diagram of the ED system, which consists of an anode, a cathode, a positively charged membrane (anion-exchange membrane) (AEM), a negatively charged membrane (cation-exchange membrane) (CEM), and two compartments for electrode rinse near the cathode and anode. The applied electrical voltage across the anode and cathode allowed the transport of positive ions within the CEM, while negative ions were transported through the AEM. Freshwater is obtained by salt transport from the diluted to the concentrated solutions. Generous et al. [53] calculated the maximum entropy generation rate within the ED desalination system using an

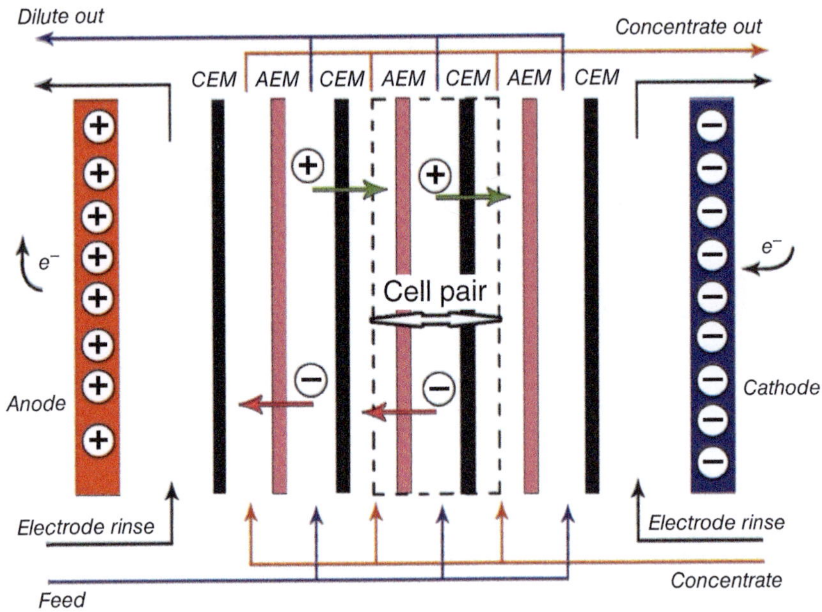

Figure 3.6 Electrodialysis schematic diagram; CEM is cation-exchange membrane, AEM is anion-exchange membrane. Source: Generous et al. [53]/with permission of Elsevier.

exergy-based model. The authors found that the maximum entropy generation contribution was in the membranes by 80%, followed by concentrated and diluted channels by 16% and membrane boundary layers by 4%. They also found that the entropy generation rate could be lowered by increasing the membrane thickness up to 1 mm. Exergy efficiency and entropy generation increased with an increase in the electric field. On the other hand, optimal exergy and current efficiency could be obtained at 0.4–0.65 mm channel width. Chehayeb and Lienhard [54] investigated the use of the Maxwell–Stefan model for ion transport phenomenon and entropy generation rate analysis owing to the viscous effect, flux of ions, and pressure losses. The impact of salinity change on entropy generation was studied. The results showed that the highest entropy generation was in the channels at low salinity; however, at high salinity, entropy generation was high in the membranes. Moreover, lowering the membrane thickness had a positive effect on entropy generation and energy consumption. In another study [55] the authors used equipartition theory for entropy generation optimization of batch ED units. Electrical energy consumption was compared at different operating conditions of constant voltage, constant entropy rate, and constant current. The results revealed that the ED process performed well at constant entropy rate generation rather than the operation with constant current. Energy consumption and cost can be lowered by minimizing the fixed expenses of the ED system.

3.2.5 Hybrid Systems

Membrane desalination processes are integrated with other techniques to decrease the vast electrical energy consumption and enhance the exergetic performance. The RO system consumes energy greater than the minimum work for 1 m^3 freshwater production. Subsequently, an inevitable energy reduction is still required for sustainable freshwater production. A novel RO system was designed by integrating reverse osmosis with pressure-retarded osmosis (RO-PRO) powered by solar energy for 5,000 m^3/day capacity [56]. Solar thermal energy and salinity-gradient energy are combined to reduce electricity consumption for desalination. The integrated system (Figure 3.7) consists of PRO to dilute feed seawater and generate mechanical energy by mixing seawater with treated wastewater. Exergy analysis revealed that the energy consumption of the integrated system decreased to 0.39 kWh/m^3, which is less than that of the RO system with 50% recovery. Additionally, the exergy analysis revealed that hydro turbine, throttling valve, pump, and RO module improvements can enhance the integrated system performance.

Also, the performance of a novel forward osmosis and multi-effect distillation (FO-MED) integrated system was investigated from the thermodynamic and economic analysis point of view to decrease MED's environmental impact [57] as shown in Figure (3.8). The FO-MED system consisted of

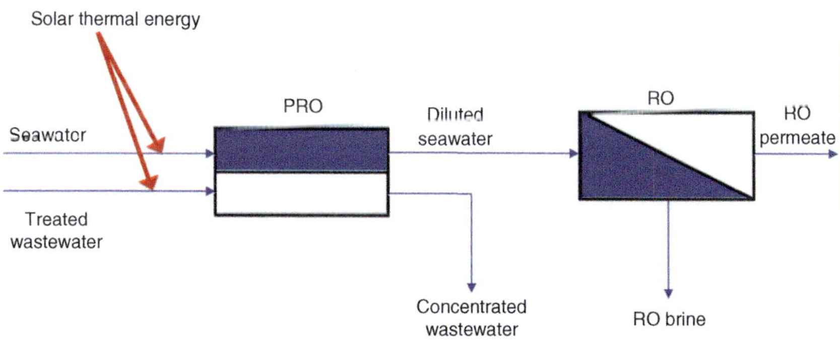

Figure 3.7 Schematic diagram of pressure-retarded osmosis (PRO) and reverse osmosis (RO) hybrid system using solar thermal energy. Source: Tong et al. [56]/with permission of Elsevier.

evaporation–condensation effects through vessels. Feed seawater was evaporated by the indirect contact between the sprayed feedwater over the evaporators and the condensed vapor inside the tubes. The first effect was heated using saturated external steam; however, the energy source of the following stages was the vapor produced in the previous effect. The feed preheating process was performed using part of the vapor produced in each vessel. The distillate was collected from each evaporator and preheater through flash boxes for additional vapor production. The vapor produced in the last stage was condensed using cold intake seawater. The rejected brine was discharged again to the sea, which affected the environment badly. Accordingly, the rejected brine from MED was utilized as a draw solution for the FO process, and the draw solution outlet was the MED feed. This integration process reduced the scaling risk by mitigation of the seawater divalent ion concentration.

The results indicated that employing FO pretreatment of 100 m^3/day and MED plant with 8 effects at 65°C can enable an increase in the heating steam temperature to 100°C and expand the number of effects to 16 while mitigating the scaling tendency.

This integration leads to a 44% reduction in specific thermal energy consumption and a 21% decline in specific heat transfer area. The saline water volume that was discharged into the sea decreased by 40%, and the water footprint was decreased by 36%. Exergy analysis indicated that the MED component has the highest exergy destruction.The cost of water produced by the FO-MED process is higher ($7.6/$m^3$) compared to the stand-alone MED process ($4.6/$m^3$). However, using economical and highly efficient membranes could reduce the cost of the product by 33%, bringing it down to $5.1/$m^3$ (Figure 3.8).

Comprehensive theoretical modeling and exergy analysis of an integrated unit of sweeping gas membrane distillation-bubble column dehumidifier (SGMD-BCD) (Figure 3.9) were performed to decrease energy consumption and cost [58]. Mass and heat equations were solved analytically for SGMD-BCD

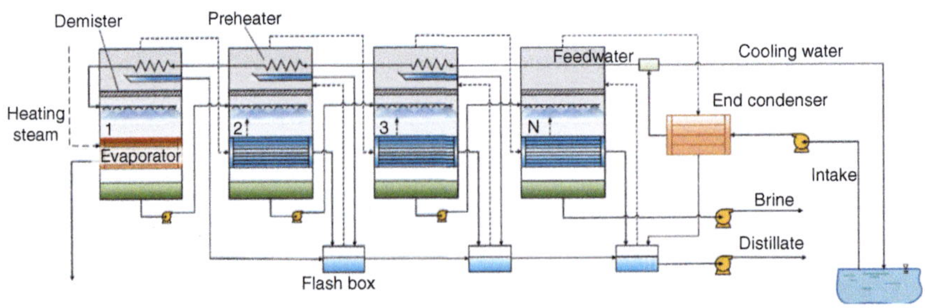

Figure 3.8 Schematic representation of forward osmosis/multi-effect distillation integrated system. Source: Ortega-Delgado et al. [57]/with permission of Elsevier.

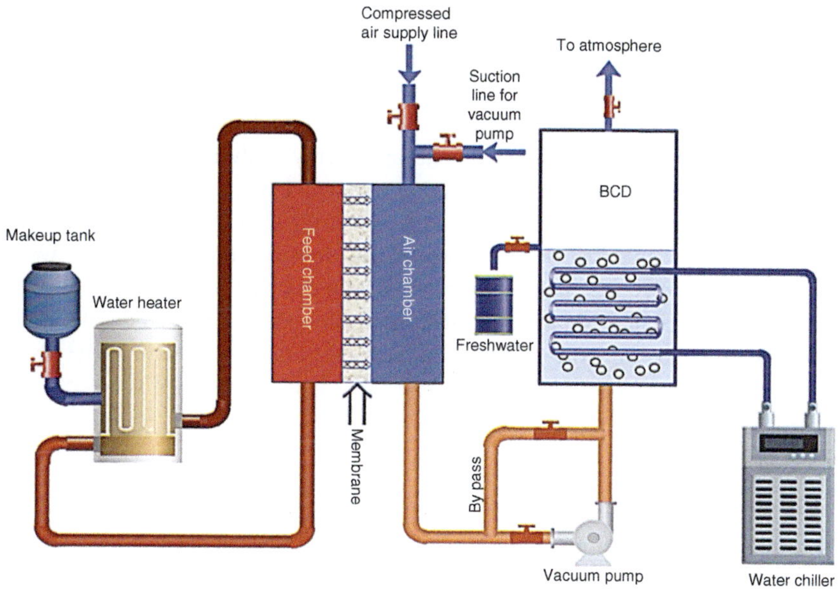

Figure 3.9 Schematic representation of vacuum-assisted sweeping air membrane distillation/bubble column dehumidifier integrated system. Source: Kotb et al. [58]/with permission of Elsevier.

system performance prediction, and the model was validated with the available experimental data in the literature [59]. The system consisted of an SGMD unit where the vapor pressure difference between the feed and sweeping gas (air) sides was the main driving force for the transport process. The discharged humid air from the SGMD unit was cooled in a BCD unit using a water chiller. The results showed that increasing the feed flow rate and temperature promoted the productivity with a lower SEC. The exergy results revealed that the exergetic efficiency of the integrated system increased by decreasing the feed flow rate and air pressure while increasing the air flow rate. The exergy destruction of the SGMD system was higher than that of the BCD unit.

Conventional and advanced exergy analyses were performed for a brackish water reverse osmosis and forward osmosis (BWRO-FO) desalination integrated system for zero-liquid discharge and decreasing the SEC [60]. The results of the conventional exergy analysis revealed that 87.06% of the total exergy destruction occurred in pumps and 6.51% in the RO membrane.

The advanced exergy analysis revealed that 1.19 kW was the unavoidable exergy destruction in the pumps. In addition, the avoidable exergy destruction could be decreased by the appropriate selection of the pumps. The BWRO-FO integrated system achieved greater than 80% water recovery and 0.43 kWh/m^3 SEC.

3.3 Conclusion

In this chapter, an overview of the exergy concept and terminologies was performed for a better understanding of the second law or exergy analysis. The equations and constraints of conventional exergy analysis were discussed. In addition, the splitting of exergy destruction into avoidable, unavoidable, endogenous, and exogenous was also stated. Previous studies on the conventional and advanced exergy analyses of membrane-based desalination technologies were reviewed to identify the main sources of exergy destruction and ways to promote exergy efficiency. Finally, the integration of membrane-based desalination (RO, FO, and ED) and thermal desalination processes (MED and HDH) was reviewed.

Nomenclature

Symbol	Description
$Ex_{d,k}^{AV}$	Avoidable exergy destruction of the k-th component [kW]
$Ex_{d,k}^{EN}$	Endogenous exergy destruction of the k-th component [kW]
$Ex_{d,k}^{Ex}$	Exogenous exergy destruction of the k-th component [kW]
$Ex_{d,k}^{UV}$	Unavoidable exergy destruction of the k-th component [kW]
$Ex_{d,k}$	Component exergy destruction [kW]
Ex_f	Fuel exergy rate [kW]
Ex_{in}	Inlet exergy rate [kW]
Ex_{ke}	Kinetic exergy rate [kW]
Ex_{out}	Outlet exergy rate [kW]
Ex_p	Product exergy rate [kW]
Ex_{po}	Potential exergy rate [kW]
ex_{CH}	Chemical exergy [kJ/kg]
ex_{PH}	Physical exergy [kJ/kg]
ex_f	Flow exergy [kJ/kg]
Ex_t	Total exergy rate [kW]
M_w	Seawater flowrate [kg/h]
P	Pressure [Pa]
S_w	Seawater salinity [g/L]
T	Absolute temperature [K]
W_i	Species mass fraction
W_{act}	Actual work [kW]
W_{min}	Minimum work for separation [kW]
η_{II}	Second law efficiency [%]

Questions

1. Differentiate between the first and the second laws of thermodynamics.
2. What is the meaning of the word "exergy"?
3. Mention and define the two main exergy types.
4. What is meant by thermodynamic analysis?
5. Differentiate between energy and exergy analyses.
6. Differentiate between conventional and advanced exergy analyses.
7. What is meant by endogenous, exogenous, avoidable, and unavoidable exergy destructions?
8. What are the definition and equation of the total exergy?
9. What are the conditions of the dead state?
10. Define exergy destruction.
11. Define fuel exergy.
12. Define product exergy.
13. What is meant by second law efficiency? And write the equation.
14. Define the minimum work of the desalination process.
15. Define the endogenous avoidable exergy destruction.
16. Define the endogenous unavoidable exergy destruction.
17. Define the exogenous avoidable exergy destruction.
18. Define the exogenous unavoidable exergy destruction.
19. State one exergy point that the reverse osmosis (RO) process for water desalination suffers from, and suggest how this exergy may be overcome.

20 What is the main exergy destruction in the forward osmosis (FO) process for water desalination?

21 What are the maximum entropy generation locations of the electrodialysis (ED) process for water desalination?

References

1. Shahrim, N.A., Abounahia, N.M., El-Sayed, A.M.A. et al. (2023). An overview on the progress in produced water desalination by membrane-based technology. *J. Water Process Eng.* 51: 103479. https://doi.org/10.1016/J.JWPE.2022.103479.
2. Nassrullah, H., Anis, S.F., Hashaikeh, R., and Hilal, N. (2020). Energy for desalination: a state-of-the-art review. *Desalination* 491: https://doi.org/10.1016/j.desal.2020.114569.
3. Dincer, I. and Cengel, Y.A. (2001). Energy, entropy and exergy concepts and their roles in thermal engineering. *Entropy* 3 (3): 116–149. www.mdpi.org/entropy/http://www.geocities.com/ibrahimdincer.
4. Abdelmoez, W., Ashour, E.A., and Sayed, N.M. (2019). Water desalination by humidification–dehumidification technology with performance evaluation using exergy analysis. *Desalin. Water Treat.* 148: 1–19. https://doi.org/10.5004/dwt.2019.23668.
5. Dinçer, I. and Rosen, M.A. (2007). *Exergy, Energy, Environment and Sustainable Development*. Oxford: Elsevier.
6. Sharqawy, M.H., Lienhard V, J.H., and Zubair, S.M. (2011). On exergy calculations of seawater with applications in desalination systems. *Int. J. Therm. Sci.* 50 (2): 187–196. https://doi.org/10.1016/J.IJTHERMALSCI.2010.09.013.
7. Rahimi-Ahar, Z. and Hatamipour, M.S. (2024). Exergy analysis of thermal desalination processes: a review. *Clean Techn. Environ. Policy* 26 (5): 1335–1362. https://doi.org/10.1007/s10098-023-02491-6.
8. Ebrahimi, M., Carriveau, R., Ting, D.S.K., and McGillis, A. (2019). Conventional and advanced exergy analysis of a grid connected underwater compressed air energy storage facility. *Appl. Energy* 242: 1198–1208. https://doi.org/10.1016/j.apenergy.2019.03.135.
9. Esmaeilion, F., Soltani, M., Nathwani, J. et al. (2023). Conventional and advanced exergy and exergoeconomic analysis of a novel multi-fueled cogeneration system with desalination. *Energy Rep.* 9: 5704–5720. https://doi.org/10.1016/j.egyr.2023.05.003.
10. Blanco-Marigorta, A.M., Lozano-Medina, A., and Marcos, J.D. (2017). The exergetic efficiency as a performance evaluation tool in reverse osmosis

desalination plants in operation. *Desalination* 413: 19–28. https://doi.org/10.1016/j.desal.2017.03.007.

11 Abdelmoez, W., Sayed, N.M., and Ashour, E.A. (2021). Energy and exergy analysis of a solar/hybrid humidification–dehumidification desalination system. *Desalin. Water Treat.* 209: 1–15. https://doi.org/10.5004/dwt.2021.26508.

12 Xia, X., Liu, Z., Wang, Z. et al. (2023). Energy, conventional and advanced exergy analysis for the organic Rankine cycle-vapor compression refrigeration combined system driven by low-grade waste heat. *Appl. Therm. Eng.* 220: https://doi.org/10.1016/j.applthermaleng.2022.119650.

13 Eshoul, N.M., Agnew, B., Al-Weshahi, M.A., and Atab, M.S. (2015). Exergy analysis of a two-pass reverse osmosis (RO) desalination unit with and without an energy recovery turbine (ERT) and pressure exchanger (PX). *Energies (Basel)* 8 (7): 6910–6925. https://doi.org/10.3390/en8076910.

14 Fellaou, S., Ruiz-Garcia, A., and Gourich, B. (2021). Enhanced exergy analysis of a full-scale brackish water reverse osmosis desalination plant. *Desalination* 506: https://doi.org/10.1016/j.desal.2021.114999.

15 Sharqawy, M.H., Zubair, S.M., and Lienhard, J.H. (2011). Second law analysis of reverse osmosis desalination plants: an alternative design using pressure retarded osmosis. *Energy* 36 (11): 6617–6626. https://doi.org/10.1016/j.energy.2011.08.056.

16 Wang, Y., Qin, G., Zhang, Y. et al. (2021). Conventional and advanced exergy analyses of an organic Rankine cycle by using the thermodynamic cycle approach. *Energy Sci. Eng.* 9 (12): 2474–2492. https://doi.org/10.1002/ese3.980.

17 Cerci, Y. (2002). Exergy analysis of a reverse osmosis desalination plant in California. *Desalination* 142 (3): 257–266. www.elsevier.com/locate/desal.

18 Romero-Ternero, V., Garcia-Rodriguez, L., and Gómez-Camacho, C. (2005). Exergy analysis of a seawater reverse osmosis plant. *Desalination* 175 (2): 197–207.

19 Aljundi, I.H. (2009). Second-law analysis of a reverse osmosis plant in Jordan. *Desalination* 239 (1–3): 207–215. https://doi.org/10.1016/j.desal.2008.03.019.

20 Sadri, S., Khoshkhoo, R.H., and Ameri, M. (2016). Multi objective optimization of reverse osmosis desalination plant with exergy approach. *J. Mech. Sci. Technol.* 30 (10): 4807–4814. https://doi.org/10.1007/s12206-016-0953-4.

21 Patel, D., Mudgal, A., Patel, V. et al. (2023). Energy, exergy, economic and environment analysis of standalone forward osmosis (FO) system for domestic wastewater treatment. *Desalination* 567: https://doi.org/10.1016/j.desal.2023.116995.

22 Lin, S., Yip, N.Y., and Elimelech, M. (2014). Direct contact membrane distillation with heat recovery: thermodynamic insights from module scale modeling. *J. Membr. Sci.* 453: 498–515. https://doi.org/10.1016/j.memsci.2013.11.016.

23 Shirazi, M.M.A., Kargari, A., Ismail, A.F., and Matsuura, T. (2016). Computational fluid dynamic (CFD) opportunities applied to the membrane distillation process: state-of-the-art and perspectives. *Desalination* 377: 73–90. https://doi.org/10.1016/j.desal.2015.09.010.

24 Jantaporn, W., Ali, A., and Aimar, P. (2017). Specific energy requirement of direct contact membrane distillation. *Chem. Eng. Res. Des.* 128: 15–26. https://doi.org/10.1016/j.cherd.2017.09.031.

25 Sayed, N.M., Noby, H., Thu, K., and El-Shazly, A.H. (2023). Improved modeling of Janus membrane considering the influence of hydrophilic layer characteristics. *J. Non-Equilib. Thermodyn.* https://doi.org/10.1515/jnet-2023-0037.

26 Yang, S., Abdalkareem Jasim, S., Bokov, D. et al. (2022). Membrane distillation technology for molecular separation: a review on the fouling, wetting and transport phenomena. *J. Mol. Liq.* 349: https://doi.org/10.1016/j.molliq.2021.118115.

27 Alessandro, F., Macedonio, F., and Drioli, E. (2023). New materials and phenomena in membrane distillation. *Chemistry (Easton)* 5 (1): 65–84. https://doi.org/10.3390/chemistry5010006.

28 Bin Abid, M., Wahab, R.A., Salam, M.A. et al. (2023). Desalination technologies, membrane distillation, and electrospinning, an overview. *Heliyon* 9 (2): e12810. https://doi.org/10.1016/j.heliyon.2023.e12810.

29 Francis, L., Ahmed, F.E., and Hilal, N. (2022). Electrospun membranes for membrane distillation: the state of play and recent advances. *Desalination* 526: https://doi.org/10.1016/j.desal.2021.115511.

30 Lawal, D.U., Antar, M.A., Ismaila, K.G. et al. (2022). Hybrid multi-stage flash (MSF) and membrane distillation (MD) desalination system for energy saving and brine minimization. *Desalination* https://doi.org/10.1016/j.desal.2022.116231.

31 Chen, Z., Li, J., Zhou, J., and Chen, X. (2023). Photothermal Janus PPy-SiO_2@PAN/F-SiO_2@PVDF-HFP membrane for high-efficient, low energy and stable desalination through solar membrane distillation. *Chem. Eng. J.* 451: https://doi.org/10.1016/j.cej.2022.138473.

32 Yang, G., Ng, D., Huang, Z. et al. (2023). Janus hollow fibre membranes with intrusion anchored structure for robust desalination and leachate treatment in direct contact membrane distillation. *Desalination* 551: 116423. https://doi.org/10.1016/j.desal.2023.116423.

33 Duong, H.C., Cooper, P., Nelemans, B. et al. (2015). Optimising thermal efficiency of direct contact membrane distillation by brine recycling for small-scale seawater desalination. *Desalination* 374: 1–9. https://doi.org/10.1016/j.desal.2015.07.009.

34 Ahmed, F.E., Lalia, B.S., Hashaikeh, R., and Hilal, N. (2020). Alternative heating techniques in membrane distillation: a review. *Desalination* 496: https://doi.org/10.1016/j.desal.2020.114713.

35 Karanikola, V., Moore, S.E., Deshmukh, A. et al. (2019). Economic performance of membrane distillation configurations in optimal solar thermal desalination systems. *Desalination* 472: https://doi.org/10.1016/j.desal.2019.114164.

36 Elmarghany, M.R., El-Shazly, A.H., Salem, M.S. et al. (2019). Thermal analysis evaluation of direct contact membrane distillation system. *Case Stud. Therm. Eng.* 13: https://doi.org/10.1016/j.csite.2018.100377.

37 Banat, F. and Jwaied, N. (2008). Exergy analysis of desalination by solar-powered membrane distillation units. *Desalination* 230 (1–3): 27–40. https://doi.org/10.1016/J.DESAL.2007.11.013.

38 Ali, E., Orfi, J., and Najib, A. (2018). Assessing the thermal efficiency of brackish water desalination by membrane distillation using exergy analysis. *Arab. J. Sci. Eng.* 43 (5): https://doi.org/10.1007/s13369-017-2831-5.

39 Miladi, R., Frikha, N., and Gabsi, S. (2017). Exergy analysis of a solar-powered vacuum membrane distillation unit using two models. *Energy* 120: 872–883. https://doi.org/10.1016/j.energy.2016.11.133.

40 Si, Z., Han, D., Gu, J. et al. (2020). Exergy analysis of a vacuum membrane distillation system integrated with mechanical vapor recompression for sulfuric acid waste treatment. *Appl. Therm. Eng.* 178: https://doi.org/10.1016/j.applthermaleng.2020.115516.

41 Woldemariam, D., Martin, A., and Santarelli, M. (2017). Exergy analysis of air-gap membrane distillation systems for water purification applications. *Appl. Sci.* 7 (3): https://doi.org/10.3390/APP7030301.

42 Deshmukh, A. and Elimelech, M. (2017). Understanding the impact of membrane properties and transport phenomena on the energetic performance of membrane distillation desalination. *J. Membr.Sci* 539: 458–474. https://doi.org/10.1016/j.memsci.2017.05.017.

43 Ali, E. (2017). Energy efficient configuration of membrane distillation units for brackish water desalination using exergy analysis. *Chem. Eng. Res. Des.* 125: 245–256. https://doi.org/10.1016/j.cherd.2017.07.020.

44 Najib, A., Orfi, J., Ali, E., and Saleh, J. (2019). Thermodynamics analysis of a direct contact membrane distillation with/without heat recovery based on experimental data. *Desalination* 466: 52–67. https://doi.org/10.1016/j.desal.2019.05.009.

45 Si, Z., Han, D., Xing, Y., and Xiang, J. (2022). Experimental and numerical study on thermodynamic characteristics of a vacuum membrane distillation system based on mechanical vapor recompression for sulfuric acid waste.

Chem. Eng. Process. Process Intensif. 174: https://doi.org/10.1016/j.cep.2022.108862.

46 Qin, Y., Zhang, H., Wang, F. et al. (2022). Integration of high-temperature PEMFC with air gap membrane distillation for simultaneous electricity generation and seawater desalination: a theoretical investigation. *Desalination* 543: https://doi.org/10.1016/j.desal.2022.116100.

47 Lai, C., Zhang, H., and Hu, Z. (2022). Performance analyses on a concentrated photovoltaics driven direct contact membrane distillation coupled system. *Desalination* 531: https://doi.org/10.1016/j.desal.2022.115695.

48 Lai, C., Zhu, H., Zhang, H., and Hu, Z. (2023). Performance analyses on a novel evacuated U-tube solar collector powered direct contact membrane distillation hybrid system for seawater desalination. *Appl. Therm. Eng.* 219: https://doi.org/10.1016/j.applthermaleng.2022.119490.

49 Lin, B. and Malmali, M. (2023). Energy and exergy analysis of multi-stage vacuum membrane distillation integrated with mechanical vapor compression. *Sep. Purif. Technol.* 306: 122568. https://doi.org/10.1016/j.seppur.2022.122568.

50 Signorato, F., Morciano, M., Bergamasco, L. et al. (2020). Exergy analysis of solar desalination systems based on passive multi-effect membrane distillation. *Energy Rep.* 6: 445–454. https://doi.org/10.1016/j.egyr.2020.02.005.

51 Luo, A. and Lior, N. (2016). Critical review of membrane distillation performance criteria. *Desalin. Water Treat.* 57 (43): 20093–20140. https://doi.org/10.1080/19443994.2016.1152637.

52 Choi, J., Choi, Y., Lee, J. et al. (2021). Exergy analysis of a direct contact membrane distillation (DCMD) system based on computational fluid dynamics (CFD). *Membranes (Basel)* 11 (7): https://doi.org/10.3390/membranes11070525.

53 Generous, M.M., Qasem, N.A.A., and Zubair, S.M. (2020). Exergy-based entropy-generation analysis of electrodialysis desalination systems. *Energy Convers. Manag.* 220: https://doi.org/10.1016/j.enconman.2020.113119.

54 Chehayeb, K.M. and Lienhard, J.H. (2017). Entropy generation analysis of electrodialysis. *Desalination* 413: 184–198. https://doi.org/10.1016/j.desal.2017.03.001.

55 Chehayeb, K.M. and Lienhard, J.H. (2019). On the electrical operation of batch electrodialysis for reduced energy consumption. *Environ. Sci. (Camb)* 5 (6): 1172–1182. https://doi.org/10.1039/c9ew00097f.

56 Tong, X., Liu, S., Chen, Y., and Crittenden, J. (2020). Thermodynamic analysis of a solar thermal facilitated membrane seawater desalination process. *J. Clean. Prod.* 256: https://doi.org/10.1016/j.jclepro.2020.120398.

57 Ortega-Delgado, B., Palenzuela, P., Altaee, A. et al. (2022). Thermo-economic assessment of forward osmosis as pretreatment to boost the performance and sustainability of multi-effect distillation for seawater desalination. *Desalination* 541: https://doi.org/10.1016/j.desal.2022.115989.

58 Kotb, M., Khalifa, A., Hussein, A. et al. (2024). Performance and exergy analysis of vacuum-assisted sweeping air membrane distillation and bubble column dehumidifier. *Appl. Therm. Eng.* 245: https://doi.org/10.1016/j.applthermaleng.2024.122800.

59 Hussein, A., Khalifa, A.E., Alawad, S.M., and Antar, M.A. (2022). Experimental investigation of membrane distillation with bubble column dehumidifier system for water desalination. *Case Stud. Therm. Eng.* 39: https://doi.org/10.1016/j.csite.2022.102383.

60 Patel, D., Mudgal, A., Patel, V. et al. (2024). Exergy analysis for enhanced performance of integrated batch reverse osmosis – forward osmosis system for brackish water treatment. *Desalination* 580: https://doi.org/10.1016/j.desal.2024.117548.

4

Water Desalination via Solar Energy and Photovoltaic

Yunqian Dai[1], Xiangyu Meng[1], Haoxuan Li[2], Ziyu Tao[1], and Xiao Shen[2]

[1] *School of Chemistry and Chemical Engineering, Southeast University, Nanjing, China*
[2] *Key Laboratory of Eco-Textiles, Jiangnan University, Wuxi, China*

4.1 Introduction and Development Outline of Solar Water Desalination

4.1.1 Introduction and Definition of the Solar Water Evaporation Process

Water, being among the richest substances found on our planet, blankets approximately 71% of the Earth's surface, and notably, more than 77% of this vast expanse constitutes seawater. Freshwater, regarded as a globally strategic resource, is crucial for industrial production, daily life, and human survival. However, challenges such as industrial expansion, population growth, climate change, and environmental pollution are escalating freshwater scarcity, impacting a wide population base. Therefore, immediate attention must be given to exploring efficient and cost-effective water resource management and purification technologies to mitigate the freshwater crisis and ensure water resources can meet the needs of future generations [1].

Currently, traditional methods like reverse osmosis (RO) and multistage flash evaporation, while effective, are highly dependent on fossil fuels, costly, and environmentally harmful [2]. Consequently, faced with the limitations of conventional methods, there is increasing interest in harnessing abundant, clean, renewable, and economically viable solar energy to drive seawater desalination [3]. Figure 4.1 depicts the global annual average direct solar irradiance, showcasing solar energy as a secure, clean, renewable, sustainable, and widely accessible energy source [5].

Solar-driven seawater desalination refers to the process of using solar energy directly on the surface of water or on light-absorbing materials to induce localized thermal effects, prompting water molecules to transition from the liquid to the gas

Water Treatment and Desalination: Membrane Separation using Renewable Energy Resources, First Edition. Edited by Mohamed Bassyouni.

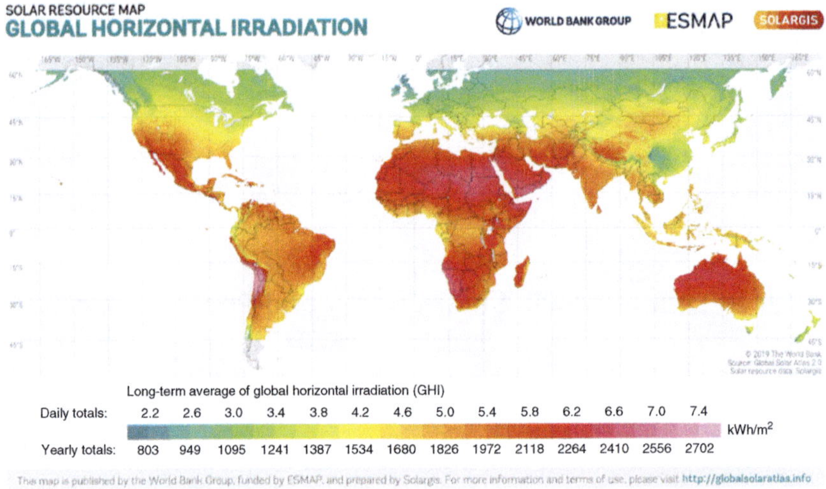

Figure 4.1 Illustration of global annual average solar heat. Source: Adapted from MacFarlane [4].

phase through evaporation to remove salts. Directly heating water sources with light results in high losses in light absorption and thermal effects, leading to low evaporation efficiency. To overcome this challenge, significant advancements have been made in solar-driven interfacial evaporation technology, which utilizes photothermal materials with photothermal conversion capabilities localized at the air/liquid interface to heat the water molecules. Using solar energy to drive seawater desalination, by the photothermal interfacial evaporation process, is an ideal strategy to simultaneously address clean water shortages and environmental concerns [6]. The core of this technology lies in the efficient absorption and rapid conversion of solar energy into heat, directly heating water molecules to transition from the liquid to the gas phase, achieving green and sustainable water treatment and purification [7]. Therefore, we will now proceed to explain these core steps in detail in the following sections.

4.1.1.1 Photon Absorption Process

When light strikes the surface of an object, it consists of a series of photons. If atoms or molecules of the material absorb these photons, they transfer energy to internal particles. In the context of light-driven water evaporation, specific light-absorbing materials (e.g., black coatings, metal nanoparticles [NPs]) are designed to efficiently absorb photons from the solar spectrum. The absorption of photons adheres to quantum mechanics principles, requiring that the photon's energy matches or exceeds the energy gap needed for electron state transitions in the material. Once absorbed, the photon energy transforms into thermal energy via photothermal conversion, leading to a rise in the material's temperature. The process of photothermal conversion is explained in the following section.

4.1.1.2 Photothermal Conversion Process

Photothermal materials are crucial components in new solar water evaporation devices. These materials are capable of absorbing solar energy under sunlight and converting a portion or all of the incident light into heat energy. Clearly, high-performance photothermal materials, characterized by their proficient photothermal conversion capabilities, are crucial for enhancing the efficacy of solar-driven water evaporation systems. Over the past decade, researchers have extensively studied materials that exhibit high absorption and conversion efficiencies within the solar spectrum. This focus aims to optimize the performance of solar energy harvesting technologies. Photothermal mechanisms can be categorized into three types [8]: plasmonic heating (Figure 4.2a), nonradiative relaxation (Figure 4.2b), and thermal vibration (Figure 4.2c), which will be detailed later.

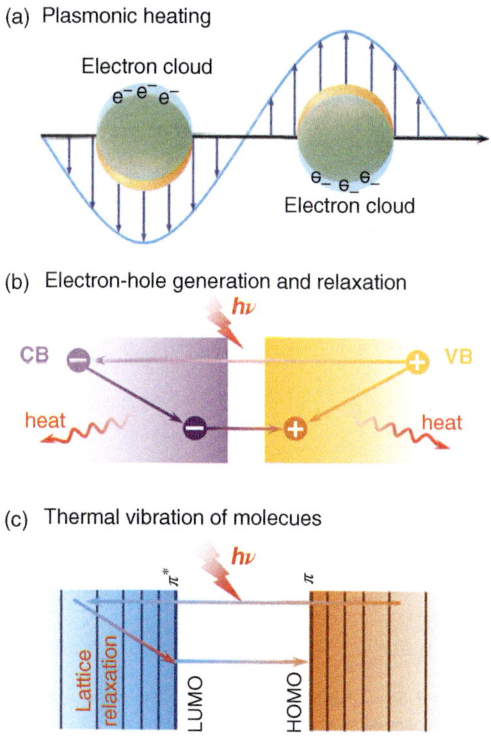

Figure 4.2 Illustration of the mechanisms of photothermal conversion. (a) Plasmonic heating. (b) Electron–hole generation and relaxation. (c) Thermal vibration. Source: Chen [8]/with permission of Elsevier.

4.1.1.3 Seawater Evaporation and Collection Process

Evaporation is a gradual process of heat absorption occurring at the surface of water or moist materials. Several elements enhance this occurrence, including the surface's absorption of radiative heat, the extent of the surface area conducive to evaporation, the surrounding air's humidity levels, and the amount of convective heat transfer [7]. At the macroscopic level, water evaporation is an intuitive phenomenon where water transitions from a liquid phase to a vapor phase and releases into the atmosphere. This process involves absorbing thermal energy from the surroundings and follows the principles of energy conservation and phase transition in physics. Figure 4.3 illustrates the processes of energy and mass transfer. Solar energy gets absorbed and transformed into localized heat at the evaporation interface through photothermal conversion by solar absorbers/evaporators. Through capillary action, large quantities of water are continuously transported to porous channels on the evaporation surface. Subsequently, the generated localized heat is converted into latent heat, driving water evaporation efficiently. The water vapor produced at the evaporation interface migrates to a condenser where it cools and transforms back into purified water [9].

From a microscopic perspective, water evaporation involves dynamic changes in the hydrogen-bonded network between water molecules. Water can be found in three distinct states: solid (i.e., ice), liquid, and gas (i.e., water vapor), each characterized by different hydrogen-bonded networks. Water molecules, being

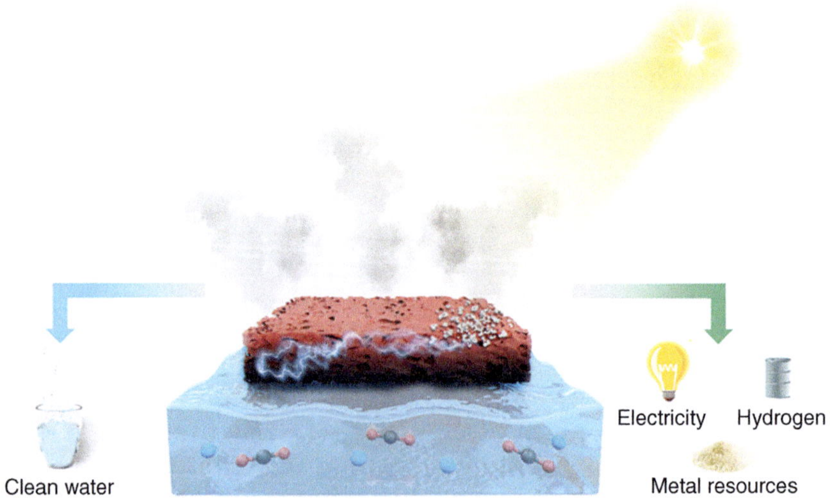

Figure 4.3 Schematic diagram of the energy and mass transfer process of interfacial solar water evaporation process. Source: Shi [9]/American Chemical Society.

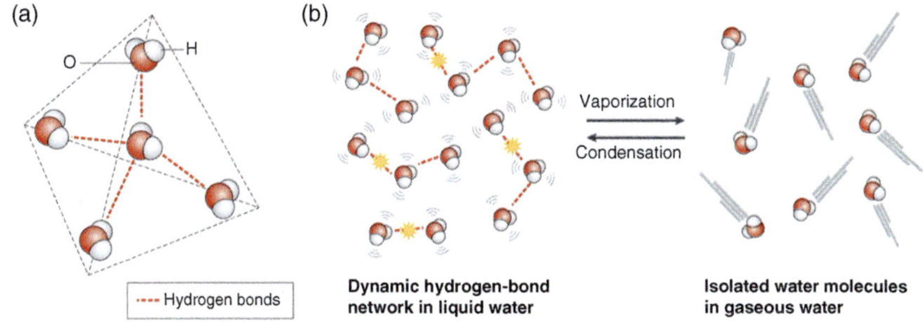

Figure 4.4 (a) Illustration of hydrogen bonds between water molecules. (b) Illustration of molecular changes in the hydrogen-bond network during vaporization and condensation. Source: Zhao [10]/Springer Nature.

polar, attract each other through hydrogen bonding due to the disparity in electronegativity between oxygen and hydrogen atoms. Each water molecule provides four sites—two O–H groups and two lone pairs of electrons on the oxygen atom—for hydrogen bonding with neighboring molecules (Figure 4.4a). When external energy is absorbed, such as heat from photothermal conversion, the vibration of water molecules intensifies, leading to frequent breaking and reformation of the hydrogen bonds. As the temperature rises, the hydrogen-bonded network loosens, allowing more water molecules to break free from the hydrogen bonds and transition from liquid to freely moving gaseous molecules (i.e., water vapor) (Figure 4.4b). Upon condensation, the dynamic hydrogen-bonded network can reform [11]. This process not only involves a change in energy states but also a fundamental rearrangement of the molecular structure, reflecting the microscopic mechanisms underlying the phase transition of water molecules during desalination [10].

4.1.2 Historical Development

Solar energy and water are two of the most abundant resources on Earth. Solar desalination has been historically used for freshwater production. Over the past decade, advancements in interface solar evaporation have significantly enhanced efficiency by approximately 90%, even as solar intensity has decreased. This improvement is primarily attributed to rapid developments in novel photothermal materials and photothermal engineering. Modern nanotechnology in materials science has driven the development of these photothermal materials, which can harness the entire solar spectrum with exceptional efficiency in converting it into heat. Here, we present a landmark modern development in photothermal water evaporation for freshwater production, structured around the evolution of materials science advancements (Figure 4.5).

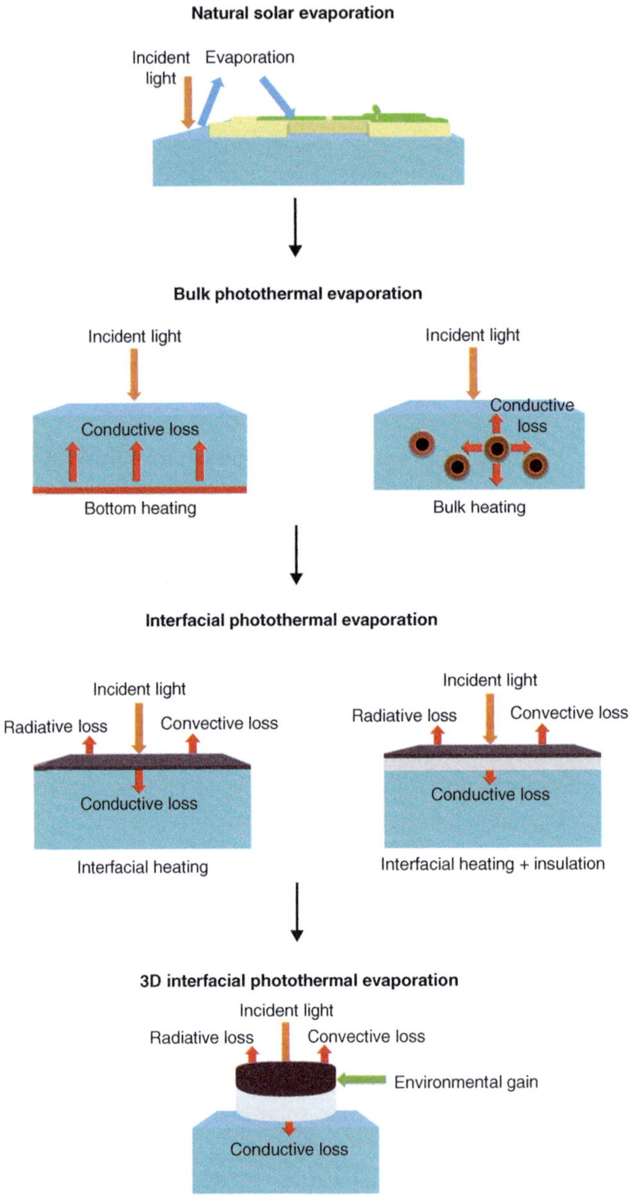

Figure 4.5 Scheme for illustrating the development of solar evaporation (produced by author).

4.1.2.1 Stage One: Direct Solar Heating

In the initial research, water was directly used as a solar absorber. However, due to water's narrow response range to the solar spectrum, the efficiency of solar utilization was not ideal. Commercially, solar radiation is reflected or concentrated by collectors and then converted into thermal energy in concentrating solar power (CSP) systems [12]. Subsequently, this thermal energy can be used to generate electricity via steam turbines, which are employed in steam generation under varying high temperatures or pressures. Commercial CSP systems include parabolic trough collectors (PTCs), solar power towers (SPTs), linear Fresnel reflectors (LFRs), and parabolic dish systems (PDSs). The PTC system consists of four main components: the absorber system, the heat exchanger system, the thermal storage system, and the power generation system [13]. This system can achieve high temperatures, effectively converting heat into electricity while reducing storage costs. In parabolic trough systems, metal pipes are enclosed within glass tubes, with vacuum or air between the two tubes to minimize heat transfer. These pipes are positioned at the focal line of the condenser, where temperatures can reach up to 400°C. In LFR technology, the receiver and reflector are fixed components, contributing to lower overall system costs. However, its conversion efficiency is lower compared to PTCs and SPTs. Over the past few decades, to enhance the optothermal performance of CSP systems, selective absorption coatings have been applied on the vacuum tubes of solar collectors. These coatings enhance light absorption across the visible- to near-infrared (NIR) spectrum of the solar spectrum while reducing thermal radiation losses in the far-infrared region. Nevertheless, the overall efficiency of these solar selective coatings is limited by the effectiveness of heat transfer to the working fluid [13]. The substantial spatial investment required for reflector mirrors makes them impractical and unsuitable for household use. Consequently, there is a pressing need to develop a new, compact, and cost-effective solar steam generation system under these circumstances.

4.1.2.2 Stage Two: Photothermal Heating

Lately, advancements in photothermal materials have led to new system designs capable of achieving efficient solar-driven evaporation without the need for complex concentrators. Photothermal materials integrated into water sources localize solar light absorption and heat conversion to the evaporation region of the material, effectively transferring heat to the surrounding water medium. Initially, efforts to enhance light absorption involved modifying the evaporation pool, known as bottom heating (Figure 4.5) [14]. For instance, lining the bottom of the evaporation pool with highly absorbent black materials to boost evaporation rates. However, under this design, the efficiency gains from enhanced evaporation are limited because sunlight needs to penetrate through the water layer, with only a fraction being absorbed by the black material. Moreover, bottom heating designs suffer from inefficient heat exchange phenomena: Heated water at the bottom of the pool conducts heat to unheated water during the upward evaporation process. Consequently, the overall temperature equilibrium of the water in the evaporation pool cannot sustainably increase.

Subsequently, with further advancements in photothermal materials, a new evaporation method called volumetric heating (or optical nanofluid heating) was developed (Figure 4.5). A previous work pioneered solar-driven water evaporation based on Au NPs dispersed uniformly in bulk water [15]. This configuration, due to its bulk water heating approach, yielded unsatisfactory water evaporation efficiency under both conventional and nonconcentrated solar irradiation. Previous work reported a water evaporation efficiency of 69% under 10-day illumination using dispersed graphene oxide, carbon black (CB), and graphene in water [16]. Another work achieved enhanced water evaporation efficiency by dispersing Au NPs and light-scattering polystyrene NPs in bulk water, achieving an efficiency of 35.6% under 35.36 W/cm^2 532 nm laser irradiation [17]. Nanofluids exhibit better stability, rheological properties, and high thermal conductivity compared to nonnanofluids. Since the 1990s, extensive experimental and theoretical studies have demonstrated that dispersing photothermal NPs in water to form nanofluids exhibits excellent photothermal performance due to sunlight absorption, localized heating, and water evaporation. The bulk water remains at lower temperatures, with vapor bubbles generated around the photothermal material floating on the water surface and releasing steam. However, bubble generation is a phase-change process that requires overcoming energy barriers, limiting evaporation efficiency despite the effective sunlight conversion by nanofluids.

4.1.2.3 Stage Three: Interfacial Photothermal Heating

In contrast to bulk water heating, the concept of interfacial heating emerged in the early 2010s and has since become a popular method for solar-driven water evaporation (Figure 4.5). Volumetric solar-driven steam generation systems, including nonequilibrium and equilibrium heating, unavoidably suffer from radiation, conduction, and convective heat losses due to the substantial working fluid being heated, which leads to reduced steam generation efficiency. Interfacial solar evaporation systems represent a promising strategy that utilizes nonsubmerged solar absorbers to localize solar heat and steam generation at the water–air interface. In these systems, thermal conversion is confined to the interface, heating only the surface water. Consequently, the temperature of the underlying bulk liquid remains at lower levels, thereby minimizing thermal losses and parasitic heat transfer to the bulk water postconversion heating. Moreover, due to efficient evaporation, the surface temperature of the absorber is lower, thereby reducing radiative and convective heat losses from the absorber surface. Through this approach, solar-to-steam conversion efficiencies exceeding 90% can be achieved. In 2011, researchers reported one of the pioneering works introducing interfacial heating in solar thermal applications with a Fe_3O_4/carbon composite material that self-floats on water surfaces [18]. Subsequently, examples such as Au films [19], self-assembled black titanium dioxide (TiO_2) films [20], and self-assembled magnetic mixed metal oxide (MMO) microsphere films emerged [21]. These interfacial heating systems provide better energy efficiency under similar conditions (especially under nonconcentrated solar radiation) compared to dispersed configurations, as they selectively heat only at the air–water interface where water evaporation occurs.

However, the presence of high-temperature zones at the air–water interface in interfacial heating systems poses a challenge of thermal conduction loss from the photothermal material to the underlying bulk water, prompting the use of thermal insulation layers to separate the photothermal layer from the bulk water (Figure 4.5). In 2014, researchers first proposed and prepared a dual-layer structure, with a foam carbon base layer designed as a deliberate thermal insulation layer [22]. In 2015, a work reported an air-filled paper thermal insulation layer supporting its top self-assembled gold (Au) film [23]. Additionally, the base layer improves the mechanical properties of the photothermal material as a substrate. Since then, the dual-layer structure has been widely adopted due to improved energy efficiency and mechanical stability. Ho's research team reported a novel elastic cellular sponge that can self-limit and store water without contacting bulk water [24]. The ultralight nitrogen-rich carbon sponge easily absorbs water and performs efficient in-situ solar evaporation. Carbon sponges feature advantageous built-in structural hierarchical porosity, interconnecting mesoporous fibers, forming elastic large-pore openings. These components and built-in structural characteristics favor broadband light absorption and isolated thermal localization. Through the capillary action of the sponge, self-contained water is absorbed and transported to permanent hotspots for evaporation, completely isolated from bulk water, achieving a high efficiency of 90%, higher than any common suspended configuration. In 2017, researchers reported two-dimensional (2D) channels. In this design, 2D channels are placed outside a single functional thermal insulation layer of polystyrene foam with closely packed pores [25, 26], increasing energy efficiency by 83–85%. Subsequently, the literature reported a one-dimensional (1D) channel design using capillary effects in porous substrates. In both 2D and 1D channel configurations, thermal losses due to heat conduction to bulk water are almost eliminated.

In addition to the above artificial structures, the transpiration of trees, relying on unique porous structures and mass/heat transfer properties, has inspired the development of wood-based interfacial evaporators. Natural woods with excellent hydrophilicity and vertically aligned microchannels effectively transport water and serve as carriers for solar absorption and water evaporation. Inspired by the ecological relationship between energy and water, Hu's team demonstrated high-efficiency solar steam generation based on natural wood [27]. A dual-layer wood-based solar steam generation device comprising radially cut wood and top carbonized surfaces was obtained through a simple heating process. Variations in the microstructures of various natural woods (poplar, pine, and coconut wood) significantly affect solar steam generation performance; efficiencies are 86.7%, 76.3%, and 46.8%, respectively, under 10 Suns. Higher evaporation efficiencies are achieved with porous wood. Due to its abundance, biocompatibility, hydrophilicity, low thermal conductivity, and high light absorption, wood with natural container structures is an excellent support layer for solar steam generation. Several studies have also reported on hybrid wood-coated solar absorbers for interfacial solar evaporation.

4.1.2.4 Stage Four: Three-Dimensional Interfacial Photothermal Heating

Next, researchers have found that, in laboratory environments, the top surfaces of three-dimensional (3D) evaporators are typically directly exposed to solar irradiation, while the side surfaces maintain temperatures at or below ambient levels. As a result, these evaporators utilize ambient temperatures to achieve passive cooling and dark evaporation from the surrounding air (Figure 4.5). Early designs of solar interface evaporators, whether 1D or 2D, inherently suffer from higher thermal conduction losses because a significant portion of their total surface area directly contacts bulk water (including immersion below).

This challenge has been mitigated by adopting 3D indirect contact systems. Here, photothermal evaporators are mounted on partially submerged thermal insulation materials used for water transport. This configuration increases the total surface area exposed to solar energy and allows for higher evaporation rates. Polypyrrole (PPy) has been integrated into the frameworks of poly(*N*-isopropylacrylamide) and alginate to form 3D hydrogels. These thermal-responsive hydrogels exhibit peak performance at a height of 4 cm, achieving a deionized water evaporation rate of 4.145 $kg/m^2/h$ under a single solar irradiation [28]. Another approach involves preparing 3D photothermal aerogels by spray-coating cotton sheets with a mixture of reduced graphene oxide and sodium alginate. Aerogels designed at heights of 9 and 14 cm achieve pure water evaporation rates of 6.1 and 7.6 $kg/m^2/h$, respectively [29]. Alternative geometric designs with vertically oriented channels have been developed by increasing the angle of solar irradiation refraction onto the evaporator surface. In one study, iron-doped carbon nanotubes (CNTs) were spray-coated onto corrugated polyvinyl alcohol (PVA)-impregnated melamine foam for a 3D evaporator utilizing two vertical channels to transport large volumes of water to horizontal channels supporting the corrugated Fe/CNT surface. This manufacturing process creates a Janus structure introducing a hydrophobic photothermal layer atop the hydrophilic foam, allowing extended operation under saline conditions and stable saline water supplies. Vertical channel heights and surface refraction angles are optimized to 5.5 cm and 60°, respectively. This design demonstrates continuous operation for up to 12 hours, with performance declining from 4.2 $kg/m^2/h$ at 3.5 wt.% salinity to 3 $kg/m^2/h$ at 21 wt.% salinity [30]. Despite reduced performance at higher salt concentrations, it remains competitive compared to other high-performance systems. 3D-printed solar interface evaporators with a volcano-like structure have been manufactured using a fused deposition modeling approach [31]. The structure utilizes black polylactic acid substrate subsequently coated with polydopamine (PDA) to enhance its hydrophilic properties. The volcano-like structure, with a height-to-diameter ratio of 1, maximizes the potential of the Marangoni effect to transport water upward due to temperature gradients. Combined with three layers of PDA, this design achieves an evaporation rate of 4.02 $kg/m^2/h$. Solar-driven 3D interface evaporators demonstrate stable and high evaporation rates under solar and passive ambient energy conditions, offering significant potential for water production and resource recovery from seawater and wastewater.

4.2 Introduction to Material and Device Design Strategies for Solar Water Evaporation

Based on the above content, we have gained insights into the background, development history, and process explanation of photothermal evaporation for water desalination and treatment. Now, we will focus on detailing the current design strategies for photothermal materials and water condenser/collector devices in photothermal evaporation technology.

4.2.1 Introduction to the Design Strategies of Photothermal Water Evaporation Materials

Photothermal conversion materials have been reported as achieving effective solar-driven evaporation, showcasing their significant roles as solar radiation receptors and converters [32]. These materials mainly include metal NPs, semiconductors, and carbon-based materials. Isoexcited metal NPs have a narrow absorption spectrum. To efficiently absorb light over a broad wavelength range, isoexcited NPs with varying sizes and shapes can be designed within highly disordered templates to enhance broadband light absorption effectively (Figure 4.6a). Semiconductors are generally employed as solar energy absorbers owing to their

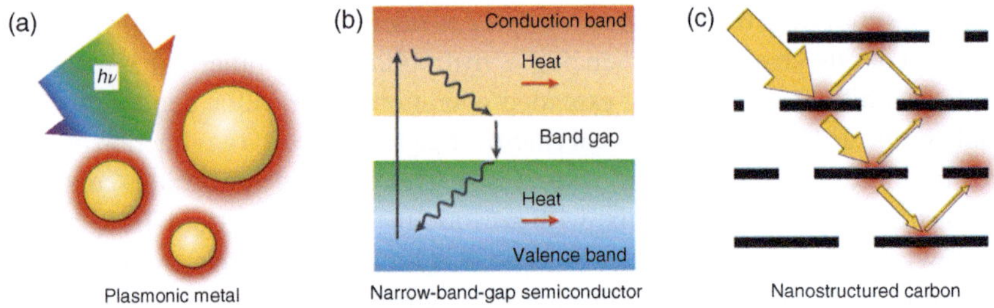

Figure 4.6 (a) A schematic illustration of plasmonic absorbers with metal nanoparticles of random sizes and distributions that can absorb light over a wide wavelength range. (b) A schematic illustration of heat generation in narrow-bandgap semiconductors. (c) A schematic illustration of a carbon nanostructure with long internal optical path lengths to enhance light absorption. Source: Zhao [11]/Springer Nature.

tunable bandgaps. Current research focuses on reducing bandgaps to enable absorption across a broader visible spectrum, thereby enhancing solar energy harvesting efficiency (Figure 4.6b). Carbon-based materials, such as amorphous carbon, CNTs, and graphene, can absorb broad-spectrum solar radiation from visible to infrared wavelengths but are limited by unfavorable surface reflections. Hence, researchers have proposed various carbon nanostructures, including oriented, porous, and layered architectures, to increase optical path lengths for light capture and utilize multiple internal reflections to minimize energy loss from reflections (Figure 4.6c). Understanding and controlling the performance of these materials hinge on exploring their underlying operational principles, strengths, and limitations. The subsequent discussion will delve into specific categorizations.

4.2.1.1 Metal Nanoparticles

Metal NPs, especially Au, silver (Ag), aluminum (Al), germanium (Ge), and Au–Ag alloys, have demonstrated excellent performance in photo-excited water evaporation due to their high light-trapping ability and efficient photothermal conversion mechanism employing the localized surface plasmon resonance (LSPR) effect. These NPs act as a solar energy-absorbing medium in the solar evaporation system [33]. The LSPR effect induces the oscillation of electrons on the NP surfaces to collectively oscillate and resonate with the incident light when irradiated by the light with a specific frequency, which efficiently traps the photon energy and translates it into thermal energy, and then released via the Joule heating effect.

The advantages of LSPR NPs are obvious. First, they exhibit ultra-high light absorption efficiency at the LSPR peak, targeting specific bands of the solar spectrum; second, by creating localized high-temperature points (hotspot effect), they achieve the instantaneous localized heating required for rapid water evaporation. In addition, when assembling LSPR units into innovative structural designs, such as the arrangement on porous anodized aluminum oxide (AAO), efficient light absorption over a wide spectral range can be endowed [34]. However, LSPR metals have several limitations, including the high cost of noble metal NPs represented by Au and Ag, potential environmental risks, corrosion, or structural degradation due to long-term exposure, which affect the long-term stability and efficiency of the photothermal materials. In addition, the frequency dependence of the LSPR effect limits the full utilization of the entire solar spectrum when using metals as photothermal materials.

When the applied electric field aligns with the dimer's axis, both heat production and electric field amplification escalate. Intriguingly, plasmonic nanostructures exhibit a distinctive photothermal isosbestic behavior, manifesting a constant temperature regardless of alterations in the light polarization state at certain wavelengths (Figure 4.7a) [40]. The isosbestic point for a specific plasmonic nanostructure appears where absorption spectra under varying polarization orientations intersect. Given the inherent limits of a solitary dimer in expanding the LSPR band, assembling and synergizing multiple NPs are necessary to augment light absorption across a wider spectral range. Linear arrangements of Al NPs have demonstrated a broadening of extinction peaks, with notable shifts toward the visible and NIR regions, as the number of particles in the assembly grows (Figure 4.7b). Random configurations of Au NPs serve as ideal absorbers, achieving near-total light absorption. The suppression of back reflection enables an average absorbance exceeding 99% across a broad spectrum from 400 nm to 10 μm. Simulation reveals temperature distributions for arrays of nine Ag NPs with adjustable spacing (Figure 4.7c). Simultaneous irradiation of numerous NPs results in uniformly high temperatures system-wide, integrating all individual temperature increments. 3D or irregular nanostructures, like those in plasmonic colloidosomes formed via assembly techniques, enhance plasmon coupling, boosting photothermal heating. A case in point is the creation of 3D dark plasmonic colloidosomes from Au nanospheres using an emulsion templating method (Figure 4.7d) [41].

Prior studies showcased silica–Au (SiO_2/Au) nanoshells for achieving effective solar-driven evaporation (Figure 4.7e) [42]. In separate research, Au NPs were layered onto an air-laid paper to fabricate a biomimetic plasmonic film exhibiting potent photothermal conversion capabilities [38]. This freestanding film displayed enhanced light absorption (up to 85%) and superior evaporation efficiency (reaching 77.8%) under solar exposure [23, 43]. Contrasted with plasmonic materials consisting of singular components, integrated photothermal structures featuring multifaceted composites or designs represent advantageous setups for amplifying photothermal performance (Figure 4.7f).

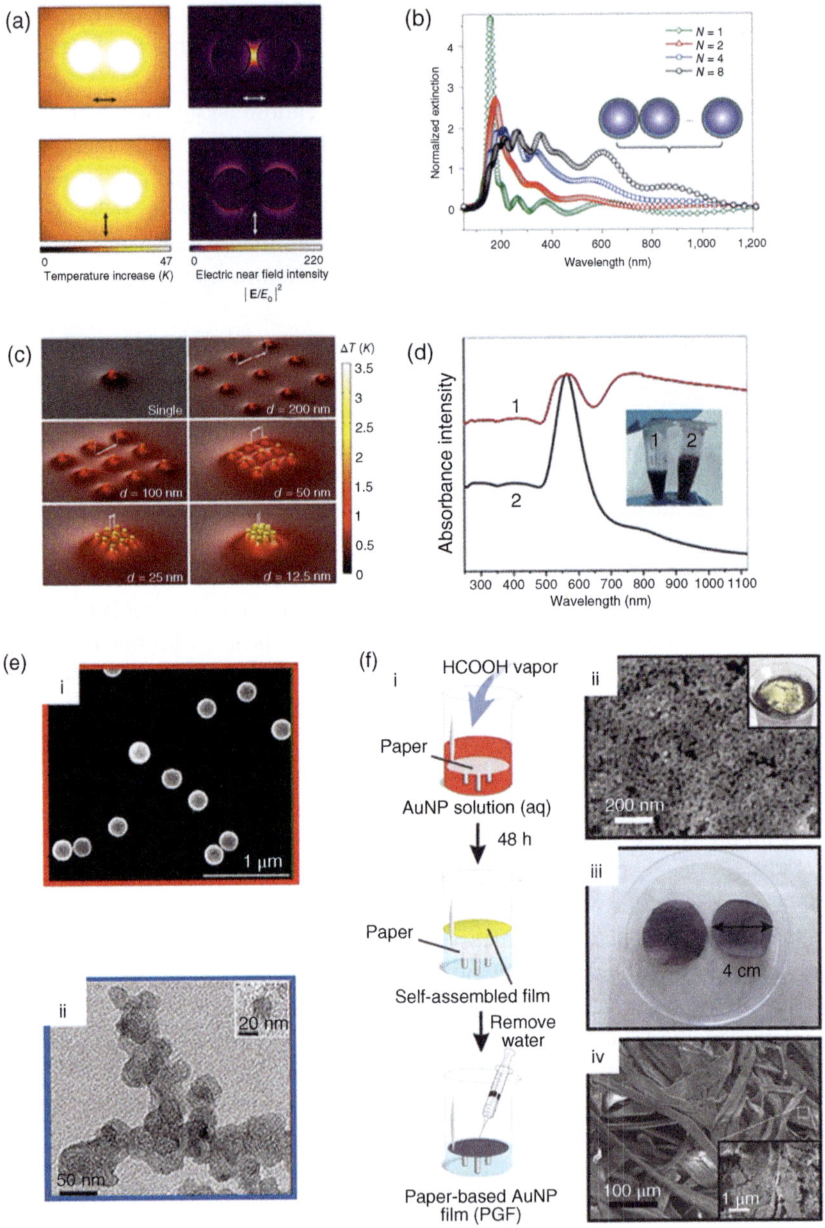
(a)
0
47
Temperature increase (K)
0
220
Electric near field intensity
|E/E0|^2
(b)
N = 1
N = 2
N = 4
N = 8
Normalized extinction
Wavelength (nm)
200
400
600
800
1,000
1,200
(c)
ΔT (K)
Single
d = 200 nm
d = 100 nm
d = 50 nm
d = 25 nm
d = 12.5 nm
3.5
3
2.5
2
1.5
1
0.5
0
(d)
Absorbance intensity
1
2
300 400 500 600 700 800 900 1000 1100
Wavelength (nm)
(e)
i
1 μm
ii
20 nm
50 nm
(f)
i
HCOOH vapor
Paper
AuNP solution (aq)
48 h
Paper
Self-assembled film
Remove water
Paper-based AuNP film (PGF)
ii
200 nm
iii
4 cm
iv
100 μm
1 μm

Figure 4.7 (a) Temperature distributions (left column) of an Au dimer under the longitudinal (top) and transverse (bottom) polarization excitation and the corresponding electric near-field enhancements (right column). Source: Reproduced with permission from Metwally [35]/American Chemical Society. (b) Dependence of the calculated normalized extinction spectra of one-dimensional (1D) Al nanoparticle chains on the particle number. Source: Zhou [36]/Springer Nature. (c) Calculated temperature maps of a single array and five arrays of nine Ag nanoparticles in a square lattice with lateral periods from 12.5 to 200 nm. Source: Reproduced with permission from Metwally [35]/ Springer Nature. (d) Measured absorption spectra of self-assembled Au colloidosomes (red) and the Au nanoparticles (black) suspended in 1-butanol. The inset shows the corresponding scanning electron microscopy (SEM) image of the Au colloidosomes. Source: [37]/with permission of John Wiley & Sons. (e) SEM and transmission electron microscopy (TEM) characterization of SiO_2/Au nanoparticles. Source: Reproduced with permission from Neumann [38]; Copyright ©2012 American Chemical Society. (f) Process of preparing paper-based Au nanoparticle films and the corresponding SEM images. Source: Reproduced with permission from Gao [39]/John Wiley & Sons.

4.2.1.2 Semiconductor Materials

At the forefront of light-driven water evaporation technology, low-cost, low-toxicity narrow-bandgap semiconductor materials are emerging as key media for solar thermal conversion [44, 45]. Upon excitation by photons possessing adequate energy, a semiconducting material generates electron–hole pairs whose energies are on a par with the bandgap. The energized electrons may relinquish their energy by emitting photons or, alternatively, transfer it to the material's lattice via nonradiative relaxation pathways. In a semiconductor, the recombination of charge carriers releasing phonons rather than photons can elevate lattice heat dissipation, consequently inducing a localized rise in temperature [46, 47].

The unique advantage of narrow-bandgap semiconductors is that their energy band structure perfectly matches the solar spectrum, allowing most photon energies to easily exceed the bandgap threshold, inducing electrons to jump from the valence band to the conduction band, and the electron–hole pairs formed, which are then rapidly released as heat through nonradiative composites, efficiently heating up the material and accelerating the evaporation of water. In contrast, the energy conversion efficiency of wide-bandgap semiconductors is limited by the tendency of electron–hole pairs to lose energy in the form of radiative composites.

Through dimensional modulation (e.g., NPs, quantum dots) and complex structural designs (e.g., 3D light-trapping systems), these materials achieve further optimization of performance while maintaining cost-effectiveness and environmental compatibility. Nevertheless, long-term stability, limitations in photothermal conversion efficiency, and large-scale fabrication challenges for certain high-performance materials remain challenges for current research. Addressing this issue, strategies such as incorporating defects or utilizing complex superstructures have been investigated to broaden the spectral response range of semiconductor materials [48]. As an illustration, hydrogen-doped

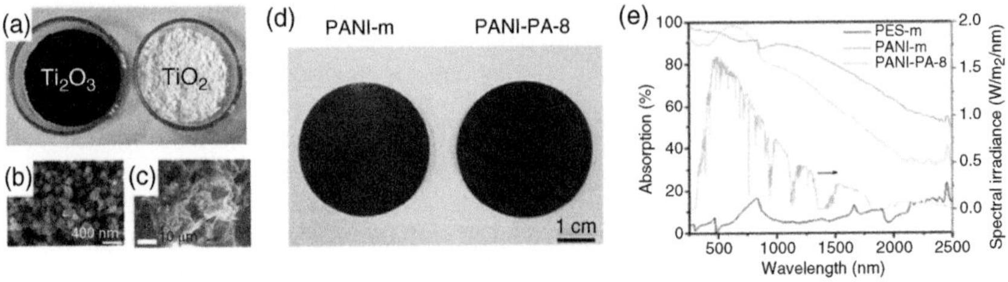

Figure 4.8 (a) Photographs of Ti_2O_3 and TiO_2 nanoparticles. Source: Reproduced with permission from Gaertner [45]/John Wiley & Sons. (b) SEM images of Cu/G nanoparticles. Source: Reproduced with permission from Yuan [49]/Elsevier. (c) Morphology and elemental distribution of prepared GA-$CuFeSe_2$. Source: Reproduced with permission from Wei [50]/Elsevier. (d) Photographs of the polyaniline-modified (PANI-m). (e) UV–vis-NIR absorption spectrum. Source: Reproduced with permission from Chen [8]/Royal Society of Chemistry.

black TiO_2 and titanium trioxide (Ti_2O_3) with engineered narrow bandgaps (Figure 4.8a) have been devised to harness the full spectrum of solar energy for the purpose of steam generation. Similarly, structures resembling sea urchins composed of copper nanodots embedded in graphene (Cu/G) (Figure 4.8b), as well as graphene integrated with GA-$CuFeSe_2$ (Figure 4.8c), have been shown to effectively absorb the entire solar spectrum. This results in a potent photothermal heating effect suitable for the generation of vapor from solar energy [45, 49, 50].

Moreover, polyaniline (PANI), a material renowned for its metallic sheen and status as a prototypical organic semiconductor, owes its distinction to the extensive linear conjugated π-electron framework within its molecular structure, facilitating the free movement and transfer of electrons. Its intrinsic photothermal conversion attributes have rendered PANI a popular choice for applications in solar-driven interfacial water evaporation [51]. Through oxidative polymerization of aniline on a polyether sulfone microfiltration-membrane (PES-m), followed by interfacial polymerization to create a polyamide (PA) layer atop the PANI-coated membrane, PANI-PA films were synthesized (Figure 4.8d). These dark blue PANI-PA films showcased over 93.5% light absorption across the 250–800-nm spectral range. Conversely, the average light absorption in the NIR spectrum (800–2500 nm) stood at 55.5% (Figure 4.8e), revealing PANI's superior absorption in the ultraviolet (UV)–visible spectrum, yet falling short of complete sunlight absorption in the NIR domain [8]. The incorporation of dopants into PANI results in heightened light absorption due to bimodal or polariton transitions. One study amalgamated $Ti_3C_2T_x$ MXene with PANI to produce composite fibers via in-situ polymerization of aniline, utilizing wet spinning and subsequently drying the fibers to yield superhydrophilic MXene/polyaniline (MPs) nonwoven fabrics.

The integration of MXene with PANI significantly bolstered the utilization efficiency in the NIR spectrum, enabling the MPs to exhibit outstanding light absorption efficiency across the entire spectral range, averaging 88.3% [52].

4.2.1.3 Carbon-Based Materials

Carbon-based materials exhibit superior light absorption features and are notably adept at generating heat through lattice vibrations. Despite the significant energy disparity between the σ and σ^* orbitals in conventional carbon bonds, such as C–C, C–O, and C–H, the loosely bound electrons in these substances can readily transition from π orbitals to π^* orbitals under low-energy photon exposure [53]. Upon encountering photons with energy corresponding to an electronic transition within the material, the π electrons ascend from their ground state (highest occupied molecular orbital [HOMO]) to a more energetic state (lowest unoccupied molecular orbital [LUMO]). Following the relaxation of these excited electrons back to the ground state via vibration–electron coupling, the surplus energy is discharged as heat. Furthermore, the conjugation or hyperconjugation of π orbitals significantly alters the electron transitions between HOMO and LUMO. As the quantity of π bonds escalates, the energy gap between these orbitals diminishes [22, 54], a phenomenon that can be leveraged to fine-tune the photothermal properties of carbon-based materials for various applications.

The advantages of carbonaceous materials are their broadband light absorption, economy, stability, ease of production on a large scale, environmental friendliness, and structural designability. However, the water transfer in carbon photothermal materials is hard to control. Recent investigations have indicated that highly hydrophilic surfaces do not necessarily equate to optimal conditions for efficient water evaporation, particularly concerning the uppermost evaporation surface. An overly hydrophilic characteristic on this surface can sometimes result in excessive water accumulation, which in turn leads to greater heat dissipation and hinders the maintenance of self-sustained solar evaporation. To counteract this limitation, Chen et al. introduced a strategy focusing on heat localization [55]. Their approach involved a dual-layer carbon architecture, where the first layer comprised a light-absorbing, hydrophilic, porous structure made from expanded graphite. Coupled with this was a second layer of insulating, hydrophilic, porous carbon foam acting as a thermal barrier. This innovative design aimed to optimize the balance between light absorption and heat retention, thereby enhancing the overall efficiency of solar-driven water evaporation (Figure 4.9a) [53]. In 2015, researchers showcased a durable hydrophobic evaporative structure by using a fluoroalkyl silane-modified PPy-coated stainless steel (SS) mesh film and demonstrated a high evaporation rate by optimizing the wettability of evaporation interfaces (Figure 4.9b) [56].

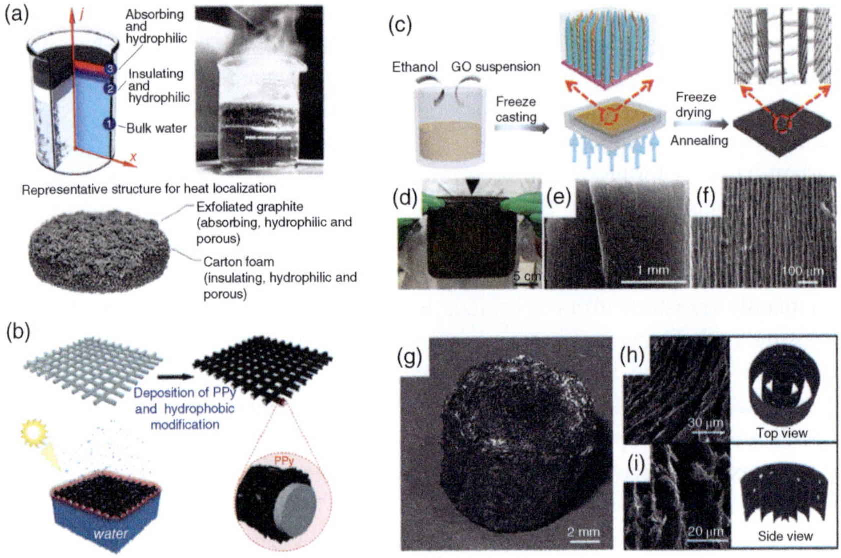

Figure 4.9 (a) A schematic illustration of a digital image of a double-layered evaporator to realize heat localization for interfacial solar steam generation, the fabrication process, and a digital photograph. Source: Reproduced with permission from Zhang [55]/Springer Nature. (b) A schematic illustration of the preparation of the light-to-heat conversion membrane for interfacial heating. Source: Lianbin Zhang et al., 2015 [56]/with permission of John Wiley & Sons. (c) A schematic of the fabrication of vertically aligned graphene sheet membranes (VA-GSM). (d) A photograph of VA-GSM. (e, f) A schematic illustration of the SEM images of VA-GSM. Source: Reproduced with permission from [57]/American Chemical Society. (g) A representative photograph of the N-RGO aerogel with a ring-like architecture. (h) The SEM images and the corresponding schemes of the ring-like N-RGO aerogel from the top view and (i) the side view. Source: Reproduced with permission from [56]/American Chemical Society.

At the initial stage of developing carbon solar evaporation materials, the unoptimized graphene channel system faces the problem of disordered sheet arrangement. The disordered channel structure hinders the effective rise of water and is not conducive to the smooth dissolution and discharge of salts, which are prone to form crystalline precipitation on the channel surface. These crystal accumulations not only block the path of water flow but also may obscure the light, further inhibiting the photothermal conversion efficiency and limiting the performance of the whole light-driven water evaporation system. In response to this question, researchers developed a graphene membrane with a vertically aligned layered structure (Figures 4.9c–f) so that the graphene channels show a regular and ordered arrangement pattern to promote the rapid evaporation of water [57]. This structural change resulted in significant performance enhancement. The ordered channels promote smoother migration of water molecules while ensuring effective salt dissolution and reducing the risk of crystalline blockage, maintaining an unobstructed optical transmission path. Furthermore, to simultaneously manage the rapid water transfer and effective solar heating, researchers assembled porous N-doped reduced graphene oxide (N-RGO) meshes in the 3D photothermal aerogel, endowed with a unique annual ring architecture (Figures 4.9g–i) [56]. Representative SEM images illustrate that the N-RGO meshes are oriented perpendicularly, creating parallel microchannels throughout the resilient aerogel, which boasts a substantial porosity of 88%. In addition to the innovative design of vertical microchannels, Dai's work incorporates a unique 3D gradient within these channels, strategically concentrating water toward the hottest areas to facilitate highly efficient interfacial evaporation. This dual approach, combining the vertical channel architecture with the intelligent gradient sizing, results in remarkable performance metrics: an elevated evaporation rate reaching 2.53 $kg/m^2/h$ and a solar-to-vapor conversion efficiency that surpasses honeycomb-like aerogels by 90.3%, highlighting a substantial 41% improvement.

4.2.1.4 Hybridized Materials

It is worth noting that Au not only exhibits photothermal effects on its own but can also be hybridized with other materials to improve photothermal efficiency or form multifunctional composites. The formation of hybrid materials is one way to achieve enhanced light absorption and photothermal heating for effective solar evaporation. Hybrid materials of various combinations, including bimetallic, metal–organic, and metal–semiconductor composites, have been developed to improve photothermal conversion efficiency. By utilizing the optical properties of individual materials, these hybrid materials acquire complementary and/or synergistic optical properties [46]. As an excellent plasma metal, Au NPs can interact with each other at a suitable distance to produce a strong resonance and are able to form hotspots at high temperatures (Figures 4.10a, b), like the formation of a localized strong thermal zone between two closely neighboring grapes in a microwave oven. Researchers localized Au NPs at pores of the graphene meshes [56]. This design ensures that water molecules transport through these pores, with the hotspots around the holes highly concentrating thermal energy onto the water channels within the pores. This precise localization of thermal energy along the transport paths of water molecules dramatically enhances evaporation efficiency and improves both mass and heat transfer through an optimized physical structure. Furthermore, by carefully tuning the distribution distances of Au NPs, it is possible to adjust and extend the material's light absorption capability across the visible spectrum. This nonuniform distribution strategy cleverly harnesses the LSPR properties of Au NPs, enabling the composite material to comprehensively absorb the solar spectrum, enhancing light absorption efficiency and significantly improving the performance of light-driven water evaporation.

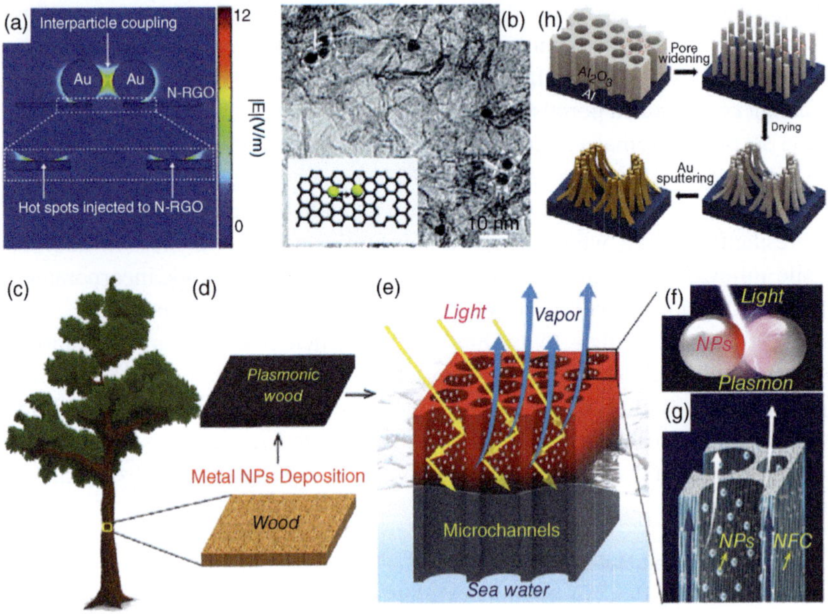

Figure 4.10 (a) Simulation results of the hot-electron distribution of Au nanoparticles around graphene nanopores. (b) TEM images showing the Au nanoparticles around the nanopores (highlighted by rectangles), with corresponding schematic illustrations. Source: Reproduced with permission from Meng [56]/Royal Society of Chemistry. (c) A tree transports water from the bottom upward and absorbs sunlight for photosynthesis. (d) After nanoparticle decoration, the natural wood is cut perpendicular to the growth direction of the tree and turns black due to the plasmonic effect of the metal nanoparticles. (e) After metal nanoparticle decoration, light can be guided into the wood lumen and be fully absorbed for steam generation. (f) Schematic of the plasmonic effect of two adjacent metal nanoparticles (NPs). (g) Zoomed-in schematic illustrating the water transport along the microchannels in wood. The cell wall is composed of many nanofibrous cellulose (NFC). Source: Zhu [58]/Advanced Energy Materials. (h) Schematic illustration of the method to fabricate black gold membranes by employing a pore-widening process in hexagonal AAO templates. Source: Bae [57]/Springer Nature/CC BY 4.0.

Similarly, Zhu has mastered the art of solar absorption, reaching nearly 99% efficiency across an expansive spectrum from 200 to 2500 nm [58]. He accomplished this by embedding a variety of plasmonic metal NPs into the 3D, mesoporous structure of natural wood, a novel material dubbed "plasmonic wood" (as shown in Figures 4.10a–d). This integration capitalizes on the wood's inherent architecture and the NPs' plasmonic attributes to significantly amplify the breadth and intensity of light absorption. Figure 4.10b captures the metamorphosis of the wood into its plasmonic variant, a visual testament to its enhanced absorptive capabilities. The diagrams in Figures 4.10c, d show the mechanism by which light interacts with the plasmonic NPs lining the wood's microchannels, a process pivotal for the conversion of solar energy into thermal energy. The even distribution of NPs throughout the wood's microchannels, highlighted in Figure 4.10e, is essential for achieving exceptional broadband absorption. The juxtaposition in Figure 4.10a between the pristine wood and the plasmonic wood underscores the dramatic shift in optical characteristics, from minimal to maximal light absorption, underscoring the profound impact of NP incorporation on the material's optical behavior.

As depicted in Figure 4.10f, a prior study elaborated a flexible, thin-film membrane through the self-assembly of collapsed nanowires. This was achieved following an appropriate enlargement of the pores in AAO templates, which had been prepared using a standard two-step anodization procedure. The resultant structure presents a meticulously organized array of hexagonal nanopores, optimally designed to amplify light absorption and the LSPR effect in the deposited metallic layers [57]. The team succeeded in generating water vapor efficiently, with a solar thermal conversion efficiency of 57% at a light intensity of $20\ kW/m^2$.

4.2.2 Introduction to the Design Strategies of Water Condensation and Collection Devices

Based on the above discussion, we have gained insights into strategies and current practices in material design for photothermal evaporation. For solar-induced interfacial steam generation (SISG) in desalination, in addition to highly efficient seawater evaporation, a vapor condensation device to assist water collection is essential. Over the past few years, growing attention has been devoted to designing desalination devices to achieve a high vapor collection rate. As a pioneering solar desalination technology, the device of traditional solar distillers serves as the foundation for guiding the condensation chamber design of solar interfacial evaporation systems. In the following section, we will introduce devices and structural strategies for freshwater or clean water condensation and collection.

4.2.2.1 Single-Stage Condensation Device

Upward Vapor Collection A traditional condensation device is essentially a confined chamber with a transparent roof that allows incident light to pass through and facilitates vapor condensation. The first condensation device for water collection in an SSG system, reported by Wang and coworkers in 2015, was composed of transparent plastic walls to separate the evaporation chamber and condensation chamber and a tilted transparent cover to enclose the device (Figure 4.11a) [59]. Upon exposure to solar radiation, the process of seawater evaporation occurs, with the resulting vapor rising to occupy the upper compartment. Positioned on the

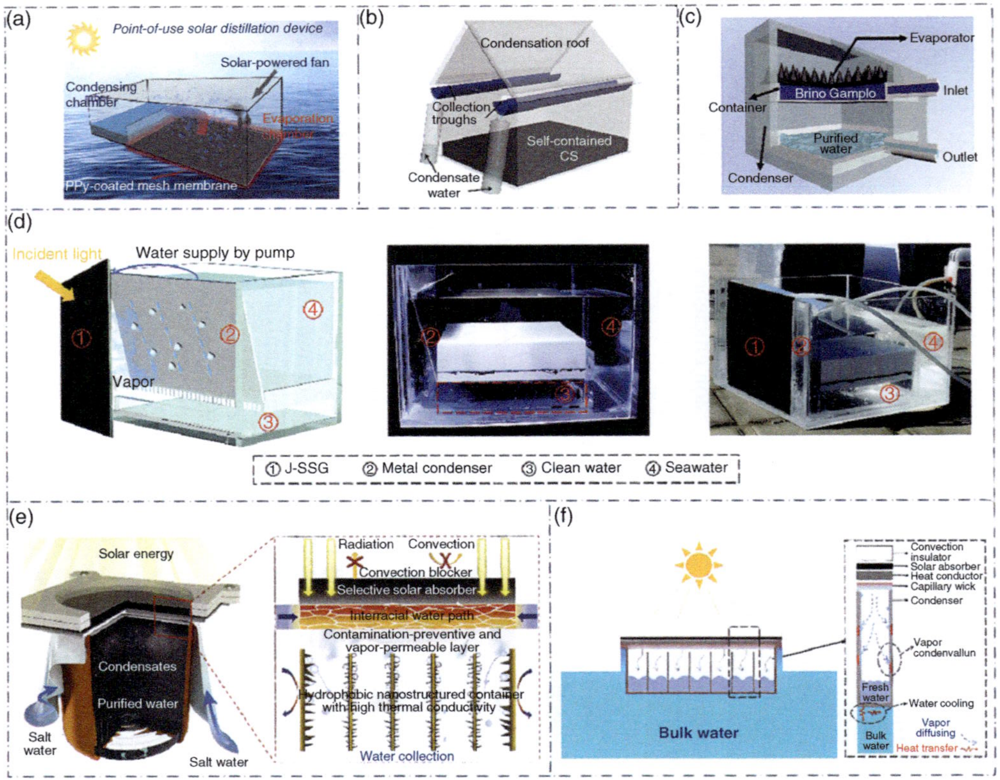
(a)
Point-of-use solar distillation device
Solar-powered fan
Condensing chamber
Evaporation chamber
PPy-coated mesh membrane
(b)
Condensation roof
Collection troughs
Self-contained CS
Condensate water
(c)
Evaporator
Brino Gamplo
Inlet
Container
Purified water
Outlet
Condenser
(d)
Incident light
Water supply by pump
Vapor
① J-SSG
② Metal condenser
③ Clean water
④ Seawater
(e)
Solar energy
Condensates
Purified water
Salt water
Salt water
Radiation
Convection
Convection blocker
Selective solar absorber
Interfacial water path
Contamination-preventive and vapor-permeable layer
Hydrophobic nanostructured container with high thermal conductivity
Water collection
(f)
Bulk water
Convection insulator
Solar absorber
Heat conductor
Capillary wick
Condenser
Vapor condenvallun
Fresh water
Water cooling
Bulk water
Vapor diffusing
Heat transfer

Figure 4.11 Single-stage condensation device. (a) An illustration of the integrated solar distillation apparatus designed for the on-site and streamlined purification of freshwater. Source: Zhang [59]/John Wiley & Sons. (b) A diagrammatic representation of a prototype solar still for residential applications. Source: Zhu [24]/John Wiley & Sons. (c) A schematic representation of the batch purification prototype, demonstrating the solar-assisted transformation of brine into distilled water, highlighting the evaporation on a 3D surface and the subsequent condensation and collection mechanisms. Source: Wu [60]/Springer Nature/CC BY 4.0. (d) A multifaceted depiction of the J-SSG solar desalination system, including an exploded-view diagram for component analysis (left), a side view with visually distinct blue water (middle), and an actual outdoor test scenario (right). Source: Reproduced with permission from Yao [61]/Royal Society of Chemistry. (e) A comprehensive depiction of the ISWP, with a structural diagram (left) detailing the layers of the device and a process schematic (right) illustrating the solar-driven purification sequence, from light absorption to vapor condensation. Source: Wang [62]/with permission of Elsevier. (f) A schematic representation of the solar desalination system's operation, highlighting the conversion of solar irradiation into thermal energy, the subsequent vaporization at the wick, and the condensation phase that releases latent heat, which is then dissipated through the grid-trough condenser. Source: Du [63]/with permission of Elsevier.

wall of the evaporation chamber is a solar-powered fan that creates a flow of air, propelling the vapor toward the condensation chamber where it is collected as freshwater. The device, positioned with its bottom in direct contact with seawater, utilized seawater as the cooling medium to facilitate liquid condensation. The recorded freshwater yield reached $0.15\,kg/m^2/h$, a significant increase of 15-fold compared to that of the condensation system devoid of photothermic evaporation technology. Later, this single-slope condensation device design was widely adopted [64–69]. To improve the light absorption of the SSG system, especially considering the variation in the incidence angle of natural sunlight, condensation devices with symmetrical surface structures have been developed [70–73]. As shown in Figure 4.11b, a potential simple and passive prototype for condensate collection was constructed with Plexiglas with high transmittance to sunlight and two collection troughs for vapor collection [24].

To save surface space, the single-slope condensation device was improved by integrating the condensing chamber directly under the photothermal evaporation tray. As shown in Figure 4.11c, a series of purification equipment, including an intake pipe, a condensing unit, a saline solution reservoir, and a discharge pipe, was assembled to emulate the actual water purification procedure [60]. The developed evaporator features an array of 3D structures, with the gaps between them filled by flat surfaces. Vapor generated from seawater continuously condenses on the top and sidewall of the condenser, and then trickles down to the bottom for collection via an outlet tube. The system is designed for continuous purification, with seawater replenishment through an inlet tube and purified water collection through an outlet tube. The proximity of the photothermal container to the condensation wall enhances the condensation efficiency. The collection rate of 1.72 $kg/m^2/h$ is lower than that of the open system (2.54 $kg/m^2/h$), which is due to the higher humidity and condensate retention in the closed system. The water quality was verified using Inductively Coupled Plasma Mass Spectrometry (ICP-MS), revealing ion concentrations well below the World Health Organization (WHO) and Environmental Protection Agency (EPA) drinking water standards, thus confirming the purification efficacy of the system. Hence, during evaporation, vapor is generated and transferred to the inner surface of the chamber to condense, where the tilted or domed roof facilitates the sliding of the condensed liquid for collection [24, 71, 74–76].

Side Vapor Condensation In the conventional condensation design mentioned before, the top cover serves as the main passage for light irradiation. As a result, sunlight may be blocked and/or scattered by the vapor and water droplets, and hence the concept of lateral condensation has been proposed [6, 61]. For example, in a work that proposed a novel solar desalination system (Figure 4.11d) [61], a dual-functional Janus-interface solar steam generator (J-SSG) was successfully implemented, featuring a distinct separation of water evaporation and solar–thermal conversion processes on the opposing sides of a thin-film device. This innovative J-SSG is capable of achieving a remarkable water evaporation rate of 2.21 $kg/m^2/h$ under standard solar irradiance, within an expansive area of 100 cm^2. Notably, a new benchmark with a high specific water productivity (SWP) of 1.95 $kg/m^2/h$ was set in a straightforward setup, with an impressive efficiency of 88%, representing the ratio of the SWP to the evaporation rate. During an outdoor evaluation in Beijing, with solar energy levels of 15 $MJ/m^2/h$, the device effortlessly yielded 10 L of purified water per square meter. The introduction of Janus-interface engineering in solar steam generators offers a groundbreaking approach and an effective solution for scalable solar-driven water generation. As shown in Figure 4.11d, the accompanying illustration presents a detailed exploded-view diagram of a solar desalination system that incorporates a J-SSG, which integrates solar absorption and conversion with the processes of water evaporation, water source storage, and vapor collection. The J-SSG is positioned vertically, with the solar absorption face oriented outward to capture sunlight and the water evaporation face facing inward toward the system. Direct sunlight irradiates the light absorber of the J-SSG, which reduces the estimated value by approximately 26%, caused by reflection and scattering from the water vapor produced during the process. This design enhancement results in about a 35% increase in sunlight utilization efficiency when compared to the previous models of solar steam generators that required transparent covers to direct light incidence. Moreover, the desalination system includes a water tank designed to hold the raw water source, such as seawater. This tank features an inclined surface crafted from a superhydrophobic metal, which is strategically positioned to face the vapor generation side of the J-SSG. This specialized surface is instrumental in the efficient condensation of water vapor back into a liquid form, thereby contributing to the overall effectiveness of the solar desalination process. Furthermore, the surface wettability of a chamber is crucial for the effective and targeted collection of water: Discrete droplets tend to form on a hydrophobic surface, while film-wise condensation occurs on a hydrophilic surface [1, 77]. Droplet condensation results in a lower heat transfer resistance than does film-wise condensation. However, the presence of water droplets inevitably scatters the incident light, declining the light energy input for evaporation [70, 78–80]. As introduced in the study, a superhydrophilic coating was applied on the inner side of the condensation chamber to improve water collection. During solar evaporation, the pristine surface without a hydrophilic coating became blurry within 30 minutes due to the formation of small water droplets. In comparison, the hydrophilic coating surface remained transparent even after six hours of continuous operation, resulting in a 37% improvement in the vapor collection rate compared with that of the pristine surface [81]. Although film-wise condensation is beneficial for reducing incident light flashes, the direction of vapor condensation does not affect the absorbance of the incident light, and the metal-sloped surface facing the vapor-generating side is superhydrophobic for better condensate collection. In the solar desalination process, seawater is directed to the rear of the photothermal surface for the purpose of evaporation. The cold seawater in the tank facilitates the quick condensation of vapor on the inclined metal surface, thereby avoiding light energy loss on the photothermal surface. The presence of condensing water droplets leads to reflection and scattering, which impact the light transmission. Benefiting from this synergistic effect, the water production rate has achieved 1.95 $kg/m^2/h$, nearly matching the evaporation rate in an open environment, which is 2.21 $kg/m^2/h$. Furthermore, the collected water has demonstrated an ion rejection rate of 99.9%.

Downward Vapor Collection All of the above condensation collection devices collect upward, which inevitably causes light-scattering problems during the collection process, further augmenting the optical loss experienced by the upper transparent cover. The top cover serves as the main passage for light irradiation, and sunlight may be blocked and/or scattered by vapor and water droplets. Therefore, to avoid light-scattering issues, the generated vapor was driven downward to condense in the bottom collector, avoiding optical loss from the transparent cover on the top. Many groups have reported novel downward vapor collection methods, which are different from the conventional vapor condensation direction methods [62, 63, 82–85].

A previous study introduced an innovative single-stage solar water purifier with an inverted design. This design features a top selective absorber and a honeycombed water condenser at the bottom, both of which are constructed with hydrophobic nanostructured copper. Sunlight passes through a convection blocker and is absorbed by the selective absorber, which converts it into heat. This heat is then transferred to evaporate water on a hydrophilic layer. The resulting vapor passes through a contamination-preventive layer that is also vapor-permeable and condenses in the condenser at the bottom. This inverted configuration minimizes optical losses caused by vapor condensation and enhances heat transfer and condensation efficiency, as illustrated in Figure 4.11e. Consequently, it achieved a remarkable water collection rate of 1.063 $kg/m^2/h$, ranking among the top performers for single-stage solar water purifiers, as reported in [82].

Li and his team developed an inverted iceberg-inspired solar water generator (ISWG), which integrates a thermoelectric generator coated with a photothermal substance, a bilayer all-fiber nonwoven fabric, and a passive cooling vapor condenser. Taking cues from the structure of an iceberg, the ISWG system is predominantly submerged beneath the water surface, with only the evaporation component exposed above the waterline. This design of the ISWG system circumvents losses in solar energy input due to vapor condensation and efficiently transfers heat to the adjacent water, facilitating a passive cooling effect and an improved vapor capture. Additionally, the thermoelectric generator leverages the temperature gradient between the photothermal and evaporation layers, employing a sandwiched arrangement for thermal conduction instead of a conventional metal plate. The ISWG is capable of achieving an impressive vapor collection rate of 1.02 $kg/m^2/h$ and an additional electrical output power of 0.47 W/m^2 under sunlight, with ambient temperatures around 42°C. This innovation could lead to the advancement of a highly efficient solar-driven water and electricity co-generation system, which is particularly beneficial for tropical island regions [62].

Du and Huang have introduced a practical approach to vapor condensation, characterized by a streamlined design, affordability, scalability for mass production, and adaptable operation. The energy efficiency of the solar–vapor–water cycle has been theoretically dissected by examining the heat and mass transfer dynamics within a solar desalination setup, offering valuable insights for enhancing vapor condensation performance. The innovative system, which integrates an advanced grid-trough condenser with a reverse heat-localized evaporator, demonstrates exceptional promise for efficient and user-friendly solar desalination solutions. Figure 4.11f describes the fundamental principle of a floating solar desalination system, which is composed of two main components: (1) A reverse-engineered heat-localized evaporation assembly, designed to optimize vapor production. This innovative structure positions vapor generation on the opposite side of solar energy absorption, enhancing efficiency. (2) A grid-trough condensation module that floats on the water's surface, utilizing water cooling to efficiently condense the vapor. Upon exposure to solar irradiation, the solar absorber captures the majority of incoming solar rays, transforming them into thermal energy. To minimize heat dissipation from the absorber to the environment, it is shielded by a convection barrier, directing heat primarily through a downward heat conductor. The solar absorber is affixed to this conductor, creating a photothermal plate. When the system is bathed in sunlight, the solar absorber soaks up solar rays, generating heat that is transferred to a capillary wick layer, promoting vaporization. The vapor then migrates to the condenser located beneath the evaporation zone. Contact with the condensing surface triggers vapor condensation, releasing latent heat, which is then conducted away into the underlying water mass via the grid-trough condenser. Under optimal conditions and one sun illumination, the system can achieve a freshwater yield of 0.393 $kg/m^2/h$ and a vapor collection efficiency of 38.8%. This performance results in a desirable equilibrium between efficiency, cost, and system complexity. Seawater trials and outdoor tests have validated the system's robustness, demonstrating its ability to float autonomously on a lake and consistently produce freshwater. Moreover, the energy efficiency of the system can potentially be enhanced through structural refinements and the incorporation of advanced materials [63].

4.2.2.2 Multistage Condensation Device

Despite these advancements in single-stage condensation devices, a gradual decrease in the rates of evaporation or condensation was observed in a continuous solar desalination process. This is because in a single-stage condensation device, the condensing process relies on the release of latent heat through energy exchange with the ambient environment. Consequently, this results in a temperature increase within the condensation chamber, thereby hindering the ongoing phase transition processes. To reuse the lost latent heat and cool the condensation chamber, multistage condensation systems have been developed [86–93].

The concept of a multistage solar evaporation-distillation system was first proposed by Zhou and coworkers in 2018 [86]. As shown in Figure 4.12a, a concise solar–thermal membrane distillation setup, characterized by three key structural elements—focused solar–thermal energy application, efficient cooling mechanisms, and latent heat reuse—has been introduced. Under open system conditions with 1 kW/m^2 solar irradiance, the system achieved a steam generation rate of 0.98 $kg/m^2/h$. Furthermore, with a two-tier configuration, the water yield is enhanced to 1.02 $kg/m^2/h$, resulting in a solar-to-water conversion efficiency of 72%. This innovative approach offers a straightforward and highly efficient method for water purification, making it particularly suitable

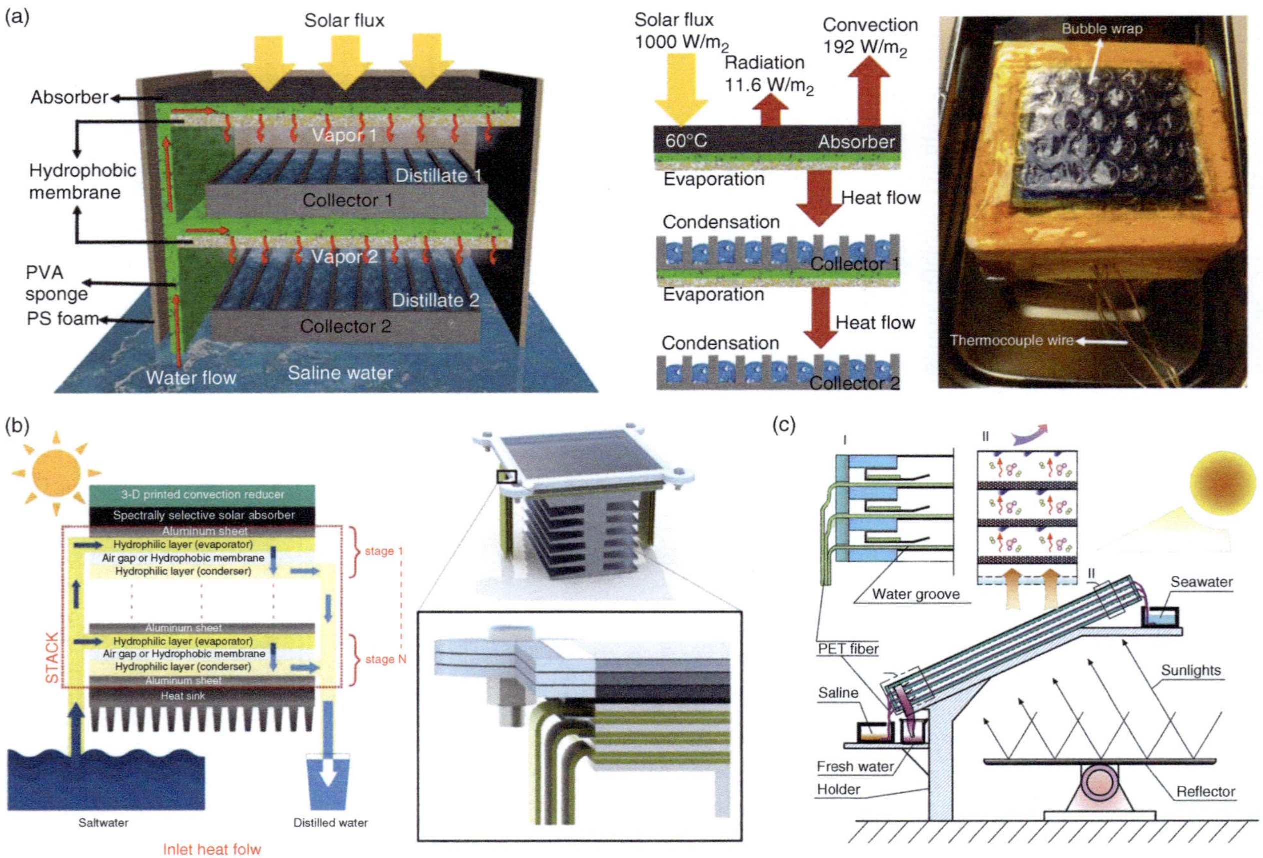
(a)
Solar flux
Absorber
Hydrophobic membrane
PVA sponge
PS foam
Vapor 1
Distillate 1
Collector 1
Vapor 2
Distillate 2
Collector 2
Water flow
Saline water
Solar flux 1000 W/m2
Radiation 11.6 W/m2
Convection 192 W/m2
60°C
Absorber
Evaporation
Heat flow
Condensation
Collector 1
Evaporation
Heat flow
Condensation
Collector 2
Bubble wrap
Thermocouple wire
(b)
3-D printed convection reducer
Spectrally selective solar absorber
Aluminum sheet
Hydrophilic layer (evaporator)
Air gap or Hydrophobic membrane
Hydrophilic layer (condenser)
Aluminum sheet
Hydrophilic layer (evaporator)
Air gap or Hydrophobic membrane
Hydrophilic layer (condenser)
Aluminum sheet
Heat sink
stage 1
stage N
STACK
Saltwater
Distilled water
Inlet heat folw
(c)
I
II
Water groove
PET fiber
Saline
Fresh water
Holder
Seawater
Sunlights
Reflector

Figure 4.12 Multistage condensation device. (a) A comprehensive depiction of the high-efficiency solar–thermal membrane distillation system, featuring a cross-sectional view of its components and water pathway (left), an energy balance and heat transfer analysis for the absorber under solar irradiance of 1,000 W/m^2 assuming a 5% thermal emittance and a temperature target of 60°C (middle), and an outdoor image of the water-harvesting prototype in action (right). Source: Xue [95]/John Wiley & Sons/CC BY 4.0. (b) A visual representation of the modular distiller, including a schematic overview (left) and an isometric view with sectional details of a three-stage prototype (right), depicting the stages of operation and the use of thermal management components to optimize performance. Source: Chiavazzo [91]/Springer Nature. (c) A comprehensive schematic of the distillation system, detailing the freshwater egress system (local graph I) and the associated heat and mass transfer processes (local graph II), providing a clear understanding of the system's operational dynamics.
Source: Ma [93]/with permission of Elsevier.

for domestic solar water purification systems. Additionally, this system relies exclusively on solar power. The solar absorber is warmed by the solar flux, creating a localized hot area. This area is passively saturated with the surrounding saline solution through a PVA sponge, which in turn produces vapor. The vapor is then directed downward to traverse a hydrophobic membrane and condense on a collector surface. The membrane itself is composed of a poly(vinylidene fluoride-*co*-hexafluoropropylene) (PVDF-HFP) nanofiber mesh positioned behind the PVA layer. The system leverages a capillary action to replace the intricate flow mechanisms typically found in conventional membrane distillation processes, eliminating the need for additional energy inputs. The low thermal conductivity of the PVA sponge also ensures minimal heat dissipation into the surrounding water [94]. The condenser in this system is strategically positioned to avoid direct contact with the membrane, which minimizes heat conduction to the surrounding water, thereby preventing unnecessary heat loss. This design ensures that the solar energy is effectively concentrated within the internal hot zone, enabling the water to reach elevated temperatures. In contrast to high-temperature vapor, bulk water remains relatively cooler, providing an ideal environment for vapor condensation. The release of latent heat to the condenser helps maintain a significant temperature gradient across the membrane, which facilitates the downward flow of vapor. By doing so, the system circumvents the issue of "temperature polarization," a common challenge in distillation processes. The solar energy input effectively counterbalances the latent heat carried away by the evaporated water and the heat released into the cooler bulk saline water, maintaining an optimal temperature difference for efficient distillation [86]. Notably, in this passive system, the water yield is inherently constrained by the condensation process. The thermal resistance caused by noncondensable gases remains a challenge and cannot be easily mitigated. Nonetheless, the system's simplicity in operation and its economic viability make it a promising solution for regions that are remote and economically disadvantaged.

Previous research introduced an inexpensive and passive multistage solar still, which incorporated effective energy utilization strategies to substantially boost the production of potable water [90]. As shown in Figure 4.12b, solar energy is captured and transformed into thermal energy on the top surface of the initial stage of the distiller. This stage features a thermal insulation system comprising three 2-mm air gaps, which are sandwiched between thin sheets of transparent linear low-density polyethylene (LLDPE). These sheets are held in place by an acrylonitrile butadiene styrene (ABS) frame, which is fabricated using 3D printing technology. Within these air gaps, convective heat transfer is minimized, promoting heat conduction as the primary mode of thermal transfer. At the base of the distiller, an Al heat sink is installed to effectively dissipate heat from the final distillation stage into the surrounding environment. It is important to note that this heat sink operates solely on natural convection, eliminating the need for an external power source. Each distillation stage is composed of two hydrophilic layers that are separated by a hydrophobic microporous membrane, without the need for any additional mechanical components. Under practical conditions, the system has shown an impressive distillate flow rate of nearly 3 $L/m^2/h$ from seawater, even at less than one solar intensity, which is a twofold improvement over that of recent passive distillation systems. Furthermore, theoretical models suggest the potential for a flow rate of up to 6 $L/m^2/h$. This sustainable system could be instrumental in addressing the freshwater requirements of remote and economically disadvantaged communities, ensuring long-term reliability.

As a potent desalination technique, multistage solar distillation, which leverages interfacial thermal localization, faces challenges related to the buildup and crystallization of salt ions. This issue complicates the achievement of scaled and enduring water generation. A recent study has put forth and investigated an entirely passive and exceptionally efficient mechanism for the discharge of concentrated brine. This mechanism is predicated on the principle of open siphon flow, offering a promising solution to the aforementioned challenges in the desalination process [92]. Nanofibers were utilized to form the core of a superhydrophilic membrane for the purpose of examining the open siphon flow phenomenon. Research has revealed that when the suction height is sufficiently high and the cross-flow length is short, the siphon flow can reach a consistent level. This consistency enables the synchronization of siphon flow rates with the heat capacities across various evaporative condensation interfaces, which is crucial for optimizing the efficiency of the desalination process. As shown in Figure 4.12c, a theoretical and experimental investigation was conducted on a three-stage solar still. When exposed to one sun irradiation, a water production rate of 238 g/h was achieved. This output represented approximately 82% of the total seawater inflow, with the remaining 18% being discharged as concentrated brine with a salinity level of 19.6%. Furthermore, during an all-day test, the system produced 940 g of water, averaging a yield ratio of 1.5, which translates to an impressive 11.99 kg/d/m^2. The findings of this study have the potential to enhance large-scale interfacial evaporation processes and contribute to more effective desalination techniques.

Desalination processes that leverage the "excess thermal energy" from photovoltaic (PV) panels offer a dual benefit of generating potable water and electricity. Such systems are particularly advantageous for remote areas that lack access to conventional power and water infrastructure. Based on theoretical frameworks, scientists have shown that it is feasible to enhance the yield of freshwater while simultaneously reducing the operating temperature of solar cells. This approach not only optimizes the performance of solar energy systems but also addresses the critical needs of off-grid communities for sustainable electricity and water resources. A cutting-edge system, the five-stage PV-membrane distillation evaporation crystallizer (PMDEC), was implemented to achieve a remarkable and consistent freshwater output of 2.45 kg/m^2/h under standard solar irradiation. This innovative setup has also contributed to a significant decrease in the solar cell temperature to 47°C, which is notably lower than the 62°C, observed in similar cells without the evaporation. This temperature reduction led to an 8% increase in the power generation efficiency of the cell. Furthermore, the PMDEC system has effectively addressed the issue of brine concentration by incorporating a bottom evaporation crystallizer. This component ensures that the brine, a byproduct of the desalination process, is fully evaporated, leading to a zero-liquid discharge outcome. This not only makes the system environmentally friendly but also optimizes the overall efficiency of the desalination and energy production processes.

Despite these advancements, the currently reported data on water collection are limited to small-scale devices monitored within a short timeframe (several hours to a day) under specific conditions (location and environmental temperature). This limitation hinders an accurate reflection of realistic freshwater production performance under various weather conditions, making it challenging to provide a stable and secure water supply solution solely relying on solar energy. Consequently, exploring alternative energy sources becomes imperative. Additionally, the fabrication cost of the condensation device is another issue that needs to be considered for an affordable drinkable water supply.

4.3 The Development of Integrating Solar Water Evaporation with Sustainable Energy Conversion Applications

Electricity is an indispensable force that drives both industrial operations and everyday activities, supplying the necessary power for manufacturing, transportation, communication infrastructure, and residential appliances. However, the prevalent methods of electricity generation are largely dependent on nonsustainable energy sources, including fossil fuels and nuclear power. This reliance leads to significant environmental challenges, such as the exacerbation of climate change, increased air pollution, and worries about the sustainability of the energy supply and the scarcity of these finite resources. To foster sustainable growth and mitigate the ecological footprint, there is an urgent need to invest in and advance renewable energy solutions such as solar, geothermal, and wind energy, thereby reducing our dependence on conventional energy sources. Commonly, hydropower manifests in various forms, harnessing mechanical, latent, and chemical energy from water. In addition, significant energy potentials within water, particularly in regard to phase changes such as evaporation, remain largely untapped and often overlooked. Approximately 35% of the solar energy absorbed by the Earth is absorbed by water, amounting to an impressive 60 petawatts (10^{15} W). Harnessing even a small fraction of this immense energy stored in water could easily meet the global energy demands of 10^{12} W. Solar water evaporation represents a spontaneous continuous phase-change process where liquid water converts to vapor by absorbing heat from the environment. Unlike other forms of mobile hydropower (e.g., conventional hydroelectricity), water in solar evaporation serves not as a direct energy carrier but as a medium for converting environmental heat into electricity.

In typical solar-driven seawater desalination systems based on evaporation, interface evaporation induces specific conditions, such as rapid vapor generation, accelerated water flow, salinity gradients in thermoelectric evaporators, high humidity environments within the device, temperature differentials between the evaporating surface and bulk water, and internal/external device environments. These conditions collectively generate additional electricity and facilitate clean water production, making integrated solar evaporation technology for distributed water and power supplies highly appealing. Here, we explore advancements in technologies that integrate solar water distillation/evaporation with other energy conversion methods including PVs, thermoelectric, hydrovaltic, mechanical power generation, and salinity-gradient powering, for generating electrical energy to meet the energy needs for seawater desalination and beyond.

4.3.1 Combining Solar Water Evaporation with Photovoltaic Power Generation

Solar power, as a plentiful and omnipresent resource, has effectively showcased its capacity in addressing the water–energy challenge in an environmentally friendly manner. The interconnection between water and energy is crucial, affecting their availability and security. Despite this, current technologies for water and energy remain in separate domains. It is imperative to concentrate on developing integrated and synergistic solutions that enhance efficiency. In a world where water scarcity is becoming more prevalent, desalination is a key strategy for overcoming water shortages, given that most of the Earth's water is found in the form of seawater. Electricity, like water, is fundamental to contemporary society. However, there exists an imbalance between the production of electricity and its demand. The use of fossil fuels for power generation, which has been the norm for more than a century, is now fraught with escalating security and environmental hazards due to the consumption of finite fossil resources and the exacerbation of pollution contributing to global warming [22].

The traditional method of water desalination through RO is characterized by its high energy consumption and low recovery rates of seawater, primarily due to the challenges of brine concentration and the restricted osmotic pressure that can be exerted [95]. In contrast, membrane desalination (MD) addresses the common issues associated with RO, including the need for lower operating pressures and the management of high-concentration brine. This technology leads to improved water recovery rates [96]. However, MD faces challenges such as temperature polarization and the significant costs associated with the electricity required for heating. The fusion of established MD technology with innovative photothermal distillation (PD) technology has given rise to the sustainable approach of photothermal membrane distillation (PMD). PMD tackles some of the lingering issues in desalination by harnessing sunlight for localized heating at the membrane's feedwater surface, offering a method to generate freshwater without the need for electricity [97]. Incorporating thermally driven processes with established expertise can mitigate the deficiencies inherent to any singular methodology.

PV systems, which rely exclusively on sunlight as their energy source, stand out among other energy technologies due to their minimal carbon emissions, advanced technological readiness, and accessible entry points. Driven by economic advantages and supportive regulatory frameworks, PVs are anticipated to capture a growing segment of the energy market [98]. While PVs hold great potential, their advancement is hindered by limitations in converting sunlight into electricity efficiently. Upon exposure to sunlight, photons with a short wavelength (above the bandgap of the semiconductor) are captured by the PV cell and transformed into electrical energy. In contrast, photons with a long wavelength (below the bandgap) are converted into heat, which constitutes over 70% of the total solar energy absorbed. Consequently, the PV panel heats up, and this rise in temperature negatively impacts the performance of the PV system, leading to an approximate 0.5% decrease in energy efficiency for each degree Celsius such that the temperature increases [99]. This situation is further exacerbated by the fact that heat not only accelerates the aging of the PV panels but also has the potential to raise local temperatures in the case of large-scale solar installations, contributing to the urban heat island effect. This phenomenon is a concern for both the performance of PVs and the surrounding environment. As a result, there is a growing interest in the industry and academia to investigate the most effective cooling methods for PV panels. Considering that semiconducting PV cells do not utilize long-wavelength photons and that short-wavelength photons are more than sufficient for photothermal applications, it makes sense to develop a solar capture system that is tailored to the specific spectrum of sunlight. This approach could optimize the use of the solar spectrum, improving the efficiency of PV systems while also addressing the issue of excess heat.

Integrating PV with PMD is indeed a key strategy for making the most of the solar spectrum by addressing the gaps in waveband utilization. Through a hybrid approach, high-energy photons are specifically harnessed by the PV cell to generate electricity. Meanwhile, the low-energy photons are harnessed and converted into thermal energy, which can be effectively utilized in the desalination process. This integration not only enhances the overall efficiency of solar energy utilization but also contributes to more sustainable and integrated solutions for energy and water challenges (Figure 4.13a). It is important to recognize that the desalination process serves a dual purpose: It generates freshwater and simultaneously provides essential cooling for the PV panels. This approach, straightforward yet efficient, tackles the intertwined challenges of energy and water with minimal environmental and land impacts. In summary, the field of PVs has seen significant advancements, characterized by a steady increase in efficiency and a decrease in associated costs [101–103]. Nonetheless, the performance of individual PV cells remains below the 30% threshold, with a significant majority (>70%) of the solar energy being wasted. Once absorbed, photons that possess energy exceeding the bandgap quickly convert to thermal energy through the process of thermalization, which moves photocarriers to the band edges [104]. Meanwhile, photons with energy below the bandgap will either contribute to heat generation if they remain within the cell or simply pass through without being harnessed for energy production (Figure 4.13b). The energy released through thermalization contributes to a rise in the temperature of the solar cell, which in turn adversely affects its PV efficiency [100].

To enhance both PV efficiency and freshwater generation, Wang and colleagues proposed an advanced hybrid system. This system is designed to incorporate

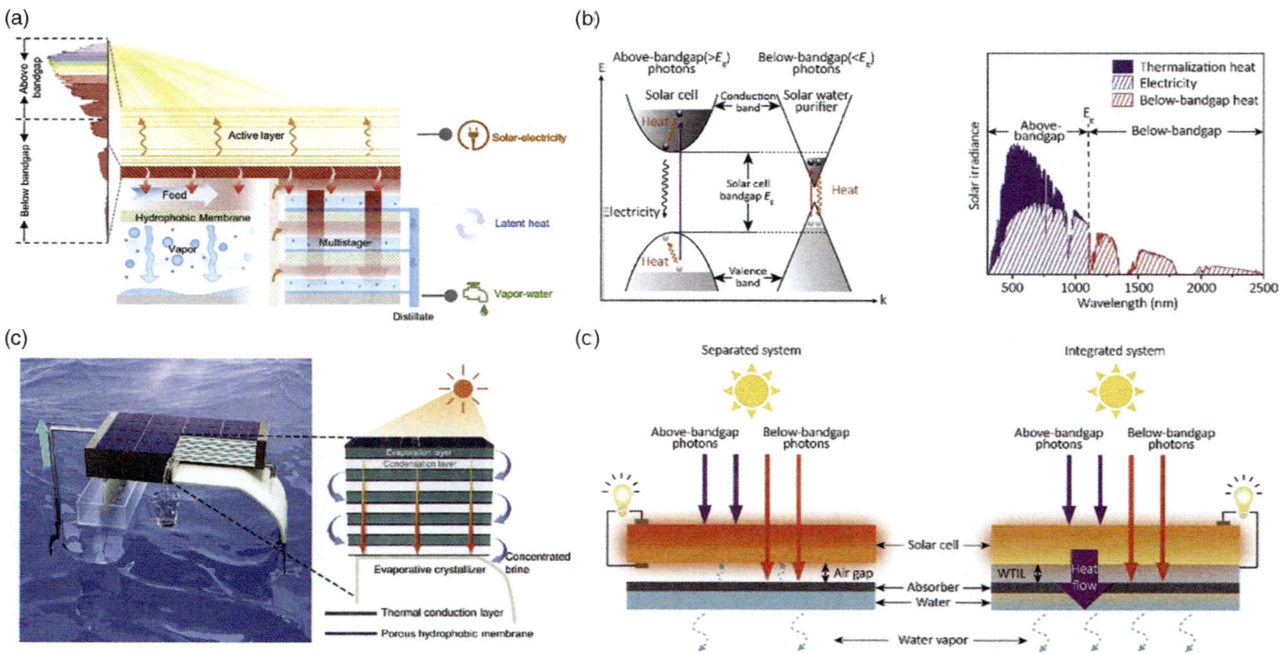

Figure 4.13 Photovoltaic power generation principles and applications. (a) Diagram depicting the strategy for harnessing the full solar spectrum. Source: Ding [95]/with permission of Elsevier. (b) A schematic depiction of the tandem solar electricity–water generator, featuring the band diagram on the left and a breakdown of solar energy utilization for dual output on the right, with distinct color coding for clarity. Source: Xu [100]/with permission of Elsevier. (c) Diagram of the hybrid photovoltaic and membrane distillation apparatus coupled with an evaporative crystallizer (PMDEC). Source: Reproduced with permission from Wang [90]. (d) A schematic depiction of the tandem solar generator, illustrating the transition from a separated to an integrated design, with a focus on the pivotal role of the water-proof thermal interconnecting layer (WTIL) in synergistically increasing both electricity and clean water production. Source: Reproduced with permission from Xu [100]/Elsevier, Reproduced with permission from [113]/ELSEVIER.

a multistage MD process alongside PV cells for optimal results [90]. As shown in Figure 4.13c, the silicon PV cell is strategically positioned above the distillation unit to capture and filter high-energy photons, while the remaining energy is harnessed to heat the feedwater in the distillation unit below. This setup represents an advance over previous hybrid systems that combine PV with distillation, as it employs an optimized cross-flow multistage membrane distiller to enhance the recycling of latent heat. Consequently, the excess heat generated by the solar cell is effectively dissipated and channeled downward through successive stages to produce potable water. The system's refined membranes and the number of distillation stages have been meticulously calibrated to achieve an optimal balance between vapor and heat transfer, significantly increasing the yield of freshwater. This is accomplished even after the PV cell has filtered out the high-energy photons. Furthermore, an evaporative crystallizer is integrated at the final condensing stage to ensure zero-liquid discharge and to function as a passive heat sink for the entire system. In addition to nearly doubling the freshwater yield, the multistage heat recycling mechanism efficiently extracts heat from the cell panel, reducing its temperature by 10°C. This temperature reduction leads to an increase in the electrical output. The practicality of this integrated water–energy technology has been confirmed through real-world testing, underscoring its significant potential in the field.

Researchers have crafted an integrated tandem solar system that performs a dual function: It generates electricity through the upper solar cell and concurrently produces clean water from the lower purification unit. This innovative system leverages the entire solar spectrum to achieve its objectives, making efficient use of all available sunlight for both power and water production [104]. A moisture-resistant thermal bridging component (MRTBC) links the upper and lower apparatuses, allowing for their collaborative operation. The well-conceived layout of the dual-function solar energy and water production system is depicted in Figure 4.13d. The system comprises two solar energy converters: a PV cell situated at the top, which generates electricity by capturing photons with energy above the bandgap, and a solar-driven water purification unit positioned beneath the PV cell, designed to produce potable water by harnessing the excess heat. A thermal interconnection layer (TIL) is strategically inserted to firmly bond these two units, facilitating the efficient conveyance of thermal energy from the upper PV cell to the lower water purification unit. The TIL plays a pivotal role in the synergistic operation of the devices, as it allows the top cell to be cooled by the water purification process beneath, which in turn enhances the cell's power conversion efficiency. Simultaneously, the water purification unit capitalizes on the otherwise wasted thermal energy from the top cell to increase the water purification capacity. To validate the practicality of the integrated solar electricity and water generation setup, an experimental prototype was constructed by integrating an infrared-transparent crystalline silicon PV cell with a solar water purifier that utilizes reduced graphene oxide (r-GO). Under sunlight with an intensity of 1.0 kW/m^2, this hybrid tandem system is capable of harnessing 74.6% of the incident solar energy, with 20.4% allocated to electricity generation and 54.2% to water vaporization. The remaining 15% of the solar energy that is not harnessed includes 11% due to optical reflection and 4% due to heat dissipation through radiation and convection, indicating potential areas for further optimization.

As shown in Figure 4.14a, the process of energy transfer, starting with solar energy as the primary source and the subsequent transformation of water energy during desalination, forms the energy flow within the solar-driven interfacial evaporation system, as referenced [109]. Earlier studies proposed a PV-MD apparatus (depicted in Figure 4.14b) that is capable of consistently generating clean water at a rate exceeding 1.64 $kg/m^2/h$ from seawater, without any degradation in its electricity generation efficiency, which remains above 11%, under the illumination of one sun [93]. The achievement of a high rate of clean water production is facilitated by the integration of a multistage membrane distillation (MSMD) system at the rear of the solar cell, which effectively recaptures the latent heat released during the condensation of water vapor at each stage of distillation. This integrated device has the potential to substantially decrease initial financial outlays by utilizing a shared land area and mounting infrastructure, thereby offering a feasible approach to repurposing an electricity generation facility from a water-consuming entity to one that generates freshwater.

In an effort to generate freshwater and simultaneously mitigate the release of greenhouse gases from power production, a study has suggested integrating a PV system with a novel desalination process that utilizes a passive, MSMD approach [105]. As shown in Figure 4.14c, the innovative desalination apparatus operates on low-temperature thermal energy and is devoid of the need for mechanical or electrical components. This system is capable of generating a maximum of 2 $L/m^2/h$ under standard solar irradiance. Additionally, the integration with the desalination process has resulted in a 4.5% enhancement in the efficiency of the PV module, which was attributed to a 9°C reduction in temperature. This research opens up possibilities for the development of compact, passive unit-based systems that can efficiently deliver energy and potable water with minimal ecological and esthetic footprints, particularly suitable for remote locations and emergency scenarios.

Previous research has developed an innovative hybrid setup that integrates a solar PV panel with a MSMD unit, enabling the concurrent production of electrical power and potable water (Figure 4.14d) [106]. In the integrated PV-MD configuration, the thermal energy produced by the PV panel is efficiently dissipated through an evaporative cooling mechanism, which lowers the operational temperature and consequently enhances the solar-to-electric

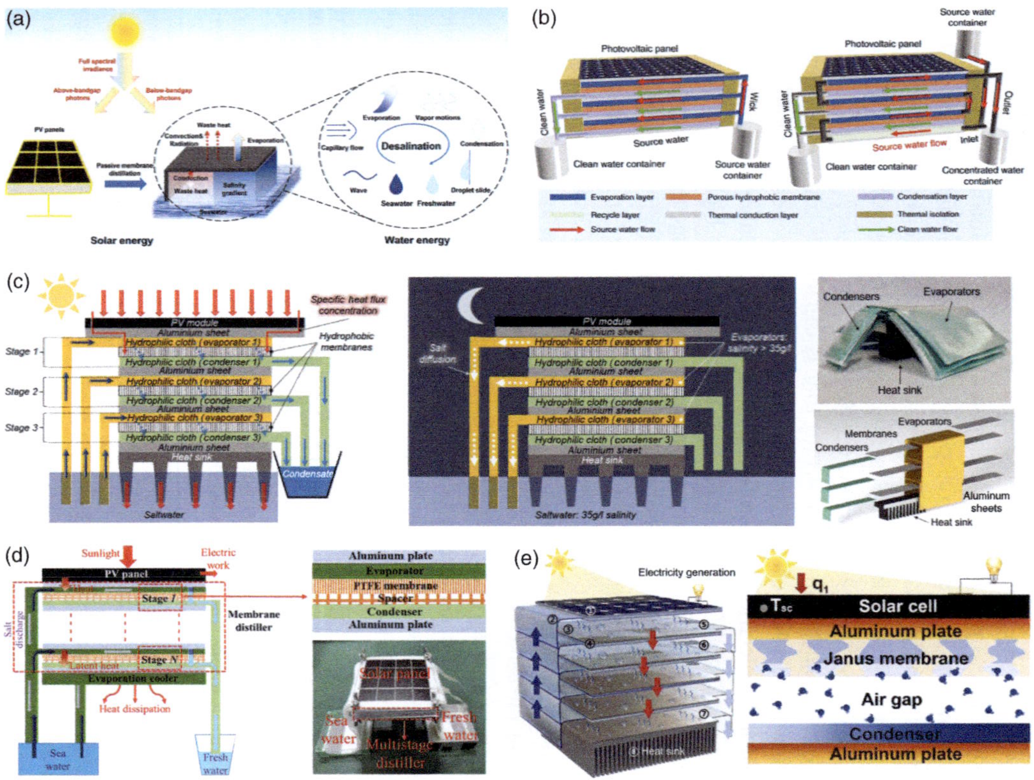
(a)
PV panels
Passive membrane distillation
Solar energy
Desalination
Evaporation
Vapor motions
Capillary flow
Condensation
Wave
Seawater
Freshwater
Droplet slide
Water energy
(b)
Photovoltaic panel
Clean water
Wick
Source water
Clean water container
Source water container
Outlet
Inlet
Source water flow
Concentrated water container
Evaporation layer
Porous hydrophobic membrane
Condensation layer
Recycle layer
Thermal conduction layer
Thermal isolation
Source water flow
Clean water flow
(c)
Specific heat flux concentration
PV module
Aluminium sheet
Hydrophobic membranes
Stage 1
Stage 2
Stage 3
Hydrophilic cloth (evaporator 1)
Hydrophilic cloth (condenser 1)
Hydrophilic cloth (evaporator 2)
Hydrophilic cloth (condenser 2)
Hydrophilic cloth (evaporator 3)
Hydrophilic cloth (condenser 3)
Heat sink
Condensate
Saltwater
Salt diffusion
Evaporators: salinity > 35g/l
Saltwater: 35g/l salinity
Condensers
Evaporators
Heat sink
Membranes
Aluminum sheets
(d)
Sunlight
Electric work
PV panel
Stage 1
Stage N
Salt discharge
Membrane distiller
Latent heat
Evaporation cooler
Heat dissipation
Sea water
Fresh water
Aluminum plate
Evaporator
PTFE membrane
Spacer
Condenser
Aluminum plate
Solar panel
Multistage distiller
(e)
Electricity generation
Heat sink
q1
Tsc
Solar cell
Aluminum plate
Janus membrane
Air gap
Condenser
Aluminum plate

Figure 4.14 Hydropower co-generation devices. (a) A diagrammatic representation depicting the pathway of energy transmission within the SDIE framework. Source: Gui [105]/with permission of Elsevier. (b) A schematic diagram showing the operation of the hybrid photovoltaic-membrane distillation system, featuring two distinct modes: (left panel) the dead-end filtration mode, where the feedwater is drawn into the evaporative section by the red-marked wick, and the resultant vapor condenses, exiting via the green-marked channel; and (right panel) the cross-flow filtration mode, characterized by the feedwater's passage into the evaporative section along the red-marked path, with the condensed liquid being discharged from the condensing section in the direction indicated by the green arrow. Source: Wang [94]/Springer Nature/CC BY 4.0. (c) An overview diagram of the photovoltaic module integrated with a passive multistage membrane distillation apparatus, highlighting: (left) the constituent elements and the dynamics of thermal and material transfer during the operational phase of daylight hours; (center) the system's self-cleaning mechanism engaged during the inactive nocturnal period; an illustrative depiction of the three-stage passive distillation unit; and a detailed disassembled perspective of the same three-stage unit. Source: Reproduced with permission from Antonetto [106]/Elsevier. (d) The fundamental configuration of a photovoltaic-membrane distillation (PV-MD) apparatus. (Left) A diagrammatic representation of the PV-MD setup, featuring directional indicators for the pathways of energy and substance exchange; an in-depth depiction of the MD assembly; and a visual snapshot of the complete system, encompassing a photovoltaic panel, a tiered distillation unit, a brine reservoir, and a freshwater receptacle. Source: Reproduced with permission from Huang [107]/Elsevier. (e) A conceptual diagram depicting the structure of a buoyant water-to-electricity conversion system, integrating a solar photovoltaic module with a membrane distillation process. Source: Liu [108]/with permission of Elsevier.

conversion efficiency. Moreover, the excess heat can be harnessed to power the MSMD process for the generation of freshwater. Under standard solar conditions, the hybrid system delivered an electrical power output of 66.6 W/m^2 with a resistance of 10 Ω, and it produced freshwater at a rate of 1.11 $kg/m^2/h$ across four distillation stages. Consequently, this system is well suited for implementation as a retrofit solution in existing PV modules, offering improvements in both PV efficiency and freshwater yield.

A recent study introduced a pioneering approach to integration by merging a floating PV array with a five-tier MD setup, facilitating the concurrent generation of clean water and electricity atop water bodies (Figure 4.14e) [107]. Mounted on the rear side of the PV cell, the MD apparatus, which incorporates a Janus membrane in its multistage design, passively regulates the temperature of the PV cell. It efficiently captures and reuses the excess thermal energy and effectively recycles the enthalpy of vaporization. This cohesive system delivers a remarkable freshwater yield of 4.14 $kg/m^2/h$ while also sustaining an elevated electrical power efficiency of 16.4% under standard solar irradiance, thereby optimizing the overall conversion of solar energy. With a daily water production capacity of 11.6 $kg/m^2/day$, this setup ranks among the most efficient solar stills under natural daylight conditions. This integrated solution sets a precedent for the co-generation of clean water and electricity with a reduced environmental impact, broadening the scope of terrestrial solar energy systems to include aquatic environments.

Capitalizing on the minimal initial investment, both financially and technically, the convergence of solar-powered PV systems with PMD holds promise for profoundly impacting the interconnection of water and energy resources. This strategy presents considerable opportunities for the synergistic and economical generation of both water and electricity. The notion of integrated water and energy production is on the horizon of becoming a common practice. Even though many of these innovative and sustainable systems are still in the experimental and developmental phase, it is imperative that we persistently drive innovation to meet the escalating demands and address the challenges posed by a changing climate.

4.3.2 Combining Solar Water Evaporation with Thermoelectric Power Generation

During solar evaporation, the temperature of the evaporator surface is commonly and typically higher than the temperature of bulk water and the surrounding environment, leading to energy loss through radiation, convection, and conduction into the environment. To mitigate this thermal energy waste, integrating thermoelectric devices into solar evaporators can harness temperature gradients between the photothermal evaporator and the bulk water, thus powering electricity. Standard thermoelectric generator units consist of *p*-type and *n*-type materials connected through conductive links. Applying a temperature gradient across these materials induces the Seebeck effect, converting thermal energy into electrical power. Ho's team incorporated thermoelectric generators at the base of solar absorbers, which were fabricated from CNT/cellulose nanocrystal (CNC) nanocomposites, to exploit the temperature difference between the solar absorber

Figure 4.15 (a) Synergistic production of water and electricity occurs through solar evaporation at interfaces and thermoelectric generation. Source: Zhu [110]/John Wiley & Sons. (b) Vaporization enthalpy storage and recycling enable the concurrent generation of clean water and thermoelectricity. Source: Li [111]/with permission of Elsevier. (c) Schematic diagram of the working mechanism of the thermoregulator-based evaporator. Source: Jiang [112]/John Wiley & Sons. (d) Schematic and operational diagrams of fibrous carbon sponge device. Schematic representation of electric potential induced by steam generation in carbon sponge. Source: Zhu [24]/John Wiley & Sons. (e) Diagram illustrating an interfacial evaporation system employing SDIE for solar-powered water purification. Voltage generation of the material after 30 minutes of immersion following multiple cycles under a 50 N load. Source: Meng [113]/John Wiley & Sons.

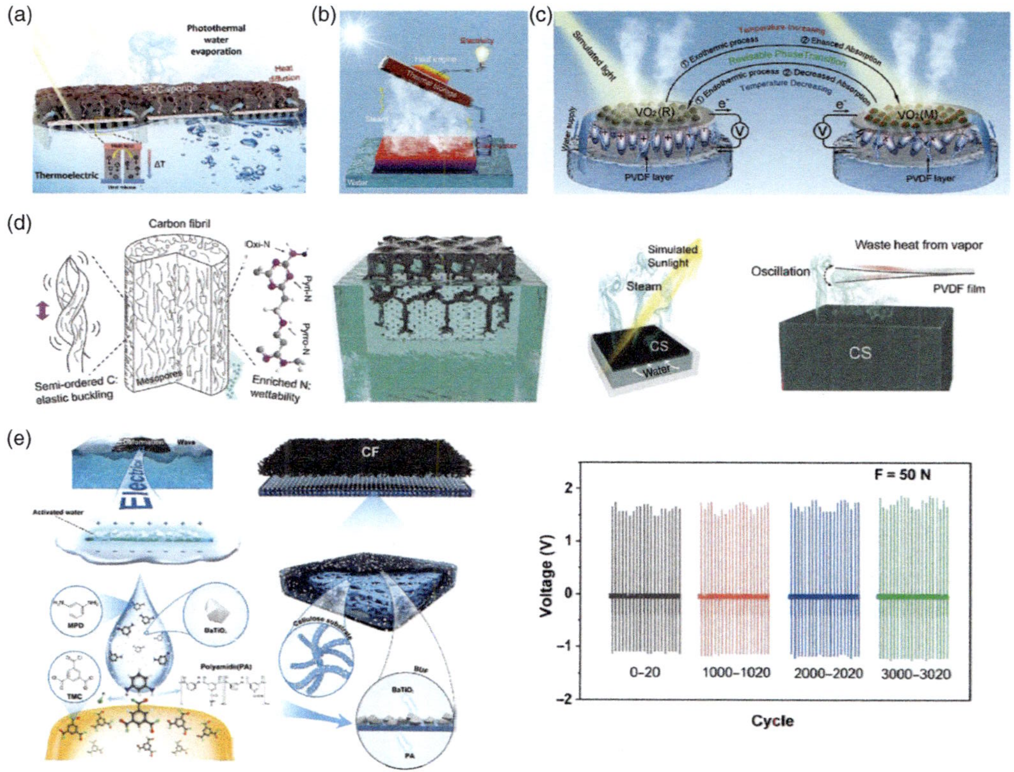
(a)
Photothermal water evaporation
Heat diffusion
Thermoelectric
ΔT
(b)
Electricity
Thermal storage
Water
(c)
Simulated light
Temperature increasing
① Exothermic process
② Enhanced Absorption
Reversible Phase Transition
① Endothermic process
② Decreased Absorption
Temperature Decreasing
VO2(R)
VO2(M)
e⁻
V
Water supply
PVDF layer
PVDF layer
(d)
Carbon fibril
Oxi-N
Pyri-N
Pyrro-N
Semi-ordered C: elastic buckling
Mesopores
Enriched N: wettability
Simulated Sunlight
Steam
CS
Water
Waste heat from vapor
Oscillation
PVDF film
CS
(e)
Wave
Electric
Activated water
MPD
TMC
BaTiO3
Polyamide(PA)
CF
Cellulose substrate
BaTiO3
PA
F = 50 N
Voltage (V)
2
1
0
−1
−2
0–20
1000–1020
2000–2020
3000–3020
Cycle

and the bulk water for generating electricity (Figure 4.15a) [110]. They achieved a maximum output power of 0.32 mW under a solar intensity of 1 Sun. Additionally, thermoelectric generators effectively suppressed thermal conduction losses, increasing the solar evaporation rate of water from 1.2 to 1.36 $kg/m^2/h$. Zhu's team installed thermoelectric generators in place of solar evaporators for real-time storage and recovering latent heat, enabling combined freshwater and electrical power production (Figure 4.15b) [111]. Using graphite and nonwoven fabric nanocomposite thin films featured high optical absorption and high thermal stability as the light absorbers, the system achieved a power density of 292.9 W/m^2 when irradiated under a light intensity of 30 Sun, possessing a solar conversion efficiency of 1.23% and a water evaporation rate of 34.8 $kg/m^2/h$. Here, the generated electricity can power small household appliances, such as fans and lights.

In contrast to thermoelectric generator mechanisms, pyroelectric materials exhibit spontaneous polarization in ferroelectric materials without any external electric field. Due to their unique thermal response, these materials can convert waste heat into electricity. When heated, dipole moments within the material lose orientation due to thermal vibrations, reducing spontaneous polarization levels attached to the material surface. Under short-circuit conditions, current flows between the material's two polar surfaces. Similarly, upon cooling, pyroelectric materials regain dipole orientation, increasing spontaneous polarization levels and generating reverse current under short-circuit conditions. Leveraging this principle, researchers designed a self-regulating temperature interface system with PVDF films and NPs doped with tungsten in vanadium dioxide (VO_2), exhibiting temperature gradients. When integrated into a photothermal interface evaporation system, it achieved pyroelectric power generation during solar evaporation (Figure 4.15c) [112]. When exposed to simulated solar irradiation of 1.1 kW/m^2, the solar evaporation system achieved self-regulating temperature production. It achieved a maximum temperature differential of 7°C, an evaporation efficiency of 71.43%, and a peak output power density of 104 $\mu W/m^2$.

4.3.3 Combining Solar Water Evaporation with Mechanical Energy Conversion for Electricity Generation

Solar evaporation-piezoelectric (SE-PZ) technology integrates interface evaporation with piezoelectric effects, representing an advanced solar energy conversion method. After liquid evaporation, the generated vapor applies pressure to piezoelectric materials, causing deformation that converts into electricity via the piezoelectric effect, thereby achieving effective solar power conversion. This technology involves crucial elements, such as solar interface evaporators, liquid mediums, piezoelectric materials, electrodes, and circuits. Steam generated through evaporation exerts pressure on contact with piezoelectric materials.

The division of positive and negative charges within the piezoelectric material generates a potential difference. This potential difference can be utilized through connected electrodes and circuitry to either power external devices or store electrical energy. The implementation of piezoelectric nanogenerators (PENGs) and triboelectric nanogenerators (TENGs) efficiently harnesses mechanical energy from steam and condensed liquid droplets to produce electricity. Ho's research team placed a layer of ferroelectric PVDF film atop nanostructured carbon sponge solar absorbers to harness thermal mechanical energy from steam for electricity generation (Figure 4.15d) [112]. Due to the thermal and kinetic energy of steam, electricity generation was achieved through pyroelectric–piezoelectric effects, where the diffusion of steam causes mechanical oscillations and temperature variations in PVDF films. The system produced an output power density of 240.7 $\mu W/m^2$ under 1 Sun irradiation and achieved a water evaporation rate of 1.39 $kg/m^2/h$. Recently, researchers integrated thin-film piezoelectric nanomembranes into solar evaporative systems using the integrated thermal management and evaporation system (TMES) technology (Figure 4.15e) [113]. The generated electricity from the evaporator activates adjacent water areas. This promotes evaporation, setting it apart from nonpiezoelectric evaporators. Additionally, it showed more than a 20% increase in efficiency compared to other evaporation systems and demonstrated durability through 3,000 repetitions.

4.3.4 Solar Water Evaporation-Induced Hydrovaltic Effect on Electricity Generation

At the solid–liquid interface, chemical functional groups on solid surfaces interact with water molecules through complex physicochemical processes. Water molecules, being polar, possess oxygen atoms with negative charges and hydrogen atoms with positive charges, capable of orientation and alignment in an electric field. As water molecules approach a solid surface, induced by the net charge on the solid surface, forces act on both ends of water molecules with positive charges and negative charges, generating a potential difference. It causes separation of charges within water molecules, an uneven distribution of positive and negative charges, forming an electrostatic field. The electrostatic field further induces orientational alignment of water molecules, where positive and negative charge portions face toward the solid surface, thus forming the hydrovaltic effect. In 2017, Guo's team pioneered a generator utilizing water evaporation on nanostructured carbon materials, reliably generating up to 1 V of continuous voltage under ambient conditions. Subsequent years saw widespread research interest in this field. As depicted in Figure 4.16a, the evaporation-induced generator (EIG) is composed of a porous CB film with CNTs deposited as electrodes at both ends [114]. One end of the EIG is immersed in water. As water evaporates from the gas–liquid interface, capillary forces within the porous carbon film move the

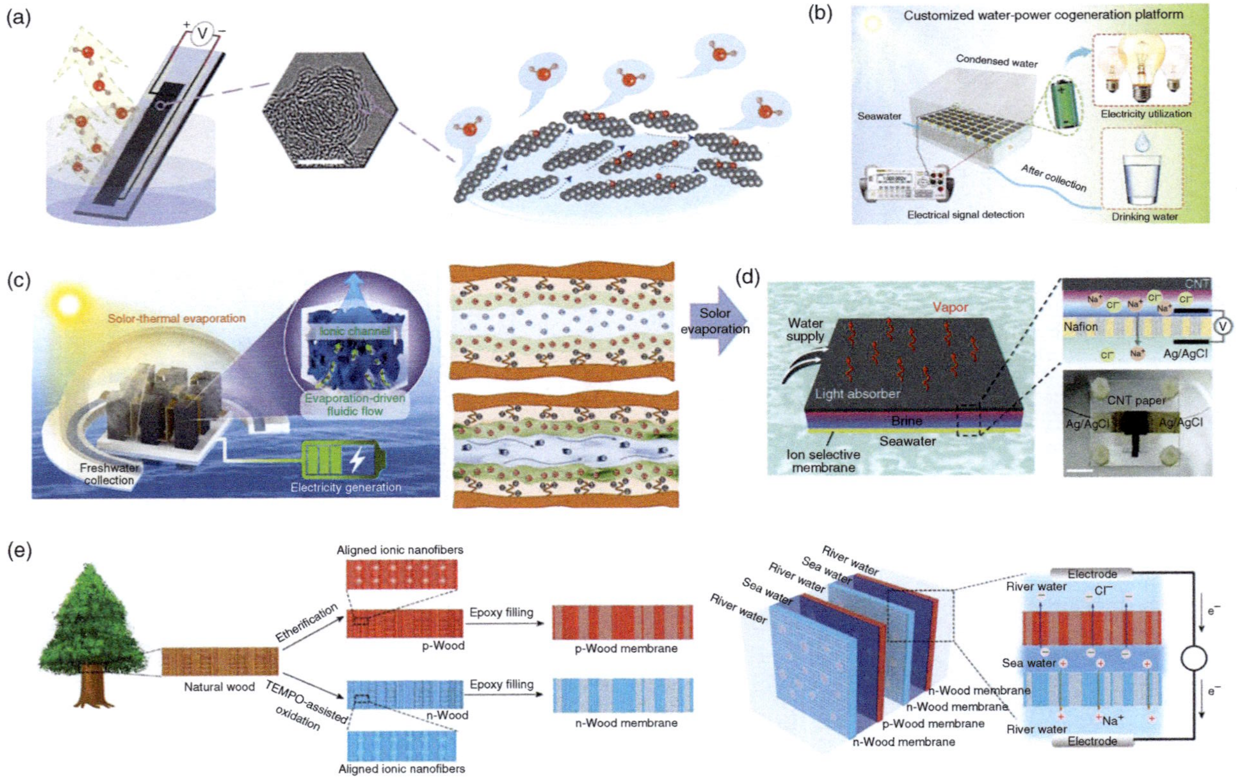
(a)
(b)
Customized water-power cogeneration platform
Condensed water
Seawater
Electricity utilization
After collection
Electrical signal detection
Drinking water
(c)
Solor-thermal evaporation
Ionic channel
Evaporation-driven fluidic flow
Freshwater collection
Electricity generation
Solor evaporation
(d)
Vapor
Water supply
Light absorber
Brine
Seawater
Ion selective membrane
CNT
Nafion
Ag/AgCl
CNT paper
(e)
Aligned ionic nanofibers
Etherification
p-Wood
Epoxy filling
p-Wood membrane
Natural wood
TEMPO-assisted oxidation
n-Wood
n-Wood membrane
River water
Sea water
n-Wood membrane
p-Wood membrane
Electrode
Cl^-
Na^+
e^-

Figure 4.16 (a) Device setups for electricity generators driven by evaporation. Source: Xue [114]/Springer Nature. (b) Schematic system design for reaching ultra-high voltages exceeding 100 V and impressive freshwater collection utilizing a thermodiffusion-enhanced hydrovoltaic generator. Source: Chen [115]/John Wiley & Sons. (c) Diagram, operational mechanism, and voltage output of 3D polymer sponge evaporators for integrated high-flux solar–thermal desalination and power generation. Source: Xu [116]/with permission of Elsevier. (d) Electricity produced by ion diffusion directed by salinity gradients. Source: Yang [117]/Royal Society of Chemistry. (e) The concept of power generation using ionized wood membranes in a salinity gradient. Manufacturing steps for ionized wood membranes. Diagram depicting the functioning of stacked ionized wood membrane pairs to harness salinity-gradient energy between freshwater and seawater for electricity production. Source: Wu [118]/John Wiley & Sons.

adjacent water molecules to compensate for water loss, causing continuous water movement within the channels of the porous carbon film. This can drive the ion-solution movement through confined pathways or porous media, generating a potential difference. In this process, dynamic mass and heat transfer combined with nanostructured materials continuously convert thermal energy from water evaporation into electrical energy. Tang's team developed enhanced solar evaporation and electricity generation using biomimetic hydrogels incorporating dedicated liquid transport channels, graphene quantum dot/MXene hybrids, and varying hydrophobic interfaces (Figure 4.16b), simulating the transpiration process of lotus stems and leaves. Synergistic thermal diffusion effects enabled an integrated evaporative hydrovaltic generator (IEHVG) to achieve an output power density of 45.6 $\mu W/cm^2$, capable of powering gadgets or charging industrial supercapacitors [115]. A comprehensive system comprising 192 IEHVG units achieved record-breaking voltages up to 105 V, with a freshwater collection rate of 2.0 $L/m^2/h$ from the seawater in sunlit outdoor environments. One work reported on a PVA sponge integrated with poly(3,4-ethylene dioxythiophene), polystyrene sulfonate, and CNT composites as a 3D dual-function solar thermal evaporator for the high-throughput desalination of seawater and electricity production (Figure 4.16c) [116]. The 3D sponge features layered porous structures and strong solar absorption, achieving a water vaporization rate of as high as 6.8 $kg/m^2/h$ with 1 Sun illumination, with an evaporation efficiency of 94.9%. Furthermore, evaporation-driven high-throughput directed flow enables seawater to move through negatively charged channels within the sponge, generating a consistent potential of 117.8 mV.

4.3.5 Combining Solar Water Evaporation with Salinity Gradients for Electricity Generation

Ongoing evaporation of seawater at the evaporation interface results in elevated salt concentrations relative to the bulk water, forming a salinity gradient that facilitates directional ion diffusion for generating electricity. The disparity in salinity between seawater and freshwater holds vast renewable energy, which can be captured during mixing and separation processes, referred to as "blue energy." Zhou's research group positioned ion-selective membranes under solar absorbers made from CNTs to leverage ion diffusion induced by salinity gradients for power generation (Figure 4.16d) [117]. They achieved a power density of $1\,W/m^2$ under 1 Sun illumination with a water evaporation rate of $1.15\,kg/m^2/h$. Following this, the team developed an integrated system combining reverse electrodialysis (RED) and thermogalvanic cells (TGCs) to generate electricity via salinity gradients (Figure 4.16d). This integrated approach minimizes the buildup of residual ions on individual TGC and RED electrodes, thereby reducing ion concentration polarization. The system attained a power density of $1.11\,W/m^2$ under 1 Sun illumination and a water evaporation rate of $1.4\,kg/m^2/h$. Previous work demonstrated ion wood membranes for this application (Figure 4.16e) [118]. They successfully developed two types of ion wood membranes: positively charged (p-wood) and negatively charged (n-wood) ion wood membranes. Due to their ion selectivity, sodium cations from seawater preferentially migrate through the n-wood membrane to the river water, while chloride anions move in the opposite direction through the p-wood membrane from seawater to river water. Effective charge separation resulted in a high voltage output of up to 9.8 V. This emerging technology proposes a cost-effective method to utilize evaporated saline water for salinity-gradient power generation, feasible even during nighttime operations, ultimately achieving zero-liquid discharge. This revision aims to enhance clarity and readability while maintaining technical accuracy and coherence in describing the innovative applications of salinity gradients for electricity production.

The above discussion highlighted several advanced solar thermal evaporation and energy conversion technologies beyond PVs, showcasing broader applications in future seawater desalination. Improvement and development directions for these integrated technologies are proposed as follows: For combined heat and power generation of clean water and electricity, there is a trade-off between sufficient steam generation, which typically lowers the surface temperature of solar absorbers, and the generation of electricity from steam and thermal gradients. Hence, there is a critical need for solar absorbers/devices with dynamic configurations to adjust evaporation and power output, meeting the diverse scenarios of water and power demands. Additionally, in electricity generation from salinity gradients, continuous and abundant steam production often leads to high-salinity gradients, promoting ion diffusion and enhancing power output. However, this could lead to salt buildup on the surface of solar absorbers, ultimately decreasing steam generation rates. Hence, it is highly advisable to incorporate antifouling solar absorbers into thermogalvanic systems to prevent salt crystallization on their surfaces, thereby maintaining high power output and adequate steam generation. For photothermal devices, enhanced management of the transport pathways of photothermal fluids is necessary to counteract ion and water molecule movement paths opposite to the direction of temperature gradients caused solely by thermal diffusion, thereby ensuring a directed and stable current flow. Furthermore, in the future, these technologies should collaboratively address issues of unstable operating voltages caused by fluctuating solar irradiance from Earth's sun, aiming to develop more industrially viable solar thermal evaporation and energy integration devices with high efficiency and stability, beyond laboratory-scale development.

4.4 Summary and Outlook

In this chapter, we have introduced the principles of photothermal water evaporation and seawater desalination, along with their development history, material and device strategies, and the integration with other solar-driven energy conversion technologies such as PVs, photo-electrochemical systems, and thermoelectrics to accelerate seawater desalination or impart new application potentials. At the end of this chapter, we propose the following directions for addressing current technological challenges.

4.4.1 Enhancing the Efficiency of Solar-to-Freshwater Conversion

Future research should focus on improving the comprehensive efficiency of each process within the system, including light absorption, photothermal conversion, water evaporation, condensation collection, and separation. Specific measures include developing high-efficiency light-absorbing materials, optimizing thermal management strategies, and designing advanced condensation collection systems to maximize the conversion efficiency from light to freshwater. Studies on photothermal materials should consider not only their light absorption capacity but also their thermal conductivity and evaporation efficiency to guarantee efficient photothermal conversion and quick water evaporation.

4.4.2 Reducing Costs

The high cost of new photothermal materials has been a major barrier to large-scale applications. Future efforts should emphasize developing low-cost, easily accessible raw materials, simplifying synthesis processes, and reducing manufacturing costs. Additionally, enhancing the adaptability of materials and devices to varying light intensities and directions and improving their resistance to high temperatures, low temperatures, high-salinity seawater, and water containing pollutants or corrosive substances will help extend the lifespan of equipment and further reduce operational costs. Exploring strategies such as waste reuse and the circular economy can also help lower material costs and achieve sustainable resource utilization.

4.4.3 Developing Smart and Automated Photothermal Evaporation Materials

By using sensor technology to monitor the temperature, humidity, salt accumulation, evaporation rate, light intensity, material structure, and chemical state changes at the evaporation interface in real time, a database of operating conditions and materials can be formed. Utilizing artificial intelligence to analyze these data can enable automatic system adjustments and optimization, early warning of potential issues, and improved system stability and efficiency. In the future, smart systems can be integrated with the Internet of Things (IoT) technology for remote monitoring and management, increasing the flexibility and responsiveness of seawater desalination systems. Additionally, intelligent algorithms and big data analysis will help identify potential optimization areas, enhancing the overall economic benefits of the system.

4.4.4 Interdisciplinary Research

Utilizing system optimization, numerical simulations, and computational fluid dynamics (CFD) tools can predict and optimize mass and heat transfer processes at the material evaporation interface and optimize system design and operational parameters. During this process, it is necessary to consider the coupling of multiple physical processes, such as photothermal conversion, heat conduction, water evaporation, airflow, and water condensation, to improve the overall energy conversion efficiency of the system. Developing real-time responsive simulation models and algorithms will enable photothermal water evaporation systems to adjust operations and structural designs based on environmental conditions and energy supply, ensuring efficient and stable operation under various natural conditions. Interdisciplinary research can also promote the rapid development of new materials and technologies, such as combining bioengineering techniques to develop biomimetic materials to enhance photothermal conversion efficiency and durability.

4.4.5 Leveraging the Unique Physical and Chemical Behaviors of Photothermal Interface Evaporation

The unique properties exhibited during photothermal interface evaporation, such as photothermal conversion, localized thermal management, high-speed gas–liquid flow, and spontaneous, controllable temperature and ion concentration gradients, can provide new application opportunities for more energy conversion and chemical reactions. For instance, by enhancing the utilization of temperature gradients, phase changes, ion gradients, interactions between water and interfaces, interactions between ions and interfaces, localized heating and evaporation cooling, and selective ion and molecular transport and separation, new energy devices and photothermal catalytic systems can be designed, promoting advancements in other disciplines. Particularly in an era of increasing energy scarcity and environmental pressure, developing multifunctional integrated systems using these characteristics is crucial.

4.4.6 Eco-Friendliness and Sustainability

The development of photothermal water evaporation technology must consider ecological friendliness and sustainability. Future research should focus more on environmental impacts, developing materials and processes that are harmless or minimally harmful to the ecological environment. For example, using bio-based or biodegradable materials to reduce environmental pollution and resource consumption. Additionally, the design of photothermal water evaporation systems should account for energy consumption and the carbon footprint, aiming to utilize renewable energy as much as possible to achieve carbon neutrality or negative carbon emissions. Developing recycling and resource recovery technologies can further reduce environmental impacts and achieve sustainable development.

4.4.7 Multifunctional Integration and Extended Applications of Photothermal Evaporation Technology

Photothermal evaporation technology is not limited to seawater desalination, it can also extend to other areas, such as wastewater treatment, pollutant separation, and concentration extraction. By integrating photothermal evaporation with other technologies like PVs, electrochemistry, and catalysis, multifunctional integrated systems can be developed to further enhance resource utilization efficiency and economic benefits. This multifunctional integration can achieve efficient resource utilization and provide flexible solutions for different application scenarios, meeting diverse needs.

4.5 Conclusion

In conclusion, the development prospects for photothermal seawater desalination technology are broad. By continuously improving efficiency, reducing costs, smart system operations, interdisciplinary research, and exploring new application fields, we are likely to achieve more efficient, economical, and intelligent seawater desalination technology in the future, contributing to solving the global water resource crisis. Further research and applications require the concerted efforts of global researchers and engineers, as well as continuous innovation and practical advancements in technology.

Questions

1 Why is solar water evaporation technology important for addressing the water and energy crisis in sustainable energy conversion applications?

2 Briefly explain the principle of solar seawater desalination.

3 How does the combination of photovoltaic (PV) technology and seawater desalination work, and what are its advantages?

4 What is the thermoelectric effect, and how is it applied in solar water evaporation systems?

5 What is "blue energy," and how can it generate electricity through salinity gradients in solar water evaporation systems?

6 How does the integration of mechanical energy conversion achieve solar water evaporation and electricity generation?

7 Discuss the challenges of solar seawater desalination in practical applications.

References

1 Zhu, L.L. (2019). Recent progress in solar-driven interfacial water evaporation: Advanced designs and applications. *Nano Energy* 57: 507.
2 Shahzad, M.W. (2022). Demystifying integrated power and desalination processes evaluation based on standard primary energy approach. *Therm. Sci. Eng. Prog.* 27: 101153.
3 Li, C.N. (2013). Solar assisted sea water desalination: a review. *Renew. Sustain. Energy Rev.* 19: 136.
4 MacFarlane, D.R. (2020). A roadmap to the ammonia economy. *Joule* 4: 1186.
5 Larsen, T.A. (2016). Emerging solutions to the water challenges of an urbanizing world. *Science* 352: 928.
6 Wu, X. (2024). Interfacial solar evaporation: from fundamental research to applications. *Adv. Mater.* 36: 2313090.
7 Dao, V.D. (2024). Nature-inspired design for high-efficiency solar-driven water evaporation. *J. Power Sources* 609: 234676.
8 Chen, C.J. (2019). Challenges and opportunities for solar evaporation. *Joule* 3: 683.
9 Shi, P.R. (2024). Cogeneration of clean water and valuable energy/resources via interfacial solar evaporation. *Nano Lett.* 24: 5673.

10 Zhao, F. (2020). Materials for solar-powered water evaporation. *Nat. Rev. Mater.* 5: 388.

11 Nagata, Y. (2015). Molecular mechanism of water evaporation. *Phys. Rev. Lett.* 115: 236102.

12 Kurkute, N. (2022). A thorough review of the existing concentrated solar power technologies and various performance enhancing techniques. *J. Therm. Anal. Calorim.* 147: 14713.

13 Deng, Z.Y. (2017). The emergence of solar thermal utilization: solar-driven steam generation. *J. Mater. Chem. A* 5: 7691.

14 Zhang, Y. (2020). Structure architecting for salt-rejecting solar interfacial desalination to achieve high-performance evaporation with in situ energy generation. *Adv. Sci.* 7: 1903478.

15 Chang, K.F. (2024). Key technology developments for solar-driven interface evaporation on structural innovation and thermal design. *Nano Energy* 132: 110369.

16 Ni, G. (2015). Volumetric solar heating of nanofluids for direct vapor generation. *Nano Energy* 17: 290.

17 Zhao, D. (2015). Enhancing localized evaporation through separated light absorbing centers and scattering centers. *Sci. Rep.* 5: 17276.

18 Zeng, Y. (2011). Solar evaporation enhancement using floating light-absorbing magnetic particles. *Energy Environ. Sci.* 4: 4074.

19 Wang, Z. (2014). Bio-inspired evaporation through plasmonic film of nanoparticles at the air-water interface. *Small* 10: 3234.

20 Zhu, G. (2016). Constructing Black Titania with unique nanocage structure for solar desalination. *ACS Appl. Mater. Interfaces* 8: 31716.

21 Chen, R. (2017). Magnetically recyclable self-assembled thin films for highly efficient water evaporation by interfacial solar heating. *RSC Adv.* 7: 19849.

22 Ghasemi, H. (2014). Solar steam generation by heat localization. *Nat. Commun.* 5: 4449.

23 Liu, Y. (2015). A bioinspired, reusable, paper-based system for high-performance large-scale evaporation. *Adv. Mater.* 27: 2768.

24 Zhu, L.L. (2018). Self-contained monolithic carbon sponges for solar-driven interfacial water evaporation distillation and electricity generation. *Adv. Energy Mater.* 8: 1702149.

25 Shi, L. (2017). Rational design of a bi-layered reduced graphene oxide film on polystyrene foam for solar-driven interfacial water evaporation. *J. Mater. Chem. A* 5: 16212.

26 Li, X. (2016). Graphene oxide-based efficient and scalable solar desalination under one sun with a confined 2D water path. *Proc. Natl. Acad. Sci. USA* 113: 13953.

27 Chen, C. (2017). Highly flexible and efficient solar steam generation device. *Adv. Mater.* 29: 1701756.

28 Lim, H.W. (2023). 3D thermoresponsive hydrogel with enhanced water uptake and active evaporation for effective interfacial solar steam generation. *Desalination* 550: 116368.

29 Zhao, X. (2020). Overcoming salt crystallization with ionic hydrogel for accelerating solar evaporation. *Desalination* 482: 114385.

30 Wang, Y. (2023). Nano-coprecipitation route synthesis of highly-efficient sub-micron (Sr,Ba)2SiO4:Eu2+ phosphors with enhanced thermal stability for microLED color conversion. *ACS Appl. Mater. Interfaces* 15: 40595.

31 Zhu, J. (2023). FDM 3D-printed volcanic-shaped structure for ultrafast solar-driven interfacial evaporation and efficient energy utilization. *Desalination* 548: 116275.

32 Liu, H.D. (2019). Interfacial solar-to-heat conversion for desalination. *Adv. Energy Mater.* 9: 1900310.

33 Liu, G.H. (2017). Solar water evaporation by black photothermal sheets. *Nano Energy* 41: 269.

34 Zhou, L. (2016). Self-assembly of highly efficient, broadband plasmonic absorbers for solar steam generation. *Sci. Adv.* 2: 1501227.

35 Metwally, K. (2017). Isosbestic thermoplasmonic nanostructures. *ACS Photonics* 4: 1544.

36 Zhou, L. (2016). 3D self-assembly of aluminium nanoparticles for plasmon-enhanced solar desalination. *Nat. Photonics* 10: 393.

37 Liu, D.L. (2015). Black gold: plasmonic colloidosomes with broadband absorption self-assembled from monodispersed gold nanospheres by using a reverse emulsion system. *Angew. Chem. Int. Ed.* 54: 9596.

38 Neumann, O. (2013). Solar vapor generation enabled by nanoparticles. *ACS Nano* 7: 42.

39 Gao, M.M. (2019). Solar absorber material and system designs for photothermal water vaporization towards clean water and energy production. *Energy Environ. Sci.* 12: 841.

40 Cui, X.M. (2023). Photothermal nanomaterials: a powerful light-to-heat converter. *Chem. Rev.* 123: 6891.

41 Chen, J.X. (2019). Space-confined seeded growth of Black Silver nanostructures for solar steam generation. *Nano Lett.* 19: 400.

42 Huang, Z.M. (2022). From materials to devices: rationally designing solar steam system for advanced applications. *Small Methods* 6: 2200835.

43 Cortés, E. (2020). Challenges in plasmonic catalysis. *ACS Nano* 14: 16202.

44 Wang, J. (2016). High-performance photothermal conversion of narrow-bandgap Ti_2O_3 nanoparticles. *Adv. Mater.* 29: 1603730.

45 Gaertner, W.W. (1961). Photothermal effect in semiconductors. *Phys. Rev.* 122: 419.

46 Yao, J.D. (2017). Alloying-assisted phonon engineering of layered $BiInSe_3$@nickel foam for efficient solar-enabled water evaporation. *Nanoscale* 9: 16396.

47 Xu, J.J. (2018). Copper nanodot-embedded graphene urchins of nearly full-spectrum solar absorption and extraordinary solar desalination. *Nano Energy* 53: 425.

48 Sun, A.Q. (2020). Super-performance photothermal conversion of 3D macrostructure graphene-$CuFeSe_2$ aerogel contributes to durable and fast clean-up of highly viscous crude oil in seawater. *Nano Energy* 70: 104511.

49 Yuan, R.X. (2015). Superhydrophobic polyaniline hollow spheres with mesoporous brain-like convex-fold shell textures. *J. Mater. Chem. A* 3: 19299.

50 Wei, X. (2022). A polyaniline nanofiber array supported ultrathin polyamide membrane for solar-driven volatile organic compound removal. *J. Mater. Chem. A* 10: 20424.

51 Ding, R.J. (2023). Achieving fast interfacial solar vapor generation and aqueous acid purification using $Ti_3C_2T_x$ MXene/PANI non-woven fabrics. *Mater. Horiz.* 10: 2262.

52 Wang, Y.F. (2020). Conjugated-polymer-based nanomaterials for photothermal therapy. *ACS Appl. Polym. Mater.* 2: 4258.

53 Yu, C.C. (2020). Polymer-based nanomaterials for noninvasive cancer photothermal therapy. *ACS Appl. Polym. Mater.* 2: 4289.

54 Meng, X.Y. (2020). Gradient-aligned Au/graphene meshes with confined heat at multiple levels for solar evaporation and anti-gravity catalytic conversion. *J. Mater. Chem. A* 8: 16570.

55 Zhang, P.P. (2017). Vertically aligned graphene sheets membrane for highly efficient solar thermal generation of clean water. *ACS Nano* 11: 5087.

56 Meng, X.Y. (2020). Gradient vertical channels within aerogels based on N-doped graphene meshes toward efficient and salt-resistant solar evaporation. *ACS Sustain. Chem. Eng.* 8: 4955.

57 Bae, K. (2015). Flexible thin-film black gold membranes with ultrabroadband plasmonic nanofocusing for efficient solar vapour generation. *Nat. Commun.* 6: 10103.

58 Zhu, M. (2018). Plasmonic wood for high-efficiency solar steam generation. *Adv. Energy Mater.* 8: 1701028.

59 Zhang, L. (2015). Hydrophobic light-to-heat conversion membranes with self-healing ability for interfacial solar heating. *Adv. Mater.* 27: 4889.

60 Wu, L. (2020). Highly efficient three-dimensional solar evaporator for high salinity desalination by localized crystallization. *Nat. Commun.* 11: 521.

61 Yao, H. (2021). Janus-interface engineering boosting solar steam towards high-efficiency water collection. *Energy Environ. Sci.* 14: 5330.

62 Wang, F. (2021). A high-performing single-stage invert-structured solar water purifier through enhanced absorption and condensation. *Joule* 5: 1602.

63 Du, C. (2022). A floating vapor condensation structure in a heat-localized solar evaporation system for facile solar desalination. *Appl. Therm. Eng.* 201: 117834.

64 Liu, Z. (2017). Extremely cost-effective and efficient solar vapor generation under nonconcentrated illumination using thermally isolated black paper. *Global Chall.* 1: 1600003.

65 Wang, Z. (2021). Highly conductive amorphous pentlandite anchored with ultrafine platinum nanoparticles for efficient pH-universal hydrogen evolution reaction. *Adv. Funct. Mater.* 31: 1.

66 Wu, X. (2018). Evaporation above a bulk water surface using an oil lamp inspired highly efficient solar-steam generation strategy. *J. Mater. Chem. A* 6: 12267.

67 Wu, X. (2019). A flexible photothermal cotton-CuS nanocage-agarose aerogel towards portable solar steam generation. *Nano Energy* 56: 708.

68 Cui, L. (2018). High rate production of clean water based on the combined photo-electro-thermal effect of graphene architecture. *Adv. Mater.* 30: 1706805.

69 Zhang, P. (2020). High throughput of clean water excluding ions, organic media, and bacteria from defect-abundant graphene aerogel under sunlight. *Nano Energy* 46: 415.

70 Ni, G. (2018). A salt-rejecting floating solar still for low-cost desalination. *Energy Environ. Sci.* 11: 1510.

71 Shi, Y. (2021). All-day fresh water harvesting by microstructured hydrogel membranes. *Nat. Commun.* 12: 2797.

72 Gong, F. (2019). Scalable, eco-friendly and ultrafast solar steam generators based on one-step melamine-derived carbon sponges toward water purification. *Nano Energy* 58: 322.

73 Zhang, P. (2018). Three-dimensional water evaporation on a macroporous vertically aligned graphene pillar array under one sun. *J. Mater. Chem. A* 6: 15303.

74 Singh, S.C. (2020). Solar-trackable super-wicking black metal panel for photothermal water sanitation. *Nat. Sustainability* 3: 938.

75 Wu, X. (2021). Confinement capillarity of thin coating for boosting solar-driven water evaporation. *Adv. Funct. Mater.* 31: 1.

76 Zhang, Q. (2019). Vertically aligned Janus MXene-based aerogels for solar desalination with high efficiency and salt resistance. *ACS Nano* 13: 13196.

77 Wang, Z. (2016). Environmental applications of interfacial materials with special wettability. *Environ. Sci. Technol.* 50: 2132.

78 Miljkovic, N. (2013). Condensation heat transfer on superhydrophobic surfaces. *MRS Bull.* 38: 397.

79 He, S. (2019). Nature-inspired salt resistant bimodal porous solar evaporator for efficient and stable water desalination. *Energy Environ. Sci.* 12: 1558.

80 Kuang, Y. (2019). A high-performance self-regenerating solar evaporator for continuous water desalination. *Adv. Mater.* 31: e1900498.

81 Wang, Y. (2020). Rational 3D coiled morphology for efficient solar-driven desalination. *Environ. Sci. Technol.* 54: 16240.

82 Javed, M. (2021). Complete system to generate clean water from a contaminated water body by a handmade flower-like light absorber. *ACS Omega* 6: 35104.

83 Jin, B. (2023). Iceberg-inspired solar water generator for enhanced thermoelectricity–freshwater synergistic production. *Chem. Eng. J.* 469: 143906.

84 Gao, C. (2023). Reversed vapor generation with Janus fabric evaporator and comprehensive thermal management for efficient interfacial solar distillation. *Chem. Eng. J.* 463: 142002.

85 Xu, J. (2020). All-day freshwater harvesting through combined solar-driven interfacial desalination and passive radiative cooling. *ACS Appl. Mater. Interfaces* 12: 47612.

86 Shao, Y. (2022). Marangoni effect drives salt crystallization away from the distillation zone for large-scale continuous solar passive desalination. *ACS Appl. Mater. Interfaces* 14: 30324.

87 Xue, G. (2018). Highly efficient water harvesting with optimized solar thermal membrane distillation device. *Global Chall.* 2: 1800001.

88 Morciano, M. (2020). Solar passive distiller with high productivity and Marangoni effect-driven salt rejection. *Energy Environ. Sci.* 13: 3646–3655.

89 Huang, L. (2020). Enhanced water yield of solar desalination by thermal concentrated multistage distiller. *Desalination* 477: 114260.

90 Wang, W. (2021). Integrated solar-driven PV cooling and seawater desalination with zero liquid discharge. *Joule* 5: 1873.

91 Chiavazzo, E. (2018). Passive solar high-yield seawater desalination by modular and low-cost distillation. *Nat. Sustainability* 1: 763.

92 Yang, L. (2021). Photovoltaic-multistage desalination of hypersaline waters for simultaneous electricity, water and salt harvesting via automatic rinsing. *Nano Energy* 87: 106163.

93 Ma, X. (2022). A small heat capacity solar distiller with extra effective discharge for brine by the siphoning of a hydrophilic membranous wick. *Desalination* 521: 115306.

94 Wang, W. (2019). Simultaneous production of fresh water and electricity via multistage solar photovoltaic membrane distillation. *Nat. Commun.* 10: 3012.

95 Ding, T. (2021). Using the sun to co-generate electricity and freshwater. *Joule* 5: 1639.
96 Lee, K.P. (2011). A review of reverse osmosis membrane materials for desalination—development to date and future potential. *J. Membr. Sci.* 370: 1.
97 Abu-Zeid, M.A. (2015). A comprehensive review of vacuum membrane distillation technique. *Desalination* 356: 1.
98 Gao, M. (2021). Photothermal membrane distillation toward solar water production. *Small Methods* 5: 2001200.
99 Nugent, D. (2014). Assessing the lifecycle greenhouse gas emissions from solar PV and wind energy: a critical meta-survey. *Energy Policy* 65: 229.
100 Xu, N. (2020). Synergistic tandem solar electricity-water generators. *Joule* 4: 347.
101 Hasanuzzaman, M. (2016). Global advancement of cooling technologies for PV systems: a review. *Sol. Energy* 137: 25.
102 Polman, A. (2016). Photovoltaic materials: present efficiencies and future challenges. *Science* 352: aad4424.
103 Green, M.A. (2018). Solar cell efficiency tables (version 52). *Prog. Photovolt. Res. Appl.* 26: 427.
104 Cariou, R. (2018). III–V-on-silicon solar cells reaching 33% photoconversion efficiency in two-terminal configuration. *Nat. Energy* 3: 326.
105 Gui, J. (2023). Hybrid solar evaporation system for water and electricity co-generation: comprehensive utilization of solar and water energy. *Nano Energy* 107: 108155.
106 Antonetto, G. (2021). Synergistic freshwater and electricity production using passive membrane distillation and waste heat recovered from camouflaged photovoltaic modules. *J. Clean. Prod.* 318: 128464.
107 Huang, L. (2020). Solar-driven co-generation of electricity and water by evaporation cooling. *Desalination* 488: 114533.
108 Liu, M. (2024). Synergistic solar-powered water-electricity generation: an integrated floating system on water. *Nano Energy* 119: 109074.
109 Radziemska, E. (2003). The effect of temperature on the power drop in crystalline silicon solar cells. *Renew. Energy* 28: 1.
110 Zhu, L.L. (2019). Shape conformal and thermal insulative organic solar absorber sponge for photothermal water evaporation and thermoelectric power generation. *Adv. Energy Mater.* 9: 1900250.
111 Li, X.Q. (2018). Storage and recycling of interfacial solar steam enthalpy. *Joule* 2: 2477.
112 Jiang, M. (2020). Bioinspired temperature regulation in interfacial evaporation. *Adv. Funct. Mater.* 30: 1910481.
113 Meng, S. (2022). A wave-driven piezoelectric solar evaporator for water purification. *Adv. Energy Mater.* 12: 2200087.

114 Xue, G. (2017). Water-evaporation-induced electricity with nanostructured carbon materials. *Nat. Nanotechnol.* 12: 317.

115 Chen, Y. (2024). Achieving ultrahigh voltage over 100 V and remarkable freshwater harvesting based on thermodiffusion enhanced hydrovoltaic generator. *Adv. Energy Mater.* 14: 2400529.

116 Xu, Y. (2022). All-in-one polymer sponge composite 3D evaporators for simultaneous high-flux solar-thermal desalination and electricity generation. *Nano Energy* 93: 106882.

117 Yang, P.H. (2017). Solar-driven simultaneous steam production and electricity generation from salinity. *Energy Environ. Sci.* 10: 1923.

118 Wu, Q.Y. (2020). Salinity-gradient power generation with ionized wood membranes. *Adv. Energy Mater.* 10: 1902590.

5

Membrane-Based Desalination Utilizing Wind and Wave Energy

Mohamed S. Salem, Mahmoud A. Shouman, Osama Abdelrehim, and Asmaa A. Khater

Mechanical Power Engineering Department, Faculty of Engineering, Mansoura University, Mansoura, Egypt

5.1 Introduction

Water scarcity is a growing global crisis, profoundly affecting sustainable development, economic growth, and social stability. Rapid industrialization, population growth, and higher living standards have significantly increased water consumption rates, exacerbating the strain on available water resources. The escalating frequency and severity of droughts further highlight the urgency of addressing this issue, particularly in regions already facing limited access to drinkable water.

Membrane desalination, which involves using semipermeable membranes to remove salts and impurities from seawater and brackish water, has emerged as a vital solution to meet the increasing demand for potable water. Historically, desalination systems have relied on fossil fuels, leading to high energy consumption and significant greenhouse gas emissions. However, the environmental impact of fossil fuels has necessitated the exploration of alternative energy sources. Renewable energy-assisted membrane desalination technologies are becoming increasingly attractive due to their potential to provide sustainable and environmentally friendly water production. By harnessing renewable resources such as solar, wind, wave, and geothermal energy, membrane desalination can be achieved with minimal ecological footprint, offering a promising path toward alleviating the global water scarcity crisis.

Water Treatment and Desalination: Membrane Separation using Renewable Energy Resources, First Edition. Edited by Mohamed Bassyouni.

5.2 Wind-Powered Membrane Desalination

Wind power technology is one of the most promising renewable energy sources as it rivals conventional energy sources and requires less capital and investment costs compared to other renewable energy sources. Large-capacity wind turbines (WTs) are economically suitable for installation in coastal areas, which frequently experience water shortages and are often characterized by abundant wind resources. Estimates suggest that harnessing 20% of the global wind energy potential could yield an annual electricity generation of 123 petawatt-hours (PWh) [1]. This creates an ideal scenario for the integration of wind energy with membrane desalination processes. Wind power's technological maturity and its ability to generate electricity in water-scarce regions make it a suitable candidate for membrane desalination applications. Unlike other renewable sources, wind energy's intermittent nature can be effectively managed by storing fresh water produced during periods of adequate power supply. Additionally, membrane desalination systems can be operated directly by WTs or by harnessing the waste heat generated by these turbines. Consequently, wind-powered membrane desalination offers a sustainable and reliable solution to meet the growing demand for fresh water in areas with limited resources, ensuring a continuous supply even as wind conditions fluctuate.

Wind energy systems can be coupled with desalination processes either indirectly through a local electricity grid or directly by connecting the WT to the desalination plant. This technology is particularly well-suited for small- to medium-scale desalination plants producing between 1 and 250 m^3/day.

The key design parameters influencing WPD plant performance include:

- **Water Demand and Plant Capacity:** Aligning water production with consumption needs.
- **Site Characteristics:** Wind turbine and desalination plant location, altitude, and site development requirements.
- **Feed Water Quality:** Salinity levels impacting desalination process efficiency.
- **Wind Conditions:** Wind speed distribution for optimal turbine performance.
- **System Configuration:** Direct or indirect coupling, energy storage, and power distribution.
- **Water Storage:** Capacity for managing fluctuations in water production and demand.
- **Energy Consumption:** Desalination unit energy efficiency and power management.
- **Membrane Performance:** Salt rejection, permeate flux, and membrane area.

5.2.1 Wind-Powered Desalination Plant Design Challenges

The integration of WTs and desalination plants presents a complex set of design challenges. The primary issues revolve around:

- **Energy Variability:** Wind energy is inherently intermittent, making it difficult to match power generation with desalination demand.
- **Energy Efficiency:** Optimizing the energy consumption of the desalination process to maximize water production per unit of wind energy.
- **System Sizing:** Determining the optimal capacity of both the WT and the desalination plant to meet water demand while considering economic factors.
- **Energy Storage:** Evaluating the necessity and effectiveness of energy storage solutions (e.g., batteries and water pumping) to mitigate the impact of wind variability.
- **Economic Feasibility:** Balancing capital and operational costs with the value of the produced water.

5.2.2 Wind-Powered Desalination Plant Design Considerations

A) **Energy Production and Consumption**
   - **Wind Turbine Capacity Factor (CF):** Accurate estimation of the capacity factor is crucial for determining energy production. The usual value is between 20% and 30%.
   - **Desalination Energy Intensity:** Selecting the most energy-efficient desalination technology is essential to optimize water production.
   - **Energy Matching:** Developing strategies to align wind energy generation with desalination demand, including demand management and flexible operation.

B) **Energy Storage**
   - **Storage Technology:** Evaluating the suitability of different storage technologies (e.g., batteries, pumped hydro, etc.) based on specific project requirements.
   - **Storage Capacity:** Determining the optimal storage size to balance energy storage costs with system reliability.
   - **Storage Efficiency:** Considering the round-trip efficiency of energy storage systems.

C) **System Integration**
   - **Control Systems:** Developing advanced control algorithms to manage energy flows and optimize system performance.

- **Grid Connection:** Assessing the feasibility and benefits of grid connection for energy exchange and backup power.
- **Water Storage:** Incorporating water storage to buffer fluctuations in water production and demand.

D) **Economic Analysis**

- **Levelized Cost of Water (LCOW):** Calculating the total cost of water production, including capital, operational, and energy costs.
- **Economic Optimization:** Identifying the optimal system configuration to minimize LCOW.
- **Financial Analysis:** Evaluating the economic viability of the project, considering factors such as revenue, subsidies, and financing options.

E) **Additional Considerations**

1) **Environmental Impact:** Assessing the potential environmental impacts of the project, including land use, water consumption, and noise pollution.
2) **Reliability and Maintainability:** Designing for high system reliability and easy maintenance.
3) **Scalability:** Considering the potential for scaling up the system to meet growing water demand.

By carefully addressing these design challenges, it is possible to develop successful and sustainable WPD systems.

5.2.3 Reverse Osmosis (RO)—Wind-Powered Desalination Systems

Wind energy, as a clean and renewable energy source, has attracted researchers as it can be used to electrically or mechanically power RO desalination systems [2]. The unpredictable behavior of wind-produced power requires the use of energy storage systems to achieve a smoother operation before connection to the RO system [3]. The conventional RO-wind system is shown in Figure 5.1. Examples of RO-wind power plants include the 80 MW wind-powered Kwinana desalination plant in Perth, Western Australia (140.000 m^3/day), and the 140 MW wind-powered Sydney (Kurnell) desalination plant (250.000 m^3/day) [4].

Wind energy is usually coupled with RO desalination systems as an electric energy plant that can be used to power the system after stabilization either by going through a storage system or by connecting the plant to the grid. As RO is mostly suitable for seawater desalination, it might be suitable to be integrated with wind energy that is convenient in coastal areas [5]. To investigate the feasibility of using wind-directed powered RO systems without any control system, Dahioui and Loudiyi [6] carried out a study using MATLAB that investigated the effect of unstable electricity produced from a direct wind system on an RO desalination system. It not only affects productivity, but this also leads to instability in the feed water pressure, which can be a direct cause of system failure

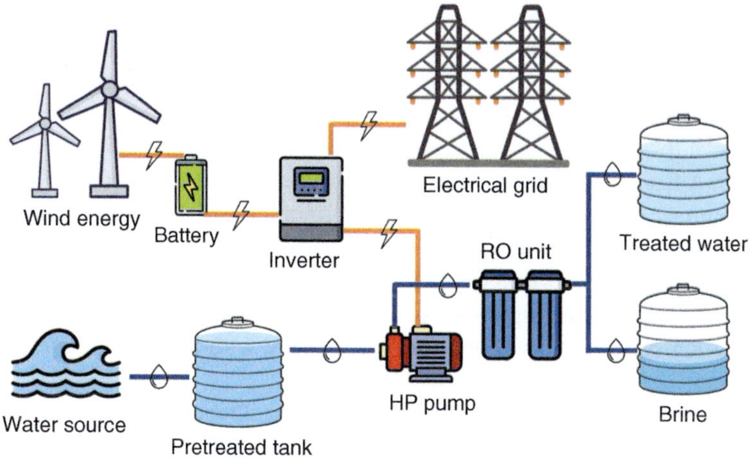

Figure 5.1 Layout of a conventional RO-wind desalination system.

due to membrane damage. The study does not hide the fact that using a stabilizing system might be feasible and beneficial given the advantage of using a clean and renewable energy source such as wind energy, especially in coastal cities.

Different studies were held to propose satisfactory solutions to eliminate the effect of produced fluctuations. The most common system employed to reduce the effect of directly using produced power is the use of batteries to store this energy. Nevertheless, these systems may need a connection to other energy sources or a backup system to avoid battery degradation due to fluctuated charging and discharging behaviors [7–9]. The RO unit on those systems is mainly wind-driven as the wind energy is used to power the unit pumps and store additional power in the battery storage unit. The stored energy is used to stabilize the system or even power the system in no-wind situations. A common component that exists in such systems is a battery charger controller that is used to avoid overcharging.

Another proposed solution to the fluctuations problem is the use of a supercapacitor bank to control the dynamic behavior of the system [10, 11]. In those systems, a controller is usually used to control the pump rotating speed, thereby adjusting flow rates and pressure. It was noticed that supercapacitors enhanced the short-term dynamic behavior of a WT, resulting in steady operation and improved performance.

A fully wind-driven RO system with two 230 kW WTs and eight RO units was utilized in the Canary Islands [12]. The system used 10 kW battery banks and inverters as part of the control unit. Depending on the supplied wind power, the number of RO modules can vary using the control system. This arrangement did not show any significant change in the quality or quantity of the average water produced per unit. To assess the reliability of using wind-driven RO systems completely off the grid, Peñate et al. [13] held a study in which an analysis of two systems with the same capacity of 1,000 m^3/day and different numbers of RO modules to achieve this capacity was conducted. The first system used only one RO module with a fixed 1,000 m^3/day capacity and the other used two 400 m^3/day capacities connected to a 200 m^3/day unit. The use of a gradual capacity configuration instead of a fixed one resulted in up to 8% water production but a higher year-long operation rate and lower excess energy. Moreover, due to flexibility in design, a gradual capacity configuration can give a lower stop time with low wind conditions. Such systems can be enhanced if integrated with an efficient energy storage system.

Another study focused on the economic and environmental effects of using a wind-powered RO desalination system [14]. In this study, a 30-kW WT was used to power a small RO unit with the target of producing 1 m^3/h of fresh water. CO_2 emission reduction was about 80 tons/year with about 3 ~ 6.5 USD/m^3 for off-grid systems and <3 USD/m^3 for on-grid systems. This study shows the potential of using wind-powered RO systems from both technical and environmental aspects. Other studies were also carried out that verify this conclusion [15–17]. A larger-scale study of a 5,000 m^3/day RO plant capacity and 10 MW wind energy was conducted by Dehmas et al. [18]. The analysis was performed considering weather conditions in Algeria. The reduction of CO_2 emissions reached 8,500 tons/year. The study also proved that the use of wind-driven RO desalination can be economically advantageous.

5.2.3.1 Autonomous Wind-Powered Desalination Systems with Backup

Remote coastal regions and island communities often lack a grid infrastructure, making stand-alone power systems essential. Wind energy, a prevalent resource in these areas, presents a viable option for electricity generation. However, the intermittent nature of wind power poses challenges for desalination processes, which typically require stable operating conditions. Power fluctuations can adversely affect the desalination equipment lifespan and performance, while interruptions can disrupt the water supply, impacting reliant communities. Consequently, the robustness and reliability of the system are paramount. Optimal component sizing is crucial to balance water and electricity production, minimize disruptions, and control costs.

To mitigate the impacts of wind power variability, backup systems such as diesel generators or energy storage solutions are commonly integrated. These systems enhance system reliability and ensure a consistent power supply for the desalination process. A summary of some of the previous stand-alone wind-powered RO desalination projects can be found in Table 5.1.

Table 5.1 Summary of some research concerning stand-alone wind-powered RO desalination plants.

References	Year	Water source	Energy storage	Capacity (m^3/d)
[19]	2020	Brackish water	Microhydraulic plant	7.2
[11]	2015	Seawater	Batteries/ supercapacitor	18
[13]	2011	Seawater	Batteries/flywheel	1,000
[20]	2010	Seawater	Hydraulic plant	4320
[21]	2008	Seawater	Batteries	3.12
[22]	2007	Seawater	Flywheel	180–1,400
[23]	2006	Seawater	Batteries	18–96
[24]	2004	Seawater	—	96
[12]	2003	Seawater	Batteries/flywheel	72–192
[25]	2001	Seawater	Batteries/microhydraulic plant	8.4
[26]	2000	Seawater	—	100
[27]	1997	Seawater	Batteries	6.2
[28]	1995	Seawater	—	5–20
[17]	1994	Brackish water	—	22–33
[29]	1981	Brackish water	—	1.5–9

Diesel Generators as Backup Power for Desalination Diesel generators offer a reliable power source for desalination plants, complementing wind energy during periods of insufficient generation. By reducing diesel fuel consumption, wind-diesel hybrid systems can lower operational costs and carbon emissions. Several studies have demonstrated the economic benefits of integrating wind power into desalination systems.

Habali and Saleh [17] compared a stand-alone diesel-powered desalination plant with a wind–diesel hybrid system, finding significant cost reductions through wind energy integration. Henderson et al. [30] explored the potential of seawater desalination using surplus wind power from a hybrid system. While the specific case study was economically unfavorable, the authors highlighted the potential of desalination for energy storage and load management in larger-scale applications.

Despite these advantages, diesel generators introduce environmental concerns due to emissions and increase operational costs through fuel dependency. Moreover, potential fuel shortages, maintenance issues, and power outages can limit water production.

Energy Storage as a Backup for Wind-Powered Desalination Energy storage is crucial for mitigating the intermittency of wind power in desalination systems. Studies have explored various storage options, including batteries, pumped hydro, and thermal storage.

Early WPD projects, such as the GKSS-Research Centre's plant on Süderoog Island [29], encountered challenges due to wind variability. Battery storage was essential to enable stable desalination operation.

Island communities like Merssinin village often rely on stand-alone systems. Integrating photovoltaics (PV) into desalination systems is a common strategy. Hybrid PV-wind-powered desalination with combined battery and pumped hydro storage can address both electricity and fresh water needs [25].

For regions with limited grid infrastructure, such as the islands of Split and Dalmatia, decentralized WPD systems are considered [26]. Energy storage is crucial to ensure reliable water supply in these locations.

To manage wind power fluctuations, desalination systems can operate under near-constant or variable conditions. Energy storage is key for achieving near-constant operation by smoothing power variations and providing backup power [19]. Modular energy storage systems, combining hydraulic and water storage, can enhance system flexibility and efficiency.

Challenges and Trade-offs in Energy Storage for Wind-Powered Desalination Maintaining constant operating conditions for desalination plants often necessitates substantial energy storage, leading to increased capital costs. Infield's study [27] highlighted the complex interplay between wind speed, battery capacity, and operating pressure in stand-alone wind-powered RO systems. Optimal system design requires careful consideration of these parameters to balance water production, energy consumption, and economic viability.

While energy storage is essential for stabilizing power supply, minimizing its size is crucial for cost reduction. Water storage can be a more economical alternative for matching supply and demand. However, achieving a constant power supply often necessitates larger energy storage systems.

Hybrid renewable energy systems, combining wind and solar power, can potentially reduce the required energy storage capacity. A prototype in Lavrio, Athens, demonstrated the feasibility of this approach [21]. Nevertheless, hybrid systems introduce complexities in energy management and control, potentially offsetting initial cost savings.

Modular Desalination Capacity for Wind-Powered Systems Modular desalination systems offer flexibility by adjusting capacity to match fluctuating wind power. However, frequent on/off cycles can negatively impact membrane performance.

Veza and coworkers [31] compared modular and fixed-capacity seawater RO (SWRO) systems, finding that modular systems better utilize wind energy but produce less water overall. Power control is crucial for maximizing water production.

Carta et al. [11] introduced a small-scale modular system with a flywheel for power smoothing. While initial tests showed no significant membrane deterioration, long-term effects require further investigation.

Optimizing modular capacity systems can improve performance and reduce costs. Combining modularity with variable operating conditions can enhance system efficiency.

Koklas and Papathanassiou [23] analyzed the impact of system components on water production and operational costs, concluding that larger desalination units with fewer start/stop cycles are more efficient. A 2:1–3:1 ratio between WT and desalination capacity is recommended.

Spyrou and Anagnostopoulos [20] investigated a hybrid wind–solar desalination system with hydraulic storage. Results indicated that hydro-storage is beneficial for larger water demands. Wind-only systems can produce water at lower costs than conventional sources.

Neris et al. [28] studied a modular system with two units, demonstrating effective power management through a resistive dump load and fluid coupling.

A comprehensive comparison of different system configurations was presented by Ali et al. [32], considering variable and fixed operation, capacity, and energy storage options. Results suggest that variable capacity systems can improve performance but may have higher initial costs.

A study by Subiela et al. [33] compared WT–battery, WT-PV, and WT–diesel systems with fixed and variable capacity desalination. Modular systems increased operation time but produced less water and had higher costs compared to variable operations.

5.2.3.2 Autonomous Wind-Powered Desalination Without Backup

To minimize capital and operational costs, stand-alone WPD systems often operate without full-scale energy storage, subjecting desalination equipment to variable conditions. For RO systems, this implies fluctuating pressure and flow rates.

Effective system design necessitates careful component sizing, identification of operational limits, robust control strategies, and thorough performance evaluation under dynamic conditions. Miranda and Infield [34] proposed a system with independently controlled pumps to optimize wind energy capture and water production. De la Nuez Pestana et al. [24] demonstrated the feasibility of operating RO membranes under variable conditions without significant damage.

Alternative approaches include the direct coupling of WTs and desalination units, as explored by Moreno and Pinilla, and the use of electrical heaters to manage surplus energy, as implemented in Baltic Sea pilot plants [35].

Optimizing system performance requires balancing water production, energy efficiency, and equipment longevity. Studies by Koklas and Papathanassiou [23], Spyrou and Anagnostopoulos [20], and others have investigated various system configurations and operational strategies, including modular desalination, hybrid power sources, and energy storage options.

While variable operation presents challenges, it can lead to cost reductions compared to systems with extensive energy storage. However, careful system design, control, and component selection are essential for ensuring optimal performance and reliability.

5.2.3.3 Direct-Driven Desalination

To circumvent energy losses associated with electrical conversion, direct-driven WPD systems mechanically couple WTs to desalination pumps or compressors. By eliminating the generator and associated electronics, these systems enhance efficiency and reliability, particularly in remote locations.

Typical configurations involve placing the pump or compressor at the tower base or within the nacelle, with power transmission achieved through gearboxes, pistons, or belt systems [36, 37]. Alternatively, the pump or compressor can be directly integrated into the nacelle, using the fluid as an energy transfer medium [38].

Direct-driven systems offer advantages in terms of performance, robustness, and reliability, while also reducing nacelle weight [39].

While electrical drive transmission systems dominate the wind energy sector, hydraulic and mechanical drive systems remain in their developmental stages, occupying a niche market. Consequently, research and prototyping of directly connected wind-driven desalination systems are limited. Existing prototypes exhibit a wide range of capacities, configurations, and turbine designs, spanning from sub-1 kW to 8 MW wind power and below 1 m^3/d to 864,000 m^3/d water production.

Reverse osmosis is the predominant desalination technology employed in these systems. Notably, a majority of recent prototypes and case studies focus on offshore applications rather than onshore. Initial tests and simulations demonstrate the feasibility of various operational strategies to mitigate wind energy fluctuations. With further technological advancements, this approach holds promise for cost-competitive desalination compared to conventional systems.

Direct-driven desalination systems could be with or without an intermediate energy storage stage.

Direct-Driven Desalination with Intermediate Energy Storage Early prototypes of wind-driven desalination systems incorporated intermediate energy storage to mitigate the intermittent nature of wind power. A pioneering system developed by Robinson et al. in 1988 [40] employed a windmill-driven pump to extract groundwater, storing it in pressurized tanks that acted as hydraulic accumulators. A solenoid valve regulated water release for desalination. Despite demonstrating feasibility for capacities exceeding 500 L/d, the system suffered from significant energy losses (52%) and low recovery rates (9–9.7%). To enhance efficiency, an energy recovery device (ERD) and advanced control systems were suggested.

Subsequent prototypes, such as the one in Coconut Island, Hawaii [41], refined the concept by using multiple solenoid valves to maintain a constant operating pressure. This enabled water pretreatment and accommodated varying feed water salinities. However, both systems were limited to brackish water due to insufficient pressure generation.

The WindDeSalter technology [38] introduced pressure accumulators within the WT tower for RO desalination processes. This configuration smoothed pressure fluctuations and enabled load management through pump and RO unit activation/deactivation based on wind availability.

Cutajar et al. [42] introduced an innovative hydropneumatic energy storage system integrated into a floating offshore wind-driven desalination plant. A tension leg platform supports the WT, which directly drives a swash plate piston pump replacing the conventional generator. Pressurized seawater is stored in a submerged chamber within the platform's gravity foundation, compressing air into an upper chamber. This process stores energy. During periods of low wind, compressed air displaces seawater, providing high-pressure feed to an onboard RO desalination unit. Produced fresh water is transported to shore, while brine undergoes energy recovery and is discharged.

The system incorporated two flow control strategies: constant feed pressure and constant flow rate, both preventing chamber overfilling or emptying. Simulations demonstrated the system's effectiveness in smoothing fluctuations, producing 33,000 m^3/d of fresh water using a 5 MW WT and 6.5 MWh hydropneumatic storage.

Direct-Driven Desalination Without Intermediate Energy Storage To circumvent the added cost, weight, and volume associated with energy storage systems, direct mechanical coupling of WTs to desalination pumps has been explored. As exemplified by Heijman et al. [36], a multi-vane WT directly drives a high-pressure pump via a drivetrain, eliminating the need for energy storage. This configuration results in variable pump speed and permeate production corresponding to wind fluctuations. While maintaining a constant recovery rate without auxiliary control, the prototype achieved a maximum recovery of 25%, lower than commercial systems. Fresh water storage is included to bridge periods of wind absence.

A similar approach is presented by Greco et al. [37], where the high-pressure pump replaces the generator in the WT nacelle. Produced water is split for desalination and electricity generation via a Pelton turbine. An isobaric ERD (iSave) enhances system efficiency. With a rated capacity of 600 m^3/d, the prototype's performance is influenced by wind speed variations. To optimize fresh water and electricity output within component constraints, an operational strategy is proposed, involving active control of blade pitch, Pelton turbine spear valve, and ERD flow rates.

Wind Energy Marine Units (WEMUs) for Seawater Desalination A relatively novel concept, the Wind Energy Marine Unit (WEMU), has been proposed for seawater desalination [39]. The WEMU is a vertical-axis WT featuring a rotating semisubmersible pontoon as its rotor. Equipped with a hydraulic transmission system comprising hydrostatic water pumps directly connected to the rotor and high-head water turbines for electricity generation, the WEMU offers two desalination configurations.

Water pumped by the WEMU is partially converted into electricity via a hydraulic turbine generator, while the remaining pressurized water is directly fed into a RO module for desalination. Excess energy in the high-pressure brine is recovered through a high-head hydraulic turbine generator to produce electricity. According to the authors, this approach holds significant promise for addressing water scarcity in regions like Northern Crimea, with the added benefit of electricity generation.

5.2.4 Electrodialysis (ED)–Wind-Powered Desalination Systems

Electrodialysis (ED) is a membrane-based separation process that exploits an applied electric potential to facilitate ion transport. In the context of saline or brackish water treatment, salt ions are selectively removed from the water phase via ion exchange membranes (IEMs) under the influence of an electric field. A fundamental ED cell comprises an IEM positioned between two electrodes. ED stacks consist of alternating cation exchange membranes (CEMs) and anion exchange membranes (AEMs) intercalated between cathode and anode electrodes. Spacers are employed to partition the membranes and create distinct fluid compartments.

CEMs and AEMs exhibit selective permeability to ions based on their respective charge. As feedwater is introduced into the ED unit and distributed across distinct compartments, an applied voltage initiates the ion separation process. This results in the formation of dilute and concentrated compartments within the system. The purified water, devoid of salts, is subsequently recovered from the dilute compartment [43].

Wind turbines could be used to provide the necessary power for the ED process either directly or indirectly.

Some of the previous investigations include the work by Veza et al. [31], who conducted a pilot-scale evaluation of a wind-powered ED desalination system in Gran Canaria as a part of the "Seawater Desalination with an Autonomous Wind Energy System (SDAWES) project," in 2002. The initiative sought to determine the optimal desalination technology for medium-scale, off-grid wind energy systems. The system employed a 72–192 m^3/day ED unit coupled with two 460 kW WTs to treat brackish water. To regulate ED current and feedwater pressure, the system incorporated DC and variable frequency drivers (VFDs), respectively, controlled by a programmable logic controller.

The electrodialysis reversal system comprised two stacks, each containing 4 hydraulic stages and 2 electrical stages with 340 cell pairs. To optimize performance, the system was equipped with DC drivers for each electrical stage (E1 and E2) in both stacks, VFDs for the feed and brine recycling pumps, and an advanced control system. The DC drivers provided a maximum power output of 38 kW (0–170 VDC) for stage E1 and 19 kW (0–170 VDC) for stage E2. The VFDs regulated the feed and brine pump speeds to control feed pressure and stack inlet differential pressure, respectively. The system was powered by two WTs with a nominal capacity of 230 kW each.

The ED unit was designed to dynamically adjust product flow based on available wind power. Operational modes included constant product quality, constant product flow rate, and variable operation. The ED system demonstrated rapid adaptability to changing operating conditions, with response times within 20 seconds.

In 2016, Malek et al. [44] designed a laboratory-scale wind-powered ED system for synthetic brackish water desalination. The system integrated a WT directly with the ED unit, eliminating the need for energy storage. This configuration was hypothesized to enhance efficiency by bypassing AC–DC conversion and charging/discharging losses. However, the system's performance was influenced by fluctuating wind conditions, leading to variable water production, especially at sub-cut-in wind speeds and high turbulence intensities. Notably, the system's specific energy consumption (SEC) remained stable at 3.36–3.38 kWh/m^3 despite wind variations.

Furthermore, the study revealed that system efficiency was not solely dependent on wind speed but was also influenced by the ED unit's load resistance. Overall, the system demonstrated reliable off-grid operation with a degree of resilience to wind fluctuations. The optimal performance of this configuration necessitated wind speeds exceeding turbine-rated values, emphasizing the critical role of site selection. Such systems are well-suited for remote regions with limited fresh water access. A key advantage is the relatively low SEC coupled with insensitivity to wind speed variations.

5.2.5 Membrane Distillation (MD) via Harnessing Waste Heat of Wind Turbines

In this technique, a circulating cooling fluid extracts heat from the components of a WT. This heat is transferred to saline water via a heat exchanger. The transferred heat is used to warm the feedwater in the direct-contact membrane distillation (DCMD) module, generating water vapor that crosses through the pores of the membrane and then condenses at the outlet of the module. The mathematical model assumes that all losses in WT components (generator and gearbox) are in the form of heat [45, 46]. This amount of heat is calculated through the following equation:

$$Q_{WH} = (1 - \eta_{generator})(1 - \eta_{gearbox})P_{avg},$$

where Q_{WH} is the waste heat generated in the WT components and $\eta_{generator}$ and $\eta_{gearbox}$ are the efficiency of the generator and gearbox, respectively.

Moreover, the generated heat is transferred to saline water by applying the energy balance equation driven from the 1st law of thermodynamics,

$$Q_{WH} = \dot{m}_{CF}(h_{CF,\,out} - h_{CF,\,in}) = \dot{m}_{sw}(h_{sw,\,in} - h_{sw,\,out}),$$

where $\dot{m}_{CF}$ and $\dot{m}_{sw}$ are the mass flow rate of the circulating cooling fluid and the mass flow rate of saline water, respectively, $h_{CF,\,out}$ and $h_{CF,\,in}$ are the enthalpies of the circulating cooling fluid at the outlet and inlet, respectively, and $h_{sw,\,in}$ and $h_{sw,\,out}$ are the enthalpies of the saline water at the inlet and outlet, respectively.

Memon et al. [46] investigated four configurations of 15 DCMD modules numerically. In the first configuration, the 15 modules were connected in series. In the second configuration, the 15 modules were connected in parallel. The third and fourth configurations were series–parallel arrangements of 3 × 5 and 5 × 3 modules, respectively. The study compared the effects of wind speed, seawater flow rate, and temperature on the daily water production, permeate flux, evaporation efficiency, and energy consumption of the four configurations. The highest daily water production was achieved with the first configuration, yielding a production rate of 1,927–10,159 kg/day depending on the wind speed and seawater flow rates. The third configuration exhibited the lowest energy consumption and the highest heat gain and evaporation efficiency.

5.3 Wave-Powered Membrane Desalination

The simultaneous presence of wave energy and saline water in one location makes wave energy an ideal approach for desalination technology. Marine energy, recognized for its stability and high energy density compared to solar and wind energy, enables direct hydraulic power take-off (PTO). This means energy can be converted directly from waves into power-producing and power-consuming devices, enhancing overall efficiency by reducing energy conversion steps and cutting material costs.

Wave energy presents a promising alternative to solar and wind energy due to its vast potential and unique characteristics. The ocean's stored energy capacity can provide up to 200% of the current global electricity demand, making wave energy a substantial resource [47]. Wave energy can be converted into electricity to power a desalination plant, as presented by Sharmila et al. [48] and Kim and Hwang [49].

Sharmila et al. [48] reviewed an Indian RO plant that operated by converting wave energy extracted from an oscillating water column (OWC) into electricity through an impulse turbine, producing 10 m^3 of water daily. Kim and Hwang [49] introduced a vacuum-enhanced air-gap membrane distillation system, which is a thermal-based desalination technique. The system primarily derives the necessary heat from a two-stage heat pump (HP), supplemented by the thermal energy of the surrounding seawater and the recovered latent heat from condensate fresh water. The two-stage HP operates using electrical energy generated by a flywheel-generator mechanism, which is powered by the heaving motion of a wave energy converter (WEC). The PTO, HP, and desalination system are all housed within a single-cylinder WEC buoy. The proposed system, named WavoWater by the authors, uses a 1-m diameter buoy that consumes 3.1 kWh of wave energy to produce 12.6 L of water.

A more efficient method involves directly pressurizing seawater through hydraulic power. This direct pressurization method eliminates multiple energy conversions from hydraulic to electric and then back to hydraulic, thus removing the need for a hydraulic turbine/motor, an electric generator, motors, and hydraulic pumps. Additionally, marine energy has a minimal land footprint, making it even more suitable for coastal regions.

Although most WECs provide adequate internal space for integrating a desalination plant [50, 51], alternative configurations may involve situating the plant on the seabed near the WECs [51]. These approaches significantly reduce the infrastructure demands of water pipelines connecting the desalination facility to the shore. They also ensure that brine disposal occurs away from the coastline, thereby minimizing environmental impact. Additionally, such designs require a minimal land area for deployment.

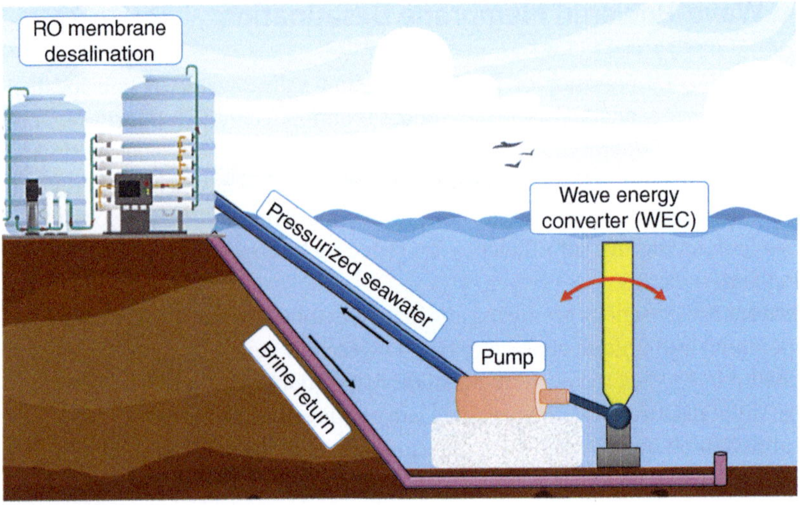

Figure 5.2 Layout of a wave-powered RO (WPRO) desalination system.

Typically, WPD plants are positioned onshore, where high-pressure seawater is transferred to the shore using pumps driven by WECs to facilitate the RO process, as depicted in Figure 5.2. By streamlining the energy transfer process, wave-powered membrane desalination becomes a more effective and resource-efficient solution for producing potable water in coastal regions. Considering that approximately 40% of the global population resides within 100 km of a coastline [52], utilizing the available energy-dense resource of ocean waves to power SWRO presents a practical solution to coastal water scarcity.

5.3.1 Integration in Ocean Environment

Desalination powered directly by pressurized seawater from wave energy is not a new concept; it was demonstrated over four decades ago with DELBUOY, developed by Hicks et al. [51] at the University of Delaware, USA, in 1976. DELBUOY was a pioneering system designed to directly convert ocean wave energy into power for an SWRO membrane-based desalination system. System commercialization was managed by ISTI Delaware, Inc., in 1985, with five full-scale sea trials and commercial installations. The DELBUOY device consists of a lightweight cylinder-shaped buoy with a shallow draft, a single-acting positive displacement submerged pump, and an anchor system, combined with a single RO membrane module. Wave-induced motions drive the pump, utilizing energy stored in rubber cable springs. Clean seawater is drawn into the pump through prefilters, and the pressurized seawater is regulated by check valves and damped via a hydraulic accumulator. The pressurized seawater is then filtered through the RO membrane, converting at a 25–30% ratio, and the produced fresh water is transported to shore via a low-pressure pipeline. A pump stroke longer than 2 m is designed to handle storm waves and tidal variations. The buoy-to-pump piston area ratio allows the system to amplify the wave-induced pressure to 5,500 kPa needed for the RO process. The system could produce an average daily production of 1.1 m^3 of potable water under conditions of wave heights ranging from 0.6 to 1.5 m, wave periods from 3 to 8 seconds, and water depths between 15 and 20 m. Deploying multiple units at a single site could achieve the desired water production rate, up to 6 m^3/day, offering flexible deployment procedures. Operating experience indicates that DELBUOY was an economical alternative to traditional RO desalination systems, especially considering the high-power costs and a five-year payback period. Johannessen et al. [53] presented another attempt to integrate an RO plant into the WECs by developing a freely floating buoy enclosing a moonpool open to the sea from below, with a disc-shaped floating body covering the upper side. Light anchor lines secure the moonpool float to maintain horizontal stability. The unit features a double-acting pressure magnifier that drives the vertical pistons for the RO pumps by utilizing the irregular motion between two floating bodies. Despite offering a novel approach to WPD, operational details and specific daily production metrics were typically not provided.

A WPD system developed by Oneka Inc. [50] can produce >500 m^3 of fresh water daily. The Oneka system involves deploying a single buoy for small-scale or an array of buoys for larger-scale desalination projects off the coast, each anchored to the ocean floor. The seawater is pressurized through the heaving movement of the buoy and is sent to the RO unit attached to the buoy. The produced fresh water is then pumped to the shore for storage or distribution. Real-time monitoring of water quality and system performance is achieved by sensors powered by solar panels. The produced brine is only 30% saltier than seawater, significantly lower than conventional RO desalination plants, ensuring no detectable change in salinity within 2–3 m of the buoy and minimal environmental impact on marine ecosystems.

5.3.2 Onshore Deployment Strategies

Folley and Whittaker [54] conducted a techno-economic analysis of an autonomous wave-powered RO (WPRO) plant, indicating that producing fresh water can be as low as £0.45/m^3. The model highlighted the advantages of such a system, including its deployment versatility and reduced environmental impact. The proposed study investigated a plant consisting of an Oyster WEC, a double-acting high-pressure pump, a standard RO membrane, and a pressure exchanger-intensifier for energy recovery. The Oyster extracts energy from the waves to directly drive the high-pressure pump. Non-return valves create a unidirectional flow, which is conditioned using a high-pressure gas-filled accumulator and a pressure relief valve to maintain stable pressure levels. Brine is fed into the Clark pump, developed by the authors [55], which serves as the pressure exchanger-intensifier, before being discharged back into the sea. In the study, the capital cost of the Oyster converter was estimated at £875,000, translating to an annualized cost of £82,000. Using 4% of capital costs for Operation and Maintenance (O&M), the annual cost was estimated at £35,000. These early-stage development costs are expected to decrease over time. The desalination plant costs are proportional to the number of RO pressure tubes, with each tube and associated equipment costing £36,000 and having an annualized cost of £3,370. O&M costs were estimated at £1,500 per tube. Results showed that a 2,500 m^3/day plant would cost £1,100,000 with annual O&M costs of £45,750. These costs were lower than traditional desalination plants due to the absence of electrical components and reduced building requirements. Despite the similar cost of water production compared to standard plants (£0.45/m^3), the autonomous WPD system has significant potential due to its lower SEC and greater deployment flexibility. However, its capital costs were higher, and it operated at an average load factor of about 62%, discarding some captured wave energy due to plant size limitations.

Magagna and Muller [56] developed an off-grid WPRO system designed to operate without electricity by utilizing wave energy to drive an RO membrane. The system comprised two main components: a high-pressure pump to generate the necessary pressure for the RO, known as the WaveCatcher, and a wave-driven pump, an OWC pump, which supplied the required water to the WaveCatcher. The high-pressure pump operated simply by accumulating water from the OWC pump in a tank attached to the end of a piston pump. The weight of the collected water activated the piston to generate the desired high pressure. The depth of the accumulating tank and the length of the piston stroke were the key parameters affecting the system's efficiency. A tested 1:6 scale model demonstrated that the WaveCatcher achieved a pressure of 0.42 MPa from an initial pressure of 2 kPa. Coupling the proposed system with an RO membrane was not investigated.

Ramudu [57] also studied an off-grid WPD system to address water scarcity in remote communities. The proposed system integrates the surge wave energy converter (SWEC) from Resolute Marine Energy, Inc., which extracts energy from ocean waves with an efficiency of 40%. This WEC features a submerged paddle connected to a shaft at the seabed, acting as a hinge that oscillates with wave motion. According to the study, an array of approximately 25 Surge WECs was predicted to produce 0.9 million m^3 of fresh water annually, sufficient for about 30,000 people. In addition to meeting critical water needs, the system was estimated to eliminate approximately 104,000 tons of CO_2 emissions annually. The estimated payback time for the system was approximately 3.5 years, showcasing its viability and economic feasibility for remote communities.

Ylänen and Lampinen [58] developed the AaltoRO, a WPRO system that combines a WaveRoller, an oscillating SWEC (OSWEC), two adaptive pressure generators (APGs), conventional RO membranes, and a hydraulic turbocharger. The system comprises two parts: the sea-based part, which includes the WaveRoller and APGs, and the onshore part, which includes the RO membranes and the turbocharger. The WaveRoller, a 25–30 m-wide and 10 m-high oscillating panel, harnesses wave energy to pressurize seawater for the RO process. The APG, designed at Aalto University, smooths pressure fluctuations, maintaining a consistent operating pressure through its adaptive valve construction and control algorithm. The pressurized seawater is transported to shore for biological treatment and then further pressurized by the hydraulic turbocharger to recover the brine's energy. The optimal pressure for the system's operation was 45 bar, lower than the usual 60–65 bar, ensuring economical operation and system longevity with a 30% recovery rate. This pressure level provides the best balance between cost (around 0.8 €/m^3) and system reliability, making it suitable for the AaltoRO and similar WPD systems.

Suchithra et al. [59] conducted a numerical analysis and designed a pilot WPD system. In the proposed design, seawater first flows through pretreatment filters, including a multimedia sand filter (MSF), an activated carbon filter, and an ultrafilter. After pretreatment, the seawater enters a double-acting cylinder driven by a flap-type OWSC. The seawater is then pumped to a hydraulic accumulator to reduce feedwater fluctuation before finally reaching the RO unit. Results showed that the designed system could operate at pressures up to 50 bar and produce 100 L of fresh water per hour with total dissolved solids less than 500 mg/L.

In 2021, Resolute Marine Energy, Inc. [60], introduced a commercial WPD system called "Wave2O," which can produce up to 4,000 m^3/day of fresh water at an estimated cost of US\$1.25/m^3. The system comprises an oscillating WEC that delivers pressurized seawater to the onshore RO unit.

Researchers at the National Renewable Energy Laboratory (NREL) designed and constructed a WPD system called the Hydraulic and Electric Reverse Osmosis Wave (HERO) [61, 62]. The HERO WEC operates in two alternating modes within the same system: electric and hydraulic. In electric mode, the WEC generates electricity and sends it to the shore via a cable to power the RO unit. In hydraulic mode, the high pressure required for desalination is achieved directly by driving a rotary pump, eliminating the need for electricity generation, electronics, and active controls, but involving an accumulator. Both modes achieve the same performance, providing 22 L/min at 620 kPa for the RO module. These two modes are achieved by alternating the drivetrain between a rotary electric generator and a rotary pump. The HERO system consists of three major units: the WEC, the power electronics, and the RO module. The WEC features a polyvinyl chloride octagonal-shaped float that encloses (1) a winch for converting linear motion into rotary motion, which is attached to (2) a clutch for retracting after each wave oscillation, and (3) a gearbox to increase the speed of the winch to drive either the rotary pump or the generator. The power electronics include a power cable, rectifier, charge controller, battery bank, inverter, and submersible pump. The RO module comprises two Spectra LB 400 units, each equipped with a Clark pump as an ERD. A comparison of the performance of the two operating modes was not available due to calm wave conditions during the hydraulic mode testing.

In a recent study, Mi et al. [63] explored a WPRO desalination system using a 1:10 scaled model in a wave tank. The system employs an OSWEC to directly drive a double-acting piston pump, which pressurizes seawater for desalination. A bladder-type accumulator smooths pressure fluctuations, and a needle valve regulates system pressure and water recovery. Optimal specific water productivity (SWP) achieved was 2.23 m^3/kWh for 3.5 g/L salinity, scaling to 0.22 m^3/kWh for 35 g/L seawater. Tuning the needle valve enhanced SWP by 17% and decreased system pressure by 24%. The study demonstrated a maximum permeate flux of 11.81 L/m^2/h and 97.74% salt rejection, highlighting the potential of wave energy for sustainable desalination. Future research should focus on system optimization and energy recovery to further improve performance.

5.3.3 Energy Recovery in Wave-Powered Desalination (WPD) Systems

As aforementioned, high-pressure seawater is pumped to the shore where desalination plants are positioned, allowing for chemical pretreatment and more space for installing ERDs, which aim to reuse energy from the pressurized brine. ERDs are an effective method to enhance the overall efficiency of the desalination system. Since desalination of seawater is an energy-intensive technology, these devices typically reduce SEC by approximately 40–60% and are essential in RO desalination plants. ERDs are classified into three categories based on the working principle as follows [64–66]:

1) **Hydraulic to Mechanical-Assisted Pumping:** This was the first energy recovery technology, dating back to prior to 1980, and is no longer used in recent RO plants due to its low efficiency. This inefficiency arises from the double energy conversion process: first, from hydraulic energy from the reject brine into mechanical energy of a rotating shaft, and then back into hydraulic energy in the feed pump. Examples of ERDs of this type include Francis turbines (reverse running pumps) and Pelton wheels.
2) **Hydraulically Driven Pumping in Series:** This type was introduced in the late 1980s when hydraulic turbochargers, commercially known as FEDCO's

hydraulic pressure boosters, were first used and more widely adopted in the 1990s. The turbocharger device consists of a pump impeller and a turbine runner on the same shaft, placed in series with the main feed pump.

3) **Hydraulically Driven Pumping in Parallel:** This type also emerged in the late 1980s with a new technology boasting efficiencies exceeding 90%, based on the "theory of work exchange," which involves the direct transfer of hydraulic energy without intermediate conversion into mechanical energy. This category includes several ERDs such as the rotary isobaric pressure exchanger, work exchangers, piston-type isobaric ERDs like the dual work exchanger energy recovery, pressure intensifiers like the simple Clark pump [67] and ENERCON pressure intensifier [68], and pressure exchanger-intensifier (modified Clark pump) [55].

The most applied types of ERDs in wave energy are turbochargers [58] and Clark pumps [54]. These devices are favored due to their high efficiency and ability to integrate effectively with WPD systems. In WPD systems, selecting the right ERDs is crucial for optimizing energy efficiency and ensuring the sustainability of the desalination process. Integrating these devices allows WPD systems to harness the kinetic energy of waves more effectively, reducing reliance on external energy sources and lowering operational costs. By leveraging the inherent energy within the pressurized brine, these devices significantly reduce the SEC of the desalination process, aligning with the global pursuit of more efficient and sustainable water treatment solutions.

5.4 Conclusion

Renewable energy technologies, such as wind and wave power, offer a promising avenue for sustainable desalination to address global water scarcity. While significant progress has been made, further advancements are essential to optimize the integration of these technologies.

Careful consideration of site-specific conditions and existing expertise is crucial for successful desalination projects. Key areas for improvement include reducing the cost and size of energy harvesting and recovery devices, as well as exploring alternative battery technologies.

To accelerate the development and deployment of renewable energy-powered desalination, increased research and development efforts, coupled with industry collaboration, are imperative. A focus on producing reliable, compact, and affordable desalination systems is essential. Additionally, knowledge dissemination and training programs will be vital in fostering wider adoption of this technology, especially in regions facing water scarcity, such as the Mediterranean, the Middle East, and Africa.

Questions

1. Explain the role of renewable energy in membrane desalination and discuss why it is considered a sustainable solution for water scarcity.

2. How do fluctuations in renewable energy sources like wind and wave affect the desalination process, and what strategies can be implemented to mitigate these effects?

3. What are the main considerations when choosing a site for wind-powered desalination plants, and how do these considerations impact plant performance?

4. Explain the principle of DCMD, and how waste heat from wind turbines can be utilized in this process.

5. Describe the key challenges in designing wind-powered desalination plants and propose potential solutions to address these challenges.

6. What is the significance of energy storage in wind-powered desalination systems, and what types of storage technologies are commonly used?

7. Discuss the process of reverse osmosis (RO) in desalination and explain how wind energy can be integrated into an RO system.

8. Describe the concept of direct-driven desalination systems. How do they differ from traditional wind-driven desalination setups?

9. How does the use of hybrid renewable energy systems (e.g., combining wind and solar) improve the reliability and efficiency of desalination plants?

10. Compare the advantages and disadvantages of wind-powered and WPD systems in terms of energy efficiency and environmental impact.

11. What are the challenges of scaling up WPD systems for large-scale use, and what are some potential solutions?

12. Analyze the design and operation of a WPD system like DELBUOY. What factors influence its performance and efficiency?

13 Explain how energy recovery devices (ERDs) enhance the efficiency of WPD systems and provide examples of commonly used ERDs.

14 What are the environmental considerations for implementing large-scale wind or WPD systems in coastal regions?

References

1 Archer, C.L. and Jacobson, M.Z. (2005). Evaluation of global wind power. *J. Geophys. Res. Atmos.* 110 (D12): https://agupubs.onlinelibrary.wiley.com/doi/epdf/10.1029/2004JD005462.

2 Rashidi, M.M., Mahariq, I., Murshid, N. et al. (2022). Applying wind energy as a clean source for reverse osmosis desalination: a comprehensive review. *Alex. Eng. J.* 61 (12): 12977–12989. https://doi.org/10.1016/j.aej.2022.06.056.

3 Nurjanah, I., Chang, T.-T., You, S.-J. et al. (2024). Reverse osmosis integrated with renewable energy as sustainable technology: a review. *Desalination* 581: 117590. https://doi.org/10.1016/j.desal.2024.117590.

4 Peñate, B. and García-Rodríguez, L. (2012). Current trends and future prospects in the design of seawater reverse osmosis desalination technology. *Desalination* 284: 1–8. https://doi.org/10.1016/j.desal.2011.09.010.

5 Ghaffour, N., Lattemann, S., Missimer, T. et al. (2014). Renewable energy-driven innovative energy-efficient desalination technologies. *Appl. Energy* 136: 1155–1165. https://doi.org/10.1016/j.apenergy.2014.03.033.

6 Dahioui, Y. and Loudiyi, K. (2013). Wind powered water desalination. In: *2013 International Renewable and Sustainable Energy Conference (IRSEC)*, 257–262. https://doi.org/10.1109/IRSEC.2013.6529659.

7 Mathioulakis, E., Belessiotis, V., and Delyannis, E. (2007). Desalination by using alternative energy: review and state-of-the-art. *Desalination* 203 (1): 346–365. https://doi.org/10.1016/j.desal.2006.03.531.

8 Lai, W., Ma, Q., Lu, H. et al. (2016). Effects of wind intermittence and fluctuation on reverse osmosis desalination process and solution strategies. *Desalination* 395: 17–27. https://doi.org/10.1016/j.desal.2016.05.019.

9 Cipollina, A., Tzen, E., Subiela, V. et al. (2015). Renewable energy desalination: performance analysis and operating data of existing RES desalination plants. *Desalin. Water Treat.* 55 (11): 3120–3140. https://doi.org/10.1080/19443994.2014.959734.

10 Park, G.L., Schäfer, A.I., and Richards, B.S. (2013). Renewable energy-powered membrane technology: supercapacitors for buffering resource fluctuations in

a wind-powered membrane system for brackish water desalination. *Renew. Energy* 50: 126–135. https://doi.org/10.1016/j.renene.2012.05.026.

11 Carta, J.A., González, J., Cabrera, P., and Subiela, V.J. (2015). Preliminary experimental analysis of a small-scale prototype SWRO desalination plant, designed for continuous adjustment of its energy consumption to the widely varying power generated by a stand-alone wind turbine. *Appl. Energy* 137: 222–239. https://doi.org/10.1016/j.apenergy.2014.09.093.

12 Carta, J.A., González, J., and Subiela, V. (2003). Operational analysis of an innovative wind powered reverse osmosis system installed in the Canary Islands. *Sol. Energy* 75 (2): 153–168. https://doi.org/10.1016/S0038-092X(03)00247-0.

13 Peñate, B., Castellano, F., Bello, A., and García-Rodríguez, L. (2011). Assessment of a stand-alone gradual capacity reverse osmosis desalination plant to adapt to wind power availability: a case study. *Energy* 36 (7): 4372–4384. https://doi.org/10.1016/j.energy.2011.04.005.

14 Gökçek, M. and Gökçek, Ö.B. (2016). Technical and economic evaluation of freshwater production from a wind-powered small-scale seawater reverse osmosis system (WP-SWRO). *Desalination* 381: 47–57. https://doi.org/10.1016/j.desal.2015.12.004.

15 Loutatidou, S., Liosis, N., Pohl, R. et al. (2017). Wind-powered desalination for strategic water storage: techno-economic assessment of concept. *Desalination* 408: 36–51. https://doi.org/10.1016/j.desal.2017.01.002.

16 Kiranoudis, C.T., Voros, N.G., and Maroulis, Z.B. (1997). Wind energy exploitation for reverse osmosis desalination plants. *Desalination* 109 (2): 195–209. https://doi.org/10.1016/S0011-9164(97)00065-9.

17 Habali, S.M. and Saleh, I.A. (1994). Design of stand-alone brackish water desalination wind energy system for Jordan. *Sol. Energy* 52 (6): 525–532. https://doi.org/10.1016/0038-092X(94)90660-2.

18 Dehmas, D.A. et al. (2011). On the use of wind energy to power reverse osmosis desalination plant: a case study from Ténès (Algeria). *Renew. Sust. Energ. Rev.* 15 (2): 956–963. https://doi.org/10.1016/j.rser.2010.11.004.

19 Ben Ali, I., Turki, M., Belhadj, J., and Roboam, X. (2020). Systemic design and energy management of a standalone battery-less PV/wind driven brackish water reverse osmosis desalination system. *Sustain Energy Technol Assess* 42: 100884.

20 Spyrou, I.D. and Anagnostopoulos, J.S. (2010). Design study of a stand-alone desalination system powered by renewable energy sources and a pumped storage unit. *Desalination* 257 (1–3): 137–149.

21 Tzen, E., Theofilloyianakos, D., and Kologios, Z. (2008). Autonomous reverse osmosis units driven by RE sources experiences and lessons learned. *Desalination* 221 (1–3): 29–36.

22 Paulsen, K. and Hensel, F. (2007). Design of an autarkic water and energy supply driven by renewable energy using commercially available components. *Desalination* 203 (1–3): 455–462.

23 Koklas, P.A. and Papathanassiou, S.A. (2006). Component sizing for an autonomous wind-driven desalination plant. *Renew. Energy* 31 (13): 2122–2139.

24 de la Nuez Pestana, I., Latorre, F.J.G., Espinoza, C.A., and Gotor, A.G. (2004). Optimization of RO desalination systems powered by renewable energies. Part I: wind energy. *Desalination* 160 (3): 293–299.

25 Manolakos, D., Papadakis, G., Papantonis, D., and Kyritsis, S. (2001). A simulation-optimisation programme for designing hybrid energy systems for supplying electricity and fresh water through desalination to remote areas: case study: the Merssini village, Donoussa island, Aegean Sea, Greece. *Energy* 26 (7): 679–704.

26 Vujčić, R. and Krneta, M. (2000). Wind-driven seawater desalination plant for agricultural development on the islands of the County of Split and Dalmatia. *Renew. Energy* 19 (1–2): 173–183.

27 Infield, D. (1997). Performance analysis of a small wind powered reverse osmosis plant. *Sol. Energy* 61 (6): 415–421.

28 Neris, A.S., Giannakopoulos, G.B., and Vovos, N.A. (1995). Autonomous wind turbine supplying a reverse osmosis desalination unit. *Wind Eng.* 325–347.

29 Petersen, G., Fries, S., Mohn, J., and Müller, A. (1981). Wind and solar powered reverse osmosis desalination units-design, start up, operating experience. *Desalination* 39: 125–135.

30 Henderson, C.R., Manwell, J.F., and McGowan, J.G. (2009). A wind/diesel hybrid system with desalination for Star Island, NH: feasibility study results. *Desalination* 237 (1–3): 318–329.

31 Veza, J., Peñate, B., and Castellano, F. (2004). Electrodialysis desalination designed for off-grid wind energy. *Desalination* 160 (3): 211–221.

32 Ali, E., Bumazza, M., Eltamaly, A. et al. (2021). Optimization of wind driven RO plant for brackish water desalination during wind speed fluctuation with and without battery. *Membranes (Basel)* 11 (2): 77.

33 Subiela, V.J., Peñate, B., and García-Rodríguez, L. (2020). Design recommendations and cost assessment for off-grid wind-powered-seawater reverse osmosis desalination with medium-size capacity. *Desalin. Water Treat.* 180: 16–36.

34 Miranda, M.S. and Infield, D. (2003). Renewable energy for desalination: a wind-powered seawater reverse-osmosis system without batteries. *Desalination* 153 (1–3): 9–16.

35 Moreno, F. and Pinilla, A. (2005). Preliminary experimental study of a small reverse osmosiswind-powered desalination plant. *Desalination* 171 (3): 257–265.

36 Heijman, S.G.J., Rabinovitch, E., Bos, F. et al. (2009). Sustainable seawater desalination: stand-alone small scale windmill and reverse osmosis system. *Desalination* 248 (1–3): 114–117.

37 Greco, F., De Bruycker, D., Velez-Isaza, A. et al. (2020). Preliminary design of a hydraulic wind turbine drive train for integrated electricity production and seawater desalination. In: *Journal of Physics: Conference Series*, 032015. IOP Publishing.

38 Witte, T., Siegfriedsen, S., and El-Allawy, M. (2003). WindDeSalter® technology direct use of wind energy for seawater desalination by vapour compression or reverse osmosis. *Desalination* 156 (1–3): 275–279.

39 Cheboxarov, V.V., Yakimovich, B.A., Abd Ali, L.M., and Al-Rufee, F.M. (2019). An offshore wind-power-based water desalination complex as a response to an emergency in water supply to Northern Crimea. *Appl. Sol. Energy* 55: 260–264.

40 Robinson, R., Ho, G., and Mathew, K. (1992). Development of a reliable low-cost reverse osmosis desalination unit for remote communities. *Desalination* 86 (1): 9–26.

41 Liu, C.C.K., Jae-Woo, P., Migita, R., and Gang, Q. (2002). Experiments of a prototype wind-driven reverse osmosis desalination system with feedback control. *Desalination* 150 (3): 277–287.

42 Cutajar, C., Sant, T., Buhagiar, D., and Farrugia, R.N. (2019). Modelling of a hybrid floating wind, energy storage and desalination unit. In: *2019 Offshore Energy and Storage Summit (OSES)*, 1–11. IEEE.

43 Curto, D., Franzitta, V., and Guercio, A. (2021). A review of the water desalination technologies. *Appl. Sci.* 11 (2): 670.

44 Malek, P., Ortiz, J.M., and Schulte-Herbrüggen, H.M.A. (2016). Decentralized desalination of brackish water using an electrodialysis system directly powered by wind energy. *Desalination* 377: 54–64.

45 Khalilzadeh, S. and Hossein Nezhad, A. (2018). Utilization of waste heat of a high-capacity wind turbine in multi effect distillation desalination: energy, exergy and thermoeconomic analysis. *Desalination* 439: 119–137. https://doi.org/10.1016/j.desal.2018.04.010.

46 Memon, S., Lee, H.-S., Kim, W.-S., and Kim, Y.-D. (2022). Parametric investigation of modular configuration of multi-stage direct contact membrane distillation powered by waste heat of wind turbine. *Desalination* 533: 115770. https://doi.org/10.1016/j.desal.2022.115770.

47 Lamb, B.S. and Lesemann, A.R. (2019). From watt-scale to megawatt-scale: advances in wave energy technologies for offshore applications. In *Offshore Technology Conference,* OTC. https://doi.org/10.4043/29600-MS.

48 Sharmila, N., Jalihal, P., Swamy, A.K., and Ravindran, M. (2004). Wave powered desalination system. *Energy* 29 (11): 1659–1672. https://doi.org/10.1016/j.energy.2004.03.099.

49 Kim, G.S. and Hwang, Y. (2022). Wave-powered and zero-discharging membrane-distillation desalination system: conceptual design and analysis. *Water (Basel)* 14 (12): 1897. https://doi.org/10.3390/w14121897.

50 Oneka: our wave-powered sustainable desalination solution [Online]. https://www.onekawater.com/sea-water-desalination-technology/ (accessed 28 July 2024).

51 Hicks, D.C., Mitcheson, G.R., Pleass, C.M., and Salevan, J.F. (1989). Delbouy: ocean wave-powered seawater reverse osmosis desalination systems. *Desalination* 73: 81–94. https://doi.org/10.1016/0011-9164(89)87006-7.

52 Steven, A., Addo, K.A., Llewellyn, G. et al. (2023). Coastal development: resilience, restoration and infrastructure requirements. In: *The Blue Compendium: From Knowledge to Action for a Sustainable Ocean Economy* (ed. J. Lubchenco and P.M. Haugan), 213–277. Cham: Springer International Publishing, https://doi.org/10.1007/978-3-031-16277-0_7.

53 Johannessen, T.B., Brandsvoll, R.A., Palmstrøm, A. et al. (2011). A system for wave driven desalination of seawater. In: *Volume 5: Ocean Space Utilization; Ocean Renewable Energy*, 467–474. ASMEDC, https://doi.org/10.1115/OMAE2011-49419.

54 Folley, M. and Whittaker, T. (2009). The cost of water from an autonomous wave-powered desalination plant. *Renew. Energy* 34 (1): 75–81. https://doi.org/10.1016/j.renene.2008.03.009.

55 Folley, M., Peñate Suarez, B., and Whittaker, T. (2008). An autonomous wave-powered desalination system. *Desalination* 220 (1): 412–421. https://doi.org/10.1016/j.desal.2007.01.044.

56 Magagna, D. and Muller, G. (2009). A wave energy driven RO stand-alone desalination system: initial design and testing. *Desalin. Water Treat.* 7 (1–3): 47–52. https://doi.org/10.5004/dwt.2009.699.

57 Ramudu, E. (2011). Ocean wave energy-driven desalination systems for off grid coastal communities in developing countries. In: *2011 IEEE Global Humanitarian Technology Conference*, 287–289. https://doi.org/10.1109/GHTC.2011.38.

58 Ylänen, M.M.M. and Lampinen, M.J. (2014). Determining optimal operating pressure for AaltoRO—a novel wave powered desalination system. *Renew. Energy* 69: 386–392. https://doi.org/10.1016/j.renene.2014.03.061.

59 Suchithra, R., Das, T.K., Rajagopalan, K. et al. (2022). Numerical modelling and design of a small-scale wave-powered desalination system. *Ocean Eng.* 256: 111419. https://doi.org/10.1016/j.oceaneng.2022.111419.

60 Wave2O™ a new wave-powered, flexible seawater desalination system for developing countries [Online]. https://solarimpulse.com/solutions-explorer/wave2o (accessed 28 July 2024).

61 Nakhai, A., McGilton, B., and Jenne, S. (2022). Summary report of HERO WEC test article for waves to water: electrical power take-off. National Renewable Energy Lab (NREL), Golden, CO (United States).
62 Jenne, S., Gore, S., Muglia, M. et al. (2024). The design, fabrication, and test program for NREL's wave-powered desalination system. National Renewable Energy Laboratory (NREL), Golden, CO (United States).
63 Mi, J., Wu, X., Capper, J. et al. (2023). Experimental investigation of a reverse osmosis desalination system directly powered by wave energy. *Appl. Energy* 343: 121194. https://doi.org/10.1016/j.apenergy.2023.121194.
64 Al-Hawaj, O.M. (2003). The work exchanger for reverse osmosis plants. *Desalination* 157 (1): 23–27. https://doi.org/10.1016/S0011-9164(03)00378-3.
65 Guirguis, M.J. (2011). Energy recovery devices in seawater reverse osmosis desalination plants with emphasis on efficiency and economical analysis of isobaric versus centrifugal devices. University of South Florida.
66 Kadaj, E. and Bosleman, R. (2018). Chapter 11 – energy recovery devices in membrane desalination processes. In: *Renewable Energy Powered Desalination Handbook* (ed. V.G. Gude), 415–444. Butterworth-Heinemann. https://doi.org/10.1016/B978-0-12-815244-7.00011-8.
67 Thomson, M., Miranda, M.S., and Infield, D. (2003). A small-scale seawater reverse-osmosis system with excellent energy efficiency over a wide operating range. *Desalination* 153 (1): 229–236. https://doi.org/10.1016/S0011-9164(02)01141-4.
68 Paulsen, K. and Hensel, F. (2005). Introduction of a new energy recovery system—optimized for the combination with renewable energy. *Desalination* 184 (1): 211–215. https://doi.org/10.1016/j.desal.2005.03.060.

6

Utilization of Geothermal Energy in Water Desalination Systems

Amr Mansi[1,2], Maggie Nasry Attia[1,2], Noran Ashraf[1,2], Hamad AlMohamadi[3,4], and Mohamed Bassyouni[1,2,3,4]

[1] *Center of Excellence for Membrane Testing and Characterization (CEMTC), Port Said University, Port Said, Egypt*
[2] *Department of Chemical Engineering, Faculty of Engineering, Port Said University, Port Said, Egypt*
[3] *Department of Chemical Engineering, Faculty of Engineering, Islamic University of Madinah, Madinah, Saudi Arabia*
[4] *Sustainability Research Center, Islamic University of Madinah, Madinah, Saudi Arabia*

6.1 Introduction

The global water shortage is becoming a more pressing issue due to population growth and climate change. Freshwater consumption is expected to increase significantly at the global level. Around 40% of the global population, lack access to safe and clean drinking water [1, 2]. Furthermore, drinking contaminated water is the primary cause of 90% of infectious diseases. Moreover, a report by the World Resources Institute predicted that more than 3.5 billion people will face water scarcity by 2025. More water evaporates from sources due to global warming, which inevitably lowers the available drinking water and increases the demand for unconventional sources of water [3]. Since this may threaten global water security in the near future, prompt actions are inevitable [4]. Desalting water is the most rational way to handle the organization of freshwater supply [2]. The oceans and the shores of land contain over 97.5% of the world's water, which is theoretically drinkable if its salinity is lowered [5]. Energy is used in all desalting processes to either remove salt or clean saltwater [6]. Desalination has drawn a lot of attention lately as one of the best solutions to the freshwater crisis issue [7]. With current technologies, desalination still demands a significant amount of non-renewable energy sources [8]. However, new research indicates that the long-term depletion of petroleum supplies may occur slightly later [9].

Water Treatment and Desalination: Membrane Separation using Renewable Energy Resources, First Edition. Edited by Mohamed Bassyouni.

As a result, the use of renewable resources, particularly geothermal desalination, for desalination has increased [10]. There are two types of geothermal energy resources in the world namely high- and low-enthalpy reservoirs [11]. The two types are classified based on temperature and geology. For instance, the lower enthalpy geothermal sources are those with temperatures below 150°C and about 1,000 m below the surface [12]. The utilization of geothermal sources has several advantages. For instance, the high-capacity factor of geothermal energy sources allows the supply of consistent and reliable heat sources for thermal desalination and reverse osmosis processes. Additionally, the technology for geothermal energy production is mature and commercially viable. The majority of the world's geothermal source temperatures fall between 70 and 90°C, making them perfect for low-temperature multi-efficient desalination (MED) desalination [13]. Power generation and other process heat operations can be conducted using high-grade sources that are above 100°C. Geothermal desalination may produce power and water products simultaneously and is a cost-effective process. From an economic perspective, by minimizing imported fossil fuels, geothermal desalination enhances both environmental sustainability and energy security [14].

The efficiency and viability of using geothermal energy for desalination are increased by technological advancements in desalination, such as enhanced heat extraction and membrane distillation (MD) [15, 16]. Thermal energy production from underground aquifers is currently a highly developed technology. It is not affected by weather whims or the changing of the seasons. The advantage offered by geothermal reservoirs to the current desalination technologies is the high-energy capacity that provides a consistent energy supply. Geothermal desalination can produce water and electricity simultaneously, saving costs. MED operated at low temperatures is optimum for the average temperature range of the majority of global geothermal sources (i.e., 70–90°C) [17].

In this chapter, the availability of geothermal sources and applicability of geothermal energy in desalination applications were assessed with actual case studies. Moreover, the future perspectives, limitations, and considerations for using geothermal energy in desalination were discussed.

6.2 Global Status of Geothermal Energy

Operational geothermal power plants are limited to only 32 countries around the globe [18]. In 2023, the combined capacity of those plants reached around 16 GW from approximately 200 geothermal fields. The growth trend of geothermal energy

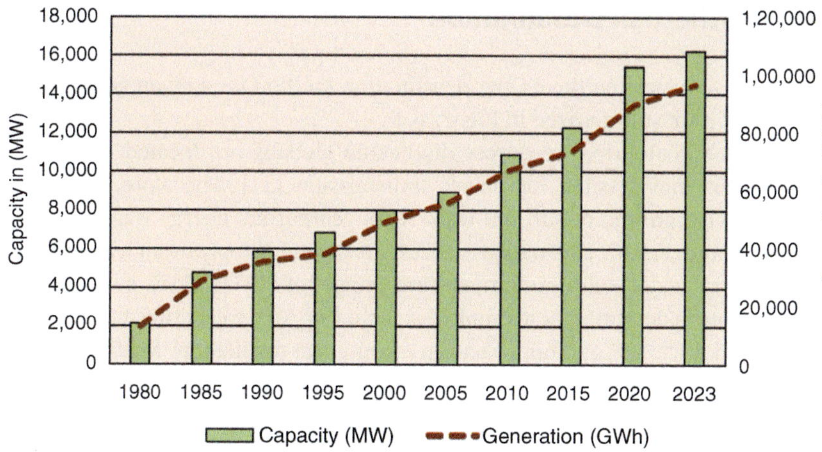

Figure 6.1 Growth trend of geothermal energy capacity and the corresponding electricity production from geothermal sources. Source: Gutiérrez-Negrín [19]/with permission of Springer Nature.

capacity and electricity generation is shown in Figure 6.1. Considering the current millennium, the geothermal energy capacity doubled between 2000 and 2023 from 7,970 to 16,320 MW. The electricity generation from geothermal sources growth was slightly lower than the growth in the geothermal industry. The electricity generation increased by 96% between 2000 and 2023 from 49,260 to 96,550 GWh [19].

The share of geothermal energy in global energy and electricity production is only 0.34% and 0.16%, respectively. Nevertheless, geothermal energy is vital in some regions where it contributes significantly to electricity production. For example, geothermal energy contributed 45% to the total energy production in Kenya, 32% in Iceland, and between 10 and 20% in Costa Rica, Nicaragua, and New Zealand.

It is important to note that geothermal energy has a superior capacity factor when compared to other renewable energy sources [20]. The capacity factor, defined as the ratio of the net energy generated by a process to the total energy available from the source, can reach up to 95% for geothermal energy if operated in a closed-loop configuration. The capacity factor for the common renewable energy sources is less than one-third of the value for geothermal energy. For instance, the capacity factors for wind, tidal, and photovoltaic cells are 30%, 30%, and 20%, respectively. A high-capacity factor motivates the scalability and investment of a technology [21].

6.3 Geothermal Desalination

The criteria for selecting the proper desalination method for a given geothermal energy source are summarized in Figure 6.2.

Most of the geothermal resources discovered globally are located in remote areas that are not feasible for power transmission [22]. Therefore, desalination of brackish ground or sub-soil water with geothermal energy was proposed as a feasible and realistic alternative to secure freshwater supply in such areas [23].

Geothermal water with a temperature range of 40–70°C is suitable for low-temperature desalination technologies including MD, solar stills, humidification–dehumidification, and high vacuum multi-effect distillation. Medium-grade sources (> 70°C) are suitable for multi-stage flash and multi-effect evaporation processes. The highest enthalpy sources (up to 200°C) can be used for cogeneration processes in large-scale desalination plants. The nature of the geothermal fluid is detrimental to the available application. The equilibrium between the dissolved salts shifts with temperature and pressure that may promote scaling. The total dissolved solids in a geothermal fluid varies in both composition and concentration depending on the nature of the source rocks and reservoir location with Sodium chloride being the most abundant salt [24]. The reported total dissolved solids range for high enthalpy geothermal sources was 2,500–81,000 ppm while for lower-grade geothermal sources the reported range was 1,100–8,200 ppm The higher salinity sources can be used for indirect heating of the water to be desalinated to avoid scaling and fouling issues. Moreover, a hybrid configuration can be adopted where the cooled geothermal fluid is further desalinated in an MD cell before being reinjected back

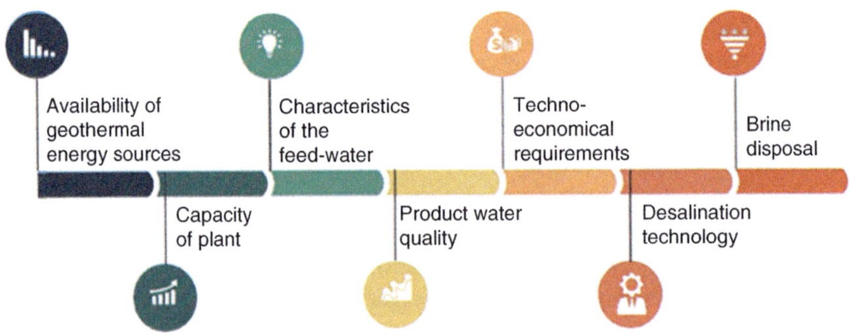

Figure 6.2 The important parameters in geothermal desalination. Source: Shelare et al. [12]/MDPI/CC BY 4.0.

to the reservoir. Indeed, this hybrid scheme has its own drawbacks because of the gradual increase in the salinity of the reinjected geothermal fluid. Therefore, precise calculations should be performed to assess the sustainability of the hybrid scheme and estimated lifetime.

6.3.1 Geothermal Energy Utilization Schemes

There are two methods of geothermal desalination (direct and indirect). The direct geothermal desalination process heats seawater directly to its boiling point in an evaporator, then transferring the steam to a condenser to produce freshwater. Indirect geothermal desalination converts geothermal energy into electricity, which is then used to power membrane-based desalination processes. The general operation scheme of geothermal power plants is that the hot fluid is brought from underground to the surface to produce heated media. The heated media (e.g., steam) is used to drive a power turbine and the cooled fluid is reinjected back into the reservoir. There are four main types of technology for producing geothermal energy, and they differ in terms of efficiency, reliability, and reliability. The four main types are shown in Figure 6.3. Figure 6.4 shows the percentage of energy produced based on the classification of geothermal power generation technology.

Dry steam is one of the oldest methods used to produce geothermal energy. This technology is applicable when the steam out of the geothermal reservoir is dry or superheated and can be used directly to produce electricity from a turbine-generator setup. The thermodynamic cycle is completed by reinjecting the condensate back into the reservoir.

Flash steam is currently the predominant method of geothermal energy production. The principle of the process is that the emerging fluid, which is at a high temperature and pressure, is flashed in a low-pressure vessel to a superheated steam feed to the turbine-generator system.

On the other hand, the binary power cycles can be operated at lower temperatures than the flash steam cycle. The hot fluid out of the geothermal reservoir exchanges heat with a fluid having a lower boiling point than water. The evaporated fluid is then directed to the turbine for energy generation [25].

The efficiency of geothermal energy power plants varies according to the technology used and according to the temperature. Other considerations include the climate, location, depth, and heat lost inside the equipment, turbines, and generators. In general, the temperature required to drive the turbine is higher than 150°C [25]. The highest estimates for the efficiency of energy converted from geothermal heat are 21%, which occurred in the Darajat in Indonesia (vapor-dominated system), the worldwide average efficiency is estimated at 12% [27].

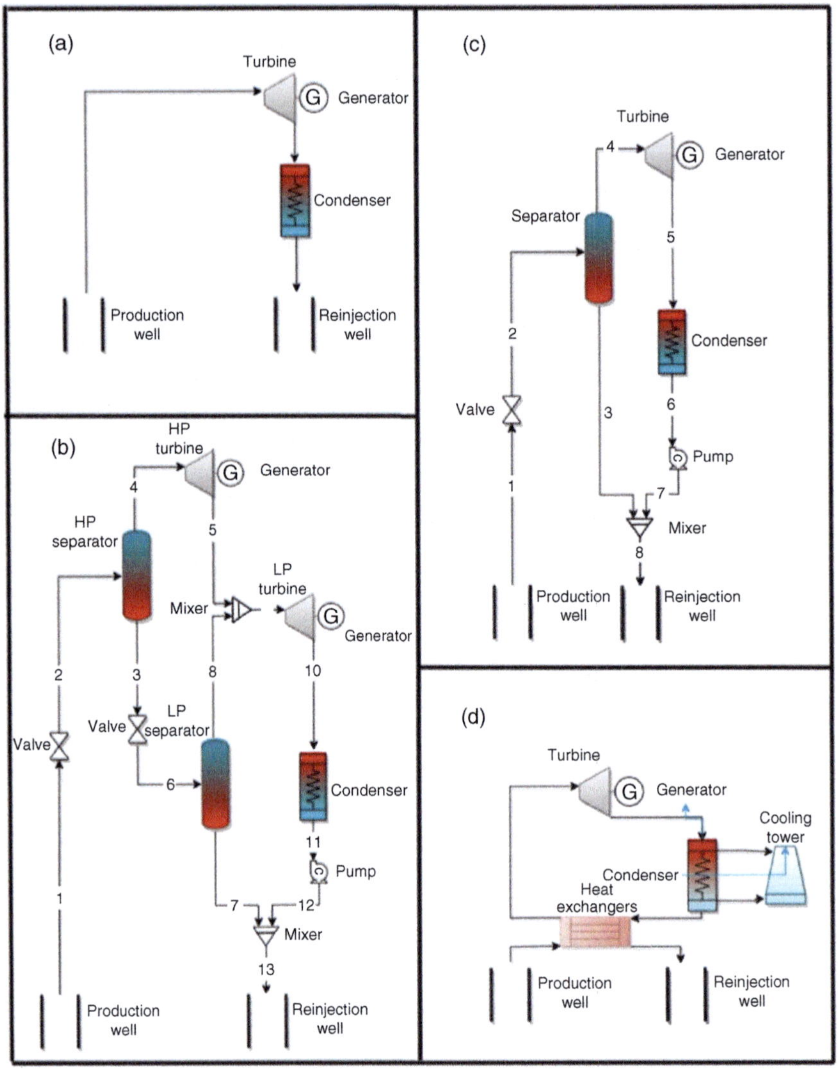

Figure 6.3 Configurations of energy production systems by geothermal energy: (a) dry steam, (b) double flash steam, (c) single flash steam, and (d) binary organic Rankine cycle. Source: Pietrasanta et al. [25]/with permission of Elsevier.

6.3.2 Geothermal Desalination for Thermophysical Systems

The traditional desalination technologies require intensive thermal energy for the separation of water from the other constituents. Coupling these technologies with geothermal desalination can minimize the energy required and reduce the

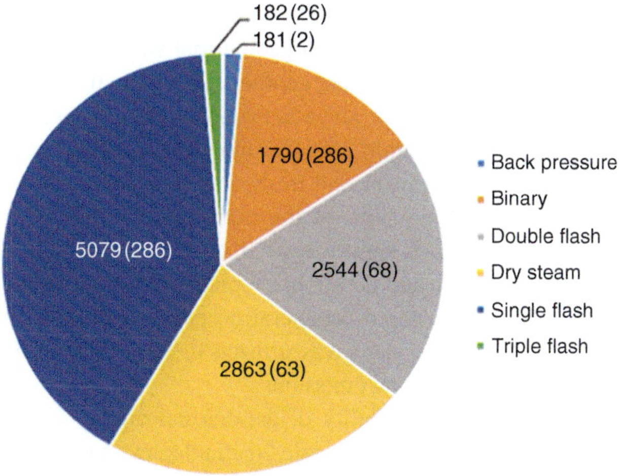

Figure 6.4 Installed geothermal plants energy capacity in MWe and number of units for different geothermal technologies. Source: Sharmin et al. [26]/with permission of Elsevier.

carbon footprint. A recent study examined the efficiency of solar stills that used brackish geothermal water in South Algeria as the feed. The enthalpy is mostly low in South Algeria's geothermal sources that have a maximum temperature of 60–70°C. The performance of a developed system that consists of a capillary film solar distiller was examined. The experimental outcomes from the distillation unit were compared to theoretical data, which showed that the heat recovery for the system was capable of generating up to 20 L/m^2/day, unlike the regular solar still that produces 5–6 L/m^2/day [28].

Previous research studied an aero-evapo-condensation process that showed potential for cooling and desalting geothermal water. The mass and heat transfer were evaluated, by using modeling and experimental data, within a temperature range of 60–90° [29].

Another research examined the heat transfer in the aero-evapo-condensation process using a geothermal spring that has a water temperature range of 70° and it also investigated the impact of hydrodynamic and thermal parameters on the performance that resulted in a good correlation between experimental and theoretical results [30].

In a feasibility study, a hybrid configuration consisting of a multi-stage flash (MSF) unit and a multi-effect boiling unit was tested using geothermal sources in Baja California, Mexico, where the geothermal heat source was available at 80°C. This study found that the ratio of desalinated water to geothermal water was about 1–14 [31].

In a previous study, A prototype of the MSF-MED configuration at a coastal desalination plant was developed to test the coastal geothermal water source at 80°. It reported a ratio of 1–5.9 for freshwater to the geothermal source. It showed that 118 m^3/day of geothermal heat flow can produce 20 m^3/day of freshwater [32].

Koroneos evaluated the potential of multi-effect evaporation coupled with geothermal energy for co-generation of power and freshwater in Nisyros island, Greece [33]. The lower temperature operation of the multi-effect evaporation decreased the overall scaling potential. A significant reduction in the environmental impacts of the power generation and desalination processes was reported compared to a traditional fossil fuel-powered cogeneration plant. Moreover, in this process, waste heat from the power plant was utilized in the desalination process eliminating associated environmental emissions.

A previous study demonstrated the feasibility of geothermal energy-driven desalination technologies in a site test. The investigated technologies, comprising MSF, vertical tube evaporator (VTE), and high-temperature electrodialysis desalination (HTED), had the common characteristic of smooth and enhanced heat transfer surfaces [34]. Results showed the successful operation of the VTE and MSF units at low and high temperatures at 87.7 and 132°C. The HTED units were operated at a three-stage and two-stage configurations at 71 and 60°C respectively.

A seawater desalination plant driven by low-enthalpy geothermal energy located on the Milos Island had been described [35]. The process design consisted of coupling a geothermal groundwater source having a temperature range of 75–90°C with multi-effect distillation. Results showed that the proposed process would cut out about 58.15 GWh of the total energy demand of the desalination plant operating at a capacity of 600–800 m^3 freshwater per day. This indicates that thermal desalination processes can benefit significantly when coupled with geothermal sources even in the lack of direct geothermal energy.

6.3.3 Geothermal Desalination for Membrane Processes

The majority of the currently operated desalination plants are based on reverse osmosis technology. Kamal reported a significant decrease in the cost of potable water can be prompted at seawater reverse-osmosis plants located in southern California [36]. Increasing the feed water temperature by 1°C can increase the productivity of the membrane by 2–3%. Most commercial reverse osmosis (RO) membranes can be operated at temperatures up to 40°, but some new membranes can tolerate higher temperatures. For example, the Backpulseable™ polypropylene tubular membranes can operate at temperatures up to 60°C. It should be noted that increasing the temperature of the feed water to RO units up to 40°C (temperature tolerance of most commercialized membranes) can increase its productivity of about 20–30%.

On the other hand, MD is a novel technology that utilizes a hydrophobic membrane for water desalination at low temperatures. The permeate flow rates in membrane desalination process are significantly influenced by the feed water temperature. Lower-temperature feeds have higher viscosities and higher resistances to penetrate the membrane pores, while higher-temperature feeds yield higher flux due to lower viscosity [37]. The energy demand of membrane processes for pumping the cold feed water during winter seasons is inevitably increased to meet the required daily production. A geothermal energy-based zero-liquid discharge setup of air-gap MD and fluidized bed crystallizer was investigated [38]. The water salinity varied between 3,000 and 35,000 ppm and the permeate flux was found to deteriorate with increasing salinity. This was due to vapor pressure reduction and the increase in concentration and temperature polarizations. The maximum reported permeate flow value was reported as 7.5 kg/m^2/h.

6.4 Future Perspectives

According to the International Energy Agency, the global geothermal energy's share in power production is forecast to escalate by the year 2050 as shown in Figure 6.5. Geothermal energy contribution to electricity production is estimated to reach 1,400 TWh/y, which represents approximately 3.5% of the global energy consumption by 2050. This contribution will minimize the carbon footprint of energy production by roughly 8×10^8 metric tons CO_2 per year. The rapid expansion of geothermal power production observed between 2015 and 2030 is

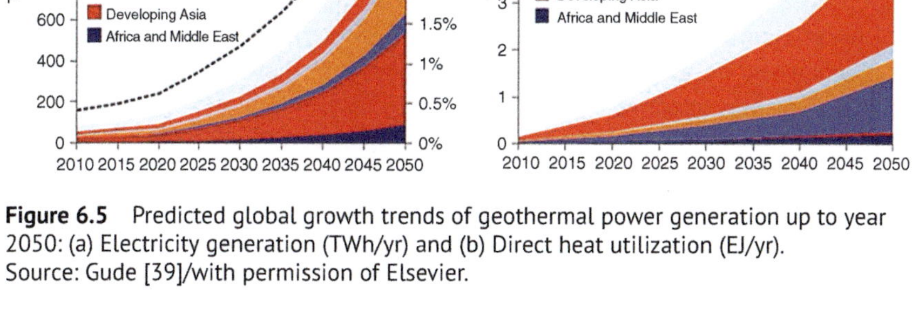

Figure 6.5 Predicted global growth trends of geothermal power generation up to year 2050: (a) Electricity generation (TWh/yr) and (b) Direct heat utilization (EJ/yr). Source: Gude [39]/with permission of Elsevier.

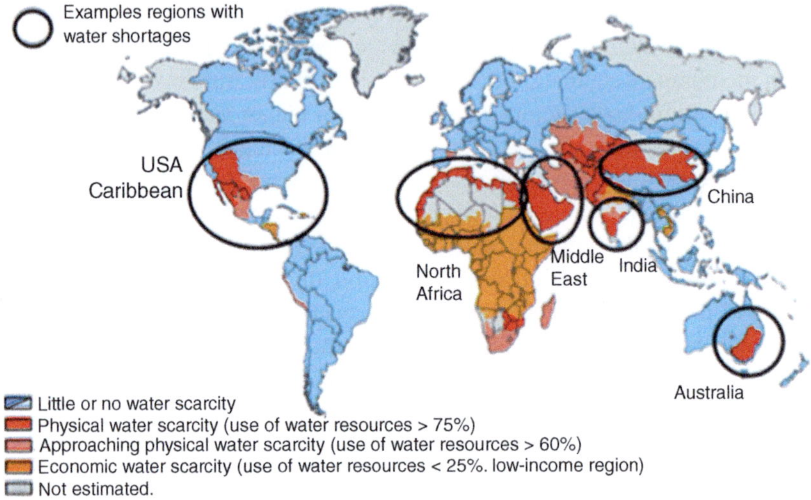

Figure 6.6 Water stress regions around the world. Source: Gude [39]/with permission of Elsevier.

mainly fueled by the rapid deployment of the conventional high-temperature wells. This development is driven by relatively attractive economics but limited to regions where such resources are available. Extensive development, research, and demonstration resources are required in the current decade to ensure geothermal developed systems become commercially feasible by 2030 [39].

Many regions around the world currently face major water stress due to economic or physical water scarcity. Physical water scarcity means the depletion of water resources due to excess need and inadequate water management. Whereas the economic water scarcity is the lack of feasible technologies or resources to produce potable water that is the case in least-developed countries with high salinity water resources. Geothermal energy can, therefore, be considered a viable solution to eliminate economic water scarcity because it is a low-cost energy source. Figure 6.6 illustrates the global water-stress regions. Interestingly, the regions striving against high water-stressed correspond with high-temperature geothermal sources [40]. The incorporation of geothermal resources with power production and desalination application feasibility in these regions is presented in the following subsections.

6.4.1 Geothermal Desalination Potential in the USA

More than 50% of the total population in the USA are located in coastal areas. The future water demand in the coastal regions cannot be met by solely

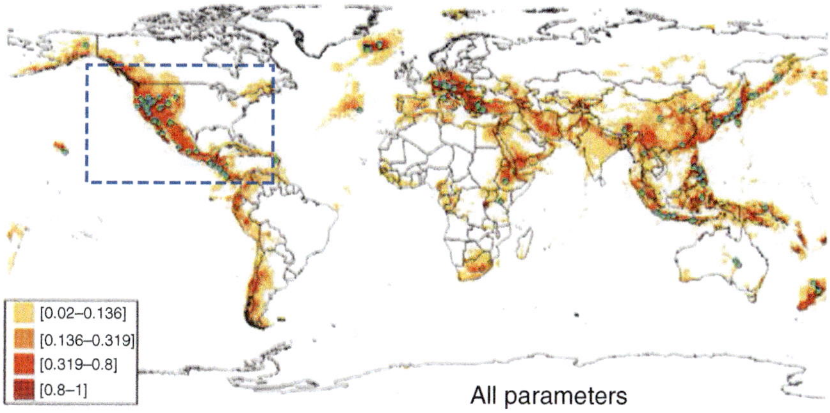

Figure 6.7 Geothermal source potential and the source temperatures in the USA. Source: Coro and Trumpy [41]/Elsevier/CC BY 4.0.

relying on the traditional surface and groundwater resources but rather include environmentally and economically sound use of diverse water resources. This can be achieved by including saline ocean water, brackish water, and the reuse of reclaimed water. This situation prevails in Texas, Florida, and California. Fortunately, most of the land area of the United States lies above potential geothermal resources. As observed in Figure 6.7, the high temperature sources are more concentrated in the western parts of the country. The abundant geothermal wells with a temperature range of 100–150°C are most suitable for desalination applications. Numerous geothermal projects are located and being constructed throughout the USA mainly in, but not limited to, western Nevada, the Geysers resource area, the Salton Sea, and Northern California. Some regions of New Mexico are also scarce in surface water and groundwater sources and the geothermal desalination is a very attractive route in these regions.

6.4.2 Geothermal Desalination Potential in the Middle East and Africa

The Middle East has a potential share of geothermal resources. For instance, substantial geothermal hotspots were discovered in the Kingdom of Saudi Arabia. The country's CO_2 emission from fossil fuels has doubled from the year 2000 till the present time reaching approximately 492 million metric tons per year. Saudi Arabia is now investigating implementing a renewable footprint for water and power sectors. The coastal region along the Red Sea has potential high and moderate enthalpy resources for electricity generation. The Arabian Gulf region, more specifically the sedimentary basins, may contain considerable low-enthalpy geothermal sources suitable for desalination [23].

Similar to Saudi Arabia, potable water production is among the major energy-intensive sectors in the United Arab Emirates. The UAE initiated plans to minimize the non-renewable energy footprint, especially for water desalination. MSF contributes to about 80% of the total desalination capacity in the UAE that requires a significant amount of energy. This technology is mainly used for handling high-salinity seawater. Cogeneration schemes are utilized to increase resource utilization efficiencies. On the other hand, MSF is a low-temperature desalination technology that motivates the use of low-enthalpy geothermal sources available across the UAE.

The availability of geothermal sources in the Sub-Saharan African region is limited to a few countries, with the potential for power generation greatest in Ethiopia and Kenya with estimated geothermal electricity production potential of 41,000 and 3,000 MWe, respectively. The two countries together have an installed total geothermal power capacity of 217 MW. Meanwhile, the United States Agency for International Development (USAID) developed a master plan to expand the currently operational Aluto Langano geothermal plant in Ethiopia to a production capacity of 5 TW by 2037.

North African countries contain a few discovered geothermal sources but their use is only limited to heating and irrigation of greenhouses.

Geothermal development in Sub-Saharan African countries is hindered by various economic and social obstacles. The elevated drilling cost is beyond the economic affordability of these countries and necessitates support from foreign countries. The utilization and potential capabilities of geothermal energy are underestimated or poorly recognized. Moreover, the limited technical and human capacity signifies the essential contribution of private investors in these areas.

6.4.3 Geothermal Desalination Potential in Australia

The locations of desalination plants for Australia along with the temperature profile of geothermal sources are illustrated in Figure 6.8. It is noted that high enthalpy geothermal sources with a temperature range between 160 and 235°C are available in some regions of the continent that are suitable for electricity generation. It is important to note that the high enthalpy geothermal source locations are far from the existing desalination plants, which means that geothermal sources are available in remote areas suitable for desalination applications. Australia has an abundance of small communities that strive against water scarcity and severe drought issues. The implementation of geothermal desalination is a resource-efficient option for these locations.

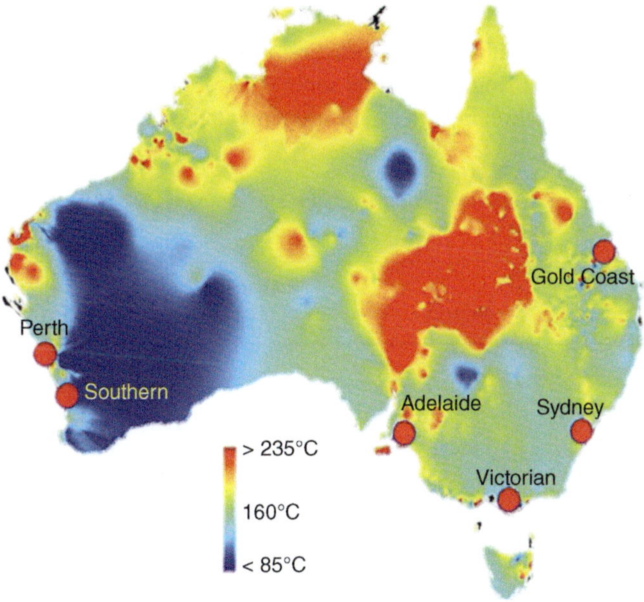

Figure 6.8 Geothermal source potential and the constructed desalination plants in Australia. Source: Gude [39]/with permission of Elsevier.

6.5 Considerations and Challenges for Geothermal Desalination

An ideal geothermal source should be approximate to the place of application [42], readily accessible, and available at moderate depths with a quality appropriate for the intended use. It is possible to summarize the primary criteria for choosing a quality geothermal resource as follows [43, 44].

i) For a cogeneration scheme, high temperatures are necessary for improved power generation efficiency.
ii) A significant amount of stored thermal energy to extend the life of the reservoir.
iii) Availability of potential energy difference between the production wells and the disposal wells to enable discharge under gravity.
iv) Generates almost neutral pH fluids for low rates of corrosion in power plants and wells.
v) High permeability to guarantee sufficient outputs from each well.
vi) A minimal chance of hydrothermal outbursts and volcanic activity.
vii) The distance from power transmission cables or loads.

The exploration for a geothermal reservoir, estimation of the reservoir's volume, thermodynamic properties, chemical characteristics, and long-term availability are the fundamental steps in using geothermal sources for desalination applications. For geothermal desalination facilities to be implemented successfully, a few

Table 6.1 Possible impacts of geothermal sources on the environment based on desalination and power production.

Impact	Severity	Probability of occurrence	Duration
Social economic problems	Low	Low	Short-term
Well blowouts	Low/medium	Low	Short-term
Land subsidence	Low/medium	Low	Long-term
Air pollutants emissions	Medium	Low	Short-term
Underground contamination	Medium	Low	Long-term
Discharged surface water	Low/medium	Medium	Short-term/ long-term
Thermal or chemical contamination	Medium/high	Medium	Short-term/ long-term
Disposal of solid waste	Medium/high	Medium	Short-term
High noise levels	Medium/high	High	Short-term

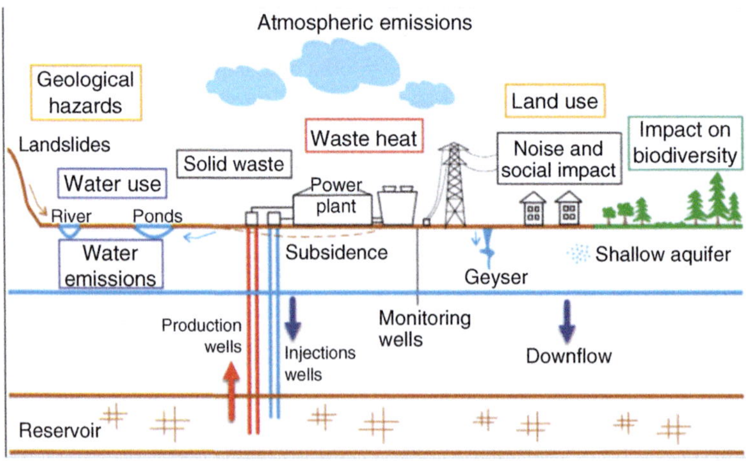

Figure 6.9 Environmental impacts associated with geothermal energy. Source: Sayed et al. [46]/with permission of Elsevier.

more crucial elements must be taken into consideration. Land use [45], geological dangers, atmospheric releases, waste heat, water footprint, and socioeconomic issues are the main concerns. Table 6.1 depicts the probability, duration, and severity of the socio-economic and environmental impacts associated with using geothermal energy in desalination. These impacts are better illustrated in Figure 6.9. Certain events and their impact may have little chance of happening. Regardless, the environmental impacts of geothermal power and water facilities are subject to a variety of environmental restrictions [39]. There may be regional variations in these regulations and criteria, however, the major environmental regulations that apply to geothermal operations are given in Table 6.2.

Some significant factors must be considered for successful implementation of geothermal desalination plants [39, 40].

Table 6.2 Major environmental regulations that apply to geothermal operations.

Laws & regulations[a)]	Surface water	Air	Geothermal fluids	Liquid waste	Solid waste	Noise	Cultural resources	Subsidence/seismicity	Biological resources
National Environmental Policy Act (NEPA)	√	√	√	√	√	√	√	√	√
Geothermal Resource Operational	√	√	√	√	√	√	√	√	√
Occupational Safety & Health Act (OSHA)		√	√	√	√	√			
Toxic Substance Control Act				√	√				
Resource Conservation and Recovery Act (RCRA)				√	√				
Federal Water Pollution Control Act (NPDES)	√			√					
Clean Air Act		√							
Safe Drinking Water Act (Underground Injection Control Regulations)				√					
Noise Control Act						√			
Endangered Species Act									√

a) The data in the table is based on the power production and desalination in the USA.

6.5.1 Land Use

Large tracts of land are needed for both the extraction of energy and the management and treatment of geothermal water in geothermal energy systems. For every MW of energy produced, 1.2–2.7 tm^2 of land are needed. Facilities responsible for the management of geothermal brines have been stated to have larger land footprints. It is important to mention that low land footprints are necessary for flash-steam facilities with high capacity [39]. Over the life time of the plant the depth of extraction and site characteristics change, so with each different scenario regular and appropriately reinforced mitigations must be used. To achieve a balanced scenario, there must generally be appropriate site allocation with respect to operational guidelines and constructions. Combining a geothermal plant with other renewable energy systems is another widely used strategy under development that seems to be the most effective in enhancing the plant's performance. The hybrid energy generation schemes are reported to decrease the land-use footprint at added efficiency [46].

6.5.2 Geological Hazards

Other considerations related to geological hazards arise from the fact that deep soil excavation, which involves substantial amounts of extracted and circulated geothermal waters, is a prerequisite for geothermal power facilities. This could have a little impact but potentially lead to seismic instability in some regions. Land subsidence could potentially result from ongoing geothermal water extraction [40]. The ground may fracture or subside due to steam or hot water coming from the subterranean geothermal well and the various ways the soil reacts to the injected fluids.

Seismic events were reported in a planned geothermal power plant at Basel. The reported earthquakes reached the Richter magnitude of 3.4 after roughly one week of the hydraulic stimulation [47].

6.5.3 Waste Heat Releases

The application of geothermal energy for power generation to drive water desalination plants yields significant amounts of waste heat. The main causes of this are the significant geothermal flows and the ineffective conversion of electricity. Waste heat emissions from geothermal power stations are significantly higher than those from other types of conventional power plants. The waste heat is often released as vapors into the atmosphere [40].

6.5.4 Atmospheric Emissions

Significant amounts of non-condensable gasses, mostly connected to flash-steam and dry steam power plants, are released by geothermal wells. These gasses include H_2S, CO_2, and CH_4. Ammonia, Radon, boron, and mercury are further trace contaminants. Various power plants release CO_2 at varying rates, Around 380–1,045 kg of CO_2 eq/MWh is produced through geothermal power plants [47]. Other reports showed that sulfur dioxide emissions can range from 159 to 1,000 mg/kWh [48].

6.5.5 Water Footprint

The water footprint is a critical parameter in geothermal applications. Water is needed for a geothermal well's whole life, from drilling to operation. The amount of water needed for completing a single geothermal well can reach up to 2,300 m^3; this value varies depending on the depth, geology, and technology. Water use for power generation using air or hybrid cooling facilities ranges from 0 to 1.5 m^3/MWh, whereas water cooling towers can use up to 17 m^3/MWh. Furthermore, the considerable brines that are discharged from these wells are frequently reinjected into the wells; nonetheless, spillage frequently happens and introduces pollutants, including heavy metals and other harmful compounds. Furthermore, geothermal brines have the drawback of having a high concentration of salt, which leads to hard-scale formation [49]. The ammonia and hydrogen sulfide in geothermal heat sources may be the primary causes of the scale formation [44].

6.5.6 Noise and Social Impacts

Noise issues involve well excavation, drilling and operations, some esthetic issues can be expected in some locations. The quantity of wells and drilling equipment being used determine the noise levels from each well and the combined impacts. The creation of jobs for the local community would have a social impact. Given that geothermal activities will need specialized labor, this benefit might be limited. Water production and geothermal energy can benefit remote communities that have limited access to other renewable energy resources and connection to the grid. The local economies can sometimes be improved when geothermal wells are used for transitory communal applications like tourism [40].

6.6 Conclusion

Despite the availability of renewable energy, harnessing its value still needs more sustainable methods. Geothermal sources provide efficient resource utilization due to their significantly high capacity factor compared with other renewable sources and are promising in desalination applications. However, the use of cogeneration or poly-generation desalination schemes is favorable since geothermal sources vary significantly in their heat content and enthalpy. The high enthalpy sources are suitable for the simultaneous production of freshwater and electricity while the lower enthalpy sources are more suitable for the thermophysical desalination technologies and/or MD process. Accordingly, statistical forecasts indicated that the global utilization of geothermal energy for industrial purposes and power generation is expected to escalate in the future. In conjunction with this development, the incorporation of geothermal energy in desalination applications should be assigned priority in water-stressed regions as a sustainable option. Nevertheless, the geothermal desalination viability should be carefully evaluated case by case because not all geothermal sources are suitable for incorporation in desalination applications. Uncertainties related to the characteristics of the geothermal reservoir should be taken into consideration in the preliminary design stage to ensure accurate estimation of the costs and expected life span of the desalination plant. Other factors such as land area, emissions, hazards, etc. must be considered for successful implementation of geothermal desalination plants.

Questions

1 State the significance of using geothermal energy in desalination compared to other renewable energy sources.

2 State the criterion for selecting the suitable application for a geothermal source based on the source temperature.

3 State three possible combinations of geothermal energy and seawater desalination technologies.

4 Discuss the difference between physical water scarcity and economic water scarcity and the role geothermal energy can play to minimize their influence.

5 Compare by flow sheeting, the difference between the dry steam cycle and the single flash steam cycle.

6 What are the primary criteria for choosing a quality geothermal resource?

7 Discuss briefly the challenges of geothermal desalination in practical applications.

References

1 UNESCO (2024) World Water Report warns of an imminent water crisis. https://www.unesco.org/en/articles/world-water-report-warns-imminent-water-crisis (accessed 1 July 2024).

2 Maddah, H.A., Alzhrani, A.S., Bassyouni, M. et al. (2018). Evaluation of various membrane filtration modules for the treatment of seawater. *Appl. Water Sci.* 8 (6): 1–13. https://doi.org/10.1007/S13201-018-0793-8/FIGURES/16.

3 Tomaszewska, B., Akkurt, G.G., Kaczmarczyk, M. et al. (2021). Utilization of renewable energy sources in desalination of geothermal water for agriculture. *Desalination* 513: 115151. https://doi.org/10.1016/J.DESAL.2021.115151.

4 Mansi, A.E., El-Marsafy, S.M., Elhenawy, Y., and Bassyouni, M. (2023). Assessing the potential and limitations of membrane-based technologies for the treatment of oilfield produced water. *Alex. Eng. J.* 68: 787–815. https://doi.org/10.1016/J.AEJ.2022.12.013.

5 Milousi, M., Pappas, A., Vouros, A.P. et al. (2022). Evaluating the technical and environmental capabilities of geothermal systems through life cycle assessment. *Energies* 15 (15): 5673. https://doi.org/10.3390/EN15155673.

6 Elhenawy, Y., Moustafa, G.H., Attia, A.M. et al. (2022). Performance enhancement of a hybrid multi effect evaporation/membrane distillation system driven by solar energy for desalination. *J. Environ. Chem. Eng.* 10 (6): 108855. https://doi.org/10.1016/J.JECE.2022.108855.

7 Elminshawy, N.A.S., Ahmed, A., Osama, A. et al. (2024). The potential of optimized floating photovoltaic system for energy production in the Northern Lakes of Egypt. *Eng. Anal. Bound. Elem.* 161: 226–246. https://doi.org/10.1016/J.ENGANABOUND.2024.01.022.

8 Alzahrani, K.M., Hee, J.L., Elsakka, M. et al. (2024). Techno-economic assessment of vertical axis wind turbine driven RO desalination with compressed air energy storage for remote communities. *Desalination* 592: 118094. https://doi.org/10.1016/J.DESAL.2024.118094.

9 Sakamoto, H., Ronquim, F.M., Martins Seckler, M., and Kulay, L. (2019). Environmental performance of effluent conditioning systems for reuse in oil refining plants: a case study in Brazil. *Energies* 12 (2): 326. https://doi.org/10.3390/EN12020326.

10 Ahmed, A., Elsakka, M., Elminshawy, N. et al. (2023). Recent advances in floating photovoltaic systems. *Chem. Rec.* 23 (12): e202300229. https://doi.org/10.1002/TCR.202300229.

11 Prus, P., Maitah, M., Borge-Diez, D. et al. (2022). The potential and development of the geothermal energy market in Poland and the Baltic states—selected aspects. *Energies* 15 (11): 4142. https://doi.org/10.3390/EN15114142.

12 Shelare, S., Kumar, R., Gajbhiye, T., and Kanchan, S. (2023). Role of geothermal energy in sustainable water desalination—a review on current status, parameters, and challenges. *Energies* 16 (6): 2901. https://doi.org/10.3390/EN16062901.

13 Elhenawy, Y., Fouad, K., Bassyouni, M., and Majozi, T. (2023). Design and performance a novel hybrid membrane distillation/humidification–dehumidification system. *Energy Convers. Manage.* 286: 117039. https://doi.org/10.1016/J.ENCONMAN.2023.117039.

14 Elsakka, M.M., Refaat, A., Alzahrani, K.M. et al. (2024). Techno-economic assessment of a novel wind-powered RO system with a compressed air energy storage for water desalination. *Energy* 311: 133296. https://doi.org/10.1016/J.ENERGY.2024.133296.

15 Chen, J., Yan, B., Xu, T., and Xia, F. (2023). Hydrochemical evolution characteristics and mechanism of groundwater funnel areas under artificial governance in Hengshui City, North China. *Ecol. Indic.* 148: 110059. https://doi.org/10.1016/J.ECOLIND.2023.110059.

16 Yudha, S.W., Tjahjono, B., and Longhurst, P. (2022). Unearthing the dynamics of Indonesia's geothermal energy development. *Energies* 15 (14): 5009. https://doi.org/10.3390/EN15145009.

17 Elhenawy, Y., Moustafa, G.H., Abdel-Hamid, S.M.S. et al. (2022). Experimental investigation of two novel arrangements of air gap membrane distillation module with heat recovery. *Energy Rep.* 8: 8563–8573. https://doi.org/10.1016/J.EGYR.2022.06.068.

18 Toth, A., Lund, J., Huttrer, G., and Fenerty, D. (2022). Current world situation for geothermal energy production, based on WGC2020+ 1 reports from 88 countries. *Oceania* 28 (236,745): 0–259.

19 Gutiérrez-Negrín, L.C.A. (2024). Evolution of worldwide geothermal power 2020–2023. *Geotherm. Energy* 12 (1): 14.

20 Li, K., Bian, H., Liu, C. et al. (2015). Comparison of geothermal with solar and wind power generation systems. *Renew. Sustain. Energy Rev.* 42: 1464–1474.

21 Ahmed, A., Alzahrani, M., Shanks, K. et al. (2022). Reliability and temperature limits of the focal spot of a primary optical component for an ultra-high concentrated photovoltaic system. *AIP Conf. Proc.* 2550 (1): https://doi.org/10.1063/5.0099091/16220440/040001_1_ONLINE.PDF.

22 Ali, E., Orfi, J., AlAnsary, H. et al. (2020). Integration of multi effect evaporation and membrane distillation desalination processes for enhanced performance and recovery ratios. *Desalination* 493: 114619. https://doi.org/10.1016/j.desal.2020.114619.

23 Prajapati, M., Shah, M., and Soni, B. (2021). A review of geothermal integrated desalination: a sustainable solution to overcome potential freshwater shortages. *J. Cleaner Prod.* 326: 129412. https://doi.org/10.1016/j.jclepro.2021.129412.

24 Finster, M., Clark, C., Schroeder, J., and Martino, L. (2015). Geothermal produced fluids: characteristics, treatment technologies, and management options. *Renew. Sustain. Energy Rev.* 50: 952–966.

25 Pietrasanta, A.M., Mussati, S.F., Aguirre, P.A. et al. (2022). Water-renewable energy nexus: optimization of geothermal energy-powered seawater desalination systems. *Renew. Energy* 196: 234–246. https://doi.org/10.1016/j.renene.2022.06.146.

26 Sharmin, T., Khan, N.R., Akram, M.S., and Ehsan, M.M. (2023). A state-of-the-art review on geothermal energy extraction, utilization, and improvement strategies: conventional, hybridized, and enhanced geothermal systems. *Int. J. Thermofluids* 18: 100323. https://doi.org/10.1016/j.ijft.2023.100323.

27 Zarrouk, S.J. and Moon, H. (2014). Efficiency of geothermal power plants: a worldwide review. *Geothermics* 51: 142–153. https://doi.org/10.1016/J.GEOTHERMICS.2013.11.001.

28 Bouchekima, B. (2003). Solar desalination plant for small size use in remote arid areas of South Algeria for the production of drinking water. *Desalination* 156 (1–3): 353–354.

29 Bourouni, K., Martin, R., Tadrist, L., and Chaibi, M.T. (1999). Heat transfer and evaporation in geothermal desalination units. *Appl. Energy* 64 (1–4): 129–147.

30 Bourouni, K., Deronzier, J.C., and Tadrist, L. (1999). Experimentation and modelling of an innovative geothermal desalination unit. *Desalination* 125 (1–3, 153): 147.

31 G. Rodrıguez, M. Rodrıguez, J. Perez, and J. Veza (1996). A systematic approach to desalination powered by solar, wind and geothermal energy sources. *Proceedings of the Mediterranean conference on renewable energy sources for water production*, pp. 20–25.

32 Gutiérrez, H. and Espíndola, S. (2010). Using low enthalpy geothermal resources to desalinate sea water and electricity production on desert areas in Mexico. *GHC Bulletin* 29: 19–24.

33 Koroneos, C. and Roumbas, G. (2012). Geothermal waters heat integration for the desalination of sea water. *Desalination Water Treat* 37 (1–3): 69–76.

34 Boegli, W.J., Suemoto, S.H., and Trompeter, K.M. (1977). Geothermal desalting at the East Mesa test site. *Desalination* 22 (1–3): 77–90.

35 Goosen, M., Mahmoudi, H., and Ghaffour, N. (2012). Overview of renewable energy technologies for freshwater production. In: *Renewable Energy Applications for Freshwater Production* (ed. J. Bundschuh and J. Hoinkis), 25–77. London, UK: CRS Press and IWA Publishing.

36 Kamal, I. (1992). Cogeneration desalination with reverse osmosis: a means of augmenting water supplies in Southern California. *Desalination* 88 (1–3): 355–369.

37 Mansi, A.E., El-Marsafy, S.M., Elhenawy, Y., and Bassyouni, M. (2024). Assessing the fouling behavior of PTFE membrane in air-gap membrane distillation against oil-in-brine stabilized emulsions. *Appl. Water Sci.* 14 (2): 35.

38 Bouguecha, S. and Dhahbi, M. (2003). Fluidised bed crystalliser and air gap membrane distillation as a solution to geothermal water desalination. *Desalination* 152 (1–3): 237–244.

39 Gude, V.G. (2016). Geothermal source potential for water desalination - current status and future perspective. *Renew. Sustain. Energy Rev.* https://doi.org/10.1016/j.rser.2015.12.186.

40 Shah, M., Prajapati, M., and Sircar, A. (2023). A comprehensive and systematic study of geothermal energy-based desalination systems for sustainable development. *Geothermics* 115: https://doi.org/10.1016/j.geothermics.2023.102829.

41 Coro, G. and Trumpy, E. (2020). Predicting geographical suitability of geothermal power plants. *J. Cleaner Prod.* 267: 121874.

42 Zheng, B., Xu, J., Ni, T., and Li, M. (2015). Geothermal energy utilization trends from a technological paradigm perspective. *Renew. Energy* 77: 430–441. https://doi.org/10.1016/J.RENENE.2014.12.035.

43 Barbier, E. (2002). Geothermal energy technology and current status: an overview. *Renew. Sustain. Energy Rev.* 6 (1–2): 3–65. https://doi.org/10.1016/S1364-0321(02)00002-3.

44 Gude, V.G. (2018). Geothermal source for water desalination—challenges and opportunities. In: *Renewable Energy Powered Desalination Handbook: Application and Thermodynamics*, 141–176. Butterworth-Heinemann. https://doi.org/10.1016/B978-0-12-815244-7.00004-0.

45 Bayer, P., Rybach, L., Blum, P., and Brauchler, R. (2013). Review on life cycle environmental effects of geothermal power generation. *Renew. Sustain. Energy Rev.* 26: 446–463. https://doi.org/10.1016/J.RSER.2013.05.039.

46 Sayed, E.T., Wilberforce, T., Elsaid, K. et al. (2021). A critical review on environmental impacts of renewable energy systems and mitigation strategies: wind, hydro, biomass and geothermal. *Sci. Total Environ.* 766: https://doi.org/10.1016/j.scitotenv.2020.144505.

47 Rahman, A., Farrok, O., and Haque, M.M. (2022). Environmental impact of renewable energy source based electrical power plants: solar, wind, hydroelectric, biomass, geothermal, tidal, ocean, and osmotic. *Renew. Sustain. Energy Rev.* https://doi.org/10.1016/j.rser.2022.112279.

48 Kagel, A., Bates, D., and Gawell, K. (2005). *A Guide to Geothermal Energy and the Environment*. Geothermal Energy Association https://doi.org/10.2172/897425.

49 Belessiotis, V. and Delyannis, E. (2001). Water shortage and renewable energies (RE) desalination—possible technological applications. *Desalination* 139 (1–3): 133–138. https://doi.org/10.1016/S0011-9164(01)00302-2.

7

Bioenergy and Biogas Utilization for Water Desalination

Naira Meky[1] *and Amor Abdelkader*[2]

[1] *Civil Engineering Department, Faculty of Engineering, Aswan University, Aswan, Egypt*
[2] *Faculty of Science and Technology, Bournemouth University, Poole, UK*

7.1 Introduction

The massive population growth, changing consumer behavior, and economic expansion are the main factors affecting the water usage rate worldwide as it has increased eightfold during the last century. As a result, major issues are expected to continue to arise. The proper freshwater quality as well as its availability are extremely crucial to ensuring a high living standard and a healthy environment. Moreover, freshwater is essential for humans' daily activities such as drinking, cooking, cleaning, and manufacturing. Therefore, recently, water security has been a vital concern for the health of humans and the environment. Not all freshwater sources are easily accessible. Groundwater makes up just over 30% of the Earth's freshwater, while icecaps and glaciers contain more than 68% of the total. Lakes, rivers, and marshes that are easily reachable contain only about 0.3% of our freshwater. It is estimated that we already use more than half of the available freshwater. Now, desalination is considered a dependable solution to ensure sufficient freshwater demand [1].

Salts and minerals removal from salt water to make it drinkable is the definition of the water desalination process. This process is particularly important in areas facing freshwater scarcity. Desalinating seawater produces freshwater but also results in a by-product called brine. While desalination is expensive due to energy consumption, it is a viable solution for providing freshwater, especially in regions with limited access to freshwater sources. The uptake and development of desalination, however, have been hampered in many countries by high desalination costs and high prices for desalinated water (which are twice or three

Water Treatment and Desalination: Membrane Separation using Renewable Energy Resources,
First Edition. Edited by Mohamed Bassyouni.

times higher than those from traditional water sources). The primary elements influencing the ultimate cost of desalinated water are production capacity, water salinity, capital and operating expenditures, and applied desalination technology. Moreover, the desalination process is energy-intensive, and the high cost of electricity has comparable economic effects to fuel or other ongoing project expenses that cannot be depreciated over time. Bioenergy-powered desalination is a solution recently investigated by researchers that might help avoid the high energy expenses of the process [2].

Recently, technology developers have focused on using renewable energy resources to power the desalination process, with a special focus on using bioenergy. Adoption of the proposed technology makes water desalination more cost-competitive as it can reduce energy costs and increase water accessibility which helps cities facing water scarcity, and finally helps us to move forward toward a more sustainable and eco-friendly future. Bioenergy is one of the green energy resources that helps the reduction of fossil fuel dependency. Bioenergy-powered desalination is a cheaper, cleaner, and eco-friendly alternative compared to conventional desalination processes powered by fossil fuels. It lowers the impact of climate change as it does not release any carbon emissions. To summarize, bioenergy-powered desalination is an affordable sustainable technology that is suggested as an excellent solution for providing freshwater to areas suffering water scarcity.

7.1.1 Objectives of the Chapter

This chapter spotlights the potential adoption of bioenergy-powered desalination technology as a novel, cost-effective, and eco-friendly water supply system. The key objectives of the chapter are as follows:

- Providing an overview of bioenergy, including its sources, biogas production, and applications, followed by an overview of conventional desalination technologies and challenges faced in conventional desalination.
- Evaluating the feasibility of using bioenergy to power desalination plants.
- Identifying the environmental and economic advantages of bioenergy-powered desalination vs. conventional methods.
- Discussing the challenges and limitations of efficiently integrating bioenergy and desalination.
- Finding research gaps, making recommendations for future studies, and suggesting ways to improve the integration of bioenergy into desalination processes.

7.1.2 Structure of the Chapter

Through consecutive six sections, Chapter 7 discusses the feasibility of combining bioenergy and water desalination and provides a road map for researchers to advance the sustainable application of this technology. The synopsis of the sections is as follows:

- Section 7.1 comprises an introduction to the chapter topic, an overview, motivations, and specific objectives in detail.
- Section 7.2 discusses the concept of bioenergy, its sources, and biogas generation and composition. In addition, it gives a summary of the biogas applications and the main operating factors affecting its production.
- Section 7.3 summarizes the conventional desalination technologies and reviews their energy, environmental, and economic considerations. Moreover, the feasibility of combining bioenergy and water desalination is covered with a special focus on using biogas as a sustainable alternative energy source providing the required heat and electricity for the desalination process.
- Section 7.4 highlights the environmental and economic gains of using bioenergy-powered water desalination systems such as carbon footprint reduction and waste management. The section also compares conventional and bioenergy-powered desalination systems from the point of view of economics.
- Section 7.5 runs through the potential limitations, challenges, and technical barriers of the practical application of bioenergy-powered desalination systems. Furthermore, it shows the recent research gaps in integrating bioenergy into desalination to make it a real solution.
- Section 7.6 highlights the significance of ongoing research and summarizes the main conclusions and recommendations to assist the practical use of bioenergy-powered desalination systems.

7.2 Overview of Bioenergy

Bioenergy means any energy product in the form of solid, liquid, or gas that is produced from biomass or biological waste materials. It is considered an effective low-cost and sustainable option to attain the recent high energy demand and reduce greenhouse gas (GHG) emissions. Biomass is any organic material that can be converted into energy (i.e., heat, electricity, and fuel) by using chemical, mechanical, biological, or enzymatic processes such as municipal solid trash, crops, forest, and agricultural residue [3]. Based on biomass and

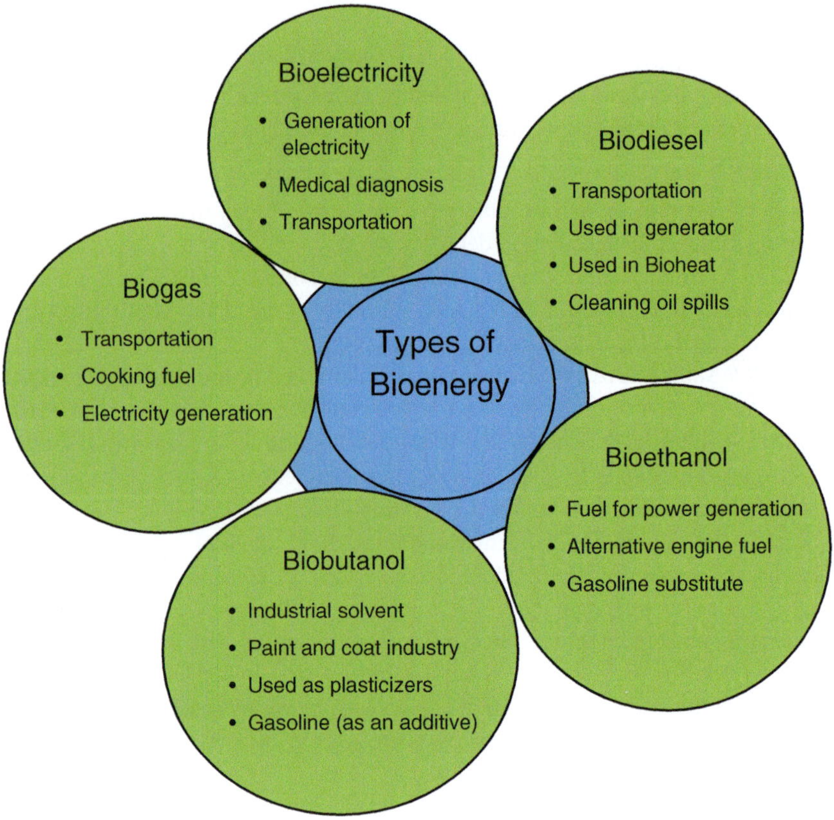

Figure 7.1 Types of bioenergy. Source: Singh et al. [4]/with permission of Elsevier.

bioresources, bioenergy can be divided into several groups. These include biogas (which includes biomethane and biohydrogen), biofuels (such as biodiesel, bioethanol, and biobutanol), and bioelectricity, as shown in Figure 7.1. According to recent research forecasts, anaerobic waste digestion and biorefineries, along with advanced bioenergy generated by biomass technology, might potentially overtake fossil fuels as the fastest-growing renewable energy source by 2030 [4]. According to [5], bioenergy provides 10% of the global energy demand.

7.2.1 Bioenergy Sources

- **Agriculture Residues**

  Agricultural crop residues include wheat straw, oat straw, rice straw, straw, sorghum stubble, and corn stover (stalks, leaves, husks, and cobs). Any form of these materials can produce energy in two ways: burned directly to produce heat energy, or biologically treated to produce biogas [6].

- **Municipal Solid Waste (MSW)**

Production of MSW has increased dramatically globally due to urbanization and increased consumption rates. It accounted for 1.3 billion metric tons in 2020 approximately and is expected to be 2.2 billion metric tons by 2025. The organic part of MSW is an excellent feedstock for bioenergy production and waste management, which proposes a dual-benefit method of treating the organic MSW which reduces the huge waste amounts buried in sanitary landfills, and biogas production [7].

- **Forestry Products**

Forestry products or lignocellulosic wastes are considered the most predominant streams; they contain cellulose, hemicelluloses, and lignin. Forest wastes are ideal biomass for bioenergy and biogas production; however, pretreatment is a crucial step for these materials to foster cellulose and hemicelluloses enzymatic conversion [8].

- **Animal Waste**

Animal waste material is not fully used as it should be; either the organic waste is burned immediately or released into the environment. After being left out in the open for an extended period, a small amount of this is utilized as fertilizer in agricultural areas. The improper management of animal feces and the careless release of waste materials into pastures, rivers, and arable fields contaminate the soil and cause degradation. The production of bioenergy from animal waste has potential benefits for the environment. Fermented fertilizer made from animal dung promotes crop and energy production while reducing adverse environmental effects [9].

- **Algae**

Algae have a great potential for bioenergy production compared to other plants due to their rapid growth, shorter life cycles, superior photosynthesis, and lack of need for fertile land. The rapid growth of algae requires heat, carbon dioxide (CO_2), and nutrients; all of them are richly available in bioenergy-producing bioreactors. Biogas production from algae technology has superior energy efficiency as an output compared to alternative methods used to produce green diesel, biodiesel, bioethanol, and hydrogen. Additionally, their energy output and sustainability can be enhanced by using anaerobic digestion (AD) [10].

7.2.2 Biogas Production

Biogas is a form of bioenergy that is produced through the biological breakdown of organic matter (i.e., food or animal waste) by microorganisms in an oxygen-free environment; this biological process is known as anaerobic digestion. The produced biogas consists of different gases including 60% methane (CH_4) and 35–40% CO_2, with trace amounts of hydrogen (H_2), ammonia (NH_3), and hydrogen sulfide (H_2S). The volume of produced biogas depends on the used waste type. Biogas can be produced purposefully as a fuel by an industrial process, or it might happen naturally. Since the biological material used to make biogas will naturally decompose, it is less harmful to the environment to capture the gases produced during this process and use them as a source of energy rather than letting them escape into the atmosphere [11].

The biogas production process is influenced by several operating factors that have been found over the years by research. Digester design, pH, temperature, carbon-to-nitrogen (C/N) ratio, organic loading rate (OLR), hydraulic retention time (HRT), and toxin presence are the primary influencing factors. All of these, however, are highly dependent on the kind of inoculum and substrate that are used. Some of these parameters are uncontrollable; thus, it is rarely feasible to maintain all of them under field settings. In addition, feedstock mixing affects the digestion efficiency; the biogas production rate increases by adjusting the method and strength of mixing. In conclusion, all these operating factors need to be appropriately optimized to maximize biogas production rates [11].

The anaerobic digestion process occurs in tightly sealed vessels known as digesters or bioreactors. Their sizes and shapes depend on the used feedstock characteristics. The main function of all bioreactors is to produce biogas from organic matter through anaerobic digestion. The produced biogas is a mixture of gases that can be used as a sustainable energy source. This function can be accomplished using different types or strategies of digesters; their design, mixing and handling methods, and digestion strategies may have a slight effect on process efficiency. The various anaerobic digester types are covered below.

- **Batch Digesters**

For this type, the feedstock is added once into tightly closed containers and left till the fermentation process is ended and the biogas production is stopped. Then, the container is emptied and cleaned, and a new batch of feedstock is added. However, some digestate is retained to serve as an inoculum for the fresh feedstock. Batch bioreactors are very simple and do not need regular maintenance. They are often used for high fiber and solid content feedstocks (ranging from 30% to 40%) and are considered suitable for small-scale fermentation. However, there are drawbacks to batch bioreactors, such as fluctuations in gas production and quality. A series of batch digesters fed one after the other is necessary to achieve a consistent biogas output. Other limitations include the significant time required to load and empty the batch digesters, biogas losses during discharge, and a maximum height for the bioreactors [12].

- **Continuous Digesters**

To enable continuous biogas production, feedstock is continually fed into continuous digesters. Continuously fed systems come in one, two, and multi-stage varieties. One-stage systems lack certain advantages of two-stage systems, such as better yields and rates of biogas production. By providing the ideal conditions for various microbes at each stage, they boost the production of methane. Multistage digesters are more costly to construct and operate than single-stage digesters, but they provide faster, more effective, and less expensive processes overall. For MSW, the methane yield in two-stage reactors is 21% higher than in single-stage reactors [13].

- **Plug Flow Digesters**

A plug flow digester uses an extended, thin horizontal tank to break down feedstock without the need for a mixing system. The tank is heated, insulated, and typically made of reinforced concrete, steel, or fiberglass. Feedstock is continuously added to the tank, pushing older materials through the digester. This creates a "plug" of material that moves through the vessel, giving the digester its name. The tank is five times longer than it is wide, allowing for efficient anaerobic digestion. In reality, the situation is more complicated, with different parts of the waste moving through the system at different speeds, and some parts possibly settling or floating in the tank. The main advantage of the plug-flow design is that it is easy to install and has low operating costs. However, it is not as effective or reliable as the fully mixed design [14].

- **Advanced Digesters**

Three main characteristics must be satisfied to construct any anaerobic digester: It must be able to constantly manage a high OLR, produce large amounts of high-quality biogas, and have a short hydraulic retention period to have a lower reactor volume. Advanced reactors that are more efficient than the previously described traditional reactors include the anaerobic sequencing batch reactor (ASBR), anaerobic baffled reactor (ABR), up-flow anaerobic sludge blanket (UASB) reactor, and anaerobic filters. The characteristics of the primary feedstock, particularly the total solids content, determine which digester type is best. In continuous stirred-tank reactors (CSTRs), feedstocks with high levels of slurry and total solids are often processed. However, anaerobic filters, fluidized bed reactors, and UASB reactors are the main systems used for the digestion of soluble organic wastes. The primary applications of co-digestion are in wet single-step procedures (like CSTR). Continuous systems produce gas at a steady state and constant rate because they are fed constantly, and the digestate residue is released at the same rate. However, only continuously supplied substrates can be used for this kind of process. If not, a semicontinuous feeding schedule that occurs multiple times a day is used. Continuous systems feature minimal capital costs, easy construction and operation, and quick acidification; however, they can have drawbacks, such as increased volatile fatty acid (VFA) production [15].

7.2.3 Anaerobic Digestion Process

This biological process stabilizes and biodegrades complex organic matter in the absence of oxygen through a group of microorganisms, resulting in the production of high-energy biogas that can replace fossil fuels. The nutrient-rich by-products of anaerobic digestion are utilized as soil supplements. Anaerobic digestion is carried out under different temperature conditions such as thermophilic, psychrophilic, and mesophilic [16]. The anaerobic digestion process comprises four fundamental processes: hydrolysis, acidogenesis, acetogenesis, and methanogenesis (Figure 7.2).

- **Hydrolysis**

Anaerobic digesters rely on hydrolysis to break down complex organic compounds into smaller components that can be utilized by bacteria. Special enzymes help convert carbohydrates, lipids, and proteins into sugars, fatty acids, and amino acids. However, some substrates like lignin and cellulose can be difficult to degrade, so enzymes are added to aid in the process. The rate of hydrolysis can

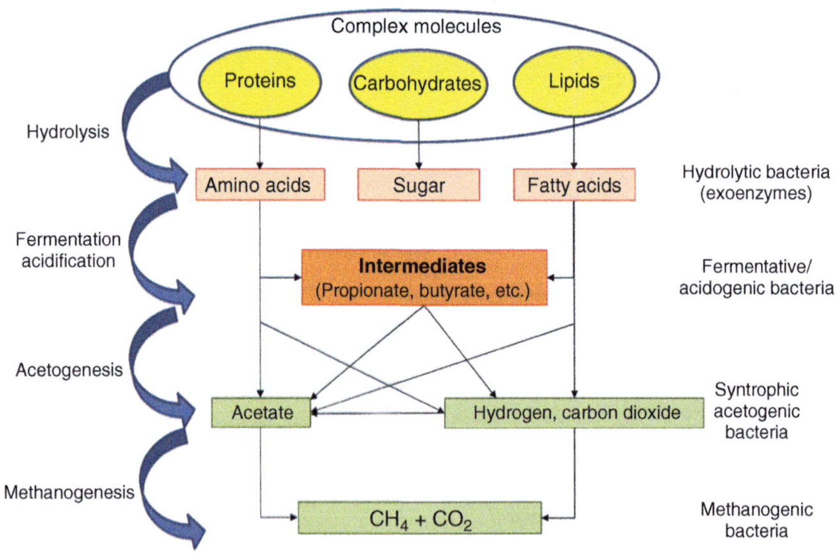

Figure 7.2 Four key stages of the anaerobic digestion process. Source: Grippi et al. [16]/MDPI/CC BY 4.0.

impact the overall process, and various waste pretreatment methods are being explored to speed it up. Generally, hydrolysis works best at temperatures between 30°C and 50°C and a pH of 5–7.

- **Acidogenesis**

  Acidogenic bacteria absorb the by-products of hydrolysis and create VFAs as well as other products like acetates, propionate, and butyrate during the acidogenesis stage of anaerobic digestion. Depending on the circumstances in the digester, these concentrations can change. The regeneration of acidogenic bacteria during acidogenesis occurs more quickly than during other phases of anaerobic digestion as it takes less than 36 hours. However, digester failure may result from VFA acidification. Anaerobic digestion can be inhibited at high concentrations by ammonia generation via deamination, and VFA production from amino acids in wastes rich in protein, such as sewage wastewaters.

- **Acetogenesis**

  Acetate bacteria transform the by-products of acidogenesis into acetic acid, hydrogen, and carbon dioxide during the acetogenesis process. Metabolism occurs after the formation of acetic acid. The process of producing acetate from hydrogen and carbon dioxide is known as chemoautotrophic acetogenesis. Acid fermentation is the term used to describe the first three stages combined. This procedure turns organic matter into a substrate for the subsequent methanogenesis process rather than removing it from the liquid phase. The products of the acid phase are transformed by acetate bacteria into acetates and hydrogen throughout this process, which methanogenic bacteria can use.

- **Methanogenesis**

  The last phase of anaerobic digestion, known as methanogenesis, is when methanogenic microbes break down available intermediates to generate methane. Since 99% of methanogenic bacteria are destroyed after 10 hours of exposure to oxygen, these microorganisms are extremely sensitive to oxygen. Methanogenic bacteria can only create methane from a restricted variety of substrates, mostly hydrogen, and acetate, but they can also use formate, methanol, and methylamines. In contrast to the early stages of anaerobic digestion, they require a higher pH and a lower redox potential, which makes laboratory cultivation difficult. They regenerate more slowly as well, taking 5–16 days; however, certain species can double in just two hours. The process of methanogenesis in batch reactors comes to an end when biogas production stops, typically after around 40 days. The extent of digestion can be assessed by measuring the number of volatile solids and the sludge's ability to dewater.

7.2.4 Factors Affecting the Anaerobic Digestion Process

The production of biogas is affected by several factors including initial pH values, HRT, mixing, inoculum characteristics and composition, feedstock composition, and temperature. Two of these factors have the highest impact on fermentation process efficiency and the production of biogas; the two variables are feedstock nature and composition, and retention time. Biogas production increases, while the retention time increases; also, the high C/N ratio in the feedstock positively affects biogas production. Additionally, fermentation process efficiency and the production of biogas are influenced by other variables including OLR, and the presence of toxins. To summarize, understanding and optimizing these operating factors could potentially increase biogas productivity; thus, it is crucial to optimize these operating factors to maximize biogas production yields and enhance the fermentation process. The main operating factors that affect biogas production and fermentation are outlined below.

- **Hydraulic Retention Time (HRT)**

  HRT means the time the input feedstock spends inside the bioreactor until it is ejected. Biogas production increases, while the retention time increases; however, the biogas production rate increases at first and then progressively decreases till the fermentation process nears the end. This means the time required to accomplish 70–80% fermentation is less than the time to reach complete fermentation. Therefore, the bioreactor must be designed with HRT guaranteeing 70–80% digestion at least. HRT often ranges from 20 to 120 days by considering other operating factors (i.e., operating temperature and digester's design). HRT is typically 40–50 days for hot regions, while it may increase to 100 days for cold regions [17].

- **Temperature**

  Temperature has a great effect on the anaerobic digestion process, and it affects microorganisms' activity and feedstock breakdown. Both fermentation process efficiency and the production of biogas increase, while temperature increases. As a result, it is important to find the optimum temperature. Previous research reported that the optimum temperature range is 25–45°C considering heat consumption inside the reactor. Other researchers found that as the temperature increased, the concentration of free ammonia (an inhibitor) also rose, leading to a decrease in biogas production. For anaerobic digestion, the two primary temperature ranges are mesophilic (35°C) and thermophilic (55°C). While thermophilic digestion produces more biogas and more effectively destroys pathogens, mesophilic digestion is slower but uses less energy. While research has also been done on atypical temperature regimes including psychrophilic digestion at −15°C to +15°C and achieved moderate biogas generation, some digesters are not heated and undergo seasonal swings in methane production [17].

- **Carbon to Nitrogen (C/N) Ratio**

In anaerobic digestion, the C/N ratio is crucial. Carbon is used as an energy source, and nitrogen promotes the growth of microorganisms. Each nutrient has the potential to be a constraint. The ideal ratio is in the range of 20–30. Because of the quick consumption of nitrogen, gas production is limited when the C/N ratio is high. On the other hand, a low C/N ratio causes ammonia to build up and raises the pH above 8.5, which is harmful to methanogenesis. To obtain the optimal C/N ratio, combining substrates with different C/N ratios is reported to be very effective. Studies demonstrated that the C/N ratio range from 30:1 to 35:1 is the optimal range due to the faster conversion of carbon compared to nitrogen during the digesting process [18].

- **pH Value**

The initial pH value inside bioreactors is crucial as methanogenic bacteria are sensitive to the acidic environment which negatively affects the biogas production. The pH value fluctuates inside the digester through the anaerobic fermentation phases. During the acetogenesis stage, the pH value decreases below 5, which affects the methanogens due to the acids' accumulation inside the reactor. The initial pH value must remain constant with the use of buffers (i.e., lime or calcium carbonate). The optimal value for methane-producing bacteria and biogas production is recommended to be neutral to slightly alkaline (6.8–8.5), while the optimal values for hydrolysis and acetogenesis are 5.5 and 6.5, respectively. The ideal range for the pH of anaerobic digestion waste has been determined by several investigations to be approximately 7 [19].

7.2.5 Biogas Composition, Properties, and Applications

Anaerobic digestion of organic waste results in the production of biogas, which is a mixture of methane, CO_2, and trace amounts of other gases. Wastewater treatment facilities, landfill gas recovery systems, and biodigesters can all produce it. Biogas can be utilized as an energy source or to produce heat and electricity, as illustrated in Figure 7.3. Biogas normally contains between 60% and 75% methane by volume. The amount of methane in biogas determines its energy content; a higher concentration of methane results in a higher calorific value. "Renewable natural gas," or biomethane, is a nearly pure form of methane (biogas containing more than 90% methane). It is made by "upgrading" biogas with a lower heating value (LHV) of about 36 MJ/m^3 and is identical to natural gas [21]. The primary applications of biogas are covered in this section, with a focus on utilizing combined heat and power (CHP) systems to generate heat and electricity.

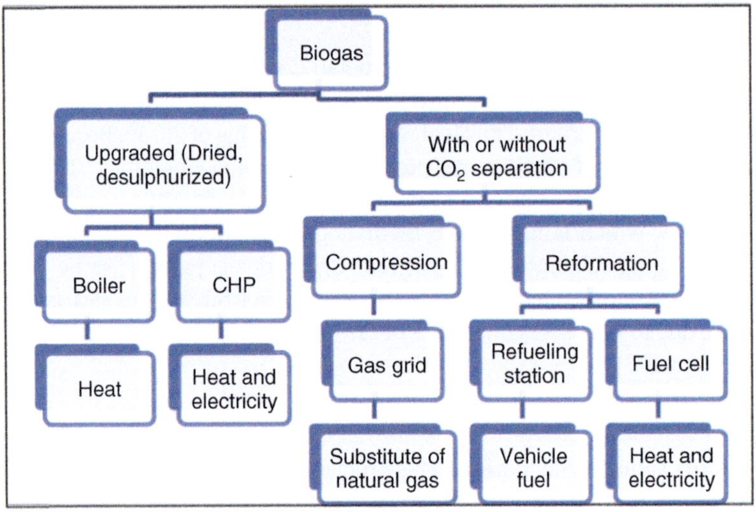

Figure 7.3 Applications of biogas. Source: Yousuf et al. [20]/with permission of Springer Nature.

7.2.5.1 Electricity Generation

Producing electricity from biogas can be a highly effective way to utilize renewable energy sources. An internal combustion engine uses biogas to produce mechanical energy. Electricity is produced by an electric generator, which is rotated by mechanical energy. This is only applicable, though, if the power generator's released heat can be put to good use both financially and environmentally. Biogas has an energy content of roughly 6 kWh/m^3, or roughly 21–23.5 MJ/m^3 on average. This means that 1 m^3 of biogas is equivalent to 0.5–0.6 L of diesel fuel. However, only about 1.7 kWh may be produced from 1 m^3 of biogas because of conversion losses. In general, larger biogas plants are more economical than smaller ones. Nonetheless, the technology of producing energy from biogas is suitable even for modest applications with power outputs between 10 and 100 kW. For biogas plants, the international market provides advanced, dependable, high-quality technology. Germany, China, Thailand, Brazil, and Kenya possess the necessary knowledge and resources to build biogas power plants. Maintaining a biogas-powered generator is comparable to maintaining one that runs on fossil fuels [22].

7.2.5.2 Heat Production

The most basic application of biogas is as a thermal energy source. Small biogas systems can supply the thermal energy needed for simple cooking and water heating in places where fuels are scarce. Biogas is another fuel that gas lighting systems can use to illuminate. The air-to-gas ratio of conventional gas burners can be readily changed to accommodate biogas. Gas burners only need a gas pressure of 8–25 mbar and must keep H_2S levels below 100 ppm to reach a dew point of 150°C; hence, there is little requirement for biogas quality. The type of digester used and the wastefeed materials in the anaerobic digester have an impact on the energy in biogas. Natural gas, which is almost entirely composed of methane, usually has two heating values per cubic meter: 35.8 MJ/m^3 and 39.8 MJ/m^3. Biogas will have a LHV of 21.5 MJ/m^3 if it is dry and contains about 60% methane. The combustion of biogas produces water vapor and carbon dioxide. More energy is captured as the water vapor condenses, raising the higher heating value (HHV). High water-content biogas has less energy potential and can lower CHP generator efficiency. To optimize biogas's energy potential, drying must be considered [23].

7.2.5.3 Combined Heat and Power (CHP) Systems

Systems that produce CHP make use of a fuel's inherent waste heat as well as its capacity to produce power. In certain CHP systems, the main output is heat, with electrical power coming in second (bottoming cycle). In some CHP systems, the waste heat is used to heat process water (topping cycle) and is predominantly produced as electrical power. Either way, using the fuel (biogas) to produce solely electricity or heat results in substantially lower efficiency than the entire (combined) efficiency of the power and heat produced and used. Gas turbines, which can produce both heat and power, have low maintenance requirements and comparable efficiency to spark-ignition engines, despite their high initial costs (micro-turbines, 25–100 kW; large turbines, >100 kW). However, most CHP applications employ internal combustion engines. To use biogas in these systems, water vapor and H_2S must be removed (to below 100 ppm). Fuel cells, with their high efficiency and low emissions, are seen as the small-scale power plants of the future to produce heat and electricity. Washington State is home to one of the biggest digester/fuel cell combinations. Located at the South Treatment Plant in Renton, Washington, the fuel cell has a daily capacity of approximately 154,000 ft^3 of biogas consumption and can generate up to 1 MW (1,000,000 W) of energy. This would power 1,000 homes, but it is being used to keep the plant running instead. Furthermore, to cut energy costs and GHG emissions, biogas CHP systems are being employed more frequently in US municipal wastewater treatment plants [24]. In the same vein, CHP systems are an essential component of the European Union's 2021–2050 carbon neutrality policy [25].

7.3 Water Desalination Techniques and Integration with Bioenergy

7.3.1 Conventional Desalination Methods

The artificial process of desalination transforms saline water into freshwater using multiple methods. While each has benefits and drawbacks, they are all helpful. The techniques can be separated into two categories: **thermal** (such as multistage flash distillation) and **membrane** (such as reverse osmosis [RO]). Distillation, or boiling and recondensing saltwater to remove salt and contaminants, is the conventional method of desalination. Reverse osmosis and distillation are the two most used desalination techniques. Most of the world's desalination capacity is now held by two technologies: reverse osmosis and multistage flash distillation (MSF).

7.3.1.1 Distillation

The process of thermal desalination, also known as distillation, is among the oldest methods for turning brackish and saltwater into drinkable water. The concepts of condensation and boiling, or evaporation, form its foundation. Heat is applied to water to cause it to evaporate. As the vapor condenses to create new water, the salt is left behind. The necessary thermal energy is now generated in waste heat boilers, steam generators, or by extracting back-pressure steam from power plant turbines. The most popular methods for thermal desalination are MSF, multiple-effect distillation (MED), evaporation via vapor compression (VC), and sun desalination of water [26].

7.3.1.2 Reverse Osmosis

A semipermeable membrane is utilized in the process of water purification called reverse osmosis to separate water molecules from other contaminants. This technology is used to supply drinking water on industrial and large scales; it applies reverse pressure to overcome osmotic pressure. RO is very effective in the removal of bacteria, contaminants, and dissolved or suspended chemical species. The purified solvent travels to the other side of the membrane, while RO retains the solute on the pressured side. The passage of molecules is determined by their relative sizes. "Selective" membranes reject large molecules, while smaller molecules, such as solvent molecules like water, are allowed to pass through. In the desalination business right now, RO is thought to be the best and most efficient membrane-based desalination technology [27].

7.3.1.3 Electrodialysis

Ions are transported via semipermeable membranes by electrodialysis (ED), a membrane-based process that uses an applied electric field. Desalination, table salt manufacturing, wine stabilization, whey demineralization, and pickling bath recovery are among the processes that use electrodialysis. Polyelectrolyte membranes are used in an electrodialysis cell to separate the permeating stream from the feed stream during the operation. The cation exchange membrane is represented by the anionic polyelectrolyte, while the anion exchange membrane is represented by the cationic polyelectrolyte. These polymers are frequently cross-linked for stability during the electrodialysis process because they are soluble in water or dispersible. Anion permeability is allowed via the anion exchange membranes, while cation permeability is allowed through the cation exchange membranes, but anion permeability is rejected [28].

7.3.2 Challenges of Conventional Desalination

Concerns with desalination plants include brine waste disposal, high energy use, and chemical contamination. Negative environmental impacts include harming marine life and contributing to climate change.

7.3.2.1 Energy Intensity

While desalination can produce an endless supply of freshwater, there are significant technical challenges. The main one is the process's enormous energy consumption. The energy needed to extract salts and other dissolved substances from water makes desalination a labor intensive procedure. It is theoretically possible to desalinate 1 m^3 of salt water with 0.86 kWh of electricity (3 kJ/kg). Depending on the process type, modern desalination plants utilize anywhere from 5 to 26 times as much as this theoretical minimum. Desalination operations must be optimized for energy efficiency through technological advancements and economies of scale. Furthermore, using a large amount of energy leads to the consumption of a large amount of fossil fuels, resulting in increased CO_2 emissions and contributing to climate change [29]

7.3.2.2 Environment Impacts

Hypersaline slurry, or brine, forms when minerals, extracted salts, and a portion of source water combine. Unlike seawater, brine has a much higher salt content, posing challenges for disposal. Typically, brine waste is disposed of in the sea, which can have negative effects on the environment. To mitigate these effects, brine can be mixed with other sources of water to reduce its saltiness or released using diffusers. Diffusers help spread brine at different locations and facilitate its mixing with ocean water. Desalination plants discharge brine at several locations. Due to its hazardously high concentration of salts and other minerals, brine waste also presents a risk to aquatic life and water quality. Due to its salinity and high density, brine waste can build up at and around disposal sites, suffocating organisms that live on the ocean floor and drastically changing coastal habitats. Furthermore, fish and other aquatic animals suffer severe harm or perish if they become stuck in the pipeline system due to water intakes designed for desalination systems. Finally, if renewable energy sources are not used for the desalination process instead of relying on fossil fuels, desalination can increase GHG emissions and accelerate climate change [30].

7.3.2.3 Economic Considerations

The final price of the produced drinkable water via desalination is highly influenced by the high construction and maintenance costs of desalination plants. This is a major challenge for the desalination sector. However, by optimizing plant design and size, reducing energy costs using renewable energy sources, and incorporating technological advancements in thermal and membrane systems, the capital and operating costs of desalination structures can be significantly reduced. The economic feasibility of desalination technology depends on the lifespan of desalination systems as well as capital and operational expenses. The price of generated water is directly related to the amount of desalinated water that is ultimately required.

7.3.3 Biogas Utilization in Desalination

The desalination process relies heavily on the burning of fossil fuels, which are finite and contribute to climate change and air pollution. Electrical energy costs represent approximately about 35–40% of the total operating costs in desalination systems. To address the issues of non-renewability and limited access to fossil fuels, renewable alternatives such as **biogas** can be employed. Biogas is a powerful sustainable and environmentally friendly alternative energy source for powering water desalination plants, particularly after upgrading. Biogas can effectively supply the necessary heat and electricity for the desalination process. By adopting the technology of utilizing biogas to power desalination systems, several environmental and economic benefits will be gained including waste management, GHG emissions reduction, and circular economy principles application. Biogas can easily be integrated with diverse energy systems and existing infrastructure. Adopting this technology ensures a cleaner, sustainable, and greener future for generations to come [31].

7.3.3.1 Energy Generation for Desalination (Heat and Electricity)

Biogas can produce the necessary heat and electricity to power water desalination plants. For thermal desalination plants that need heat to evaporate salty water, biogas is burned and produces heat that can power the thermal processes. On the other hand, biogas can be utilized to generate electricity; it can be used to power combustion engines that convert biogas to mechanical

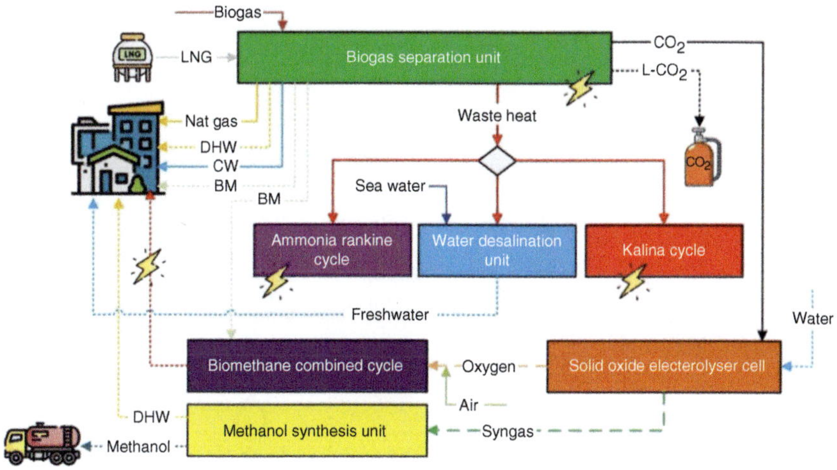

Figure 7.4 Diagram illustrating an innovative method for producing electricity and heat from biogas and using that heat to desalinate seawater to create freshwater. Source: Song et al. [32]/with permission of Elsevier.

energy to power electric generators and produce electricity. This is very efficient for desalination plants that require a lot of electricity such as RO and ED systems. Biogas-powered water desalination technology achieves a more affordable, sustainable, eco-friendly desalination process; it ultimately lowers both high energy costs and the carbon footprint of conventional desalination systems.

A study conducted by Song et al. [32] explored a new method of using biogas to generate heat and electricity required to power a seawater desalination system (Figure 7.4). The study demonstrated that using biogas has significant promise for desalination applications in terms of economic and environmental performance. Based on the study findings, using the biogas approach exhibits an energy efficiency of 91% and CO_2 emission reduction to −0.1041 kg/kWh.

7.3.3.2 Combined and Hybrid Systems for Desalination

Recently, cutting-edge research examined innovative hybrid water desalination systems that combine several renewable energy sources (i.e., biogas, solar, or wind energy) to power the desalination plants. This approach was found to increase desalination performance efficiency and energy supply and output while reducing energy costs and carbon footprints, with a reduced environmental impact as a result. On the other hand, CHP systems can help maximize energy output by converting biogas to generate both electricity and heat. In conclusion, innovative hybrid desalination systems that combine biogas with other renewable energy sources provide a dependable, sustainable, and environmentally friendly desalination operation.

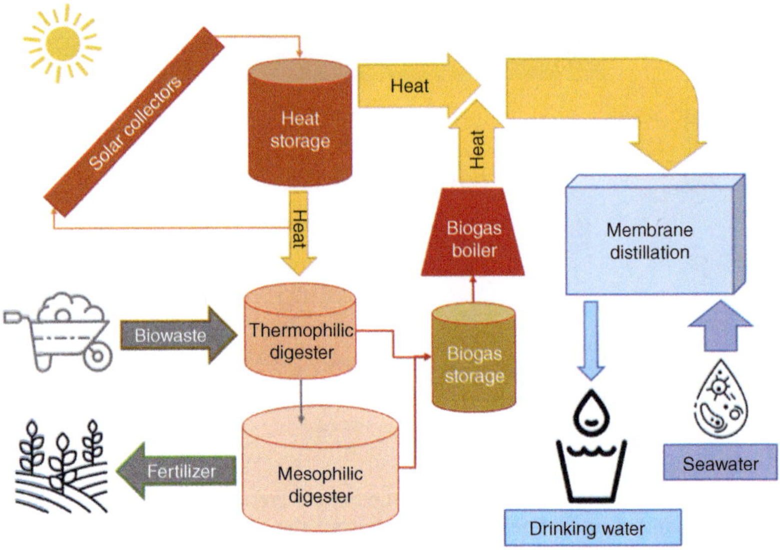

Figure 7.5 An innovative membrane water desalination pilot plant powered by a hybrid solar-biogas energy source. Source: Ref. [33]/Port Said University.

For example, Figure 7.5 shows an innovative membrane water desalination pilot plant powered by a hybrid solar–biogas energy source, which was the subject of a project completed by Port Said University in Egypt. The project is a feasibility study for a solar and biogas-powered water desalination plant. Specifically, using biogas from anaerobic digestion as a backup to solar energy replaces solar energy during overcast days and overnight. Moreover, one of the project's great goals is to manufacture units with an Egyptian industrial enterprise. As a result of the partnership, an integrated desalination system will be created that can be sold and set up in underdeveloped and rural areas such as Sinai, Nile Delta, and Upper Egypt. These devices will supply these communities with inexpensive freshwater and biogas for home use [33].

7.4 Environmental and Economic Aspects

As mentioned earlier, desalination has some challenges, such as high energy requirements, high costs, and negative environmental impacts. One proposed solution is to use biogas in desalination processes, which is believed to be an effective way to address these issues. In this section, the financial and environmental benefits of using biogas to power water desalination plants will be highlighted.

7.4.1 Sustainability of Bioenergy in Desalination

7.4.1.1 Reduction in Carbon Footprint

Desalination mainly depends on fossil fuels to generate the necessary heat and electricity, which increases GHG emissions and accelerates climate change, as mentioned in Section 3.2.2. Transition to using renewable energy sources for the heat and electricity needed to desalinate water is the effective solution to this issue. This change will help decrease carbon emissions, minimize environmental impact, and mitigate climate change in the end. Biogas-powered desalination systems were reported to reduce carbon emissions by up to 70% [5].

7.4.1.2 Waste Management Benefits

Biogas is high in biomethane and can be utilized to produce energy. It is produced by converting organic waste and leftovers into a gas mixture by the anaerobic fermentation process. This technology offers a circular economy, waste management, and disposal site demand reduction, and offers a renewable energy source.

7.4.2 Economic Analysis

The transition to using biogas as a renewable energy source instead of fossil fuels to power the desalination process is one of the efforts being made to lower desalinated water prices. Using biogas reduces desalination costs compared to traditional desalination processes. This topic is examined in detail below by contrasting traditional desalination units with biogas-powered desalination units.

- **Initial Capital Costs**

  The initial capital costs required for desalination unit construction include the cost of land, infrastructure, selected desalination technology, and project size. These costs may be influenced by project geography and location; for example, coastal areas need more complex engineering requirements than other places. It will not, however, be more expensive than a conventional desalination unit or a desalination unit fueled by biogas; the price will remain the same. The initial capital expenses for biogas-producing plants are already quite high. When we compare biogas-powered desalination units to conventional fossil fuel-powered units, it is clear that the biogas-powered units have significantly higher initial capital costs. This is because fossil fuel-powered units benefit from pre-existing infrastructure and economies of scale.

- **Operational and Energy Costs**

  During the operating phase, desalination facilities incur both fixed and variable operating expenses. Labor, insurance, upkeep, and regulatory compliance are examples of fixed costs. Chemicals, energy costs, and additional labor are examples of variable costs that change depending on the amount of production. Desalination units powered by biogas have lower operating costs compared to traditional desalination systems fueled by fossil fuels. First, the cost of input energy from fossil fuels is high, while biogas can be produced from free or low-cost waste. Second, biogas plants have extremely low operating expenses. Bioenergy-powered desalination systems were reported to save operational costs by up to 30% [34].

In terms of energy costs, several factors influence energy consumption, including technology, size, fuel price fluctuations, and efficiency. Biogas-powered desalination units have lower energy costs compared to traditional units due to reduced waste feedstock expenses and lower electricity costs. In addition, it is believed to have a more stable energy cost over time when compared to the consistently increasing price of fossil fuels.

In conclusion, in terms of cost, desalination techniques powered by bioenergy have higher initial capital costs. However, they are a competitive and sustainable alternative due to their lower operational and energy costs.

7.5 Challenges, Limitations, and Future Research Directions

7.5.1 Technical Challenges and Limitations

The development of currently used bioenergy production systems involves the optimization of operating conditions, with an emphasis on key elements. Maintaining optimal parameters such as pH, temperature, and nutritional balance, while optimizing the conversion of biogas into energy, is crucial for the practical use of desalination units powered by biogas. These factors collectively regulate the biogas output to maintain efficiency and continuity of operation. Integrated systems with biogas face a significant hurdle in maintaining their viability: current infrastructure integration. Given that many desalination machines were already constructed to run on fossil fuels, it appears to be quite challenging. To ensure that the switch to biogas is compatible with the current operating procedures, adjustments must be made to the infrastructure, control systems, and input energy systems. Finally, the practical feasibility of adopting desalination powered by biogas technology heavily depends on regulatory frameworks, funding, regional environmental laws, and government support [35].

7.5.2 Research Gaps and Needs

The aforementioned technological obstacles and constraints need to be thoroughly investigated to discover efficient solutions for supporting and advancing desalination systems powered by bioenergy. The suggested research gaps that require more investigation are highlighted in the ensuing paragraphs.

- **Feedstock availability and variability** significantly affect the rates and composition of the produced biogas. Additional research is needed to find the optimal mixture of feedstocks to maximize the produced biogas yields. Moreover, feedstock availability around the desalination plants is another crucial factor. The sustainability and scalability of biogas-powered desalination technology can be improved by considering these two feedstock-related factors [36].
- **Optimizing anaerobic digestion** helps increase biogas production yield, energy supply, and fermentation process stability while preserving the steady performance of desalination systems. Thus, highlighting the optimal operating parameters including pH, temperature, C/N ratio, feedstock selection, and hydraulic HRT is a very important topic to be researched.
- **Biogas purification and upgrading** of the produced biogas before using it are extremely important to enhance the quality and energy conversion. Other components of biogas other than biomethane (i.e., moisture, CO_2, and H_2S) need to be eliminated as they negatively impact the machinery and operating efficiency of desalination units. Consequently, the development of biogas upgrading and purification technologies such as membrane separation, pressure swing adsorption (PSA), and chemical scrubbing needs more investigation and research [37].
- **Effective integration of bioenergy with desalination technologies** needs a deep understanding and knowledge of two key points to ensure the practical and efficient adoption of this technology without any drawbacks: (1) the mechanism of advanced systems capable of continuously utilizing the produced biogas to generate heat and electricity for powering desalination units and (2) the energy requirements and operational features of desalination processes.
- **Life cycle assessment (LCA)** to assess the potential environmental impact of biogas-powered desalination systems is seen as very crucial to ensure approach sustainability. LCA research is needed in terms of waste management, intake issues, climate change, and brine disposal.
- **Economic analysis** significantly impacts the feasibility and scalability of biogas-powered water desalination systems. Therefore, capital, operating, and energy costs as well as market circumstances need extensive investigations.
- **Enhancing public awareness** about bioenergy-powered desalination technology can help improve the engagement of the community and stakeholders to adopt this approach. Consequently, public awareness is considered a crucial research gap. The related research must highlight the economic and environmental benefits of integrating bioenergy and desalination.

7.6 Conclusion and Recommendations

- It is both practical and possible to combine water desalination technology with bioenergy, particularly biogas. Furthermore, the most feasible and environmentally friendly technique with the largest biogas production and the least negative effect on the environment is anaerobic digestion-based biogas generation.
- By employing alternate renewable energy sources in place of fossil fuels, this technology can address several pressing environmental issues, such as waste management, reducing carbon emissions, and promoting the principles of the circular economy.
- Desalination driven by biogas will undoubtedly reduce the issue of energy intensity in conventional desalination methods by lowering reliance on fossil fuels and substituting them with environmentally benign, renewable biogas energy that may be converted to either heat or electricity, or both, in the case of CHP.
- Biogas-powered desalination systems have a higher initial capital cost than conventional methods, but they are less expensive to operate, maintain, and provide energy. Additionally, they manage waste, which lowers carbon emissions and mitigates climate change.
- Practical application of biogas-powered desalination systems is currently hampered by a few issues and constraints. Future studies and research are required to address these issues to improve system performance and implement cutting-edge, creative integrated biogas-desalination technology.
- Government assistance, financing, and environmental legislation are required to motivate stakeholders and communities to use this technology. Furthermore, raising public knowledge of the economic and environmental advantages of combining desalination with bioenergy enhances the long-term viability and scalability of biogas-powered desalination plants.

In conclusion, the integration of biogas and water desalination technology offers viable and workable answers to the problems of existing drinking water availability and its impending scarcity. It also tackles the issue of the energy sector's detrimental effects on climate change and its reliance on fossil fuels. However, several issues and difficulties need to be solved before this technology can be used practically. To overcome these obstacles, a great deal of research is required, and community and governmental backing is crucial for successfully implementing biogas-powered desalination projects.

List of Abbreviations

ABR	Anaerobic baffled reactor
AD	Anaerobic digestion
ASBR	Anaerobic sequencing batch reactor
C/N	Carbon-to-Nitrogen ratio
CH_4	Methane
CO_2	Carbon dioxide
ED	Electrodialysis
GHG	Greenhouse gas
H_2	Hydrogen
H_2S	Hydrogen sulfide
HRT	Hydraulic retention time
LCA	Life cycle assessment
LHV	Lower heating value
MED	Multiple effect distillation
MSF	Multiple stage flash distillation
MSW	Municipal solid waste
NH_3	Ammonia
OLR	Organic loading rate
PSA	Pressure swing adsorption
RO	Reverse osmosis
UASB	Upflow anaerobic sludge blanket
VC	Vapor compression
VFAs	Volatile fatty acids

Questions

1 What factors contribute to the increasing global demand for freshwater, and why is desalination considered a vital solution?

2 How does bioenergy, particularly biogas, address the challenges of conventional desalination methods?

3 What are the main sources of bioenergy, and how do they contribute to sustainable energy production?

4 What are the primary methods of biogas production, and how do operating factors influence its efficiency?

5 What are the most commonly used desalination methods, and what are their advantages and limitations?

6 What are the economic and environmental advantages of integrating biogas with desalination systems?

7 What are the main challenges and future research directions for biogas-powered desalination systems?

References

1 Li, Z., Xingtao, X., Sheng, X. et al. (2021). Solar-powered sustainable water production: state-of-the-art technologies for sunlight–energy-water nexus. *ACS Nano* 15 (8): 12535–12566. https://doi.org/10.1021/acsnano.1c01590.

2 Goren, A.Y., Dincer, I., and Khalvati, A. (2024). Development of a unique integrated bioreactor for simultaneous desalination and bioenergy and biohydrogen production. *Chem. Eng. J.* 486: 150312. https://doi.org/10.1016/j.cej.2024.150312.

3 Gielen, D., Boshell, F., Saygin, D. et al. (2019). The role of renewable energy in the global energy transformation. *Energy Strat. Rev.* 24: 38–50. https://doi.org/10.1016/j.esr.2019.01.006.

4 Singh, S., Sarkar, P., and Dutta, K. (2022). Bioenergy: an overview of bioenergy as a sustainable and renewable source of energy. In: *Bioprospecting of Microbial Diversity* (ed. P. Verma and M.P. Shah), 483–502. Elsevier. https://doi.org/10.1016/B978-0-323-90958-7.00006-6.

5 International Renewable Energy Agency (IRENA). (2021). World Energy Transitions Outlook: 1.5°C Pathway. https://www.irena.org/publications/2021/Jun/World-Energy-Transitions-Outlook.

6 Balali, Y. and Stegen, S. (2021). Review of energy storage systems for vehicles based on technology, environmental impacts, and costs. *Renew. Sustain. Energy Rev.* 135: 110185. https://doi.org/10.1016/j.rser.2020.110185.

7 Dehkordi, S.M., Noorbakhsh, M., Jahromi, A.R.T. et al. (2020). Investigation of biogas production potential from mechanical separated municipal solid waste as an approach for developing countries (case study: Isfahan-Iran). *Renew. Sustain. Energy Rev.* 119: 109586. https://doi.org/10.1016/j.rser.2019.109586.

8 Aslanzadeh, S., Berg, A., Taherzadeh, M.J., and Horváth, I.S. (2014). Biogas production from N-methylmorpholine-N-oxide (NMMO) pretreated

forest residues. *Appl. Biochem. Biotechnol.* 172: 2998–3008. https://doi.org/10.1007/s12010-014-0747-z.

9 Ardebili, S.M. and Safieddin. (2020). Green electricity generation potential from biogas produced by anaerobic digestion of farm animal waste and agriculture residues in Iran. *Renew. Energy* 154: 29–37. https://doi.org/10.1016/j.renene.2020.02.102.

10 Yaashikaa, P.R., Keerthana Devi, M., and Senthil Kumar, P. (2022). Algal biofuels: technological perspective on cultivation, fuel extraction and engineering genetic pathway for enhancing productivity. *Fuel* 320: 123814. https://doi.org/10.1016/j.fuel.2022.123814.

11 Atelge, M.R., Krisa, D., Kumar, G. et al. (2020). Biogas production from organic waste: recent progress and perspectives. *Waste Biomass Valori.* 11: 1019–1040. https://doi.org/10.1007/s12649-018-00546-0.

12 Li, Y., Park, S.Y., and Zhu, J. (2011). Solid-state anaerobic digestion for methane production from organic waste. *Renew. Sustain. Energy Rev.* 15 (1): 821–826. https://doi.org/10.1016/j.rser.2010.07.042.

13 Mao, C., Feng, Y., Wang, X., and Ren, G. (2015). Review on research achievements of biogas from anaerobic digestion. *Renew. Sustain. Energy Rev.* 45: 540–555. https://doi.org/10.1016/j.rser.2015.02.032.

14 Lansing, S., Martin, J.F., Botero, R.B. et al. (2010). Methane production in low-cost, unheated, plug-flow digesters treating swine manure and used cooking grease. *Bioresour. Technol.* 101 (12): 4362–4370. https://doi.org/10.1016/j.biortech.2010.01.100.

15 Karki, R., Wachiranon Chuenchart, K.C., Surendra, S.S. et al. (2021). Anaerobic co-digestion: current status and perspectives. *Bioresour. Technol.* 330: 125001. https://doi.org/10.1016/j.biortech.2021.125001.

16 Grippi, D., Clemente, R., and Pilar Bernal, M. (2020). Chemical and bioenergetic characterization of biofuels from plant biomass: perspectives for Southern Europe. *Appl. Sci.* 10 (10): 3571. https://doi.org/10.3390/app10103571.

17 Nie, E., He, P., Zhang, H. et al. (2021). How does temperature regulate anaerobic digestion? *Renew. Sustain. Energy Rev.* 150: 111453. https://doi.org/10.1016/j.rser.2021.111453.

18 Choi, Y., Ryu, J., and Lee, S.R. (2020). Influence of carbon type and carbon to nitrogen ratio on the biochemical methane potential, pH, and ammonia nitrogen in anaerobic digestion. *J. Anim. Sci. Technol.* 62 (1): 74. https://doi.org/10.5187/jast.2020.62.1.74.

19 Induchoodan, T.G., Haq, I., and Kalamdhad, A.S. (2022). Factors affecting anaerobic digestion for biogas production: a review. *Adv. Org. Waste Mgmt.* 223–233. https://doi.org/10.1016/B978-0-323-85792-5.00020-4.

20 Yousuf, A., Khan, M.R., Pirozzi, D. et al. (2017). Economic and market value of biogas technology. In: *Waste Biomass Management–A Holistic Approach*

(ed. L. Singh and V.C. Kalia), 137–158. Springer. https://doi.org/10.1007/978-3-319-49595-8_7.

21 Abubakar, A.M. (2022). Biodigester and feedstock type: characteristic, selection, and global biogas production. *J Eng Res Sci* 1 (2): 170–187. https://dx.doi.org/10.55708/js0103018.

22 Venus, T.E., Strauss, F., Venus, T.J., and Sauer, J. (2021). Understanding stakeholder preferences for future biogas development in Germany. *Land Use Policy* 109: 105704. https://doi.org/10.1016/j.landusepol.2021.105704.

23 Dalpaz, R., Konrad, O., da Silva Cyrne, C.C. et al. (2020). Using biogas for energy cogeneration: an analysis of electric and thermal energy generation from agro-industrial waste. *Sustainable Energy Technol. Assess.* 40: 100774. https://doi.org/10.1016/j.seta.2020.100774.

24 Environmental Protection Agency (EPA). (2020). Biogas Opportunities Roadmap Progress Report. U.S. Environmental Protection Agency. https://www.epa.gov/sites/default/files/2015-12/documents/biogas-roadmap-progress-report.pdf.

25 European Commission. (2021). Fit for 55: delivering the EU's 2030 climate target on the way to climate neutrality. European Commission. https://ec.europa.eu/commission/presscorner/detail/en/IP_21_3541.

26 Yao, M., Tijing, L.D., Naidu, G. et al. (2020). A review of membrane wettability for the treatment of saline water deploying membrane distillation. *Desalination* 479: 114312. https://doi.org/10.1016/j.desal.2020.114312.

27 Hailemariam, R.H., Woo, Y.C., Damtie, M.M. et al. (2020). Reverse osmosis membrane fabrication and modification technologies and future trends: a review. *Adv. Colloid Interface Sci.* 276: 102100. https://doi.org/10.1016/j.cis.2019.102100.

28 Mir, N. and Bicer, Y. (2021). Integration of electrodialysis with renewable energy sources for sustainable freshwater production: a review. *J. Environ. Manage.* 289: 112496. https://doi.org/10.1016/j.jenvman.2021.112496.

29 Ghaffour, N., Missimer, T.M., and Amy, G.L. (2013). Technical review and evaluation of the economics of water desalination: current and future challenges for better water supply sustainability. *Desalination* 309: 197–207. https://doi.org/10.1016/j.desal.2012.10.015.

30 Elsaid, K., Kamil, M., Sayed, E.T. et al. (2020). Environmental impact of desalination technologies: a review. *Sci. Total Environ.* 748: 141528. https://doi.org/10.1016/j.scitotenv.2020.141528.

31 Morote, Á.-F., Rico, A.-M., and Moltó, E. (2017). Critical review of desalination in Spain: a resource for the future? *Geogr. Res.* 55 (4): 412–423. https://doi.org/10.1111/1745-5871.12232.

32 Song, Y., Ahmad, S.F., Houran, M.A. et al. (2023). Multi-variable study of a novel multigeneration system using biogas separation unit and LNG cold energy utilization, producing electricity, cooling, heat, fresh water, liquid CO_2, biomethane, and methanol. *Process Saf. Environ. Prot.* 180: 616–638. https://doi.org/10.1016/j.psep.2023.10.023.

33 Port Said University. (2022). Hybrid membrane water desalination—Port Said University. https://psu.edu.eg/en/hybrid-membrane-water-desalination/ (accessed 13 January 2025).

34 International Desalination Association (IDA). (2021). The state of desalination and water reuse: an IDA and GWI Sourcebook. IDA Global Connections. https://idadesal.org/publications/2021-the-state-of-desalination-and-water-reuse/.

35 Paranjpe, A., Saxena, S., and Jain, P. (2023). A review on performance improvement of anaerobic digestion using co-digestion of food waste and sewage sludge. *J. Environ. Manage.* 338: 117733. https://doi.org/10.1016/j.jenvman.2023.117733.

36 Koutsiantzi, C., Mitrakas, M., Zouboulis, A. et al. (2022). Evaluation of polymeric membranes' performance during laboratory-scale experiments, regarding the CO_2 separation from CH_4. *Chemosphere* 299: 134224. https://doi.org/10.1016/j.chemosphere.2022.134224.

37 Khan, M.U., Lee, J.T.E., Bashir, M.A. et al. (2021). Current status of biogas upgrading for direct biomethane use: a review. *Renew. Sustain. Energy Rev.* 149: 111343. https://doi.org/10.1016/j.rser.2021.111343.

8

Emerging Desalination Techniques Powered by Waste Heat

Jesús A. Claudio-Rizo, Denis A. Cabrera-Munguía, Tirso E. Flores-Guía, and Lucía F. Cano-Salazar

Department of Advanced Materials, Faculty of Chemical Sciences, Autonomous University of Coahuila, Saltillo, Coahuila, Mexico

8.1 Introduction

The twenty-first century has witnessed the coming together of booming populations, urbanization, and climate change, which put immense pressure on freshwater supplies across the globe. The number of individuals suffering from water scarcity has reached a mark of almost two billion people today and is expected to increase due to the unreliability of traditional water sources. One way out of it is desalination—change seawater into fresh drinking water. However, usually conventional desalination methods use a lot of energy, are expensive, and are unfriendly to the environment. A combination of these factors makes it necessary to make use of innovative ways that will improve our lives by desalting using waste heat. This strategy not only points toward an abundant future for renewable freshwater but also demonstrates how recycling and resource optimization can be transformative. Despite its significance as an energy source, waste heat, which results from industrial processes like power production as well as natural events, remains one untapped resource in many situations. While there are many other competing uses for waste heat and hence often it is overlooked despite its potential capacity for being an excellent source of power (Figure 8.1).

As the population of the world explodes and freshwater sources dwindle, the development of sustainable or efficient desalination methods is increasingly critical. Although they are effective, conventional desalination processes often rely heavily on high energy usage, making them less environmentally friendly and economically viable. In a period where climate change, population growth, and industrial demands increasingly strain freshwater availability, finding efficient and green methods for desalination is crucial. Common methods of desalination

Water Treatment and Desalination: Membrane Separation using Renewable Energy Resources, First Edition. Edited by Mohamed Bassyouni.

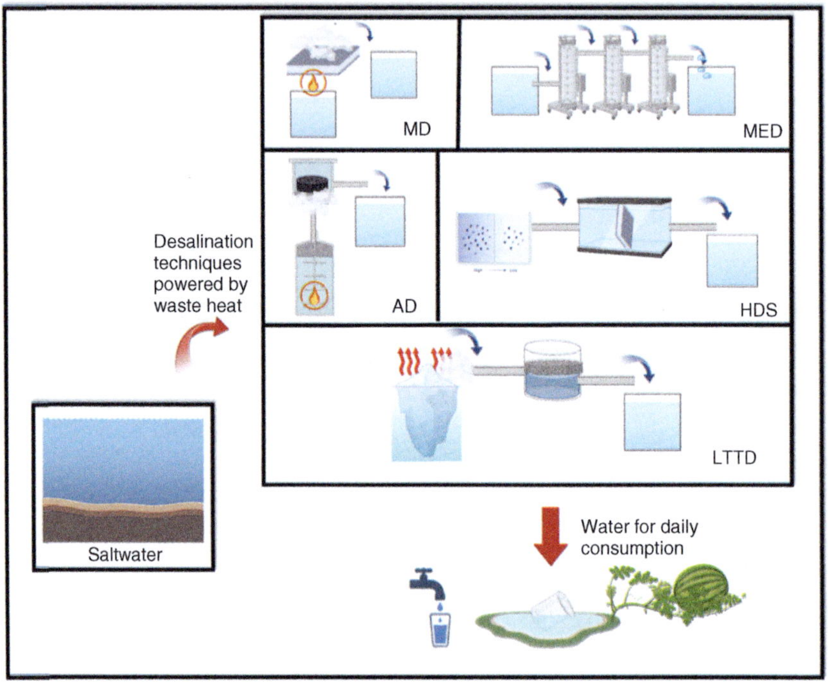

Figure 8.1 General diagram showing the new waste heat-powered desalination methods.

include thermal distillation and reverse osmosis (RO). These processes have typically required large amounts of energy as well as raised environmental concerns. There is hope in emerging waste heat-driven desalination technologies that can convert industrial effluents into a valuable water resource. Thus, when coupled with other conventional means of seawater treatment, these new techniques will create room for the security and enhancement of our water supply, leading to better living standards for various populations globally. Nevertheless, innovation is now moving toward converting waste heat—a by-product from industrial operations or power generation—into an important source of water to be used in the process of desalting it through an unorthodox approach. Among these novel techniques propelled by waste heat, membrane distillation (MD), multi-effect distillation (MED), adsorption desalination (AD), hybrid desalination systems (HDSs), and low-temperature thermal desalination (LTTD) are at the vanguard of sustainable freshwater production (Figure 8.1).

MD is an advanced technology that utilizes low-grade waste heat to propel the desalination process. It functions by having a temperature gradient between its membrane with the warm side heated from waste heat sources (Figure 8.2).

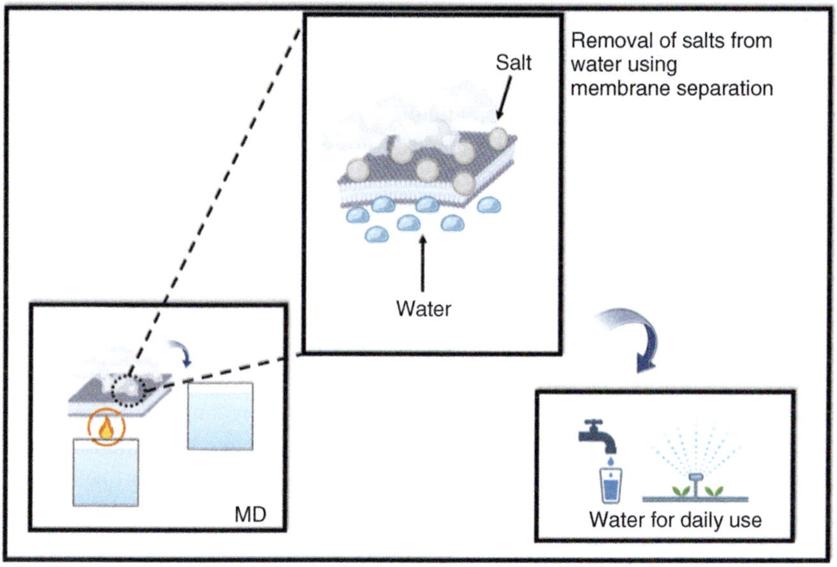

Figure 8.2 Using membranes and the MD process, salts are removed from water to produce water fit for human consumption.

Consequently, water vapor passes through this membrane due to the difference in temperature condensing on the cooler side. Due to its low energy requirements, it can efficiently desalinate saline water, including brine, emanating from traditional desalination plants, thereby reducing environmental discharge. For instance, MD is modular and scalable, making it ideal for use in decentralized and remote areas such as those within coastal communities and islands as well. Another point is that MD is integrable into existing waste heat sources; hence, it is versatile for sustainable applications of desalting tools; MD adopts a temperature gradient across a selectively permeable membrane induced by low-grade waste heat, allowing water vapor passage while retaining salts and impurities. This process needs only a supply of heat at temperatures as low as 60°C, which is usually rejected by industries in large quantities. The other thing is that MD can be deployed in decentralized systems where people live far away from urban centers and do not have access to clean potable water [1].

The use of small-scale residual cold by MD diminishes the need for extra energy to run it, thus reducing costs and having less effects on the environment. Portable MD systems are fully scalable, allowing them to easily adapt to varying circumstances, making them suitable for decentralized and remote applications. Figure 8.2 shows that they effectively remove salts, producing high-purity water. However, fouling and scaling can occur in MD membranes while dealing with high-saline feed waters, demanding routine servicing especially where highly saline feeds are used. In addition to this, a disadvantage of MD systems over common methods such as RO is their capability to operate at lower throughputs, which may impede large-scale deployment.

MED is an old method made new by exploiting the heat from waste to derive more energy from it and reduce the cost of operation. MED works on an arrangement of stages (or effects) in which seawater is heated and evaporated one after another, with each stage operating at a progressively lower pressure as it takes its energy from waste heat. The steam generated in one level heats up the next stage, thus obtaining maximum use of power (Figure 8.3). MED offers a compelling solution for large-scale desalination because it can harness heat wasted by industries such as power plants or refineries. MED is a strong method that involves several stages where water is evaporated and then condensed, with each step being fed by the vapor coming out of the previous stage. With the inclusion of waste heat, MED operates on higher efficiencies for energy. Lower operational costs are achieved through multiple-stage heat reuse in MED, making desalinated water affordable and available to many people. This greater effectiveness may result in reduced water expenses borne by societies, allowing them to allocate more resources into building education systems, improving healthcare delivery mechanisms or infrastructure projects. By engaging many steps in recycling heating sources, for instance, MED achieves efficient utility usage far better than

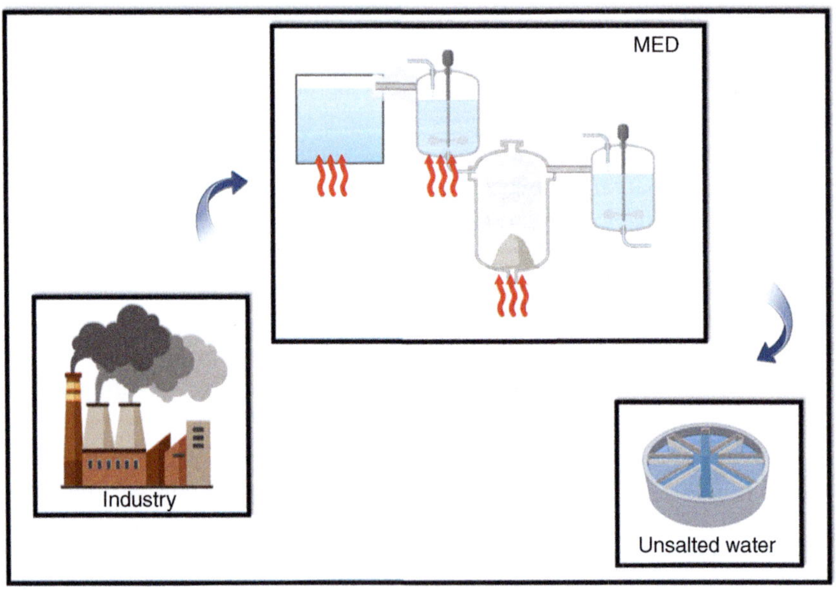

Figure 8.3 MED's adaptability in treating wastewater with industrial residual energy.

other methods, hence reducing operational expenditures [2]. In essence, this means that there can be no fixed rules concerning issues such as user training requirements and how long certain types of users will take to acquire proficiency levels needed for effective model use, because these things will depend on a wide range of factors specific to individual software companies themselves. The feed waters used in MED comprise wastewater treatment effluents, high-salinity brackish waters, and industrial wastes whose qualities may fluctuate considerably across different locations. The initial cost of MED systems can be substantial, which can be a hindrance to small communities or underdeveloped areas. MED installations usually require more space than other desalination technologies, and this may pose a limitation if the area is densely populated [3].

AD is a new technology that uses waste heat at low temperature to make freshwater by absorption. These AD systems have adsorbent materials like silica gel and zeolites to capture the water vapor from saline solutions. The substances discharge this liquid when heated by industrial waste heat, condensing it into freshwater (Figure 8.4). It is an interesting feature of AD that it can work at low temperatures up to 50°C, making it well-suited for use with poor quality low-grade waste heat [4]. Designed with modularity in mind, AD is easily integrated into existing industrial practices as an adaptable and energy-efficient desalination solution. In turn, the AD absorbs the vapor from saline media which, after being heated by a lower-temperature waste matter, turns into clean water. As opposed to lots of other

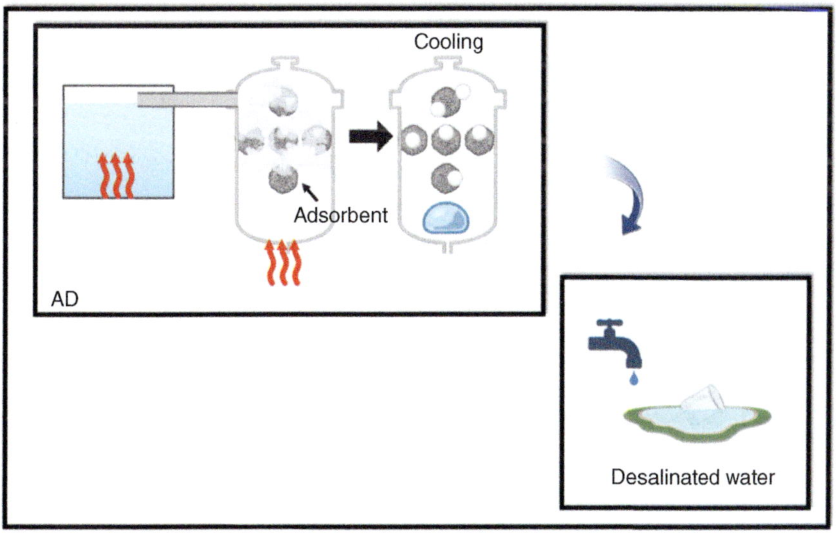

Figure 8.4 Illustration of AD water desalination utilizing cooling and adsorbent materials.

methods, AD processes are capable of running at lower temperatures; hence, they are adjustable to various types of wasted heat sources occurring in different climates and conditions. The availability of reliable freshwater contributes directly to better public health through a decrease in incidences of waterborne diseases and improved sanitation practices. Because they operate at lower temperatures, AD systems can be used with many different sources of exhausted heat.

The design of AD systems can be easily integrated into already existing industrial set-ups and expanded to meet different water demands. It is possible for AD systems to function with a very minimal energy level which improves their sustainability and reduces operational costs. The adsorbent materials employed in AD can degrade with time, necessitating periodic replacement and increasing maintenance costs. AD systems may have lower water production capacities than other technologies, thus posing difficulties in meeting increased water demands [5].

Also, HDSs which constitute a combination of traditional technologies are important with regard to the efficiency of the desalination process (Figure 8.5). The technology can be adjusted to cater for specific requirements linked to water quality fluctuations and energy availability. This attribute makes it even more attractive in areas where resources are unevenly distributed or there is

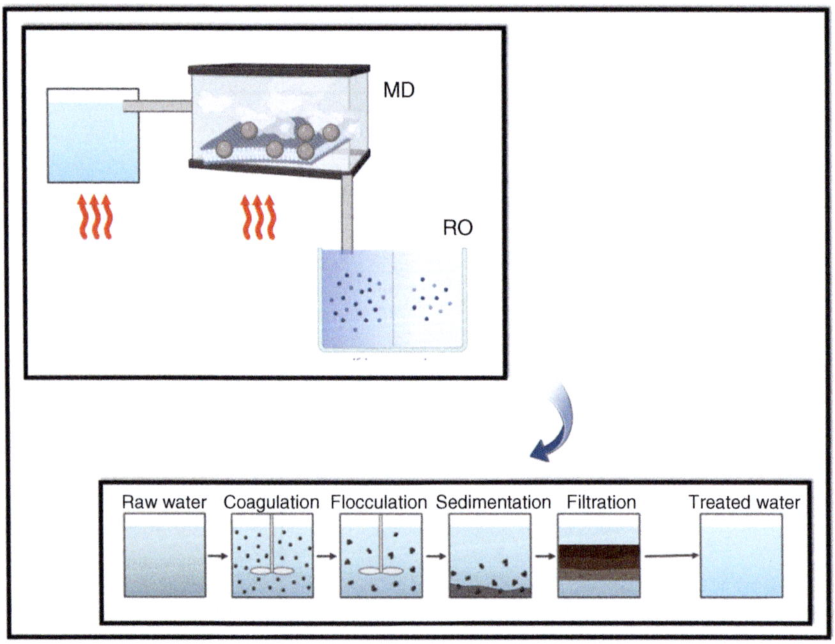

Figure 8.5 HDS method for treating salt water and obtaining potable water.

high variability in demand. Additionally, hybrid systems can also adapt even when different waste qualities are observed and if waste heat is available so that resources are optimally used and developed sustainably. It promotes the development of resilient water infrastructure required to support rapid urban population growth and economic development. Through this optimization, waste heat usage by hybrid systems has led to a reduction in energy consumption during different stages of the process, hence improving efficiency.

Advantageously, the flexibility within hybrid systems allows them to suit any environmental conditions as well as various operations through customization of hybrid systems depending on feed-water qualities and waste heat source specificities. In other words, combining multiple technologies could enhance reliability or robustness and increase production capacity for these regions' systems, at least on paper: Hybrid System Design in terms of Reliability & Robustness [6]. However, such projects may impose heavy initial costs putting strain on budgets [6].

Heat gradients present in the sea are beneficial in LTTD processes (Figure 8.6). These heat gradients can be used for water condensation without using assisted energy sources, resulting in low energy demand systems, representing a sustainable alternative [7].

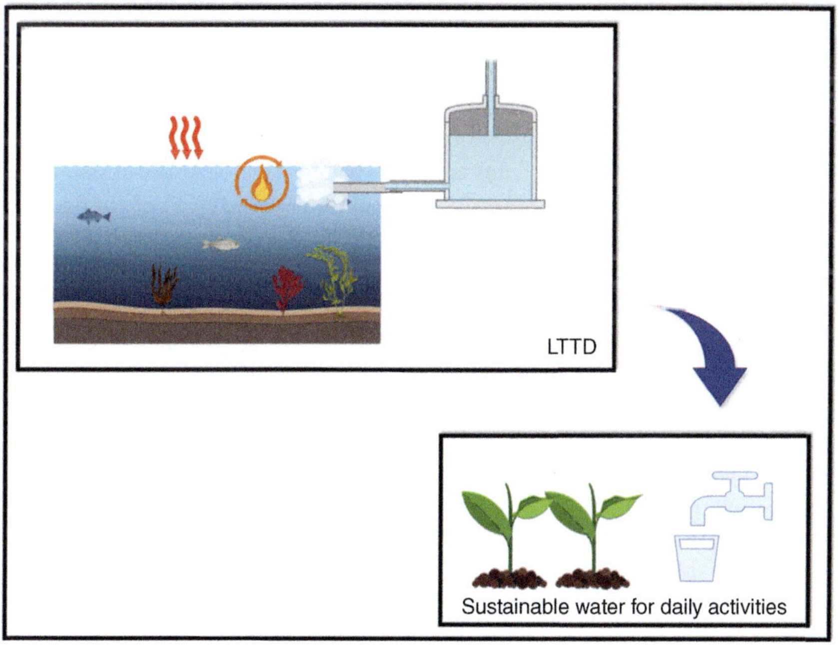

Figure 8.6 Emerging LTTD technique for the desalination of seawater and obtaining sustainable water.

This makes LTTD attractive economically as it helps in cutting down on operational expenses. Therefore, because of this reliance on nature's own temperature fluctuations, the technology has limited applicability in terms of geographical scope. The capacity of such systems might also be below other means of desalinating water, making them less effective under high-demand conditions. For better water production, traditional approaches such as RO and multistage flash distillation (MSF) could be integrated with new desalination strategies driven by waste heat. This can make systems more efficient, using less energy while increasing the range of desalination technologies that are applicable. Hybrid systems employing waste heat for pretreatment or secondary desalination steps require significantly less total energy compared to conventional stand-alone alternatives.

Desalination by-product waste heat can be used to enhance the quality of human life internationally. This is possible just by changing an ignored waste product into something beneficial—creating a sustainable, ample freshwater supply that supports healthy communities, resilient ecosystems, and strong economies. Public health depends on access to clean water because it is a basic requirement. The need for reliable drinking water supplies even in slum areas has been met with an answer in the form of desalination powered by waste heat, which ultimately reduces diseases and leads to a reduction in death rates. Water becomes affordable when desalination costs go down due to waste heat reutilization, hence freeing other vital funds for education or medication.

There is a very pressing need to ensure adequate and reliable sources of freshwater as the world grapples with escalating water scarcity driven by climate change, population growth, and unsustainable water usage. Seawater desalination is an essential method for converting the voluminous reserves of the world's oceans into drinkable freshwater. But traditional desalination procedures are often energy-intensive and environmentally challenging. At present, heat-driven desalination technologies in general and those using waste heat in particular offer a promising way toward achieving global sustainability goals (Figure 8.7). These new technologies may efficiently produce drinking water from saltwater while not consuming too much power. This chapter examines how these developing approaches can be aligned with and contribute to the United Nations' Sustainable Development Goals (SDGs), addressing current issues related to seawater desalination. MD and MED, which are two heat-powered techniques, make use of waste heat or low-grade thermal energy to generate freshwater from saline liquids. Such technology could be installed in arid regions where there is little water, providing people with an uninterrupted supply of high-quality drinking water especially near coastal areas or in arid zones. The optimization of energy consumption through the integration of waste heat recovery into heat-powered desalination supports this target [8].

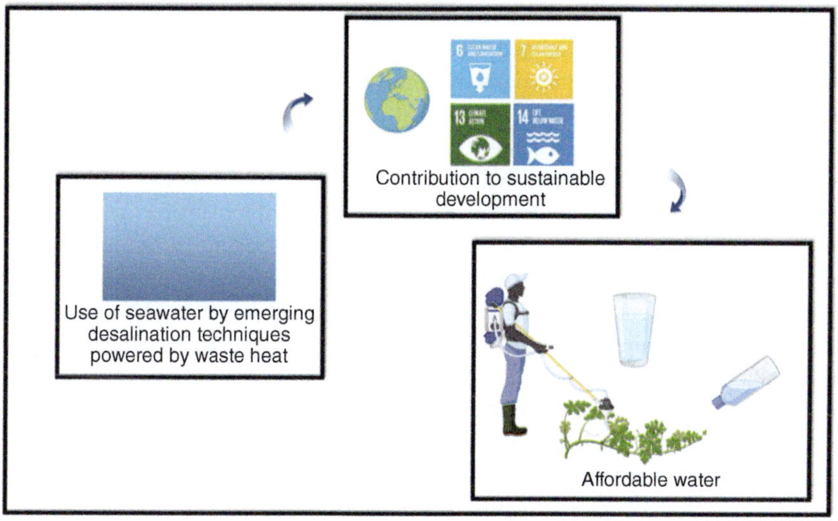

Figure 8.7 New methods of desalination driven by waste heat are important approaches for the worldwide growth of water-related sustainability.

Infrastructure and technological development alongside supportive policies and incentives are required to scale these technologies beyond pilot plants to large-scale operations. Dealing with brine concentrate by-product is one of the biggest environmental challenges related to desalination. Innovative techniques are necessary for minimizing brine, treating it properly, and reusing it ecologically. For sustainable exploitation of ocean resources, intake and discharge practices by desalination plants should not harm marine life at all costs. The combination of emerging desalination approaches with existing water treatment and distribution structures is complicated. To be able to do this, there must be careful planning as well as technology adaptation; hence, stakeholder collaboration has been put in place for promotion purposes so that everything goes according to plan to make them more effective and sustainable, the customization of desalination solutions to match local climate, economy, and environment is mandatory.

To lower the initial cost barriers and encourage investments in thermal-powered desalination technologies, governments and international organizations can give financial incentives and subsidies. For promoting waste heat utilization as well as sustainable management of the by-products of desalination for adoption of these technologies, it is critical to have supportive regulatory frameworks. Research and development investment still plays an important role in boosting the efficiency, scalability, and cost-effectiveness of heat-powered desalination methods. This technological revolution will rely on a collaborative effort among academic researchers, industrialists, and governments. The deployment of new technology can be hastened if public agencies partner with private companies; this allows them to share related risks with rewards for technological improvement. Public awareness campaigns that educate communities about the benefits and challenges associated with using heat-powered desalination can help gain support from the public for being involved. Sustainable water use is important, along with conservation efforts, which are made known through such campaigns. By engaging local stakeholders in planning and implementing desalination projects, solutions can be crafted in a way that suits specific community needs or conditions. Eco-friendly water treatment systems such as advanced brine handling units or energy-saving techniques should therefore be developed and employed for sustainable desalination processes. To improve the sustainability and resilience of these operations, adaptive management approaches are needed to allow for continuous monitoring coupled with adjustments where necessary.

A revolutionary opportunity has been created by combining emerging heat-driven desalination techniques with sustainable practices. More so, these modern ways reduce energy requirements and running costs as well as address environmental concerns. Consequently, the use of waste heat produced from industrial activities or power plants is an alternative to traditional desalinating processes that contributes to reduced carbon emissions and a healthier planet. Thus, while aiming to quench the thirst for clean water due to rising demand, innovative approaches are significant in ensuring this state is maintained for future generations. Having overcome these challenges through partnership, innovation and supportive policies will lead us toward achieving our global sustainability goals where everybody can access clean water. Taking up the task of adopting waste heat-powered desalination solutions offers some glimmer of hope in times of increasing water scarcity. Besides solving the immediate need for sustainable water resources, these creative solutions reflect the circular economy and resource optimization ideas.

When we invest in and upscale these technologies, we make a future where clean water is everywhere but not rare and where innovation benefits everyone in society and improves our way of living. The innovative landscape of desalination powered by waste heat is increasingly showing more sustainable solutions economically to global water problems. This integration of emerging methods for desalination that are powered by waste heat with existing processes opens a transformational window to address the global water crisis. These new approaches entail several benefits which range from better energy efficiency, environmental sustainability, water quality upgrade, and operational resilience improvement. While issues such as system complexity and upfront capital costs persist, the potential gains far outweigh the obstacles. Combining waste heat with advanced desalination techniques creates a trustworthy source of freshwater that raises living standard for worldwide communities. It must be noted that this forward-looking method does not only cater for immediate water needs but also lays the ground for a future, where attaining sanitation and quality drinking water will become an unconditional right of all people around the globe.

Different technologies such as MD, MED, AD, hybrid systems, and LTTD offer their own set of unique benefits that can change garbage into something precious. Not only do these techniques continue to advance and grow together but they also represent the relationship between technology and environmental conservation. By embracing them, we are getting closer to a future where all people will have access to clean water that would rely on the sustainable use of energy and resources. This chapter will delve deeply into physiochemicals and engineering principles used in current desalination techniques powered by waste heat. Furthermore, it will focus on the major application of these new technologies aiming to provide viable water for various human activities.

8.2 Multi-Effect Distillation (MED) and Membrane Distillation (MD)

In particular, when combined with waste heat recovery and renewable energy, desalination has the potential to become an affordable and environmentally beneficial option. By harnessing diverse renewable sources—like wind, hydro, tidal, and wave power for electricity, and solar, geothermal, and biomass for both thermal and electrical energy—we can greatly enhance the sustainability of desalination processes [9].

Incorporating waste heat, which is typically categorized by its temperature range, further optimizes the process [10]. Various mechanical systems, such as heat exchangers, furnace regenerators, heat pumps, boilers, passive air preheaters, and thermodynamic cycles, can be deployed to capture and utilize this waste heat, driving desalination in a more energy-efficient and sustainable direction [10].

Desalination technologies can be broadly categorized into two main types: phase change (thermal) and non-phase change (membrane-based) processes. Phase change: Thermal desalination techniques involve heating water to create a phase change, where it evaporates, leaving salts and impurities behind. The resulting high-purity freshwater vapor is then condensed under high pressure in cooling pipes. This method includes several advanced processes:

- **MED**: Utilizes multiple stages of evaporation to maximize efficiency.
- **Multistage Flash (MSF) Distillation**: Involves rapidly evaporating water at reduced pressures across multiple stages.
- **Mechanical Vapor Compression (MVC)**: Compresses and recycles the vapor for condensation.
- **Thermal Vapor Compression (TVC)**: Uses steam jet ejectors to compress and recycle the vapor.

Non-phase change (membrane-based desalination): Membrane desalination relies on advanced filtration techniques where semipermeable membranes with varying pore sizes remove dissolved salts, pollutants, and microorganisms from water. This process often requires significant energy or pressure to force water through the membranes. Key techniques include the following:

- **RO**: Utilizes high pressure to push water through a semipermeable membrane, filtering out salts and impurities.
- **MD**: Employs a thermal gradient across a membrane to facilitate vapor transport and condensation.
- **Microfiltration (MF)**: Uses membranes with larger pores to filter out bacteria and suspended solids.
- **Ultrafiltration (UF)**: Targets smaller particles, including viruses and colloids.
- **Nanofiltration (NF)**: Focuses on removing small organic molecules and certain dissolved salts.
- **Membrane Bioreactor (MB)**: Integrates biological treatment with membrane filtration for wastewater recycling.
- **Electrodialysis (ED)**: Applies an electric field to drive ions through ion-selective membranes, separating salts from water.

By leveraging these innovative desalination techniques, we can effectively address water scarcity issues while optimizing energy consumption and minimizing environmental impact [9, 10].

8.2.1 Multiple-Effect Distillation

8.2.1.1 Principle and Operating Conditions

The primary thermal desalination processes widely used commercially are MSF, MED, and MVC [11]. MED was initially implemented with a capacity of 500 m^3/day [10]. An MED plant comprises multiple single evaporators (effects) arranged in series with a main condenser, as well as various feed water pre-heaters and flashing chambers, where water is heated to temperatures up to 110°C [12]. The maximum brine temperature is capped at 120°C due to calcium sulfate scaling, while the minimum temperature is determined by the available temperature in the wave [9].

Steam is produced in a set of boilers or steam turbines using fossil fuels, waste heat, and/or renewable energy sources. This steam is applied in the first evaporation stage, and the vapor generated in the first evaporation effect is then used in the second stage, continuing in this manner until the temperature drops to 30–40°C. Most industrial MED desalination plants take the vapor generated in the final stage, compress it to the desired temperature, and reuse it in the first evaporation stage. Saline water serves as a cooling fluid for the distillate (freshwater) condenser, and it is preheated using heat recovery exchangers supplied by the steam produced in each effect chamber before being fed into the evaporation stages [9]. Consequently, MED is considered the most promising thermal desalination method due to its lower operating costs and energy consumption, making it more efficient than MSF [10].

8.2.1.2 MED Configurations

MED systems can be configured in several ways to optimize their efficiency and performance. These configurations include forward feed (FF), backward feed (BF), and parallel cross-feed (PCF), each offering unique operational benefits (Figure 8.8):

- **FF**: In this configuration, both the feed water and steam flow in the same direction through the series of evaporation stages, creating a cocurrent arrangement. This method is straightforward and is effective in applications where the simplicity of flow direction is advantageous.
- **BF**: This approach contrasts with FF by having the feed water and vapor flow in opposite directions, establishing a counter-current flow. This configuration can improve the efficiency of heat recovery as the coldest feed water meets the coldest condensate, enhancing the overall thermal gradient.
- **PCF**: This configuration introduces fresh feed water at each stage and extracts concentrated brine from each effect. This approach is more complex but excels in performance. PCF maximizes energy efficiency and minimizes specific heat consumption by leveraging the optimal use of thermal energy across all stages, leading to higher output and reduced operational costs.

Because of its better performance and lower specific heat consumption among these configurations, PCF is frequently chosen [12]; this makes it perfect for high-demand desalination operations where efficiency is crucial.

8.2.1.3 Limitations and MED Economics

However, though it is highly effective, MED poses certain challenges in terms of environmental footprints and operational costs. Around 70% of the seawater that MED systems treat as brine is discharged back into the ocean. The elevated salinity level affects the flora and fauna within the area, resulting in a loss of biodiversity and change in marine life behavior [12].

It is common for the distillate and brine from the last effect of MED to be released at temperatures between 35 and 40°C. This thermal discharge raises seawater temperature, forcing marine animals to struggle as their ecological balances are altered [12]. Desalination energy requirements can increase due to high salinity levels and varying temperatures of seawater, thereby increasing costs. On a per-unit basis, larger plants have lower water production costs compared to smaller installations due to economies of scale. Costs associated with chemicals required for pretreatment and maintenance add to operating expenses. The cost also depends on how far the seawater intake is from the desalting plant

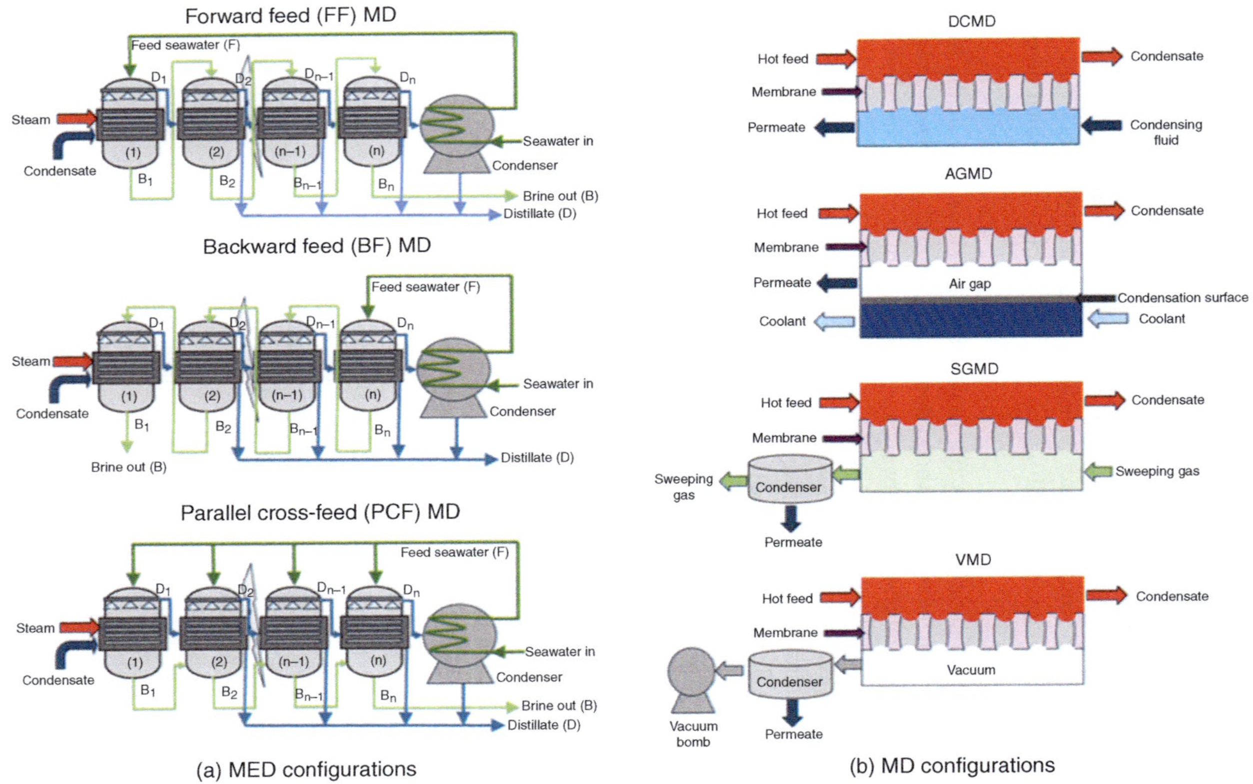

Figure 8.8 The most typical arrangement of the desalination processes (a) MED and (b) MD.

because pumping infrastructure would have to be built for water transportation. Typically, thermal energy used for seawater evaporation accounts for a significant proportion of these costs.

The cost reduction is urgently needed when it comes to optimizing heat usage [13].

Some of these solutions have been developed to solve the challenges and make MED systems more efficient such as:

- **Energy Saving Through Reducing Extraction Vapor**: This can be achieved by lowering the amount of steam extracted which will lower operating costs.
- **Renewable Energy and Waste Heat Integration**: The use of renewable sources of energy in addition to recovering waste heat from industrial processes or other power production can provide low-cost thermal inputs [10, 11].
- **MSF Distillation or TVC Plus MED**: By combining multiple technologies, the hybrid systems enhance overall system efficiency, hence cutting down on costs; such hybrid configurations are able to achieve higher water removal rates with lower production costs [10].
- **New Configurations and Operational Adjustments**: Moreover, Table 8.1 illustrates various types of single and hybrid configurations while also proposing a number of different performance-enhancing parameters that could bring further savings. For example, this involves designing new ways for integrating MED with other thermal methods as well as optimizing available thermal energy.

By taking care of these environmental issues and economic points, MED technology becomes sustainable and economically viable, hence playing its part in meeting global water demands while minimizing ecological impacts.

8.2.2 Membrane Distillation

8.2.2.1 Principle and Operating Conditions

The method of operation is called membrane distillation. Unlike conventional thermal approaches that operate at atmospheric pressure, multistage flash (MSF) and multiple effect distillation (MED) techniques have rather moderate demands and can tolerate high feed salinity in some cases [26, 27].

Operations are as follows:

a) **Formation of a Temperature Gradient**: One part of a hydrophobic membrane contacts the hot saline feeding solution [26, 27].
b) **Vapor Transportation**: Heat is transferred from the bulk hot solution to the surface of the membrane by means of the boundary layer.
c) **Cold Side Condensation**: The vapor recondenses into liquid form after it has come into contact with the cold side of the membrane [28].

Table 8.1 Single and hybridized MED and MD units for water desalination.

Water desalination technique	Remarks	Water price ($/m³)	References
MED	• Utilizing low-temperature waste heat from exhaust flue gases	1.13–2.9	[14]
MED	• An optimal number of effects of 17	0.614	[15]
MED-TVC	• A zero liquid discharge • Use of solar energy • Salt recovery • Profit 5.71 $/m^3$	Not presented	[16]
MED-TVC	• Utilization of low-grade heat from flue gas • Water production costs decrease with the optimization of thermal vapor compression • The lowest water production costs were achieved with five effects for MED and four effects for TVC, respectively	2.84–3.13	[17]
MED-evaporative crystallization	• A zero liquid discharge • Forward feed MED • Industrial waste heat and solar energy can be implemented	4.17	[18]
MED-RO	• Cogeneration of water in a power plant • Cooling water in MED provided by the reject water on RO	0.81 (MED) 0.58 (RO) when are coupled	[19]
MED-AD	• Heat recovery from residual brines • Use of solar energy	Not presented	[20]
MED-MD	• Parallel feed configuration of MED-MD yields the best performance • Preheating of the MED feed by the MD process • Treating and reduction of brines	1.84–2.05	[21]
MD	• Configuration: Water gap membrane distillation (WGMD) • The natural convection in the water gap is critical • A maximum efficiency of 93 with a feed at 90°C	Not presented	[22]

Table 8.1 (Continued)

Water desalination technique	Remarks	Water price ($\$/m^3$)	References
MD	• Use of solar energy • Water purification and energy generation • Design suitable for residential and commercial buildings for potable and domestic hot water	10	[23]
MD	• Brackish water • Combination of solar thermal energy with a waste heat source • Renewable energy reduces the water cost production	1.6	[24]
MSF-WGMD	• MSF and WGMD were operated with three stages • The decrease of electric energy consumption is affected by an increase in the feed temperature	Not presented	[25]

The importance of MD technology lies in using these concepts to achieve low energy desalination and hence reducing operational expenses which make it the most appropriate technology for green water solutions. Interestingly, this technique has several advantages; it works well under normal air pressure within the system; there is also no need for high temperatures or vacuum conditions that are essential in conventional thermal desalination processes [26]. On the other hand, unlike some other types of desalination that are less effective with feed waters having very high salinity, MD can keep its performance at a high level [26, 27]. By applying the inherent advantages of the MD process, water desalination can be performed efficiently with less energy consumption and negative impact on the environment. Therefore, it is preferred to be used in emerging water applications like small-scale rural community supply or industrial uses [27].

Typically, the ideal working temperature for MDs is 40°C; a significant increase in this temperature improves the partial vapor flux. To enhance efficiency and to cause turbulence, spacers are usually employed in module channels [26, 28]. Therefore, these variables should be maximized to make MD systems for desalination more dependable and efficient and that has increased its interest as a sustainable water treatment method. In addition, there are hybrid MD technologies such as BR-MD (bioreactors), ED-MD (electrodialysis), EC-MD (electrocoagulation), MDC (crystallization), AD-MD (adsorption desalination), MSF/MED-MD (thermal processes), and FO-MD (forward osmosis) [26]. Recent research on several kinds of single and combination MD is shown in Table 8.1.

8.2.2.2 MD Configurations

Various configurations of MD technology are available to meet specific operational requirements and applications (Figure 8.9). DCMD, AGMD, SGMD, and VMD are the four major categories. Each one has its own advantages for use in different types of desalination and purification processes.

The simplest setup is the DCMD, frequently employed in small-scale laboratory experiments. One side of a hydrophobic membrane is in direct contact with the hot saline feed, while the opposite side is in contact with the cold distillate. This direct contact allows the immediate condensation of water vapor into coolant on the distillate side.

- **Operating Conditions**: The feed temperature ranges from ambient up to 90°C, whereas the cold solution is typically kept at around 20°C [26].
- **DCMD Advantages**: The simple design works well for experimental setups or other narrow industrial purposes such as arsenic removal, crystallization, and dye effluent treatment [27, 30].
- **Drawbacks**: It has high conductive heat loss through direct contact that may adversely affect energy efficiency, thereby reducing it.

AGMD is a development on the DCMD process, and unlike in it, a gap exists between the membrane and the distillate collector. In this setup, water vapor permeates through the membrane and the air gap before condensing onto a cold surface.

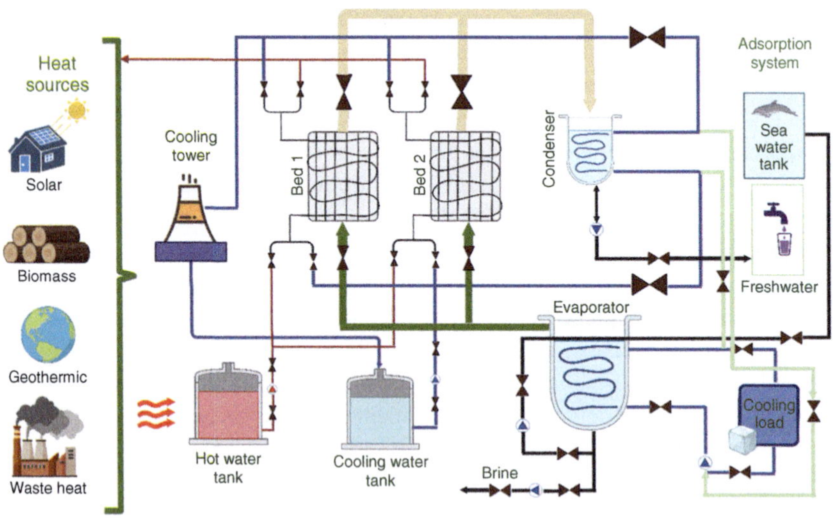

Figure 8.9 Adsorption system for cooling and desalination applications. Source: Alsaman et al. [29]/Springer Nature/CC BY 4.0.

- **Operating Principles**: Due to its presence as an additional thermal barrier, the air gap reduces heat conduction. With respect to the distillate side, there is no direct contact with liquid which lessens concerns over wetting of the membranes.
- **Applications**: Some of these include water desalination, fruit juice concentration, separating azeotropic mixtures, and removing volatile organic compounds from solutions [30, 31].
- **Benefits**: The AGMD design helps in the separation of volatile substances while reducing chances of wetting on the distillate side, making it applicable for different liquid separations.

In the SGMD method, cold inert gas flows over the distilled product side that sweeps away water vapor from the membrane element toward the module outlet where it condenses.

- **Mechanism**: The flow through this vacuum gap acts as an insulator which decreases thermal polarization and loss of heat. The collected vapor gets condensed outside of the membrane module, thus preventing cooling within the system [26, 31].
- **Applications**: It can also be used in the desalination of brackish water, distillation of azeotropic mixtures, wastewater treatment, and removing volatile organic compounds [30, 31].
- **Advantages**: This enables SGMD to minimize thermal polarization that prevents wetting on the distillate side, and this helps to make it more robust.

By choosing the correct MD configuration, operators can customize the desalination process as per their individual needs while ensuring that they balance between efficiency, cost, and application requirements. Hence, each configuration has its own operational benefits, making MD a versatile tool for modern-day desalination processes and separation. This exercise should be based on these rules.

8.2.2.3 MD Module Types

Various module configurations are employed by MD to optimize its performance for different uses. The major features of these configuration are their flat sheet, tubular, and spiral wound types, which have particular advantages and disadvantages with regard to design, operational aspects, and scalability.

Flat sheet modules: These modules are commonly used in small laboratories due to the ease of their use and operation.

- **Structure and Fabrication**: A hydrophobic membrane is placed between two rigid frames which may be made of plastic or metal. This method has made construction and maintenance very simple for the flat sheet module [26].

- **Convenience**: Membranes can be easily replaced, while the whole module is easy to clean, a fact that makes it ideal for small-scale pilot plants as well as laboratory applications.
- **Constraints**: On the other hand, compared to other types of modules, effective membrane areas on this type are relatively low. For this reason, scaling up or increased capacity, rendering it ineffective in commercial applications, is limited [31].
- **Applications**: In research and development scenarios where rapid manipulation of membrane surfaces takes place or membranes need constant changing, the best option would be flat sheet modules.

Tubular modules: They consist of hollow fiber membranes arranged in a shell tube to maximize the effective membrane area, making them useful in commercial processes.

- **High Packing Density**: The hollow fibers facilitate high packing density, adding to the effective membrane area for desalination. It makes tubular modules efficient and therefore suitable even for larger-scale operations [26, 31].
- **Durability and Maintenance**: Tubular modules require special sealing to withstand thermal expansion that ensures long-term durability. Cleaning and replacement of individual fibers may prove challenging. Replacing damaged fibers within the bundle is difficult and often necessitates complete module replacement if repairs are required [31].
- **Applications**: Tubular modules find application in industries or commercial entities with high membrane areas and strong performance needs.

Spiral wound modules: Spiral wound modules comprise multiple layers of membranes wrapped on a central collection pipe located inside a tubular housing.

- **High Area-to-Volume Ratio**: This design has a very high effective surface area, allowing spiral wound modules to be highly efficient as well as space-saving. They balance operational handling, permeation rate, and packing density effectively [26].
- **Operational Efficiency**: Spiral wound modules allow easy handling and offer better control over fouling which is important in achieving consistent performance during desalination processes [31].
- **Flexibility**: Besides being used in traditional desalination methods such as reverse osmosis, spiral wound modules are likewise employed in commercial MD settings because they can be easily scaled up or down.
- **Use Cases**: These are mostly found in large-scale desalination plants and industrial settings where space is a major concern and high flow rates are required.

By selecting the right module design, it is possible to adapt MD systems to different operation requirements which range from laboratory-scale research to industrial scale. Every configuration provided possesses its own unique benefits that can help to improve the efficiency and flexibility of membrane distillation technology.

8.2.2.4 Membrane Classification

Membrane properties in MD are critical in determining the efficiency and effectiveness of the desalting process. Membranes with low thermal conductivity and high hydrophobic characteristics ensure stable separation processes, prevent extensive heat losses, and protect against membrane wetting. These membranes use commonly available materials such as polytetrafluoroethylene (PTFE), polyvinylidene fluoride (PVDF), polypropylene (PP), or polydimethylsiloxane (PDMS). PTFE is, however, the most preferred among these due to its exceptional resistance to both thermal and chemical factors [26–28, 30–32].

Mechanism of Wetting: When liquid enters the membrane pores causing reduced flux or compromised salt rejection, it is referred to as wetting. The more hydrophobic a membrane is, the more resistant it becomes to wetting. Another important factor is liquid entry pressure (LEP) which is the minimum pressure required for a liquid to pass through the pores of a membrane [28].

Types of membranes based on wettability and fouling:

a) Superhydrophobic membranes:
   Characteristics: These membranes have apparent water contact angles greater than 150° exhibiting high resistance to wetting with remarkable self-cleaning properties. As a result, they are highly efficient in maintaining stable flux during long-term MD operations [28].
   Applications: Ideal for processes that require wetting resistance and long-term performance.
b) Omniphobic membranes:
   Features: They are made of materials with lower surface tension than superhydrophobic membranes [26].
   Challenges: The manufacture of omniphobic membranes can be difficult and often uses toxic substances such as fluorine, which pose potential health risks, particularly in potable water production [28].
   Applications: They are suitable for environments exposed to a wide range of organic contaminants.
c) Janus membranes:
   Unique Structure: These membranes have dual wettability properties on opposite sides that can be controlled by altering their surface morphology, charge, and material composition [28].

Advantages: Janus membranes exhibit excellent resistance against both wetting and fouling, ensuring stable performance under harsh conditions [27, 32].
Applications: Their versatility is a key attribute as they find use in various industrial applications where the membrane's two faces encounter different operational conditions.

Membrane fabrication techniques: Different methods may be involved in making the membrane with each method having an impact on its structural as well as functional aspects.

- **Lamination**:
  In this case, one layers materials to create a composite membrane that has different distinct functional layers.
- **Phase Inversion**:
  It involves transforming a polymer solution into a porous membrane structure frequently used to develop the active layer of the membrane.
- **Electrospinning**:
  This approach creates nanofibers with rugged surfaces, big pore diameters, and high porosity. Electrospun membranes are popular for being extremely superhydrophobic and superoleophilic as a result of incorporating hydrophobic particles like PTFE, PDMS, carbon nanotubes, and modified silica. However, their mechanical strength is often lower than expected which is a problem in long-term large-scale applications [32].

Types of membrane materials:

a) Materials for a nanocomposite
   Structure: The hydrophobic particles added to these electrospun nanofibers increase their hydrophobicity and surface roughness.
   Limitations: However, there is the challenge of low strength and difficulty in scaling up the process of electrospinning, making these membranes unsuitable for small-scale desalination [33].
   Improvements: Some of these include SiO_2, TiO_2, or other organic–inorganic nanoparticles such as MWCNTs, graphene that could be incorporated into polymer membranes to enhance them [32].
b) Commercial polymer membranes with modified surfaces:
   Commonly Used Polymers: Surface modification of PVDF enhances MD performance.
   Modification Techniques: An immersion membrane is placed in a titania or silica nanoparticle slurry, deposited onto an adhesive layer, and subsequently fluorinated. This process induces superhydrophobicity or omniphobicity while reducing pore sizes, thereby altering permeability [28].

Applications: Various processes of desalination and separation can benefit from better surface characteristics which provide improved MD performance.

c) Synthesized polymer substrates with modified surfaces:
   Characteristics: These are primarily designed for treating highly saline/oil-contaminated liquids like emulsified oily wastewater and surfactant-laden oilfield-generated water [28].
   Applications: When traditional membranes fail in complex wastewater treatment situations, it is perfect.
d) Changing the surface of inorganic membranes:
   Materials: Usual sources of these include ceramic materials like glass, alumina, and titanium dioxide as well as silica.
   Properties: There is very little effect on them from chemicals and high temperatures. Such a membrane would be chemically stable enough that it could withstand severe environments and heat. In this way, a ceramic membrane can be made hydrophilic or superhydrophobic by means of the fluorination process [28].
   Hence, the decrease in energy efficiency of MD processes is because of the need for ceramic membranes which require temperatures higher than 800°C during curing [28].
   Applications: It can be used where advanced wastewater treatment needs do not meet common membranes.

These novel materials and manufacturing procedures permit MD customization for various applications starting from large-scale desalination plants to lab studies.

8.2.2.5 Fouling and Cleaning of Membranes

Fouling is a very big problem in MD which affects the membrane performance and operational efficiency. This happens when foulants (particles, salts, organic matter, and microorganisms) interact with the membrane surface. In these interactions, there are also van der Waals forces, electrical double layer forces, and charge-related interactions between these foulants and the membrane.

Mechanisms and impacts of fouling:

- **Van der Waals Interactions**:
  These forces are weak attractive forces that exist between molecules or particles on one hand, and on the other hand between them and the surface of the membrane. They play a part in the initial deposition of fouling substances.
- **Electrical Double-Layer Forces**:
  These arise from the presence of electrical charges at the interaction zone or interface between the solution and membranes.

- **Charge Interactions**:
  Fouling behavior also depends on charges that distinguish foulants from membrane surfaces, enabling how particles adhere to it.

Impact on MD performance: The deposition of foulants on the membrane surface can diminish the temperature gradient crucial for the MD process. This drop in temperature difference reduces the driving force for vapor transport, thereby decreasing the overall performance and efficiency of the MD system [27].

Strategies to mitigate fouling:

- **Enhancing Membrane Hydrophobicity**:
  Improving the hydrophilicity of a membrane can prevent liquid from entering into its pores, thus decreasing the fouling occurrence.
- **Optimizing Pore Size**:
  Using membranes with relatively big pore sizes may help preserve a flow rate and reduce the chances of pore fouling by pollutants.
- **Maintaining Low Operating Pressure**:
  It is possible to reduce the formation of a compacted fouling layer on the membrane by operating at lower pressures, which will also prevent foulant compaction onto it.

Types of fouling:

a) Inorganic fouling:
   - **Characteristics**: Resulting from deposits such as hard salt precipitates.
   - **Triggers**: Temperature differences and steam generation cause precipitation of these salts on the surface of a membrane [27, 32, 33].

b) Organic fouling:
   - **Characteristics**: Caused by colloidal organic matter accumulation.
   - **Sources**: Natural organic matter in feed water, industrial effluents, or biological metabolites contribute to organic fouling [9, 12, 13].

c) Biofouling or biological fouling:

- **Properties**: The attachment and growth of microorganisms on the membrane surface leading to biofilm formation can significantly hinder MD performance. **Causes**: The presence of bacteria, algae, and other microorganisms in the feed water [27, 32, 33] triggers biofouling.

Cleaning Membrane Techniques:

a) Chemical cleaning
   - Chemical cleaning which is usually done every few months depending on the membrane life span as well as operating conditions (can by definition only restore some fraction of its original performance).

- Frequent chemical cleaning may result into such membrane side effects like wetting, degradation, and the development of residual deposits that can serve as sites for further fouling [32].

b) Cleaners:
- **Chlorine and Acids**: Also, they can contaminate the reject stream, thereby increasing disposal costs, but are effective at removing scales like calcium carbonate as well as iron oxides.
- **Alkaline Solutions**: Ideal for scrubbing off organic foulants.
- **Sodium Hypochlorite (NaClO)**: Although it has a demerit of oxidizing the polymer structure, causing pore deformation when used frequently to remove biofoulants, thus damaging the membrane.
- **Surfactants**: This form micellar structures around hydrophobic organics like foulants [33].

c) Physical cleaning methods:
- **Air Backwashing and Aeration**: They are techniques used to dislodge foulants off the membrane surface with the help of air bubbles or jets.
- **Vibration/Ultrasound**: Vibration or ultrasonic waves applied to the membrane to remove foulants through mechanical agitation.
- **Water Flushing**: High-pressure water is used to flush away foulants from membrane surfaces.

Efficiency and Limitations: That notwithstanding, as compared to chemicals, physical cleaning methods are less effective, but they do not involve toxic substances and may be gentler on the membrane. In the MD process, however, this requires additional energy input that enhances overall efficiency [33].

Efficient fouling management in MD systems calls for combining strategies of enhancing membrane properties, optimizing operational conditions, and applying suitable cleaning techniques. This can lead to significantly increased sustainability and cost-effectiveness in desalination because longer-lasting and more efficient MD membranes can be achieved by addressing both prevention and remediation of fouling.

8.2.2.6 Pretreatment for the Feed

MD is an adaptable technology used in desalination and wastewater treatment, offering advantages over conventional techniques. However, the fouling of membranes still remains a significant challenge that affects performance and service life. This implies that feed water must be properly pretreated to prevent fouling and extend membrane life. As a result, effective pretreatment can significantly reduce the amount of foulants entering the MD system, increasing total operational stability and efficiency.

The following types are common pretreatment procedures:

a) Absorption:
   Pollutants are captured on the surfaces of materials with high surface area, which attract specific impurities.
   Benefits:
   - High efficacy in removing trace levels of pollutants.
   - Some adsorbents can be chosen for the selective elimination of certain contaminants.
b) Fluctuation and coagulation:
   Coagulation follows where water's colloids and suspended particles become destabilized due to the addition of coagulants (most often metal salts or polymers). Flocculation happens when these small aggregates combine to form bigger flocs that can be easily separated through sedimentation or filtration.
   - Often used for removing colloids, suspended solids, and organic matters from water.
   - Doing this increases the efficiency of pressure-driven processes and decreases fouling in MD, thus reducing the particulate load [33].
   Advantages:
   - Works well in highly organically contaminated, turbid waters.
   - Efficient downstream filtration and membrane distillation (MD) strategies have been put in place.
   Challenges:
   - The cost of coagulation-flocculation may be very high and generate lots of sludge that needs to be disposed of.
   - Careful adjustment of coagulant dosing rates and pH is necessary for the process to work as intended.
c) Prior to filtering:
   The prefilter is used to remove sedimentary solids and larger particles before water enters the MD system through various filter media.
   - These include cartridge filters, MF, UF, or NF membranes.
   - Frequently used in water treatment plants for prevention of overfouling on subsequent membranes.
   - On the other hand, while NF can also take away finer impurities such as divalent ions and dissolved organic matter, MF/UF are excellent at eliminating particles including bacteria or certain viruses [33].
   Benefits:
   - It also helps stabilize the MD membrane by decreasing fouling load on it.
   - Fractionates some specific feed water components based on size exclusion theories.

- Pretreatment solution that remains stable all its life requiring only simple operation without skilled manpower.

Problems:

- To keep performance at its optimum level, routine maintenance including the replacement of filter cartridges is needed.
- Furthermore, the UF and NF sophisticated membrane filtration systems are very expensive to buy and run.
- Optimization and integration:

Integration of these pretreatment methods can optimize MD system operation as well as prolonging its lifespan. Utilized together, these techniques can facilitate safeguarding the MD membrane against diverse foulants and pollutants.

Combining filtration and adsorption:

- Prefiltration addresses particle and colloidal debris while adsorption eliminates dissolved organic matter and heavy metals.
- Typically, this blend enhances the stability and efficiency of the MD process while minimizing fouling load generally.

Filtration after flocculation and coagulation:

- A larger part comprises coagulation-flocculation for aggregating bigger particles that would be removed through fine filtering (MF or UF), hence reducing both suspended solids and colloids effectively.
- In this way, highly turbid feed waters or those loaded with biological materials are treated most efficiently by this method.

These feed waters have uniquely formatted pretreatment: Depending on the properties of the feed water, these methods can be altered. For example, for extremely saline industrial effluents, a combination of adsorption to remove organic contaminants and NF to reduce salt concentration before entering the MD process may prove beneficial. Proper pretreatment is essential in ensuring that MD systems operate economically and last a long time. Adsorption, coagulation-flocculation, and pre-filtration are techniques employed in this respect to decrease the fouling potential of feed water. As a result of this, there will be an improved membrane performance as well as longer operational life expectancy. By combining these tailor-made pretreatment processes for diverse waters, fouling could be managed effectively, hence optimizing the MD process.

8.3 Adsorption Desalination and Hybrid Desalination Systems

The planet will conform to the UN SDGs by 2030 that advocate for pure water [34]. This has been done by the scientific community and others. Desalination is one approach to mitigate the problem of water scarcity in arid zones and coastal areas where there is a limited supply of freshwater [35]. The process through which seawater with high salt concentration becomes suitable for human consumption, irrigation, or industrial use after being converted into freshwater is referred to as desalination [36]. The technology additionally eases pressure on water supplies due to population growth and global warming, thus minimizing overdependence on overexploited freshwater resources and securing clean drinking water throughout the country at all times [37]. This part focuses on two approaches of desalination.

8.3.1 Adsorption Desalination

Many ways of desalination exist, one of which is the process of adsorption. Desalination uses seawater (water with a high salt content) and brackish water (it has more salts than freshwater but less than that in seawater). This physical process involves adhesion occurring between a solid phase material (adsorbent) and adsorbate (the substance to be removed) [38–40]. The efficiency of the adsorption depends on several factors like pore size, presence of functional groups, point of zero charge (PZC), and specific surface area (SSA), among others. Furthermore, it is worth noting that the cost factor for the adsorbents plays an important role as well [39–42]. Adsorption proceeds from adsorption chillers [43] which initially had problems such as low performance and energy use. Since then, researchers have been working to enhance the properties of such absorbents by increasing capacity, thereby making AD viable.

8.3.1.1 Mechanics of Adsorption Desalination

AD operates in two primary stages:

- **Evaporation-Adsorption**: During this phase, water evaporates, and the resulting vapor is adsorbed by the adsorbent.
- **Desorption-Condensation**: Here, the adsorbed water is released (desorbed) from the adsorbent and then condensed into liquid form.

Despite following the basic principles of conventional adsorption, AD involves more complex processes. A standard AD system typically comprises two or more beds, a condenser, and an evaporator [44, 45]. Advances in research have led to increasingly sophisticated designs, often featuring multiple evaporators or stages [29, 46, 47].

8.3.1.2 Energy and Heat Management in AD

During the desalination and cooling process in AD systems, heat generated from adsorption and condensation is dissipated using cooling water supplied through heat exchangers. The energy needed for evaporation comes from an external source [29]. In systems with heat recovery, the heat from condensation can be recycled for evaporation, improving energy efficiency. Additionally, the energy required for AD operations can be sourced from various renewable and sustainable sources such as biomass, solar energy, geothermal energy, and waste heat [29].

8.3.1.3 System Overview

A typical two-bed AD system consists of a condenser, an evaporator, a cooling tower, cooling and heating tanks, a desalinated water tank, water pumps, and valves (Figure 8.9). Heat transfer lines, illustrated in green, facilitate the movement of heat within the system [29].

8.3.1.4 Research on Adsorption-Based Water Desalination

Moreover, these two methods are energy-intensive and operationally difficult. Consequently, scientists were compelled to explore the applicability of adsorption which is one of the simplest ways for seawater desalination.

8.3.1.5 Desalinating Using Carbon Nanotubes

The unique surface properties and small size of carbon nanotubes (CNTs) make them a potential option for seawater desalination [48, 49]. They are hydrophobic and have large accessible areas so as to eliminate impurities from saline water and retain salts [50]. However, the direct application of CNTs poses practical limitations especially in the case of large scale seawater treatment as it requires not only costly processing, but also development and procurement [50, 51]. In order to overcome this limitation, researchers attempted embedding CNTs into membranes or matrices [50, 51].

CNTs are more manageable and efficient when used in a matrix or membrane. These composite materials extract salt using CNT's electrostatic interaction, making them appropriate choices for desalination. By enhancing the efficiency and scalability of adsorption, this method also eases the use of CNTs. Our attention is on these state-of-the-art materials combined with their integration into useful systems to provide adsorption-based desalination technologies that consume less energy than traditional methods while being easier to operate.

In 2013, Yang et al. [52] carried out a study on the adsorption capacity of plasma modified ultralong carbon nanotubes (UCNTs) for water desalination. They coated UCNTs on porous substrates composed of mixed cellulose ester (MCE). The composite device has demonstrated significant success in desalinating low to medium salt concentration waters. More investigation and testing of this system using seawater are required, as indicated by the scientists [52].

The giant pore volumes found in this activated carbon make it an attractive alternative for CDI electrodes, resulting in a substantial enhancement in desalination efficiency, according to Zhang et al. [53]. Askalany et al. [54] in 2020 proposed a two-stage ejector cycle method aimed at increasing the production rate and performance of desalinated water. The best half-cycle period required by this novel system is 400 seconds, with which freshwater output can increase up to 23 m^3 per ton of silica gel every day. With a COP value equal to 1.64, it could be three times more efficient than some existing AD systems.

The data shows that a two-stage ejector cycle is an innovative invention in the desalination industry, providing a more efficient and cost-effective alternative to the traditional AD technologies. In 2014, Mitra et al. [55] carried out simulation experiments on a four-bed single-stage system combined with silica gel adsorption desalination. It was found that for this system to operate effectively, both the desalination and cooling capacity need to be maximized, with an ideal average cycle time of 600–900 seconds. In addition, COP increased as the cycle times rose. This also reduces energy consumption by approximately 25%. This implies that cheaper and eco-friendly desalination processes can be realized.

8.3.1.6 Natural Zeolites and Innovative Adsorbent Materials in Desalination

The adsorbent characteristics of natural zeolites for desalination have been extensively studied [56–58]. For their higher ability to exchange ions and selectivity for different ions, these minerals are now being recognized in water desalination. A remarkable study by Wajima et al. explored the capability of an Iizaka deposit from Fukushima as a natural zeolite for desalination [59]. They achieved an 81% NaCl removal during batch tests. To illustrate that their desalination process could be practically applied, they employed this treated water in the growth of radishes over a ten-day period. Although they observed that the water quality was suitable for use in agriculture, they realized that it would need further treatment before being fit for human or industrial consumption.

For desalinating seawater, Wibowo et al. [60] investigated the use of natural zeolite from Sukabumi, Bandung, Indonesia. In addition, they discovered that zeolite's adsorption capacity could be significantly improved through thermal activation. To do this experiment, they heated the zeolite for three hours at 225°C. The researcher carried out a study using different masses of thermally activated zeolite in a batch process with 100 mL saline water from the Java Sea containing an initial salt content of 35 ppt. The outcome revealed a satisfying result, whereby only 7.5 g of zeolite powder removed 3.2 ppt, equivalent to approximately a 91% desalination rate.

8.3.1.7 Enhanced Adsorption Capacity with Metal–Organic Networks of the ZIF Family

Metal–organic frameworks (MOFs) based on imidazolate zeolites (ZIFs) have been incorporated into polymeric materials to promote the adsorption capacity of metal ions. In this sense, Li et al. [61] generated a composite material based on polydopamine (PDA) with this type of nanostructured material, evidencing potential properties for application in desalination.

8.3.1.8 Advances in Adsorbent Composites and Innovations in Desalination Systems

Baiju et al. [62] investigated desalination using three different types of adsorbent composites, focusing on silica gel, MOFs, and polyvinylpyrrolidone (PVP) as a binder. The pore size of these composites was determined to be less than 2 nm, making them microporous materials. This type of composite material demonstrated excellent performance in terms of mechanics and thermal stability, resulting in materials with potential application in desalination.

8.3.1.9 Theoretical Study on Heat Transfer Enhancement

Askalany et al. [63] introduced an extraordinary concept that could enhance the heat transfer coefficient of the adsorbent bed to make ADs more effective. In this study, they did an experiment in which they replaced 30% of the adsorptive base with a hypothetical aluminum species mixed with silica gel.

Due to enhanced rates of heat exchange, daily freshwater production would be expected to rise considerably. The system can manufacture up to 17 m^3 of silica gel per ton per day in the absence of recuperative heat. It is interesting that it may become as high as 42 m^3/day when recovery is applied. These breakthroughs might lead to a fourfold increase in classical adsorption desalination technologies and revolutionize the industry process.

8.3.1.10 Capacitive Deionization Technology

The project that was carried out in 2023 [64] demonstrates how the structure of different adsorbent materials is connected to their properties. It was therefore decided to define this feature in industry as a desalination coefficient or capacity (CDI). CDI depends on this nature of the material under study, which determines the electrodes that are formed using various materials having an optimum CDI for demineralization purposes through water electrolysis through electrochemical measurement methods.

8.3.1.11 Vermiculite as a Promising 2D Material

Among the promising materials for CDI and other adsorption processes is vermiculite (VMT). Upon heating, VMT transforms into a porous material with excellent adsorption properties [65, 66]. Recently, Zhao et al. [66] investigated the potential of a $MgCl_2$/VMT composite for water vapor adsorption from the atmosphere, comparing it to other materials. The composite with 31% salt content outperformed others, particularly in its sensitivity to humidity. The adsorption capacity of $MgCl_2$/VMT increased up to 2.5 times as humidity rose from

50% to 80%. Remarkably, this composite demonstrated durability, maintaining an average adsorption capacity of 0.3275 g/g over 10 adsorption–desorption cycles. These characteristics suggest that $MgCl_2$/VMT could be highly effective in desalination applications.

8.3.1.12 Mechanisms of Water Vapor Adsorption and Capillary Adsorption

The study of water vapor adsorption in porous materials like VMT points to their potential in desalination through capillary adsorption. This process leverages the macroscopic concept of capillary-induced fluid transport at the nanoscale, where the adsorption and movement of water are driven by capillary forces within the porous structure [68, 69].

Hwang et al. concurrently studied VMT and In Silica (MPS) using a simulator [70]. A complete kinetic study of this research was performed, and isothermal models were used to compare various pore sizes of MPS.

Hwang et al. revealed that MPS had an adsorption capacity of 0.57 kg/kg, signifying its high potential for application in desalination systems operated by adsorption. The encouraging outcomes obtained from both VMT and MPS in the course of adsorption underscore their possibility of significantly changing desalting techniques.

8.3.1.13 Enhancements in Adsorption-Cooling and Desalination Systems

A versatile approach to the current water shortage with a sustainable approach and less demanding application conditions is the adsorption-cooling and desalination (ACD) process. The use of activated carbon (Max) represents a great advantage in this approach. By combining activated carbon with ammonium carbonate, it has been possible to desalinate around 13 cubic meters of water per ton of this adsorbent material by an ACD process [29]. This highlights Max as a potent adsorbent material for enhancing ACD systems. AD supports global efforts to mitigate water scarcity while reducing reliance on conventional, energy-intensive desalination processes. Utilizing renewable energy sources in these systems aligns with SDGs, offering a greener alternative for water production.

8.3.2 Hybrid Desalination Systems

HDSs have emerged as a pivotal innovation in the search for sustainable and efficient methods to produce potable water, particularly in regions facing acute water scarcity [35]. These systems integrate various water treatment technologies, such as RO, MD, and advanced adsorption techniques, to enhance the overall effectiveness of the desalination process [71, 72]. By leveraging the synergies among multiple technologies, hybrid systems achieve superior efficiency in removing salts and contaminants from water, adapting flexibly to different water sources and varying feed water quality [73].

8.3.2.1 Benefits of Hybrid Desalination Systems

- **Energy Savings**: The combination of desalination processes reduces energy use since the residual energy of each process is used [74, 75]. It has been widely reported that the use of MD is associated with energy efficiency in desalination [76].
- **Better Water Quality**: The combination of several desalination techniques results in water with an improved degree of purity since various contaminants are removed by each technique used [77].
- **Process Optimization**: Hybrid systems can be applied in different application scenarios from seawater and river water, taking advantage of the energy and infrastructure that characterize these systems. This can be leveraged for current water scarcity, contributing to sustainable solutions for water use [34, 73, 74].

8.3.2.2 Challenges and Considerations

There are certain teething problems relating to HDSs, yet they have advantages.

- **Technical Complexity**: The development and maintenance of a system that brings together several technologies in one necessitates highly skilled experience in technology [75]. As complex as it may be, the situation can lead to increased costs of running which will need training and updating of skills among personnel.
- **Exorbitant Installation Costs**: Hybrid systems require a lot of upfront capital due to the necessity to combine numerous technologies [75].
- **Handling Waste and By-Products**: Hybrid systems frequently produce multiple waste streams and by-products. Since different emissions require special handling procedures and disposal mechanisms under environmental regulations, control is more complicated unlike emissions from monotechnological systems [75].

8.3.2.3 Advantages of Hybrid Desalination Systems

HDSs are significantly more efficient, cost-effective, and environmentally friendly than traditional methods. This means that such systems recover more water and consume less energy as well. As a result, they are sustainable in the long run [77–80]. They can be used for different sources of water and operating conditions that will make them suitable for many parts of the world [76].

Hybrid systems' performance is also boosted by advanced technologies such as nanotechnology, MOFs, and graphene membranes to enhance the overall effectiveness of the process used in removing contaminants, including chemicals, from seawater or saline solutions [81, 82].

There are, however, challenges faced in terms of their operational costs and energy consumption [74]. The latest developments in science concentrate on improving efficiency to reduce costs and minimize negative environmental consequences. Key areas of improvement include:

- **Energy Sources**: Exploring renewable energy sources.
- **Materials**: Developing advanced materials with higher durability and efficiency to optimize desalination processes.
- **Chemicals**: Researching eco-friendly chemicals and treatment methods to mitigate environmental impacts during desalination operations.

Investing in HDSs not only addresses water scarcity effectively but also promotes sustainability and resilience in water resource management. By leveraging advanced technologies and optimizing operational strategies, these systems play a crucial role in ensuring reliable access to clean water while minimizing ecological footprint.

8.3.2.4 Energy Sources Used in Hybrid Desalination Systems

The pioneers in establishing desalination sites are definitely the countries of the Middle East (United Arab Emirates, Saudi Arabia, and the Persian Gulf), largely because they are located in oil-rich areas, allowing the plants to use thermal energy for their operation. Approximately 50% of desalination plants are situated in this region due to their evident problems with freshwater availability [83]. However, there are other countries where this technology is gaining significant importance, such as the United States of America, Spain, Israel, Algeria, India, Australia, and China. As mentioned earlier, one of the disadvantages of desalination plants is the high energy consumption and the use of fossil fuels. For this reason, hybrid methods have been developed that include the use of renewable energy for these purposes. Table 8.2 shows four types of energy sources used, as well as the desalination methods that have been adapted to these sources.

8.3.2.5 Advances in Hybrid Desalination Systems Utilizing Renewable Energy

Chaouki et al. explored a hybrid desalination system combining adsorption and multiple effect distillation (MEDAD) powered by solar energy. This study aimed to reduce operational costs and enhance productivity in seawater desalination using renewable energy. Key parameters investigated included system stages, additional desalination stages, and heat recovery from residual brines. Results indicated a 2.6-fold increase in system efficiency and a 57% reduction in specific energy consumption due to the adsorption stage addition. The freshwater production rate increased by 14%, with a 12.8% improvement in performance ratio, highlighting the competitiveness of this system for seawater desalination [20].

Bamufleh et al. [84] proposed an optimization approach for a hybrid system combining MED and MD. The system focused on treating brine from MED through MD to reduce brine discharge and enhance freshwater recovery.

Table 8.2 Different desalination methods with renewable energy resources.

Renewable energy resources	Geothermal	Heat	Multistage flash	
			MED	
			VC	
		Electricity	RO	
			MD	
			Electrodialysis (EDy)	
	Solar	**Photovoltaic**	RO	
			MD	
			EDy	
		Solar thermal	Electricity	RO
				MD
				EDy
			Indirect process	MSF
				MED
				VC
	Wind	**Electricity**	RO	
			EDy	
	Renewable energy	**Electricity**	RO	

Additionally, excess heat generated was utilized to lower cooling and heating loads, integrating industrial facility efficiencies into the desalination process.

González-Bravo et al. [85] evaluated various energy sources including fossil fuels, biofuels, and solar energy for desalination processes. Their findings underscored favorable cost-benefit ratios, environmental benefits, and significant energy savings achieved through integrated heating and cooling technologies.

8.3.2.6 Innovations in Hybrid Desalination Systems

Rajesh et al. [86] conducted a study in Vellore, India, where the system effectively utilized waste heat from the panel to preheat saline water for the system, enhancing freshwater yield. Brine from the system was reused to improve freshwater extraction in the system. Experimental results showed an electrical efficiency of the panel 0.4% higher than a reference panel. The hybrid system achieved an average freshwater production of 6,400 to 7,450 mL/m^2 with overall efficiencies between 61.17% and 67.71%, exergy efficiencies from 17.16% to 17.87%, and a unit water cost ranging from 0.09 to 0.082 USD per liter [86].

Abhishek et al. proposed a hybrid system combining AD and open humidification–dehumidification (OHD) processes to enhance water capacity compared to conventional AD systems. The AD-OHD system was evaluated by conducting the various operating conditions using an OHD chamber and a single-bed AD. Its precise switching on and off periods, for example, demonstrated its best performance with a desalinated water productivity rate of 1.12 L/h and GOR of 0.78 which is achievable only at this mode. In this way, the data provided in this study showed that AD-OHD is better than traditional systems (AD—80%, OHD—50%) in terms of desalinated water production [87].

These studies show how hybrids can provide new ways to create more effective desalination plants. That is why, instead of simply increasing freshwater production rates, using PVT panels in combination with advanced process schemes, for instance, AD-OHD, also improves overall system efficiencies as well as cost effectiveness in addressing water scarcity challenges.

AD efficiency was enhanced by blending AD with new desalination technologies in the study conducted by Baiju et al. [88]. The design of a decentralized hybrid system, which involves the application of TED and TECs on the cold side of traditional AD, is the outcome. This changes how AD can be done at higher temperatures and pressures than older ones that generally operated only between 35°C and 65°C (evaporator temperature) where the heat side of the TEC couples to the AD evaporator.

Water generation in different modes and the performance coefficient were also considered when developing the small-scale hybrid system (AD-TED). Under both humidified and non-humidified conditions, water production rates were equal at 1.05 and 1.15 L/h, respectively, for closed-flow mode operation. From this, it would appear that even an open or closed flow configuration does not affect performance [88].

The world's freshwater resources are endangered by overextraction, pollution, and climate change; hence, there is a need for more research on adsorption and HDSs. Some of the key advantages of adsorption desalination include reduced energy usage, low costs, and ability to operate at lower temperatures, making it a better alternative to the widely used reverse osmosis technologies.

To improve flexibility as well as efficiency in water purification, MD or solar desalination can be combined with adsorption methods in hybrid systems. As such, these options have potential for modification depending on various temperature zones and geographical regions regarding availability of local resources. This calls for sufficient studies and funds toward addressing the global problem of water shortages. The contemporary technique of desalination offers promise for ecological scalability while ensuring material affluence toward saving mankind's existence on Earth through sustainable distribution of scalable green solutions, thereby promising that potable water will not be a luxury item for future generations anywhere in the world.

8.4 Low-Temperature Thermal Desalination (LTTD)

Some 40% of the world's population is currently impacted by water scarcity, a number expected to increase due to rising temperatures, rapid population growth, urbanization, and changing consumption patterns. As previously emphasized in this chapter, desalination stands out as one of the most promising solutions to mitigate water scarcity. Desalination involves the removal of salts and minerals from saline water, transforming it into potable water. The classification of saline water is based on its total dissolved solid (TDS) content, with seawater typically containing the highest levels, depending on the source [89]. Currently, the most widely used desalination processes include distillation, RO, and electrodialysis. However, a diverse array of desalination techniques exists, necessitating a clear classification. Desalination systems can be categorized based on the energy source they utilize—thermal, mechanical, electrical, or chemical—or the specific method employed for desalination, such as evaporation-condensation, filtration, and crystallization processes [90, 91].

Globally, RO and thermal distillation processes—specifically MED and MSF desalinations are the most prevalent desalination methods. RO typically consumes about 5 kWh/m^3, whereas thermal processes like MED and MSF require between 10 and 15 kWh/m^3 [92]. Despite the higher energy consumption, thermal desalination processes remain viable due to their reliable performance and minimal need for extensive pretreatment. In thermal desalination, saline water is heated and evaporated upon contact with a hot surface, leaving behind the salts. The water vapor is then condensed in cooling pipes to produce freshwater [12].

One of the biggest problems with these techniques is that they usually require high operating temperatures, causing large heat losses and lowering system efficiency. However, LTTD provides a better energy-saving option. In LTTD, the evaporation energy comes from seawater, making it more sustainable. India has been in the forefront in coming up with LTTD technology for desalination processes. Thermocline-driven desalination or alternatively vacuum distillation or flash evaporation at nearly ambient temperatures and low pressure as compared to those levels reduces water's energy conversion requirements from liquid to gaseous states [93]. This reduces the total amount of electricity required by such systems, making them particularly suited for remote areas where both power and freshwater are valuable [94].

To get it to the point where it can be used in industry, however, some of its drawbacks will need to be dealt with. Nevertheless, LTTD systems during this process still have a purity level that is higher than other desalination processes. What makes them different from other desalination methods is their ability to operate at near-ambient temperature conditions, which minimizes heat loss, saves energy, and increases output while consuming less fuel if they are coupled with renewable energy sources. One example is the Ocean Thermal Energy Conversion (OTEC) integration [95]. The heat generated from desalination powers a turbine for electricity generation [96]. In terms of thermal desalination procedures, flash evaporation and vacuum distillation are worth mentioning because they utilize two separate sources of heat. On the other hand, flash evaporation could be used in large-scale production due to its capacity to produce large water volumes.

LTTD harnesses the temperature gradient between two bodies of water to efficiently produce freshwater. In this process, warmer surface water is evaporated at low pressures—below the saturated vapor pressure of water—and the resulting vapor is condensed using cooler deep-sea water. This method not only yields high-quality freshwater but also contributes to marine life by enriching the returned deep water, which is high in minerals and plankton, thus creating potential breeding grounds for various marine species [97, 98]. Figure 8.10 illustrates the primary components of LTTD technology. It shows the dual piping system for extracting hot and cold water, pumps for water transport and pressure maintenance, a vacuum pump, an evaporation chamber equipped with a demister, and a condenser with a freshwater outlet [99, 100].

The operational process begins with hot surface water being pressurized under atmospheric conditions and passing through the vacuum pump. By working at low pressures and utilizing a modest temperature gradient, the vapor pressure is elevated, which reduces the boiling point of the water, allowing it to evaporate.

In the vacuum chamber, surface seawater passes through the evaporator using heat exchangers. Here, the seawater evaporates at the bottom of the chamber, undergoes a phase change, and moves as vapor to the condensation chamber. The cooling effect provided by the deep-sea water, pumped from lower depths, condenses the vapor back into freshwater [101].

8.4.1 Challenges and Innovations in Implementing LTTD Technology

Implementing LTTD technology presents several challenges, primarily due to the need for long high-density polyethylene (HDPE) pipes to extract cold water from significant ocean depths. Suitable temperatures for this process are typically found at depths ranging from 350 m to nearly 1,000 m, depending on the location. These pipelines can either be buried in the seabed or supported on elevated structures. Burying requires extensive underwater dredging, while trestle support demands massive substructures, both of which are costly undertakings. To address these complexities, Prakash and collaborators conducted comprehensive analyses on

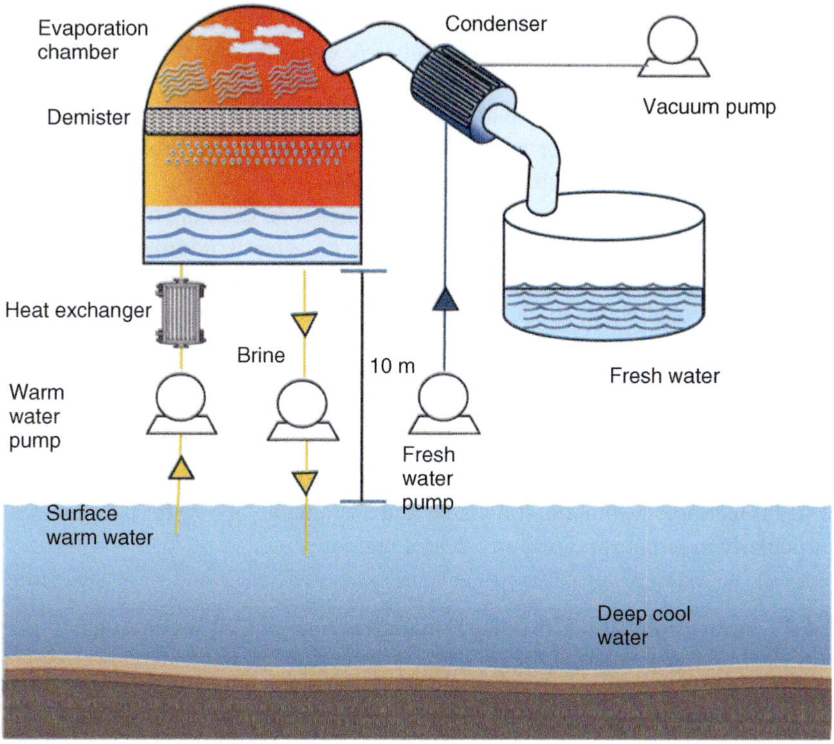

Figure 8.10 General design of the LTTD process.

various sections of the pipeline under different oceanic conditions. Their study focused on the impact of varying flow speeds and axial tensions, especially considering the vortex induced vibration (VIV) phenomenon, which can significantly affect the structural integrity of the pipes [102]. HDPE pipes are favored in these applications due to their tightness against leaks, ease of installation, and favorable thermal conductivity. However, they are susceptible to fatigue damage caused by oscillations from marine currents. To mitigate these oscillations, steel weights are often used at strategic points along the pipe. Yet, these measures alone do not adequately address the in-line VIV.

In response, Saravanan explored an innovative solution involving helical strakes. A comparative analysis between numerical simulations and a physical pipeline prototype demonstrated that partial-length helical strakes significantly suppress VIV responses more effectively than steel clamps placed at two different regions. This innovative approach substantially increases the lifespan of the pipes, making them more reliable for LTTD operations [103].

Another operational challenge is the initial pumping of cold water from the depths to the condenser. Additionally, constructing infrastructure on fragile seabeds can be problematic. Innovations such as floating platforms and ship-supported offshore pilot plants have been developed to overcome these difficulties. These floating or mobile units provide flexible and robust solutions where traditional fixed structures are not feasible.

LTTD technology has had a transformative impact on the islands of the Indian Ocean. The implementation of this technology has greatly improved the health and well-being of residents by providing a reliable source of freshwater. Currently, several operational plants in the region showcase the versatility and effectiveness of LTTD. These include laboratory and experimental facilities, a pilot plant on Kavaratti Island, a floating plant off the mainland, and technology demonstration sites at thermal power stations in South India, such as the land-based plant at the North Chennai Thermal Power Station [97, 98]. These advancements highlight how innovative engineering solutions can address the inherent challenges of LTTD technology, making it a viable and sustainable option for desalination, particularly in remote and resource-scarce regions.

8.4.2 Innovations and Efficiency Enhancements in LTTD Technology

In a groundbreaking study, Abraham et al. [7] harnessed the temperature gradient between warm surface water and cooler deep marine water to develop a model desalination plant. This pilot plant was designed to demonstrate the feasibility of LTTD technology. Remarkably, during its initial trials, the plant exceeded expectations by producing approximately 20% more freshwater than projected. It also achieved a competitive production cost of about 0.2 US cents per liter. These promising outcomes led to the successful implementation of a floating desalination plant model in 2007, which operated for over 20 days for experimental evaluations.

The focus of these evaluations was on optimizing critical components of the system. Key areas of investigation included the vacuum system, the orientation of nozzles within the flash chamber, and the dimensions of the flash chamber itself. Each of these factors plays a crucial role in enhancing energy efficiency and overall system performance [104]. The studies underscored the versatility of the thermocline-driven system, not just for desalination but also for potential applications in energy generation, aquaculture, and air-conditioning.

8.4.2.1 Key Factors Influencing LTTD Efficiency

Several factors influence the efficiency of freshwater production in LTTD systems. These include the vacuum load, operating pressure, deaeration process, mass flow rate, and the size and shape of the condenser. Each factor impacts different stages of the desalination process. Understanding and optimizing these parameters are essential for improving system performance.

From an energy perspective, the two most significant factors are the water-cooling and vacuum systems, which together account for approximately 60% of the total energy consumption in an LTTD system. Maintaining optimal operating conditions in the vacuum system is particularly critical. The effectiveness of the vacuum system is influenced by several factors:

- **Water Vapor Escape**: Approximately 60.4% of the vacuum load comes from water vapor escaping the condenser.
- **Non-Condensable Gases**: These gases contribute about 37.9% to the vacuum load.
- **Air Leaks**: Air leaks, primarily through flanges, gaskets, and other instruments, account for about 1.5% of the vacuum load [105].

Improving the efficiency of LTTD systems requires lowering these vacuum loads. Innovations in minimizing vapor escape, managing non-condensable gases, and sealing air leaks can significantly improve the overall performance and energy efficiency of the desalination process.

8.4.3 Experimental Findings and Design Improvements

The experimental studies conducted on the floating desalination plant highlighted several design improvements. For instance, optimizing the orientation and size of nozzles in the flash chamber can significantly reduce energy consumption and enhance water production. Additionally, careful management of the vacuum system, including controlling air leaks and non-condensable gases, is vital for maintaining efficient operation.

The research conducted by Abraham et al. demonstrates the potential of thermocline-driven systems beyond just water desalination. By efficiently utilizing the natural temperature gradients in ocean waters, these systems can also be adapted for sustainable energy generation, supporting aquaculture, and providing air-conditioning. This multifaceted approach positions LTTD technology as a versatile and sustainable solution for addressing various challenges in remote and resource-scarce regions [104, 105].

8.4.3.1 Enhancements and Innovations in LTTD Systems

In some advanced LTTD models, the combustion chamber is ingeniously positioned about 10 m above sea level. This design leverages natural atmospheric pressure differences to create a vacuum in the upper sections of the feed and extraction columns for desalinated water. By utilizing this natural vacuum, these systems can operate without the need for mechanical pumping after the initial start-up phase.

However, traditional thermal desalination systems, especially those utilizing low-quality energy, often rely on mechanical vacuum pumps throughout the process. These pumps are crucial for pressurizing seawater and maintaining the vacuum conditions necessary for efficient operation. A jet pump is commonly used to remove non-condensable gases (NCs) such as nitrogen, oxygen, and argon, which are by-products of the desalination process and can impede heat transfer in the condenser, thereby increasing the energy demands of the system and operational costs [106, 107].

8.4.3.2 Reducing Energy Consumption in LTTD Systems

Vacuum maintenance usually needs so much energy, thus motivating current attempts to invent alternative methods for lowering energy usage. One such recent attempt is permeate gap membrane distillation (PGMD). In this case, a hydrophobic membrane separates vapor from seawater. The system architecture includes a region with clean water between the membrane and the condensation plate. On one side of the membrane, hot seawater flows, while on the other side, there is cool water along the surface of the condensation plate. As a result, vacuum pumps are not required and it operates at low pressures, hence saving 22 kWh/m^3 as compared to 2.5 kWh/m^3 that PGMD consumes [108].

8.4.3.3 Environmental and Ecological Considerations

When choosing between desalination methods, it is important to consider, not just energy consumption but also the environmental and ecological implications. A few of the benefits of LTTD over RO can be deduced from comparative studies. Unlike the RO treatment process, LTTD systems do not result in chemical pollution and only have minor influence on ornamental fish as well as other wildlife species existing within marine environments. Furthermore, unlike ROs, they cause no harm to marine life and require less training to operate, thus making them suitable for most regions [109].

8.4.3.4 Optimizing Water Quality in LTTD Systems

LTTD plants; water quality improvement methods have been explored by Balaji and his associates [110] with a focus on the salinity variations of produced water due to tidal changes. There were two main modifications that their research aimed at achieving in the combustion chamber configuration: altering the geometry of the spout pipe and adjusting the elevation of the demister. They noted that increasing the height of the demister significantly improved water quality. Raising it from 2.25 to 2.7 m reduces the entrainment of saltwater droplets into the condenser, hence promoting their coalescence instead, thus stopping them from passing through into the condenser and contaminating the freshwater [110].

8.4.4 Advanced Mathematical Models and Heat Transfer Optimization in LTTD

The understanding and optimization of the evaporation and condensation processes have a major impact on the efficiency and performance of LTTD systems. To estimate steam output and validate experimental data, numerous mathematical models have been created to predict the instantaneous rates of seawater droplet evaporation.

8.4.4.1 Evaporation Rate Models

In a thermal fluidic model, Karmakar et al. [111] constructed an LTTD system that included a flash drum and a vacuum spray nozzle system. The authors' findings showed that the application of spray nozzles can highly increase the performance of the device by reducing droplet size and pressure. Verification of this model with experimental results from existing literature was found to be in good agreement with the experimental data. In these systems, increasing seawater temperature by 10–20°C above ambient conditions causes the evaporation of water from the evaporating chamber into the condenser section. The condenser is meant to allow heat transfer for cooling the vapor back into a liquid state.

8.4.4.2 Heat Transfer in Condensers

Balaji et al. [110] investigated the impact of the surface area of the condenser on the efficiency of the heat transfer. They compared experimental data with mathematical models generated by the Heat Transfer Research Institute (HTRI) software version 6.0 and the Kern mathematical model. Key parameters analyzed included the overall heat transfer coefficient and the coefficients for both the tube and shell sides.

8.4.4.3 Heat Exchanger Design and Optimization

Heat exchangers play a vital role in enabling the LTTD system to transfer heat across liquids. These shell and tube heat exchangers will commonly warm briny water before it goes into the chamber for combustion. However, because of the vibrations, these heat exchangers suffer from corrosion and excessive bending strains. In optimizing its design, one must take care about such elements like fluid-elastic instability, the appearance of a vortex, and the acoustic resonance of the tube bundle [111]. Karmakar et al. have proposed that in order to improve the efficiency of possible implementation using solar energy as a driver for heating seawater temperatures from 25°C up to 65°C prior to being fed into the vaporizer [111].

8.4.4.4 Enhancing Condenser Efficiency

To optimize the design of shell and tube condensers, Elakkiyadasan et al. [112] considered critical factors. Their research showed that increasing the water flow rate leads to increased heat transfer coefficient values. From this relationship, it is possible to deduce that increasing flow velocity minimizes pressure drop and enhances condenser performance. However, it also requires larger pumps for which higher costs and power consumption are associated. Hence, there should be a trade-off between maximizing heat transfer efficiency and minimizing energy usage.

8.4.5 Integrating Renewable Energy and Industrial Heat Sources in LTTD

LTTD technology leverages the natural temperature gradient between the layers of ocean to produce freshwater. However, beyond these natural gradients, LTTD can be combined with heat sources from industrial processes or supported by renewable energy technologies to enhance its efficiency and application scope.

8.4.5.1 Utilizing Waste Heat from Industrial Processes

LTTD systems can capitalize on waste heat from industrial processes, which often discharge water at high temperatures. By harnessing this thermal energy, it is possible to create the necessary temperature differential for desalination. Table 8.3 outlines several examples where LTTD is augmented by technologies such as spray desalination, flash desalination, and vapor desalination, demonstrating how combining these methods can improve desalination efficiency and output.

8.4.5.2 Solar Energy Integration

One of the most prominent renewable energy sources for thermal desalination is solar energy. Solar energy is advantageous because it provides both the heat and electricity required for desalination processes. Given that LTTD operates at lower temperatures, it aligns well with the intermittent nature of solar power as it minimizes energy losses and utilizes the residual heat without generating greenhouse gas emissions. To address the intermittency of solar energy, thermal energy storage (TES) systems are incorporated into LTTD setups.

Types of TES systems:

- **Sensitive Heat Storage**: Stores energy by altering the temperature of the storage medium. This method is commonly used in LTTD processes, with operational temperatures for desalination typically ranging between 60 and 80°C [120].
- **Latent Heat Storage**: Involves storing energy during phase changes, such as from solid to liquid.
- **Thermochemical Storage**: Uses chemical reactions to store and release energy.

Among these, sensitive heat storage is the most widely used due to its simplicity and efficiency in storing thermal energy.

Table 8.3 Technologies utilizing low temperatures for water desalination.

Technology	Method	Objective	Conclusion	References
Spray assisted	Mathematical modeling	Spray-assisted LTTD systems are thoroughly analyzed and optimized across different operational parameters.	A MED outperforms a single stage, achieving superior thermal efficiency.	[113]
Low-temperature thermal by thermal energy storage	Mathematical model	LTTD system—thermal energy with storage tank based on the solar collector.	Production of 100 L/d of freshwater.	[114]
Flash vaporization	Pilot plant and mathematical model	Flash vaporization experimentally feasible.	Presenting a low-energy desalination approach.	[115]
Vapor diffusion	Mathematical model and experimental test rig	Freshwater yield variations under different conditions.	Good theoretical predictions.	[116]
Vacuum desalination	Prototype	Enhance the performance of a condenser tube in desalination by incorporating fins on its outer surface.	Demonstrates a substantial impact on condensate production rates.	[117]
Directional solvent extraction (DSE)	Computational framework	Assess the techno-economic viability of using fatty acid and ionic liquid.	Optimizing the process and conducting rapid sensitivity analysis.	[118]
Low-temperature electrolysis	Software	Evaluate an innovative energy to generate thermal energy.	Utilization of solar energy via parabolic trough solar collectors and thermal storage tanks.	[119]

8.4.5.3 Case Studies and Prototypes

Gude developed a prototype that operates within a temperature range of 40–50°C, suitable for low-demand thermal energy sources like solar power or industrial waste heat. This system exhibited an energy consumption of 0.96 kWh/kg and an efficiency of 69%, producing around 0.31 kg/h of freshwater. While this may seem modest, it represents a significant improvement over single-stage solar desalination technologies, which consume about 1.4 kWh/kg. The specific energy requirement of the prototype of Gude could drop to 0.6 kWh/kg under optimal conditions, making it an efficient solution compared to conventional methods. Increasing the evaporation temperature can further enhance water production, albeit with a rise in energy consumption [114].

Askalany and colleagues [63] proposed using waste heat from internal combustion engines to power a desalination system. They explored the impacts of various parameters, including evaporator and condenser temperatures, vacuum pressure, and flow rates. Their findings indicated that lower vacuum pressures and condenser temperatures (from 28°C to 20°C) significantly boost freshwater production [63].

8.4.6 Advantages and Future Trends in LTTD

- **Relying on Renewable Energy**: By using solar and ocean renewable energy, these systems become more sustainable and less dependent on fossil fuels.
- **Ease of Operation**: These systems have less complicated operational requirements because they have simpler designs, easier maintenance needs, and lower chemical pretreatment needs.
- **Benefits for the Environment**: LTTD uses fewer chemicals and has reduced energy requirements thus reducing maintenance concerns as well as environmental problems. Furthermore, integration with membrane technology may enable safe disposal of brine into the ocean, hence reducing its impact on the environment.
- **Reduction of Thermal Pollution**: Combining LTTD with industrial heat sources can directly stop hot water flow into the ocean, thereby protecting marine ecosystems from industrial heat pollution.
- Additionally, these systems combine LTTD with renewable forms of energy and waste industrial heat, providing a safe, sustainable answer to water scarcity especially in places rich in thermal resources but poor in freshwater [116–119].

8.5 Perspectives and Conclusion

In this chapter, we argue that there is a glimmer of hope for finding sustainable water solutions to SDGs through new desalination methods that leverage waste heat. Examples are MD, MED, AD, HDS, and LTTD which exploit different advanced industrial wastes, solar energy, and geothermal sources to produce freshwater efficiently and in an eco-friendly manner. Freshwater can be produced from industrial waste heat using the invention, which was previously considered as a waste product, making it possible for everyone who needs cheap and safe drinking water to have access to it. The technologies are significant because they provide both ecological soundness and resource productivity [121, 122]. Additionally, a part of the energy lost during processing was reclaimed as waste heat back into the environment, thus helping decrease thermal pollution and carbon emissions related to efforts aimed at mitigating climate change. Combining already-existing industrial processes with water desalination also promotes sustainable industrial practices by maximizing resource utilization and reducing environmental impact. Nonetheless, there are several crucial concerns that need to be addressed so as to fully realize the potential of waste heat-powered desalination systems. For this interdisciplinary scientific challenge to be surpassed, some technical barriers such as recovery efficiency, system dependability, and integration issues should be investigated.

Material science advancements, studies in thermodynamics, and optimization of processes are key factors that enhance the whole system's performance while reducing the expenses associated with its operations, hence making it useful in different industries. Moreover, these technologies' use is determined by social and economic aspects. Their adoption depends on the cost efficiency, regulation, and public opinion. The right balance between scientific progress and socioeconomic viability should be struck for widespread use and impacts to occur. In conclusion, this means waste heat-powered desalination is one of several SDG-enabling solutions for water scarcity. As such, there is a need to look at issues like future research directions, stakeholder collaborations, and innovations that would help overcome these hurdles and fully tap into this technology's prospects for sustainable global water security goals, hence tackling sustainability issues too [123–125].

Questions

1. What is the primary focus of emerging desalination techniques powered by waste heat?

2. How does membrane distillation (MD) work in the context of waste heat-powered desalination?

3. What distinguishes multi-effect distillation (MED) in desalination processes?

4 Explain the principle behind adsorption desalination (AD) and its reliance on waste heat.

5 What are hybrid desalination systems (HDSs) and how do they optimize energy efficiency?

6 How does low-temperature thermal desalination (LTTD) utilize waste heat for desalination?

7 What practical implications do waste heat-powered desalination techniques have in addressing water scarcity?

8 Why is continued research and development crucial for emerging desalination technologies?

9 How do emerging desalination techniques address the challenge of energy consumption?

10 What geographical relevance do waste heat-powered desalination techniques have?

11 How do emerging desalination techniques powered by waste heat contribute to achieving the United Nations Sustainable Development Goals (SDGs)?

References

1 Mejia-Mendez, D.L., Castel, C., Lemaitre, C., and Favre, E. (2018). Membrane distillation (MD) processes for water desalination applications. Can dense selfstanding membranes compete with microporous hydrophobic materials? *Chem. Eng. Sci.* 188 (12): 84–96.
2 Dessouky, S., Oh, J.H., Yang, M. et al. (2012). *Pavement repair strategies for selected distresses in FM roadways*. Texas: Texas Transportation Institute.
3 Al-Mutaz, I.S. and Wazeer, I. (2015). Current status and future directions of MED-TVC desalination technology. *Desalin. Water Treat.* 55 (1): 1–9.
4 Ge, H., Jensen, P.D., and Batstone, D.J. (2011). Temperature phased anaerobic digestion increases apparent hydrolysis rate for waste activated sludge. *Water Research* 45 (4): 1597–1606.
5 Riaz, N., Sultan, M., Miyazaki, T. et al. (2021). A review of recent advances in adsorption desalination technologies. *Int. Commun. Heat Mass Transfer* 128: 105594.

6 Sztekler, K., Kalawa, W., Bujok, T. et al. (2024). Hybrid desalination system for Baltic Sea water: a preliminary study. *Desalination* 574: 117269.
7 Abraham, R. and Singh, T.R. (2006). Thermocline-driven desalination: the technology and its potential. *Int. J. Nucl. Desalin.* 2 (2): 109–116.
8 Bhagwati, A., Shah, M., and Prajapati, M. (2023). Emerging technologies to sustainability: a comprehensive study on solar desalination for sustainable development. *Sustainable Manuf. Serv. Econ.* 2: 100007.
9 Curto, D., Franzitta, V., and Guercio, A. (2021). A review of the water desalination technologies. *Appl. Sci.* 11 (2): 670.
10 Elsaid, K., Taha Sayed, E., Yousef, B.A.A. et al. (2020). Recent progress on the utilization of waste heat for desalination: a review. *Energy Convers. Manag.* 221: 113105.
11 Almasoudi, S. and Jamoussi, B. (2024). Desalination technologies and their environmental impacts: a review. *Sustainable Chem. One World* 1: 100002.
12 Feria-Díaz, J.J., López-Méndez, M.C., Rodríguez-Miranda, J.P. et al. (2021). Commercial thermal technologies for desalination of water from renewable energies: a state of the art review. *Processes* 9: 262.
13 Baniasad Askari, I. and Ameri, M. (2021). A techno-economic review of multi effect desalination systems integrated with different solar thermal sources. *Appl. Therm. Eng.* 185: 116323.
14 Goodarzi, S., Jahanshahi Javaran, E., Rahnama, M., and Ahmadi, M. (2019). Techno-economic evaluation of a multi effect distillation system driven by low-temperature waste heat from exhaust flue gases. *Desalination* 460: 64–80.
15 Al-hotmani, O.M.A., Al-Obaidi, M.A., Filippini, G. et al. (2020). Optimisation of multi effect distillation based desalination system for minimum production cost for freshwater via repetitive simulation. *Comput. Chem. Eng.* 135: 106710.
16 Panagopoulos, A. (2020). Techno-economic evaluation of a solar multi-effect distillation/thermal vapor compression hybrid system for brine treatment and salt recovery. *Chem. Eng. Process. Process Intensif.* 152: 107934.
17 Xue, Y., Du, X., and Yang, L. (2021). Optimization of parallel/cross flow multi-effect desalination system with flue gas pre-heating considering thermo-vapor compression coupling. *Desalination* 506: 114998.
18 Chen, Q., Burhan, M., Shahzad, M.W. et al. (2021). A zero liquid discharge system integrating multi-effect distillation and evaporative crystallization for desalination brine treatment. *Desalination* 502: 114928.
19 Emamdoost, N., Jafarian, A., and Kouhikamali, R. (2020). Implementing multiple-effect distillation and reverse osmosis thermal coupling to improve desalination process performance in combined water and power plants. *Energy Convers. Manag.* 221: 113176.
20 Ghenai, C., Kabakebji, D., Douba, I., and Yassin, A. (2021). Performance analysis and optimization of hybrid multi-effect distillation adsorption

desalination system powered with solar thermal energy for high salinity sea water. *Energy* 215 (B): 119212.

21 Ali, E., Orfi, J., AlAnsary, H. et al. (2021). Cost analysis of multiple effect evaporation and membrane distillation hybrid desalination system. *Desalination* 517: 115258.

22 Alawad, S.M. and Khalifa, A.E. (2019). Analysis of water gap membrane distillation process for water desalination. *Desalination* 470: 114088.

23 Li, Q., Beier, L.J., Tan, J. et al. (2019). An integrated, solar-driven membrane distillation system for water purification and energy generation. *Appl. Energy* 237: 534–548.

24 Usman, H.S., Touati, K., and Rahaman, M.S. (2021). An economic evaluation of renewable energy-powered membrane distillation for desalination of brackish water. *Renew. Energy* 169: 1294–1304.

25 Khalifa, A.E. and Alawad, S.M. (2018). Air gap and water gap multistage membrane distillation for water desalination. *Desalination* 437: 175–183.

26 Ghaffour, N., Soukane, S., Lee, J.G. et al. (2019). Membrane distillation hybrids for water production and energy efficiency enhancement: a critical review. *Appl. Energy* 254: 113698.

27 Lee, W.J., Ng, Z.C., Hubadillah, S.K. et al. (2020). Fouling mitigation in forward osmosis and membrane distillation for desalination. *Desalination* 480: 114338.

28 Yao, M., Tijing, L.D., Naidu, G. et al. (2020). A review of membrane wettability for the treatment of saline water deploying membrane distillation. *Desalination* 479: 114312.

29 Alsaman, A.S., Ahmed, M.S., Ibrahim, E.M.M. et al. (2023). Experimental investigation of porous carbon for cooling and desalination applications. *NPJ Clean Water* 6 (1): 4.

30 Abid, M.B., Wahab, R.A., Salam, M.A. et al. (2023). Desalination technologies, membrane distillation, and electrospinning, an overview. *Heliyon* 9 (2): e12810.

31 González, D., Amigo, J., and Suárez, F. (2017). Membrane distillation: perspectives for sustainable and improved desalination. *Renew. Sust. Energ. Rev.* 80: 238–259.

32 Ravi, J., Othman, M.H.D., Matsuura, T. et al. (2020). Polymeric membranes for desalination using membrane distillation: a review. *Desalination* 490: 114530.

33 Abdel-Karim, A., Leaper, S., Skuse, C. et al. (2021). Membrane cleaning and pretreatments in membrane distillation – a review. *Chem. Eng. J.* 422: 129696.

34 Naciones Unidas (2018). *La Agenda 2030 y los Objetivos de Desarrollo Sostenible: Una Oportunidad para América Latina y el Caribe*. Santiago.

35 Alsaman, A.S., Askalany, A.A., Harby, K., and Ahmed, M.S. (2016). A state of the art of hybrid adsorption desalination-cooling systems. *Renew. Sust. Energ. Rev.* 58: 692–703.

36 Burn, S., Hoang, M., Zarzo, D. et al. (2015). Desalination techniques - a review of the opportunities for desalination in agriculture. *Desalination* 364: 2–16.

37 Inamuddin and Khan, A. (2021). Sustainable materials and systems for water desalination. In: *Advances in Science, Technology & Innovation* (ed. Inamuddin and A. Khan). Cham: Springer International Publishing.

38 Vlckova, B., Matejka, P., Simonova, J. et al. (1993). Surface-enhanced resonance Raman spectra of free base 5,10,15,20-tetrakis(4-carboxyphenyl) porphyrin and its silver complex in systems with silver colloid: direct adsorption in comparison to adsorption via molecular spacer. *J. Phys. Chem.* 97 (38): 9719–9729.

39 Pourhakkak, P., Taghizadeh, A., Taghizadeh, M. et al. (2021). Fundamentals of adsorption technology. *Interface Sci. Technol.* 33: 1–70.

40 Duncan, H. and Lasia, A. (2008). Separation of hydrogen adsorption and absorption on Pd thin films. *Electrochim. Acta* 53 (23): 6845–6850.

41 Mahmoodi, N.M., Oveisi, M., Taghizadeh, A., and Taghizadeh, M. (2020). Synthesis of pearl necklace-like ZIF-8@chitosan/PVA nanofiber with synergistic effect for recycling aqueous dye removal. *Carbohydr. Polym.* 227: 115364.

42 Mahmoodi, N.M., Oveisi, M., Taghizadeh, A., and Taghizadeh, M. (2019). Novel magnetic amine functionalized carbon nanotube/metal-organic framework nanocomposites: from green ultrasound-assisted synthesis to detailed selective pollutant removal modelling from binary systems. *J. Hazard. Mater.* 368: 746–759.

43 Thimmaiah, P.C., Sharafian, A., Rouhani, M. et al. (2017). Evaluation of low-pressure flooded evaporator performance for adsorption chillers. *Energy* 122: 144–158.

44 Ali, E.S., Askalany, A.A., and Zohir, A.E. (2020). Innovative employing of salt hydration with adsorption to enhance performance of desalination and heat transformation systems. *Appl. Therm. Eng.* 179: 115614.

45 Thu, K., Saha, B.B., Chua, K.J., and Ng, K.C. (2016). Performance investigation of a waste heat-driven 3-bed 2-evaporator adsorption cycle for cooling and desalination. *Int. J. Heat Mass Transf.* 101: 1111–1122.

46 Mitra, S., Kumar, P., Srinivasan, K., and Dutta, P. (2015). Performance evaluation of a two-stage silica gel + water adsorption based cooling-cum-desalination system. *Int. J. Refrig.* 58: 186–198.

47 Ali, E.S., Muhammad-Asfahan, H., Sultan, M., and Askalany, A.A. (2021). A novel ejectors integration with two-stages adsorption desalination: away to scavenge the ambient energy. *Sustain Energy Technol Assess* 48: 101658.

48 Majumder, M., Chopra, N., Andrews, R., and Hinds, B.J. (2005). Nanoscale hydrodynamics: enhanced flow in carbon nanotubes. *Nature* 438 (7064): 44.

49 Hummer, G., Rasaiah, J.C., and Noworyta, J.P. (2001). Water conduction through the hydrophobic channel of a carbon nanotube. *Nature* 414 (6860): 188–190.

50 Porada, S., Sales, B.B., Hamelers, H.V.M., and Biesheuvel, P.M. (2012). Water desalination with wires. *J. Phys. Chem. Lett.* 3 (12): 1613–1618.

51 Li, H., Pan, L., Zhang, Y. et al. (2010). Kinetics and thermodynamics study for electrosorption of NaCl onto carbon nanotubes and carbon nanofibers electrodes. *Chem. Phys. Lett.* 485 (1-3): 161–166.

52 Yang, H.Y., Han, Z.J., Yu, S.F. et al. (2013). Carbon nanotube membranes with ultrahigh specific adsorption capacity for water desalination and purification. *Nat. Commun.* 4: 3220.

53 Zhang, Z., Zhang, Y., Jiang, C. et al. (2022). Highly efficient capacitive desalination for brackish water using super activated carbon with ultra-high pore volume. *Desalination* 529: 115653.

54 Askalany, A., Ali, E.S., and Mohammed, R.H. (2020). A novel cycle for adsorption desalination system with two stages-ejector for higher water production and efficiency. *Desalination* 496: 114753.

55 Mitra, S., Srinivasan, K., Kumar, P. et al. (2014). Solar driven adsorption desalination system. *Energy Procedia* 49: 2261–2269.

56 Guo, X.-F., Ji, Z.-Y., Yuan, J.-S. et al. (2016). Recovery of K^+ from concentrates from brackish and seawater desalination with modified clinoptilolite. *Desalin. Water Treat.* 57 (15): 6829–6837.

57 Xue, Z., Li, Z., Ma, J. et al. (2014). Effective removal of Mg^{2+} and Ca^{2+} ions by mesoporous LTA zeolite. *Desalination* 341: 10–18.

58 Aghakhani, A., Mousavi, S.F., Mostafazadeh-Fard, B. et al. (2011). Application of some combined adsorbents to remove salinity parameters from drainage water. *Desalination* 275 (1-3): 217–223.

59 Wajima, T., Shimizu, T., Yamato, T., and Ikegami, Y. (2010). Removal of NaCl from seawater using natural zeolite. *Toxicol. Environ. Chem.* 92 (1): 21–26.

60 Wibowo, E., Sutisna, R., and M., Murniati, R., Khairurrijal, and Abdullah, M. (2017). Utilization of natural zeolite as sorbent material for seawater desalination. *Procedia Engineering* 170: 8–13.

61 Li, X., Zhang, H., Yang, X. et al. (2024). Mesoporous dopamine-modified leaf-like zeolitic imidazolate frameworks derived carbon for efficient capacitive deionization. *J. Colloid Interface Sci.* 654 (A): 559–567.

62 Baiju, V., Abhishek, P., Priya, K.L., and Mohammed, S.P.A. (2022). Performance investigation of silica gel based consolidated composite adsorbents effective for adsorption desalination systems. *Mater. Today Commun.* 32: 104015.
63 Askalany, A.A., El-Sayed, M.A., and El-Sayed, M.A. (2021). Effect of metal additives in the bed on the performance parameters of silica gel-based adsorption desalination systems. *Energies* 14 (21): 7226.
64 Aristov, Y.I., Restuccia, G., Tokarev, M.M. et al. (2000). Selective water sorbents for multiple applications. 11. $CaCl_2$ confined to expanded vermiculite. *React. Kinet. Catal. Lett.* 71 (2): 377–384.
65 Wan, Y., Fan, Y., Dan, J. et al. (2019). A review of recent advances in two-dimensional natural clay vermiculite-based nanomaterials. *Mater. Res. Express* 6 (10): 102002.
66 Zhao, H., Huang, T., Liu, T. et al. (2021). Synthesis of $MgCl_2$/vermiculite and its water vapor adsorption-desorption performance. *Int. J. Energy Res.* 45 (15): 21375–21389.
67 Alsaman, A.S., Maher, H., Ghazy, M. et al. (2024). 2D materials for adsorption desalination applications: a state of the art. *Therm. Sci. Eng. Prog.* 49: 102455.
68 Acquaroli, L.N., Urteaga, R., Berli, C.L.A., and Koropecki, R.R. (2011). Capillary filling in nanostructured porous silicon. *Langmuir* 27 (5): 2067 2072.
69 Hulikal Chakrapani, T. and den Otter, W.K. (2020). Capillary imbibition of binary fluid mixtures in nanochannels. *Langmuir* 36 (42): 12712–12722.
70 Hwang, J., Kataoka, S., Endo, A., and Daiguji, H. (2015). Adsorption and desorption of water in two-dimensional hexagonal mesoporous silica with different pore dimensions. *J. Phys. Chem. C* 119 (46): 26171–26182.
71 Mitra, S., Thu, K., Saha, B.B., and Dutta, P. (2017). Performance evaluation and determination of minimum desorption temperature of a two-stage air cooled silica gel/water adsorption system. *Appl. Energy* 206: 507–518.
72 Alazab, A.A., Qasem, N.A.A., and Baaqeel, H. (2023). Performance evaluation of a novel integrated adsorption desalination system with direct contact membrane distillation plant. *Desalination* 552: 116441.
73 Kashyout, A.E.B., Hassan, A., Hassan, G. et al. (2021). Hybrid renewable energy/hybrid desalination potentials for remote areas: selected cases studied in Egypt. *RSC Adv.* 11 (22): 13201–13219.
74 Esmaeilion, F. (2020). Hybrid renewable energy systems for desalination. *Appl Water Sci* 10: 84.
75 Feria-Díaz, J.J., Correa-Mahecha, F., López-Méndez, M.C. et al. (2021). Recent desalination technologies by hybridization and integration with reverse osmosis: a review. *Water* 13 (10): 1369.

76 Padrón, I., Avila, D., Marichal, G.N., and Rodríguez, J.A. (2019). Assessment of hybrid renewable energy systems to supplied energy to autonomous desalination systems in two islands of the canary archipelago. *Renew. Sust. Energ. Rev.* 101: 221–230.

77 Harby, K., Emad, M., Benghanem, M. et al. (2024). Reverse osmosis hybridization with other desalination techniques: an overview and opportunities. *Desalination* 581: 117600.

78 Alardhi, S.M., Ali, N.S., Cata Saady, N.M. et al. (2024). Separation techniques in different configurations of hybrid systems via synergetic adsorption and membrane processes for water treatment: a review. *J. Ind. Eng. Chem.* 130: 91–104.

79 Abdelaziz, G.B., El-Said, E.M.S., Dahab, M.A. et al. (2021). Hybrid solar desalination systems review. In: *Energy Sources Part A: Recovery, Utilization and Environmental Effects*, 1–31.

80 Ghazaie, S.H., Sadeghi, K., Sokolova, E. et al. (2020). Comparative analysis of hybrid desalination technologies powered by SMR. *Energies* 13 (19): 5006.

81 Youssef, P.G., Dakkama, H., Mahmoud, S.M., and AL-Dadah, R.K. (2017). Experimental investigation of adsorption water desalination/cooling system using CPO-27Ni MOF. *Desalination* 404: 192–199.

82 Elsayed, E., AL-Dadah, R., Mahmoud, S. et al. (2020). Experimental testing of aluminium fumarate MOF for adsorption desalination. *Desalination* 475: 114170.

83 Jones, E., Qadir, M., van Vliet, M.T.H. et al. (2019). The state of desalination and brine production: A global outlook. *Sci. Total Environ.* 657: 1343–1356.

84 Bamufleh, H., Abdelhady, F., Baaqeel, H.M., and El-Halwagi, M.M. (2017). Optimization of multi-effect distillation with brine treatment via membrane distillation and process heat integration. *Desalination* 408: 110–118.

85 González-Bravo, R., Ponce-Ortega, J.M., and El-Halwagi, M.M. (2017). Optimal design of water desalination systems involving waste heat recovery. *Ind. Eng. Chem. Res.* 56 (7): 1834–1847.

86 Rajesh, S. and Chiranjeevi, C. (2024). Experimental investigations on PVT-integrated hybrid desalination system: energy, exergy, and economic analysis. *Energy Sources, Part A* 46 (1): 3793–3806.

87 Abhishek, P. and Baiju, V. (2024). Experimental analysis on hybrid adsorption and open air cycle humidification dehumidification (AD–OHD) desalination system. *Desalination* 572: 117130.

88 Baiju, V., Abhishek, P., and Harikrishnan, S. (2024). Performance investigations of hybrid adsorption and thermo electric dehumidification desalination system. *Energy Convers. Manag.* 300: 117912.

89 Micale, G., Cipollina, A., and Rizzuti, L. (2009). Seawater desalination for freshwater production. In: *Green Energy and Technology* (ed. H. Machrafi), 1–15. Springer Science and Business Media Deutschland GmbH.

90 Alkaisi, A., Mossad, R., and Sharifian-Barforoush, A. (2017). A review of the water desalination systems integrated with renewable energy. In: *Energy Procedia* (ed. A.E. Childress, M. Duke, N. Ghaffour, et al.), 268–274. Elsevier Ltd.

91 Al-Sahali, M. and Ettouney, H. (2007). Developments in thermal desalination processes: design, energy, and costing aspects. *Desalination* 214 (1–3): 227–240.

92 Ettouney, H. (2009). Conventional thermal processes. In: *Green Energy and Technology* (ed. H. Machrafi), 17–40. Springer Science and Business Media Deutschland GmbH.

93 Silva, A.J., Methot, R.L., Panich, M. et al. (1998) Thermocline driven desalination: status for cape verde. *IEEE Oceanic Engineering Society. OCEANS'98*. Conference Proceedings (Cat. No.98CH36259), Nice, France, 1998, 2, 983–987.

94 Chen, Q., Burhan, M., Li, Y., and Ng, K.C. (2021). An ocean thermocline desalination system using the direct spray method. *Desalination* 520: 115373.

95 Natarajan, S.K., Suraparaju, S.K., and Elavarasan, R.M. (2022). A review on low-temperature thermal desalination approach. *Environ. Sci. Pollut. Res.* 29 (22): 32443–32466.

96 Guduru, S., Bajaj, P., and Gonsalves, O.N. (2020). India's low temperature thermal desalination technology: Water diplomacy with small island developing states in the indo-pacific region. *Marit. Aff.* 16 (2): 30–45.

97 Jalihal, P. and Venkatesan, R. (2019). Advanced desalination technologies. In: *Sustainable Water and Wastewater Processing*, 93–131. Elsevier.

98 Kathiroli, S., Jalihal, P., and Sistla, P.V.S. (2008). Barge mounted low temperature thermal desalination plant. In: *The Proceedings of the Eighteenth (2008) International Offshore and Polar Engineering Conference*: Vancouver: International Society of Offshore and Polar Engineers.

99 Saad, M., Ahmed, M., and Morcos, V.M. (2011). Performance analysis of a vacuum desalination system. *Proceedings of the ASME 2011 International Mechanical Engineering Congress and Exposition*, 1, 421–425.

100 Sistla, P.V.S., Venkatesan, G., Jalihal, P., and Kathiroli, S. (2009). Low temperature thermal desalination plants. In: *The Proceedings of the Eighth (2009) ISOPE Ocean Mining (& Gas Hydrates) Symposium*, Chennai, India.

101 Gude, V.G., Nirmalakhandan, N., and Deng, S. (2011). Sustainable low temperature desalination: a case for renewable energy. *J. Renewable Sustainable Energy* 3 (4): 043108.

102 Arkadu, J.P., Kiran S., Srinivasula, R.A. et al. (2014) Vortex induced vibration analysis of intake pipes in shallow waters for low temperature thermal desalination plants. In: *Conference: International Conference on Innovative Technologies and Management for Water Security.*

103 Saravanan, R., Bhattacharyya, S.K., Ramanamurthy, M.V.R., and Panneer Selvam, R. (2023). Benchmarking of VIV numerical analysis with prototype response for fatigue assessment of inverse catenary coldwater pipelines. *SN Appl. Sci.* 5: 94.

104 Abraham, R. (2007). Experimental studies on a desalination plant using ocean temperature difference. *Int. J. Nucl. Desalin.* 2 (4).

105 Balaji, D., Abraham, R., and Ramana Murthy, M.V. (2016). Experimental study on the vacuum load of low-temperature thermal desalination plant. *Desalin. Water Treat.* 57 (55): 26830–26844.

106 Yuan, G., Zhang, L., Zhang, H., and Wang, Z. (2011). Numerical and experimental investigation of performance of the liquid-gas and liquid jet pumps in desalination systems. *Desalination* 276 (1-3): 89–95.

107 Venkatesan, G., Prakash Kumar, L.S.S., and Jalihal, P. (2020). Influence of non-condensable gases on performance of the low temperature thermal desalination plants. *Curr. Sci.* 118 (11): 1718–1724.

108 Nayar, K.G., Swaminathan, J., Warsinger, D.M., and Lienhard V.J.H. (2015). Performance limits and opportunities for low temperature thermal desalination. In: *Proceedings of the 2015 Indian Water Week*, New Delhi, India.

109 Venkatesan, R. (2014). Comparison between LTTD and RO process of sea-water desalination: an integrated economic, environmental and ecological framework. *Curr. Sci.* 106 (3): 378–386.

110 Balaji, D., Abraham, R., Phani Kumar, S.V.S., and Ramana Murthy, M.V. (2015). Experimental and comparative study of a sea water-cooled surface condenser of LTTD plant with HTRI and Kern method. *Desalin. Water Treat.* 57 (41): 19540–19556.

111 Karmakar, A., Patra, A.K., Saha, S., and Jana, S.K. (2024). Design and simulation of a low-temperature thermal desalination system. *Chem. Eng. Technol.* 47 (1): 160–174.

112 Elakkiyadasan, R., Kumar, P.M., Subramanian, M. et al. (2021). Optimization of shell and tube condenser for low temperature thermal desalination plant. *E3S Web Conf.* 309: 01011.

113 Chen, Q., Kum, J.M., and Chua, K.J. (2018). A spray assisted low-temperature desalination technology. In: *Emerging Technologies for Sustainable Desalination Handbook* (ed. G. Gude), 255–284. Elsevier.

114 Gude, V.G. and Nirmalakhandan, N. (2009). Desalination at low temperatures and low pressures. *Desalination* 244 (1-3): 239–247.

115 Muthunayagam, A.E., Ramamurthi, K., and Paden, J.R. (2005). Low temperature flash vaporization for desalination. *Desalination* 180 (1-3): 25–32.

116 Muthunayagam, A.E., Ramamurthi, K., and Paden, J.R. (2005). Modelling and experiments on vaporization of saline water at low temperatures and reduced pressures. *Appl. Therm. Eng.* 25 (5-6): 941–952.

117 Setyawan, E.Y., Nakhoda, Y.I., Siagian, P., and Napitupulu, R.A.M. (2020) Geometry fin performance in vacuum desalination system condenser tubes using low temperature. In: *IOP Conference Series: Materials Science and Engineering*, Institute of Physics Publishing.

118 Garciadiego, A., Luo, T., and Dowling, A.W. (2021). Molecular design targets and optimization of low-temperature thermal desalination systems. *Desalination* 504: 114941.

119 Delpisheh, M., Haghghi, M.A., Athari, H., and Mehrpooya, M. (2021). Desalinated water and hydrogen generation from seawater via a desalination unit and a low temperature electrolysis using a novel solar-based setup. *Int. J. Hydrog. Energy* 46 (10): 7211–7229.

120 Gude, V.G., Nirmalakhandan, N., Deng, S., and Maganti, A. (2012). Low temperature desalination using solar collectors augmented by thermal energy storage. *Appl. Energy* 91 (1): 466–474.

121 Angelakis, A.N., Valipour, M., Choo, K.-H. et al. (2021). Desalination: from ancient to present and future. *Water* 13: 2222.

122 Ibrahim, A.G.M., Rashad, A.M., and Dincer, I. (2017). Exergoeconomic analysis for cost optimization of a solar distillation system. *Sol. Energy* 151: 22–32.

123 Alaei-Shahmirzadi, M.A., Hosseini, S.S., Luo, J., and Ortiz, I. (2018). Significance, evolution and recent advances in adsorption technology, materials and processes for desalination, water softening and salt removal. *J. Environ. Manag.* 215: 324–344.

124 Donald B Broughton, Clarence G Gerhold (1984). Continuous sorption process employing fixed bed of sorbent and moving inlets and outlets, US2985589A.

125 Reddy Pilli, S., Arkadu, J.P., Sistla, P., and Ramana Murthy, M.V. (2014) Vibration analysis of shell and tube heat exchanger for low temperature thermal desalination plant. In: *International Conference on Innovative Technologies and Management for Water Security.*

9

Minerals Recovery from Seawater by Using Membrane Crystallization

Ali Bahadar[1] and Mohamed Helmy Abdel-Aziz[1,2]

[1]*Chemical and Materials Engineering Department, King Abdulaziz University, Raabigh, Saudi Arabia*
[2]*Chemical Engineering Department, Faculty of Engineering, Alexandria University, Alexandria, Egypt*

9.1 Introduction

Seawater comprises diverse valued minerals such as potassium, magnesium, calcium, lithium, strontium, and bromine, all of which have extensive applications including agriculture, pharmaceuticals, cosmetics, and seawater desalination [1, 2]. Potassium chloride, vital in fertilizers, improves crop yield, whereas potassium carbonate is used to produce glass and soaps. Strontium carbonate, known for its use in fireworks, makes vibrant visual displays. Lithium carbonate is critical for manufacturing lightweight metal alloys used in aerospace and electronics, and magnesium minerals play an important role in producing high-strength, lightweight materials for automotive and aerospace applications.

The high solubility of these minerals in seawater facilitates their extraction, mostly in desalination processes where they can be concentrated. This method provides environmental and economic advantages, reducing the necessity for land-based mining and minimizing ecological damage. Extracting minerals from seawater provides an alternative source of essential materials along with contributing to energy conservation and environmental sustainability by lowering carbon emissions and protecting natural habitats [3].

Despite seawater's abundance, traditional extraction methods are often incompetent due to the low concentration of target minerals. Technologies like ion exchange, adsorption, and solvent extraction, though commonly used, suffer from issues such as low selectivity and slow kinetics, making large-scale recovery from seawater economically challenging [4]. Membrane crystallization (MCr), a novel technique, shows promise by eliminating these restrictions [5]. This process permits the selective crystallization of minerals under controlled conditions, making it a potentially more efficient solution for seawater mineral recovery.

Water Treatment and Desalination: Membrane Separation using Renewable Energy Resources,
First Edition. Edited by Mohamed Bassyouni.

9.1.1 Importance of Minerals and Their Applications

Minerals are crucial to modern society, with applications ranging from agriculture to electronics. Seawater, which contains a wide array of dissolved minerals, represents a promising supply for meeting rising industrial demands. Key minerals like lithium, potassium, and magnesium are particularly required for their roles in areas such as energy production, electronics, and water purification [6].

Seawater, though rich in minerals, has extraction challenges because of the low concentrations of valuable ions [5]. This requires advanced recovery technologies like membrane crystallization, which offers higher efficiency by manipulating the solubility and precipitation behavior of minerals. The ability to recover minerals such as lithium and precious metals from seawater offers a sustainable substitute for traditional mining.

9.1.2 Challenges in Traditional Mineral Extraction Methods

The increasing request for minerals has strained traditional mining operations, which face environmental, physical, and economic limitations. As terrestrial mineral reserves become depleted, other sources such as seawater are attracting attention. However, conventional extraction technologies are often ineffective due to seawater's low mineral concentrations. Processes such as ion exchange and solvent extraction suffer from poor efficiency and high energy requirements, hindering large-scale application [7].

Membrane crystallization emerges as a promising solution to these challenges. By applying thermodynamically stable processes under controlled conditions, it allows the selective recovery of minerals from seawater. This technology not only mitigates the restrictions of conventional methods but also offers a pathway for the large-scale recovery of valuable minerals like lithium, potassium, and magnesium from seawater [8].

9.1.3 Potential of Seawater for Mineral Recovery

The uniform composition of seawater globally makes it a viable resource for mineral extraction [9]. Despite previous challenges in recovering minerals like sodium and chloride, recent developments in technology offer new chances to make benefits from seawater's rich mineral reserves [10].

Efforts are now focused on developing ways to recover not only essential minerals but also to address environmental concerns, such as the removal of hazardous materials like heavy metals and radioactive elements [11]. The recovery of valuable minerals from seawater, coupled with advancements in membrane technology, holds great potential for sustainable resource management.

9.2 Membrane Crystallization: Principles and Mechanisms

Membrane crystallization is a promising chemical separation technique that combines the multi-discipline features of materials science, process control, and the chemical industry. It may engineer the crystallization process in terms of microscale mass transfer and interfacial process control. Due to its remarkable features and revolutionary potential in theoretical studies of special membrane materials, distinct membrane processes, and scientific research tools related to membrane crystallization, it has attracted significant attention from researchers in recent years. Indeed, progress in the research on membrane crystallization has enabled membrane crystallization to progress from theoretical concepts to practical industrial applications. In particular, the membrane-based technology in enhancing nucleation and growth offers several advantages regarding precision in process control, high efficiency of micromixing, controllable heterogeneous nucleation, and production of high-quality crystal products [12].

9.2.1 Crystallization

The process of crystallization involves forming solid crystals from a liquid solution in order to separate it into solid and liquid phases. It has typically been used with conventional equipment like crystallizers or evaporators for the separation and purification of industrial streams. Because of their advantages over conventional technology in terms of versatility and low energy consumption, membrane technologies are emerging as a possible alternative to conventional crystallizers [8].

The method of crystallization, which yields clean water and recoverable salts by removing dissolved salts and other contaminants from wastewater, is based on the solid crystal formation from a solution [13, 14]. Crystallization involves the removal of the solute from the liquid solution to transform it into a pure solid crystalline phase. This allows several valuable and useful compounds to be recovered from wastewater, including NaCl, Na_2CO_3, ammonium, phosphates, and the removal of heavy metals, and the recovery of water itself by the application of crystallization techniques [15–21].

9.2.2 Membrane-assisted Crystallization (MCr)

Membrane crystallization processes' regulating mechanisms fall into one of two general categories: (1) Crystal growth processes directly involving the membrane; (2). Nucleation initiation processes where the membrane induces concentration up to the supersaturation ratio [22]. The two main strategies are as follows: (1) Membranes for crystal formation and (2) Concentrative crystallization.

One of the most common uses of membrane crystallization is in crystal growth membranes, where the membrane surface serves as a growth site for heterogeneous nuclei in addition to concentrating the feed solution to reach supersaturation. Applications of membrane crystallization technology in different industries are presented in Figure 9.1. A temperature-induced partial vapor gradient, an osmotically regulated vapor pressure difference, or a transmembrane pressure difference can all represent significant driving forces. Its central component is a porous hydrophobic membrane that, because of restrictions placed on solvent flow when it is liquid, functions as a permselective barrier to perform solvent evaporation. The surface accelerates the kinetics toward the crystal state, which in turn promotes quicker nucleation. This happens as a result of the decrease in Gibbs free energy, which favors the transition to a stable state with lower energy. As a result, the kinetically regulated regime gives way to the thermodynamically controlled regime, enlarging the metastable zone. The achievement of the supersaturation state is also favorable to this [23].

Therefore, the solvent selectively evaporates through the porous membrane surface under the effect of the driving force. As a result, the feed gradually becomes concentrated in the direction of the supersaturation condition. In a heterogeneous model of crystallization, this is in attenuation with simultaneous assimilation of the concentrated solute molecules on the membrane surface shifting toward the formation of crystal nuclei [24]. The utilization of membrane crystallization in a direct method has been implemented in various domains, including biomolecule manufacturing, active pharmaceutical ingredient preparation, and selective molecular synthesis. In addition to producing a narrow crystal size distribution (CSD), this method has been shown to produce the target polymorph with increased selectivity. Studies using static membrane crystallizers and forced circulation have demonstrated the synthesis of L-glutamic acid, polymorphs, and glycine (selected products) in response to hydrodynamic condition changes [25].

A diffusion-type growth model is commonly used to grow crystals on the membrane surface. However, solitary crystal formation follows this more precisely than clustered crystal growth. The population growth model, which shows a sigmoidal pattern as the rates of nucleation and crystallization increase with an increase in supersaturation, can be used to explain the growth of the crystal assembly [26, 27].

Concentrative crystallization is another membrane crystallization method in which the solution is dewatered using the membrane. In this case, crystal formation does not directly occur on the membrane surface. The aggregation of potential nuclei in the bulk feed begins, depending on the saturation level attained. Over time, this gets increasingly concentrated. Concentrative crystallization can be achieved using standard pressure-driven membrane processes like Ultrafiltration (UF), Nanofiltration (NF), and Reverse Osmosis (RO) [28].

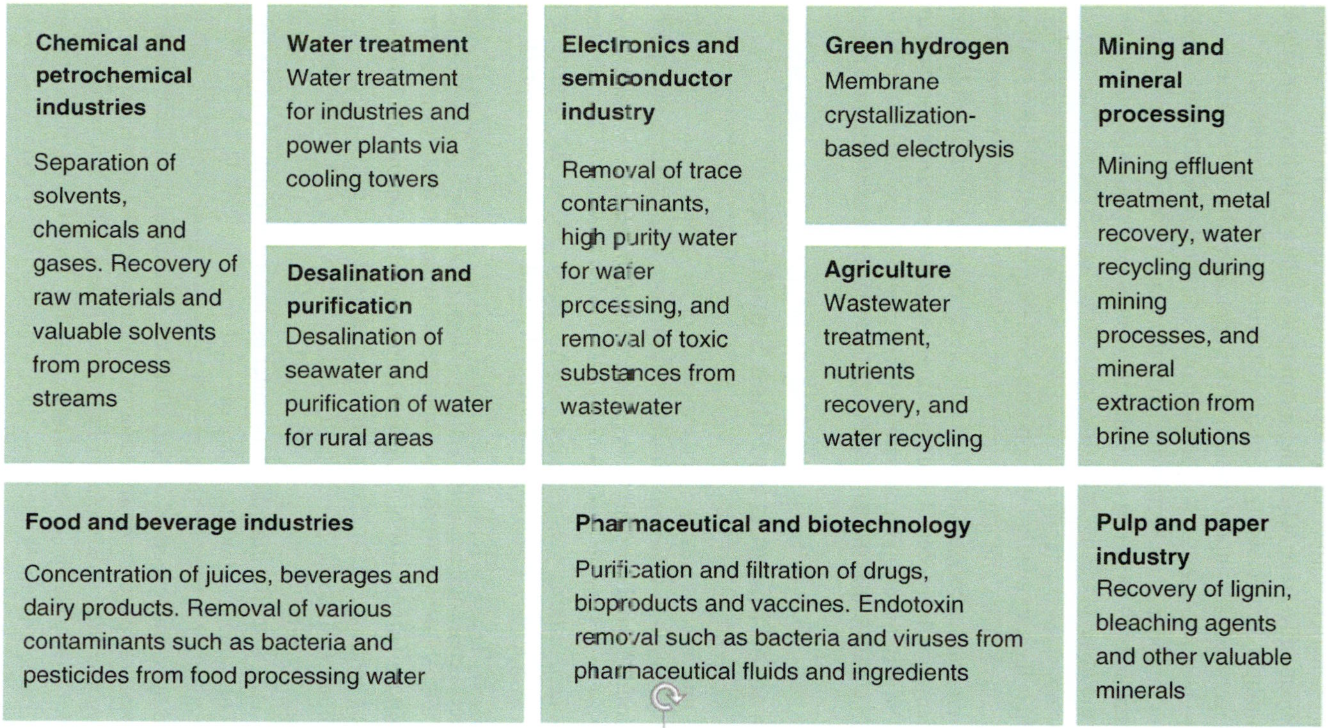

Figure 9.1 Applications of membrane crystallization technology in different industries.

The primary benefit is in the relative avoidance of concentration polarization, an issue that frequently arises when the membrane surface is utilized directly for crystal formation. The flexibility of integration with traditional membrane-based separation techniques is another benefit of this approach. The concentrate may be subjected to heat treatment, sonic precipitation, chemical precipitation, or other processes to promote crystal formation once the required level of concentration has been reached [29, 30]. The solution experiences nearly minimal feed losses and/or evaporative losses because the membrane technique is employed for separation.

Researchers have shown that the concentrative form of crystallization produces both competitive fluxes and improved process yields [31]. Another benefit of this method is its compact design. This allows for the execution of many modules based on the demands of the process. Other membrane-based crystallizers also have this property. The system design is comparatively simpler in this procedure since there is less of a need to modify the flat sheet and/or hollow fiber configuration design.

9.3 Targeted Minerals and Their Significance

Many vital minerals that are rare and expensive in land-based form can be found in notable amounts in seawater. However, at present, mining from the water is limited to a small number of minerals—those with high concentrations. Seawater mining is becoming an increasingly attractive option because of the problems that land-based mining sectors have recently faced due to the depletion of high-grade ores, sustainable water and energy demand, and environmental issues [1]. The four most concentrated metal ions in seawater—Na, Mg, Ca, and K—are currently recovered commercially due to their high concentration. These metallic elements, which are present in the highest concentration, have been extracted commercially as carbonates, sulfates, and chlorides, whilst magnesium has been extracted as hydroxide [32]. There are other cations as well; however, in far smaller amounts: silicon (Si), boron (B), and strontium (Sr). Furthermore, with regard to the so-called energy shift, lithium (Li), which has become strategically important as an essential ingredient for batteries, has a concentration of 0.17 ppm [33, 34]. The concentration of refractory and noble metals is too low to be accurately quantified. Because their market values are significantly less than the capital and operating expenses of extraction, low quantities of mineral elements have not been recovered from saltwater. With the advent of numerous seawater desalination facilities, this scenario has, nevertheless, altered in recent times. The need for freshwater has significantly expanded due to industrialization and rapid population growth. While seawater is available in vast quantities, it needs to be desalinated because its 33–37 g/L dissolved salt concentration is too high for agriculture, industry, or drinking. This has led to the global development of desalination plants, which use seawater RO technology to create freshwater [35].

Numerous minerals are produced as byproducts in the depleted brine developed during the seawater extraction process. Not only would the cost of producing water reduce significantly if these minerals could be economically recovered, but the pollution issues related to the disposal of brine would also be reduced. For instance, the market value of Na, Ca, Mg, and K recovered from the rejected brine of a Saudi Arabian desalination plant is predicted to be almost $US18 billion annually [36].

Due to the advantages of lower environmental risks and resource recovery, the scientific community is paying increasing attention to desalination facilities' ability to recover minerals from rejected brines and seawater [1]. The disposal of waste brines, which have a higher salinity concentration than feed seawater, produced by existing desalination systems, presents significant environmental issues. Consequently, the process of desalinating saltwater, which extracts rich minerals, highlights the move toward a circular economic paradigm. By using the brine byproduct strategically, which was previously regarded as waste, environmental concerns are reduced, and a more sustainable approach to resource management is encouraged. Economic diversification and technological advancement in a variety of industrial areas, from energy storage to agriculture, are facilitated by the economically feasible recovery of minerals. Moreover, including mineral recovery in desalination procedures offers an environmentally friendly tactic. Through reducing the negative environmental effects of careless brine discharge, the strategy is in line with sustainable waste management techniques [37].

A recent review study by Sharkh et al. [9] evaluated the products of seawater desalination that are most likely to be economically viable in the medium future and the procedures and technologies that are most important to enable commercial production in an economic environment. According to their research, in order to obtain a concentration increase of this kind when directly mining seawater, the energy required for the operation of seawater desalination plants would need to be injected into the system. When compared to the terrestrial mining of the same chemicals, the extraction of mineral products from desalination plant concentrate may be advantageous. The ocean's nearly limitless size, its consistent composition, its enormous ability to dilute treated waste streams, and the mining operation's steady and fixed footprint are a few examples.

Kumar et al. [38] highlighted how desalination concentrate as a source of minerals becomes increasingly commercially and ecologically feasible as technology for seawater brine mining advances. The concentration of minerals in the concentrate increases along with their market price, and this increases the economic benefit of mineral extraction. When it comes to this, mining compounds containing elements like Mg, Na, Ca, K, Sr, Li, Br, B, and Rb may be economically viable to harvest from concentrate if appropriate techniques for brine concentration and extraction are created [10]. From an economic perspective, it is necessary to compare the revenue potential—which is dependent on changes in the market price of commodities—with the cost of extraction. Brine extraction has the added advantage of reducing brine amounts, making it less invasive on the environment than conventional mining. Many products, such as uranium, magnesium, potassium salts, sodium hydroxide, chlorine, bromine, and chlorine, have been deemed likely to be commercially viable [38]. These products are currently or have traditionally been generated economically from saltwater. The summary of concentrations of different elements in seawater is presented in Table 9.1.

Shahmansouri et al. [42] conducted an economic analysis of previous, ongoing, or suggested extraction strategies for the manufacture of sodium chloride, bromine, magnesium hydroxide, and sodium hydroxide; chlorine and sodium hydroxide; magnesia; and sodium hydroxide in another study. As per the results of the analysis, extraction could yield a profit after around ten years, provided that high-purity commodities can be produced, except for bromine. According to a sensitivity study, commodity pricing, which is influenced by factors such as

Table 9.1 Concentration of different elements in seawater.

Elements in seawater	Concentration (ppm)	CO_2 emissions (kg/ton)	Recovery	References
Chlorine	19,400	10–15	—	[39]
Calcium	400	8–20	Up to 90%	[40]
Potassium	392	6–15	Up to 85%	[40]
Magnesium	1,290	10–25	Up to 95%	[39–41]
Sodium	10,800	5–10	Up to 99%	[40]
Sulphate	2,708	12–18	—	[39–41]
Strontium	8.1	—	—	[40, 41]
Barium	0.021	15–20	—	[40]
Copper	0.0009	15–20	—	[40]
Lithium	0.17	20–25	50–70%	[39–41]
Manganese	0.0004	15–20	—	[40]
Nickel	0.0066	18–22	—	[39–41]
Uranium	0.0033	22–25	—	[40, 41]
Rubidium	0.12	—	—	[40]
Zinc	0.005	—	—	[40, 41]

product purity, market volatility, and operating and maintenance expenses, has a significant impact on how profitable extraction is. The viability of manufacturing commodities from desalination concentrate for profit is contingent upon various criteria, including but not limited to proximity to a buyer, extraction efficiency, product purity and uniformity, safety, material handling, storage and transportation. This research suggests that extraction from desalination concentrate is unlikely to significantly improve the economics of desalination unless concentrate disposal costs are reduced considerably due to the marginally attractive economics of extraction and significant uncertainties associated with producing commodities. Nonetheless, their research focused more on the cost-benefit analysis of specific minerals than it did on the differences in mineral extraction techniques [42].

The economic viability of extracting salt concentrate from desalination facilities' rejected brine water is investigated in a relatively recent study by del Villar et al. [43]. In recent times, there has been a notable surge in the pricing of specific elements due to their high concentration and high demand from current industrial applications. The markets exhibit significant volatility, suggesting a persistent and ongoing increasing trend in specific elements found in the saline reject concentrates obtained from desalination plants. Although this behavior makes it difficult to predict the future value of these materials, it does enable us to anticipate a rising prospective revenue stream. The economic worth of the elements that were recovered from the reject flows of desalination facilities in Spain, which included some elements including boron, calcium, strontium, magnesium, sodium, lithium, rubidium, and gallium, was estimated by the study to be between 13.4 and 29.8 billion euros annually. If we simply consider the most valuable components, which make up 0.1% of the entire volume, the annual potential revenue would be approximately 3.8 billion euros. Because of their concentration in the salt rejects, the remaining components, however less valued, have potential. Of the eight elements that were chosen, the high concentration of sodium—whose price ranges from €0.67/kg to €2.02/kg—in the reject water thus indicates the element with the highest potential for profit [43].

9.4 Membrane Crystallization Process Design

Membrane crystallization can be used to extract lithium, sodium, magnesium, barium, and strontium, among other necessary salts. The application of membrane crystallization is driven by the need to manufacture, separate, and purify the elements in wastewater and saltwater into solid crystalline form in many industries, technologies, and scientific domains. Many of the materials that are used in commercial products today, such as fine chemicals, pigments, personal care

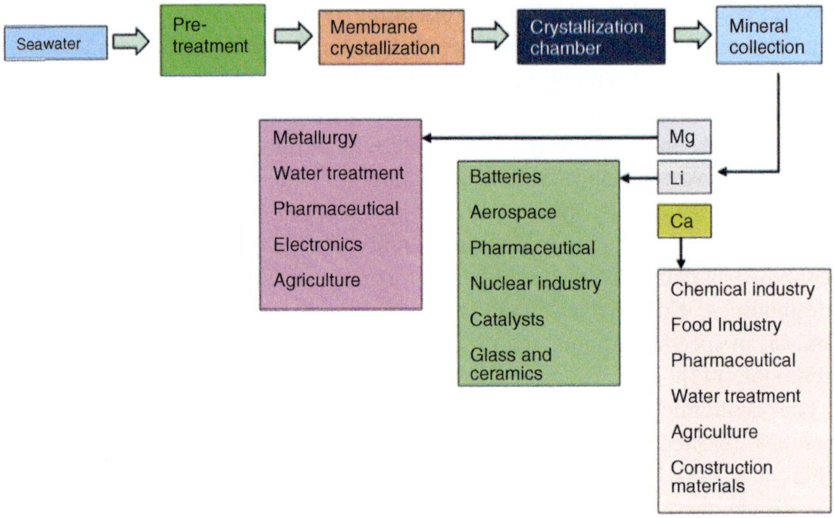

Figure 9.2 General flowchart for membrane crystallization process, mineral recovery and applications.

items, hygiene goods, cosmetics, and several food additives, are produced as crystalline powders, which enhance the stability of the products [44]. Applications can include the creation of technology, like photonic crystals and organic semiconductors. In contrast, the systematic investigation of novel medications with sources generated from crystal structures could lead to advancements in medicine [45]. The membrane crystallization process, mineral recovery, and applications are depicted in Figure 9.2.

9.4.1 Membrane Distillation Crystallization (MDCr)

High-concentrated brines and wastewater released from industrial operations can currently be used to collect freshwater and minerals using the MDCr technology, which has shown great promise [31, 46]. The MDCr is a hybrid process that combines classical crystallization with membrane distillation to achieve controlled process performance [22]. In contrast to traditional crystallization procedures, MDCr regulates the rate of solution supersaturation, particle nucleation, and growth [47, 48]. As a result, homogeneous size distributions and high-quality particles are produced. Moreover, enhanced process separation results in a high mineral extraction capacity. It is interesting to note that process intensification is moving toward cleaner, more efficient technologies that utilize solar energy [49].

MDCr recovers minerals through a process of supersaturating the feed solution and allowing clean water to evaporate across the membrane [49, 50]. The MDCr ensures environmental preservation and a circular economy by promoting zero-liquid discharge (ZLD) due to the recovery of minerals from the feed [50, 51]. MDCr has been effectively tested for the recovery of a variety of mineral salts, such as calcium and magnesium salts and sodium chloride.

Membrane distillation (MD) and membrane crystallization (MCr) share a common principle. Heat and mass transfer concepts underpin the MD/MCr function. To aid in the separation process, they use a temperature gradient across a hydrophobic membrane. In its role as a selective barrier, the membrane allows only vapor molecules to flow through from the hotter feed side to the cooler permeate side, blocking the passage of liquid and non-volatile solutes [52]. Each pore has a vapor/liquid interface because the hydrophobic polymeric membrane stops water from penetrating the pores. Water diffuses through the pores, evaporates at the membrane's interface on the heated side (retentate), and condenses on the cold side (distillate). In MDCr, a hydrophobic porous hollow fiber membrane module works to prevent salt deposition and/or accumulation inside the MDCr plant by tightly controlling salt supersaturation during crystallization in a circulating crystallizer and recovery in a crystal recovery system [53, 54]. The different design configurations of MDCr are depicted in Figure 9.3. The procedure causes the solution to become supersaturated, which leads to the creation of a metastable

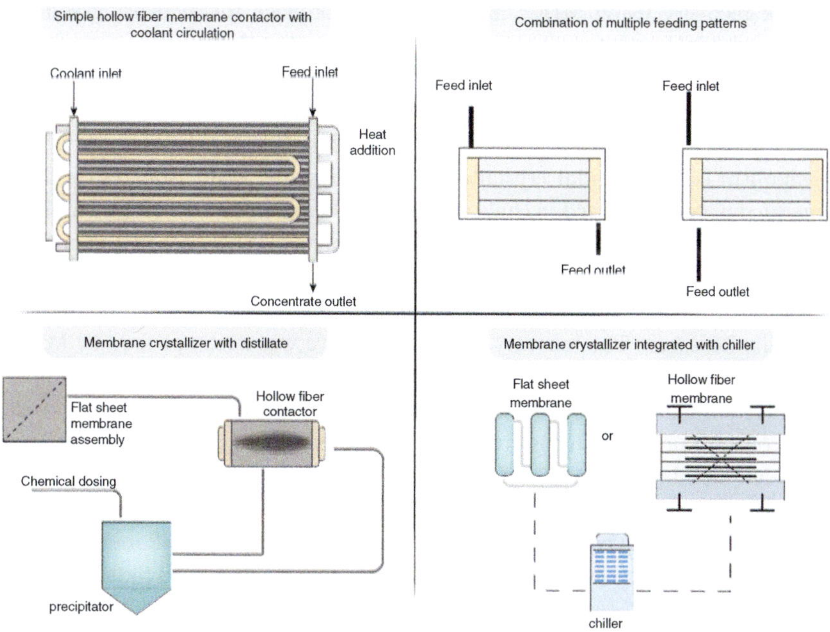

Figure 9.3 Different design configurations of MDCr.

state that facilitates crystal nucleation and growth [55, 56]. Because MDCr allows for ideal control over the supersaturation level, it is a desirable approach for brine concentration. Additionally, in comparison to other methods of solid separation, like cooling or evaporative crystallization, it can yield crystals of a superior caliber [57]. There are four basic MD/MCr arrangements, and each has distinct flow and contact patterns: Vacuum MD/MCr, Air Gap MD/MCr, Sweeping Gas MD/MCr, and Direct Contact MD/MCr. These designs provide tailored solutions for particular separation challenges by providing a range of alternatives that balance energy economy, distillate purity, and process intensity [45, 58, 59].

MD offers many benefits for crystallization, including the ability to operate at high concentrations, better fluid distribution, heterogeneous nucleation, high surface-to-volume ratios, the ability to separate nucleation from growth, simple solvent removal control for controlled supersaturation, total rejection of nonvolatile solutes, and the potential for low operating temperatures [60, 61]. Membrane distillation-crystallization has a number of benefits, but it also has certain disadvantages. The primary drawback stems from the membrane itself; it increases the resistance to mass transfer, which reduces the water flux through the membrane [62], thus lowering the rate of evaporation. The scaling phenomenon on top of the membrane, which can result in a decrease in flux or even total membrane blockage, is another significant disadvantage.

Brine recovery using MDCr has been commonplace for recovering sodium chloride, calcium sulfate, and similar chemicals [63, 64]. The recovery of salts from the concentrated feedstocks occurs after MD is performed at a temperature range of 50–80°C using hollow fiber membrane contactors comprised of hydrophobic materials like polyvinylidene fluoride (PVDF). Utilizing MD as a mechanism of preferential crystallization, other saline water sources, such as generated water connected to shale gas and coal bed methane development, have also shown competitive salt recovery. Researchers have experimentally examined a recovery factor of 37% NaCl from produced water [46]. This translates to 16.4 kg of NaCl per m^3 of produced water that was obtained from the Kuwait Institute of Scientific Research. The recovery mentioned above was accomplished by MDCr utilizing a modular hollow fiber architecture with a total surface area of 0.1 m^2. It is interesting to note that MD processes have the operational flexibility to work with concentrated feedstocks from a variety of sources, such as concentrated brine. NF membrane rejects and concentrates RO [65, 66]. Studies on comparative resource recovery have shown that MD is more effective than RO brines at selectively separating barium, strontium, and magnesium from NF retentate. For monovalent ions such as lithium, the opposite tendency is applied [40].

Although it has been demonstrated that MDCr can generate high-quality crystals, this ability is only applicable to the principal salts found in brines and saltwater at high concentrations. It has been suggested that only at high water recovery rates may MDCr be able to selectively fractionalize important minerals present in low quantities in brine and saline seawater. In these circumstances, supersaturated brines may cause scaling development, particularly in gypsum ($CaSO_4{\cdot}2H_2O$), which is thermally insoluble [67]. This usually causes operational disruptions for the MDC. For example, in order to avoid gypsum scaling on the membrane's surface, Ca was removed as $CaCO_3$ by adding Na_2CO_3 before the recovery of NaCl and $MgSO_4$ by MDCr [53]. $Na_2CO_3/NaHCO_3$ was created by Drioli et al. through a reaction between CO_2 and NaOH [55]. A membrane contactor device was used to provide this function. In order to avoid $Mg(OH)_2$ precipitation, which obstructs the production of MgSO4, the pH of the crystallizing solution was brought to 5 by adding HCl. Moreover, the performance of MD and MDCr would be limited at high water recovery due to the notable influence of polarization and higher barrier to vapor transport within the membrane pores [53].

9.4.2 Osmotic Membrane Crystallization (OMC)

Another cutting-edge method of crystallization is osmotic membrane crystallization (OMC), where the primary driving factor for separation is the concentration gradient [22]. In OMC, the feed and draw solutions are separated by a semi-permeable membrane. The feed solution, which needs to be concentrated through crystallization, is placed on one side of the membrane. To create a more considerable osmotic pressure than that of the feed concentration, a synthetic solution with a higher concentration is used as the draw solution on the other side. This setup generates a concentration gradient across the membrane, causing the solvent to move from the feed side (higher concentration) to the draw solution side (lower concentration). Ideally, no external pressure is needed for the solvent flow; however, a small amount of pressure—between 1 and 1.5 bars—may be applied to expedite the process [68, 69]. OMC is especially suitable for concentrating solutions where temperature-induced denaturation is a concern, such as in food processing, pharmaceutical synthesis, or biomolecule purification because it requires only low external pressure and operates at ambient temperature [70]. Variations of this process include pressure-driven methods for concentration or selective recovery of compounds from RO concentrates [71, 72].

OMC can utilize either membranes designed for direct crystal formation or those for concentrative crystallization, with both scenarios employing a semi-permeable membrane. The membranes can be hydrophobic or hydrophilic, allowing selective solvent transport in the aqueous state or selective vapor transport. The hydrophobic forward osmotic asymmetric membranes used in OMC typically feature a nonwoven support layer and a densely packed active layer. RO membranes have also been used in osmotic crystallization research. The critical difference between RO and forward osmosis (FO) membranes is their support layer thickness; RO membranes are thicker due to the high pressures involved, which affects concentration polarization dynamics and mass transport boundary layer formation. Early membranes were made from materials such as cellulose acetate, cellulose triacetate, and polyamide (PA), with an active layer thickness of approximately 0.1–1 μm, supported by a porous polyester layer around 100–200 μm thick [73–75].

A novel process, osmotic membrane distillation crystallization (OMDC), employs a non-selective hydrophobic microporous membrane to bring two liquids into contact [76]. Water evaporates from the feed side to the osmotic side due to concentration differences, creating a vapor pressure gradient that drives the process. OMDC offers advantages over traditional distillation and crystallization methods, including higher mass transfer with smaller equipment and lower energy consumption [58, 62]. The partial pressure gradient, rather than additional pressure, drives the process, reducing equipment costs and improving process safety. If available, residual heat or renewable energy sources can further decrease overall costs and environmental impact [58, 77, 78]. Additionally, using polymer materials in OMDC equipment helps reduce or eliminate erosion issues [8, 31, 79–81].

A recent study by Alvarez et al. proposed OMDC as a novel method for crystallizing sodium bicarbonate ($NaHCO_3$) [82]. In this process, $NaHCO_3$ is produced from soda ash and carbon dioxide (CO_2). The purity and shape of the $NaHCO_3$ crystals determine their suitability for various applications in the food, pharmaceutical, and agricultural sectors. Traditional crystallizers require the introduction of CO_2 to the tank environment for $NaHCO_3$ to crystallize [83]. The study investigated the optimal operating conditions for OMDC, including feed and osmotic velocities, feed concentration, and temperature. It was observed that higher feed and osmotic velocities increased the likelihood of membrane wetting. Additionally, temperature affected crystal size and purity, with higher temperatures resulting in more extensive and purer crystals [82].

9.4.3 Concentrative Crystallization

Membrane-based separation approaches that integrate multiple unit operations have the potential to serve as viable alternatives to traditional processes like fractionation. Among these, RO is one of the most widely used membrane technologies employed in desalination in both industrial and residential settings. However, a significant challenge remains: the management and disposal of the highly concentrated saline reject streams generated by RO desalination plants. This is where membrane-based crystallization methods come into play [22].

Membrane crystallization techniques offer a promising solution for brine and chemical recovery from RO concentrates. This approach not only addresses the issue of concentrate disposal but also provides value-added recovery, thus encouraging the process. Three main methods for RO-based crystallization are: freestanding RO, a combination of NF and RO, and RO integrated with chemical precipitation. Concentrating feed streams using RO has proven effective. The concentration of the feed solution to its supersaturation level can initiate nucleation, and seeding can further enhance kinetics up until the secondary nucleation stage [22].

Membrane crystallization can also be effectively combined with other technologies, such as microfiltration (MF) [84, 85], NF [31, 55], FO [86], and RO [87], to expand its range of applications. The relatively simple design and operational flexibility of membrane crystallization make it suitable for integration with these technologies. Compared to other separation methods, membrane crystallization is less affected by solution concentration, particularly in high-concentration systems [31]. For instance, coupling NF and RO with membrane crystallization can further concentrate the high-salinity brine, potentially increasing the water recovery rate to 99.9% and reducing the need to discharge high-salt wastewater into the environment. Additionally, combining membrane crystallization with RO can significantly reduce the membrane area required for crystal recovery [87].

RO/NF membranes, which are among the most commonly available types, are used in membrane crystallization processes. Both hollow fiber and flat sheet membrane constructions are available for industrial and residential applications. The most commonly used polymers include thin-film composites and cellulose acetate. Other materials, such as cellulose diacetate and cellulose triacetate, are produced through phase inversion and feature an anisotropic structure with a thin upper layer on a porous sub-layer [88]. Polyamide membranes, known for their increased selectivity, are also used alongside cellulose acetate [89].

Despite the potential of RO-based membrane crystallization systems, several obstacles need to be addressed for widespread commercial application. These include low confidence in scale-up, shorter membrane shelf life, and limited research on process integration. Successful commercialization will require careful consideration of trade-offs and cost-benefit analyses, including operational costs, installation expenses, building materials, and recurring consumables, all while maintaining efficiency, yield, energy consumption, and overall sustainability [22].

9.4.4 Antisolvent Crystallization

One of the most intriguing areas of research in advanced separation processes is the integration of antisolvent techniques with membrane crystallization processes. Membranes provide precise control over the crystal system and antisolvent dosage, and they enhance mixing by offering a high surface area of contact. This integration serves two main functions: (1) membranes used in crystal growth [90], and (2) surfaces on which liquid films can form [91]. This approach has significant potential across various industries, including polymer coating and pharmaceuticals.

Antisolvent membrane crystallization can play a crucial role in enhancing mixing efficiency, controlling product size distribution, and increasing the surface area of contact in drug delivery systems and related process developments. For example, a simple hollow fiber membrane could offer a practical and sustainable method for producing nanosuspensions. Porous microfiltration membranes have demonstrated optimal micromixing for synthesizing the model drug indomethacin under various operating conditions [92]. In one study, a hollow fiber porous membrane unit configured in a shell-and-tube arrangement was used. In this setup, the antisolvent (water) was circulated through the tube side, while the polymeric suspension was circulated through the shell side. This arrangement facilitated homogeneous precipitation of the polymer as a coating on the particle surface due to the controlled and rapid mixing [93].

Comparative performance tests of antisolvent crystallization systems without membrane separation have highlighted issues such as irregular crystal morphology, wide particle size variation, and inadequate contact surfaces due to poor mixing. Integrating antisolvent dosing with a membrane-based crystallizer helps mitigate these disadvantages by improving mixing efficiency and product consistency [91].

9.5 Membrane Materials and Their Performance

At present, around 20 polymers are utilized to support the production of commercial membranes. These include polyethersulfone (PES), polysulfone (PSF), polypropylene (PP), polyacrylonitrile (PAN), polyimide (PI), PA, polytetrafluoroethylene (PTFE) and PVDF [94].

9.5.1 Polysulfone (PSF)

PSF membranes are renowned for their broad chemical compatibility, resilience to high temperatures, and toughness. Due to their durability, they are frequently utilized in ultrafiltration. A microporous membrane has an open, spongy structure with a carefully regulated network of minuscule pores, ranging from 0.03 to 10 μm in diameter [95]. These pores may also be nanoscale and may be helpful in mineral recovery membrane crystallization. The hydrophilicity of PSF membranes significantly impacts their performance [96]. Common polymeric materials include PP, polyphenylene ether-nitrile (PPENK), polycarbonate (PC), PAN, PES, PVDF, PI, and PTFE. PSF is a preferred option in membrane creation due to its accessibility, chemical stability, ease of modification, high mechanical strength, surface charge, and broad pH and temperature operating range [97]. PSF also exhibits low cytotoxicity and offers intrinsic biocompatibility because it contains aryl or alkyl sulfone groups [98]. Its high natural content of aromatic groups further enhances its exceptional biocompatibility [99]. The strong mechanical properties of PSF, including a tensile strength between 70 and 83 MPa, have made it a favored choice for mineral recovery membranes [100].

Modifying PSF membranes is an excellent way to introduce hydrophilic functional groups and enhance membrane performance [101]. Membrane modification has become a critical area in the field, with PSF membranes often being altered via coating, chemical grafting, physical mixing, and layer-by-layer techniques to improve their functionality [102]. The performance of different types of membranes used in membrane crystallization is depicted in Table 9.2.

9.5.2 Polyacrylonitrile (PAN)

PAN is a crucial membrane material for ultrafiltration. PAN membranes offer strong mechanical properties, are hydrophilic, and are resistant to a variety of chemicals. They are frequently employed in applications requiring low fouling and high flux. Since PAN has favorable qualities such as good mechanical properties, solvent resistance, significant chemical stability, and lower cost compared to other polymers like PES and PSF, it has become one of the most widely used polymers in membrane fabrication [105, 106].

One of PAN's disadvantages is membrane fouling, which limits the use and advancement of PAN ultrafiltration membranes due to declining performance after fouling. There are numerous ways to enhance the hydrophilicity of PAN membranes, such as adding polymer layers or unique functional groups to the membrane surface [107].

9.5.3 Polyvinylidene Fluoride (PVDF)

PVDF is well known for its ability to withstand a variety of substances, such as bases, oxidizing agents, acids, and organic solvents. This makes it ideal for applications involving frequent exposure to harsh chemical conditions, such as mineral recovery and membrane crystallization. PVDF membranes are also valued for their high mechanical strength and hydrophobic qualities [108].

PVDF is a semicrystalline polymer with a repeating unit of $-(CH_2CF_2)_n-$. It possesses strong mechanical properties, good thermal stability, chemical resistance, and excellent aging resistance, all of which are critical for the real-world use of separation membranes. Additionally, PVDF exhibits good processability in the fabrication of tubular, hollow fiber, and flat sheet membranes. Dimethyl

Table 9.2 Performance of different types of membranes used in membrane crystallization.

Type	Material	Selectivity (%)	Membrane pore size	Thermal stability	Fouling resistance	Applications	References
Ultrafiltration	PAN, polysulfone	90–95	2–100 nm	140°C	High	Removal of fine particles and organic materials	[5, 46, 49, 60, 103, 104]
Microfiltration	PVDF, PP, zirconia	70–90	100–500 nm	120°C	Medium	Separation of large particles	[5, 46, 49, 60, 103, 104]
Reverse osmosis	Thin film composites, polyamide, PP	99.9	<1 nm	150°C	Ultra high	Desalination and purification of water	[5, 46, 49, 60, 103, 104]
Nanofiltration	Polyamide, PP and ceramics	95–99	1–2 nm	200°C	High	Desalination and ion separation	[5, 46, 49, 60, 103, 104]
Electrodialysis	Ion-exchange membranes	90–98	Variable	100°C	Low to medium	Ion separation	[5, 46, 49, 60, 103, 104]

formamide (DMF), *N*-methyl-2-pyrrolidone (NMP), and *N*,*N*-dimethyl acetamide (DMAc) are some common solvents in which PVDF is soluble [109].

PVDF membranes are used in various fields, including biofuel recovery, lithium-ion battery separators, ion exchange processes, microfiltration, ultrafiltration, membrane bioreactors (MBR), membrane distillation, gas separation and stripping, and the removal of pollutants like boron, volatile organic compounds, and ammonia from water. Moreover, PVDF materials are employed in the production of many commercial membranes that are available today. However, two significant issues restrict PVDF membrane applications: wetting in membrane contactors and fouling in water treatment and mineral recovery. Fouling and wetting reduce membrane efficiency and performance, increase operating costs, and can lead to failure. Hydrophilic or hydrophobic modifications to PVDF membranes have garnered significant interest due to their ability to enhance resistance to fouling and wetting. Surface modifications, such as hydrophilic polymer grafting or plasma treatment, can significantly improve fouling resistance, extend membrane life, and reduce maintenance needs [110].

9.5.4 Polypropylene (PP)

Mineral recovery and membrane crystallization are two common membrane technologies that utilize PP, a thermoplastic polymer. Due to its properties, PP is well-suited for these applications, particularly in harsh chemical environments. It is highly resistant to a variety of substances, including acids, bases, and organic solvents. This makes it ideal for aggressive solution processes, such as mineral recovery, where the membrane is exposed to strong chemicals [94].

The linear hydrocarbon polymer PP is characterized by its repeating $-CH_2CH(CH_3)-$ units. Polypropylene can exist in different forms: pure amorphous polymer (atactic) or semi-crystalline (atactic, isotactic, or syndiotactic) [111].

PP membranes can function effectively at temperatures as high as 90–100°C. While this range is not the highest among polymers, it is sufficient for many crystallization processes, which often occur at lower temperatures. Additionally, PP has excellent mechanical qualities, such as flexibility and tensile strength, enabling it to withstand the pressures required for membrane filtration without significant deformation. Due to its hydrophobic nature, PP can be advantageous in processes involving non-aqueous phases. However, this characteristic can also lead to fouling, especially when handling hydrophobic substances. Modifications can be made to reduce the likelihood of fouling [94].

9.5.5 Polyamide (PA)

Nylon, also referred to as polyamide, is a versatile polymer used in membrane applications such as mineral recovery and membrane crystallization. Its unique combination of properties hydrophilicity, chemical resistance, and mechanical strength makes it an excellent material for operations requiring durability and precision in separation under harsh conditions [112]. Polyamide thin-film composite (TFC) membranes have been highly successful, largely due to the formation of strong, ultra-thin polyamide films and the composite design, which allows for the independent optimization of each layer. Polyamide composites quickly outperformed their cellulosic counterparts in most applications, including organic molecule separations, mineral recovery, and gas separation, due to their remarkable performance and scalable manufacturing process [113, 114].

Polyamide films typically feature "ridge-and-valley" structures, resulting in surface roughness on the order of 100 nm. A rougher polyamide surface is believed to increase water permeability by expanding the membrane's surface area [115]. Polyamide membranes exhibit strong chemical resistance, particularly in basic and mildly acidic conditions. However, their use in extremely harsh chemical environments may be limited due to sensitivity to strong acids, bases, and oxidizing agents [116].

Unlike many other polymers, polyamide is naturally hydrophilic, which reduces fouling in aqueous environments. This hydrophilicity allows PA membranes to maintain higher flux rates and decrease the frequency of cleaning cycles, making them especially useful in water-based processes [114]. PA membranes are frequently used in membrane crystallization due to their ability to separate solutes based on size and molecular interactions. This selectivity is crucial in applications such as mineral recovery, pharmaceuticals, and specialty chemicals, where high-purity crystals are needed. Polyamide membranes' ability to selectively separate ions and minerals from complex mixtures enables the recovery of valuable metals like lithium, gold, and rare earth elements. This selectivity is also advantageous in nanofiltration and ion exchange processes [60, 117].

9.6 Operating Parameters and Process Optimization

In the field of membrane crystallization, process control is essential to maximize crystal output and ensure high quality. By considering the membrane's chemical-physical properties along with process variables such as temperature, concentration, and flow rate, it may be possible to adjust the solvent evaporation rate and the rate at which supersaturation occurs. The result is better control over the rate at which crystals form and grow. Consequently, membrane crystallization allows for the production of highly precise crystal morphologies and structures compared to standard crystallization methods [44].

It is also important to note that the surface composition of voided membranes can absorb dissolved molecules. This capturing process leads to localized concentration, which induces supersaturation and eventual nucleation. These conditions are critical for crystal formation as they promote uniformity and nucleation, which would otherwise be insufficient [90]. Optimization is necessary to prevent unwanted crystallization inside the module and tubing, which is essential for a functional MDCr process. Choosing the proper membrane distillation (MD) and crystallization operating conditions is vital. These factors include process temperature, solution supersaturation, flow rates, and crystallization time. Additionally, crystal size distribution is influenced by temperature and flow rate. A deeper understanding of these factors provides insight into the MDCr process, ensuring optimal performance while achieving zero liquid discharge to the environment [47].

9.6.1 Temperature

Temperature is a critical factor in membrane crystallization, as an increase in feed temperature can boost crystallization flux. As the feed temperature increases, vapor pressure rises, heightening the driving force across the membrane. Attia et al. [118] assessed the effect of temperature using synthetic electrospun PVDF, superhydrophobic alumina, and commercial PVDF membranes in a comparative air-gap membrane distillation (AGMD) process. They established a direct correlation between permeate flux and feed temperature [118]. Similarly, Chen et al. [119] observed that a higher temperature gradient positively affects the diffusion coefficient, leading to increased vapor flux. However, higher temperatures can also affect membrane stability [119]. Each system requires an optimal temperature, as the heat capacities of solutions vary with salt type and concentration [49].

9.6.2 Solution Supersaturation

The concentration of the feed solution eventually reaches its critical saturation due to the slow movement of water vapor from the feed stream into the distillate. A further increase in feed supersaturation facilitates the recovery of crystal salts from the crystallization reactor [22]. Feed concentration also alters flux performance by lowering vapor pressure and increasing temperature polarization [120, 121]. The rise in salt concentration reduces water vapor pressure, leading to a drop in flow. Moreover, concentration polarization intensifies as salt concentration increases, reducing both mass and heat transfer. Higher feed concentrations also heighten the risk of membrane fouling [120, 121], and the formation of crystals on the membrane surface can lead to pore blockage.

9.6.3 Impact of Feed Flow, Recirculation Rates, and Air Gap

The relationship between feed flow rate and permeate flux in membrane distillation has been studied. Chen et al. [119] discovered that decreasing temperature polarization and thermal boundary layer thickness results in an increase in permeate flux when the flow rate is increased. This was further supported by Naidu et al. [122], who observed that the permeate flux increased from 8.4 to 9.6 kg/m^{2}h when the flow rate was increased from 50 to 100 L/h. Heat and mass transfer are significantly improved when the membrane distillation process is run under convective diffusion conditions, particularly in transitional or turbulent areas. This is important for the process's economic viability [123]. Furthermore, through increasing heat transfer and decreasing boundary layer resistance, a high recirculation rate also helps to increase flux performance. Temperature polarization is greatly impacted by changes in recirculation rate, as noted by Martínez-Díez et al. [121], which further increases permeate flux. The air gap is important; a smaller distance improves flux and raises the temperature gradient, which increases permeating flux [49].

9.6.4 Membrane Characteristics

Hydrophobic microporous membranes, typically with pore diameters between 100 nm and 1 μm, are used in MD due to their non-wetting characteristics, which allow vapor passage while retaining liquid. An optimal pore size is essential to balance permeate quality and flux performance. Although larger pores may increase flux, they can reduce contaminant rejection. In addition to pore size and porosity, membrane thickness is a crucial property. Thinner membranes enhance mass and heat transfer, reducing resistance and increasing permeate flux [124].

Polymeric membranes such as PVDF, PP, and PTFE are commonly used in MD for medical dye extraction. PTFE is favored for its exceptional hydrophobicity, thermal stability, and chemical resistance, while modified PVDF membranes exhibit excellent contaminant rejection and performance. Although ceramic membranes are known for their mechanical, thermal, and chemical durability, their inherent hydrophilicity requires hydrophobic treatment for effective use in MD.

9.6.5 Fouling

Fouling in MDCr is a common issue that impairs process efficiency when microbiological, colloidal, organic, or inorganic contaminants accumulate on the membrane's surface or inside its pores. The consequences of fouling include blockages, decreased water flux, reduced salt rejection, higher operating costs, and diminished membrane hydrophobicity, which exacerbates membrane wetting and further decreases mineral salt rejection. Permeability flow reduction in MDCr is typically caused by either partial or complete membrane wetting. Complete wetting allows salt ions to pass through the membrane, degrading water quality.

Table 9.3 Parameter optimization of membrane crystallization and effect on mineral recovery.

Parameter	Range	Effect on mineral recovery	Prevention techniques	References
Feed concentration	35–150 g/L	High concentrations improve the nucleation process but with increased fouling	Supersaturation should be maintained with the controlled concentration gradient	[8, 22, 57, 125]
Feed temperature	Up to 100°C	Crystallization and solubility are dependent on temperature. High temperature leads to better mass transfer processes.	An optimum temperature should be maintained for enhanced solubility and reduced premature scaling	[8, 22, 57, 60, 125, 126]
Flow rate	5–100 L/h	A high flow rate may damage membrane surfaces leading to reduced fouling but can affect crystal size.	The flow rate should be optimized to reduce fouling.	[22, 57, 60, 125, 126]
Pressure	Up to 10 bar	Pressure controls flux and affects the crystal formation. High pressure means more compact crystals	The flux and crystal quality should be balanced with optimized pressure.	[22, 57, 125, 126]
pH	6–8	Influences ion solubility and membrane stability	Adjustment of pH to maintain membrane integrity with optimized solubility	[8, 22, 57, 60, 125, 126]

Fouling is influenced by hydrodynamic conditions, membrane properties, and feed solution characteristics, with scaling from sparingly soluble salts such as calcium carbonate and sulfate being particularly problematic. Fouling can manifest as non-porous, increasing hydraulic and thermal resistance, or porous, leading to thermal resistance. Addressing these issues is crucial for maintaining high MDCr process efficiency [47]. The parameter optimization of membrane crystallization and effect on mineral recovery are presented in Table 9.3.

9.7 Challenges and Mitigation Strategies

Future research projects may address the challenges related to the scalability and practical deployment of membrane crystallization technologies. Several criteria must be considered, including technological viability on a larger scale, system integration, process design, and economic feasibility [44]. Due to its low-grade thermal energy usage, membrane crystallization has been explored as a potential technology platform for solar desalination and wastewater treatment with high salinity. However, membrane crystallization still faces significant obstacles, such as fouling and pore wetting, which can lead to reduced permeation flow. Developing membranes with exceptional anti-wetting and scaling properties is, therefore, a promising approach to addressing these issues. Figure 9.4 provides the major challenges and opportunities for membrane crystallization.

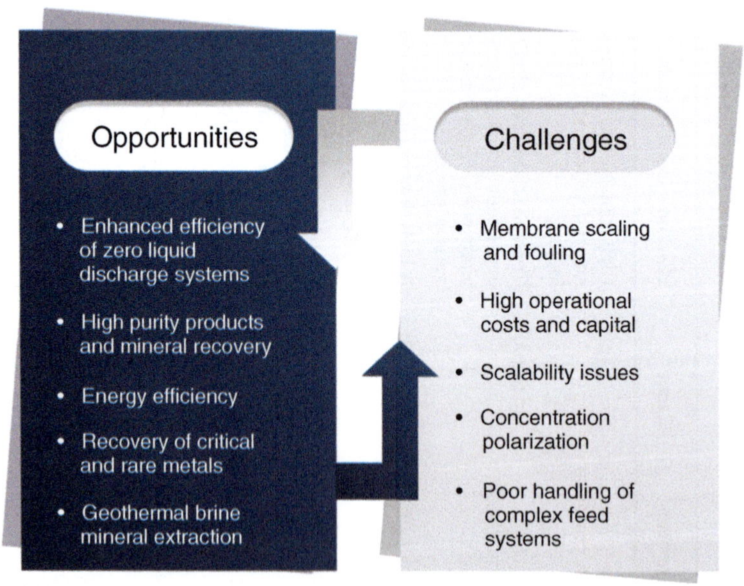

Figure 9.4 Challenges and opportunities for MCr.

According to Tong et al. [127], organic fouling and mineral scaling severely limit the efficiency of membrane desalination processes. When foulants and scalants interact, combined fouling and scaling behavior becomes more complex and realistic compared to individual fouling or scaling [127]. The quality of feedwater, operating conditions, and membrane characteristics all affect membrane fouling. Research aims to optimize conditions and modify the membrane surface to improve flux performance. Membrane fouling, which can be organic, biofouling, inorganic, or colloidal, results from the interaction of feed fluid particles with the membrane surface. However, a comprehensive understanding of membrane fouling remains lacking, as most studies investigate different causes and often use artificial solutions that do not accurately represent real-world scenarios. More mathematical simulations and models, such as computational fluid dynamics (CFD), are needed to predict and understand membrane fouling in MD systems, considering variables such as operating conditions, feed composition, and membrane characteristics. Most studies focus on inorganic fouling, with preventing inorganic foulants from becoming supersaturated being the most effective mitigation strategy. In addition to complex membrane cleaning techniques, such as using acidic solutions, which may shorten the membrane's lifespan, adjusting feed temperature and pH can also be beneficial. Organic fouling can be managed using cleaning techniques like backwashing or NaOH treatments, as polysaccharides are less likely to foul. Tough biological fouling, which often accompanies organic fouling, is particularly challenging to minimize without pre-treatment. While chlorination and high feed temperatures may help address biofouling, predicting the most effective approaches is difficult due to data limitations. Colloidal fouling is less studied, and completely removing colloids from the MD feed solution remains challenging, despite pre-treatment methods like ultrafiltration and microfiltration helping manage colloid levels [128].

In a study [129] on the effectiveness of a bench-scale MDCr method, Zhang et al. focused on how operating parameters affected crystal quality. They found that although permeate fluxes at 40°C and 50°C were consistent for all solutions, aquarium and raw salt solutions at 60°C showed a notable decrease compared to a laboratory NaCl solution. This decrease in flux was attributed to impurities, specifically calcium and magnesium salts, which can precipitate on the membrane surface at elevated temperatures, obstructing the pores. The study concluded that contaminants in salt solutions might cause increased variability in permeate flux at higher temperatures, emphasizing the requirement to address fouling and scaling issues for optimizing the MDCr process and improving its potential for large-scale applications [129].

Nthunya et al. [130], documented several issues with flat sheet membranes, including non-uniform crystal size distribution, fouling, and flux degradation. They suggested that using hollow fiber direct contact membrane distillation crystallization for mineral recovery from acid mine drainage could address these problems [130]. Although membrane-based techniques for lithium recovery are efficient, they are not energy-efficient, particularly when dealing with low-concentration brines. Lithium recovery via membrane technology is effective but energy-intensive and requires optimization. $LiMn_2O_4$ (LMO) batteries, although having high capacity and environmental benefits, face stability issues due to Mn^{3+} dissolution. Modified LMO materials, such as Ni-doped LMO, exhibit improved charge density and cyclic stability [131]. Optimization is crucial for viability, as even effective methods like electrodialysis face significant energy challenges [6]. Compared to pressure-driven membrane technologies that have been scaled up for practical applications, MD is a promising low-grade thermal energy technique for hypersaline brine from various industries. However, MD must overcome several significant technological issues that could compromise performance or cause process failure, including membrane fouling, mineral scaling, and pore wetting. Effective strategies for addressing scaling and fouling include membrane pretreatment, feedwater conditioning, and cleaning. Innovations such as pulse flow, intermittent back purging, and micro/nanobubbles further improve MD system performance. Future study should concentrate on comprehending the shared effects of various breakdown mechanisms using complex feed solutions. This method could bridge the gap between basic science and practical use, making MD a reliable method for treating a range of hypersaline brines [132].

9.8 Case Studies and Applications

The successful application of membrane crystallization for crystallization from generated water relies heavily on the availability of appropriate membranes and an understanding of related heat and mass transport phenomena. The material, shape, and surface properties of the membrane significantly impact process performance and stability [133, 134]. Furthermore, research on protein molecule crystallization has shown that used membranes can also affect crystallization kinetics, polymorphs, and crystal properties [135].

One study [46], explored the potential of membrane crystallization for recovering freshwater and minerals from microfiltered oilfield-produced water. This research investigated the impact of process variables, such as feed inlet temperature, feed flow rate, and membrane material, on crystal properties and transport phenomena. Both lab and semi-pilot scale tests demonstrated the potential of membrane crystallization for salt recovery and treated generated water. Microscopy, scanning electron microscopy (SEM), energy dispersive X-ray (EDX), and X-ray diffraction (XRD) analyses confirmed that the recovered crystals were sodium chloride with very high purity (>99.9%). Experimental results indicated that 16.4 kg of NaCl could be recovered per cubic meter of produced water, with a recovery factor of 37% [46].

Christie et al. [136] investigated membrane distillation–crystallization for creating carbonate minerals from carbon dioxide-loaded amine solutions, waste products such as fly ash, and desalination brines. They conducted trials with feed solutions dosed with 0.18 M Ca^{2+} and Mg^{2+} after being loaded with 5–15% CO_2 and heated to 40–50°C. Membranes with higher roughness and lower surface energy accelerated mineralization by 20% due to increased vapor flux. While membrane wetting tolerance improved by 96.2% at lower operating temperatures, the crystal growth rate decreased by 48.3% under the same conditions. Using sweeping gas membrane distillation improved membrane wetting tolerance by 37.5% and reduced the mineralization rate by 71.6% [136].

Bouchrit et al. [137] studied membrane crystallization in concentrated wastewater to produce high-quality water and sodium sulfate (Na_2SO_4) crystals. They kept permeate and feed flow rates constant at 2 L/h and conducted trials at a permeating temperature of 25°C. The recovery ratio reached 80%, with a Na_2SO_4 crystal yield of around 100 kg/m^3. The flow data showed cycles of decreasing and increasing, caused by phenomena in membrane crystallization that raise concentration to supersaturation and initiate crystallization. Scaling on the membrane surface contributed to a hyperbolic trend, with the highest cycle magnitude observed at $T_f = 70$°C, having a ΔJ of 7.81 kg/m^2 h. The seedling, growing at a rate of 1 g/h, helped control crystal growth, and SEM findings indicated that the heating-seeding combination could extend the membrane's lifespan. The method's environmental benefits were highlighted by the well-controlled, orthorhombic thenardite crystals and near-zero brine discharge [137].

Another study examined the membrane crystallization technique for recovering lithium carbonate (Li_2CO_3) crystals from artificial brine solutions. Using XRD and various microscopy techniques, the study analyzed the effects of salts (NaCl, KCl, and LiCl), feed temperature (40–60°C), and flow rate (0.81–1.3 L/min) on Li_2CO_3 crystallization in a binary solution. Results showed that higher feed temperatures led to larger and faster crystal formation. The addition of NaCl and KCl decreased the average crystal size while rising feed temperatures increased the average crystal size from 9.59 to 30.51 μm. Despite salt additions or operational parameters, the crystal morphology remained needle-like. LiCl accelerated crystallization due to the common-ion effect, but NaCl and KCl reduced crystal formation by increasing Li_2CO_3 solubility. These findings are important for further studies on recovering Li_2CO_3 from brine streams using membrane crystallization and other technologies [138].

A related study produced pure water and salt crystals from concentrated single and mixed solutions using tube membrane distillation integrated with crystallization. The research found that while saturation increased, membrane flux decreased dramatically but remained steady at 2.61 kg·m^2/h. The water yield had low conductivity (<10 μs/cm), and membrane rejection was unaffected by high quantities of zinc and nickel. The SO_4^{2-} common-ion effect promoted co-crystallization effectively. The study suggested that MD combined with crystallization is a promising technique for handling highly concentrated heavy metal solutions due to the membrane's exceptional fouling resistance [139].

Martinez et al. [140] explored using membrane crystallization to recover the crystalline form of the active medicinal drug G-1 from waste aqueous solutions. Their study found that G-1 needed to crystallize at a low feed flow rate of 5.0 mL/min. Vitrofural, a blend of G-1 and polyethylene glycol (PEG), allowed for higher concentration operations, although PEG decreased transmembrane flow by interfering with mass transfer. SEM images demonstrated that the crystallized G-1's surface shape matched that of the commercial product, indicating that membrane crystallization is a viable method for obtaining pharmaceutical chemicals like G-1 from waste streams [140].

Nthunya et al. [130] recently examined MDCr for recovering water and valuable minerals from acid mine drainage. At a feed inlet temperature of 52°C, high-quality freshwater was recovered with a 99.9% salt rejection rate and a permeate flux of 2.32 kg/m^2/h. Feed temperatures were varied to study their effects on the mineral crystal growth rate, size distribution, and shape. Sodium sulfate, dihydrogen sulfate, potassium iron oxide, manganese fluoride, and nickel oxide hydroxide were the primary minerals collected. Lower temperatures (34°C) produced orthorhombic structures, while higher temperatures (41°C and 52°C) encouraged co-crystallization. Crystal size ranged from 3.91 ± 0.97 to 14.04 ± 3.31 μm, with a restricted range for low recovery factors and a broader range as recovery progressed. Higher temperatures increased crystal formation rates, indicating the potential for economically recovering minerals and water. However, selective recovery of precious minerals might require additional units [130].

9.9 Economic Viability and Sustainability

Membrane crystallizers generally operate at ambient pressure and low temperatures (below 60°C), which reduces associated energy costs. The high purity, well-controlled nucleation, and growth achieved with membrane crystallization also minimize the need for extensive post-treatment of minerals. In contrast, conventional mining methods use physical, chemical, and sometimes thermal techniques to extract minerals from brines or ores. These methods often consume significant amounts of energy and can have adverse environmental impacts [40].

In applications where space efficiency, energy consumption, and process control are essential, membrane crystallization offers a more compact, efficient, and potentially cost-effective alternative to traditional methods. Conventional techniques, although reliable, may require more space and could be more expensive to operate and maintain [141]. MD, a novel method involving the creation of a temperature difference across a microporous hydrophobic membrane, provides several advantages over traditional procedures. These include the use of waste-grade energy, the ability to concentrate solutions to saturation, and the complete rejection of non-volatiles. The energy efficiency of MD makes it particularly attractive for processing produced water [142]. However, traditional treatment techniques have their own drawbacks, such as the use of hazardous chemicals, high treatment costs, large treatment footprints, lengthy retention times, and secondary pollution [143]. These methods also include the adsorption of dissolved organics on materials like activated carbon, organoclay, copolymers, zeolite, and resins, as well as dissolved air precipitation, chemical precipitation, chemical oxidation, and photocatalytic treatment. The analysis of different parameters for mineral extraction through membrane crystallization is summarized in Table 9.4.

Recovering nutrients plays a crucial role in sustainability and economic viability. Kim et al. [148] found that shale gas-produced water (SGPW) presents

Table 9.4 Analysis of different parameters for mineral extraction through membrane crystallization.

Parameters	Effect on membrane crystallization	Usage	References
Energy consumption	Lower energy consumption by optimizing membrane properties	100–500 kWh/ton	[28, 40, 46, 144, 145]
Water usage	Minimum water usage	10–100 L/kg of minerals	[22, 144, 146]
CO_2	Low emissions with minimum environmental impact	5–50 kg/ton of minerals	[22, 28, 144]
Efficiency	High efficiency but more research is required for scalability issues	80–90%	[22, 46, 144]
Cost	Economically feasible for recovery of minerals	$200–1,000/ton of recovered minerals	[22, 46, 144, 146, 147]
Scalability	Emerging technologies with research ongoing on scalability	–	[46, 144, 146]

significant technical challenges due to its high concentration of pollutants. MDCr has shown considerable potential for managing such wastewater by efficiently recovering both water and minerals and achieving near-ZLD conditions. The study demonstrated that under optimal conditions, MDCr could achieve an 84% recovery rate with a specific energy consumption as low as 28.2 kWh/m^3, recovering 2.72 kg/m^2 per day of solid product recovery (SPR). These findings highlight MDCr's capability to manage SGPW in a sustainable and energy-efficient manner [148]. While other methods like mechanical vapor compression and thermal evaporation have been used for SGPW treatment, they are often hindered by high costs, large footprints, and harmful chemicals. Recent investigations into membrane technologies offer a more cost-effective and practical alternative [148, 149].

Membrane technology also scales effectively, especially when processing waste streams to recover resources. Traditional mineral extraction methods often face higher costs due to the need for sophisticated infrastructure and environmental controls. High-purity products produced by membrane technology can command higher market prices compared to traditional methods, where purity may vary and additional refining processes could further increase costs [141].

One significant challenge in the industrialization of membrane crystallization is the energy-intensive nature of the initial process. This often requires various pretreatment techniques, such as membrane filtering, thermal evaporation, crystallization, and electrochemical procedures. Energy consumption can be influenced by factors such as the configuration of the MDCr system, the operational environment, and the characteristics of the brine being treated. To optimize energy use, comprehensive process simulations, energy audits, and optimization studies are necessary [5].

Despite the initial investment required for specialized technology, membrane crystallization offers long-term cost advantages, particularly in energy efficiency and environmental fulfillment. It is particularly useful in applications requiring compact operations and high-purity products. Conventional mineral extraction techniques, while reliable and scalable, often come with higher capital, operating, and environmental costs, making them less desirable in contexts where efficiency and sustainability are paramount.

9.10 Conclusion

In conclusion, in the form of a membrane, the process of mineral recovery could not be easier but cost-effective from seawater, thus, solving two problems in one shot, ecology and resource. Through crystal-clear brine material production in the absence of liquid discharge, this technology benefits nature more compared to the conventional methods. Membrane crystallization is not only a support to the circular economy by converting brine to become resources but it also contributes to the power of the process through its feature of scalability, energetic efficiency, and low cost-effectiveness. If membrane technologies and operating practices continue to evolve, then this practice would be the mainstay of the industries from energy storage to the pharmaceutical industry, thus, it will be a new way to recover scarce minerals and use the renewable resources wisely.

Questions

1. Which minerals are recoverable from seawater using membrane crystallization?
2. Why is membrane crystallization preferred over traditional methods for seawater mineral recovery?
3. How does membrane crystallization enhance the recovery of lithium from seawater?
4. What role does supersaturation play in membrane crystallization?
5. How do hydrophobic membranes aid in membrane crystallization?
6. What are the economic implications of using membrane crystallization for mineral extraction from seawater?
7. What is the environmental impact of membrane crystallization compared to traditional mining?
8. How does the membrane material affect the efficiency of crystallization?
9. What challenges are associated with membrane crystallization?
10. What is the purpose of antisolvent membrane crystallization?
11. How does membrane distillation crystallization (MDCr) differ from traditional crystallization?
12. What minerals are economically viable to recover from seawater desalination brine?

13 How can membrane crystallization contribute to zero-liquid discharge (ZLD)?

14 What is the significance of feed temperature in membrane crystallization?

15 Why are polymeric membranes like PVDF and PP preferred in membrane crystallization?

16 How does feed flow rate influence crystallization in membrane processes?

17 What are the main applications of the minerals recovered from seawater?

18 How is membrane crystallization used in pharmaceuticals?

19 What are the advantages of using membrane crystallization in brine management?

20 How does the air gap distance in membrane crystallization impact mineral recovery?

References

1 Loganathan, P., Naidu, G., and Vigneswaran, S. (2017). Mining valuable minerals from seawater: a critical review. *Environ. Sci.: Water Res. Technol.* 3 (1): 37–53. https://doi.org/10.1039/C6EW00268D.

2 Diallo, M.S., Kotte, M.R., and Cho, M. (2015). Mining critical metals and elements from seawater: opportunities and challenges. *Environ. Sci. Technol.* 49 (16): 9390–9399. https://doi.org/10.1021/acs.est.5b00463.

3 Rankin, W.J. (2011). *Minerals, Metals and Sustainability: Meeting Future Material Needs*. CSIRO publishing.

4 Kazi, O.A., Chen, W., Eatman, J.G. et al. (2023). Material design strategies for recovery of critical resources from water. *Adv. Mater.* 35 (36): 2300913.

5 Zhang, X., Koirala, R., Pramanik, B. et al. (2023). Challenges and advancements in membrane distillation crystallization for industrial applications. *Environ. Res.* 234: 116577.

6 Khalil, A., Mohammed, S., Hashaikeh, R., and Hilal, N. (2022). Lithium recovery from brine: recent developments and challenges. *Desalination* 528: 115611.

7 McLellan, B.C., Yamasue, E., Tezuka, T. et al. (2016). Critical minerals and energy–impacts and limitations of moving to unconventional resources. *Resources* 5 (2): 19.

8 Salmón, I.R. and Luis, P. (2018). Membrane crystallization via membrane distillation. *Chem. Eng. Process. Process Intensif.* 123: 258–271.

9 Sharkh, B.A., Al Amoudi, A.A., Fellows, C.L. et al. (2022). Seawater desalination concentrate—a new frontier for sustainable mining of valuable minerals. *NPJ Clean Water* 5 (1): 9.

10 Ortiz-Albo, P., Torres-Ortega, S., Gonzalez Prieto, M. et al. (2019). Techno-economic feasibility analysis for minor elements valorization from desalination concentrates. *Sep. Purif. Rev.* 48 (3): 220–241.

11 Bello, A.S., Zouari, N., Da'ana, D.A. et al. (2021). An overview of brine management: emerging desalination technologies, life cycle assessment, and metal recovery methodologies. *J. Environ. Manag.* 288: 112358.

12 Jiang, X., Niu, Y., Du, S., and He, G. (2022). Membrane crystallization: engineering the crystallization via microscale interfacial technology. *Chem. Eng. Res. Des.* 178: 454–465. https://doi.org/10.1016/j.cherd.2021.12.042.

13 Lu, H., Wang, J., Wang, T. et al. (2017). Crystallization techniques in wastewater treatment: an overview of applications. *Chemosphere* 173: 474–484. https://doi.org/10.1016/j.chemosphere.2017.01.070.

14 Flatscher, S., Bharati, V.P., and Dwivedi, A. (2023). Crystallization in water and used water purification. In: *Handbook of Water and Used Water Purification* (ed. J. Lahnsteiner), 1–24. Cham: Springer International Publishing. https://doi.org/10.1007/978-3-319-66382-1_8-1.

15 Xu, H., Zhou, Q., Wang, G., and Wang, J. (2010). Process for changing phenolic nitrate effluent into useful material by extracting, evaporating, crystallizing and coupling. Chinese Patent No. CN101654305-A. https://patents.google.com/patent/CN101654305A/en?oq=CN101654305-A.

16 Luo, X., Li, X., Wei, C. et al. (2022). Recovery of NaCl and Na_2SO_4 from high salinity brine by purification and evaporation. *Desalination* 530: 115631. https://doi.org/10.1016/j.desal.2022.115631.

17 Van Spronsen, J., Pascual, M.R., Genceli, F.E. et al. (2010). Eutectic freeze crystallization from the ternary Na_2CO_3–$NaHCO_3$–H_2O system. *Chem. Eng. Res. Des.* 88 (9): 1259–1263. https://doi.org/10.1016/j.cherd.2009.09.012.

18 Huang, H., Xiao, D., Pang, R. et al. (2014). Simultaneous removal of nutrients from simulated swine wastewater by adsorption of modified zeolite combined with struvite crystallization. *Chem. Eng. J.* 256: 431–438. https://doi.org/10.1016/j.cej.2014.07.023.

19 Rahman, M.M., Mohd, M.A., Salleh, U. et al. (2014). Production of slow release crystal fertilizer from wastewaters through struvite crystallization—A review. *Arab. J. Chem.* 7 (1): 139–155. https://doi.org/10.1016/j.arabjc.2013.10.007.

20 Lee, C.-I. and Yang, W.-F. (2005). Heavy metal removal from aqueous solution in sequential fluidized-bed reactors. *Environ. Technol.* 26 (12): 1345–1354. https://doi.org/10.1080/09593332608618613.

21 Myasnikov, S.K., Tikhonov, A.Y., Chipryakova, A.P., and Kulov, N.N. (2016). Removal of heavy metal ions from water by an combined sorption–crystallization process using activated clays. *Theor. Found. Chem. Eng.* 50 (4): 366–382. https://doi.org/10.1134/S0040579516040436.

22 Das, P., Dutta, S., and Singh, K.K. (2021). Insights into membrane crystallization: a sustainable tool for value added product recovery from effluent streams. *Sep. Purif. Technol.* 257: 117666. https://doi.org/10.1016/j.seppur.2020.117666.

23 Di Profio, G., Tucci, S., Curcio, E., and Drioli, E. (2007). Selective glycine polymorph crystallization by using microporous membranes. *Cryst. Growth Des.* 7 (3): 526–530. https://doi.org/10.1021/cg0605990.

24 Polino, M., Portugal, C.A.M., Di Profio, G. et al. (2019). Protein crystallization by membrane-assisted technology. *Cryst. Growth Des.* 19 (8): 4871–4883. https://doi.org/10.1021/acs.cgd.9b00223.

25 Di Profio, G., Curcio, E., Ferraro, S. et al. (2009). Effect of supersaturation control and heterogeneous nucleation on porous membrane surfaces in the crystallization of L -glutamic acid polymorphs. *Cryst. Growth Des.* 9 (5): 2179–2186. https://doi.org/10.1021/cg800838b.

26 Bosetti, L. and Mazzotti, M. (2020). Population balance modeling of growth and secondary nucleation by attrition and ripening. *Cryst. Growth Des.* 20 (1): 307–319. https://doi.org/10.1021/acs.cgd.9b01240.

27 Omar, H.M. and Rohani, S. (2017). Crystal population balance formulation and solution methods: a review. *Cryst. Growth Des.* 17 (7): 4028–4041. https://doi.org/10.1021/acs.cgd.7b00645.

28 Das, P., Dutta, S., Singh, K.K.K., and Maity, S. (2019). Energy saving integrated membrane crystallization: a sustainable technology solution. *Sep. Purif. Technol.* 228: 115722. https://doi.org/10.1016/j.seppur.2019.115722.

29 Sonawane, S.H., Shirsath, S.R., Khanna, P.K. et al. (2008). An innovative method for effective micro-mixing of CO_2 gas during synthesis of nano-calcite crystal using sonochemical carbonization. *Chem. Eng. J.* 143 (1–3): 308–313. https://doi.org/10.1016/j.cej.2008.05.030.

30 Zhao, C., Ge, R., Zhen, Y. et al. (2019). A hybrid process of coprecipitation-induced crystallization-capacitive deionization-ion exchange process for heavy metals removal from hypersaline ternary precursor wastewater. *Chem. Eng. J.* 378: 122136. https://doi.org/10.1016/j.cej.2019.122136.

31 Drioli, E., Di Profio, G., and Curcio, E. (2012). Progress in membrane crystallization. *Curr. Opin. Chem. Eng.* 1 (2): 178–182. https://doi.org/10.1016/j.coche.2012.03.005.

32 Bardi, U. (2010). Extracting minerals from seawater: an energy analysis. *Sustainability* 2 (4): 980–992. https://doi.org/10.3390/su2040980.

33 Turekian, K.K. (1977). The fate of metals in the oceans. *Geochim. Cosmochim. Acta* 41 (8): 1139–1144. https://doi.org/10.1016/0016-7037(77)90109-0.

34 Mends, E.A. and Chu, P. (2023). Lithium extraction from unconventional aqueous resources—a review on recent technological development for seawater and geothermal brines. *J. Environ. Chem. Eng.* 11 (5): 110710. https://doi.org/10.1016/j.jece.2023.110710.

35 Abdulsalam, A., Idris, A., Mohamed, T.A., and Ahsan, A. (2017). An integrated technique using solar and evaporation ponds for effective brine disposal management. *Int. J. Sustain. Energy* 36 (9): 914–925. https://doi.org/10.1080/14786451.2015.1135923.

36 Greenlee, L.F., Lawler, D.F., Freeman, B.D. et al. (2009). Reverse osmosis desalination: water sources, technology, and today's challenges. *Water Res.* 43 (9): 2317–2348. https://doi.org/10.1016/j.watres.2009.03.010.

37 Sabour, S.M.J.S. and Ghorashi, B. (2024). A comprehensive review of major water desalination techniques and mineral extraction from saline water. *Sep. Purif. Technol.* 349: 127913. https://doi.org/10.1016/j.seppur.2024.127913.

38 Kumar, A., Naidu, G., Fukuda, H. et al. (2021). Metals recovery from seawater desalination brines: technologies, opportunities, and challenges. *ACS Sustain. Chem. Eng.* 9 (23): 7704–7712. https://doi.org/10.1021/acssuschemeng.1c00785.

39 Ameen, F., Stagner, J.A., and Ting, D.S.-K. (2018). The carbon footprint and environmental impact assessment of desalination. *Int. J. Environ. Stud.* 75 (1): 45–58. https://doi.org/10.1080/00207233.2017.1389567.

40 Quist-Jensen, C.A., Macedonio, F., and Drioli, E. (2016). Integrated membrane desalination systems with membrane crystallization units for resource recovery: a new approach for mining from the sea. *Crystals* 6 (4): https://doi.org/10.3390/cryst6040036.

41 Wang, Y., Morosuk, T., and Cao, W. (2024). Carbon footprint of seawater desalination technologies: a review. *J. Energy Resour. Technol.* 146 (8): 080801. https://doi.org/10.1115/1.4065251.

42 Shahmansouri, A., Min, J., Jin, L., and Bellona, C. (2015). Feasibility of extracting valuable minerals from desalination concentrate: a comprehensive literature review. *J. Clean. Prod.* 100: 4–16. https://doi.org/10.1016/j.jclepro.2015.03.031.

43 Del Villar, A., Melgarejo, J., García-López, M. et al. (2023). The economic value of the extracted elements from brine concentrates of Spanish desalination plants. *Desalination* 560: 116678. https://doi.org/10.1016/j.desal.2023.116678.

44 Frappa, M., Macedonio, F., and Drioli, E. (2023). Membrane-assisted crystallization. *JRR* 1: 1018–1018. https://doi.org/10.61186/jrr.2308.1018.

45 Curcio, E. and Drioli, E. (2005). Membrane distillation and related operations—a review. *Sep. Purif. Rev.* 34 (1): 35–86. https://doi.org/10.1081/SPM-200054951.

46 Ali, A., Quist-Jensen, C., Macedonio, F., and Drioli, E. (2015). Application of membrane crystallization for minerals' recovery from produced water. *Membranes* 5 (4): 772–792. https://doi.org/10.3390/membranes5040772.

47 Chimanlal, I., Nthunya, L.N., Quist-Jensen, C., and Richards, H. (2022). Membrane distillation crystallization for water and mineral recovery: the occurrence of fouling and its control during wastewater treatment. *Front. Chem. Eng.* 4: 1066027. https://doi.org/10.3389/fceng.2022.1066027.

48 Qu, M., You, S., and Wang, L. (2021). Insights into nucleation and growth kinetics in seeded vacuum membrane distillation crystallization. *J. Membr. Sci.* 620: 118813. https://doi.org/10.1016/j.memsci.2020.118813.

49 Pramanik, B.K., Thangavadivel, K., Shu, L., and Jegatheesan, V. (2016). A critical review of membrane crystallization for the purification of water and recovery of minerals. *Rev. Environ. Sci. Biotechnol.* 15 (3): 411–439. https://doi.org/10.1007/s11157-016-9403-0.

50 Choi, Y., Naidu, G., Nghiem, L.D. et al. (2019). Membrane distillation crystallization for brine mining and zero liquid discharge: opportunities, challenges, and recent progress. *Environ. Sci.: Water Res. Technol.* 5 (7): 1202–1221. https://doi.org/10.1039/C9EW00157C.

51 Nthunya, L.N., Bopape, M.F., Mahlangu, O.T. et al. (2022). Fouling, performance and cost analysis of membrane-based water desalination technologies: a critical review. *J. Environ. Manag.* 301: 113922. https://doi.org/10.1016/j.jenvman.2021.113922.

52 Shirazi, M.M.A., Kirkebæk, B.S., Ali, A., and Quist-Jensen, C.A. (2024). Membrane crystallization for resource recovery from brine towards sustainability and circular economy. *Science Talks* 11: 100367. https://doi.org/10.1016/j.sctalk.2024.100367.

53 Macedonio, F. and Drioli, E. (2010). Hydrophobic membranes for salts recovery from desalination plants. *Desalin. Water Treat.* 18 (1–3): 224–234. https://doi.org/10.5004/dwt.2010.1775.

54 Charcosset, C., Kieffer, R., Mangin, D., and Puel, F. (2010). Coupling between membrane processes and crystallization operations. *Ind. Eng. Chem. Res.* 49 (12): 5489–5495. https://doi.org/10.1021/ie901824x.

55 Drioli, E., Curcio, E., Criscuoli, A., and Profio, G.D. (2004). Integrated system for recovery of $CaCO_3$, NaCl and $MgSO_4 \cdot 7H_2O$ from nanofiltration retentate. *J. Membr. Sci.* 239 (1): 27–38. https://doi.org/10.1016/j.memsci.2003.09.028.

56 Drioli, E., Criscuoli, A., and Curcio, E. (2002). Integrated membrane operations for seawater desalination. *Desalination* 147 (1–3): 77–81. https://doi.org/10.1016/S0011-9164(02)00579-9.

57 Chen, G., Lu, Y., Krantz, W.B. et al. (2014). Optimization of operating conditions for a continuous membrane distillation crystallization process with zero salty water discharge. *J. Membr. Sci.* 450: 1–11. https://doi.org/10.1016/j.memsci.2013.08.034.

58 Alkhudhiri, A., Darwish, N., and Hilal, N. (2012). Membrane distillation: a comprehensive review. *Desalination* 287: 2–18. https://doi.org/10.1016/j.desal.2011.08.027.

59 Shirazi, M.M.A., Kargari, A., Ismail, A.F., and Matsuura, T. (2016). Computational fluid dynamic (CFD) opportunities applied to the membrane distillation process: state-of-the-art and perspectives. *Desalination* 377: 73–90. https://doi.org/10.1016/j.desal.2015.09.010.

60 Chabanon, E., Mangin, D., and Charcosset, C. (2016). Membranes and crystallization processes: state of the art and prospects. *J. Membr. Sci.* 509: 57–67. https://doi.org/10.1016/j.memsci.2016.02.051.

61 Xiong, Y.Q., Peng, T., and Shi, Z.F. (2014). Research progress and enhancement of MDC process. *Adv. Mater. Res.* 955–959: 2530–2533. https://doi.org/10.4028/www.scientific.net/AMR.955-959.2530.

62 Luis, P. (2018). *Fundamental Modeling of Membrane Systems: Membrane and Process Performance*. Elsevier.

63 Kesieme, U.K., Milne, N., Cheng, C.Y. et al. (2014). Recovery of water and acid from leach solutions using direct contact membrane distillation. *Water Sci. Technol.* 69 (4): 868–875. https://doi.org/10.2166/wst.2013.788.

64 Edwie, F. and Chung, T.-S. (2012). Development of hollow fiber membranes for water and salt recovery from highly concentrated brine via direct contact membrane distillation and crystallization. *J. Membr. Sci.* 421–422: 111–123. https://doi.org/10.1016/j.memsci.2012.07.001.

65 Hickenbottom, K.L. and Cath, T.Y. (2014). Sustainable operation of membrane distillation for enhancement of mineral recovery from hypersaline solutions. *J. Membr. Sci.* 454: 426–435. https://doi.org/10.1016/j.memsci.2013.12.043.

66 Nariyoshi, Y.N., Pantoja, C.E., and Seckler, M.M. (2016). Evaluation of sodium chloride crystallization in membrane distillation crystallization applied to water desalination. *Braz. J. Chem. Eng.* 33 (3): 675–690. https://doi.org/10.1590/0104-6632.20160333s20150133.

67 Naidu, G., Jeong, S., Vigneswaran, S. et al. (2016). A review on fouling of membrane distillation. *Desalin. Water Treat.* 57 (22): 10052–10076. https://doi.org/10.1080/19443994.2015.1040271.

68 Jamil, S., Jeong, S., and Vigneswaran, S. (2016). Application of pressure assisted forward osmosis for water purification and reuse of reverse osmosis concentrate from a water reclamation plant. *Sep. Purif. Technol.* 171: 182–190. https://doi.org/10.1016/j.seppur.2016.07.036.

69 Lutchmiah, K., Verliefde, A.R.D., Roest, K. et al. (2014). Forward osmosis for application in wastewater treatment: a review. *Water Res.* 58: 179–197. https://doi.org/10.1016/j.watres.2014.03.045.

70 Profio, G.D., Curcio, E., Cassetta, A. et al. (2003). Membrane crystallization of lysozyme: kinetic aspects. *J. Cryst. Growth* 257 (3–4): 359–369. https://doi.org/10.1016/S0022-0248(03)01462-3.

71 Benecke, J., Rozova, J., and Ernst, M. (2018). Anti-scale effects of select organic macromolecules on gypsum bulk and surface crystallization during reverse osmosis desalination. *Sep. Purif. Technol.* 198: 68–78. https://doi.org/10.1016/j.seppur.2016.11.068.

72 Li, W., Van Der Bruggen, B., and Luis, P. (2014). Integration of reverse osmosis and membrane crystallization for sodium sulphate recovery. *Chem. Eng. Process. Process Intensif.* 85: 57–68. https://doi.org/10.1016/j.cep.2014.08.003.

73 Tian, M. (2015). Design and fabrication of nanofibrous thin film composite membranes by electrospinning for osmotically driven membrane processes. Nanyang Technological University. https://doi.org/10.32657/10356/65646.

74 Cath, T.Y., Drewes, J.E., Lundin, C.D., and Hancock, N.T. (2010). Forward osmosis—reverse osmosis process offers a novel hybrid solution for water purification and reuse. *IDA J. Desal. Water Reuse* 2 (4): 16–20. https://doi.org/10.1179/ida.2010.2.4.16.

75 Dutta, S. and Nath, K. (2018). Feasibility of forward osmosis using ultra low pressure RO membrane and Glauber salt as draw solute for wastewater treatment. *J. Environ. Chem. Eng.* 6 (4): 5635–5644. https://doi.org/10.1016/j.jece.2018.08.037.

76 Drioli, E., Curcio, E., and Di Profio, G. (2005). State of the art and recent progresses in membrane contactors. *Chem. Eng. Res. Des.* 83 (3): 223–233. https://doi.org/10.1205/cherd.04203.

77 Nagy, E. (2019). Membrane distillation. In: *Basic Equations of Mass Transport Through a Membrane Layer* (ed. E. Nagy), 483–496. Elsevier. https://doi.org/10.1016/B978-0-12-813722-2.00019-4.

78 Macedonio, F., Ali, A., and Drioli, E. (2017). 3.10 membrane distillation and osmotic distillation. In: *Comprehensive Membrane Science and Engineering* (ed. E. Drioli, L. Giorno, and E. Fontananova), 282–296. Elsevier. https://doi.org/10.1016/B978-0-12-409547-2.12221-8.

79 Di Profio, G., Curcio, E., and Drioli, E. (2010). Membrane crystallization technology. In: *Comprehensive Membrane Science and Engineering* (ed. E. Drioli and L. Giorno), 21–46. Elsevier. https://doi.org/10.1016/B978-0-08-093250-7.00018-9.

80 Kramer, H.J.M., Anisi, F., Eral, B., and Stankiewicz, A.I. (2017). 4.1 membrane crystallization technology and process intensification. In: *Comprehensive Membrane Science and Engineering* (ed. E. Drioli, L. Giorno, and E. Fontananova), 1–7. Elsevier. https://doi.org/10.1016/B978-0-12-409547-2.12660-5.

81 Saberi, A., Goharrizi, A.S., and Ghader, S. (2009). Precipitation kinetics of sodium bicarbonate in an industrial bubble column crystallizer. *Cryst. Res. Technol.* 44 (2): 159–166. https://doi.org/10.1002/crat.200800429.

82 Alvarez, M.G., Sefidi, V.S., Beguin, M. et al. (2022). Osmotic membrane distillation crystallization of $NaHCO_3$. *Energies* 15 (7): 2682. https://doi.org/10.3390/en15072682.

83 Zhu, Y., Haut, B., Halloin, V., and Delplancke-Ogletree, M.-P. (2005). Investigation of crystallization kinetics of sodium bicarbonate in a continuous stirred tank crystallizer. *J. Cryst. Growth* 282 (1–2): 220–227. https://doi.org/10.1016/j.jcrysgro.2005.05.004.

84 Ali, A., Quist-Jensen, C.A., Drioli, E., and Macedonio, F. (2018). Evaluation of integrated microfiltration and membrane distillation/crystallization processes for produced water treatment. *Desalination* 434: 161–168. https://doi.org/10.1016/j.desal.2017.11.035.

85 Macedonio, F., Ali, A., Poerio, T. et al. (2014). Direct contact membrane distillation for treatment of oilfield produced water. *Sep. Purif. Technol.* 126: 69–81. https://doi.org/10.1016/j.seppur.2014.02.004.

86 Ali, A., Quist-Jensen, C.A., Jørgensen, M.K. et al. (2021). A review of membrane crystallization, forward osmosis and membrane capacitive deionization for liquid mining. *Resour. Conserv. Recycl.* 168: 105273. https://doi.org/10.1016/j.resconrec.2020.105273.

87 Macedonio, F., Curcio, E., and Drioli, E. (2007). Integrated membrane systems for seawater desalination: energetic and exergetic analysis, economic evaluation, experimental study. *Desalination* 203 (1–3): 260–276. https://doi.org/10.1016/j.desal.2006.02.021.

88 Shaulsky, E., Karanikola, V., Straub, A.P. et al. (2019). Asymmetric membranes for membrane distillation and thermo-osmotic energy conversion. *Desalination* 452: 141–148. https://doi.org/10.1016/j.desal.2018.11.005.

89 Echaide-Górriz, C., Zapata, J.A., Etxeberría-Benavides, M. et al. (2020). Polyamide/MOF bilayered thin film composite hollow fiber membranes

with tuned MOF thickness for water nanofiltration. *Sep. Purif. Technol.* 236: 116265. https://doi.org/10.1016/j.seppur.2019.116265.

90 Curcio, E., Profio, G.D., and Drioli, E. (2003). A new membrane-based crystallization technique: tests on lysozyme. *J. Cryst. Growth* 247 (1–2): 166–176. https://doi.org/10.1016/S0022-0248(02)01794-3.

91 Tuo, L., Ruan, X., Xiao, W. et al. (2019). A novel hollow fiber membrane-assisted antisolvent crystallization for enhanced mass transfer process control. *AICHE J.* 65 (2): 734–744. https://doi.org/10.1002/aic.16438.

92 Fern, J.C.W., Ohsaki, S., Watano, S., and Pfeffer, R. (2018). Continuous synthesis of nano-drug particles by antisolvent crystallization using a porous hollow-fiber membrane module. *Int. J. Pharm.* 543 (1–2): 139–150. https://doi.org/10.1016/j.ijpharm.2018.03.025.

93 Chen, D., Singh, D., Sirkar, K.K., and Pfeffer, R. (2015). Porous hollow fiber membrane-based continuous technique of polymer coating on submicron and nanoparticles via antisolvent crystallization. *Ind. Eng. Chem. Res.* 54 (19): 5237–5245. https://doi.org/10.1021/acs.iecr.5b00155.

94 Himma, N.F., Anisah, S., Prasetya, N., and Wenten, I.G. (2016). Advances in preparation, modification, and application of polypropylene membrane. *J. Polym. Eng.* 36 (4): 329–362. https://doi.org/10.1515/polyeng-2015-0112.

95 Banerjee, P., Das, R., Das, P., and Mukhopadhyay, A. (2018). Membrane technology. In: *Carbon Nanotubes for Clean Water* (ed. R. Das), 127–150. Cham: Springer International Publishing. https://doi.org/10.1007/978-3-319-95603-9_6.

96 Fathizadeh, M., Aroujalian, A., and Raisi, A. (2012). Effect of lag time in interfacial polymerization on polyamide composite membrane with different hydrophilic sub layers. *Desalination* 284: 32–41. https://doi.org/10.1016/j.desal.2011.08.034.

97 Kang, Y., Obaid, M., Jang, J. et al. (2018). Novel sulfonated graphene oxide incorporated polysulfone nanocomposite membranes for enhanced-performance in ultrafiltration process. *Chemosphere* 207: 581–589. https://doi.org/10.1016/j.chemosphere.2018.05.141.

98 Lee, A., Elam, J.W., and Darling, S.B. (2016). Membrane materials for water purification: design, development, and application. *Environ. Sci.: Water Res. Technol.* 2 (1): 17–42. https://doi.org/10.1039/C5EW00159E.

99 Koga, Y., Fujieda, H., Meguro, H. et al. (2018). Biocompatibility of polysulfone hemodialysis membranes and its mechanisms: involvement of fibrinogen and its integrin receptors in activation of platelets and neutrophils. *Artif. Organs* 42 (9): https://doi.org/10.1111/aor.13268.

100 Pouresmaeel-Selkjani, P., Jahanshahi, M., and Peyravi, M. (2017). Mechanical, thermal, and morphological properties of nanoporous reinforced polysulfone membranes. *High Perform. Polym.* 29 (7): 759–771. https://doi.org/10.1177/0954008316656742.

101 Pagidi, A., Saranya, R., Arthanareeswaran, G. et al. (2014). Enhanced oil–water separation using polysulfone membranes modified with polymeric additives. *Desalination* 344: 280–288. https://doi.org/10.1016/j.desal.2014.03.033.

102 Mamah, S.C. et al. (2021). Recent development in modification of polysulfone membrane for water treatment application. *J Water Process Eng.* 40: 101835. https://doi.org/10.1016/j.jwpe.2020.101835.

103 Olufade, A.O. and Simonson, C.J. (2018). Characterization of the evolution of crystallization fouling in membranes. *ACS Omega* 3 (12): 17188–17198. https://doi.org/10.1021/acsomega.8b01058.

104 Xue, W., Cheng, G., Wu, Y. et al. (2020). Mineral particles fouling analysis and cleaning in seawater reverse osmosis desalination. *Desalin. Water Treat.* 198: 71–79. https://doi.org/10.5004/dwt.2020.26052.

105 Nunes, S.P., Culfaz-Emecen, P.Z., Ramon, G.Z. et al. (2020). Thinking the future of membranes: perspectives for advanced and new membrane materials and manufacturing processes. *J. Membr. Sci.* 598: 117761. https://doi.org/10.1016/j.memsci.2019.117761.

106 Asadi, A., Nazari, S., Gholami, F., and Dolatshah, M. (2023). Facile and effectual surface modification of polyacrylonitrile (PAN)-based ultrafiltration membranes via manipulating the synergistic interaction of dopamine and enzymes. *J. Water Process Eng.* 52: 103562. https://doi.org/10.1016/j.jwpe.2023.103562.

107 Ding, S.G., Cheng, X.Q., Jiang, Z.X. et al. (2015). Pore morphology control and hydrophilicity of polyacrylonitrile ultrafiltration membranes. *J. Appl. Polym. Sci.* 132 (20): 41991. https://doi.org/10.1002/app.41991.

108 Saxena, P. and Shukla, P. (2021). A comprehensive review on fundamental properties and applications of poly(vinylidene fluoride) (PVDF). *Adv. Compos. Hybrid Mater.* 4 (1): 8–26. https://doi.org/10.1007/s42114-021-00217-0.

109 Liu, F., Hashim, N.A., Liu, Y. et al. (2011). Progress in the production and modification of PVDF membranes. *J. Membr. Sci.* 375 (1–2): 1–27. https://doi.org/10.1016/j.memsci.2011.03.014.

110 Kang, G. and Cao, Y. (2014). Application and modification of poly(vinylidene fluoride) (PVDF) membranes—a review. *J. Membr. Sci.* 463: 145–165. https://doi.org/10.1016/j.memsci.2014.03.055.

111 Mark, J.E. (ed.) (2007). *Physical Properties of Polymers Handbook*. New York, NY: Springer https://doi.org/10.1007/978-0-387-69002-5.

112 Pervaiz, M., Faruq, M., Jawaid, M., and Sain, M. (2017). Polyamides: developments and applications towards next-generation engineered plastics. *Curr. Org. Synth.* 14 (2): 146–155. https://doi.org/10.2174/1570179413666160831112159.

113 Lee, K.P., Arnot, T.C., and Mattia, D. (2011). A review of reverse osmosis membrane materials for desalination—development to date and future potential. *J. Membr. Sci.* 370 (1–2): 1–22. https://doi.org/10.1016/j.memsci.2010.12.036.

114 Freger, V. and Ramon, G.Z. (2021). Polyamide desalination membranes: formation, structure, and properties. *Prog. Polym. Sci.* 122: 101451. https://doi.org/10.1016/j.progpolymsci.2021.101451.

115 Ma, X., Yang, Z., Yao, Z. et al. (2019). Tuning roughness features of thin film composite polyamide membranes for simultaneously enhanced permeability, selectivity and anti-fouling performance. *J. Colloid Interface Sci.* 540: 382–388. https://doi.org/10.1016/j.jcis.2019.01.033.

116 Shao, S., Zeng, F., Long, L. et al. (2022). Nanofiltration membranes with crumpled polyamide films: a critical review on mechanisms, performances, and environmental applications. *Environ. Sci. Technol.* 56 (18): 12811–12827. https://doi.org/10.1021/acs.est.2c04736.

117 Jeon, S. and Patel, R. (2022). Recovery of valuable minerals from sea water by membrane separation and adsorption process: a review. *Membr. J* 32 (1): 13–22. https://doi.org/10.14579/MEMBRANE_JOURNAL.2022.32.1.13.

118 Attia, H., Osman, M.S., Johnson, D.J. et al. (2017). Modelling of air gap membrane distillation and its application in heavy metals removal. *Desalination* 424: 27–36. https://doi.org/10.1016/j.desal.2017.09.027.

119 Chen, T.-C., Ho, C.-D., and Yeh, H.-M. (2009). Theoretical modeling and experimental analysis of direct contact membrane distillation. *J. Membr. Sci.* 330 (1–2): 279–287. https://doi.org/10.1016/j.memsci.2008.12.063.

120 Martínez, L. (2004). Comparison of membrane distillation performance using different feeds. *Desalination* 168: 359–365. https://doi.org/10.1016/j.desal.2004.07.022.

121 Martínez-Díez, L., Florido-Díaz, F.J., and Vázquez-González, M.I. (1999). Study of evaporation efficiency in membrane distillation. *Desalination* 126 (1–3): 193–198. https://doi.org/10.1016/S0011-9164(99)00174-5.

122 Naidu, G., Choi, Y., Jeong, S. et al. (2014). Experiments and modeling of a vacuum membrane distillation for high saline water. *J. Ind. Eng. Chem.* 20 (4): 2174–2183. https://doi.org/10.1016/j.jiec.2013.09.048.

123 Liu, H., Jacob, K.I., and Wang, Y. (2024). Scaling in membrane distillation (MD): current state of art insight on mechanisms and Membrane design. *Desalination* 581: 117539. https://doi.org/10.1016/j.desal.2024.117539.

124 Alobaidani, S., Curcio, E., Macedonio, F. et al. (2008). Potential of membrane distillation in seawater desalination: thermal efficiency, sensitivity study and cost estimation. *J. Membr. Sci.* 323 (1): 85–98. https://doi.org/10.1016/j.memsci.2008.06.006.

125 Sparenberg, M.-C., Chergaoui, S., Sefidi, V.S., and Luis, P. (2021). Crystallization control via membrane distillation-crystallization: a review. *Desalination* 519: 115315.

126 Anisi, F., Thomas, K.M., and Kramer, H.J. (2017). Membrane-assisted crystallization: membrane characterization, modelling and experiments. *Chem. Eng. Sci.* 158: 277–286.

127 Tong, T., Liu, X., Li, T. et al. (2023). A tale of two foulants: the coupling of organic fouling and mineral scaling in membrane desalination. *Environ. Sci. Technol.* 57 (18): 7129–7149. https://doi.org/10.1021/acs.est.3c00414.

128 Alkhatib, A., Ayari, M.A., and Hawari, A.H. (2021). Fouling mitigation strategies for different foulants in membrane distillation. *Chem. Eng. Process. Process Intensif.* 167: 108517. https://doi.org/10.1016/j.cep.2021.108517.

129 Zhang, X., Koirala, R., Pramanik, B. et al. (2024). Performance of membrane distillation assisted crystallization and crystal characteristics for resource recovery from desalination brine. *Desalination* 574: 117244. https://doi.org/10.1016/j.desal.2023.117244.

130 Nthunya, L.N., Pinier, J., Ali, A. et al. (2024). Valorization of acid mine drainage into potable water and valuable minerals through membrane distillation crystallization. *Sep. Purif. Technol.* 334: 126084. https://doi.org/10.1016/j.seppur.2023.126084.

131 Shang, X., Hu, B., Nie, P. et al. (2021). $LiNi_{0.5}Mn_{1.5}O_4$-based hybrid capacitive deionization for highly selective adsorption of lithium from brine. *Sep. Purif. Technol.* 258: 118009. https://doi.org/10.1016/j.seppur.2020.118009.

132 Horseman, T., Yin, Y., Christie, K.S. et al. (2021). Wetting, scaling, and fouling in membrane distillation: state-of-the-art insights on fundamental mechanisms and mitigation strategies. *ACS EST Eng.* 1 (1): 117–140. https://doi.org/10.1021/acsestengg.0c00025.

133 Guillen-Burrieza, E., Thomas, R., Mansoor, B. et al. (2013). Effect of dry-out on the fouling of PVDF and PTFE membranes under conditions simulating intermittent seawater membrane distillation (SWMD). *J. Membr. Sci.* 438: 126–139. https://doi.org/10.1016/j.memsci.2013.03.014.

134 Gryta, M. (2007). Influence of polypropylene membrane surface porosity on the performance of membrane distillation process. *J. Membr. Sci.* 287 (1): 67–78. https://doi.org/10.1016/j.memsci.2006.10.011.

135 Di Profio, G., Tucci, S., Curcio, E., and Drioli, E. (2007). Controlling polymorphism with membrane-based crystallizers: application to form I and II of paracetamol. *Chem. Mater.* 19 (10): 2386–2388. https://doi.org/10.1021/cm0701005.

136 Christie, K.S.S., McGaughey, A., McBride, S.A. et al. (2023). Membrane distillation–crystallization for sustainable carbon utilization and storage.

Environ. Sci. Technol. 57 (43): 16628–16640. https://doi.org/10.1021/acs.est.3c04450.

137 Bouchrit, R., Boubakri, A., Mosbahi, T. et al. (2017). Membrane crystallization for mineral recovery from saline solution: study case Na_2 SO_4 crystals. *Desalination* 412: 1–12. https://doi.org/10.1016/j.desal.2017.02.021.

138 Shirazi, M.M.A., Ali, A., and Quist-Jensen, C.A. (2024). Membrane crystallization for recovery of lithium carbonate crystals: study on process parameters and salts effect for Li_2CO_3-NaCl-KCl-LiCl solutions. *Desalination* 571: 117109. https://doi.org/10.1016/j.desal.2023.117109.

139 Lou, X.-Y., Xu, Z., Bai, A.-P. et al. (2020). Separation and recycling of concentrated heavy metal wastewater by tube membrane distillation integrated with crystallization. *Membranes* 10 (1): 19. https://doi.org/10.3390/membranes10010019.

140 Martínez, M.B., Jullok, N., Negrín, Z.R. et al. (2014). Membrane crystallization for the recovery of a pharmaceutical compound from waste streams. *Chem. Eng. Res. Des.* 92 (2): 264–272. https://doi.org/10.1016/j.cherd.2013.07.029.

141 Sparenberg, M.-C., Ruiz Salmón, I., and Luis, P. (2020). Economic evaluation of salt recovery from wastewater via membrane distillation-crystallization. *Sep. Purif. Technol.* 235: 116075. https://doi.org/10.1016/j.seppur.2019.116075.

142 Shaffer, D.L., Arias Chavez, L.H., Ben-Sasson, M. et al. (2013). Desalination and reuse of high-salinity shale gas produced water: drivers, technologies, and future directions. *Environ. Sci. Technol.* 47 (17): 9569–9583. https://doi.org/10.1021/es401966e.

143 Fakhru'l-Razi, A., Pendashteh, A., Abdullah, L.C. et al. (2009). Review of technologies for oil and gas produced water treatment. *J. Hazard. Mater.* 170 (2–3): 530–551. https://doi.org/10.1016/j.jhazmat.2009.05.044.

144 Shen, L., Dang, M., and Han, X. (2023). Recent advances in membrane crystallization. *CrystEngComm* 25 (17): 2503–2517.

145 Tufa, R.A., Noviello, Y., Di Profio, G. et al. (2019). Integrated membrane distillation-reverse electrodialysis system for energy-efficient seawater desalination. *Appl. Energy* 253: 113551.

146 Kang, Z., Guo, H., Fan, L. et al. (2021). Scalable crystalline porous membranes: current state and perspectives. *Chem. Soc. Rev.* 50 (3): 1913–1944.

147 Poirier, K., Al Mhanna, N., and Patchigolla, K. (2022). Techno-economic analysis of brine treatment by multi-crystallization separation process for zero liquid discharge. *Separations* 9 (10): 295. https://doi.org/10.3390/separations9100295.

148 Kim, J., Kim, J., and Hong, S. (2018). Recovery of water and minerals from shale gas produced water by membrane distillation crystallization. *Water Res.* 129: 447–459. https://doi.org/10.1016/j.watres.2017.11.017.

149 Alkhudhiri, A., Darwish, N., and Hilal, N. (2013). Produced water treatment: application of air gap membrane distillation. *Desalination* 309: 46–51. https://doi.org/10.1016/j.desal.2012.09.017.

10

Integrated Hybrid Systems for Water Desalination

Heba Abdallah[1] and Eman Sobhy[2]

[1] *Chemical Engineering Department, Engineering and Renewable Energy Research Institute, National Research Centre, Dokki, Giza, Egypt*
[2] *Water Pollution Research Department, Environmental and Climate Change Research Institute, National Research Centre, Dokki, Giza, Egypt*

10.1 Introduction

Due to the scarcity of water and trends toward finding alternative means to make the water unfit for human use suitable for human consumption, energy-saving sources have been looking for ways to apply water desalination regarding the cost of energy. Due to the increasing interest in preserving the environment and the use of environmental materials and natural energy sources, a trend has been to use renewable energy, although there are many limitations such as lack of space, large land requirements, and the renewable energy facilities investment costs. Therefore, desalination technologies have been linked to renewable energy to reduce the cost of desalination processes although renewable energy is still considered expensive [1, 2].

The produced freshwater reached 120 million m^3/day in 2020 all over the world, with 40 million m^3/day in the Middle East. The disadvantage of any desalination system is the consumption of high energy. Also, the consumption was combined with pollution by emission gasses [3]. Table 10.1 provides an overview comparison for some desalination technologies in energy requirements.

Using recovery devices in desalination plants reduces the energy consumption of the plants. Fossil fuels are the main source of the energy for desalination systems and the source of environmental pollution. To reduce the consumption of energy in water desalting processes that used fossil fuels, some conditions can be applied such as using waste heat, or renewable energy. Also, utilizing thermodynamic cycles and configurations to be highly efficient, the enhancement of the system

Water Treatment and Desalination: Membrane Separation using Renewable Energy Resources, First Edition. Edited by Mohamed Bassyouni.

Table 10.1 Energy consumption for some desalination processes.

Desalination type	MSF	MED-TVC	MED	MVC	RO	ED
Typical unit size (m^3/d)	50,000–70,000	10,000–35,000	5,000–15,000	100–2,500	24,000	—
Electrical energy consumption (kWh/m^3)	4–6	1.5–2.5	1.5–2.5	7–12	3–7	2.6–5.5
Thermal energy consumption (kWh/m^3)	190–390	145–390	230–390	none	none	none
Electrical equivalent for thermal energy (kWh/m^3)	9.5–19.5	9.5–25.5	5–8.5	none	none	none
Total energy consumption (kWh/m^3)	13.5–25.5	11–28	6.5–11	7–12	3–7	2.6–5.5

Source: Alkaisia et al. [3]/with permission of Elsevier.

performance produced water quality, and reduced environmental pollution [4–6]. However, the discharge of a high volume of brine creates dangerous problems for marine ecosystems [2].

Thermal desalination represents half of the desalination market, while desalination by membranes represents the other half of the market. Thermal desalination technologies depend on saline water heating and then collecting the condensed vapor called the distillate. These technologies are mostly used in seawater desalination, categorized into three processes: multi-stage flash distillation (MSF), vapor compression distillation (VCD), and multi-effect distillation (MED) [7, 8].

The famous technologies that depend on the membrane in water desalination are reverse osmosis (RO) and electrodialysis (ED) processes. RO processes are based on the passing of water through semi-permeable membranes using high-pressure pumps (HPPs), which consume high energy. It was found that to desalinate 1,000 m^3/day to freshwater, 10,000 tons/year of oil is required [9].

10.2 Desalination by Membrane Technology

10.2.1 Reverse Osmosis (RO)

Nowadays, RO technology is considered a famous technology in the world but it needs a high-pressure pump to produce high water purity. Pressure ranges for desalinating salty water are 50–80 bar and brackish water is 10–20 bar [10]. Before RO units there are pretreatment stages, which depend on the source and the quality of the feeding water. Pretreatment stages extend the membrane lifespan by eliminating fouling. Improvements in the membrane's permeability and salt rejection capabilities can lower membrane costs, which in turn impacts the cost of the RO units and, ultimately, the desalination plant [2].

Although the RO method is most popular in Europe, Australia, and the USA, it still uses a lot of energy—roughly 3.4 kWh/m^3, while the wastewater treatment energy consumption is about kWh/m^3. The environmental pollution that emits from RO desalination units can reach 954 tons/day of CO_2 for units of 250,000 m^3/day productivity. Using an energy recovery device (ERD) in desalination plants of 5,000 m^3/day saves around 34–42% energy and if the plants have a capacity of more than 5,000 m^3/day, the energy saves around 55–65% [6].

10.2.2 Electrodialysis and Electrodialysis Reversal (ED/EDR)

ED and electrodialysis reverse (EDR) are the voltage-driven process. It needs the current power supply to produce electrical potential, which is the influencing factor in separating salts from water based on positive and negative charges of ions. The ED system is composed of sequences of membranes located on the opposite side of the electrodes (cathode and anode). The positive charge ions transfer toward the cathode, which passes the cationic exchange membrane (CEM) and is rejected by the anionic exchange membrane (AEM) [11]. The anions transfer to the anode, pass the AEM, and are rejected by the CEM. There are two compartments in this method, one compartment for concentration and the other one for dilution. Figure 10.1 illustrates the principle of ED and reverse electrodialysis (RED) processes. EDR is based on the periodic reversal of polarity, two to four times each hour, to effectively minimize the scaling and fouling of membranes, which leads to operating for huge water retrievals [13].

10.2.3 Membrane Distillation (MD)

Membrane distillation (MD) is a thermally pressure-driven membrane process. The MD system is based on passing the saline solution after heating it to the hydrophobic membrane, which permits water vapor to permeate inside the pores and prevent liquid passing. Due to the use of a cooling system, the water vapor condenses on the other side of the membrane [14, 15]. MD could use other types of energy such as solar energy and waste energy from any enterprise as a substitute energy source for heating. Moreover, it might be used in conjunction with other desalination technologies to cut down on energy and running expenses. Four distinct kinds of MD could be used: direct contact

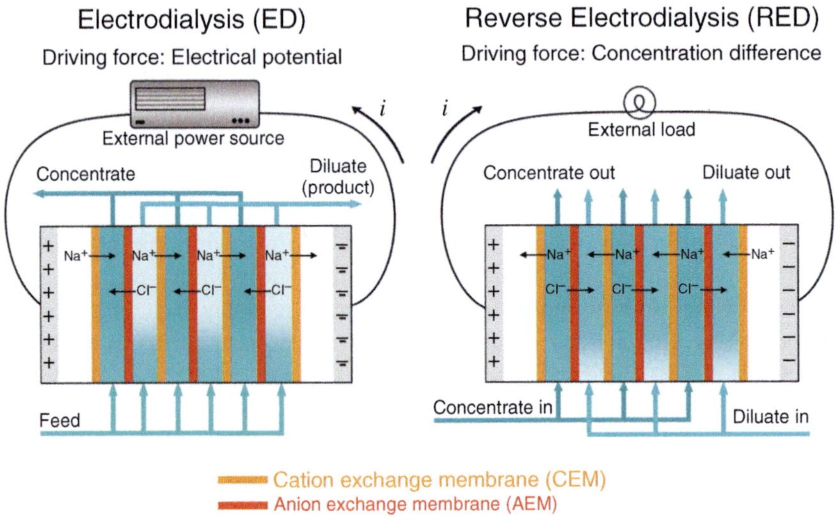

Figure 10.1 Electrodialysis and reverse electrodialysis techniques. Source: Tedesco et al. [12]/with permission of Elsevier.

membrane distillation (DCMD), vacuum membrane distillation (VMD), sweep gas membrane distillation (SGMD), and air gap membrane distillation (AGMD). For instance, a small-scale VMD process producing 10 kg/m^{2}h of permeate flux at constant temperature and seven various pressure and flow rate combinations utilizes 1.7 kWh/m^3 of energy. The SE was adjusted for additional operational settings, ranging from 1.70 to 3.96 kWh/m^3. High flux rates are produced by using high vacuum pressure, which boosts energy-efficient behavior [16]. The AGMD system requires no additional high-pressure pump to operate because it operates at atmospheric pressure, which results in relatively inexpensive electricity and heat costs. Additionally, AGMD can utilize low-strength polymer fittings and pipes that can tolerate little.

Some 50% of the capital cost of DCMD comes from the membrane cost and 30% of capital costs come from operation and maintenance costs. Using a low-grade source of heat or industrial waste heat can lead to the reduction of the price cost of freshwater as shown in Table 10.2 [17].

There are two methods by which heat is transported via the membrane matrix in MD: conduction heat that represents 20–50% of the heat loss because there is no induced flux through the membrane and latent heat of vaporization according to mass transfer through the membrane. According to that process, two heat recovery systems of MD rely on lowering operating costs: both internal and external heat recovery [18]. Usually, DCMD is used in conjunction with external heat exchangers to provide ***external heat recovery***, where the hot permeated water (distillate)

Table 10.2 Energy sources and water cost for different systems.

System	Source of energy	Water cost ($/m^3)
MD	Normal available energy	1.17
MD	Low-grade energy source	0.64
MD	Industrial waste heat	0.26
Nanofiltration + RO	With energy recovery device	0.56
Nanofiltration + RO	Without energy recovery device	0.8
RO + MD	Normal available energy	1.25

Source: Bappy et al. [17] IEOM Society International.

was exploited to heat the feed, allowing the MD module to perform heat recovery. By using this method, the production costs can be reduced to about \$2/m^3, in comparison with RO plant with 5 m^3/hour of productivity. For instance, 66% of latent heat recovery in a DCMD process is achieved by combining an integrated membrane system with a heat exchanger [18].

AGMD is an example of ***internal heat recovery***. Because of the internal latent heat recovery mechanism, AGMD often represents an interior condensation surface. The preheated feed was in contact with the membrane when it was fed seawater in a counter-direction to the condensate plate where the vapor at the air gap absorbs the latent heat of condensation. In a low-cost module like the Memstill® method, which combines latent heat recovery with AGMD, the module has the counter-current flow, the brine was utilized to recover the latent heat. It is made up of many vertical, hydrophobic, microporous membranes that have parallel cooling walls on both sides of the module that are spaced apart by air gaps as shown in Figure 10.2.

In counter-current mode the feeding water flows through the evaporator tube of the membrane after passing through a heater and dense hydrophilic tube walls that have an air gap between them. The vapor diffuses across the hydrophobic, microporous membrane that makes up the evaporator tube wall. The heat is recovered when the produced vapor condenses in the condenser tube, where the vapor pressure difference is caused by the difference in temperature between feed and cold side tubes, which brush the vapor through the membrane pores in the evaporator tube and air gap. An extensive air gap can yield energy efficiency of 85–90% [18].

In order to contrast RO and MD compared to RO, MD has reduced hardware expenses since it does not need HPP. The membrane and condenser partitions, however, drive up the costs of MD modules; in contrast, Memstill's costs are approximately 0.17/m^3, whereas ROs are \$0.25–0.35/m^3.

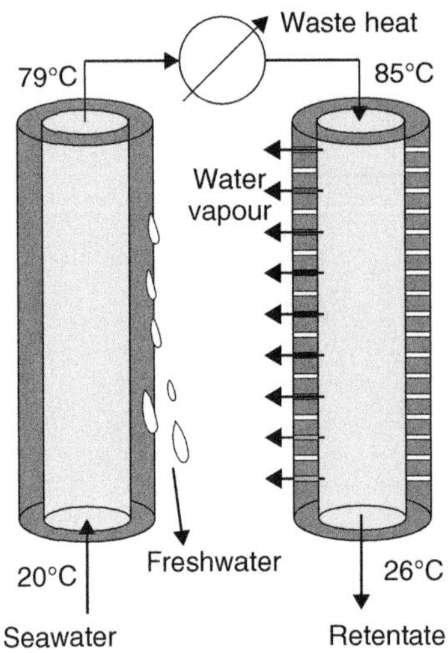

Figure 10.2 Principle of Memstill process. Source: Zhang et al. [18]/with permission of Elsevier.

10.2.4 Other Membrane Processes

RO membranes for desalination represent 60–90% of the world market share, but the limitation of the RO process and thermal desalination process has led to research for other membrane technologies [19]. The forward osmosis (FO) technique is a technology based on differences in salinity gradients. It provides high water recovery and low power consumption. Furthermore, the FO membrane offers draw solution recovery, reverse salt diffusion, and reduced membrane regeneration requirements. The FO system consists of a selective membrane and two side liquids, feed that is considered freshwater and the other side is a draw solute, which mostly is hypersaline (brine) water. According to the osmotic pressure, the freshwater migrates to the draw solute to dilute it. The primary drawbacks of the FO process are high production costs for treated water since many FO membrane packages are needed to overcome concentration polarization, which lowers permeate flux and ultimately results in water recovery [20]. Pressure retards osmosis (PRO) techniques reduce the energy consumption of desalination due to facilitating the diluted solution migration through membrane osmosis from the feeding side to the brine stream under pressure. This technique can produce a high hydraulic pressure, which powers hydro-turbines through power generation caused by the difference in osmotic pressure around two membrane sides.

A few simulation systems were constructed close to the river and the sea, and the findings showed that the energy usage was 0.72 kWh/m^3. The SEC was lowered to 0.256 kWh/m^3 using the river's water because producing power requires a lot of water [19, 21]. The capacitive deionization (CDI) process involves passing saline water between two porous carbon electrodes through an electrosorption ion process. The ions migrate to the electrodes by their charges after a constant current is given to the electrodes to create an electrical potential. This technology is an appealing choice for desalination because it uses a lot less energy than other conventional techniques [22, 23].

10.3 Hybrid Energy Sources for Water Desalination

10.3.1 High-Pressure Recovery of Brine Using Energy Recovery Device

High-pressure feeding water is needed in desalting water systems such as RO. Salinity of water, the process configuration, operating pressure, and water recovery are the parameters that determine the SEC for the process. Using high pressure leads to high energy that goes to waste in the form of high-pressure remaining brine.

Using an ERD can minimize energy loss during desalination. By extracting the residual energy from the RO system's concentrate and returning it to the feed, the ERD works to increase energy efficiency. Many ways of pumping can be applied while using ERD such as [24–26]:

Pumping from Hydraulic to Mechanical Assistance (such as a Pelton Wheel) as shown in Figure 10.3a. The apparatus is reliant on a turbine coupled to a motor and a pump. Depending on the pump being utilized, the turbines can be either Francis or Pelton wheel turbines. This process doubles the energy conversion of the brine by transforming its hydraulic energy into mechanical energy that is used to rotate the shaft. Next, the shaft's mechanical energy is converted to produce the feed hydraulic energy [25]. A case study is the fifteen RO trains at Al Jubail desalination plant in Saudi Arabia with a feed flow rate of 723 m^3/hour and working temperature range from 17 to 35°C and a pressure range of 62.4–84.2 bar.

The Pelton wheel turbine is the energy recovery technology that is being utilized. At 71.5 bar, the ERD rate is 485 m^3/hour, resulting in an overall efficiency of 81.85%. The plant has an average SEC of 7.42 kWh/m^3, and each train has an average SEC difference from 6.64to 8.74 kWh/m^3, meaning that 22.63% of the energy was saved via ERD.

The Madinat Yanbu Al-Sinaiyah SWRO desalination plant, which consists of 15 ERD trains, is another example study. While some of them run at a reject flow of 660 m^3 per hour, others run at a reject flow of 750 m^3 per hour. The feed has a salinity of approximately 42,000 mg/L. using ERD reduces the plant's average SEC from 5.56 to 4.67 kWh/m^3.

Hydraulically Driven Pumping in Series (TurboCharger) Figure 10.3b indicates this way of using ERDs, It has a casing containing an impeller, a

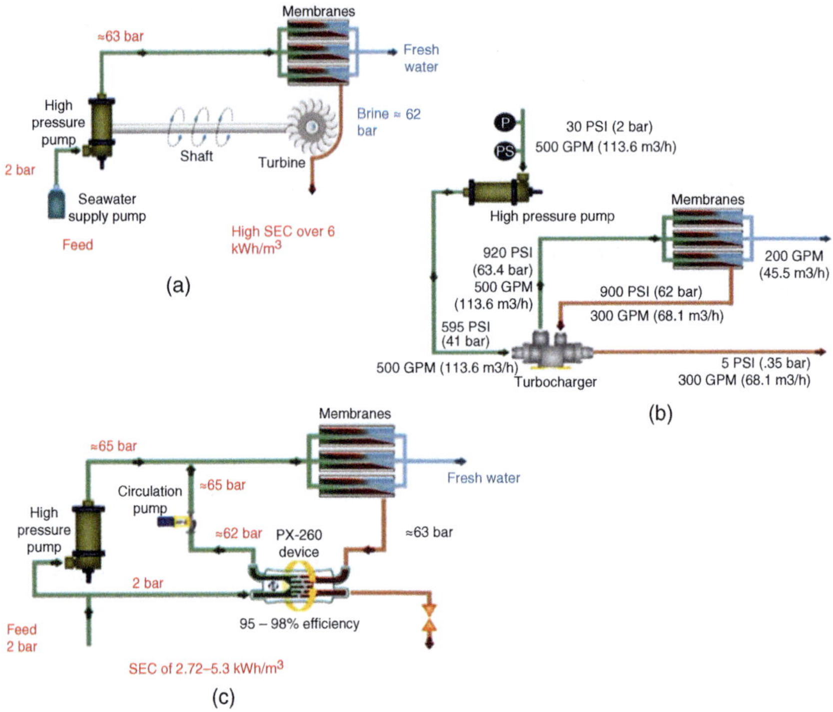

Figure 10.3 (a) Pelton wheel, (b) Turbocharger, and (c) Parallel pumping with hydraulic drive.

turbine, and a runner, they are connected to the shaft and they are in series with the pump of feed. Examples of these types of ERDs are the hydraulic pressure booster HPB, the Pelton-drive pump, and the hydraulic turbocharger.

This model of ERD can be used in small or mid-size desalination plants but it was limited because of its size, so the systems failed due to the lack of the operation efficiency. The case study was using a turbocharger in the Atlantic Ocean SWRO plant that is composed of four trains, to produce 2,500 m^3/day capacity with a recovery of 40%.

Using an ERD, the system's overall energy usage is reduced to 4.18 kWh/m^3 after increasing the feed pressure by the turbocharger to 82.76 bar, where the HPP uses 3.83 kWh/m^3 of energy overall while saving 8% of that energy [24, 25].

Parallel Pumping with Hydraulic Drive As illustrated in Figure 10.3c, this is regarded as the third class of ERD. In order to force the amount of feed to equal the amount of permeate, it uses pumping to the feed, where ERD takes place in parallel with the pump [26].

The third class of ERD provides 90% energy efficiency by exchanging brine hydraulic energy to feed hydraulic energy. There are commonly two types in this class: Dual work exchanger (DWEER) and a pressure exchanger (PX) [27]. DWEER's design uses a piston to divide the feed and reject, reducing the amount of intermixing between the two items.

A 500 m^3/hour Grand Cayman plant in Spain, with a salinity of 37,000 mg/L, provides SEC of 3.00 kWh/m^3, was the case study that employed DWEER. By employing DWEER, the plant's capacity increased to 1,699 m^3/day and SEC reduced to 2.22 kWh/m^3.

The famous and recent ERD is ERI (Energy Recovery, Inc) PX—Energy Recovery. The PX device is responsible for transforming a high-pressure fluid stream into a low-pressure stream. The PX device is made from ceramic that provides stability, and corrosion resistance. The brine pressure is used to pressurize the feed water. Setting the brine flow rate at a slightly higher flow rate lubricates the rotor, and a small amount of feeding water helps to facilitate the lubrication process.

Thus, a part of the brine flow is supplied to the membranes by the PX device, and a small amount of feedwater flow is supplied to the permeate portion by the high-pressure pump, which also has room for a tiny amount of lubricating flow, as shown in Figure 10.4. The plant in Western Australia (Perth Sea Water plant) serves as the case study for the use of PX. The plant can produce 143,000 m^3 per day. The brine stream from the RO unit flows through the twelve arrays

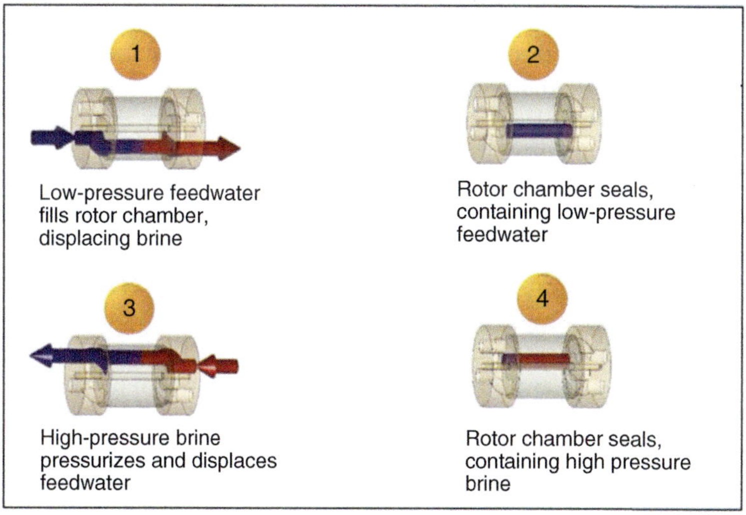

Figure 10.4 The PX device at work.

of sixteen ERD models as the PX 220 provides a capacity of 800 m^3 per hour. Vertical booster pumps (BPs) are used for RO modules to increase this pressure, where the facility uses between 3.2 and 3.5 kWh/m^3 in total. However, the RO unit's energy usage dropped to 1.4 kWh/m^3 [25].

High-Efficiency Pumps RO systems, operation are based on HPPs and BPs, so the efficiency of these pumps is the main factor for SEC in the RO system. Increasing the pump size can increase its efficiency.

The plant can be composed of several pumps, so the number of pumps can be determined according to the plant production capacity [5]. Using large and high-efficiency centrifugal HPP to aid the multiple trains of RO can save energy. However, Multistage centrifugal pumps become more efficient as pump flow increases. In desalination plants, the typical individual pump's efficiency ranges from 80% to 83%. However, the efficiency of the HPP should rise to 85% if the RO unit with two trains works by one high-pressure pump [6].

10.3.2 Integrated Renewable Energy with Membrane Technology

Different shapes of solar collectors are combined with desalination units such as flat plates, vacuum tubes, or central concentrators. Most common solar-desalination units can be powered by photovoltaic (PV) panels, where the solar energy is converted into electricity, but the backup batteries should be used for stabilization. Table 10.3 indicates the comparison between integrated desalination systems with renewable energy.

The disadvantage of the solar system is that the required energy should be stored during daylight, so the desalination unit needs a huge storage area to operate continuously [29]. Another disadvantage is that the desert places drop the dust on the collectors, which leads to a reduction in the heat capacity. Photovoltaic systems can provide 110–120 kWh/m^2 of energy collector, so the production of 1 m^3/day of treated water needs 8 kWh/m^3 that can be produced from a collector's area of 26.5–28 m^2.

Low-, medium-, and high-temperature solar collectors are the three different types available. The heat generated from high-temperature collectors like Parabolic trough collectors can be used to produce electricity or heating in RO desalination systems or thermal desalination systems respectively [29]. Using Parabolic trough collectors and a solar tower can produce power approximately from 5 to 200 MW [2].

The usefulness of the high wind potential in the coastal area can provide promising approaches in the desalination of seawater. The research indicates that using a wind turbine of a size of 6 to 30 kW in combination with a desalination unit can

Table 10.3 Comparison between integrated desalination systems with renewable energy sources.

Process	Design capacity (m^3/d)	Energy demand (kWh/m^3)	Water cost (USD/m^3)
PV-RO	<100	Electrical: 0.5–1.5 (BW); 45 (SW)	6.5–9.1 (BW); 11.7–15.6 (SW)
Wind-RO	50–2,000	Electrical: 0.5–1.5 (BW); 4–5 (SW)	Units under 100 m^3/d : 3.9–6.5 (BW); 6.5–9.1 (SW); Over 1,000 m^3/d
PV-EDR	<100	Electrical: 3–4; 0.6–1 (brackish water)	10.4–11.7; 3–16
Wind-MVC	<100	10–14 Electrical for seawater	5.2–7.8
Solar Stills	<0.1	Solar passive	1.3–6.5
Geothermal-MD	20,000	66.03 kW/kg/h	0.5
Solar-Multieffect Humidification	1–100	31.1 kWh/m^3	2.6–6.5

Source: Rabiee et al. [28]/with permission of Elsevier.

produce a high price of freshwater. So, recent research studied the combination of hybrid renewable energy systems like wind and solar. Some models were designed to combine the MED system with the MVC system and renewable energy (wind turbine + solar) to produce 100 m^3/hour. The need for water and the energy needed for desalination were taken into account in the Greek Islands. The initiative generated treated water for major islands at 1.45 euros/m^3 and small islands at 2.6 euros/m^3, given that the current cost of water in these regions is between 7 and 9 euros/m^3. However, in the south of Tunisia, the hybrid system from PV, wind, and hydrogen used the generated electricity from wind and solar sources, where the excess has been used as power for the electrolysis (EL) unit to produce hydrogen, as shown in Figure 10.5. If the system is exposed to any shortage in the input power, the fuel cell will provide the required power. The ability to work in hard conditions and the low cost of maintenance are factors, for which hydrogen was chosen. This plant produces 193.6 m^3/hour of freshwater. 14,400 people may access freshwater from this system in the winter thanks to its yearly water supply.

However, in the southern region of Tunisia, a hybrid system of PV, wind, and hydrogen has utilized the electricity produced by these sources, with any surplus power going to power the EL unit that produces hydrogen, as shown in

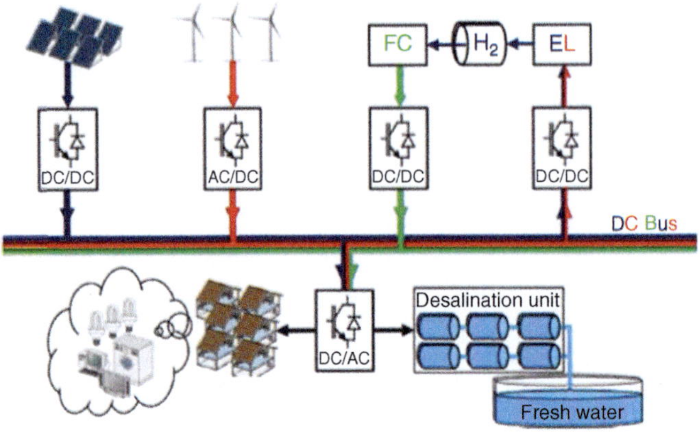

Figure 10.5 Integrated wind, PV, and hydrogen systems for desalination. Source: Smaoui et al. [30]/with permission of Elsevier.

Figure 10.5. If the system is exposed to any shortage in the input power, the fuel cell will provide the required power. The ability to work in hard conditions and low cost of maintenance are factors, for which hydrogen was chosen. This plant produces 193.6 m^3/hour of freshwater.

10.4 Desalination Using Nuclear Energy

Desalination using nuclear energy should be placed in coastal areas due to the requirement of using a large amount of water for the cooling systems. The desalination cost using nuclear energy is close to the cost of desalination using energy produced from fossil fuel. There are many nuclear plants combined with desalination plants. The first desalination plant built based on power generated from the nuclear plant was in India. It is based on a combination of MSF-RO systems. The treated water production cost usin.g nuclear energy was from 0.4 to 1.8 $/$m^3$ according to the type of nuclear reactor that is used and the plant desalination system [2].

Many hybrid desalination systems were combined with nuclear energy, while the cost depends on the type of nuclear reactor, the nuclear heating reactor (NHR), can be used to produce power for hybrid M with RO desalination system. Also, it can be used in MED alone or hybrid VC (vapor compression) with MED systems. However, in China, INET, the Tsinghua University Institute of Nuclear Energy Technology developed a MED water desalination plant that was coupled with a nuclear heating reactor (200 MW). They then constructed plants in the Shandong Peninsula in China based on feasibility studies, using two different processes: VC/MED, which produced 120,000 m^3/day, and MED, which produced 160,000 m^3/day [2].

10.5 Hybrid Desalination Systems

Most desalination technologies depend on external sources to generate energy, whether fuel or renewable energy, but there are membrane-specific methods that can produce energy, illustrated in this section.

10.5.1 Hybrid Reverse Osmosis, Membrane Distillation, and Reverse Electrodialysis

Integrated membrane operations offer several merits such as productivity, control management, ability to scale up, and appropriateness for integration [31]. Figure 10.6 indicates the integrated membrane system as a case study using the combination of RO/MD/RED. In this research, after pretreatment, (0.5 M NaCl) seawater enters the RO stage. The concentrate is returned to PX, and the treated fluid is pumped with high pressure to the RO unit. After passing through a Booster Pump (BP) and meeting the HP stream, the generated water from PX is fed back into the RO unit. Two streams—the brine stream and the permeate stream—are generated from RO. The brine with a flow rate of 500 m^3/hour is sent to PX, and its output is directed toward a heater before arriving at DCMD at a temperature of between 40 and 60°C. The retentate from DCMD was fed to RED, also the pretreated seawater was used on the other side of RED to generate electricity [32].

RO single-stage electrical energy consumption was 2.3 kWh/m^3 without using PX and after using PX it was 1.54 kWh/m^3. The electrical energy consumption

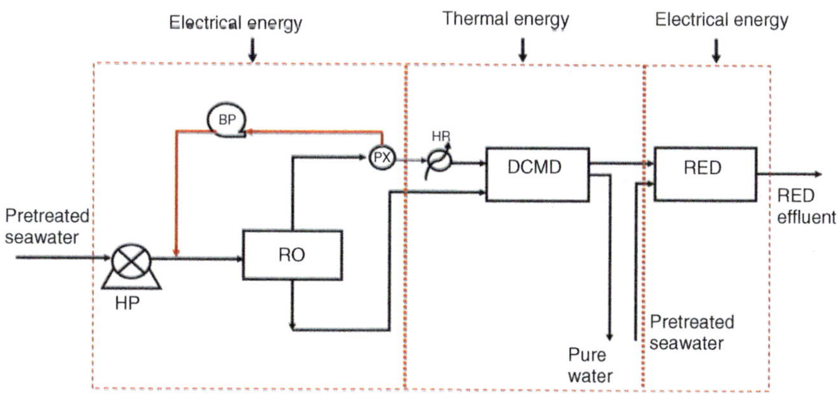

Figure 10.6 Scheme of RO/DCMD/RED integrated processes.

was reduced by 13.3% to the consumption of 1.3 kWh/m^3. However, increasing the brine concentration provides no sign of electric energy consumption after using DCMD but the thermal energy consumption was observed to be 4.2 kWh/m^3 at feeding temperature 60°C. Integration with RED using high-concentration brine from DCMD produces the lowest energetic point at a temperature of 40°C that was 2.9 kWh/m^3, with a reduction of 23% compared with 4.23 kWh/m^3 without integration with RED [33].

10.5.2 Integrated Membrane System Between RO and PRO

Introducing the PRO in the desalination process reduces the brine as a waste and reduces the energy demand and brine environmental problem. O'Toole et al. have reported the experimental results from the integrated system of RO and PRO units. This system provides a power density of 1.1–2.3 W/m^2, but this system can lower the energy required for desalination to less than 1 kWh/m^3 [34]. Another study was performed by the combination of PRO and MD in a closed-loop, where the reject from the MD system was used to extract freshwater from the feeding side of PRO, as shown in Figure 10.7. This integrated system provides high recovery, enhances power production, reduces membrane fouling, and decreases the impact on the environment [35].

The Global MVP Project in Korea (Membrane distillation, Valuable source recovery, Program) is one of the RO/MD/PRO case studies. This project aims to reduce the brine concentration and reduce energy consumption. The brine from RO enters MD as a feed. The brine produced from it enters as a draw solution to PRO, while the treated wastewater is used as the feed solution of PRO. The

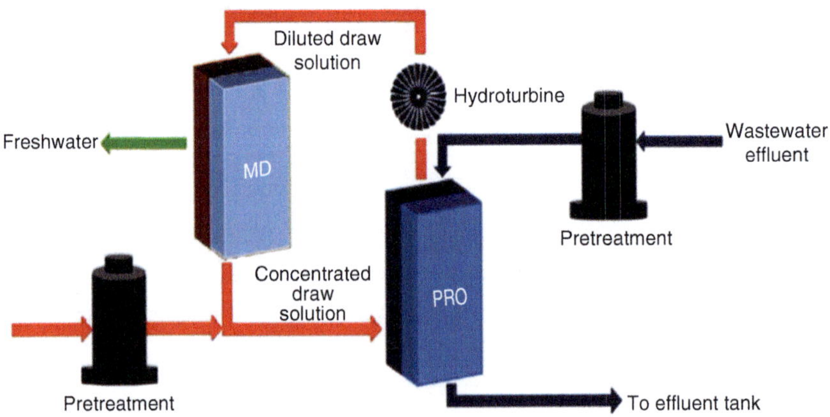

Figure 10.7 Schematic drawing of the integration of MD with PRO.

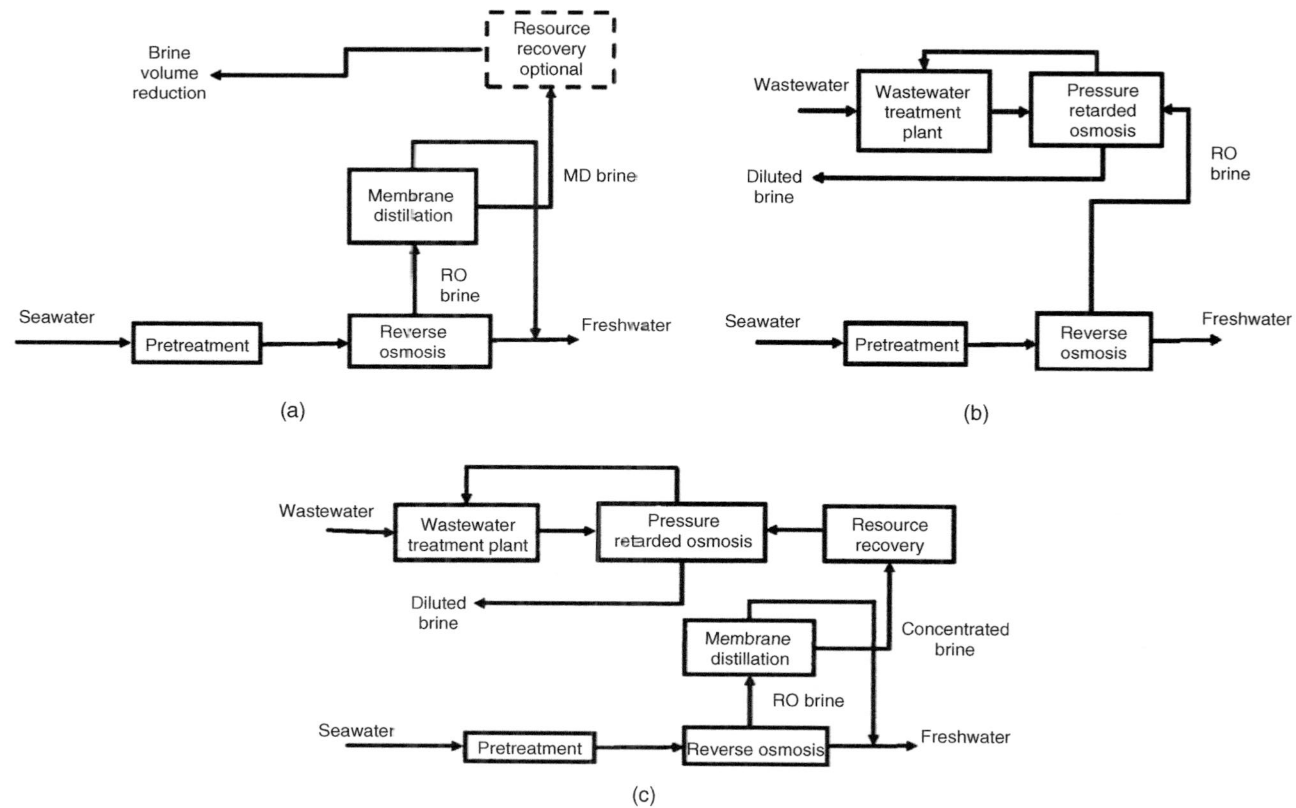

Figure 10.8 Schematics represents hybrid integrated systems.

Table 10.4 Estimated cost for 100 m^3/d plants with the different hybrid systems.

Processes	SWRO	SWRO + MD	SWRO + PRO	SWRO + MD + PRO
Initial capital cost ($/m^3.d)	115.4	115.7	166.6	144.9
Operation cost ($/m^3)	1.26	1.04	1.23	1.07
SEC (kWh/m^3)	3.3	2.8	2.87	2.68

Source: Lee et al. [38]/with permission of Elsevier.

brine of RO, in this case, decreases over 30%. In this case, the total efficiency of the plant is improved compared with the desalination unit built on an RO plant alone. The productivity of this plant was 1,000 m^3/day freshwater, the MD system just produced 400 m^3/day and the power density that was saved from the PRO system was 5 W/m^2 [36]. Figure 10.8 illustrates the schematics of the Global MVP process for water and energy recovery from RO brine: (a) using MD, (b) using PRO, and (c) using MD/PRO; these three hybrid systems reduce the electricity cost [37]. The membrane cost also affects the process economically due to the duration of membrane life and replacement time for them. The price of membrane replacement is calculated within the maintenance cost.

Therefore, the energy consumption has a great impact on operating the stations, and the membrane replacement represents a small part of the cost. Table 10.4 indicates the estimated cost for 100 m^3/day plants with different hybrid systems [38].

10.5.3 Hybrid System of Capacitive Deionization with Pressure Retard Osmosis

CDI depends on the deionization of water by applying low electrical current over porous carbon electrodes. Adsorption and desorption are two cycles that are performed in CDI [39]. After the electric current passes the cell, the adsorption step is done by adsorbing the salt ions according to their charges into the cathode and the anode and producing freshwater. After the saturation of electrodes, the desorption step begins to make regeneration for electrodes, then the brine is produced. Multistages of CDI can increase the productivity of freshwater and the concentrate solution that is produced from the desorption step can be transferred to the energy generation process such as PRO [40]. The hybrid system of PRO and CDI is illustrated in Figure 10.9. PRO is a process that depends on the gradient phenomena of osmatic pressure in the membrane, which can generate power. The concentrated stream from CDI is pressurized to PRO, to draw water from the low side concentration. This dilution creates hydraulic pressure. This pressure can run a turbine by generating energy. The high salinity stream from CDI batteries must feed the PRO process in order to produce energy during the integration of the

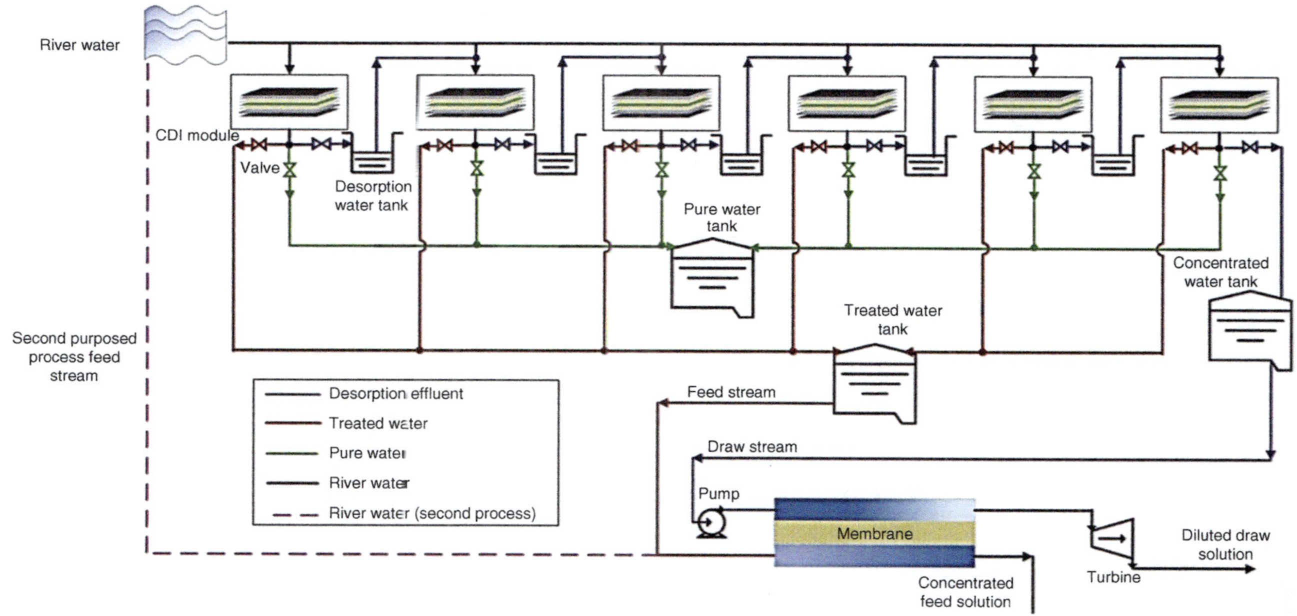

Figure 10.9 Drawing for CDI modules electrochemically integratec with PRO Source: Saleem et al. [39]/with permission of Elsevier.

CDI/PRO process [23, 39]. Because of different salinity gradients between high and low concentrations, respectively, the concentrated produced water from desorption (desorption current 15A) and the high-purity water produced from adsorption (adsorption current 3A) were employed in this system. The obtained recovery was 98.3%, while the power density of PRO was 8 W/m^2. Accordingly, the hybrid system of PRO/CDI reduced the SEC to 0.67 kJ/L, while the produced water energy consumption was 2.2 KJ/L with a pressure difference of 27.3 bar [39].

10.6 Hybrid Water Desalination System for ZLD

The zero liquid discharge (ZLD) method's objective is to separate all waste liquids from the process to save the environment by reducing wastewater and producing clean water that can be reused. This method used an evaporator and crystallizer in combination with an integrated membrane process for recovery enhancement, improving the efficiency, and decreasing the cost of the plant. Preconcentration technology provides high recovery but it is not enough to reach the brine to the saturation level, so the evaporator/crystallizer technique is required to convert the brine to the solid form. Combination of MD with MVC then crystallizer consumes energy demand equal to 1,153.5 kWh/100 m^3 brine that is lower than MVC with crystallizer that consumes 1,986 kWh/100 m^3 [41]. In China, Yangmei Taihua ZLD Project was implemented in early 2017 for Taiyuan Chemical New Materials Co., Ltd., which is a chemical company that has multiple wastewater streams. The design was based on membrane technology, where pretreatment step was carried out using microfiltration membrane and ion exchange for polishing to replace the conventional pretreatment steps [42]. A self-cleaning membrane design is applied for optimal filtration performance and reliable operation. The solution for these problems was carried out by Oasys Water partnered with Beijing Water, based on membrane brine concentrator (MBC), where the primary step depended on reducing the volume of wastewater to reuse it, as shown in Figure 10.10. The MBC is an FO-based technology that uses a highly effective draw solution to create a natural osmotic gradient against a strong and stable semi-permeable membrane for water drawing. The concentrated brine produced from MBC went to a crystallizer system to finalize the ZLD process. In the MBC system, pretreated water flows to the RO unit before entering the FO membrane system. 1200 m^3/day of wastewater as a feed is fed into the pretreatment system. The produced water from the RO membrane unit is 320 m^3/day that is fed into the FO membrane system. The brine from RO enters the FO process as a feed where it dilutes a high-strength draw solution as water is recovered from the wastewater stream. The recovery system is used to separate the draw solutes (DS) and recovered water, where DS passes in a closed-loop recovery and recondensing. The produced water from RO is less than 100 TDS and

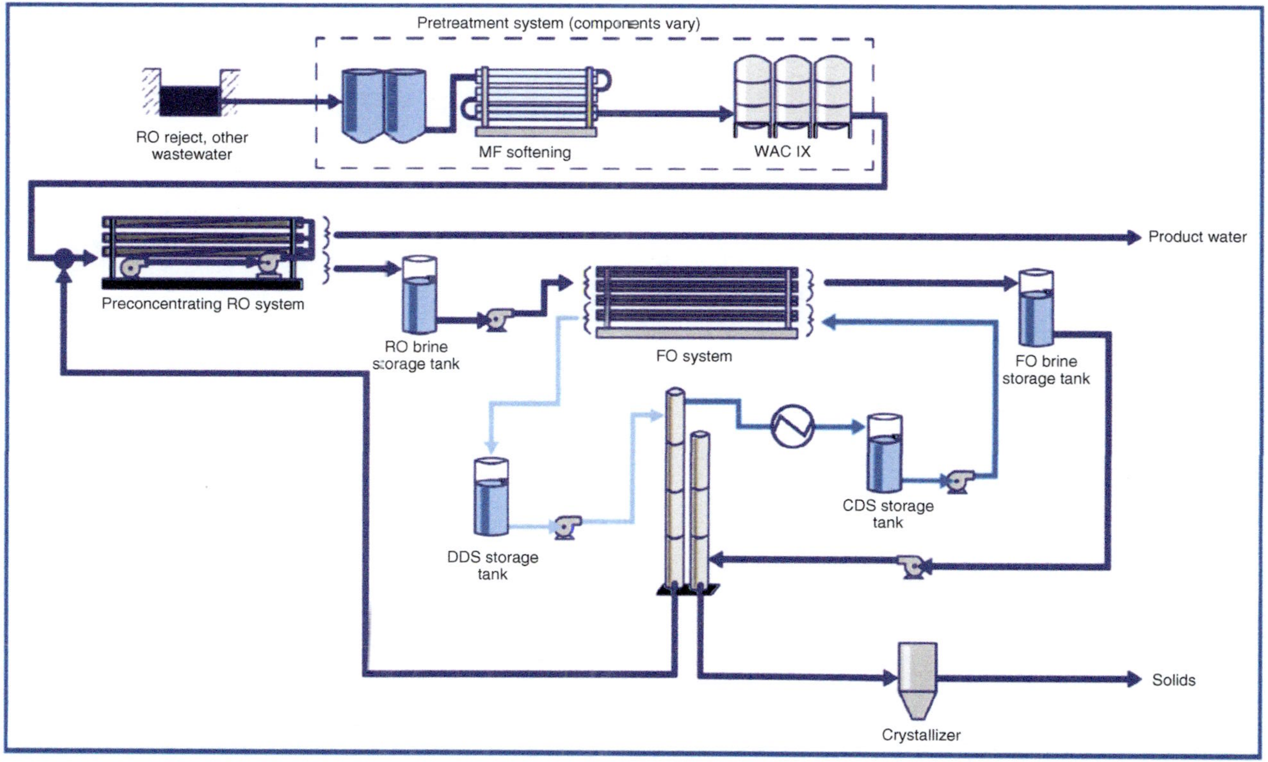

Figure 10.10 Schematic diagram for ZLD process in Yangmei Taihua, China.

Table 10.5 Advantages and disadvantages of hybrid technologies with ZLD.

Technology	Advantages	Disadvantages
MVC	Salinity >200,000 ppm	High cost
RO	Efficient energy	Fouling and scaling
MD	Salinity >200,000 ppm	Low flux and recovery
FO	High salinity, low power	Low flux
ED/EDR	Salinity >100,000 ppm	Consumes high energy

Source: Tong and Elimelech [43]/American Chemical Society/CC BY 4.0.

is used as makeup for the steam generator for the plant. Concentrated brine goes to the crystallizer as a feed at 80 m^3/day and greater than 240,000 mg/L TDS. The crystallizer produced about 90% of ZLD solid. A FO solution can be designed to match the exact composition of the effluent and remove 70–90% more clean water.

Table 10.5 illustrates the merits and disadvantages of Hybrid ZLD processes. Hybrid desalination technologies with ZLD can achieve high brine concentration using a crystallizer and then evaporator [44].

Freeze desalination (FD), MD, and crystallizer system were developed by Lu et al., to achieve a recovery of 100% for water and salt (see Figure 10.11). The FD unit received the saltwater, processed it into ice, and separated the freshwater from the brine. After reaching 80°C, the brine solution was supplied to the MD unit [46]. The feed was concentrated in MD due to the water loss, and after that, it was transported to a crystallizer as a concentrated brine [47]. 810 and 830 kWh/m^3

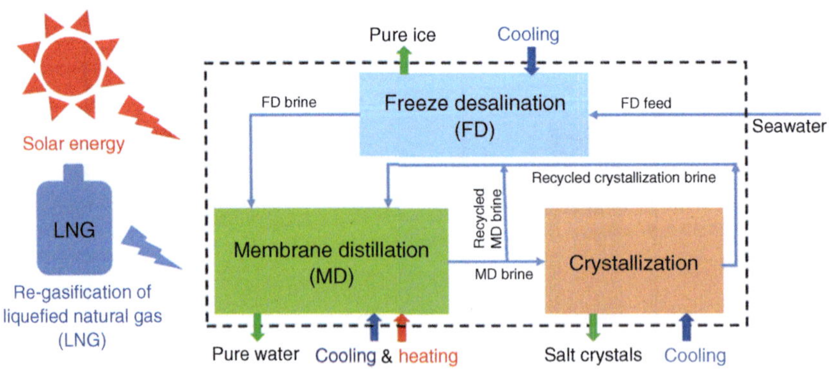

Figure 10.11 Schematic of integrated freeze desalination, crystallization and membrane distillation systems. Source: Ahmed et al. [45]/with permission of Elsevier.

were required for heating and cooling in this system. Because this device uses a lot of energy, its cost can be decreased by using solar energy.

This case study was carried out by integrating solar energy with LNG regasification for cooling. This process can use 72,000 g of seawater per day to meet the ZLD aim and produce salt of 2,520 g/day and water of 69,480 g/day. FD provided water recovery of 30%, and 70% from the DCMD unit. The solar panels were used to generate electricity. The location of the re-gasification of LNG was in Singapore, which produced the cooling energy. The re-gasification adsorbs 1,040 kJ/kg, where 207 kg LNG can produce the cooling energy for the system [46].

Minimum liquid discharge (MLD) techniques often serve as an alternative to preconcentration steps to reduce the overall cost of ZLD applications. it can provide a substantial economic advantage and water savings 95–98% water recovery through MLD drives, a reduction in discharge costs, a reduction or prevention of regulatory fines, and increased water supply in water-stressed regions with overall sustainability of municipal and industrial plants. The minimal liquid discharge process yields water recovery of around 86.15% while consuming 8.25 kWh/m^3 [48].

10.7 Case Study: China's Mega-Desalination Plant

In China, two massive desalination projects were launched in 2012. The first project was the 100,000 m^3/day Tianjin Dagang Desalination Plant, which uses ultrafiltration membrane technology to pretreat feed that is obtained from the Bo Sea. 250,000 m^3/day of pretreated water are produced by the pretreatment system. In order to create 12,500 m^3/day of desalinated water, eight separate RO trains receive the pretreated water from seawater intake pumps.

The energy recovery system, circulation pump, membrane rack, and single main high-pressure pump (Figure 10.12a) are all connected to the system. Each train in the system is intended to operate at 12.5% to 100% capacity [49].

The second project was the 100,000 m^3/day drinking water-producing Qingdao Desalination Plant in Eastern China with six train membrane system. This plant used ultrafiltration membranes to treat up to 250,000 m^3/day from the Jiao Zhou Gulf. The pretreated water is fed to the ERD and its three main HPP. Each pump provides service for two trains of the membrane, each of which is connected to a circulation pump and an energy recovery system of its own, as shown in Figure 10.12b.

A PX device, a rotary-type isobaric device, lowers the electricity utilized by the plants' HPP by up to 60% by transferring high brine pressure to conserve low feed of seawater pressure when they directly contact within the PX motor, the pressure

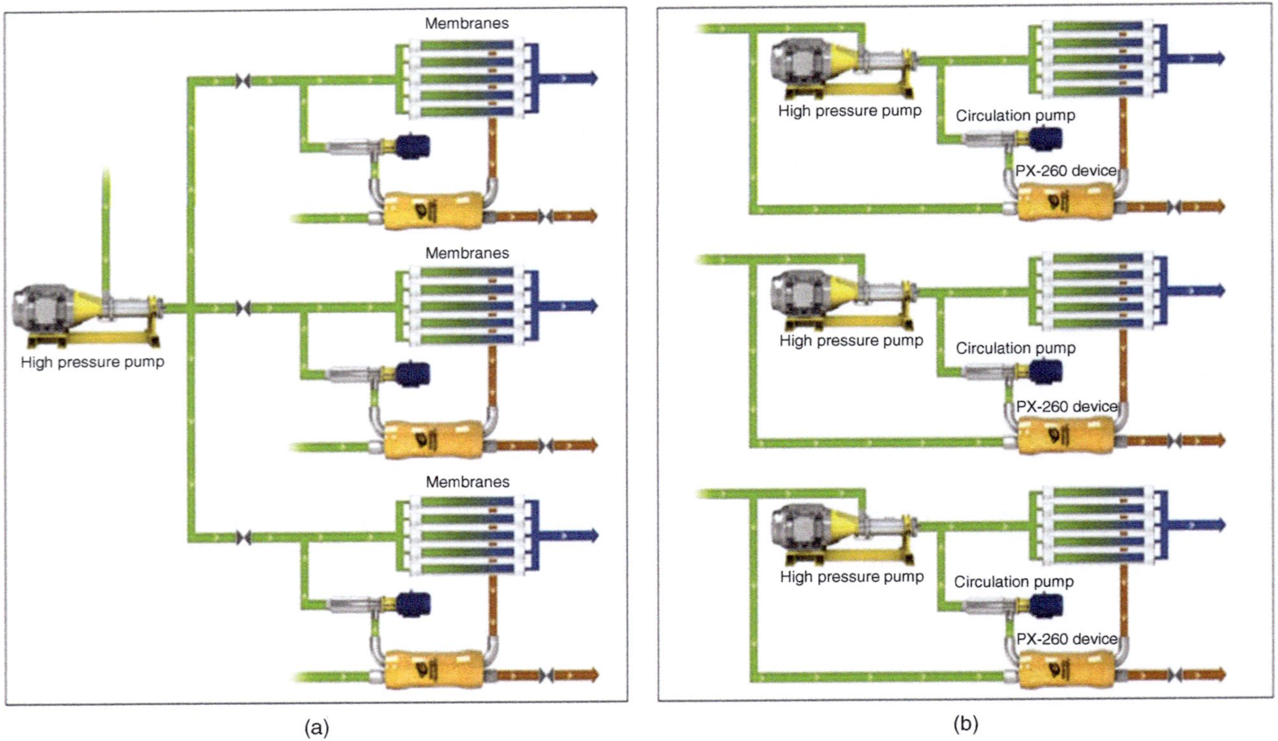

Figure 10.12 Semi-Pressure center design, individual train design.

moves from the concentrate to the seawater [49]. 144 PX-220 ERDs are working in the Tianjin plant; eight seawater RO trains are serviced by 18 PX units. A total of 102 PX-260 devices are used in the Qingdao plant; 17 PX units serve six trains.

Mixing reduces the long rotor ducts and tiny channels as well as the brief contact time (less than 0.05 seconds) between the concentrate and seawater streams.

In comparison to unused ERDs, isobaric ERDs can cut a plant's power consumption by up to 60%, and in comparison to other ERD technologies, by just 15–20%. This can result in annual savings for plants like the Tianjin and Qingdao factories of up to $7,500,000–$8,000,000 each.

The developed membranes with higher flux rates, lower pressures, larger surface areas, and high salt rejection rates were used in these desalination plants. With good overall membrane recoveries in the 40–45% range [49].

10.8 Conclusion

Recently, due to shortages in drinking water resources because of climate change, the increasing population growth in urban areas, and the need for potable water in remote areas, the trend has been directed toward desalinating seawater with different technologies. These technologies consume high energy to produce the required water quality. Significantly methods and technologies have been researched to reduce the cost of producing treated water by reducing energy consumption. However, the most common desalination technologies that are used in the world are RO and MFD technologies, and many studies have been performed to increase the efficiency of these techniques and reduce their plant costs. Integration between solar and wind turbines reduced the price of freshwater from 9 to 1.45 Euros/m^3 as mentioned in the case study of the Greek Islands. The following are the most important outcomes from this state of this art:

- Desalination technologies are too expensive due to high energy consumption, so suggesting innovative ways by development in membrane structure to produce high flux, using ERDs, using hybrid systems or integration with renewable energy, especially in remote areas can reduce the costs, making the desalination process more affordable and economical.
- RO is the popular technique in desalination, it represents 69% of desalination plants. Reduction in energy consumption can be achieved using ERDs.
- Integration systems using RO and renewable energy decreases the price of produced water, especially in remote areas.
- Hybrid renewable energies in desalination processes can provide economic and environmental benefits, especially using solar energy with wind energy.
- PRO is considered a renewable power generation technology. Introducing the PRO in the desalination process minimizes environmental problems by

reducing the brine waste. A hybrid system with PRO and RO reduces the brine of the RO to 30%.
- Incorporation of RO and ZLD improves energy and cost efficiencies.
- Hybrid membrane systems or hybrid desalination technologies decrease the cost of treated water and energy consumption.

Questions

1 What is the difference between desalination using the RO system and the electrodialysis system?

2 What are the methods that represent the heat transfer in the membrane distillation system?

3 What is the difference between RO, FO, and PRO.

4 Explain how the energy recovery device works (PX).

5 Explain how the integrated system can reduce the energy consumption of desalination plants.

References

1 Kucera, J. (2019). *Desalination: Water from Water*, 2e, 2–12. New York: Wiley.
2 Esmaeilion, F. (2020). Hybrid renewable energy systems for desalination. *Appl. Water Sci.* 10 (84): 1–47.
3 Alkaisia, A., Mossad, R., and Sharifian-Barforousha, A. (2017). A review of the water desalination systems integrated with renewable energy. *Energy Procedia* 110: 268–274.
4 Wakeel, M., Chen, B., Hayat, T. et al. (2016). Energy consumption for water use cycles in different countries: a review. *Appl. Energy* 178: 868–885.
5 Voutchkov, N. (2018). Energy use for membrane seawater desalination: current status and trends. *Desalination* 431: 2–14.
6 Kim, J., Park, K., Yang, D.R., and Hong, S. (2019). A comprehensive review of energy consumption of seawater reverse osmosis desalination plants. *Appl. Energy* 254: 1136–1152.
7 Marzooq, M., Alsabbagh, M., and Al-Zubari, W. (2018). Energy consumption in the municipal water supply sector in the Kingdom of Bahrain. *Computional, Water, Energy, and Environ. Eng.* 7 (3): 95–110.

8 Mansouri, M.T., Amidpour, M., and Ponce-Ortega, J. (2020). Optimization of the integrated power and desalination plant with algal cultivation system compromising the energy-water-environment nexus. *Sustain. Energy Technol. Assess.* 42: 1008–1079.

9 Tal, A. (2018). Addressing desalination's carbon footprint: the Israeli experience. *Water Secur.* 10: 197.

10 Mauter, M.S. and Fiske, P.S. (2020). Desalination for a circular water economy. *Energy Environ. Sci.* 13: 31–80.

11 Gurreri, L., Tamburini, A., Cipollina, A., and Micale, G. (2020). Electrodialysis applications in wastewater treatment for environmental protection and resources recovery: a systematic review on progress and perspectives. *Membranes* 10 (146): 1–93.

12 Tedesco, M., Hamelers, H.V.M., and Biesheuvel, P.M. (2016). Nernst-Planck transport theory for (reverse) electrodialysis: I. Effect of co-ion transport through the membranes. *J. Membr. Sci.* 510: 370–381.

13 Sajjad, A., Bin Mohd Yunus, M.Y., Azoddein, A.M. et al. (2020). Electrodialysis desalination for water and wastewater: a review. *Chem. Eng. J.* 380: 1222–1231.

14 El-Zanati, E., Khedr, M., El-Gendi, A. et al. (2018). Heat and mass transfer characteristics in vacuum membrane distillation for water desalination. *Desalin. Water Treat.* 132: 52–62.

15 Hardikar, M., Marquez, I., and Achill, A. (2020). Emerging investigator series: membrane distillation and high salinity: analysis and implications. *Environ. Sci.: Water Res. Technol.* 6: 15–38.

16 Ramezanianpour, M. and Sivakumar, M. (2015). Energy evaluation and treatment efficiency of vacuum membrane distillation for brackish water desalination. *J. Water Reuse Desalin.* 52: 119–131.

17 Bappy M.J. P., Bahar R., Ariff T. F. (2016). Low energy and lowcost freshwater production by membrane distillation. *Proceedings of the 2016 International Conference of Industrial Engineering and Operations Managements*, Kuala Lumpur, Malaysia (March 8–10, 2016).

18 Zhang, Y., Peng, Y., Ji, S. et al. (2015). Review of thermal efficiency and heat recycling in membrane distillation processes. *Desalination* 367: 223–239.

19 Roy, S. and Ragunath, S. (2018). Emerging membrane technologies for water and energy sustainability: future prospects, constrains and challenges. *Energies* 11: 29–97.

20 Wang, Y.N., Goh, K., Li, X. et al. (2018). Membranes and processes for forward osmosis-based desalination: recent advances and future prospects. *Desalination* 434: 81–99.

21 Lee, C., Chae, S.H., Yang, E. et al. (2020). A comprehensive review of the feasibility of pressure retarded osmosis: recent technological advances and industrial efforts towards commercialization. *Desalination* 491: 114501.

22 Chong, L.G., Chen, P.A., Huang, J.Y. et al. (2018). Capacitive deionization of a RO brackish water by AC/graphene composite electrodes. *Chemosphere* 191: 296–301.

23 Maheshwari, K. and Agrawa, M. (2020). Advances in capacitive deionization as an effective technique for reverse osmosis reject stream treatment. *J. Environ. Chem. Eng.* 8 (6): 104–413.

24 Kadaj, E. and Bosleman, R. (2018). Chapter (11): Energy recovery devices in membrane desalination processes. In: *Renewable Energy Powered Desalination Handbook, Application and Thermodynamics*, 415–444. Mathew Deans.

25 Nafey, A.S., Sharaf, M.A., and García-Rodríguez, L. (2010). Thermo-economic analysis of a combined solar organic Rankine cycle-reverse osmosis desalination process with different energy recovery configurations. *Desalination* 261: 138–147.

26 Huang, B., Pu, K., Wu, P. et al. (2020). Design, selection and application of energy recovery device in seawater desalination: a review. *Energies* 13: 41–50.

27 Sani, A.E. (2019). Design and synchronizing of Pelton turbine with centrifugal pump in RO package. *Energy* 172: 787–793.

28 Rabiee, H., Khalilpour, K.R., Betts, J.M., and Tapper, N. (2019). Chapter (13): Energy-water nexus: renewable-integrated hybridized desalination systems. In: *Polygeneration with Polystorage for Chemical and Energy Hubs* (ed. K.R. Khalilpour), 409–458. Elsevier.

29 Busch, M., Mickols, B. (2021). Economics of desalination—reducing costs by lowering energy use. https://www.waterworld.com/international/desalination/article/16200772/economics-of-desalination-reducing-costs-by-lowering-energy-use

30 Smaoui, M., Abdelkaf, A., and Krichen, L. (2015). Optimal sizing of standalone photovoltaic/wind/hydrogen hybrid system supplying a desalination unit. *Sol. Energy* 120: 263–276.

31 Ali, A., Tufa, R.A., Macedonio, F. et al. (2018). Membrane technology in renewable-energy-driven desalination. *Renew. Sust. Energ. Rev.* 81: 1–21.

32 Tufa, R.A., Noviello, Y., Di Profio, G. et al. (2019). Integrated membrane distillation-reverse electrodialysis system for energy-efficient seawater desalination. *Appl. Energy* 253: 113–551.

33 Tufa, R.A., Pawlowski, S., Veerman, J. et al. (2018). Progress and prospects in reverse electrodialysis for salinity gradient energy conversion and storage. *Appl. Energy* 225: 290–331.

34 Han, G., Zuo, J., Wan, C., and Chung, T. (2015). Hybrid pressure retarded osmosis—membrane distillation (PRO—MD) process for osmotic power and clean water generation. *Environ. Sci.: Water Res. Technol.* 1: 507–515.

35 O'Toole, G., Jones, L., Coutinho, C. et al. (2016). River-to-sea pressure retarded osmosis: resource utilization in a full-scale facility. *Desalination* 389: 39–51.

36 Xu, S., Liu, Y., Wang, Y. et al. (2017). Influential analysis of concentration polarization on water flux and power density in PRO process: modeling and experiments. *Desalination* 412: 39–48.

37 Choi, Y., Kim, S.H., and Lee, S. (2017). Comparison of performance and economics of reverse osmosis, membrane distillation, and pressure retarded osmosis hybrid systems. *Desalin. Water Treat.* 77: 19–29.

38 Lee, S., Choib, J., Park, Y. et al. (2019). Hybrid desalination processes for beneficial use of reverse osmosis brine: current status and future prospects. *Desalination* 454: 104–111.

39 Saleem, M.W., Im, B., and Kim, W.S. (2018). Electrochemical CDI integration with PRO process for water desalination and energy production: concept, simulation, and performance evaluation. *J. Electroanal. Chem.* 822: 134–143.

40 Hemmatifar, A., Palko, J.W., Stadermann, M., and Santiago, J.G. (2016). Energy breakdown in, capacitive deionization. *Water Res.* 104: 303–311.

41 Yaqub, M. and Lee, W. (2019). Zero-liquid discharge (ZLD) technology for resource recovery from wastewater: a review. *Sci. Total Environ.* 681: 551–563.

42 Oasys Water Inc. (2017). Case Study: Changxing power plant debuts the world's first forward osmosis-based zero liquid discharge application. https://www.wateronline.com/doc/changxing-power-plant-debuts-the-world-s-first-forward-osmosis-based-zero-liquid-discharge-application-0001.

43 Tong, T.Z. and Elimelech, M. (2016). The global rise of zero liquid discharge for wastewater management: drivers, technologies, and future directions. *Environ. Sci. Technol.* 50 (13): 6846–6855

44 Charisidis, C. (2018). *Brine Zero Liquid of Charge Fundametals and Design: A Guide to the Basic Conceptualization of the ZLD/MLD Process Design and the Relative Technologies Involved*, 1e, 15–25. Lenntech.

45 Ahmed, F.E., Hashaikeh, R., and Hilal, N. (2020). Hybrid technologies: the future of energy-efficient desalination – A review. *Desalination* 495: 1146–1159.

46 Lu, K.J., Cheng, Z.L., Chang, J. et al. (2019). Design of zero liquid discharge desalination (ZLDD) systems consisting of freeze desalination, membrane distillation, and crystallization powered by green energies. *Desalination* 458: 66–75.

47 Luo, L., Chang, J., and Chung, T.S. (2017). Cooling crystallization of sodium chloride via hollow fiber devices to convert waste concentrated brines to useful products. *Ind. Eng. Chem. Res.* 56: 10183–10192.

48 Togo, N., Nakagawa, K., Shintani, T. et al. (2019). Osmotically assisted reverse osmosis utilizing hollow fiber membrane module for concentration process. *Ind. Eng. Chem. Res.* 58 (16): 6721–6729.

49 Clemente, R. (2013). China's Mega-Desalination Plant Experience, A white paper on the long-term energy efficiency of China's largest desalination plants. Energy Recovery Inc.

11

Pressure-Retarded Osmosis for Blue Energy Generation

Giti Nouri, Sara Pakdaman*, and Catherine N. Mulligan*

Department of Building, Civil and Environmental Engineering, Concordia University, Quebec, Montreal, Canada

11.1 Introduction to Pressure-Retarded Osmosis

An immense strain has been imposed on Earth's natural resources, particularly on energy because of the worldwide growing population, rapid urbanization, and accelerated economic growth. Although there are advances in technology, most of the world's energy production still depends on fossil fuels [1]. Since fossil fuel resources are diminishing at a concerning rate, this reliance on nonrenewable resources is unsustainable [2]. Moreover, fossil fuels release pollutants, contributing to notable environmental degradation [3]. There is a growing interest in sustainable energy systems as a response to these upcoming challenges. To provide effective alternatives to the energy crisis, technological improvements are being actively carried out globally. Society has gradually accepted the importance of energy efficiency, establishing a change toward more attentive consumption of energy. This transition has prompted interest in alternative, renewable energies that are clean, cost-effective, and able to satisfy the increasing global energy needs [4]. Recently, salinity-gradient energy has attracted rising attention, which is produced from the release of free energy when waters with varied salinities combine, as at the conflux of rivers and seas. It is predicted that the potential harnessable power from global river discharge into the oceans exceeds 1 TW, supplying a noticeable portion of the world's energy demand. Additionally, saline water bodies such as the Great Salt Lake and the Dead Sea provide further opportunities for generating energy when mixed with less saline water bodies [5].

*Author Contributions: G.N. and S.P. contributed equally to this work

Water Treatment and Desalination: Membrane Separation using Renewable Energy Resources, First Edition. Edited by Mohamed Bassyouni.

Engineered processes are substantial to convert the available energy into usable forms. For this, pressure-retarded osmosis (PRO) has been developed as an efficient method among the various technologies. A semi-permeable membrane is applied in this process to allow freshwater permeation into highly concentrated seawater, leading to a pressure rise on the seawater side which can further be utilized to spin a turbine and produce electricity. The generated energy is derived from the Gibbs free energy of mixing, a nonreversible and thermodynamically favorable process that increases the system's entropy while offering a renewable energy source. Membrane technologies, including PRO, have demonstrated considerable potential in deriving energy from salinity gradients with numerous benefits, such as less operational complexity, capacity for large-scale installation, and the ability to treat water while recovering valuable materials. The membranes employed in these processes are carefully chosen based on chemical and mechanical characteristics, selectivity, and permeability. Ongoing research continues to enhance membrane technology, making it more efficient and flexible for different uses [6] and ideal for commercialization. PRO also exhibits unique opportunities for incorporation with current systems, reverse osmosis (RO). When integrated with RO, PRO can use RO brine concentrate as either a feed or draw solution, thereby diminishing waste and enhancing the whole process's efficiency [7].

11.1.1 Historical Development and Evolution

The idea of harnessing osmotic pressure as a type of energy, termed salinity-gradient energy, was initially introduced by Pattle in 1954 by developing a technique to exploit this energy by utilizing ion-exchange membranes to create alternating layers of freshwater and saltwater. This configuration enabled a controlled ionic flux inside the stack, which was thereafter transformed into an electrical current by electrodes connected to an external circuit, further facilitating energy extraction. Nollet was first to explain the osmosis phenomenon in 1748 [8]. Despite this early innovation laying principal work, considerable advancements in salinity-gradient energy remained limited until the oil crisis of 1973, renewing interest in other energy sources [9]. In 1973, Sidney Loeb conducted experimental implementations of PRO at Ben-Gurion University, to assess the viability of producing energy from saline water. In 1975, Loeb and Norman introduced the term PRO and patented a PRO heat engine. The performance of early PRO systems was restricted by the presence of suitable membranes despite these innovations [9, 10]. Loeb continued his research into the late 1990s and early 2000s, using the theoretical analyses carried out in 1998 and 2001 to analyze the potential for energy generation through PRO in hypersaline environments. During the 1970s and 1980s, additional studies took place, with Wick and Isaac evaluating the usage of salt domes for salinity-gradient energy, while Lee et al. developed models showing that brines, rather than seawater, might render PRO economically viable [2]. The interest in PRO decreased until the early 2000s when developments in membrane technology renewed the field. Some of the initial challenges, such as issues related to mechanical stability and salt rejection, were mitigated through the introduction of thinner, more permeable forward osmosis (FO) membranes. Nonetheless, the commercialization of osmotic power still faces noticeable challenges, including membrane fouling, optimized operational parameters, and the regeneration of draw solutions [7]. The first pilot plant to harness power from the PRO idea was built by Statkraft in Norway. This facility was designed to produce 10 kW of electricity, but it achieved only 5 kW due to the membrane's constraints. Now, Japan's MT Water Project and Korea's GMVP project have continued to evaluate the hybridization of PRO with other desalination technologies to improve energy efficiency and minimize environmental impacts [7, 9, 11].

11.1.2 Principles of Operation

In PRO, the salinity differentiation among the two solutions on either side of the membrane generates the osmotic pressure. Water migrates from the feed solution with a low salinity to the draw solution with a high salinity to balance the concentration levels even in the presence of an applied hydraulic pressure (ΔP) which must be less than the osmotic pressure variation to generate energy. Under standard conditions, the osmotic pressure dissimilarity between riverwater and seawater corresponds to a hydrostatic head of approximately 231 m, or around 23 atm. This significant pressure difference enables a PRO power plant to generate up to 1 MW/m^{3}s of freshwater processed through the system, assuming the high efficiency of the membranes and pressure exchangers [8]. A typical PRO process is depicted in Figure 11.1, elucidating the central mechanism of power generation. The pressurized brackish water is divided into two streams: one stream is directed toward a turbine for producing electrical power, while the other is routed via a pressure exchanger, helping in pressurizing the incoming seawater and thus sustaining the circulation of the system. The efficiency and power output of the PRO process are affected by several critical variables, such as the difference in salinity between the low- and high-concentration solutions, the membrane properties, and the applied pressure. The interplay of these factors governs the overall system performance [8, 11].

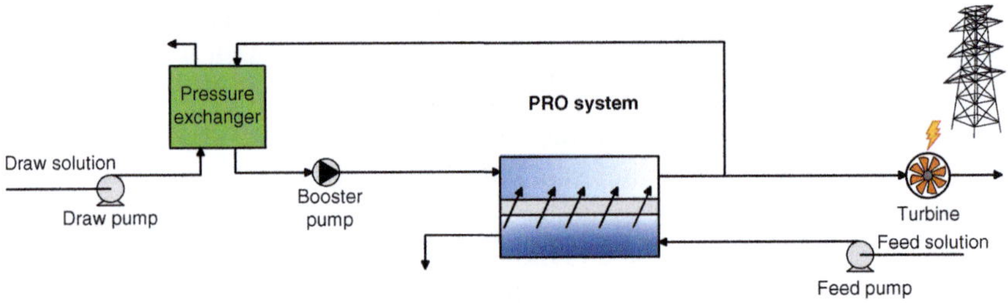

Figure 11.1 A typical process of pressure-retarded osmosis. Source: Bajraktari et al. [11]/with permission of Elsevier.

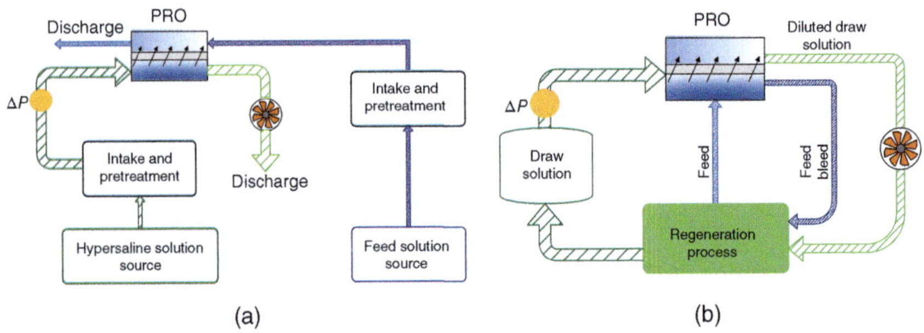

Figure 11.2 PRO's configurations: (a) open-loop and (b) closed-loop. Source: Zadeh et al. [12]/with permission of Elsevier.

11.1.3 Configurations of Pressure-Retarded Osmosis Systems

There are two primary configurations of PRO systems, namely open-loop pressure-retarded osmosis (OLPRO) and closed-loop pressure-retarded osmosis (CLPRO) (Figure 11.2). In an OLPRO system, freshwater and saline water are mixed, with both streams being discharged after the osmotic process. In contrast, in a CLPRO system, streams from the membrane modules are regenerated and recirculated into the system. CLPRO offers the advantage of choosing the draw solution with greater precision, allowing for customization based on the energy resources at specific locations. OLPRO systems incur higher costs because of the treatment of feed and draw solutions and the energy required for transporting streams. In addition, membrane fouling is increased in an OLPRO system, causing higher membrane maintenance costs. The geographical location of OLPRO plants is limited to areas near inlet streams, limiting their applicability. Conversely, CLPRO systems provide greater operational flexibility and a diminished footprint, though osmotic heat engines must be located near low-grade heat sources for effective recovery of the diluted draw solution.

CLPRO systems are environmentally advantageous since they mitigate the need for discharge, unlike OLPRO systems which often involve the discharge of outlet streams, especially when using natural sources. Overall, CLPRO systems are generally considered more flexible, practical, and efficient for PRO applications [12].

11.1.4 Factors Influencing Power Density in Pressure-Retarded Osmosis Systems

In PRO operations, the maximum power density is inherently associated with the permeability of pure water through the membrane. However, the real power density in practical PRO systems often falls short of theoretical values due to several factors, such as the loss of effective osmotic pressure from internal and external concentration polarization (ICP and ECP) and reverse solute flux. ICP happens due to localized concentration buildup within the membrane, while ECP occurs on the membrane's exterior surface. Both forms of polarization, in conjunction with membrane fouling, significantly affect the efficiency of PRO systems. ICP impairs PRO performance by increasing the concentration of salts at the interface within the membrane's active support layer, thereby impacting power density. The structural composition of the membrane is necessary for mitigating ICP, particularly in pressurized PRO operations where mechanical strength is vital to endure pressure differentials. Experimental evidence shows that membranes with a polyester-woven mesh integrated within a porous polysulfone support layer effectively decrease ICP while preserving structural integrity.

ECP occurs at the interface of the active layer and solution, posing a substantial constraint on the system's performance, especially at higher water fluxes. The inherent permeability and selectivity in membranes lead to increased water permeability, exacerbating the negative effects of reverse salt flux and ICP. Furthermore, as the PRO process progresses, the accumulation of draw salt in the feed solution reduces the effective pressure differential across the membrane, thereby diminishing power density. The decline in membrane selectivity under pressurized conditions intensifies this decline in osmotic power density. Despite the challenge of balancing membrane pressure tolerance with a high osmotic flux in PRO processes, the advancement of membranes with high power density remains critical for optimizing power generation. Two main types of membranes have been extensively utilized in PRO systems: integrated skinned membranes made from cellulose acetate (CA) and cellulose triacetate (CTA) and thin-film composite (TFC) membranes with selective layers over porous supports [2, 13].

11.1.5 Free Energy of Mixing

Gibbs free energy is calculated as follows:

$$\Delta_{mix}G = \Delta H_{mix} - T\Delta S_{mix} \tag{11.1}$$

where ΔH_{mix} and ΔS_{mix} are the enthalpy and the mixing entropy, respectively. For ideal solutions where $\Delta H_{mix} = 0$, the combination of concentrated and dilute solutions leads to a change in free energy driven solely by the entropy of mixing:

$$\begin{aligned}\Delta_{mix}G &= \Delta G_b - (\Delta G_c + \Delta G_d)\\ &= -(n_c + n_d)T\Delta_{mix}S_b + (n_c T\Delta_{mix}S_c + n_d T\Delta_{mix}S_d)\end{aligned} \tag{11.2}$$

where C, d, and b represent the concentrated, dilute, and brackish solutions, respectively. n and T are the quantity of moles and temperature, respectively. Additionally, $\Delta_{mix}S$ is the molar entropy related to the mixture of the total molar entropy of the electrolyte, as expressed by:

$$\Delta_{mix}S = -R\sum_i x_i \ln x_i \tag{11.3}$$

where R and x_i are the gas constant and the mole fraction of each component, respectively [8].

11.1.6 The Flux of Water and Salt Across a Pressure-Retarded Osmosis Membrane

Water permeation (J_w) through a membrane can be obtained by the following equation:

$$J_w = A(\Delta\pi - \Delta P) = A(\pi_{Draw} - \pi_{Feed} - \Delta P) \tag{11.4}$$

where A refers to the coefficient of water permeability, $\Delta\pi$ represents the osmotic pressure differentiation, and ΔP is the transmembrane pressure variation. π_{Draw} and π_{Feed} denote the draw and feed solutions' bulk osmotic pressures, respectively. This equation assumes ideal conditions that the membrane shows perfect selectivity and the hydrodynamics in the channels of the draw and feed are optimal, meaning that the concentrations at the surface of the membrane are equivalent to the bulk concentrations [8].

Achieving a high-power density in a PRO system per unit area of the membrane (W/m^2) is necessary for its economic viability, which can be expressed through the following relationship:

$$W = J_w\Delta P = A(\Delta\pi - \Delta P)\Delta P \tag{11.5}$$

$$J_w = -D_w\frac{dc_w}{dx} \tag{11.6}$$

where W and J_w are the power density and water flux, respectively. D_w and c_w are the diffusion coefficients of water and its concentration, respectively, and x is the coordinate to the membrane surface perpendicularly. The maximum power density can be achieved as follows [2].

$$W_{max} = A\frac{\Delta\pi^2}{4} \tag{11.7}$$

11.1.7 Actual Energy Production of Pressure-Retarded Osmosis

The performance of a PRO membrane is often compromised under real-world conditions. ICP arises when the support layer interfaces with the feed solution, leading to salt accumulation at the feed solution–membrane boundary and thus reducing the osmotic pressure difference. On the draw side, ECP can then decrease the osmotic pressure difference, thereby diminishing the driving force for water flux. Under practical conditions, the water flux (J_w) can be calculated as:

$$J_w = A(\pi_{D,m} - \pi_{icp} - \Delta\pi_{PRO}) = A(\Delta\pi_m - \Delta\pi_{PRO}) \tag{11.8}$$

where $\pi_{D,m}$ and π_{icp} denote the osmotic pressures by considering the ICP effects and the actual driving osmotic pressure difference ($\Delta\pi_{PRO}$), respectively. The effective osmotic pressure difference ($\Delta\pi_{eff}$) between the sides of the membrane can be calculated as:

$$\Delta\pi_{eff} = (\pi_D - \pi_s)\exp\left(-\frac{J_w}{k_D}\right) - (\pi_F + \pi_s)\exp(J_w k)\exp\left(\frac{J_w}{k_f}\right) \tag{11.9}$$

where K is the coefficient of the solute diffusion and k_D and k_f are the coefficients of the mass transfer on the draw and feed sides, respectively. π_s represents the osmotic pressure loss on the draw solution side because of salt diffusion:

$$\pi_s = \frac{B}{A}\left[1 + \frac{A\Delta\pi_{PRO}}{J_w}\right] \tag{11.10}$$

The relationship between J_w and salt flux (J_s) is given by:

$$\frac{J_s}{J_w} = \frac{\pi_s}{iRT} \tag{11.11}$$

where B and ΔP are the permeability of salt and the applied pressure on the draw side, respectively. The average water flux ($\overline{J_w}$) can be defined by averaging the flux at multiple points along the membrane:

$$\overline{J_w} = \sum_{j=1}^{N} J_{w_j} \tag{11.12}$$

This average flux is necessary for the accurate sizing of PRO systems and modeling of energy production, as it must account for real-world conditions such as concentration polarization (CP) and reverse salt flux. The PRO process involves different operations before and after the module of the PRO, including pretreatment of wastewater, pretreatment and pressurization of the draw solution, intake of wastewater, conversion of energy, and discharge of bleed. The average power density, $\overline{PD}$, and the actual PRO energy production, (SE_{PRO}^{Act}), are calculated as follows:

$$\overline{PD} = \overline{J_w} \Delta\pi_{PRO} \tag{11.13}$$

$$SE_{PRO}^{Act} = \eta_{TB} \left(\frac{\Delta P_{Act}}{\overline{Q}_{ww} + \overline{Q}_b} \right) \Delta Q_{PRO}^{Act} \tag{11.14}$$

where A_m and ΔQ_{PRO}^{Act} are the total membrane surface area of PRO and the PRO permeate flow, respectively. η_{TB} is the efficiency of the turbine, and A_m is expressed as the expected water flux:

$$A_m = \frac{\Delta Q_{PRO}^{Act}}{\overline{J_w}} \tag{11.15}$$

11.1.8 Maximum Energy Production by Pressure-Retarded Osmosis

Under ideal conditions, the maximum energy in a PRO system is acquired when the applied pressure is half of the osmotic pressure differentiation ($\Delta P = \pi/2$). Nevertheless of the configuration of the flow, the generated energy (SE_{PRO}) can be given as follows:

$$SE_{PRO} = \frac{\Delta P \Delta Q}{(Q_b - Q_1) + Q_{ww}} \tag{11.16}$$

where ΔQ is the flow that traverses the PRO membrane from the feed (Q_{ww}) to the draw side ($Q_b - Q_1$). To maximize energy recovery, the feed and draw solutions are set equal arbitrarily, leading to a feed flow fraction of $\emptyset = \frac{Q_{ww}}{(Q_b - Q_1 + Q_{ww})} = 0.5$. The PRO energy production can then be calculated on the basis of the osmotic pressure variation ($\Delta\pi$) and the dilution factor (DF) as follows:

$$SE_{PRO} = \frac{\Delta P}{2} DF \tag{11.17}$$

$$DF = \frac{\Delta Q}{Q_{ww}} \tag{11.18}$$

The maximum water flow across the membrane is expressed by the following equation:

$$\Delta Q_{max} = \frac{C_b - C_{ww}}{C_b + C_{ww}} Q_{ww} \tag{11.19}$$

where C_b and C_{ww} represent the draw and feed solution concentrations, respectively. For full-scale applications, it is crucial to consider the dilution effects that occur as permeate flows via the PRO membrane, as these effects can substantially reduce flux performance in comparison with small-scale systems. The maximum extractable energy (SE_{PRO}^{MAX}) can be expressed under ideal conditions, without considering the CP effects and reverse salt flux (RSF) [1].

$$SE_{PRO}^{MAX} = \frac{\Delta P}{2}\left(\frac{C_b - C_{ww}}{C_b + C_{ww}}\right) \tag{11.20}$$

11.2 Fundamentals of Pressure-Retarded Osmosis

11.2.1 Osmotic Pressure and Water Flux

Osmotic pressure is the force that must be exerted to stop water from moving across the membrane by osmosis. This arises from differences in the solute concentration between two solutions separated by such a barrier, and the greater the difference in the amounts of solute, the higher the osmotic pressure. Water molecules passively move through osmosis from an area with less dissolved solutes (higher water potential) to another region with more dissolved solutes (lower water potential) until equilibrium is reached. Such a motion gives rise to water flux through the membrane. Regarding aqueous solution, the osmotic pressure is estimated by Van't Hoff's formula [14]:

$$\pi = RT\sum iM \tag{11.21}$$

This equation uses universal gas constant (R), which equals 0.083145 L.bar/mol K; temperature T which is expressed in Kelvin; the factor of Van't Hoff (i) that each ion has its own value; and finally, the molarity (M) for that particular ion. For instance, NaCl concentrations are about 3.0–4.0% in seawater, making the osmotic pressure at these conditions to be within 25–33 bars at a temperature of 25°C. Also, rejected brines concentrated from RO desalination centers have high osmotic pressures and hypersaline waters of salty lakes like the Great Salt Lake in the USA. For instance, brines from RO desalination centers have a salt content between 6 and 7%, which translates into osmotic pressures around 50–59 bars. The Great Salt Lake is a salty lake with an average salinity of about 24%, thus resulting in osmotic pressures estimated between 200 and 290 bars. On that note, freshwater has almost zero osmotic pressure by contrast.

The water flux (Jw) can be computed by using the following formula:

$$Jw = A(\Delta\pi - \Delta P) \tag{11.22}$$

where Jw is the water flux, A represents the membrane's water permeability coefficient, ΔP refers to the applied differences in hydraulic pressure, and $\Delta\pi$ stands for the osmotic pressure differential between the two sides of a membrane.

Furthermore, in practical applications in the world, water flux deteriorates due to ICP, ECP, and backdraw solute flux [15].

If there is no external pressure ($\Delta P = 0$), there will be no flow of liquid through a membrane except for osmosis. However, for instance, in RO where water is forced against its natural gradient, external pressure is usually applied to get better rates. The need to realize how water flux and osmotic pressure impact the design of effective membranes for osmotic systems is inevitable. The performance of these systems relies on several factors such as differences in the solute concentration across the membrane and operation conditions, including temperature. Thus, the perfect management and optimization of these factors lead to increased yields when it comes to production rate but low energy consumption during osmotic processes.

11.2.2 Comparison with Different Osmosis-Based Technologies

Water treatment and desalination have increasingly become important in the use of osmosis, and it has evolved into various technologies, including PRO, RO, and FO. Moreover, though all these are founded on osmosis principles, they function differently in diverse contexts. In FO, water moves from a water that is less concentrated and is the feed solution to more concentrated water that is the draw solution spontaneously without applying force from outside as it uses the natural osmotic pressure difference. This makes FO potentially more energy-saving than RO. The areas where FO is applied include food and beverage processing, wastewater treatment, and pretreatment in desalination. Despite its promising commercial viability, this technology faces challenges such as regeneration of the draw solution and membrane fouling, which must be solved for it to be improved commercially. RO can be considered the most widely used osmotically driven technology, employed to desalinate and purify water. It is done by driving water within a semi-permeable membrane by applying hydraulic pressure above the osmotic pressure, leaving the solute behind. RO membranes are designed to reject a variety of contaminants, including salts, microorganisms, and organic molecules. PRO is a process where water transfers from a feed solution with low salinity to a draw solution with elevated salinity under applied pressure that is lower than the difference of osmotic pressure; it can generate osmotic power which can be transformed into electrical or mechanical energy. Therefore, PRO has the capability to contribute to renewable energy production from gradients in salinity, such as those occurring between freshwater and seawater in estuaries. PRO's major limitations include issues related to membrane performance, efficiencies in power conversion, and scale-up issues [16]. Figure 11.3 illustrates four potential osmotic processes that occur when saline water and pure water are separated by a semi-permeable membrane. In the first one, that is FO, as shown in

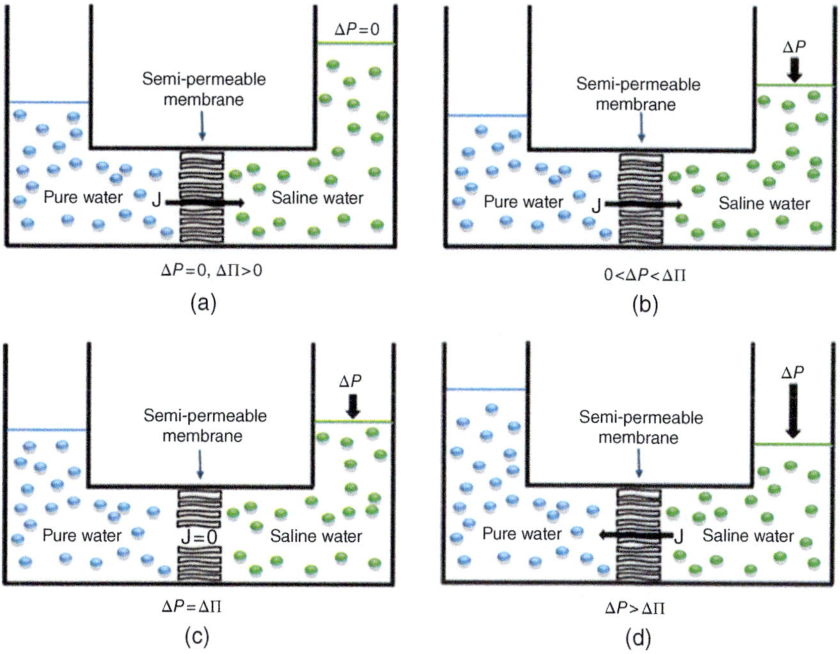

Figure 11.3 Diagrammatic depictions of four osmotic processes. Source: Helfer et al. [5]/with permission of Elsevier.

Figure 11.3a, water transfers within the membrane (J) because of the difference in osmotic pressure between the draw and feed solutions. Figure 11.3b demonstrates the PRO phenomenon. During the PRO process, the feed solution that has a lower concentration moves toward the side of the draw solution with higher concentration due to the higher $\Delta\pi$ than the hydrostatic pressure differential. This concept forms the basis for generating osmotic power. For continuous power generation, the feed solution ensures a steady flow within the membrane. Simultaneously, the saline water side must maintain consistent pressure and a fixed concentration. As a result, the volume flow increases on the side of saline water, and this further flow is able to produce energy. Figure 11.3c indicates the state that is known as osmotic equilibrium. It happens when water moves from the side with a low concentration to the side with a high concentration. Moreover, the hydrostatic pressure on the solution with higher concentration gradually rises. In the end, the osmotic flow (J) will stop. Mathematically, this occurs when ΔP equals to $\Delta\pi$. Figure 11.3d illustrates the fourth osmotic phenomenon which is RO and it occurs when ΔP exceeds $\Delta\pi$. Here, water flows from concentrated water to freshwater, leading to a negative flux. Therefore, in this phenomenon, water transfers in the opposite direction of the natural osmotic process, and due to this, it is called RO.

Moreover, this principle is utilized in advanced plants of seawater desalination. Seawater is subjected to pressure beyond its osmotic pressure, forcing it to move within a semi-permeable membrane, while the membrane rejects the salts, and only freshwater can pass through, which can then be safely consumed [17].

Each of the FO, RO, and PRO membranes serves distinct purposes and has unique properties appropriate to their specific applications. FO membranes use a concentration gradient to draw water across the membrane, requiring a highly hydrophilic and selectively permeable design to maximize water flux while rejecting solutes. In contrast, RO membranes operate under high hydraulic pressure to force water to pass through a semi-permeable barrier, necessitating robust structural integrity to withstand the pressure while ensuring high solute rejection rates. PRO membranes, designed for generating energy, combine features of both FO and RO membranes. They need the hydrophilic and selective characteristics of FO membranes to optimize the permeability of water, along with the durability and pressure resistance of RO membranes to efficiently harness osmotic pressure for power production.

11.2.3 Water Transport and Theory of Rejection

Water transport within a membrane is an important and fundamental phenomenon in different filtration or separation processes, including FO, RO, and PRO. Driving forces are different, such as different concentrations and applied pressure. In addition, membrane performance impacts the efficiency of the movement of water molecules in osmosis processes. In PRO, the process causes a prolonged fluid flow on the side of membrane permeate, leading to a diffuse flow from the surface of the membrane into the phase of the bulk liquid and maintaining the concentration of the solute higher near the surface of the membrane compared to the bulk side. According to the law of mass conservation, the sum total of both convective and diffusive flows must equal on the high-concentration side of the membrane. Thus, if salt permeates into the bulk fluid, the concentration at the external membrane interface can be nonzero, influenced by the bulk salt concentration and the coefficient of permeability. Rejection theory in PRO focuses on the membrane's ability to selectively permit water passage while effectively rejecting solutes, and this is the selective separation that keeps the osmotic pressure difference constant and drives the transport of water. The percentage difference between the concentration of solutes in the feed solution and permeate is called the rejection rate, which is effective in determining the rejection efficiency. Moreover, selectivity of the membrane, size of the solute and its charge, and CP are influential parameters in this theory.

Among the membranes, those that have a high selectivity for water over solutes are ideal for PRO because they ensure that primarily molecules of water pass through the membrane while retaining most of the solutes. Those solutes with larger size or with specific charges are rejected more effectively due to the size exclusion and electrostatic interactions. Consequently, CP can decrease the efficiency of rejection, and designing a proper system with an effective operational strategy to minimize its impacts on system performance is necessary [17, 18].

11.2.4 Phenomenon of Concentration Polarization

In PRO processes, CP happens when salt permeates near the surface of the membrane, and it can be categorized into internal and external. This phenomenon affects the concentration gradient and the performance of the membrane. Additionally, CP plays a crucial role in declining water flux and power density during osmotic processes [3]. It particularly occurs in processes that use composite membranes that are asymmetric, where a dense selective layer of thin polyamide is placed above a porous membrane support layer. As shown in Figure 11.4, ECP occurs when solute particles accumulate on the side of the dense layer, while ICP happens at the interface between the membrane

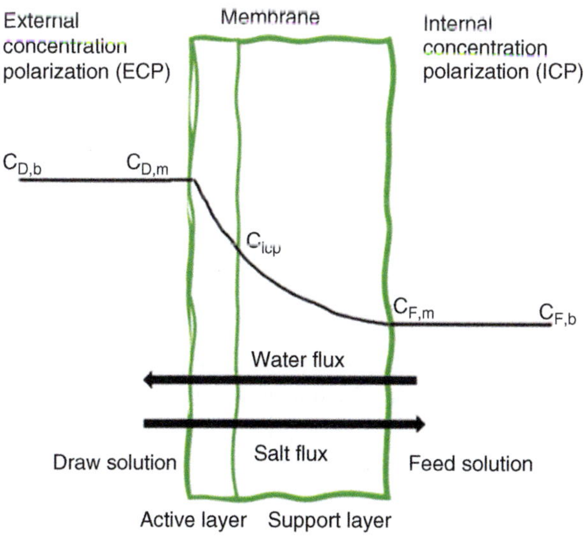

Figure 11.4 Impacts of external and internal concentration polarization on an asymmetric membrane's concentration profile during the process of pressure-retarded osmosis. Source: Adapted from Yip and Elimelech [20].

support and the selective layer [19]. ECP can be either dilutive or concentrative. When water moves from the feed solution at the bulk to the feed solution at the membrane surface (at concentrations [$C_{F,b}$] and [$C_{F,m}$], respectively), the concentration at the surface of the support layer will increase, and this results in concentrative ECP. Conversely, from draw concentration in the bulk solution ($C_{D,b}$) to the draw concentration at the membrane surface ($C_{D,m}$), the concentration at the surface of the active layer decreases, and this leads to the dilutive ECP. Shifting from the variation in bulk concentrations ($C_{D,b}$–$C_{F,b}$) to the variation in concentrations across the active layer of the membrane ($C_{D,m}$–C_{icp}) happens when ECP and ICP occur concurrently. One key factor influencing CP in osmotic operations is the membrane's orientation. In osmosis operations, the active layer can be oriented facing the feed solution or facing the draw solution. For the PRO process, facing the draw solution is typically used, whereas other osmotic operations like FO generally operate with the active layer facing the feed solution [4].

11.2.5 Fouling Challenges in Osmosis Processes

Fouling of a membrane in osmosis processes refers to the accumulation of unwanted substances on the surface of the membrane or inside its pores. This phenomenon is a significant challenge in membrane-based technologies for water filtration or generation of energy. Moreover, fouling can decrease permeate flux, deteriorate permeate quality, increase operational expenses, and minimize the durability of the membrane [20]. Membrane fouling happens due to several reasons including chemical, physical, and biological reactions among organic and inorganic substances. This results in a decline of the flux of permeate with a rise in operational costs. Thus, to make PRO commercially viable, effectively reducing membrane fouling is an important issue [21]. In PRO systems, several parameters can influence membrane fouling, such as the type of salinity gradients used, membrane types, membrane orientation, and type of draw solution. Choi et al. [7] suggested that pretreatment is an effective method to address fouling issues [22]. In another research study, several methods were employed to mitigate membrane fouling in PRO systems. Osmotic backwash, which reverses water permeation to dislodge foulants, achieved a partial recovery of about 61.3% in water flux, demonstrating its effectiveness in removing them from the membrane's porous support layer, though it was less effective at the active support interface. Pretreatment of influent streams to eliminate organic substances prevents fouling but adds energy and chemical costs. Fouling decreased projected membrane power density by about 25–26%, but osmotic backwashing restored it to approximately 85% of the original value, highlighting the possibilities of these strategies to enhance PRO systems' functionality and performance [20].

Wan et al. [23] investigated the fouling behaviors of hollow-fiber membranes used in PRO systems. This study aimed to understand how fouling affects the flux of water and osmotic power generation when employing wastewater retentate as a low-concentration solution and TFC seawater brine as a solution with a high concentration. They realized that fouling on the porous membrane substrates significantly reduced the power density from $21\,W/m^2$ with deionized water to $4.55\,W/m^2$ with real wastewater retentate. To mitigate this issue, they used pretreatment methods including ultrafiltration and nanofiltration, which enhanced the power density to 6.6 and $8.9\,W/m^2$, correspondingly. Therefore, appropriate pretreatment can effectively enhance the PRO's performance by reducing fouling on the membrane's surface [23]. There are also a large number of studies that introduce novel approaches to fouling in membrane technology through chemical modification. For example, in a study specifically designed to generate osmotic power from municipal wastewater, 2-methacryloyloxyethyl phosphorylcholine zwitterionic was used to prepare anti-fouling hollow-fiber membranes with a coating of the polyethersulfone hollow fibers with polydopamine. Fouling of the membrane significantly reduced, which led to the improvement of the membrane's performance during the PRO process [2].

In addition, another study used (3-aminopropyl) trimethoxy silane to enhance the anti-fouling characteristics of TFC membranes and improve their performance in PRO applications. The modification process involved grafting sulfonated hyperbranched polyglycerol onto the surface of the TFC membrane, and this grafting was made easier by using (3-aminopropyl) trimethoxy silane, which played a significant part in ensuring the effective attachment of the sulfonated hyperbranched polyglycerol. Finally, the results showed that using this material and method was very effective in enhancing the fouling resistance and power density of the membrane [2]. Moreover, a high concentration of the draw solution causes an increase in water flux, while, on the other hand, it can increase the chance of the accumulation of particles, including organic, inorganic, or biological ones, on the membrane. Different types of draw solutions have various effects on the fouling of the membrane and its performance. For example, divalent cations containing draw solutions such as calcium or magnesium cause more severe fouling than monovalent cations, such as sodium. The importance of selecting an appropriate draw solution is determined by this difference in order to reduce fouling issues and maintain efficient performance of the PRO process. In addition, other operational factors include the flow rates of the draw and feed solutions, their temperatures, and the specific membrane materials affect fouling issues during the osmosis processes. Therefore, by creating a proper balance between the mentioned parameters without raising the operating costs, the PRO procedure can be optimized, and the undesirable effects of membrane fouling can be mitigated [7].

11.3 Membrane Technology in the Pressure-Retarded Osmosis System

11.3.1 Development of Membranes for the Pressure-Retarded Osmosis Process

Preparing and using suitable and highly effective membranes form a crucial step in the PRO application. It was considered neither practical nor economically viable to advance the commercialization and development of PRO without these suitable membranes [24]. Two kinds of membranes are commonly utilized in most PRO systems. The first ones are CA- or CTA-synthesized membranes and the second is the TFC membrane which has a nonporous active layer on its porous support layer [25]. Early PRO studies employed cellulose-based membranes because they were commercially available, were hydrophilic, and had good mechanical strength. Moreover, a CA flat-sheet membrane was made specifically for desalination purposes [26]. Membranes that have a base of CA have some benefits including high hydrophilicity, which can enhance the water flux, and appropriate mechanical strength [27]. However, they have their own drawbacks, including low tolerance to pH changes and membranes that are produced by hydration technology innovations are costly [28]. However, the TFC membrane compared with CA-based membranes has higher water flux, and the TFC membrane is designed specifically for the generation of power in the PRO system. TFC membranes offer higher water permeability, mechanical strength, and resistance to a wide variety of the pH of the feed solution. Therefore, numerous researchers have investigated the use of TFC membranes for PRO applications that have more advantages over other kinds of membranes [25]. In order to improve PRO technology, it is important to enhance the membrane's performance since there are no effective membranes. For this reason, the polyamide layer was created by an interfacial polymerization method using hollow-fiber and flat-sheet membranes. As a result, it is crucial to reach the optimal properties of the membrane, such as its water permeability, salt rejection, mechanical strength, and structural properties. Moreover, the ideal balance between these parameters and active layer properties is also important for achieving higher osmotic power. However, the effectiveness of the membrane is limited by the support layer parameter that is a structural parameter, and it means a lower structural value can lead to higher power generation [29]. Several studies have been conducted on membrane development, and many are being applied in industrial settings. Moreover, novel nanomaterials and more effective methods have led to continuous improvements in membrane properties and performance. Compared to the earliest membranes used for osmotic processes, current membranes exhibit significant advancements. PRO membrane design, based on various mathematical simulations, has specifically focused on characteristics such as enhancement of water flux and anti-fouling properties by improving the hydrophilicity, effective salt rejection, and high mechanical strength, while the membrane can resist under higher hydraulic pressure during osmosis processes. The development of suitable PRO membranes has enabled the achievement of more osmotic power generation and represents the feasibility of using novel membranes in PRO for broader applications.

11.3.2 Flat-Sheet Membranes

Producing flat-sheet membranes is simple, and altering their characteristics with different additives and manufacturing settings is not too difficult. Several investigations have explored finding ways to improve their structural strength and permeability. For instance, Han et al. [30] investigated how posttreatment can affect the manufacturing of TFC 200 flat-sheet membrane's characteristics. NaOCl concentrations were used in the second and third batches ranging from 200 to 600 ppm. After three batches were made and exposed to various conditions of posttreatment, the higher concentrations of NaOCl were shown to increase water permeability. Additionally, reduced power density for the first and third batches was the consequence of a decrease in salt rejection as posttreatment progressed. Moreover, there are several studies that found water flux of the membranes increased when a selective polyamide layer was enhanced with nanomaterials using an interfacial polymerization method, leading to higher energy generation. Li and Chung [31] enhanced the performance of the membrane by mixing *p*-xylenediamine into a solution of m-phenylenediamine during interfacial polymerization, and this led to the higher water flux and power density of the membrane. In 2013, researchers used an electrospinning technique for synthesizing membranes, and they realized this technique increased membrane porosity and decreased structural parameters. Polyacrylonitrile was used as the polymer and support layer, and a power density of $12.2\,W/m^2$ was achieved [32]. Another study investigated surfactants that can enhance the formation of the interfacial polymerized layer of TFC membranes because of their amphiphilic nature [33, 34]. Cui et al. [35] optimized the reaction of the interfacial polymerization and applied posttreatment techniques to enhance the membrane's performance. They obtained a power density of $18.09\,W/m^2$ at an applied pressure of 22 bars by adding sodium dodecyl sulfate and applying posttreatment with *N*, *N*-dimethylformamide, where 1 M NaCl and deionized (DI) water were the draw and feed solutions, correspondingly.

11.3.3 Hollow-Fiber Membranes

Hollow-fiber membranes are another type of membrane used in the PRO system, characterized by their cylindrical shape and porous structure. Moreover, they are synthesized by using spinning techniques that make the materials separate into different phases, including nonsolvent-induced phase separation. The hollow-fiber membrane compared with the TFC membrane has a larger surface area that can lead to the higher efficiency of the filtration system. This membrane has a porous structure, and this allows selective passage of certain substances. In addition, the hollow-fiber membrane is well known due to its scalability, flexibility, mechanical strength, and efficiency in both industrial and laboratory applications [36, 37]. In a study investigating how changing the flow rate of freshwater influences the performance of a PRO system with a hollow-fiber membrane, it was found that at a slow flow rate, the main problem is the loss of freshwater and CP, and at higher flow rates, the issues are related to solute concentration and salt leakage. In the end, they suggested an ideal flow rate that can decrease the CP issue and increase the PRO performance [38]. As fouling is the main challenge in generating osmotic power, in a study, the dry-jet wet spinning technique was used for synthesizing a hollow-fiber membrane with a polyamide selective layer, to mitigate the fouling problem through amino-silane grafting in PRO applications. According to the results, it was found that mechanical strength, hydrophilicity, water flux, and power density improve when the fouling issue is addressed effectively [39]. In another study, polyethersulfone (PES) hollow fiber was used, while the seawater brine and wastewater brine were used as feed solutions. It was investigated that when wastewater brine was used as the feed solution, more and rapid fouling occurred compared with using seawater brine. To mitigate this issue, they adjusted the pH and applied pretreatment methods [40].

11.3.4 Roles of Nanomaterials in Membrane Modification for the Pressure-Retarded Osmosis Process

In several studies, researchers have focused on adding various nanomaterials to find appropriate membranes for osmotic processes and enhance the performance of membranes [41]. In 2007, the first thin-film nanocomposite (TFN) was introduced and used in the process of RO for water filtrations, for which the main reasons are enhanced anti-fouling property, selectivity, more resistance to different chemicals, and higher mechanical strength during osmosis processes. Nanomaterials can be integrated within the polyamide layer or support layer of the membrane to enhance membrane properties. Therefore, to achieve higher separation performance and optimize both the polyamide selective layer and the porous support layer, it is crucial to make novel TFN membranes due to their advantages specifically for the PRO system [42, 43]. Other reasons for investigating novel TFN membranes are to improve their cost-efficiency and decrease the energy consumption while maintaining their mechanical strength and increasing the water permeability and salt rejection [44]. Various nanomaterials have been added recently to polymer-based membranes to minimize the structural parameters and exceptional anti-biofouling properties while maintaining higher water flux under high applied hydraulic pressure in the PRO system [4]. Nanomaterials like zeolites are used for synthesizing membranes due to their structural properties and higher potential in separation processes. In a study, TFN was synthesized using zeolites (0.02–0.4 wt./v%) in the polyamide rejection layer to improve FO performance. The results showed that, compared to the TFC membrane, zeolite in TFN enhanced the permeability and flux of water with an optimal 0.1 wt./v% loading, and this is because of the porous structure of the zeolite [45]. In another study, the TFN membrane, with a nanocomposite substrate containing 0.4% by the weight of zeolite, indicated a water flux of 33.1 L/m^2h that was approximately 50% more than that of the membrane without zeolite when 10 mM NaCl was used as the feed solution and 2 M NaCl was used as the draw solution [46]. Shadravan et al. [47] explored a TFN membrane performance using zeolite nanomaterials as an additive for PRO applications while nanomaterials were used in the polyamide layer, under different test conditions.

The results presented that the TFN membrane achieved higher osmotic power and water flux in comparison with the TFC membrane. Silicon dioxide nanoparticles have a porous structure, active surface properties, and a high capacity of adsorption that make them well suited for fabricating various membranes employed in the treatment of water operations. Research was conducted to examine the incorporation of silica nanoparticles into polyamide TFN membranes at concentrations of 0.01, 0.05, and 0.1% w/v, and in this study, it was discovered that the introduction of silica nanoparticles had a great impact on the membrane hydrophilicity because their contact angle decreased from 82° to 44°. Water flux through the TFN membrane almost doubled compared with the TFC membrane, with the maximum value being equal to $36 \pm 2\ L/m^2\ h$. The membrane that contained 0.05% w/v of silicone dioxide demonstrated the higher rejection of salt at approximately $90 \pm 3\%$, attributed to effective pore blocking. Increased surface roughness, observed in atomic force microscopy images, also contributed to improved membrane performance. Although higher concentrations of silica generally enhance performance, excessive loading (0.1% w/v) leads to nanoparticle agglomeration, slightly reducing effectiveness [48]. Titanium dioxide (TiO_2) nanoparticles, especially in the anatase form, have been of interest to different industries. TiO_2 is a well-studied photocatalyst and anti-bacterial agent used in many areas including civil engineering where it has also been utilized in making self-cleaning nano paint. In this research, the solvothermal synthesis of amino-functionalized nanoporous TiO_2 and its influence on the TFC membrane was studied. TFC membranes were modified with TiO_2 nanoparticles at various concentrations, namely 0.01, 0.05, and 0.1% w/v, within the layer of polyamide. The membrane hydrophilicity, water permeability, and anti-fouling property were improved by inserting TiO_2 nanoparticles into the membrane itself. According to the results, adding just 0.05% w/v TiO_2 achieved a maximum water flux of 26.4 L/m^{2}h with high anti-fouling performance [49]. Carbon quantum dots (CQDs) are an interesting type of carbon nanoparticles, approximately 10 nm in diameter, that offer a green alternative to toxic metal-based quantum dots. In a recent study, researchers investigated TFC-PES membranes for osmotic power generation, which were reinforced by citric acid-generated CQDs. On the surface of the polydopamine-immobilized membrane, these CQDs had an average size of 3.2 nm. The modified membranes displayed better fouling resistance and higher anti-bacterial effect than unmodified ones. CQD-modified membranes demonstrated higher electrical power density and increased recovery of water after the backwashing within PRO experiments. Compared to unmodified membranes, the mean power densities for the CQD-modified membranes were 11.0 and 8.8 W/m^2 for the TFC membrane, respectively, which shows there is a significant scope for improved efficiency in the harvesting of energy applications [50]. Carbon nanotubes (CNTs) have gained extensive attention in different fields like environmental engineering due to their outstanding mechanical and electrical properties which are highly useful for engineering purposes. In a study, to mitigate the fouling issue of the membrane, functionalized carbon nanotubes (fCNTs) were added to the support layer for PRO applications.

It was found that the TFN membrane with 0.5 wt% fCNTs achieved the desired anti-fouling properties due to its surface roughness and better hydrophilicity compared with TFC membranes. In addition, the results showed a 35% decrease in the attachment of bacteria compared to the bare TFC membrane. Overall, the study concluded that adding fCNTs to the membrane support layer enhanced membrane performance and the anti-biofouling properties of the membrane and made it an appropriate membrane for PRO applications [51]. Goh et al. [52] designed hollow-fiber membranes and added fCNTs into the polyethylenimine-polyamide imide as the base materials for FO applications. The results demonstrated that water permeability increased in the hollow-fiber membranes with fCNTs by 44% compared with TFC membranes without fCNTs. In addition, water flux increased by 29% with a small increase in solute flux, which shows the higher performance of TFN membranes compared with TFC membranes. Moreover, they highlighted that TFN membranes are scalable and emphasized the benefits of using CNTs that are commercially available for the improvement of membrane performance through water filtration processes. Nanotechnology has been advanced greatly by nanomaterials in the capacity of membrane fabrication, leading to increased membrane efficiency and performance, as reported in the research. However, the application of nanomaterials has had positive impacts on membrane technologies, including enhanced hydrophilicity, improved mechanical and thermal stability, and increased clean water flux and permeability. In addition to these, membranes have also benefited from the vigorous incorporation of nanomaterials in terms of endurance, and they have also acquired anti-microbial resistance as well as fouling resistance over an extended period of time. However, it effectively rejects salts for both FO and PRO systems.

11.4 Energy, Economic, and Environmental Considerations for Pressure-Retarded Osmosis

11.4.1 Energy Assessment

Determining the specific enthalpy (h) at the inlet and outlet of the turbine is vital for evaluating the potential energy production of a full-scale PRO. Figure 11.5 shows the typical components of a PRO plant, including the feed and draw pumps and energy recovery devices such as pressure exchangers, booster pumps, and the turbine. Considering the data derived from the Gibbs free energy basis for seawater, equations (11.23) to (11.25) are applicable to determine specific enthalpy. For pressures other than atmospheric pressure (p_0), specific enthalpy ($h\,[T, p, C]$) can be obtained from equation (11.25). The ideal power output (GP_{id}) and power density (PD_{id}) of the PRO system are calculated using equations (11.26), (11.27), and (11.28), respectively.

$$h(T, p_0, C) = h_w - C\,(b_1 + b_2 w_s + b_3 w_s^2 + b_4 w_s^3 + b_5 T + b_6 T^2 + b_1 T^3 + b_8 w_s T + b_9 w_s^2 T + b_{10} w_s T^2) \tag{11.23}$$

$$h_w = 141.355 + 4202.070T - 0.535T^2 + 0.004T^3 \tag{11.24}$$

$$b_1 = -2.348 \times 10^4,\ b_2 = 3.152 \times 10^5,\ b_3 = 2.803 \times 10^6$$
$$b_4 = -1.446 \times 10^7,\ b_5 = 7.826 \times 10^3,\ b_6 = -4.417 \times 10^1$$
$$b_1 = 2.139 \times 10^{-1},\ b_8 = -1.991 \times 10^4,\ b_9 = 2.778 \times 10^4$$
$$b_{10} = 9.728 \times 10^1$$

$$h\,(T, p, C) = h(T, p_0, C) + \vartheta\,(P - p_0) \tag{11.25}$$

$$P_{TB} = \eta_{TB}\dot{m}_{TB}\,(h_{TB,in} - h_{TB,out}) \tag{11.26}$$

$$\dot{m}_{TB} = Q_p \rho_{D,out} \tag{11.27}$$

$$PD_{id} = \frac{P_{TB}}{nS_m} \tag{11.28}$$

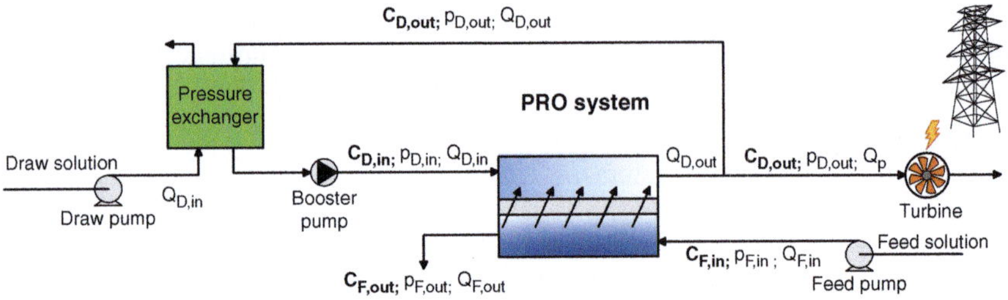

Figure 11.5 Pressure-retarded osmosis process. Source: Ruiz-García et al. [53]/Elsevier/CC BY 4.0.

where ϑ is the specific volume, the reverse of ρ. η_{TB} is the turbine performance, and n is the number of spiral-wound membrane modules in the PRO system [53].

11.4.2 Cost Contributors

Several factors contribute substantially to the overall costs while evaluating the energy, economic, and environmental considerations of a PRO. Engineering and legal expenses, along with pretreatment and transportation fees, are among the primary cost contributors, while the costs related to material disposal are minimal and almost negligible. However, labor and indirect costs significantly affect the breakdown of other operational and maintenance expenses. Pretreatment has a crucial role in increasing capital costs and decreasing the net power output of the PRO system. Transmission costs show a considerable financial obstacle, as life cycle cost analysis has shown that intake and outtake processes can lead to fundamental cost contributions. The additional expenses are heightened by the fact that the wastewater treatment plants are not located at the same places with the PRO, further driving up fees. In contrast, the end-of-life phase incurs far lower expenses, contributing a negligible amount to the total fees. This minimal effect is primarily because of the assumption that high-density polyethylene pipes will remain underground, thus decreasing landfill tipping expenses as the weight of the pipes is excluded. Regarding the distribution of other operational and maintenance expenses, indirect costs and labor as well as costs associated with the replacement of membranes emerge as the highest contributors. While frequent cleaning is essential to maintain the performance of PRO, it adds extra fees because of the consumption of the chemicals. Since the PRO system generates energy, the net energy expenses are negative, leading to cost savings. Therefore, while comparing capital expenses, net energy costs, other operational and maintenance costs, and end-of-life fees, capital costs manifest as the highest contributor, followed by other operational and maintenance expenses [54].

11.4.3 Environmental Sustainability of the Pressure-Retarded Osmosis System

PRO system outflow has a considerable environmental impact, including terrestrial, human toxicity, marine ecotoxicity, and eutrophication. Additionally, the PRO system equipment contributes remarkably to fossil fuel and ozone depletion. Conversely, components like the membrane, pressure exchangers, excavation, filtration, membrane replacement, and landfill processes have relatively marginal impacts as opposed to other input parameters. Energy-related factors, such as filtration energy, energy losses, transmission, and generated energy, affect all environmental impact indicators. The energy mix utilized in the PRO system predominantly comprises natural gas, nuclear power, and coal, leading to various emissions. Climate change is particularly affected by the release of CO_2, CH_4, and fossil methane. Moreover, terrestrial eutrophication and acidification are influenced by sulfate, nitrate, and phosphate emissions. The NO_x, NH_3, and SO_2 emissions contribute to the formation of fine particulate matter and impact the indicators of toxicity. Notably, the PRO system produces more energy than it consumes, leading to a net negative energy impact. This negative value suggests that the energy by the PRO diminishes the overall environmental effect, particularly while functioning alongside a desalination plant.

However, since the discharge of treated wastewater contains nitrogen, ammonia, orthophosphate, and phosphorus, the PRO outflow exacerbates the eutrophication of marine and freshwater. The outflow also has an impact on toxicity indicators due to the presence of metals such as magnesium, copper, and iron in the brine and treated wastewater. The equipment, particularly pumps and high-density polyethylene pipes among the components of the PRO system, emerge as the fundamental contributor to fossil fuel and ozone depletion. The production of pumps significantly contributes to ozone depletion due to the use of substances, such as chloroform, chlorofluorocarbons, and Halon 121. The turbine is defined as the largest contributor to ozone depletion where the focus is on the installation phase due to the absence of many pumps. The membrane elements have minimal effect across most indicators, except for a noticeable contribution to ozone depletion, mainly owing to the sodium hydroxide input containing chlorofluorocarbons and hydrogen chlorides from the manufacturing process. Due to the high-density polyethylene pipes, leaving only a small percentage of materials, like membranes and the pressure exchangers to be disposed of in landfills, the effects of the landfilling process are minimal [54].

11.4.3.1 Comparative Analysis of Greenhouse Gas Emissions

The carbon footprint is a crucial indicator to evaluate the environmental sustainability of renewable energies. PRO stands out as the only technology with net-zero carbon emissions among various ones. Other renewable energies, such as photovoltaics (PVs) and wind turbines, have been correlated with significant greenhouse gas emissions. Studies evaluating the environmental sustainability of wind turbines across different scales, medium-scale and large-scale, have reported net emissions of greenhouse gases ranging from 0.015 to 0.038 kg CO_2 equivalent/kWh. Other analyses of wind technology have led to average net emissions of greenhouse gases of 0.0090–0.011 kg CO_2 equivalent/kWh. Likewise, mono-crystalline silicon PVs have been revealed to produce net emissions of greenhouse gases ranging from 0.029 to 0.045 kg CO_2 equivalent/kWh. The relatively high emissions related to PV systems are largely due to the phase of manufacturing, involving steel and aluminum frame fabrication, glass production, and the creation of silicon solar cells. Moreover, the fuel mix employed in producing PV modules considerably leads to the overall carbon footprint.

For wind turbines, the key emissions of greenhouse gas contributors are in the manufacturing and transportation stages. The production of the tower and the construction of the nacelle need considerable energy, contributing to significant emissions of greenhouse gases. Stainless steel plate production releases large amounts of greenhouse gases, attributable to the significant material requirements during manufacturing and transportation. Furthermore, turbine production needs energy and external support from more powerful engines, contributing to greenhouse gas emissions. Renewable energies are challenged by their intermittency, requiring energy storage systems, which currently have low efficiency, high fees, and greenhouse gas emissions. In contrast, PRO shows the lowest net emissions of greenhouse gases as it does not need an energy storage system or extensive use of rare earth elements. One of the substantial contributors of greenhouse gases for PRO is the production of high-density polyethylene pipes; however, their production compared to steel generates fewer greenhouse gas emissions. Therefore, the overall net emissions of greenhouse gases in the PRO system are lower than those of PV and wind technologies, which depend heavily on steel [54].

11.4.4 Life Cycle Assessment and Life Cycle Cost Analysis of the Pressure-Retarded Osmosis System

It is necessary to assess PRO's environmental and economic sustainability to evaluate its competitiveness compared to other renewable energies. The assessment is classified into four key stages: installation and materials, operation, maintenance, and end of life. The transportation stage has been excluded from the evaluation, as its impact is minimal in comparison with the other stages. The system boundaries and the inventory for each stage are depicted in Figure 11.6.

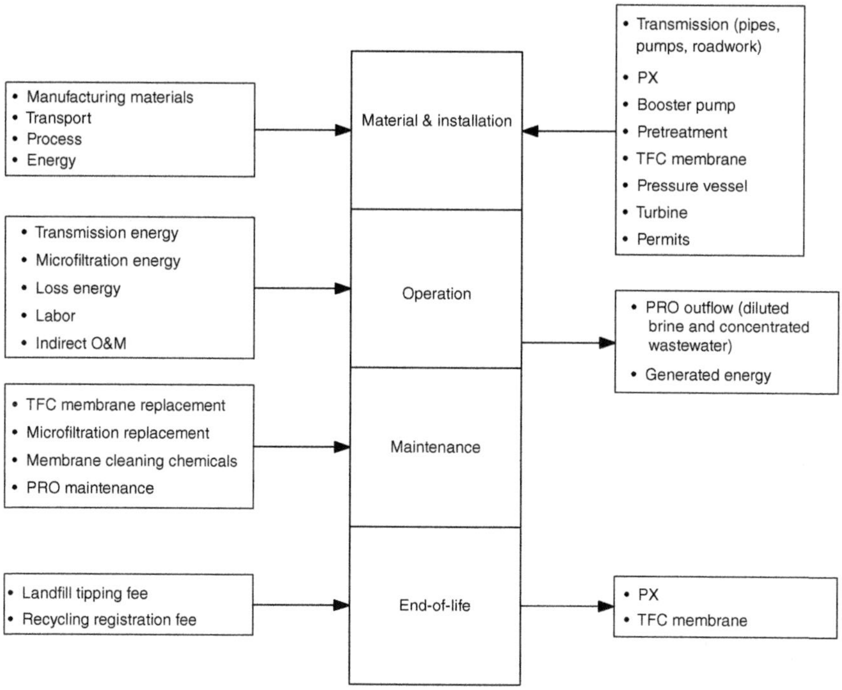

Figure 11.6 Outline of system boundaries and inventory aspects for each stage. Source: Mashrafi et al. [54]/with permission of Elsevier.

The net energy, operation, maintenance, and end-of-life costs were converted to a net present value (NPV) for the life cycle cost analysis:

$$P_v = F_v \left(\frac{1}{(1+i)^n} \right) \tag{11.29}$$

where F_v refers to the cost in year 0, i denotes the rate of discount, and n is the year. NPV was achieved by the summation of the net present value of capital, net energy, operation, maintenance, and end-of-life expenses:

$$NPV = Cap_{cost} + TNE_{cost} + O\&M_{cost} + EOL_{cost} \tag{11.30}$$

The NPV was then split by the total energy produced by the PRO during the lifespan to compare the competitiveness of the various power generation technologies. The cost is given as

$$LCOE = \frac{NPV}{G.E} \tag{11.31}$$

where G.E denotes the total net energy produced during its lifespan. The results from the life cycle cost analysis were denoted as the levelized cost of energy (LCOE) [54].

11.5 Pressure-Retarded Osmosis Hybrid Systems

Despite the sharing of essential similarities by PRO and traditional desalination processes, their fundamental goals diverge. Conventional desalination primarily seeks to produce freshwater, while PRO focuses on enhancing the efficiency of energy. This distinction has contributed to the advancement of PRO hybrid systems, integrating PRO with conventional desalination methods to optimize the usage of energy. The seawater RO-PRO system has garnered significant attention among these hybrid systems. The widespread deployment of seawater RO technology facilitates the effective modeling and simulation of seawater RO-PRO systems, leading to the initiation of pilot plants in countries, such as Japan, Singapore, and South Korea. In contrast, other hybrid configurations, like membrane distillation (MD)-PRO, have not advanced beyond the conceptual stage. Their limited commercial availability prohibits the comprehensive assessment of their performance and practical applications.

11.5.1 Reverse Osmosis-Pressure-Retarded Osmosis Hybrid System

This system demonstrates a diverse and adaptive technology capable of switching between two operational modes: (2) RO mode, which produces water with a high quality from wastewater treatment plant effluents, and (B2) PRO mode, which leverages the salinity gradient between wastewater and seawater desalination brines to generate energy. This dual mechanism enables the system to optimize its performance as a reply to fluctuating energy costs, using low-cost energy for water production and energy generation when economically advantageous. The value of the permeate must meet two critical criteria for the hybrid RO-PRO system to be feasible: (1) it must be sufficiently low to generate economically viable electricity in PRO mode, and (2) it must be sufficiently high to justify the production of water in RO mode. The system is most effectively used as an improvement to an existing PRO setup, especially in scenarios with a limited need for purified wastewater treatment plant effluents. Nevertheless, the system is predicted to prioritize purified wastewater effluent production based on the relative need for water versus energy. As depicted in Figure 11.7, the system can change seamlessly between RO and PRO modes depending on economic conditions, targeting minimization of operational costs. When there is sufficient freshwater, the system operates in PRO mode mixing freshwater and seawater to generate energy, while the system changes to RO mode to desalinate seawater when there is limited freshwater [55].

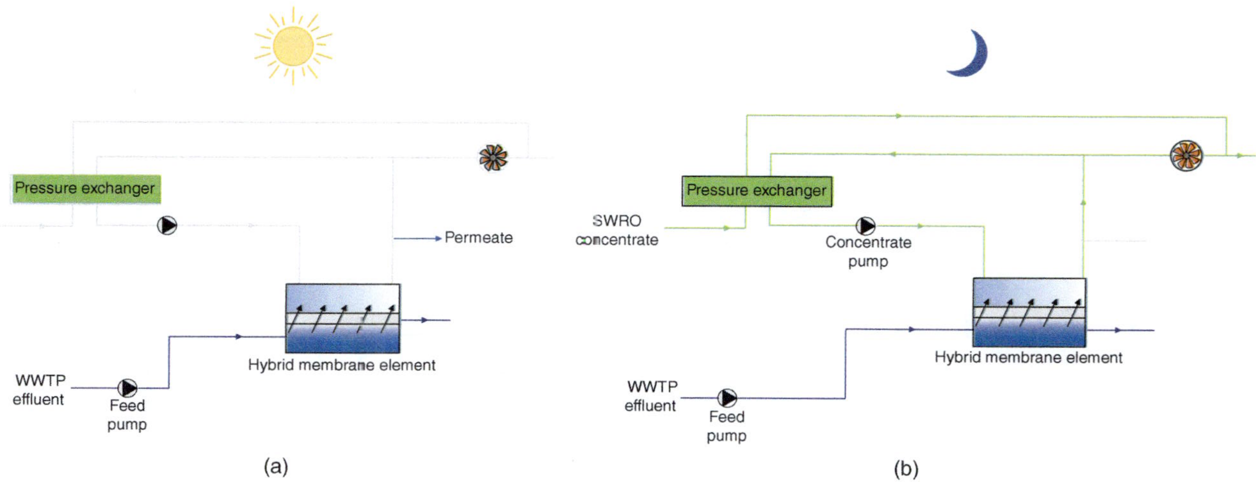

Figure 11.7 The reverse osmosis-pressure-retarded osmosis hybrid system: (A) pressure-retarded osmosis mode, and (B) reverse osmosis mode. Source: Blankert et al. [55]/with permission of Elsevier.

11.5.2 Membrane Distillation-Pressure-Retarded Osmosis Hybrid System

MD is a thermally driven separating process that heats the feed stream until partial evaporation happens. The resulting vapor passes through a porous, hydrophobic membrane and condenses on the side of colder permeate, resulting in a concentrated feed stream and high-quality permeate water, rendering it a valuable technique for concurrently recovering draw solution and producing clean water in osmotic membrane processes. The integration of FO with MD has already demonstrated significant promise in seawater desalination, industrial wastewater reuse, and wastewater reclamation. Building on this progress, hybridizing MD with PRO has been proposed as a sustainable method for generating renewable osmotic power while producing clean water. In the MD-PRO process (Figure 11.8), seawater brine, a byproduct of RO desalination, is first subjected to MD. This step concentrates the draw solution and increases its temperature while producing clean water. The consequent high-temperature, concentrated draw solution is then introduced into the PRO unit, where it interacts with low-salinity wastewater or wastewater retentate on the feed side of the PRO membrane. The differentiation in osmotic pressure promotes the diluted draw solution via a turbine, generating energy, while the concentrated wastewater, now diminished in volume, can be then processed in wastewater treatment facilities. On the other hand, the system can function by delivering high-salinity seawater brine to the draw solution side of the PRO membrane, with wastewater or wastewater retentate acting as the feed solution. The diluted

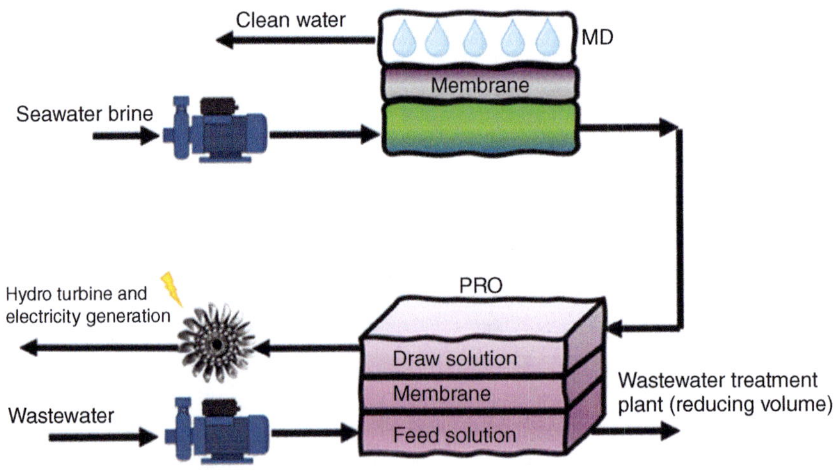

Figure 11.8 Schematic diagram of the membrane distillation-pressure-retarded osmosis process. Source: Einarsson and Wu [56]/with permission of Elsevier.

draw solution is further concentrated on the feed side of the MD process before being recycled back into the PRO process, where it is combined with incoming seawater brine. This configuration enables ongoing clean water production on the permeate side of the MD process [56].

11.5.3 Membrane Bioreactor-Pressure-Retarded Osmosis Hybrid System

This system incorporates PRO into membrane bioreactor technology for desalination and energy generation. Like the RO-PRO hybrid system, seawater is pressurized and utilized as the draw solution within the PRO system. The seawater with high salinity is diluted, producing freshwater and energy from part of the diluted draw solution. The diluted seawater then flows through a second stage of FO, where it further dilutes a particularly designed draw solution. Freshwater is extracted from this diluted draw solution, which improves the efficiency of the system. This hybrid approach incorporates the advantages of PRO and membrane bioreactor technologies to optimize the production of water and the generation of energy [7].

11.5.4 Liquid-Phase Ion-Stripping-Pressure-Retarded Osmosis Hybrid System

In a hybrid PRO system with a liquid-phase ion-stripping process, liquid-phase ion-stripping generates a salinity gradient by a thermal cycle integrated with an organic solvent which facilitates the ions' rejection, leading to freshwater production and brine concentration. The freshwater and brine then serve as the feed and draw solutions, respectively, for the PRO process. The hybrid system has demonstrated high efficiency of energy, effective extraction of solvents, and noticeable heat recovery [2].

11.5.5 Thermosiphon-Pressure-Retarded Osmosis Powered by Solar Energy

Thermosiphon systems employ natural circulation driven by buoyancy, exploiting differences of density within a thermal gradient to facilitate the generation of power. In these systems, the heated fluid rises above a colder one, creating a natural circulation, improving upward heat transfer by a collector. Recently, unlike conventional systems which depend on temperature gradients, a novel system known as brinesiphon operates by salinity differences. As shown in Figure 11.9, a thermosiphon system is hybridized with the PRO to harness the generated osmotic pressure by PRO and drive the thermosiphon circulation, thus improving fluid management and energy recovery [56].

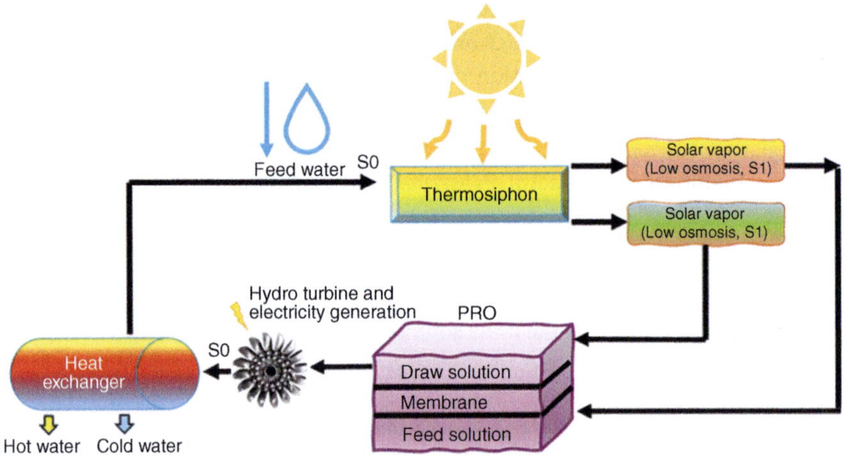

Figure 11.9 Schematic representation of thermosiphon-pressure-retarded osmosis. Source: Einarsson and Wu [56]/with permission of Elsevier.

11.5.6 Solar Pond-Pressure-Retarded Osmosis Hybrid System

Solar ponds are engineered collectors developed to capture and store solar thermal energy by utilizing a salinity gradient to retain heat in their lower layers. The pond's upper layer, exposed to solar radiation, is sustained at ambient temperature and low salinity. Below this, a nonconvective zone forms, where temperature and salinity rise with depth. The bottom layer, known as the lower convection zone, reaches temperatures between 50 and 90°C and shows the highest salinity and density. The higher density of this layer hinders it from combining with the middle layer despite solar rays penetrating and heating the bottom layer through convection, thus preserving thermal energy over extended periods. The collected thermal energy by solar ponds is typically leveraged using an organic Rankine cycle heat engine or a thermoelectric generator to generate electricity. Solar ponds offer several benefits, such as lower installation costs, reduced footprint, and the capability to function as energy storage systems. However, limitations including low quality of energy and loss of energy have driven efforts to enhance thermal efficiency and generation of electricity by combining solar ponds with other systems. Considerable integrations include thermosiphons, multi-stage desalination systems, MD, FO, and PRO. In a solar pond-PRO system (Figure 11.10), saline water from the solar pond produces two distinct streams with various salinities: low salinity in the surface layer and high salinity in the bottom layer. Thermal energy is extracted from the bottom layer of the solar pond with high temperature. The low-salinity water is employed as the draw solution in the PRO system, where it draws permeate from a freshwater feed. After passing through a turbine

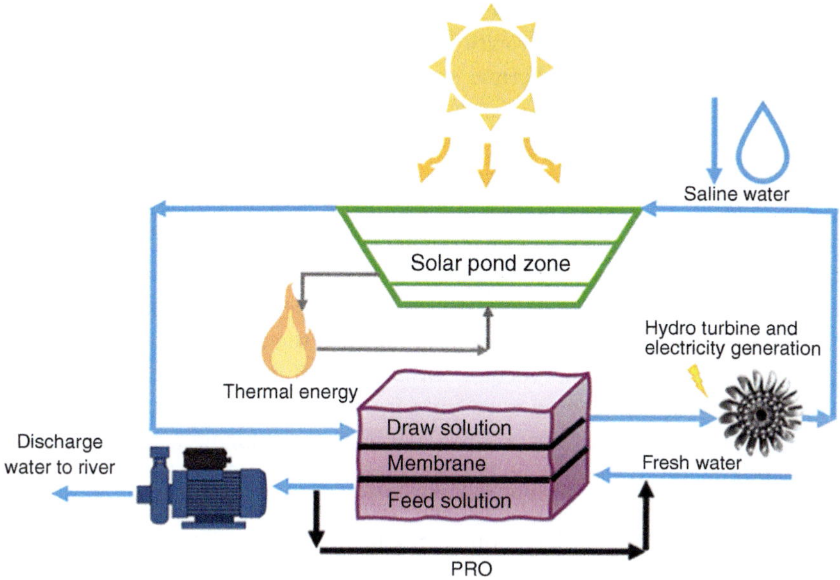

Figure 11.10 Schematic representation of solar pond-PRO. Source: Einarsson and Wu [56]/with permission of Elsevier.

to generate electricity, the diluted draw solution is recirculated back to the solar pond [56].

11.5.7 Geothermal Pressure-Retarded Osmosis Hybrid System

This system harnesses geothermal brine with high salinity within an osmotic power setup combined with geothermal heat generation. First, the geothermal brine flows through a heat exchanger in the geothermal plant, where it is preheated before entering the PRO process. The PRO system uses various feed sources to enhance its efficiency. The geothermal heat plant can provide district heating, using geothermal energy for heating purposes. Simultaneously, the produced electricity from the PRO process can be employed for residential needs or to power the geothermal plant. This dual-purpose system not only generates renewable energy but additionally supports geothermal heating, which offers a sustainable alternative for energy and heating requirements [2].

11.5.8 Multigeneration Pressure-Retarded Osmosis System

The multigeneration PRO system is a mix of multiple subsystems to harness renewable energies and generate freshwater, electrical energy, hydrogen, etc.

For example, in a multigeneration PRO system, PRO is incorporated into various subsystems. A solar subsystem with a heliostat, PV, and a heat transfer system utilizes a concentrated PV/thermal system to produce electricity and thermal energy. Wind turbines supplement the solar energy supply to provide electricity to refrigeration compressors, electrolysis units, and pumps during periods of low solar energy. A thermal energy storage subsystem stores thermal energy from the concentrated PV/thermal system using a phase-change material to smooth fluctuations out in energy supply and guarantee continuous system operation. The multistage flash desalination system uses thermal energy from the concentrated PV/thermal system and thermal energy storage units for heating seawater to a top brine temperature and flash it in multiple stages. The resulting vapor is condensed for freshwater production, with the remaining brine being discharged. An electrolysis system is used to produce hydrogen from freshwater by splitting water into hydrogen and oxygen. Proton exchange membrane fuel cells convert hydrogen's chemical energy into electricity via an electrochemical reaction with oxygen during peak demand periods. The vapor compression refrigeration system uses low-temperature refrigeration cycles to produce cooling. PRO separates solutions with various salinities to generate pressurized flow from the feed to the draw solution to drive a turbine and produce electricity [57].

Another PRO multigeneration system uses various renewable energies through the integration of various subsystems. Parabolic trough solar collectors capture solar thermal energy and convert it into power by the organic Rankine cycle subsystem and are complemented by the Kalina cycle subsystem to recover waste heat from the organic Rankine cycle to produce additional power. A single-effect absorption refrigeration cycle employs ammonia-water as the working fluid to produce cooling. FO produces freshwater and PRO generates power density. Hot and cold storage tanks guarantee reliable operation during periods without solar radiation [58]. In a sophisticated multigeneration system, a solar chimney is used to capture solar radiation and then heat the air beneath it and create buoyancy to drive a wind turbine for generating electricity. A multistage flash distillation system uses the thermal energy generated from the storage of solar chimneys to heat seawater and then generate freshwater. RO processes the brine from this system to extract more freshwater and enhance the capability of the system for the production of water. PRO is used to generate energy because of the different salinity gradients between brine and seawater. Pumped hydropower serves as a reliable source of energy by using the energy of freshwater that is held in a tank for generating electricity. A wind turbine that is independent of the presence of solar properties could generate electricity consistently. Consequently, the combination of all these technologies suggests an efficient option for having both pure water and electricity, which can mitigate the issue of inconsistent renewable energies and enhance the overall performance and sustainability of the system [57].

11.6 Pressure-Retarded Osmosis Applications

PRO has gained considerable attention as an emerging technology for sustainable energy generation, with its applications in various sectors, especially in the fields of water treatment, renewable energy generation, and industrial processes [10]. A collaborative study initiative involving teams from Australia, the USA, and Qatar is analyzing PRO's application for enhanced oil recovery and the treatment of generated water from conventional oil extraction processes (Figure 11.11). Seawater or brine from desalination plants serves as the feed solution, and the hypersaline characteristics of generated water render it an effective draw solution. This ongoing study emphasizes three critical areas: (1) identifying the pretreatment requirements for generated water to be utilized as a draw solution in PRO, (2) designing membranes that can tolerate the high pressures essential to achieve optimal power densities from hypersaline generated water, and (3) combining PRO with waterflooding, a technique widely utilized in large-scale petroleum industries. In the waterflooding process, produced water is re-injected into the oil reservoir to replace oil, thus improving production at the wellhead. The integration of

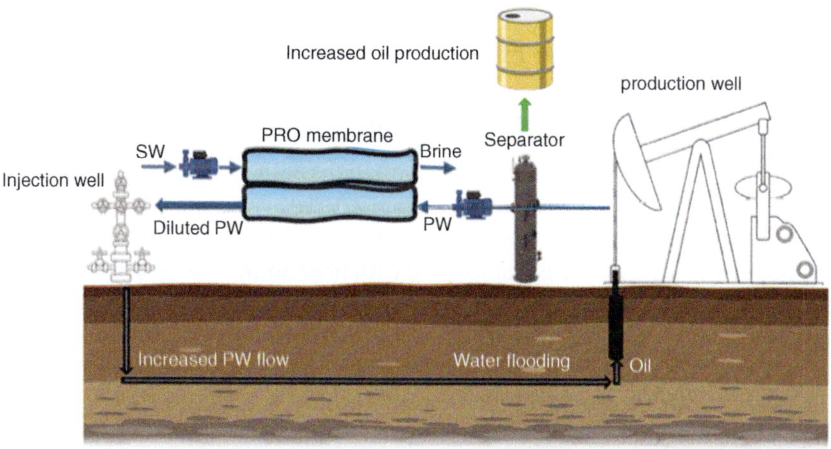

Figure 11.11 Pressure-retarded osmosis integrated with waterflooding for enhanced oil recovery where SW is seawater and PW is produced water. Source: Adapted from Gonzales et al. [2].

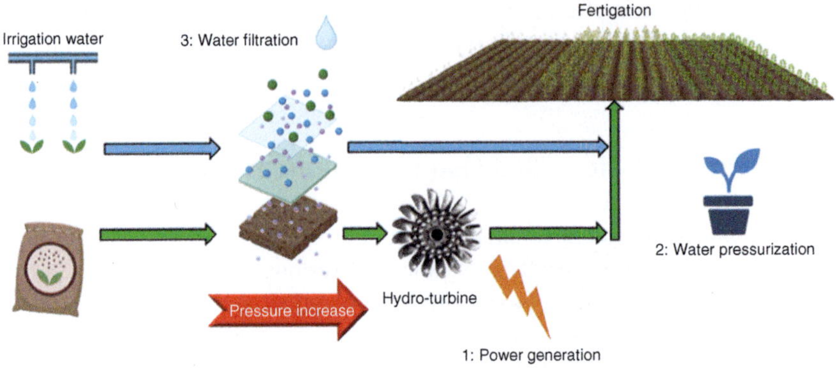

Figure 11.12 Green pressure-retarded osmosis. Source: Adapted from Gonzales et al. [2].

PRO into this process provides multiple advantages, such as the reuse of diluted generated water, reduced consumption of energy for pumping, minimized waste streams, enhanced injectivity, energy production, and increased oil production [2].

Green PRO, a specialized application of PRO, uses agricultural fertilizers as the draw solution. This innovative process seeks to simultaneously generate power, treat water, and pressurize irrigation water. As depicted in Figure 11.12, irrigation water, like river water, serves as the feed solution, while both pure and combined agricultural fertilizers are employed as the draw solutions [2].

11.7 Challenges and Future Directions

11.7.1 Technical Challenges and Solutions

Researchers have realized that there are limitations in the real-world applications of PRO, and the primary challenge facing it today is its practical feasibility. These limitations impact the maximum energy that can be effectively generated. Key issues include energy costs associated with pretreatment, pressurization, and the delivery of the system, all of which reduce the productivity and viability of the process. Consequently, practical hurdles should be addressed to develop the complete potential of PRO for sustainable generation of energy. Straub et al. [59] identified that the feasibility of PRO is significantly challenged by energy losses and operational costs, and inefficiencies within the membrane module, including CP and reverse salt flux, and inefficiencies in pressure exchangers and turbines drastically decrease the quantity of energy that can be achieved. Moreover, high-energy inputs are required for pumping source water, pretreatment to remove contaminants, and maintaining crossflow velocities in the membrane channels, often surpassing the energy output, especially with low-salinity gradients like river and seawater. In addition, the low energy density that can be obtained by combining water from sea and rivers (0.256 kWh/m^3) necessitates processing large volumes of water to generate meaningful energy, reducing the process's efficiency and economic viability, and technological limitations further complicate the situation. The current membrane technology does not yield high efficiency at the required pressures for PRO with high salinity gradients, as membranes must endure high hydraulic pressures while minimizing ICP by maintaining low structural parameters. Moreover, most experimental research is done on a modest scale, and their results do not always translate well to full-scale systems, posing a significant challenge in performance metrics such as power density and specific energy between small-scale tests and large-scale applications. Larger diameters are required for membranes with low power density to compensate for their limited energy-harnessing capabilities, which in turn increases capital costs. Superior membrane performance for PRO is anticipated to lower energy costs by producing more energy. By enhancing osmotic power, extending membrane lifespan, and reducing costs of producing energy, improved PRO membranes can boost the revenue potential for PRO demonstrations. The challenges preventing PRO from widespread adoption and commercialization have remained unchanged over the recent years. Consequently, recent research has shifted from pilot-scale implementations to discovering specialized uses and creating hybrid PRO processes. These hybrid processes seek to find appropriate applications for PRO that do not necessitate expensive infrastructure or high energy consumption for pumping and pressurization, or to use the energy extracted from PRO to meet the energy required for other processes.

11.7.2 Large-Scale Implementation

The progress of PRO is very crucial at a large scale due to its potential in enhancing the economic and environmental aspects of desalination. Global programs interested in this subject need to develop cost-effective techniques for desalinating water and managing brine discharge from hybrid seawater RO plants that create environmental challenges. The development of PRO on a large scale reflects what happens in laboratories and has two main divisions, which are based on the kind of draw solution being utilized, seawater, and seawater RO brine. Seawater is an easily accessible natural draw solution, while seawater RO brine, a byproduct from the process, offers innovative solutions to recycle waste materials dealing with issues regarding both resource utilization and waste management. In this context, a seawater RO-PRO hybrid system developed for more than two years, the pilot plant with 240 m^3/day production capacity, was built and run to determine whether this hybrid system is feasible and how it performs. The goal of this research was also to find the long-term operation stability, energy consumption, and membrane fouling behavior of the system. The results of the experimental plants revealed that specific energy consumption in the seawater RO process decreased by approximately 20% because of the incorporation of PRO, and this is attributed to the effective recovery of energy from high-saline seawater RO.

Moreover, by decreasing the concentration of high-saline water, its environmental impact was reduced. Additionally, during the experimental tests, it became more evident that pretreatment of wastewater is essential to prevent or mitigate fouling of the membrane during the PRO process. Consequently, this pilot study presents that seawater RO in PRO hybrid systems offers a method to enhance the sustainability and economic feasibility of seawater desalination [60]. In addition, in another study, researchers investigated ways to enhance the practical application of the PRO system on a large scale. In this case, to optimize the performance of PRO on a large scale, they incorporated maximum power point tracking (MPPT) into model-based controller systems. The results revealed that the integration of MPPT and model-based controller systems mitigated fundamental challenges regarding environmental fluctuations and increased the efficiency and stability of PRO on a large scale. For example, simulations demonstrated that this system can lead to a reduction of rise time, and this can reduce the experienced power oscillations. Moreover, this is very important as large-scale operations can be affected by different temperatures or changes in the salinity of bulk production. Additionally, overall system adaptability and effectiveness improved by the model-based controller component by predicting initial ideal pressures and facilitated a faster convergence of the MPPT to the maximum power point [61].

The application of a TFC membrane for generating osmotic power is investigated by another research study with the gradient of salinity energy between the wastewater of RO retentate and seawater RO brine as feed and draw solutions, respectively, in the PRO system. This research has shown that PES hollow-fiber membranes, when coated with a TFC layer via interfacial polymerization, can be scaled from laboratory to industry [62]. Moreover, in order to examine the scalability of PRO membranes for harvesting osmotic energy from specific brines, including those produced from oil and gas wells, a study suggested a comprehensive modeling tool. This tool can predict the performance of PRO in different setups which were developed by using an intricate mass transfer model and the equation of the state for electrolyte solutions. The authors have shown that there is a significant difference in nonideal behavior in the electrolytic solution upon its capacity for scaling up microfiltration systems by proving that Van't Hoff, an ideal electrolyte theory, does not work well at higher salinities. Another section discussing the optimization of multistage membrane arrangements indicated that strategic pressure adjustments, along with the distribution of membrane area across stages, enhanced energy yield significantly. These results imply the necessity for complex simulation tools in designing and optimizing PRO processes such as those applicable to high-salinity cases, while also supporting the incorporation of RO units into existing desalination plants to reduce fossil fuel consumption and associated environmental costs to make desalination more sustainable on economic grounds [63].

11.8 Conclusion

PRO is at the forefront of renewable energy technologies, offering a persuasive option for blue energy generation using natural salinity gradients between high-salinity and low-salinity solutions. The progress and refinement of membranes, the core of PRO systems, are essential to address the technical challenges that currently limit the efficiency and scalability of this technology. As explained in this chapter, the principles of osmosis, developments in membrane technology, and the hybridization of PRO with other desalination processes as well as renewable energies underscore its fundamental potential. The economic and environmental considerations highlight the vast applicability of PRO in confronting water–energy nexus challenges. Looking forward, the successful employment of PRO in large-scale applications will depend on continuous research and innovation aimed at optimizing the performance of the system, addressing fouling limitations, and ensuring economic feasibility. The roadmap for future progress in PRO includes technical advancements and strategic considerations for global implementation. To conclude, while challenges remain, the future of PRO is promising as a renewable energy with a substantial role in a sustainable and resilient future of energy.

Questions

1. Explain how energy can be generated by the PRO process.
2. What are the key benefits of using PRO for generating energy in comparison with other conventional energy sources?
3. Why are advancements in membrane technology crucial for improving PRO's efficiency?
4. What is the effect of fouling on PRO's performance?
5. What are the economic factors to make PRO commercially viable?
6. What are some of the challenges of energy generation on a large scale for PRO technology?

References

1 Touati, K. and Rahaman, M.S. (2020). Viability of pressure-retarded osmosis for harvesting energy from salinity gradients. *Renew. Sustain. Energy. Rev.* 131: 109999. https://doi.org/10.1016/j.rser.2020.109999.

2 Gonzales, R.R., Ahmed, A.W., Samer, A. et al. (2021). Salinity gradient energy generation by pressure retarded osmosis: a review. *Desalination* 500: 114841. https://doi.org/10.1016/j.desal.2020.114841.

3 Tawalbeh, M., Al-Othman, A., Abdelwahab, N. et al. (2021). Recent developments in pressure retarded osmosis for desalination and power generation. *Renew. Sustain. Energy. Rev.* 138: 110492. https://doi.org/10.1016/j.rser.2020.110492.

4 AlZainati, N., Saleem, H., Altaee, A. et al. (2021). Pressure retarded osmosis: advancement, challenges and potential. *J. Water Process Eng.* 40: 101950. https://doi.org/10.1016/j.jwpe.2021.101950.

5 Helfer, F., Lemckert, C., and Anissimov, Y.G. (2014). Osmotic power with pressure retarded osmosis: theory, performance and trends – a review. *J. Membr. Sci.* 453: 337–358. https://doi.org/10.1016/j.memsci.2013.10.053.

6 Sharma, M., Das, P.P., Chakraborty, A., and Purkait, M.K. (2022). Clean energy from salinity gradients using pressure retarded osmosis and reverse electrodialysis: a review. *Sustain. Energy. Technol. Assess.* 49: 101687. https://doi.org/10.1016/j.seta.2021.101687.

7 Altaee, A. and Sharif, A. (2015). Pressure retarded osmosis: advancement in the process applications for power generation and desalination. *Desalination* 356: 31–46. https://doi.org/10.1016/j.desal.2014.09.028.

8 Touati, K. and Tadeo, F. (2017). Green energy generation by pressure retarded osmosis: state of the art and technical advancement—review. *Int. J. Green Energy* 14 (4): 337–360. https://doi.org/10.1080/15435075.2016.1255633.

9 Micale, G., Cipollina, A., and Tamburini, A. (2016). 1 - Salinity gradient energy. In: *Sustainable Energy from Salinity Gradients* (ed. A. Cipollina and G. Micale), 1–17. Woodhead Publishing. https://doi.org/10.1016/B978-0-08-100312-1.00001-8.

10 Achilli, A. and Childress, A.E. (2010). Pressure retarded osmosis: from the vision of Sidney Loeb to the first prototype installation — review. *Desalination* 261 (3): 205–211. https://doi.org/10.1016/j.desal.2010.06.017.

11 Bajraktari, N., Hélix-Nielsen, C., and Madsen, H.T. (2017). Pressure retarded osmosis from hypersaline sources — a review. *Desalination* 413: 65–85. https://doi.org/10.1016/j.desal.2017.02.017.

12 Zadeh, A.E., Touati, K., Mulligan, C.N. et al. (2022). Closed-loop pressure retarded osmosis draw solutions and their regeneration processes: a review.

Renew. Sustain. Energy Rev. 159: 112191. https://doi.org/10.1016/j.rser.2022.112191.

13 Lee, C., Chae, S.H., Yang, E. et al. (2020). A comprehensive review of the feasibility of pressure retarded osmosis: recent technological advances and industrial efforts towards commercialization. *Desalination* 491: 114501. https://doi.org/10.1016/j.desal.2020.114501.

14 Sarp, S., Li, Z., and Saththasivam, J. (2016). Pressure retarded osmosis (PRO): past experiences, current developments, and future prospects. *Desalination* 389: 2–14. https://doi.org/10.1016/j.desal.2015.12.008.

15 Al-Anzi, B.S. and Thomas, A. (2018). One-dimensional analytical modeling of pressure-retarded osmosis in a parallel flow configuration for the desalination industry in the state of Kuwait. *Sustainability* 10 (4): https://doi.org/10.3390/su10041288.

16 Abdelkader, B.A. and Sharqawy, M.H. (2022). Challenges facing pressure retarded osmosis commercialization: a short review. *Energies* https://doi.org/10.3390/en15197325.

17 Lee, K.L., Baker, R.W., and Lonsdale, H.K. (1981). Membranes for power generation by pressure-retarded osmosis. *J. Membr. Sci.* 8: 141–471.

18 Sherwood, T., Brian, P.L.T., and Fisher, R.E. (1967). Desalination by reverse osmosis. *Ind. Eng. Chem. Fundam.* 6 (1): 2–12. https://doi.org/10.1021/i160021a001.

19 McCutcheon, J.R. and Elimelech, M. (2006). Influence of concentrative and dilutive internal concentration polarization on flux behavior in forward osmosis. *J. Membr. Sci.* 284 (1–2): 237–247. https://doi.org/10.1016/j.memsci.2006.07.049.

20 Yip, N.Y. and Elimelech, M. (2013). Influence of natural organic matter fouling and osmotic backwash on pressure retarded osmosis energy production from natural salinity gradients. *Environ. Sci. Technol.* 47 (21): 12607–12616. https://doi.org/10.1021/es403207m.

21 Emadzadeh, D. and Kruczek, B. (2020). Mini review on the effects of concentration polarization in forward osmosis and pressure-retarded osmosis processes. In: *Proceedings of the International Conference on Fluid Flow and Thermal Science (ICFFTS'20)*. Avestia Publishing. https://doi.org/10.11159/icffts20.129.

22 Choi, Y., Vigneswaran, S., and Lee, S. (2016). Evaluation of fouling potential and power density in pressure retarded osmosis (PRO) by fouling index. *Desalination* 389: 215–223. https://doi.org/10.1016/j.desal.2016.01.011.

23 Wan, C.F. and Chung, T.S. (2015). Osmotic power generation by pressure retarded osmosis using seawater brine as the draw solution and wastewater retentate as the feed. *J. Membr. Sci.* 479: 148–158. https://doi.org/10.1016/j.memsci.2014.12.036.

24 She, Q., Jin, X., and Tang, C.Y. (2012). Osmotic power production from salinity gradient resource by pressure retarded osmosis: effects of operating conditions and reverse solute diffusion. *J. Membr. Sci.* 401–402: 262–273. https://doi.org/10.1016/j.memsci.2012.02.014.

25 Gonzales, R.R., Park, M.J., Bae, T.H. et al. (2019). Melamine-based covalent organic framework-incorporated thin film nanocomposite membrane for enhanced osmotic power generation. *Desalination* 459: 10–19. https://doi.org/10.1016/j.desal.2019.02.013.

26 Suwaileh, W.A., Johnson, D.J., Sarp, S., and Hilal, N. (2018). Advances in forward osmosis membranes: altering the sub-layer structure via recent fabrication and chemical modification approaches. *Desalination* https://doi.org/10.1016/j.desal.2018.01.035.

27 Wang, X., Huang, Z., Li, L. et al. (2012). Energy generation from osmotic pressure difference between the low and high salinity water by pressure retarded osmosis. *J. Technol. Innov. Renew. Energy* 1: 122–130.

28 Altaee, A., Zaragoza, G., and Sharif, A. (2014). Pressure retarded osmosis for power generation and seawater desalination: performance analysis. *Desalination* 344: 108–115. https://doi.org/10.1016/j.desal.2014.03.022.

29 Yip, N.Y. and Elimelech, M. (2011). Performance limiting effects in power generation from salinity gradients by pressure retarded osmosis. *Environ. Sci. Technol.* 45 (23): 10273–10282. https://doi.org/10.1021/es203197e.

30 Han, G., Zhang, S., Li, X., and Chung, T.S. (2013). High performance thin film composite pressure retarded osmosis (PRO) membranes for renewable salinity-gradient energy generation. *J. Membr. Sci.* 440: 108–121. https://doi.org/10.1016/j.memsci.2013.04.001.

31 Li, X. and Chung, T.S. (2013). Effects of free volume in thin-film composite membranes on osmotic power generation. *AIChE J.* 59 (12): 4749–4761. https://doi.org/10.1002/aic.14217.

32 Song, X., Liu, Z., and Sun, D.D. (2013). Energy recovery from concentrated seawater brine by thin-film nanofiber composite pressure retarded osmosis membranes with high power density. *Energy Environ. Sci.* 6 (4): 1199–1210. https://doi.org/10.1039/c3ee23349a.

33 Mansourpanah, Y., Madaeni, S.S., and Rahimpour, A. (2009). Fabrication and development of interfacial polymerized thin-film composite nanofiltration membrane using different surfactants in organic phase; study of morphology and performance. *J. Membr. Sci.* 343 (1–2): 219–228. https://doi.org/10.1016/j.memsci.2009.07.033.

34 Kim, I.C., Jeong, B.R., Kim, S.J., and Lee, K.H. (2013). Preparation of high flux thin film composite polyamide membrane: the effect of alkyl phosphate additives during interfacial polymerization. *Desalination* 308: 111–114. https://doi.org/10.1016/j.desal.2012.08.001.

35 Cui, Y., Liu, X.Y., and Chung, T.S. (2014). Enhanced osmotic energy generation from salinity gradients by modifying thin film composite membranes. *Chem. Eng. J.* 242: 195–203. https://doi.org/10.1016/j.cej.2013.12.078.

36 Peng, N., Widjojo, N., Sukitpaneenit, P. et al. (2012). Evolution of polymeric hollow fibers as sustainable technologies: past, present, and future. *Prog. Polym. Sci.* https://doi.org/10.1016/j.progpolymsci.2012.01.001.

37 Sivertsen, E., Holt, T., Thelin, W., and Brekke, G. (2013). Pressure retarded osmosis efficiency for different hollow fibre membrane module flow configurations. *Desalination* 312: 107–123. https://doi.org/10.1016/j.desal.2012.11.019.

38 Terashima, S., Hayashi, H., Okumura, T. et al. (2016). Permeation characteristics of fresh water in hollow fiber membrane module for pressure retarded osmosis. *Mech. Eng. J.* 3 (6): 16-00416-16–00416. https://doi.org/10.1299/mej.16-00416.

39 Zhang, L., She, Q., Wang, R. et al. (2016). Unique roles of aminosilane in developing anti-fouling thin film composite (TFC) membranes for pressure retarded osmosis (PRO). *Desalination* 389: 119–128. https://doi.org/10.1016/j.desal.2015.12.024.

40 Han, G., Zhou, J., Wan, C. et al. (2016). Investigations of inorganic and organic fouling behaviors, antifouling and cleaning strategies for pressure retarded osmosis (PRO) membrane using seawater desalination brine and wastewater. *Water Res.* 103: 264–275. https://doi.org/10.1016/j.watres.2016.07.040.

41 Yip, N.Y. and Elimelech, M. (2014). Comparison of energy efficiency and power density in pressure retarded osmosis and reverse electrodialysis. *Environ. Sci. Technol.* 48 (18): 11002–11012. https://doi.org/10.1021/es5029316.

42 Zhao, D.L., Japip, S., Zhang, Y. et al. (2020). Emerging thin-film nanocomposite (TFN) membranes for reverse osmosis: a review. *Water Res.* https://doi.org/10.1016/j.watres.2020.115557.

43 Saleem, H., Saud, A., Munira, N. et al. (2022). Improved forward osmosis performance of thin film composite membranes with graphene quantum dots derived from eucalyptus tree leaves. *Nanomaterials* 12 (19): https://doi.org/10.3390/nano12193519.

44 Nambi Krishnan, J., Venkatachalam, K.R., Ghosh, O. et al. (2022). Review of thin film nanocomposite membranes and their applications in desalination. *Front. Chem.* https://doi.org/10.3389/fchem.2022.781372.

45 Ma, N., Wei, J., Liao, R., and Tang, C.Y. (2012). Zeolite-polyamide thin film nanocomposite membranes: towards enhanced performance for forward osmosis. *J. Membr. Sci.* 405–406: 149–157. https://doi.org/10.1016/j.memsci.2012.03.002.

46 Salehi, T.M., Peyravi, M., Jahanshahi, M. et al. (2018). Impacts of zeolite nanoparticles on substrate properties of thin film nanocomposite membranes for engineered osmosis. *J. Nanopart. Res.* 20 (4): https://doi.org/10.1007/s11051-018-4154-1.

47 Shadravan, A., Amani, M., and Jantrania, A. (2022). Feasibility of thin film nanocomposite membranes for clean energy using pressure retarded osmosis and reverse electrodialysis. *Energy Nexus* 7: https://doi.org/10.1016/j.nexus.2022.100141.

48 Niksefat, N., Jahanshahi, M., and Rahimpour, A. (2014). The effect of SiO_2 nanoparticles on morphology and performance of thin film composite membranes for forward osmosis application. *Desalination* 343: 140–146. https://doi.org/10.1016/j.desal.2014.03.031.

49 Emadzadeh, D., Ghanbari, M., Lau, W.J. et al. (2016). Solvothermal synthesis of nanoporous TiO_2: the impact on thin-film composite membranes for engineered osmosis application. *Nanotechnology* 27 (34): https://doi.org/10.1088/0957-4484/27/34/345702.

50 Zhao, D.L., Das, S., and Chung, T.S. (2017). Carbon quantum dots grafted antifouling membranes for osmotic power generation via pressure-retarded osmosis process. *Environ. Sci. Technol.* 51 (23): 14016–14023. https://doi.org/10.1021/acs.est.7b04190.

51 Kim, Y., Yang, E., Park, H., and Choi, H. (2020). Anti-biofouling effect of a thin film nanocomposite membrane with a functionalized-carbon-nanotube-blended polymeric support for the pressure-retarded osmosis process. *RSC Adv.* 10 (10): 5697–5703. https://doi.org/10.1039/c9ra08870a.

52 Goh, K., Setiawan, L., Wei, L. et al. (2013). Fabrication of novel functionalized multi-walled carbon nanotube immobilized hollow fiber membranes for enhanced performance in forward osmosis process. *J. Membr. Sci.* 446: 244–254. https://doi.org/10.1016/j.memsci.2013.06.022.

53 Ruiz-García, A., Tadeo, F., and Nuez, I. (2022). Simulation tool for full-scale PRO systems using SWMMs. *Desalination* 541: 116025. https://doi.org/10.1016/j.desal.2022.116025.

54 Mashrafi, S.A.L., Diaz-Elsayed, N., Benjamin, J. et al. (2022). An environmental and economic sustainability assessment of a pressure retarded osmosis system. *Desalination* 537: 115869. https://doi.org/10.1016/j.desal.2022.115869.

55 Blankert, B., Kim, Y., Vrouwenvelder, H., and Ghaffour, N. (2020). Facultative hybrid RO-PRO concept to improve economic performance of PRO: feasibility and maximizing efficiency. *Desalination* 478: 114268. https://doi.org/10.1016/j.desal.2019.114268.

56 Einarsson, S.J. and Wu, B. (2021). Thermal associated pressure-retarded osmosis processes for energy production: a review. *Sci. Total Environ.* 757: 143731. https://doi.org/10.1016/j.scitotenv.2020.143731.

57 Sezer, N. and Koç, M. (2019). Development and performance assessment of a new integrated solar, wind, and osmotic power system for multigeneration, based on thermodynamic principles. *Energy Convers. Manage.* 188: 94–111. https://doi.org/10.1016/j.enconman.2019.03.051.

58 Moosazadeh, M., Tariq, S., Safder, U., and Yoo, C.K. (2023). Techno-economic feasibility and environmental impact evaluation of a hybrid solar thermal membrane-based power desalination system. *Energy* 278: https://doi.org/10.1016/j.energy.2023.127923.

59 Straub, A.P., Deshmukh, A., and Elimelech, M. (2016). Pressure-retarded osmosis for power generation from salinity gradients: is it viable? *Energy Environ. Sci.* https://doi.org/10.1039/c5ee02985f.

60 Lee, S., Park, T., Park, Y.G. et al. (2020). Toward scale-up of seawater reverse osmosis (SWRO) – pressure retarded osmosis (PRO) hybrid system: a case study of a 240 m^3/day pilot plant. *Desalination* 491: https://doi.org/10.1016/j.desal.2020.114429.

61 He, W., Wang, Y., and Shaheed, M.H. (2015). Maximum power point tracking (MPPT) of a scale-up pressure retarded osmosis (PRO) osmotic power plant. *Appl. Energy* 158: 584–596. https://doi.org/10.1016/j.apenergy.2015.08.059.

62 Low, J.H., Zhang, J., Li, W.P. et al. (2023). Industrial scale thin-film composite membrane modules for salinity-gradient energy harvesting through pressure retarded osmosis. *Desalination* 548: https://doi.org/10.1016/j.desal.2022.116217.

63 Manzoor, H., Selam, M.A., Rahman, F.B.A. et al. (2020). A tool for assessing the scalability of pressure-retarded osmosis (PRO) membranes. *Renew. Energy* 149: 987–999. https://doi.org/10.1016/j.renene.2019.10.098.

12

Modeling and Simulation of Water Desalination Systems Using TRNSYS

Abdelfatah Marni Sandid[1], *Essam Rabea Ibrahim Mahmoud*[2], *Mohamed Bassyouni*[3,4,5], *and Yasser Elhenawy*[5,6,7]

[1] *Center of Research in Mechanics (CRM), BP N73B, Freres Ferrad, Ain El Bey, Constantine, Algeria*
[2] *Department of Mechanical Engineering, Faculty of Engineering, Islamic University of Madinah, Madinah, Saudi Arabia*
[3] *Department of Chemical Engineering, Faculty of Engineering, Islamic University of Madinah, Madinah, Saudi Arabia*
[4] *Department of Chemical Engineering, Faculty of Engineering, Port Said University, Port Said, Egypt*
[5] *Center of Excellence for Membrane Testing and Characterization (CEMTC), Port Said University, Port Said, Egypt*
[6] *Department of Mechanical Power Engineering, Faculty of Engineering, Port Said University, Port Said, Egypt*
[7] *School of Chemical and Metallurgical Engineering, University of the Witwatersrand, Johannesburg, South Africa*

12.1 Introduction

Water scarcity is a critical and growing global challenge driven by factors, such as population growth, urbanization, and climate change. Freshwater resources are becoming increasingly limited, necessitating innovative solutions to ensure a sustainable supply of potable water. Desalination, the process of removing salts and other impurities from seawater or brackish water, has emerged as a viable solution to address water scarcity. Among the various desalination methods, thermal and membrane-based processes are prominent due to their reliability and effectiveness. However, optimizing these processes for efficiency and sustainability requires advanced modeling and simulation tools [1].

Solar membrane distillation (SMD) is increasingly seen as a viable water supply option in decentralized areas. Flat-plate collectors (FPCs) are often studied as a thermal source for membrane distillation (MD) due to their ability to absorb low-grade thermal energy. Other methods, such as solar gradient ponds and

Water Treatment and Desalination: Membrane Separation using Renewable Energy Resources, First Edition. Edited by Mohamed Bassyouni.

solar stills [2], have also been applied in various MD systems. By combining conventional desalination systems with solar collectors, it is possible to create solar desalination systems tailored to specific operational scales and temperature needs [3]. For instance, one study proposed an integrated solar MD system using vacuum glass tubes for efficient energy transfer alongside a tubular MD module [4]. This system achieved a maximum thermal efficiency of 78%, with energy efficiency ranging between 4% and 5% during peak summer hours, demonstrating its capability to supply the necessary thermal energy for MD processes and its potential for sustainable water production.

Further studies have focused on solar combined membrane distillation (SCMD) systems for household applications [5]. One such system was found to produce 250 L/day of hot water and 20 L/day of distilled water. It was observed that integrating solar domestic hot water (SDHW) systems with SMD systems improved collector efficiencies and increased the solar fraction throughout the year, with a small increase in average distillate production. Additionally, research into optimizing water production costs using commercial low-temperature solar collectors across various configurations and locations found that the cost of solar energy supply is heavily dependent on the specific energy demands of desalination systems [6]. Optimizing the costs of components like membranes and collectors can make solar MD systems more feasible, and reducing heat losses while carefully planning installations based on location and raw water conditions is crucial for enhancing system performance [7].

Practical applications have also shown the viability of small-scale systems, such as those integrating spiral-wound direct-contact membrane distillation (DCMD) modules with flat-plate solar thermal collectors [8]. These systems have been able to generate over 140 kg of distillate per day under real weather conditions, with thermal efficiency varying from 0 to 0.5 depending on the time of day. Another study developed and tested a solar thermal system in the UAE to provide cold potable water and hot domestic water. Using FPCs, the system was simulated with Transient System Simulation Tool (TRNSYS) software, and economic analyses revealed a payback period as short as 6.75 years. The system also offered significant economic and environmental benefits, including the reduction of 109 tons of CO_2 emissions annually [9].

In addition to these applications, further research has utilized TRNSYS to model large-cassette MD modules for the co-production of pure water and hot water in urban households [5]. A tested SCMD system demonstrated the ability to simultaneously produce 20 L/day of pure water and 250 L/day of hot water without auxiliary heating, with only a 6% decrease in distillate productivity and a doubling of domestic hot water production. Experiments with FPCs and spiral-wound MD modules further demonstrated that the feed flow rate significantly influences module performance due to changes in bulk temperature differences and hydrodynamics [10]. These systems achieved distillate flux rates of 10–25 kg/h, with specific energy consumption ranging from 130 to 207 kWh/m^3. Simulations of a solar thermal-driven desalination plant in Freiburg, Germany, with FPCs and spiral-wound MD modules indicated that a compact system with less than 6 m^2 of collector area and no heat storage could distill 120–160 L of water during a summer day, highlighting the importance of simplicity, long maintenance-free operation, and high-quality potable water output for successful implementation.

The objectives of this chapter are to model thermal desalination systems using TRNSYS software, with a focus on air-gap membrane distillation (AGMD) and flat-plate solar collectors across various climatic conditions. The research will examine the performance of different heat exchanger fluids (water, ethylene glycol, and a water–ethylene glycol mixture) to determine their effects on thermal performance and distillate water production. By analyzing the impact of different climates, specifically in hot (Adrar), warm (Port Said), and cold (Whitehorse) regions, the study aims to assess how local conditions influence system performance and distillate output. Additionally, it seeks to quantify the effects of heat exchanger efficiency and fluid heat capacity on distillate production and overall system performance, ultimately highlighting the importance of fluid selection and local climate in optimizing desalination systems.

12.2 Materials and Methods

This section details the setup and simulation processes used in TRNSYS to model different desalination systems.

12.2.1 Air-Gap Membrane Distillation

The lab-scale AGMD desalination pilot plant is designed and installed in Port Said City, Egypt (31°16′N 32°18′E). This setup includes the primary loops for the feed solution and heating and cooling sources, with permeates collected in a distillate tank. It utilizes an Aquastill© (AS14A 3L) multi-envelope, spiral-wound AGMD membrane module with a surface area of 14.4 m^2. The high-temperature water feed channels are marked in red, while the coolant channels are blue. The coolant channels act as condenser plates, with an air gap and spacer between each coolant and hot water feed channel, where the distillate water gets collected and is expelled. The module's six membrane envelopes are spirally wound into a cylindrical shape, with countercurrent hot water feed and coolant flows. The hot water enters from the module's center and exits at the periphery, whereas the coolant flows in the opposite direction. Saline water, prepared to match saltwater concentration, is pumped through a filter and into a flat-plate heat

Figure 12.1 Experimental setup for AGMD. Source: Elhenawy et al. [11]/with permission of Elsevier.

exchanger before being divided into the AGMD modules via management valves and flow meters. Pressure sensors ensure that the feedwater pressure stays below the membrane's maximum allowable pressure (Figure 12.1) [11, 12].

The AGMD module's key specifications include its construction from polyethylene (PE) and a total net membrane surface area of 14.4 m^2. The module has a diameter of 0.4 m, an envelope width of 0.4 m, and an envelope length of 3 m. The flow channel thickness is 2 mm, while the air-gap thickness is 1 mm. The membrane sheet features a porosity of 85%, a tortuosity of 1.56, and a thermal conductivity of 1.35 W/m K. The thickness of the membrane sheet is 76 μm, with a mean pore size diameter of 0.3 μm. The module has a packing density of 115 m^2/m^3 and includes 6 evaporator (hot feed) channels, 6 condenser (coolant) channels, and 12 distillate channels. Spacer characteristics are as follows: a channel thickness of 2.01 mm, a spacer filament thickness of 1.005 mm, a spacer angle of 60°, and a filament length of 4.17 m (Figure 12.2).

12.2.2 Solar Collectors

Solar collectors are critical components in solar thermal systems, designed to capture and convert solar energy into usable heat. This section delves into the characteristics and performance of FPCs [12].

FPCs are among the most widely used solar collectors. They consist of a flat, insulated box with a transparent cover, an absorber plate, and a heat transfer fluid that circulates through the collector. The absorber plate, typically made of copper

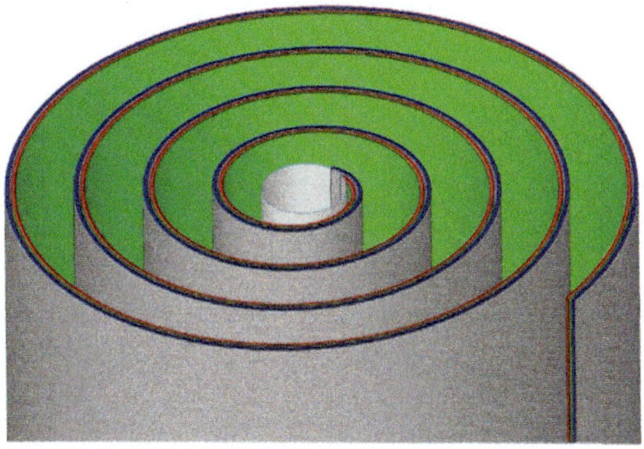

Figure 12.2 Detailed cross-sectional view of an AGMD module. Source: Marni-Sandid et al. [12]/with permission of Elsevier.

or aluminum, is coated with a selective surface that maximizes the absorption of solar radiation and minimizes heat loss, often painted black to improve absorption. The transparent cover, usually made of tempered glass or durable plastic, reduces heat loss while allowing solar radiation to penetrate through. The insulated box, located at the back and sides of the collector and usually made from fiberglass or polyurethane, minimizes heat loss to the environment. The heat transfer fluid circulates through pipes attached to the absorber plate, transferring heat to the storage tank or heat exchanger. In the current setup, the system includes six flat-plate solar collectors, each with an area of 12 m^2, arranged in series. After passing through these collectors, the heated water flows through heat exchangers to transfer heat to the AGMD unit. Additionally, a pump circulates the water within the system. The structure of a FPC is illustrated in Figure 12.3.

FPCs have the following specifications: Each collector features an absorber area of 2.10 m^2 and an aperture area of 2.16 m^2, with dimensions of 1892 mm × 1204 mm × 99 mm and a gross area of 2.28 m^2. They operate at a maximum pressure of 9 bars and can achieve a stagnation temperature of 189°C. The thermal capacity of the collectors is 10 330 kJ/($m^2 \cdot K^2$), and they have an incidence angle of 50°. The instantaneous efficiency based on the aperture area is 0.713. In terms of performance, FPCs typically have a thermal efficiency ranging from 60% to 80%, depending on the quality of materials and design. They are effective at temperatures ranging from 30°C to 80°C, though their performance decreases as the temperature difference between the collector and ambient air increases. Standard FPCs usually measure about 1.5–2 m in length and 1–1.2 m in width and are mounted at an angle of 30°–45° to optimize solar gain based on geographic location. These collectors are generally more cost-effective and simpler to install, making them suitable for moderate climates. A detailed evaluation of these solar collectors allows for the selection of the most appropriate technology based on specific application needs, climatic conditions, and budget considerations.

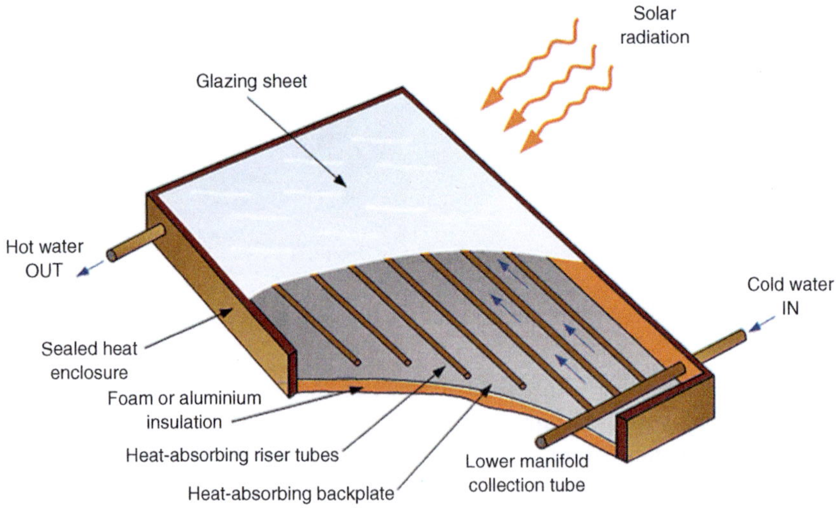

Figure 12.3 Flat-plate collector structure. Source: Shamsul et al. [13]/MDPI/CC BY 4.0.

12.2.3 Thermal Desalination Systems Using Different Heat Exchanger Fluids

Thermal desalination systems operate by utilizing heat to evaporate seawater, separating the water vapor from dissolved salts and other impurities. The water vapor is then condensed to produce freshwater. The efficiency and performance of thermal desalination systems are significantly influenced by the choice of heat exchanger fluids. This section examines the use of three different heat exchanger fluids:water, ethylene glycol, and a water–ethylene glycol mixture in thermal desalination systems.

12.2.3.1 Water as a Heat Exchanger Fluid

Water is the most commonly used heat exchanger fluid in thermal desalination systems due to its excellent thermal properties. It has a high specific heat capacity (4.18 J/g°C), meaning it can absorb and store a large amount of heat energy, making it an efficient medium for transferring thermal energy. Water is also readily available and cost-effective. However, its use in high-temperature applications can lead to issues such as scaling and corrosion in the heat exchanger and other system components. To mitigate these issues, water treatment and conditioning are often required to maintain system performance and longevity (Table 12.1) [14].

Table 12.1 Characteristics of water.

Characteristics	Water
Specific heat capacity (J/g°C)	4.18
Boiling point (°C)	100 (at 1 atm pressure)
Freezing point (°C)	0
Thermal conductivity (W/mK)	0.6
Density (kg/m^3)	1000 (at 4°C)

12.2.3.2 Ethylene Glycol as a Heat Exchanger Fluid

Ethylene glycol is another fluid used in heat exchangers for thermal desalination systems. It is known for its lower freezing point compared to water, making it suitable for systems operating in colder environments. Ethylene glycol also has good thermal conductivity (0.258 W/mK) and a relatively high boiling point (197.3°C), allowing it to effectively transfer heat at higher temperatures without evaporating. However, ethylene glycol is more expensive than water and poses environmental and health risks if leaked or improperly handled. Its higher viscosity also means it requires more energy to pump through the system, which can increase operational costs (Table 12.2) [15]

12.2.3.3 Water–Ethylene Glycol Mixture as a Heat Exchanger Fluid

Combining water and ethylene glycol creates a mixture that leverages the advantages of both fluids while mitigating some of their individual drawbacks. Typically, a 50/50 mixture of water and ethylene glycol is used, providing a balance of high specific heat capacity and lower freezing point. This mixture can

Table 12.2 Characteristics of ethylene glycol.

Characteristics	Ethylene glycol
Specific heat capacity (J/g°C)	2.43
Boiling point (°C)	197.3
Freezing point (°C)	12.9 (pure ethylene glycol)
Thermal conductivity (W/mK)	0.258
Density (kg/m^3)	1113 (at 20°C)

Table 12.3 Characteristics of the 50/50 water–ethylene glycol mixture.

Characteristics	50/50 water–ethylene glycol mixture
Specific heat capacity (J/g°C)	3.23
Boiling point (°C)	~108
Freezing point (°C)	−37
Thermal conductivity (W/mK)	0.401
Density (kg/m^3)	1065 (at 20°C)

effectively transfer heat across a wide range of temperatures, making it versatile for various thermal desalination applications. The addition of ethylene glycol reduces the risk of freezing in colder climates and decreases the likelihood of scaling and corrosion compared to using water alone. However, the mixture still retains some of the higher viscosity characteristics of ethylene glycol, requiring careful consideration of pump specifications and energy consumption (Table 12.3) [16].

12.2.3.4 Comparative Analysis of Heat Exchanger Fluids

The choice of heat exchanger fluids depends on various factors, including the operating environment, temperature ranges, and specific desalination process requirements. Water is highly efficient and cost-effective but may require extensive treatment to prevent scaling and corrosion. Ethylene glycol offers lower freezing points and high thermal conductivity but at a higher cost and with environmental considerations. The water–ethylene glycol mixture provides a balanced solution, offering versatility and reduced maintenance issues, though it comes with increased pumping energy requirements.

Using TRNSYS, simulations of thermal desalination systems with each of these fluids can help identify the optimal fluid for specific conditions. By analyzing system performance, efficiency, and operational costs, it is possible to determine the most suitable heat exchanger fluid for different desalination scenarios. Such simulations contribute to the development of more efficient and sustainable thermal desalination systems, ensuring a reliable supply of freshwater while minimizing environmental impact and operational costs.

Through detailed simulations and comparative analysis, it is possible to optimize thermal desalination systems for specific operational environments and requirements, ensuring enhanced efficiency and sustainability.

12.3 Weather Characteristics of Different Locations

The performance of solar collectors is significantly influenced by local weather conditions, which affect their efficiency in capturing and utilizing solar energy. This analysis examines the impact of weather on solar collectors in three distinct locations: Adrar (Algeria), Port Said (Egypt), and Whitehorse (Canada), based on data collected on July 29, 2018. Table 12.4 provides a comparative analysis of meteorological conditions in these locations.

In Adrar, Algeria, known for its hot desert climate with minimal rainfall, solar irradiance is notably high at 1032 W/m^2. The average temperature is around 34°C, with peak temperatures reaching up to 45°C. Humidity is very low at approximately 15%. These conditions lead to substantial energy production due to the high solar irradiance, though the extreme temperatures may challenge efficiency [17]. Port Said, Egypt, features a warm semi-arid climate with hot summers and mild winters. Solar irradiance here is moderate at 878 W/m^2. Average temperatures are around 29°C, with peak temperatures reaching 35°C, and humidity is moderate at about 60%. While this climate supports effective solar energy capture, the moderate humidity could somewhat impact collector efficiency [18].

In contrast, Whitehorse, Canada, has a cold temperate climate with cold winters and mild summers. Solar irradiance is relatively low at 532 W/m^2. Average temperatures are around 11°C, with peak temperatures up to 15°C, and humidity is high, ranging from 70% to 80%. The low irradiance and high humidity present significant challenges for solar collectors, leading to reduced energy production and efficiency [19].

The substantial climatic differences between these locations highlight the need for tailored approaches when designing and operating solar collector systems. The high irradiance and extreme temperatures in Adrar, moderate conditions in Port Said, and low irradiance with high humidity in Whitehorse all impact the operational efficiency and design considerations for solar collectors. Understanding these effects is crucial for optimizing performance and ensuring effective solar energy utilization in varying environments.

Table 12.4 Weather characteristics of different locations.

Parameter	Adrar, Algeria	Port Said, Egypt	Whitehorse, Canada
Solar irradiance (kW/m^2)	1.8	1.6	1.2
Average temperature (°C)	32	24	5
Peak temperature (°C)	45	35	15
Humidity (%)	15	60	70–80

12.4 TRNSYS Modeling of Desalination Systems

TRNSYS offers robust models for various components in desalination systems, including FPCs, and AGMD systems [13].

The FPC model in TRNSYS is represented by Type 1, which simulates a standard collector with single glazing and a metal absorber plate. This model is designed to capture solar radiation and convert it into thermal energy effectively. The performance of the collector is evaluated through parameters such as heat transfer coefficients and thermal efficiency, which are critical for determining the overall effectiveness of the solar thermal system.

In addition to collectors, TRNSYS provides models for other system components [20]:

- **Heat exchanger models**: These models, including flat-plate designs, facilitate the transfer of thermal energy between different fluids. They are crucial for optimizing energy exchange within the system, ensuring efficient thermal management.
- **Pump models**: These are used to simulate the circulation of fluids within the system, which is essential for maintaining proper fluid movement and overall system efficiency. Accurate modeling of pump performance ensures that fluid dynamics are well represented, contributing to the system's overall effectiveness.
- **Converter unit models**: These models represent devices that adjust electrical outputs to meet the specific requirements of the system. They play a critical role in ensuring that electrical energy is appropriately managed and utilized within the system.
- **Weather models**: TRNSYS includes models that simulate environmental conditions, such as solar radiation, temperature, and wind speed, for various locations. These models are integral for predicting system performance under different climatic conditions and for tailoring system design to specific environmental factors.

For AGMD, TRNSYS models can simulate the thermal desalination process using hydrophobic membranes. The performance of AGMD systems is evaluated using equations [10] for heat transfer, vapor flux, and condensation rates, integrating parameters, such as membrane properties, feed temperature, and ambient conditions.

As shown in Figure 12.4, these models collectively enable the simulation and optimization of desalination systems, providing valuable insights into their performance and efficiency under various operational conditions.

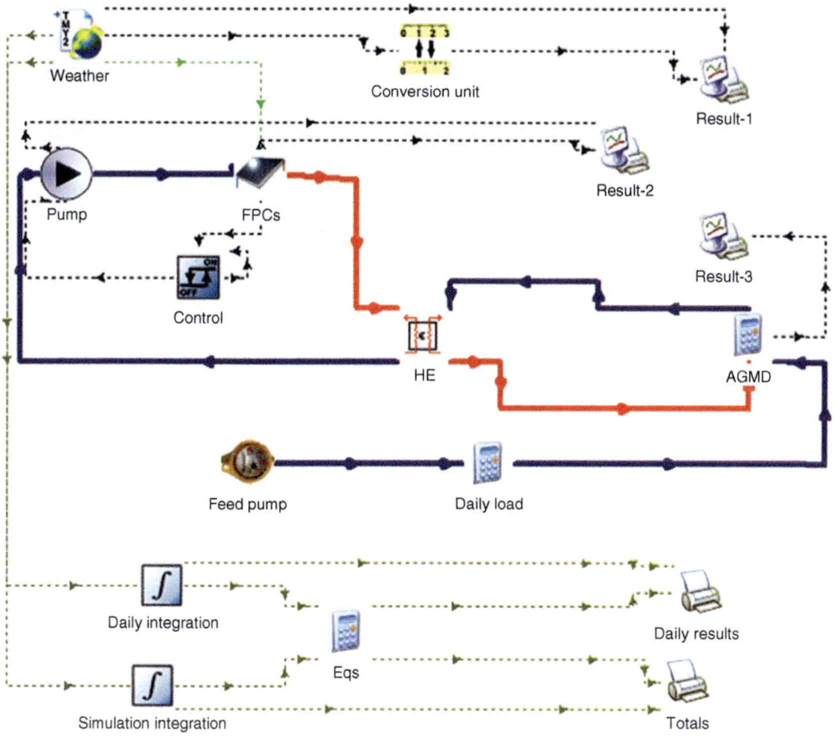

Figure 12.4 TRNSYS simulation model for integrated AGMD and FPC systems. Source: Shamsul et al. [13]/MDPI/CC BY 4.0.

12.5 Results and Discussion

12.5.1 Three-Dimensional Visualization of Air-Gap Membrane Distillation Parameters

Figure 12.5 presents a three-dimensional (3D) visualization of the relationship between feed and cooling water temperatures, flow rates, and permeate flux in an AGMD model for water desalination. Based on experimental runs, the data provide specific values for feed and cooling water temperatures, flow rates, and permeate flux. The feedwater temperature ranged from 34.76°C to 35.11°C, averaging approximately 34.91°C, while the hot feedwater temperature varied significantly from 45.1°C to 64.91°C. Hot and cold water flow rates were fixed at 600 L/h, while permeate flux exhibited the widest range, from

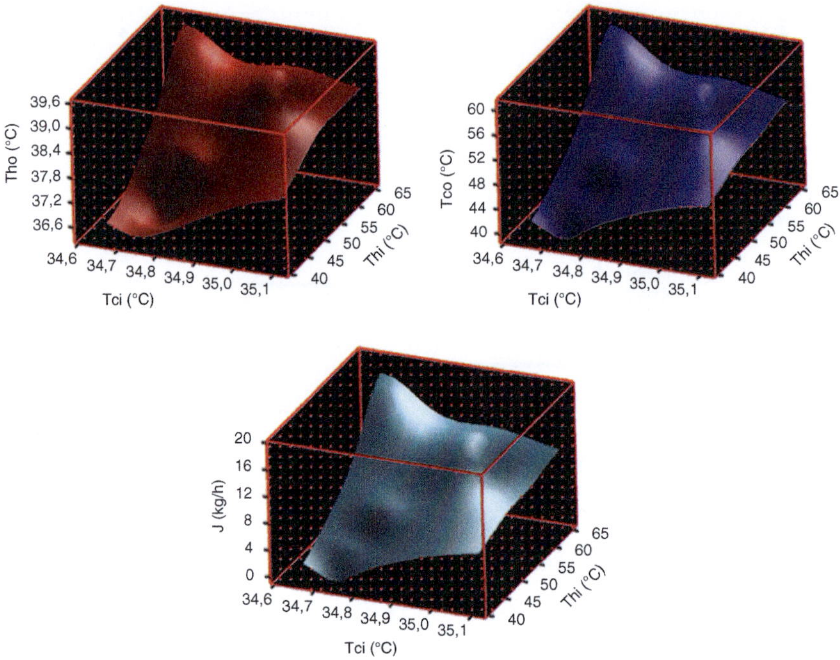

Figure 12.5 3D surface plots of permeate flux and hot and cold feedwater temperatures of the proposed AGMD model.

3.04 kg/h to 18.01 kg/h. Furthermore, higher feed and hot feedwater temperatures correlated with increased permeate flux. Figure 12.5 offers valuable insights into the relationship between various parameters and permeate flux in an AGMD model. Based on the parameters of our AGMD model, the 3D surface plots of Tco (°C), Tho (°C), and J (kg/h) are depicted in Figure 12. 5.

12.5.2 Comparative Analysis of Meteorological Parameters

Figure 12.6 presents a comparative analysis of meteorological conditions in (1) Adrar, (2) Port Said, and (3) Whitehorse on July 29. Global radiation, a critical factor for solar energy harvesting, exhibited substantial disparities. Adrar recorded the highest levels, reaching 1032 W/m^2, which is significantly greater than the 878 W/m^2 measured in Port Said and the considerably lower 532 W/m^2 observed in Whitehorse. These variations directly influence the potential solar energy input for the thermal collector system. Ambient temperature also demonstrated

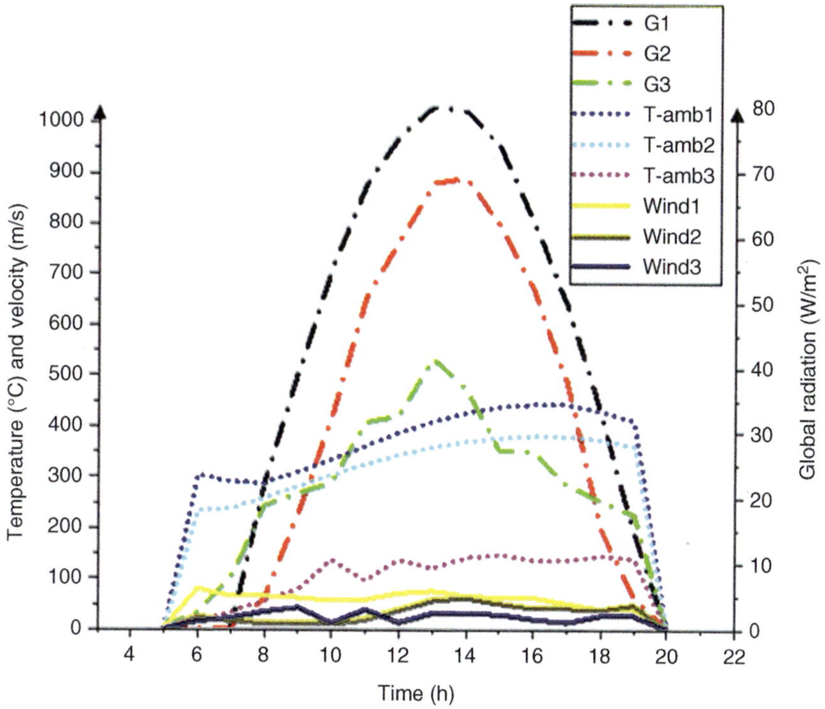

Figure 12.6 Comparative analysis of meteorological parameters across diverse climatic regions in (1) Adrar, (2) Port Said, and (3) Whitehorse.

marked differences across the three locations. Adrar experienced the highest temperature at 34°C, while Port Said recorded a more moderate 29°C. In contrast, Whitehorse encountered significantly colder conditions with a temperature of only 11°C. These temperature gradients will undoubtedly impact the thermal performance of the collector and consequently the effectiveness of the MD process. While wind velocity can affect system performance, in this study, its variations were less pronounced compared to global radiation and ambient temperature. Adrar experienced a wind speed of 5.6 m/s, slightly higher than Port Said's 4.8 m/s and Whitehorse's 2.9 m/s. Although wind can influence heat and mass transfer, its impact is anticipated to be less critical in this context. The presented information underscores the substantial climatic differences between the three locations, emphasizing the need for a tailored approach when designing and operating a thermal collector system for MD water desalination. A comprehensive analysis, incorporating additional performance indicators and statistical methodologies, is essential to fully understand the implications of these meteorological conditions on system performance and productivity.

12.5.3 Comparative Distillate Water, Inlet and Outlet Temperatures under Different Climatic Conditions

Figure 12.7 provides a comprehensive analysis of the effects of varying climates: (1) hot in Adrar, (2) warm in Port Said, and (3) cold in Whitehorse—on the inlet and outlet temperatures of collectors throughout the day on July 29. The data, representing a thermal system model for MD aimed at water desalination, reveal significant temperature variations across these diverse climates. The highest outlet temperature observed in the hot climate is 76.4°C at 14:00, while the warm climate peaks at 66.6°C at 13:00 and the cold climate reaches a maximum of 50.6°C at the same time. In terms of inlet temperatures, the hot climate registers the highest at 42.6°C at 14:00, contrasting sharply with the lowest temperature of 2.2°C recorded in the cold climate at 7:00. These findings highlight the substantial influence of climatic conditions on the thermal performance of the collectors, with hotter climates enabling higher temperature outputs. The differences between inlet and outlet temperatures across these climates also underscore the varying capacities for heat absorption and retention, which are critical for understanding the thermal dynamics of MD systems in diverse environmental settings.

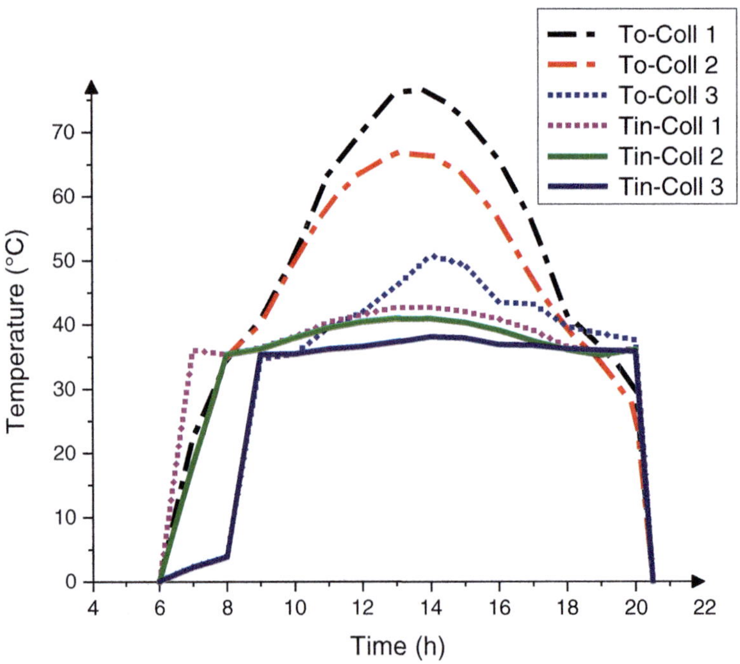

Figure 12.7 Comparative inlet and outlet temperatures of collectors under different climatic conditions.

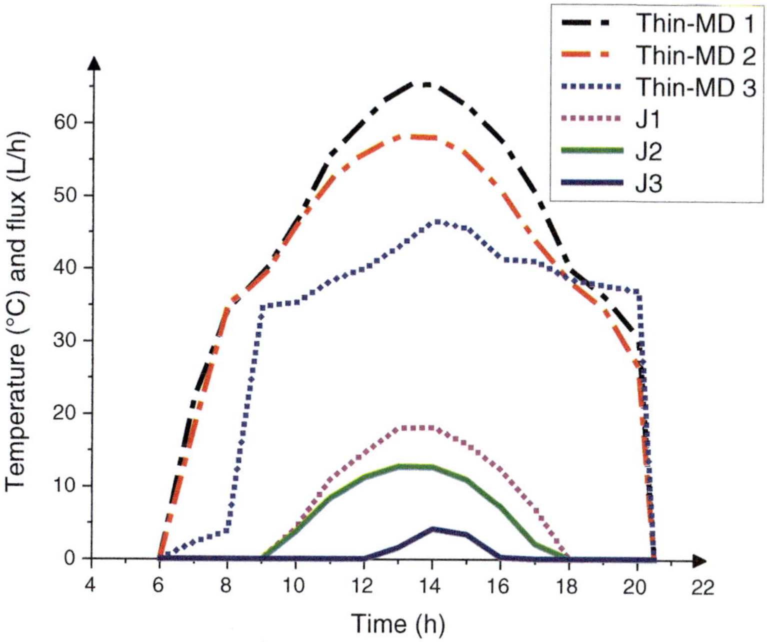

Figure 12.8 Hourly inlet temperature and distillate water production for MD under different climatic conditions.

This analysis provides valuable insights into the design and operational strategies of desalination systems tailored to specific climatic conditions.

The results presented in Figure 12.8 illustrate the impact of varying climatic conditions on the performance of an MD system for water desalination. The system was subjected to three distinct climates: (1) hot (Adrar), (2) warm (Port Said), and (3) cold (Whitehorse), and the corresponding inlet temperature and distillate water output were recorded throughout the day of July 29. The results indicated a clear correlation between climatic conditions and system performance. In the hot climate of Adrar, the maximum inlet temperature reached 65.4°C, leading to a maximum distillate water production of 18.08 L/m^2. This represents a 43.8% increase in distillate production compared to the warm climate of Port Said. Conversely, in the cold climate of Whitehorse, the maximum inlet temperature was only 46.6°C, resulting in a maximum distillate water production of 4.12 L/m^2. This represents a 67.5% decrease in distillate production compared to the hot climate of Adrar. The warm climate of Port Said demonstrated intermediate performance, with a maximum inlet temperature of 58.1°C and a maximum distillate water production of 12.74 L/m^2. These results highlight the significant influence of climatic factors on the performance of MD systems for water desalination. Further analysis and optimization of the system under different climatic conditions are necessary to enhance its output and suitability for various regions.

12.5.4 Temperature Variations in Thermal Collector Systems

Figure 12.9 presents temperature variations within a thermal collector system designed for MD water desalination using different heat capacity fluids:water, ethylene glycol, and a water–ethylene glycol mixture. The figure illustrates the influence of heat capacity on both inlet and outlet temperatures of the heat capacity fluids within the system. The higher values of collector inlet and outlet temperatures at (1) hot in Adrar, (2) warm in Port Said, (3) cold in Whitehorse were recorded on July 29. Generally, higher heat capacities correspond to lower temperatures for both inlet and outlet fluids compared to lower heat capacities. For instance, the highest heat capacity value (4.19) consistently yields lower temperatures than the lowest value (2.43) across all temperature measurements.

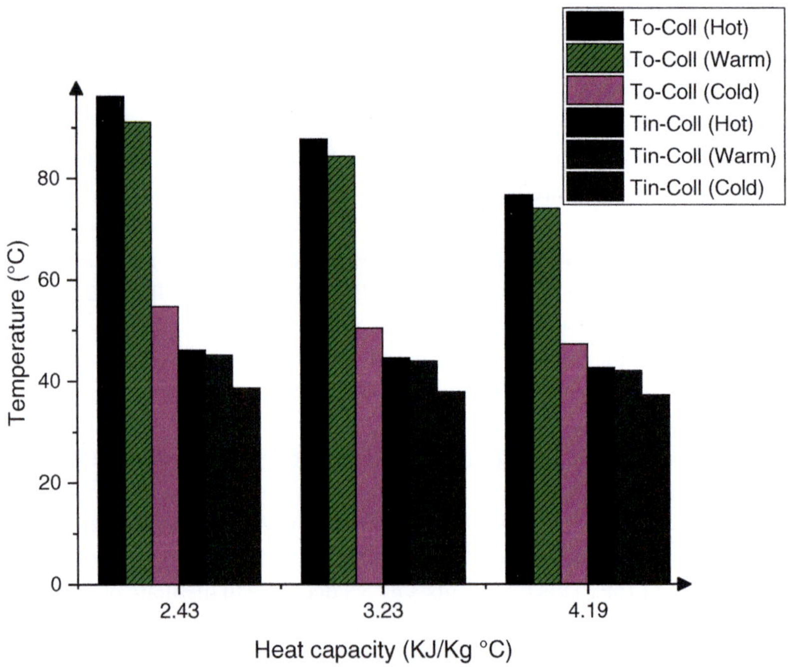

Figure 12.9 Effect of heat capacity fluids on collector temperatures under different climatic conditions.

A comparative analysis reveals significant percentage decreases in outlet temperatures for the highest heat capacity compared to the lowest: temperature 1 decreased by 21.7%, temperature 2 by 22.3%, and temperature 3 by 27.7%. This indicates a more pronounced temperature reduction at higher outlet temperatures. Additionally, a comparative analysis shows a significant 18.16% decrease in the average outlet temperature for the highest heat capacity fluid (water–ethylene glycol mixture) compared to the lowest (water). Inlet temperatures also decreased with increasing heat capacity: 7.54% for inlet temperature 1 (hottest), 6.81% for inlet temperature 2 (warm), and 3.62% for inlet temperature 3 (coldest). These findings indicate a strong correlation between increasing heat capacity and temperature reduction at both the inlet and outlet of the collector system, with a more pronounced effect on hotter fluids.

12.5.5 Influence of Heat Capacity Fluids on Distillate Production

Figure 12.10 presents the influence of varying heat capacity fluids and inlet temperatures on distillate water production within an MD system. Three different heat capacity fluids were tested:water (4.19 J/gK), ethylene glycol (2.43 J/gK), and

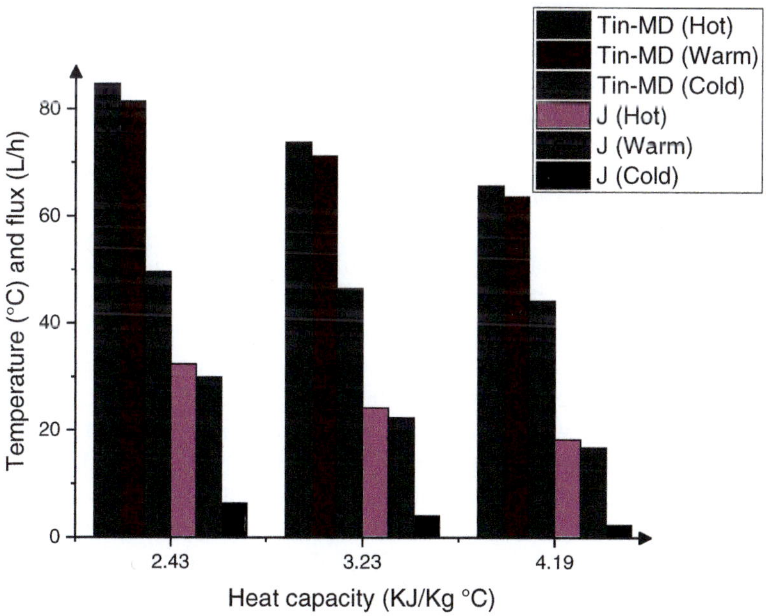

Figure 12.10 Performance evaluation of inlet temperatures and flux of an MD system under varying conditions and heat capacity fluids.

a water–ethylene glycol mixture (3.23 J/gK). The system was evaluated under three climatic conditions: hot (Adrar), warm (Port Said), and cold (Whitehorse), on July 29. The results indicated a clear correlation between heat capacity fluid, inlet temperature, and distillate water production. Higher inlet temperatures consistently led to increased distillate volume across all heat capacity fluids and locations. Ethylene glycol, with the lowest heat capacity, exhibited the highest distillate output, followed by the water–ethylene glycol mixture and then water.

The performance differences between fluids were more pronounced in hotter climates. For instance, in Adrar, the distillate volume for ethylene glycol was 32.35 L/m^2h, compared to 24.25 L/m^2h for the mixture and 18.26 L/m^2h for water. In contrast, these differences were less pronounced in Whitehorse, with distillate volumes of 68, 47, and 25.3 L/m^2h, respectively. This suggests that the heat capacity fluid's influence on distillate production is more significant in warmer environments. To quantify the performance variations, comparative percentages can be calculated. For example, in Adrar, ethylene glycol produced approximately 58% more distillate than water, while the mixture yielded a 73% increase. These percentages highlight the substantial impact of heat capacity fluid selection, particularly in hot climates. These findings indicate that employing fluids with lower heat capacities, such as ethylene glycol, can significantly enhance distillate production in MD systems, especially in regions with high ambient temperatures.

12.5.6 Effect of Heat Exchanger Efficiency on Membrane Distillation Performance

Figure 12.11 presents the influence of heat exchanger efficiency on inlet temperature, distillate water production, and overall MD performance using ethylene glycol (2.43) in the hot climate of Adrar on July 29. As shown in Figure 12.6, there is a clear correlation between heat exchanger efficiency and distillate water production across varying inlet temperatures, with a consistent increase in production as efficiency rises. For instance, at an inlet temperature of 51.27°C, elevating heat exchanger efficiency from 20% to 90% resulted in a substantial 433% surge in distillate water production, from 7.63 L/m^2h to 40.63 L/m^2h. While higher inlet temperatures generally correlate with increased distillate water production, this trend lessens at higher efficiencies, emphasizing the critical role of heat exchanger efficiency in maximizing output, particularly under elevated inlet temperature conditions. Specifically, increasing efficiency from 20% to 40% yields a 121% increase in distillate water production, while further increases of 20% in efficiency result in progressively smaller gains: 64% from 40% to 60%, 32% from 60% to 80%, and 12% from 80% to 90%. These findings underscore the critical role of heat exchanger efficiency in MD performance under

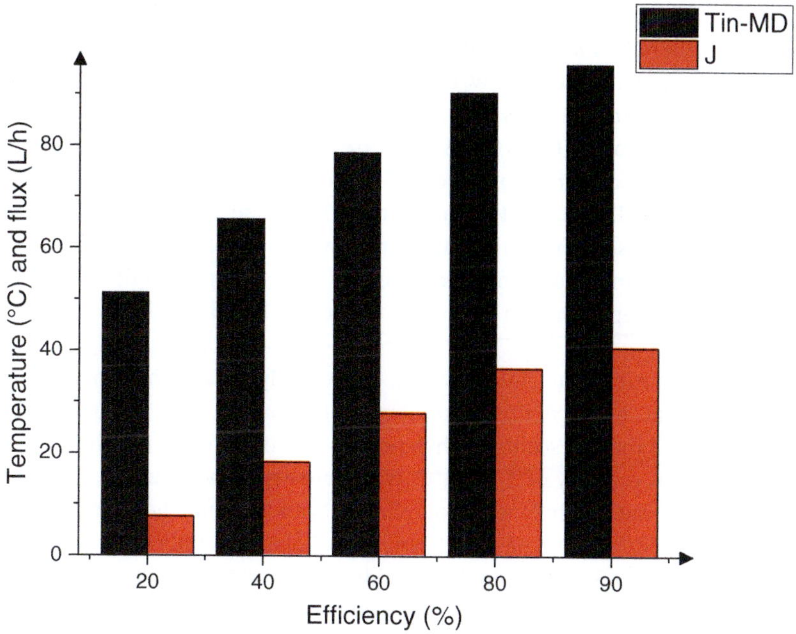

Figure 12.11 Effect of heat exchanger efficiency on membrane distillation performance under varying inlet temperatures.

hot climatic conditions. By optimizing heat exchanger efficiency, it is possible to significantly enhance distillate water production, thereby improving the overall efficiency and sustainability of water desalination systems.

12.6 Conclusion

This chapter examined various aspects of desalination systems and their interactions with environmental conditions and operational parameters, focusing on an AGMD system setup in Port Said, Egypt, and solar collectors under different climatic conditions. The AGMD system featured an Aquastill© (AS14A 3L) membrane module with a surface area of 14.4 m^2. Performance was influenced by feedwater temperatures ranging from 34.76°C to 35.11°C and hot feedwater temperatures between 45.1°C and 64.91°C, with permeate flux varying from 3.04 kg/h to 18.01 kg/h. The results showed that higher feed and hot feedwater temperatures correlated with increased permeate flux, with the highest flux observed at 18.01 kg/h, highlighting the importance of maintaining optimal thermal conditions for effective desalination.

Solar collectors tested in Adrar, Port Said, and Whitehorse showed significant performance differences due to local climatic conditions. On July 29, Adrar had the highest solar radiation at 1032 W/m^2 and a peak temperature of 34°C, leading to superior performance. Port Said had lower solar radiation (878 W/m^2) and a peak temperature of 29°C, while Whitehorse had the lowest solar radiation (532 W/m^2) and a peak temperature of 11°C. Despite wind velocity variations in Adrar at 5.6 m/s, Port Said at 4.8 m/s, and Whitehorse at 2.9 m/s, the main impact came from solar radiation and ambient temperature. The observed performance variations underscore the necessity to design solar collectors with regional-specific adaptations to optimize efficiency based on local climatic conditions.

Analysis of heat exchanger fluids—water, ethylene glycol, and their mixture—demonstrated that water, with a specific heat capacity of 4.18 J/g°C, was cost-effective but prone to scaling. Ethylene glycol, with a lower freezing point and a specific heat capacity of 2.43 J/g°C, was more expensive and had environmental risks. The water–ethylene glycol mixture, with a specific heat capacity of 3.23 J/g°C and a freezing point of −37°C, provided a balanced solution but required more pumping energy. This highlights the trade-offs between fluid properties and their impacts on system performance, suggesting that fluid selection should be carefully considered based on the operational and environmental context.

TRNSYS simulations showed that increasing heat exchanger efficiency from 20% to 90% led to a 433% increase in distillate water production, from 7.63 L/m^2h to 40.63 L/m^2h. This significant improvement in production underscores the critical impact of efficiency on the system output. Additionally, the effect of climatic conditions was notable, with Adrar achieving a maximum distillate water production of 18.08 L/m^2, compared to Port Said's 12.74 L/m^2 and Whitehorse's 4.12 L/m^2. This highlights the need to optimize system design and efficiency to enhance desalination performance in varying environments.

Looking to the future, these findings suggest that improvements in desalination technologies can significantly impact global water management. By enhancing system performance and adapting designs to local climatic conditions, desalination systems can become more sustainable and cost-effective, addressing water scarcity more effectively. Utilizing region-specific data and optimizing fluid choices will boost system reliability and effectiveness. Advances in technology, materials, and simulation models will be key to overcoming water challenges and making desalination a practical solution for providing clean water across diverse environments.

Nomenclature

AGMD	Air-gap membrane distillation
AS14A 3L	Model designation for Aquastill© multi-envelope, spiral-wound AGMD membrane module
FPC	Flat-plate collector
C_p (J/kg K)	Specific heat capacity of the fluid
λ (W/m K)	Thermal conductivity of the membrane
ϕ (%)	Porosity of the membrane
τ	Tortuosity of the membrane
d (μm)	Thickness of the membrane sheet
p (kg/m^3)	Density of the fluid
ΔT (°C)	Temperature difference
Q (kJ)	Thermal capacity
η	Efficiency of the collector
h (m^2 K/W)	Heat transfer coefficient
V (m^3)	Volume of the fluid
m (kg/s)	Mass flow rate
G (kW/m^2)	Global radiance
d (m)	Diameter of the module
w (m)	Width of the envelope
l (m)	Length of the envelope
PE	Polyethylene (membrane material)
TRNSYS	Transient System Simulation Tool

Questions

1 What software was used to model the thermal desalination systems in this study?

2 Which desalination method was the focus of this study?

3 How does climate affect distillate production in air-gap membrane distillation (AGMD) systems?

4 What impact does ethylene glycol have on AGMD system performance?

5 How does the efficiency of the heat exchanger influence distillate production?

6 What membrane surface area was used in the AGMD pilot plant?

7 Which locations were analyzed for solar collector performance?

8 What was the maximum distillate production recorded in Adrar?

9 How does a water–ethylene glycol mixture compare to water alone as a heat exchanger fluid?

10 What is the main conclusion of the study?

References

1 Marni-Sandid, A., Nehari, D., and Nehari, T. (2022). Effective study of operating parameters on the membrane distillation processes using various materials for seawater desalination. *Membr. Water Treat.* 13 (5): 235–243. https://doi.org/10.12989/mwt.2022.13.5.235.
2 Guillén-Burrieza, E., Alarcón, D.C., Palenzuela, P., and Zaragoza, G. (2015). Techno-economic assessment of a pilot scale plant for solar desalination based on existing plate and frame MD technology. *Desalination* 374: 70–80.
3 Sharon, H. and Reddy, K.S. (2015). A review of solar energy driven desalination technologies. *Renewable Sustainable Energy Rev.* 41: 1080–1118.
4 Shafieian, A. and Khiadani, M. (2019). A novel solar-driven direct contact membrane-based water desalination system. *Energy Convers. Manage.* 199: 112055.
5 Kumar, N.T.U. & Martin, A.R. (2017) 'Co-production performance evaluation of a novel solar combi system for simultaneous pure water and hot water supply in urban households of UAE', *Energies*, 10, 481 https://doi.org/10.3390/en10040481.
6 Horta, P., Zaragoza, G., and Alarcón-Padilla, D.C. (2015). Assessment of the use of solar thermal collectors for desalination. *Desalin. Water Treat.* 55: 2856–2867.
7 Saffarini, R.B., Summers, E.K., Arafat, H.A., and Lienhard, V.J.H. (2012). Economic evaluation of stand-alone solar powered membrane distillation systems. *Desalination* 229: 55–62.

8 Duong, H.C., Xia, L., Ma, Z. et al. (2017). Assessing the performance of solar thermal driven membrane distillation for seawater desalination by computer simulation. *J. Membr. Sci.* 542: 133–142. https://doi.org/10.1016/j.memsci.2017.08.007.

9 Mohan, G., Kumar, U., Pokhrel, M.K., and Martin, A. (2016). A novel solar thermal polygeneration system for sustainable production of cooling, clean water and domestic hot water in United Arab Emirates: dynamic simulation and economic evaluation. *Appl.Energy.* 167: 173–188. https://doi.org/10.1016/j.apenergy.2015.10.116.

10 Koschikowski, J., Wieghaus, M., and Rommel, M. (2003). Solar thermal driven desalination plants based on membrane distillation. *Desalination* 156 (1–3): 295–304. https://doi.org/10.1016/s0011-9164(03)00360-6.

11 Elhenawy, Y., Moustafa, G.H., Abdel-Hamid, S.M.S. et al. (2022). Experimental investigation of two novel arrangements of air gap membrane distillation module with heat recovery. *Energy. Rep.* 8: 8563–8573.

12 Marni-Sandid, A., Bassyouni, M., Nehari, D., and Elhenawy, Y. (2021). Experimental and simulation study of multichannel air gap membrane distillation process with two types of solar collectors. *Energy. Convers. Manage.* 243: 114431. https://doi.org/10.1016/j.enconman.2021.114431.

13 Azha, S., Ikmal, N., Hussin, H. et al. (2020). Thermal performance enhancement in flat plate solar collector solar water heater: a review. *Processes* 8 (7): 756.

14 Abu-Zeid, M.A.-R., Bassyouni, M., Fouad, Y. et al. (2023). Experimental and simulation study of solar-powered air-gap membrane distillation technology for water desalination. *Membranes* 13: 821. https://doi.org/10.3390/membranes13100821.

15 Le Ba, T., Várady, Z.I., Lukács, I.E., Molnár, J., Balczár, I.A. & Wongwises, S. (2020). Experimental investigation of rheological properties and thermal conductivity of SiO_2–P2^5 TiO_2 hybrid nanofluids. *J. Therm. Anal. Calorim.* 146, 493–507 https://doi.org/10.1007/s10973-020-10022-4

16 Djetel-Gothe, S., Lanzetta, F., and Bégot, S. (2020). Second law analysis for the experimental performances of a cold heat exchanger of a Stirling refrigeration machine. *Entropy* 22: 215. https://doi.org/10.3390/e22020215.

17 Marif, Y., Bechki, D., Zerrouki, M. et al. (2019). Estimation of atmospheric turbidity over Adrar city in Algeria. *J. King Saud Univ. Sci.* 31 (2): 143–149. https://doi.org/10.1016/j.jksus.2017.06.002.

18 Elhenawy, Y., Bassyouni, M., Fouad, K. et al. (2023). Experimental and numerical simulation of solar membrane distillation and humidification-dehumidification water desalination system. *Renewable Energy* 215: 118915. https://doi.org/10.1016/j.renene.2023.118915.

19 Rowland, E.L., Fresco, N., Reid, D., and Cooke, H.A. (2016). Examining climate-biome ("cliome") shifts for Yukon and its protected areas. *Global Ecol. Conserv.* 8: 1–17. https://doi.org/10.1016/j.gecco.2016.07.006.

20 Marni-Sandid, A., Nehari, T., and Nehari, D. (2021). Simulation study of an air-gap membrane distillation system for seawater desalination using solar energy. *Desalin. Water Treat.* 229: 40–51. https://doi.org/10.5004/dwt.2021.27394.

13

Economic and Environmental Aspects

Moataz El-Bagoury[1], Dina Aboelela[1], Mohammed Alyoubi[2], and Shereen M. S. Abdel-Hamid[1,3]

[1] *Faculty of Energy and Environmental Engineering, The British University in Egypt (BUE), El-Sherouk City, Egypt*
[2] *Department of Design and Engineering, Faculty of Science and Technology, Bournemouth University, Poole, UK*
[3] *Department of Basic Sciences, Innovation University, Tenth of Ramadan City, Egypt*

13.1 Introduction

Membrane technologies have been in high demand, where recently China has an overall demand estimated to be valued at $4.8 billion, which represents around 15% of the total global demand [1]. Therefore, the implementation of such projects requires an in-depth understanding of the economic and environmental aspects when designing such projects on an industrial scale. In addition, the sustainability factor is the most highly prioritized objective in any project, and this results in the urge toward developing membranes that have the capability of being produced using renewable energy (RE), are made of sustainable materials, as well as present an economic and environmental performance [2].

This chapter aims to highlight the two most important aspects that should be considered in any membrane water treatment (WT) facility, being the economic and environmental aspects. In order to erect and operate a membrane WT facility, it is essential for business developers to consider key economic aspects, such as the profit revenue, the internal rate of return (IRR), the net present value (NPV), and the payback period (PBP). The life cycle assessment (LCA) is also important as it takes into consideration the lifetime of operation and the overall economic value added, as well as the environmental impact, which is the other aspect that must be considered. The environmental aspect must be evaluated, in terms of the waste

Water Treatment and Desalination: Membrane Separation using Renewable Energy Resources, First Edition. Edited by Mohamed Bassyouni.

generated, the energy consumed, emissions discharged, and how such parameters that negatively impact the environment must be monitored and managed in an environmentally friendly manner. Generally, the sustainability of membrane technologies is the focus of this chapter, considering the economic development using environmentally friendly methods.

In this chapter, the economic aspect will be explained, considering capital costs such as equipment and other fixed costs, operating costs such as waste disposal, energy and chemical costs, the economic viability by assessing the IRR, PBP, and NPV, as well as additional techniques used in analyzing the economic viability, such as cost–benefit analysis (CBA) and LCA. The environmental aspect will be described in terms of the waste generated and how it can be dealt with, energy consumption and optimization, and sustainable methods and operations. The integration of RE with the membrane-based WT process will be discussed, and their outcomes will be assessed and will be subject to a comparative study with the conventional WT facilities as well deducing the overall benefits of RE integration.

13.2 Economic Aspects

Membrane technologies of several variants have been applied on an industrial scale and have been economically assessed, in which the costs incurred are divided into two main categories: capital costs (capital expenditures [CapEx]) and operating costs (operating expenses [OpEx]) [3].

13.2.1 Capital Costs

Capital costs are mainly composed of the procurement of equipment for the membrane facility. A study has been conducted on a direct-contact membrane distillation (DCMD) system, with an integrated nanofiltration system, with the aim of treating groundwater, in which a techno-economic assessment (TEA) that included the capital costs (C_c) was further divided into several compartments, which were then summed using equation (13.1). The costs of storage tanks (K_{tf}) and pumps (K_p), as well as the cost of pipes and valves (K_{pv}) which included the pipeline, and all types of valves needed for the flow of permeates and concentrates within the treatment process were considered. The valves required instrumentation and control costs (K_{ic}) to be encountered as well as other miscellaneous costs (K_{mi}) have been added. The main integrated membrane system cost, as well as its constituents which mainly include the module cost, was also considered within the total capital costs, all denoted as K_m. In many industrial projects, the installation cost (K_i) usually represents a considerate portion of the total fixed costs, but in this case, it seems that the miscellaneous costs had the highest

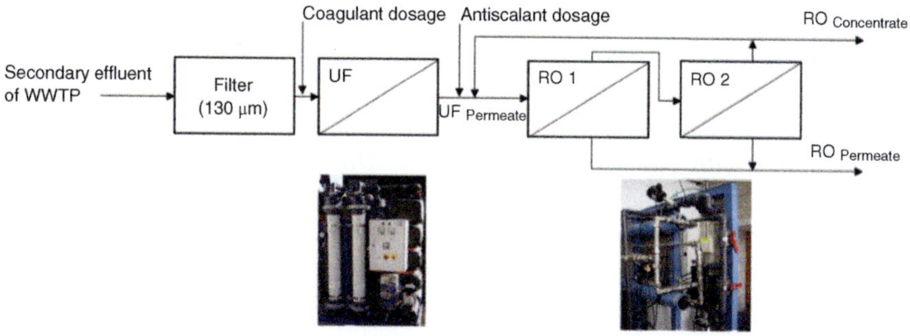

Figure 13.1 Scheme of the advanced UF-RO treatment applied to the secondary effluent of the municipal wastewater treatment plant. Source: Pérez et al. [5]/with permission of Elsevier.

proportion of contribution to the total capital costs, amounting to \$1,348,100 out of the total capital cost of \$3,236,900, for the case of a feed flow rate of 83.3 m^3/h of groundwater, with a 3,427 cm^2 membrane active surface and 70% recovery ratio, in which the economic assessment for the nanofiltration process was conducted based on the study by Elboughdiri et al. [4].

$$Cc = Kpv + Kic + Ktf + Kmi + Kp + Km + ki \tag{13.1}$$

Another setup for a membrane-based wastewater purification process for industrial reuse was made, consisting of an advanced ultrafiltration and reverse osmosis (UF-RO) treatment, as displayed in Figure 13.1, calculating the capital costs under different scenarios, which were compared to each other in terms of their membrane area. Prior to the comparison, three scenarios for the ultrafiltration (UF) process step and reverse osmosis (RO) process step were simulated at feed flow rates of 2.5 m^3/h, 5 m^3/h, and 20 m^3/h. The UF membrane areas with 80 m^2, 160 m^2, and 640 m^2 reached CapEx of €44,302, €62,220, and €122,727, respectively. As for the RO process step, another three scenarios in relation to the previous scenarios set for UF were simulated, where the RO membrane areas of 15.8, 31.6, and 126.4 m^2 possessed capital costs of €9,875, €14,161, and €29,118, respectively [5].

13.2.2 Operational Costs

Operational expenses (OpEx) may arise from various cost elements and the contribution of each cost is in relation to the application and processing within the WT facility. This theory has been practically showcased by a study article, which assessed the contribution of different operating costs for several types of water including demineralized water (for industrial applications), potable water (for domestic use), and irrigation water (for agriculture) which resulted from effluent streams from membrane-based advanced treatment processes. The main operating costs studied were of energy, chemicals, waste disposal, equipment replacement, and labor. In most cases, the cost of waste management was the highest contributor with a portion range of 42–47%. This was the norm except for one case of the irrigation water, in which the highest OpEx contributor was not waste management but was in fact labor cost with a significant 57% contribution [6].

One must take into consideration that operating expenses vary with the variant of the membrane technology being implemented, as well as the level of scalability of the overall plant. Another techno-economic analysis study was conducted for a vacuum membrane distillation (VMD) for seawater desalination, in which the operation costs in $/L were analyzed as a function of the plant size which was represented as the distilled water flow rate in L/h. The power required for feed pumps and vacuum pumps and nominal electricity charges were added alongside the miscellaneous and maintenance charges, which represented around 10–22% of the total OpEx. As the OpEx represent a function of distilled water flow rate, the relation is inversely proportional at an exponential rate, in which the flow rate increases from 1 to 100,000 L, cutting the operating cost from almost $45/L to $1/L, making it commercially viable. After a break-even analysis was carried out, it was concluded that the plant is considered commercial at the 100,000 L/h distilled water scale, which requires approximately 100,000,000 L of seawater to achieve the targeted permeate flux [7].

13.2.3 Economic Viability

Economic viability refers to the feasibility of the projects in the economy on an industrial scale. One must note that membrane technologies offer lucrative reductions in overall costs. A study has shown that when comparing membrane technologies to other types of WT, such as the five multi-effect evaporation with thermal vapor re-compression process, it has been deduced that there was an overall 75% reduction in the operating costs. Although mechanical vapor recompression may possess lower OpEx, the CapEx of such a process are relatively higher than those of RO membrane technologies [8]. The three main parameters in assessing a project's economic viability would definitely be the NPV, IRR, and PBP. A recent economic study was conducted on a 20-m^3/h plant capacity with a 39-L/m^2h permeate flow rate and a recovery rate of 83%. The total CapEx and OpEx were calculated to be €432,500 and €1.16/m^3, respectively. The process was optimized to make savings up to €291,900/year, which is about €2.09/m^3. Overall, the NPV was estimated at €751,400, a 48.6% increase in the NPV, with an IRR of 29.4% and a relative short PBP of approximately 3.3 years [9]. Higher IRRs may be achieved, which is of great interest to any business. For instance, a study conducted on desalination and natural WT using membrane processes deduced

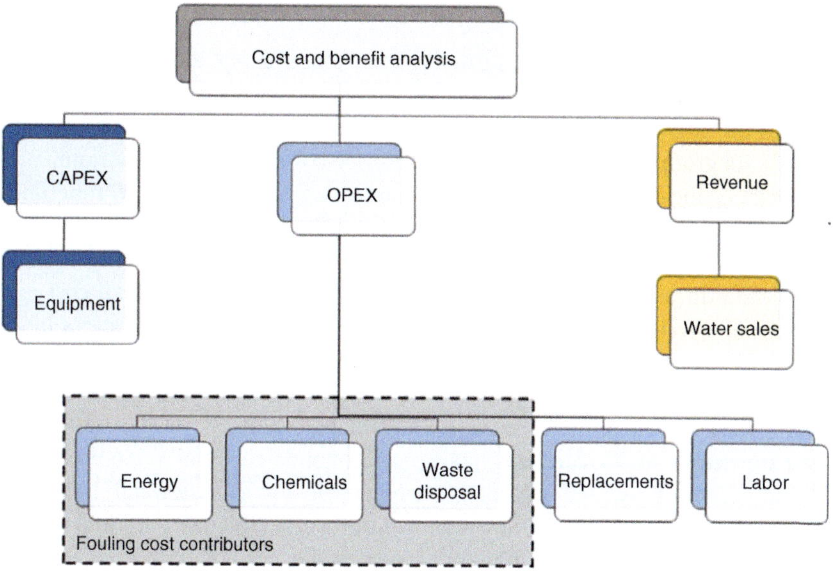

Figure 13.2 Cost factors (CAPEX, OPEX) and benefits (revenues) included in the CBA calculations. Source: Kehrein et al. [6]/IWA Publishing/CC BY 4.0.

that with an integrated system of RO with the use of wind energy as its energy source, a total capital investment cost of €196,000 and a profitability index of 1.3794 may result in high IRRs, reaching up to 225% [10].

Another factor to assess the economic viability of membrane technologies for WT facilities is the CBA, or the cost–benefit ratio (CBR), in which all components and costs, as shown in Figure 13.2, which depicts the methodology that can be followed for the calculation of CBA, are displayed [6]. CBA is divided into three main sections, in which costs are included within the CapEx and OpEx, while the main benefit comes from the third section and that is the revenue from water sales. The revenue from water sales mainly depends on the type of water being sold. A study showed that the sale of demi water, drinking water, and irrigation water came at a price of €1.10/m^3, €0.93/m^3, and €0.19/m^3, respectively [6].

One must also note that not only the application of membrane technologies for WT is economically viable, but the membranes utilized can be recycled in a recycling plant to produce other valuable products with very promising IRRs. A study on an RO membrane recycling plant was conducted to assess the financial mechanism of the plant based on CapEx, OpEx, and income for five years. It was concluded that an IRR of 73% was achieved, considering that the recycling plant was a pyrolysis plant named Plastics Energy, based in Europe, with its main revenues coming from the sales of pyrolysis oil and char [11].

13.2.4 Life Cycle Cost Assessments

Life cycle cost assessments (LCCAs) can be made using various techniques. Based on the literature available for WT projects, several methodologies were utilized for the assessment, in which life cycle costing (LCC) is categorized into three main groups, being conventional life cycle costing (CLCC), environmental life cycle costing (ELCC), and social life cycle costing (SLCC). Several WT methods and processes employing LCC methodologies will be discussed in the following section.

For the utilization of CLCC, several papers have analyzed the total cost and cost-effectiveness of such WT facilities. A study included several factors in its CLCC, by using the 3E+S model, considering the environmental, economic, energy, and social assessment. In China, CLCC was utilized to assess the economic aspect in terms of waste management, where the LCC results were for a timeframe of 30 years and had a net economic result of ¥70.62/ton [12]. In another study, two configurations were made for a wastewater treatment (WWT) facility in which the aim was to improve sludge handling and sludge recovery. The main point of difference between the two configurations was the addition of thickeners before the mixing chamber and the anaerobic digestion unit. CLCC was utilized for the calculation of the acquisition costs, or also known as the CapEx, at a timeframe of 20 years and an annual discount rate of 10%, in which the energy and polymer consumption and the OpEx were determined. It was deduced that the configuration with no thickeners had CapEx of €13,192,514, while the addition of thickeners raised the acquisition costs to €32,966,284. The energy and polymer consumption in both configurations remained identical, but one must consider the fact that the dynamic thickeners in the second configuration caused a major contribution to the increase in CapEx [13]. In another study, three models for wastewater management systems such as the septic-tank and tanker (STT), conventional gravity sewers (CGSs), and small-bare sewers (SBSs) were assessed, based on a design utilizing a lifecycle approach. CLCC methodology included the life cycle inventory (LCI) analysis which included the sewer network, the septic tank, and WWT process production, construction, and operation for the three models. After the LCI and CLCC models were applied, the three models CGSs, STT, and SBSs possessed total capital costs of $44,243,100, $16,710,400, and $31,308,900, respectively, and the 40-year present value increased to $12,532,300, while the implementation of the alternative SBS strategy caused 40% reduction in the overall collection and wastewater treatment plant (WWTP) system [14].

ELCC is another type of LCC in which the environmental load and energy consumption are significant factors that are considered within the overall

LCC [15]. A study conducted in Thailand assessed the environmental impact and cost-effectiveness of centralized municipal WWTPs with different schemes and different nutrient removal processes. The LCC model employed in the study included LCI for substance removal efficiency, life cycle impact assessment (LCIA) and sensitivity analysis for environmental impacts and sensitivity, and life cycle land use assessment (LLCA) for cost-effective results. The study deduced that the total costs of WWTPs with nutrient removal processes were significantly lower than those that do not have nutrient removal processes, with the vertical-loop reactor activated sludge being the most cost-effective scheme, having the lowest environmental impacts [16]. Another study conducted a comparative analysis between two aeration systems in WWTPs, in which the comparison was made between a surface aeration system and a blasting aeration system. The ELCC used included every stage of the aeration system, and thus included the floor area, equipment cost, operational cost, and disposal cost, with consideration of the environmental load. The ELCC was made over a 20-year life cycle in which the results for both the systems showed that the cost heavily depended on energy consumption, while during the maintenance period, the blasting aeration system was affected by several costs which included the equipment maintenance cost, maintenance cost, and economic losses caused by water loss and environmental loading. The blasting aeration system was merely more cost-effective, as it saved around ¥60,000 in annual costs in comparison to the surface aeration system [17]. Another study utilized the ELCC, in a different manner, to compare the cost of the first category integrated wastewater discharge (ICIWD) to the zero-liquid discharge (ZLD) policy set for the coal chemical industries in China. The study included a different methodology for life cycle water utilization, dividing the LCC into inputs of agents, utilities, natural water, and raw coal, the research boundary which included the WWTP and the discharge as well as the ZLD, and the outputs of effluents, byproducts, and wastes. The study concluded that the LCC of ICIWD was at an average of $0.94/ton water, but in order to reach ZLD, the cost had risen up to $1.07/ton water [18].

The third type of LCC is the social life cycle costing (SLCC), which is concerned with the social impacts that come with the products throughout their life cycle. A study conducted to assess the integration of external costs such as societal costs within the LCC of municipal WWTP used a model, called the LC^3 model, which included the running costs and the nonrunning costs, which was then input into LCI. One of the external costs calculated was the social cost of carbon (SCC) with an average of $32/ton CO_2 eq. and an annual discount rate of 3%. After analyzing the total SCC, the sum of average air emission impacts was deduced to be in a range of 5–8% of the total costs of the WWTP [19].

Table 13.1 Overview of the case study's economic factors.

Item	Unit
Capital costs	
Piping and valves	
Instruments and controls	
Tanks and frames	
Miscellaneous	
Pumps	
Membranes	
Pressure vessels	
Total CAPEX	€432,500
Operating costs	
Membrane replacement	
Energy	
Chemicals	
Concentrate disposal	
Maintenance	
Pretreatment	
Total OPEX	€1.16/m^3
Economic viability	
NPV	€751,400
IRR	29.4%
PBP	3.3 yr.

After a complete overview of all the economic factors, Table 13.1 is constructed as a demonstration of all the most common economic factors that can be found within any WWTP, whether membrane-based or any other type of process [9].

13.3 Environmental Aspects

The environmental aspects of any project must be critically assessed. In the case of membrane technologies, one must assess the environmental impact of the waste generated in terms of concentrate disposal, energy consumed by the WWTP, and the water consumption by the plant. Environmental impact assessments (EIAs) can be reviewed for such an analysis [20], and the sustainability of WWTPs will be further discussed.

13.3.1 Water Treatment

One of the beneficial environmental aspects of membrane technology, or the other treatment methods, is the removal of contaminants from water. A study has shown the effectiveness of integrating a membrane technology within the treatment of industrial effluents for the production of drinking water. Reverse osmosis (RO) was integrated into the treatment process as follows: sedimentation, followed by activated carbon, ion exchange using resins, and RO. The resulted water was of drinkable standards in which the ion exchange using resins reaching breakthrough after at 11–15 hours with 2.2–5 L effluent treatment. The fouling of RO was improved in which a high flux of water at 20–18 L/h/m^2 maintained reversible fouling dominance at a feed flow rate of 0.5–0.7 L/h.

Another study showed the effect of and enhancement in the performance of WT with a membrane, in which an integrated hybrid membrane system with enhanced water treatment (IHMS-EWT) was introduced owing to the high cost and ineffectiveness of current WT processes. IHMS-EWT showed interesting results when compared to current WT technologies, with an improvement of 98.2% in water purification, 94.2% efficiency rate, and an increase in cost-effectiveness by 82.4% [21].

Membranes synthesized on the nanoscale have shown great potential as well in WWT, which have been presented in a study on the synthesis of electrospun nanofiber membranes that have been fabricated by photocatalytic metal oxide nanoparticles of zinc oxide (ZnO) or silver-zinc oxide (Ag-ZnO) into cellulose acetate (CA)/polyvinylpyrrolidone (PVP) nanofibers using electrospinning. The composite membrane showed great results in oil and water emulsion separation, with 95% efficiency. The study concluded that the membrane can be used not only in oil/water separation but also in the separation of dye from textile wastewater, malachite green adsorption, and photocatalytic degradation of methylene blue dye, also the membrane possessed antibacterial properties. Overall, the membrane possessed low energy consumption requirements, making it an effective material for wastewater remediation [22].

13.3.2 Energy Consumption

Membrane technologies can be considered most sustainable when they have the highest efficiencies at the lowest energy consumption rates possible. A study aimed at optimizing energy reduction used electro-enhanced peroxymonosulfate (PMS) activation (E-REM-PMS), a perovskite $LaCoO_3$-functionalized Ti_4O_7 reactive electrochemical membrane (REM), to overcome restrictions in interference resistance, reusability, and WWT efficiency. At a constant rate of 39.69 per minute, it was able to remove 100% of carbamazepine (CBZ) in 233 and 23 times the electro-reactive electrochemical membrane (E-REM) system without PMS and the REM-PMS system without electricity, respectively. The enhancement of electro-activation on PMS was identified by in-situ Fourier transform infrared (FTIR), and the differences as well as the rationality of the mechanisms under three adsorption conformations were further explored by direct electron transfer (DET) analysis. Community water fluoridation (CWF) generated by treating the pharmaceutical wastewater continuously described 100% removal of CBZ and 60–70% mineralization of total organic carbon (TOC) at an ultra-high-water flux (642.86 L/m^2h) and a minimum energy consumption of 0.01 kWh/m^3. This work provided a low-cost and efficient WWT solution through a highly efficient and fouling-resistant proposed REM [23].

A comparative study concluded that for hardness control, capacitive deionization (CDI) is much more efficient than typical water softening techniques. However, there is a lack of comprehensive studies on pilot-scale CDI systems with water softening techniques. Operational-scale studies are important for determining the behavior of the desalination process, energy requirements, and selectivity of divalent cations for large-scale practical use. In this work, the electrochemical performance and energy consumption of a pilot-scale CDI technique were studied to enhance the chance of softening water. The working variables that are being studied include the voltage applied across the fibers (200–300 V or 1.0–1.5 V per pair), feed concentration (550–1250 mg/L), recovery ratio (50–80%), flow rate (3–5 L/min), and adsorption/desorption time (80–180 seconds). The total dissolved solids (TDS) removal ratio has the favorable circumstances of high voltage, low feed concentration, low recovery ratio, high flow rate, and long adsorption/desorption time. The advantages of energy efficiency, evaluated in TDS removal per unit energy consumption, are fast flow rate, long adsorption/desorption time, low voltage (LV), low input concentration, and low recovery ratio. Furthermore, different operating circumstances were assessed based on how they affected the removal of Ca^{2+} over Na^+ and how it affected energy efficiency. When compared to nonselective systems (i.e., selectivity coefficient = 1), the pilot CDI system was shown to achieve up to 45% energy-saving efficiency, with a Ca/Na selectivity coefficient of roughly 2.4 [24].

Further research focused on the novel and creative way of using gap water circulation in permeate-gap membrane distillation (PGMD) systems as an effort to present an energy-efficient water desalination system to tackle the problem of energy efficiency in the desalination industry. Here heat-mass transfer modeling of a counterflow multistage was discussed and the same was analyzed through investigation. Research analysis has confirmed that gap water circulation improved water productivity and reduced energy when compared to conventional PGMD configurations. In this study, the impact of this innovation was evaluated with a systematic approach that involved altering flow chamber configurations and analyzing working parameters. Intake flow rates at 10,000 m^3/d seawater, intake temperature and feed injection temperature, and number of stages were also varied. It was also realized that for different conditions, a counterflow design with more stages provided better feed recovery rates and lower energy consumption. Lower specific thermal energy consumption (STEC) value of the order of 92 kWh/m^3 and higher flux values of 28 kg/m^2/hour signified better mixing and turbulence due to the application of the proposed modules. Analyzing membrane distillation (MD) systems with system-based conventional PGMD systems, the authors elucidated ways in which the flow of the gap water enhanced the energy efficiency and production of MD systems significantly [25].

13.3.3 Waste Treatment

The waste generated in membrane WWTPs is mainly present in the concentrate streams, and this waste must be treated to mitigate the negative environmental impacts and avoid the increasing human and health risk concerns. Regarding the degradation of 26 pharmaceutically active compounds (PhACs) in an MD surface water concentrate, this work has investigated the performance of ultraviolet (UV), UV/hydrogen peroxide (H_2O_2), and ozonation treatments. Also, after reviewing their exposure impact on human health risks and potentials for genotoxicity as well as being a carcinogen after exposure to acute and chronic toxicity, >99% of PhACs were removed by MD with a recovery rate of 80%. In addition to solids and other ions, MD was able to reject PhACs. Concerning the removal of PhACs in the concentrate, an improvement by a factor of up to 10,000 was observed for UV/H_2O_2 and ozonation as the dosage of H_2O_2 rose and the flow rate of ozone increased, respectively. The three compounds that could not be persuaded to reduce at a faster rate than heat are prednisone, atorvastatin, and clarithromycin, the three most insoluble PhACs. Furthermore, in those processes known collectively as the advanced oxidative processes (AOPs), the color of the effluent may be significantly lowered, with total nitrogen and TOC levels. Lastly, there was a stager toxicological risk with respect to the surface water concentration. Investigating the acute and long-term effects which were reduced by the AOPs in concentrate treatment, the dose of H_2O_2 and O_3 used in the UV/H_2O_2 and ozonation processes was elevated [26].

Another study states that a better way is inland desalination particularly using the RO process which would provide additional potable water supplies. Nonetheless, the amount of the concentrated brine that is generated as a result of the procedure is relatively large. Concentration disposal can accept up to one-third of the cost of desalination and wastewater resources in deserts. The use of desalination facilities in water-scarce areas is hampered by conventional disposal systems which are fettered by characteristics, such as geographic, hydrological, weather, and legal conditions. Also, only few of these systems give much consideration to resource recovery. Another approach that reduces the amount of trash content while at the same time increasing the amount of water recovery for beneficial reuse is MD. Finally, for desalination of its concentrate stream and to allow near ZLD, MD employs thermal energy gradients. Furthermore, as shown in Section 13.2, heat recovery is included in a few MD arrangements; therefore, MD is an energy-efficient option. Air-gap membrane distillation (AGMD) for RO concentrate management was subjected to a TEA and contrasted with three traditional concentrate management systems: the three principal methods of concentration-crystallization, deep-well injection (DWI), and the final disposal by evaporation ponds. These results imply that when operating conditions are optimized and low-grade heat is available, AGMD is a competitive concentrate management system that produces water at \$0.9/m^3 as opposed to DWI at \$1.09/m^3 followed by evaporation ponds at \$1.47/m^3 and concentrators at \$6.2/m^3. The use of desalination facilities in water-scarce areas is hampered by conventional discharge systems which are bound by physical features (including geographical and geological), climatic conditions, and legal issues. Also, none of these systems focus on the recovery of resources. Another technology capable of lowering the trash volume but at the same time boosting water recovery for beneficial use is MD. MD is used to desalinate the concentrate stream and to produce nearly minimal liquid discharge by using thermal energy gradients [27].

The treatment of wastewater reverse osmosis concentration (WWROC) emanating from wastewater treatment plants (WRPs) was investigated through MD. The MD tests were performed at 85% of the water recovery of WWROC where a DCMD offered high-quality permeate water (10–15 μS/cm, 99% ion rejection) at a moderate input temperature of 55°C with only 13–15% flux loss. The MD membrane was treated with higher concentrations of WWROC, which led to the deposition of common calcium carbonate ($CaCO_3$). The flow drop of the DCMD process was not significantly affected by the low salinity and weak adherence of $CaCO_3$ on the membrane. The reduction in membrane hydrophobicity that accompanied the use of the MD membrane (the water contact angle reduced to 70% of the virgin membrane value) was a result of low-molecular-weight organic adhesion where a high organic content in WWROC with the values of 58–60 mg/L was notable. The pretreatment of WWROC with granular activated carbon (GAC) decreased the organic loading by 46–50% and removed a range of hydrophobic and hydrophilic micropollutants. This enhanced the possibility of water reuse and ensured that MD produced water of high quality and devoid of micropollutants. Concerning the aspects of selective ion precipitation, the most appropriate was brought to focus the MD-concentrated WWROC and in the case of WRPs, it minimized the liquid discharge nearly to zero [28].

13.3.4 Environmental Impact Assessment

EIA, as shown in Figure 13.3, is important for membrane WT plants. These plants are kept safe from unsustainable operations by the appraisal of expanded influence on ecosystems, water resources, and communities by EIA. This assessment defines risk areas, for example, discharge of chemicals or energy use, and directs their management. Concerning the plant design and operation, EIA also helps in reducing the environmental impact of the project while maintaining WT

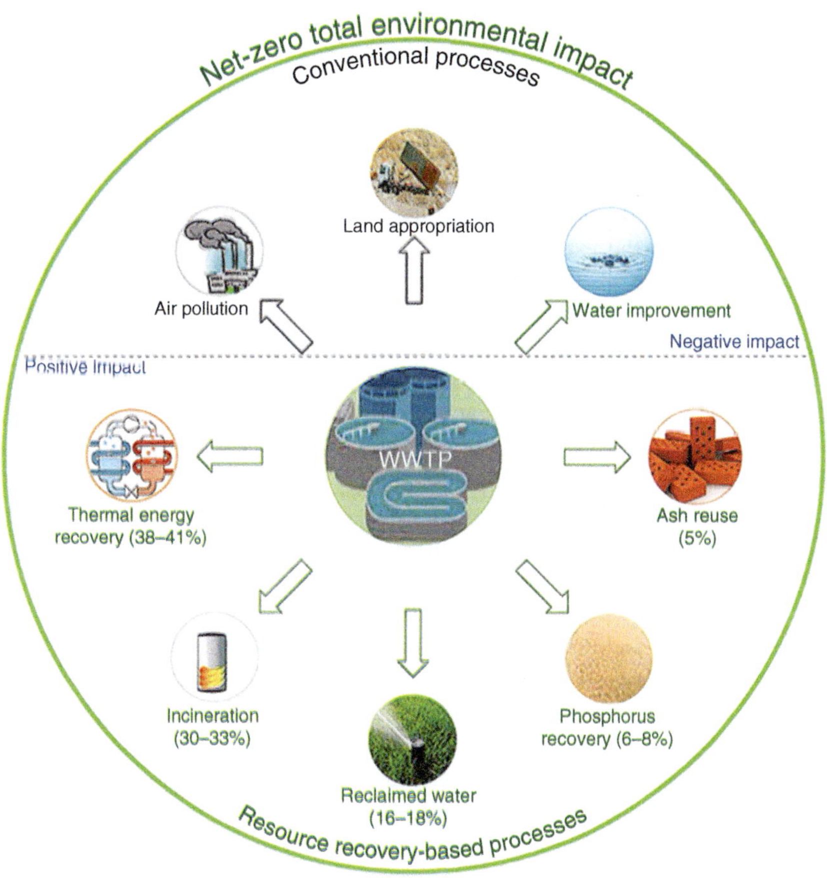

Figure 13.3 Environmental impact assessment factors for WWTP. Source: Hao et al. [29] with permission of Elsevier.

needs as well as preserving the environment. A study stated that conventional WWTPs are designed to decrease water pollution and replenish wastewater but at the same time discharge greenhouse gasses into the atmosphere and consume energy and materials. However, the harm caused by WWT to the environment may be reduced through the recovery of resources and energy with routines, such as anaerobic digestion and water treatment. More LCAs are being employed to estimate the environmental impacts of WWTPs and make suggestions for their enhancement. Applications that can result in zero operating impact on the total environment of WWTPs need to be developed for resource recovery applications that are in question. As a result, this chapter outlines and analyzes a very efficient approach to energy and resource recovery, especially for thermal energy. Based on this approach of LCA methodology, the environmental impact of a conventional WWTP was contrasted with the one described in this plan, inclusive of energy and resource recovery. Choosing the regional context of interest was a particularly important step in the LCA process. While using a Chinese scenario in this case, the data collected are of a qualitative nature and can be used cross-culturally. The outcomes, to the extent, affirmatively revealed that, unlike the new plan which adopted the resource/energy recovery-based procedures, the traditional WWTP does not make the environment better. Among other factors, thermal energy recovery alone has made a percentage contribution of approximately 40%, and therefore more research and focus should be directed toward it since it has produced a vigorous study of net-zero LCA analysis [29].

The LCA in the present work was to assess the impacts of WWT with microalgae, employing an integrated membrane bioreactor system. A functional unit of 1 kilogram of microalgal wastewater was used in this study, after pollutants' reduction by a UF membrane produced from recycled polyethylene terephthalate waste. In the meantime, the study's applied system boundary was identified using two scenarios: Cradle to gate in which water from the wastewater was only treated, and cradle to cradle in which the treated water was used to grow microalgae once again. This was clearly understood from the interaction effects of various environmental factors and hotspots identified in the different phases of the WWT process, where membrane treatment had the most impact on the environment, followed by microalgal harvesting and culture. Analyzing the results by the category of environmental effect, it was possible to determine which of the categories has a direct water-related impact—freshwater ecotoxicity, freshwater eutrophication, and marine ecotoxicity. To the best of the authors' knowledge, the following results are fairly impressive, given the fact that the results of this study were compared with those of the existing literature reporting on the LCA of WWTs using membranes. The global warming potential, which is the outstanding measure of all the environmentally impactful assessments, was only 2.94×10^{-4} kg CO_2 eq. Overall, the results of this study can be used to make the right decisions concerning the optimization of microalgal WWT and subsequent reutilization guided by the quest for sustainability [30].

13.4 Renewable Energy Integration

With recent environmental and sustainable aims, it is therefore important that RE sources be included in membrane WT plants for efficiency. With solar and wind energy, for example, plants can decrease their use of fossil energy, cut down on emissions, and reduce their power expenses. All of this is done for the benefit of the environment but also in order to preserve a water supply system that is less dependent on energy prices and interruptions within conventional power grids.

13.4.1 Renewable Energy Costs

Costs are determined by factors such as scale, source, and government policies that may impact costs through rewards and subsidies. The cost of home photovoltaic (PV) solar power is currently 0.16–0.25 $/kWh, which is generally a little more than the 0.11–0.35 $/kWh cost of conventional grid electricity in Europe. On the other hand, "grid parity," where solar power costs the same or less than the retail price of electricity from typical large-scale generators, has already been attained with PV in several countries, such as Cyprus and Italy [31, 32]. These costs are extremely sensitive to policies and subsidies in either direction; nonetheless, based on the scarcity of natural resources for energy fuel production, there is every likelihood that the market price of energy fuels from such resources will increase. On the other hand, one might expect a decrease in the cost of PV as RE because of improvements in technology and as markets grow and installed capacities increase while production costs decrease. A very recent study indicates that levelized cost of energy (LCOE) of Kenya's connected PV system was 0.21 $/kWh. This is way below the cost of the country's expensive conventional power plants such as the gas turbines and diesel generators, which make up a vast proportion of the nation's electricity generation [33].

If there is a local grid supply, then it is easy to approximate the cost of extending to it since the cost includes the labor cost, number of poles, transformers, and LV cable. On the other hand, long-distance connections, that are most likely to be developed, may be developed as part of a larger program for the extension of electrification for the provision of power to several communities, centers, or homes in a rural setting. The cost of electricity and availability and reliability of energy all play major roles that directly influence the amount of initial capital outlay in developing small-scale WT systems, and these are major expenses today.

13.4.2 Technical Challenges and Solutions

Purification of wastewater has emerged as a potential substitute for seawater desalination in the fields of industrial makeup, agricultural irrigation, and indirect potable water production. To meet water authorities' quality criteria, membrane technologies can be used in various configurations for wastewater purification and reuse [34]. It is still difficult to achieve membrane selectivity while producing high-quality products. Moreover, the membrane technique may be a useful tool for removing greenhouse emissions from industrial plant stacks and capturing CO_2. To handle the extremely high gas flows in the CO_2 collection scenario, very high gas permeance is a crucial necessity. Membrane processes are widely used, clean, eco-friendly separation technologies that need less energy than traditional methods. However, present membrane technologies are far from environmentally friendly scenarios because of the use of potentially toxic chemicals and nondegradable polymers in the manufacturing of membranes, modules, and spacers. The purported cleanliness of these membrane processes is therefore highly dubious. Furthermore, a comprehensive assessment of the impact of membrane and module production on the development of microplastics, environmental contamination, global warming, and human health is necessary. As such, careful consideration of the next generation of membrane processes which aim to be cleaner and more sustainable is imperative.

Population explosion is believed to be responsible for the upsurge in energy consumption around the world. Accordingly, the consumption of fossil fuels is expected to lead to air pollution and global warming through CO_2 emissions. The global problems with the demand for clean energy and environmental issues may be resolved by devising other RE resources. Salinity-gradient power (SGP), or the so-called blue energy, is a prospective substitute for these problems. SGP is also able to convert the chemical potential of two solutions with different salinities when the two solutions are mixed. It is considered a clean as well as a RE source with a very low carbon footprint because it does not release any poisonous gasses. Ion-exchange membranes in generating RE appear to address these challenges [35].

New forms of membranes using nanopores have been developed in the recent past specifically for harvesting blue energy from the salinity gradient. In the systematic study presented by Wang et al. [36], several process scales of nanopore-based power generation were expertly reviewed. They looked into the specific procedures by which energy had been harnessed and the effects that the pore conductance and selectivity had on power making.

13.4.3 Economic Benefits

According to technical specifications, MD, a rather new technology, offers a way for providing drinkable water in hot and remote zones. While a number of researchers have focused on the attempt to fine-tune the MD operating conditions, scant attention has been given to a crucial facet of the application of any new technology, the economics of implementation, a factor that can quickly bring the most promising idea to a halt. The costs of implementing an MD system to desalinate brackish water for drinking water production are analyzed systematically in this work. All three variables, plant size, waste heat use, and membrane permeability, are investigated individually and through their interactions. Save for employee training, this study reveals that waste heat recovery, the use of solar energy, and facility size reduction impact the total cost of purified water and, therefore, water price (WP). In the same way, recovery of waste heat may be utilized in the operation of MD to cut substantially the price of water and solar–thermal sources. For instance, an MD system, which incorporates waste heat and solar–thermal energy and hitherto accounts for 40% and 60% of its total energy requirements, respectively, will have decreased the cost of water from a merger cost of a stand-alone MD, which is floored by grid electricity of $6.80/m^3$ to under $1.6/m^3$. Finally, to assess the maximum cost savings achievable through an MD process, a sensitivity analysis was performed. The study demonstrated that there is potential to reduce the cost of making MD water by increasing the membrane permeability, using more solar–thermal and waste heat energy in the desalination plants, even if there will be the need to downscale the desalination plant capacity [37].

The cogeneration of electricity and clean water for remote regions can be possibly implemented by incorporating the photovoltaic hybrid thermal system (PVT) and DCMD in a residential building, as shown in Figure 13.4. The goal of this particular study was to introduce a microgrid for interconnections between PVT, DCMD, and a residential house in an attempt to supply clean water as well as partially meet the power requirement of the residential building. The behavior of the system was quantitatively modeled using a model that was developed and proved. Techniques of optimization and artificial neural networks (ANNs) were used. The performance of the proposed system for a number of design and operational conditions and under Riyadh–Saudi Arabia weather conditions was evaluated. The case study analysis shows that the temperature of the brackish water at PVT, capital and installation costs, and targeted water productivity are functions that plan the optimum system efficiency. It has a specific cost of water (SCW) of 23.6 $/m^3$ and requires PVT prices and power prices of 25% RE penetration. Therefore, the suggested system fulfills a quarter of the electric requirement of the residential building, while the other quarter is drawn from the grid. In addition, the proposed strategy reduced the gross amount of yearly CO_2 emissions of a single building by 4,300 kg. The findings of this project will contribute to the understanding of the application of innovative clean energy and water technologies, including PVT and DCMD, in a residential building [38].

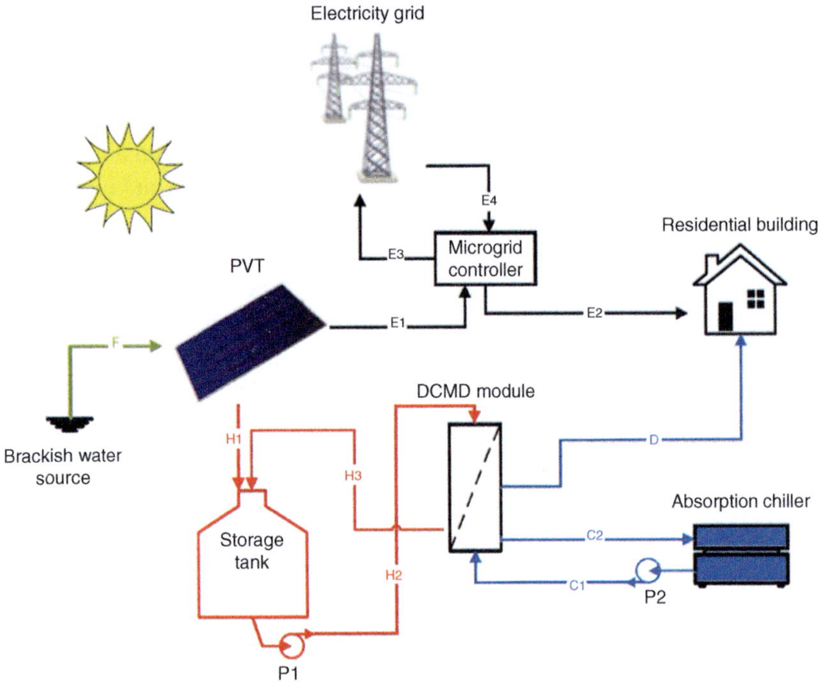

Figure 13.4 Conceptual diagram of the proposed system. Source: Almehmadi et al. [38]/MDPI/CC BY 4.0.

13.5 Comparative Study

When integrating RE within WT systems, some researchers have made critical comparative studies in order to analyze and depict the most suitable RE system for WWTPs. For instance, this research endeavors to simulate and investigate possible approaches to improve the overall efficiency, including implementation, of a brackish water RO system that is RE-operated and battery-free under simulated climatic conditions. The systems rely on PV panels, wind turbines, or both and are assessed at two flow rates (50 m^3/d and 100 m^3/d). In order to connect the RO unit to RE sources, four possibilities are taken into consideration: low; complex brackish water tank high; variable-frequency drive; and the two the last one. The next three possibilities are analyzed based on the optimized results of the first case, and then the simple method is optimized for the lowest levelized cost of water (LCOW). The approximate results show that the hybrid system can provide water with a higher available volume compared to the two separate systems of 13,500 m^3/year and 24,900 m^3/year. Nevertheless, the PV-based system is better overall in specific energy use (1.08 kWh/m^3 for both volumes) and unused energy percentage (0.32% and 0.28% for 50 and 100 m^3 devices/d, respectively) due to PV's more uniform characteristics and fewer fluctuations. Additionally, the actual economic performance of the systems is boosted by the proposed scenarios. Scenarios with enhancement to scenario four depict PV-based systems to exhibit improved efficacies while turning out to have the least LCOW worth of 1.09 \$/m^3 [39].

Another paper examines the way in which green hydrogen generation can be made to meet its freshwater requirement in sustainable energy generations. It proposes four schemes, in which proton-exchange membrane (PEM)-based water electrolysis is integrated with RE-powered desalination processes. These configurations can be run either as continuous or cyclical operations. To determine green hydrogen potential, this research evaluates the capital cost (CapEx) of integrating solar or wind power reverse osmosis (SWRO) and multi-effect distillation (MED) desalination processes. It considers the prospects of renewable power, capabilities of energy storage equipment, and efficacy of various desalination processes, particularly where these factors influence cost drivers and project economics. The least-cost SWRO system is the one that powers itself using wind energy on an intermittent basis, thus reducing the capital cost by 46%. The outcomes reveal that synergies between green desalination and green hydrogen production applications present useful gains [40].

Advanced RE-driven water desalination technologies mainly focus on innovation, in which the emphasis on the necessity of using solar energy is underscored particularly as a critical input in the multiple processes of desalination that include forward osmosis, humidification-dehumidification, MD, and combined process systems. Some of the purposes of this study were to evaluate recent advances in RE systems, especially solar systems, in addressing the water desalination function. Furthermore, the study aimed to assess the efficiency and potential improvement of integrating hybrid desalination solutions and also to outline the research challenges, limitations, and suggestions within these processes. The practical utilization of useful energy with integrated efficacy, as well as cost-efficiency, has been analyzed in numerous experiments. The evaluation also reveals that with the hybrid desalination systems, energy use may be reduced. For instance, the integration of MD and forward osmosis shows a decline in energy ratio from 0.89 to 0.64. In addition to this, a combination of the humidification-dehumidification process with RO is proven to reduce energy consumption by 50% more than the conventional RO system, coupled with an increased production of freshwater by as much as 38%. Nevertheless, there are challenges incumbent on the improvement of integration with hybrid systems to focused solar power panels, such as PV thermal and concentrated PV thermal panels [41].

Biohydrogen generation, particularly by biogas steam methane reforming (BGSMR), forms the focus of this paper and the dynamics of hydrogen production. In the context of the Portuguese system, it compares the environmental footprint of this solution to a commercial electrolysis and hydrogen synthesis through natural gas steam methane reforming (NGSMR). To provide a more detailed comparison, the solutions are expressed in terms of waste valorization rate; energy efficiency indicators; water intensity, i.e., the volume of water consumed per unit of the end product; or the proposed carbon footprint. As an application of water recycling, the assessment gauges the impact of different off-gas combustion scenarios as well as characteristic operational parameters. The fact is that it is almost impossible to obtain much information regarding the actual reforming process, and therefore the study utilizes the method of process simulation, known as the DWSIM chemical process simulator. For comparison, data were taken from water electrolyzers that are offered for sale to customers commercially. The hydrogen generated through water electrolysis is only feasible if the power system is solely reliant on renewable power generation, more so, when measured in carbon footprint. Thus, despite the possession of nearly negligible carbon footprints, biogas systems are a good fit now and in the future. Another study provides valuable knowledge of hydrogen production and shows how it may be a sustainable solution in different power system situations [42].

13.6 Conclusion

This chapter has exhaustively discussed the major economic and environmental factors that must be taken into consideration while setting up and running a membrane WT plant. It was pointed out that the process of business development involves the identification of key economic ingredients including the profit revenue, the IRR, the NPV, and the PBP so that such projects can be evaluated for financial feasibility. Moreover, the LCA offers considerable application of the assessment of the position and impact on the long-run economic and environmental benefits of membrane technologies. On an environmental note, the chapter has also outlined the strategic management of waste generation, energy utilization, and emission management in an environmentally friendly manner. The environment should not be degraded during the process, hence the need to employ sustainable processes and operations. Thus, membrane-based WT integrated with RE sourcing holds a lot of potential for enhancing sustainability. This chapter has provided a quantitative analysis of the benefits of incorporating RE into a conventional WT plant. The assessment of membrane WT facilities shows that the economic and environmental impacts should be collectively considered during the deployment of WT plants. Therefore, if these aspects are well understood and dealt with, it becomes easy for there to be long-term environmentally responsive and profitable solutions that support water security.

Questions

1 What are the types of costs within any membrane water treatment (WT) facility?

2 What are the economic factors that are used to determine a membrane WT project's viability?

3 What are the factors within a cost–benefit analysis?

4 What are the environmental impact assessment factors for a wastewater treatment plant?

5 What are the potential challenges when applying renewable energy on a membrane system?

6 What are the key differences between renewable and conventional membrane technology-based WT plants?

References

1 Zheng, X., Zhang, Z., Yu, D. et al. (2015). Overview of membrane technology applications for industrial wastewater treatment in China to increase water supply. *Resour. Conserv. Recycl.* 105: 1–10. https://doi.org/10.1016/j.resconrec.2015.09.012.
2 Khanzada, N.K., Al-Juboori, R.A., Khatri, M. et al. (2024). Sustainability in membrane technology: membrane recycling and fabrication using recycled waste. *Membranes (Basel).* 14 (2): https://doi.org/10.3390/membranes14020052.
3 Fu, Z., Zhou, Z., Liu, Z. et al. (2024). Feasibility and challenges of low-cost ceramic membranes in water treatment applications. *Desalin. Water Treat.* 320: 100739. https://doi.org/10.1016/j.dwt.2024.100739.
4 Elboughdiri, N., Cosma, R., Amari, A. et al. (2024). Evaluating the feasibility of direct contact membrane distillation and nanofiltration in ground water treatment through a techno-economic analysis. *Appl. Water Sci.* 14 (6): 1–21. https://doi.org/10.1007/s13201-024-02185-4.
5 Pérez, G., Gómez, P., Ortiz, I., and Urtiaga, A. (2022). Techno-economic assessment of a membrane-based wastewater reclamation process. *Desalination* 522: https://doi.org/10.1016/j.desal.2021.115409.
6 Kehrein, P., Jafari, M., Slagt, M. et al. (2021). A techno-economic analysis of membrane-based advanced treatment processes for the reuse of municipal wastewater. *Water Reuse* 11 (4): 705–725. https://doi.org/10.2166/wrd.2021.016.

7 Idrees, H., Ali, S., Sajid, M. et al. (2023). Techno-economic analysis of vacuum membrane distillation for seawater desalination. *Membranes (Basel).* 13 (3): https://doi.org/10.3390/membranes13030339.

8 Jevons, K. and Awe, M. (2010). Economic benefits of membrane technology vs. evaporator. *Desalination* 250 (3): 961–963. https://doi.org/10.1016/j.desal.2009.09.081.

9 Suárez, A., Fernández, P., Ramón Iglesias, J. et al. (2015). Cost assessment of membrane processes: a practical example in the dairy wastewater reclamation by reverse osmosis. *J. Memb. Sci.* 493: 389–402. https://doi.org/10.1016/j.memsci.2015.04.065.

10 Charcosset, C. (2022). Classical and recent developments of membrane processes for desalination and natural water treatment. *Membranes (Basel).* 12 (3): https://doi.org/10.3390/membranes12030267.

11 Tavares, T., Leon, F., Vaswani, J. et al. (2022). Study for recycling water treatment membranes and components towards a circular economy—case of Macaronesia area. *Membranes (Basel).* 12 (10): https://doi.org/10.3390/membranes12100970.

12 Zhou, Z., Chi, Y., Dong, J. et al. (2019). Model development of sustainability assessment from a life cycle perspective: a case study on waste management systems in China. *J. Clean. Prod.* 210: 1005–1014. https://doi.org/10.1016/j.jclepro.2018.11.074.

13 Roldán, M., Bouzas, A., Seco, A. et al. (2020). An integral approach to sludge handling in a WWTP operated for EBPR aiming phosphorus recovery: simulation of alternatives, LCA and LCC analyses. *Water Res.* 175: https://doi.org/10.1016/j.watres.2020.115647.

14 Abdallah, M., Shanableh, A., Elshazly, D., and Feroz, S. (2020). Techno-economic and environmental assessment of wastewater management systems: life cycle approach. *Environ. Impact Assess. Rev.* 82: 106378. https://doi.org/10.1016/j.eiar.2020.106378.

15 Gulotta, T.M., Salomone, R., Mondello, G. et al. (2023). Life cycle assessment and environmental life cycle costing of a unitised regenerative fuel cell stack. *Sci. Total Environ.* 901: 166007. https://doi.org/10.1016/j.scitotenv.2023.166007.

16 Talang, R.P.N., Sirivithayapakorn, S., and Polruang, S. (2020). Environmental impacts and cost-effectiveness of Thailand's centralized municipal wastewater treatment plants with different nutrient removal processes. *J. Clean. Prod.* 256: 120433. https://doi.org/10.1016/j.jclepro.2020.120433.

17 Liai, C., Hongxun, H.O.U., Weibiao, F.E.I., and Eryan, Z. (2017). Comparative research to surface aeration and blasting aeration system based on LCC theory. *IOP Conf. Ser.: Earth Environ. Sci* 63 (1): https://doi.org/10.1088/1755-1315/63/1/012029.

18 Chen, B., Yang, S., Cao, Q., and Qian, Y. (2020). Life cycle economic assessment of coal chemical wastewater treatment facing the 'zero liquid discharge' industrial water policies in China: discharge or reuse? *Energy Policy* 137: 111107. doi: 10.1016/j.enpol.2019.111107.

19 Theregowda, R.B., Vidic, R., Landis, A.E. et al. (2016). Integrating external costs with life cycle costs of emissions from tertiary treatment of municipal wastewater for reuse in cooling systems. *J. Cleaner Prod.* 112: 4733–4740. https://doi.org/10.1016/j.jclepro.2015.06.020.

20 Morrison-Saunders, A., Nykiel, A., and Atkins, N. (2024). Understanding the impact of environmental impact assessment research on policy and practice. *Environ. Impact Assess. Rev.* 104: 107334. doi: 10.1016/j.eiar.2023.107334.

21 Alluhaidan, A.S., Prabu, P., Basheer, S. et al. (2024). Integrated hybrid membrane system for enhanced water treatment and desalination for environmental preservation. *Desalin. Water Treat.* 320: 100722. https://doi.org/10.1016/j.dwt.2024.100722.

22 Chaithra, K.P., Varghese, A., Vinod, T.P., and Sunaja Devi, K.R. (2024). Multifunctional electrospun membranes incorporated with metal oxide nanoparticles, cellulose acetate, and polyvinylpyrrolidone for wastewater treatment: oil/water separation, dye adsorption, and dye degradation. *Chem. Eng. J.* 499: 156049. https://doi.org/10.1016/j.cej.2024.156049.

23 Li, W., Xiao, R., Lin, H. et al. (2022). Electro-activation of peroxymonosulfate by a graphene oxide/iron oxide nanoparticle-doped Ti_4O_7 ceramic membrane: mechanism of singlet oxygen generation in the removal of 1,4-dioxane. *J. Hazard. Mater.* 424: 127342. https://doi.org/10.1016/j.jhazmat.2021.127342.

24 Yoon, H., Jeon, S., Min, T. et al. (2024). Pilot-scale capacitive deionization for water softening: performance, energy consumption, and ion selectivity. *J. Environ. Chem. Eng.* 114259. https://doi.org/10.1016/J.JECE.2024.114259.

25 Al Hariri, A.H. and Khalifa, A.E. (2025). Heat recovery of multistage circulated permeate gap membrane distillation for energy-efficient water production. *Sep. Purif. Technol.* 352: 128284. https://doi.org/10.1016/j.seppur.2024.128284.

26 Alvim, C.B., Moreira, V.R., Lebron, Y.A.R. et al. (2020). Comparison of UV, UV/H_2O_2 and ozonation processes for the treatment of membrane distillation concentrate from surface water treatment: PhACs removal and environmental and human health risk assessment. *Chem. Eng. J.* 397: 125482. https://doi.org/10.1016/j.cej.2020.125482.

27 Felix, V., Hardikar, M., and Hickenbottom, K.L. (2024). Concentrate circularity: a comparative techno-economic analysis of membrane distillation and conventional inland concentrate management technologies. *Desalination* 574: 117213. https://doi.org/10.1016/j.desal.2023.117213.

28 Naidu, G., Jeong, S., Choi, Y., and Vigneswaran, S. (2017). Membrane distillation for wastewater reverse osmosis concentrate treatment with water reuse potential. *J. Memb. Sci.* 524: 565–575. https://doi.org/10.1016/j.memsci.2016.11.068.

29 Hao, X., Wang, X., Liu, R. et al. (2019). Environmental impacts of resource recovery from wastewater treatment plants. *Water Res.* 160: 268–277. https://doi.org/10.1016/j.watres.2019.05.068.

30 Rawindran, H., Khoo, K.S., Ethiraj, B. et al. (2024). Environmental impact assessment via life cycle analysis on ultrafiltration membrane fabricated from polyethylene terephthalate waste to treat microalgal cultivation wastewater for reusability. *Environ. Res.* 251 (P2): 118687. https://doi.org/10.1016/j.envres.2024.118687.

31 Fokaides, P.A. and Kylili, A. (2014). Towards grid parity in insular energy systems: the case of photovoltaics (PV) in Cyprus. *Energy Policy* 65: 223–228. https://doi.org/10.1016/j.enpol.2013.10.045.

32 Ameli, N. and Kammen, D.M. (2014). Innovations in financing that drive cost parity for long-term electricity sustainability: an assessment of Italy, Europe's fastest growing solar photovoltaic market. *Energy Sustain. Dev.* 19 (1): 130–137. https://doi.org/10.1016/j.esd.2014.01.001.

33 Ondraczek, J. (2014). Are we there yet? improving solar PV economics and power planning in developing countries: the case of Kenya. *Renew. Sustain. Energy Rev.* 30: 604–615. https://doi.org/10.1016/j.rser.2013.10.010.

34 Li, N.N., Fane, A.G., Ho, W.S.W., and Matsuur, T. (eds.) (2008). 29.3.1.1 All-polymeric membranes—immobilized imidazole system. In: *Advanced Membrane Technology Applications*, 757. Wiley. https://www.wiley.com/en-us/Advanced+Membrane+Technology+and+Applications-p-9780471731672.

35 Othman, N.H., Kabay, N., and Guler, E. (2022). Principles of reverse electrodialysis and development of integrated-based system for power generation and water treatment: a review. *Rev. Chem. Eng.* 38 (8): 921–958. https://doi.org/10.1515/revce-2020-0070.

36 Wang, L., Wang, Z., Patel, S.K. et al. (2021). Nanopore-based power generation from salinity gradient: why it is not viable. *ACS Nano* 15 (3): 4093–4107. https://doi.org/10.1021/acsnano.0c08628.

37 Usman, H.S., Touati, K., and Rahaman, M.S. (2021). An economic evaluation of renewable energy-powered membrane distillation for desalination of brackish water. *Renew. Energy* 169: 1294–1304. https://doi.org/10.1016/j.renene.2021.01.087.

38 Almehmadi, F.A., Najib, A., Ali, E. et al. (2022). Sustainable approach of generating water and energy: techno-economic analysis of a hybrid solar photoactive thermal system coupled with direct contact membrane distillation for water purification and electricity generation. *Appl. Sci.* 12 (18): https://doi.org/10.3390/app12189137.

39 Sohrabi, A., Meratizaman, M., and Liu, S. (2024). Comparative analysis of integrating standalone renewable energy sources with brackish water reverse osmosis plants: technical and economic perspectives. *Desalination* 571: 117106. https://doi.org/10.1016/j.desal.2023.117106.

40 Arunachalam, M., Yoo, Y., Al-Ghamdi, A.S. et al. (2024). Integrating green hydrogen production with renewable energy-powered desalination: an analysis of CAPEX implications and operational strategies. *Int. J. Hydrogen Energy* 84: 344–355. https://doi.org/10.1016/j.ijhydene.2024.08.250.

41 Tashtoush, B., Alyahya, W., Al Ghadi, M. et al. (2023). Renewable energy integration in water desalination: state-of-the-art review and comparative analysis. *Appl. Energy* 352: 121950. https://doi.org/10.1016/j.apenergy.2023.121950.

42 Bento, C., Lopes, T.F., Rodrigues, P. et al. (2024). Biogas reforming as a sustainable solution for hydrogen production: comparative environmental metrics with steam-methane reforming and water electrolysis in the Portuguese context. *Int. J. Hydrogen Energy* 66: 661–675. https://doi.org/10.1016/j.ijhydene.2024.04.113.

14

Water Treatment by Microbial Fuel Cell System

Ilse Alejandra Ortiz Mora[1], Patricia Balderas Hernández[1], Gabriela Roa Morales[1], and Eduardo Campos Medina[2]

[1]*Facultad de Química UAEMéx, CCIQS UAEM-UNAM, Toluca, Mexico*
[2]*Facultad de Planeación Urbana y Regional UAEMéx, Colonia Universidad, Toluca, Mexico*

14.1 Introduction

A global problem of main interest nowadays is the increase in energy demand [1]. This is because the world still relies greatly on fossil fuels as one of the primary sources of energy generation, which intensifies greenhouse gases like carbon dioxide and generates air pollution while contributing to global warming. In addition, there is also the inconvenience of fossil fuels being a nonrenewable resource, which means that they are being depleted as they are being used [2].

Another worldwide challenge is wastewater treatment and disposal because it contributes to the water crisis and environmental pollution due to the high toxicity of the substances they contain (Vallero), leading to drinking water shortage as a direct consequence of pollution and demand, among other factors [3]. This situation requires the development of sustainable and energy-efficient wastewater treatments that can contribute to the decontamination and recuperation of bodies of water. An example of these types of strategies can be using microbial fuel cells (MFCs), which are considered an emerging technology that uses different organic fuel sources to generate electricity by oxidizing substrates using microorganisms while treating the pollutants within the fuel source, like wastewater [2].

Using wastewater as fuel in an MFC is possible because it contains organic substrates that can be degraded by some types of microorganisms, making it a good source of energy generation. Hence, the energy crisis and environmental water pollution can be addressed using an MFC.

Water Treatment and Desalination: Membrane Separation using Renewable Energy Resources, First Edition. Edited by Mohamed Bassyouni.

14.1.1 History and Development of Microbial Fuel Cells

The history of MFCs is composed of significant scientific discoveries and technological advancements through the years, beginning with the early observations of electricity generation by bacteria or other microorganisms, leading to the development of MFC systems that could be scalable and efficient for specific purposes. Nowadays, it is certain to say that the field has evolved considerably because MFCs hold promise for a variety of applications, including wastewater treatment, the production of renewable energy, and the possibility of environmental monitoring; these factors position MFCs as a key technology for a sustainable future. To fully understand the real impact that MFCs have had over the years, this chapter provides a quick overview of the history and development of MFCs.

14.1.1.1 The First Observation

In the year 1911, a researcher named M. C. Potter, who was a professor of botany at the University of Durham in England, made the first recorded observation of electricity generation by bacteria. With his experiments, he demonstrated that microorganisms could produce a fair amount of electricity while breaking down organic compounds within the substrate, laying the groundwork for the concept of MFCs and opening the field to future applications [4].

14.1.1.2 Mid-Twentieth Century

In the 1960s, the initial developments of MFCs occurred when the researcher Bernard Cohen used Potter's work as a basis to develop the first MFC that was able to generate a measurable amount of electricity. Cohen's research focused on the biochemistry involved in the electron transfer mechanism from bacteria to the electrodes in the chambers [5].

14.1.1.3 Understanding Microbial Fuel Cells

During the 1970s, two scientists named Harold L. Cook and Bernard J. Baker conducted further experiments into MFCs; they improved the understanding of the electron transfer processes and the types of bacteria that could be used in MFCs due to their ability to produce electricity [6].

14.1.1.4 Technological Progress During the Late Twentieth Century

In the 1980s, the research focused more on improving the materials used to build MFCs instead of focusing on the microorganisms within the cells; several materials were studied for their properties and capability of usage as electrodes and membranes, or to optimize the cell and enhance the efficiency and durability of MFCs. Later, during the same period, the focus changed to the microorganisms within the substrates used in MFCs, discovering the electrogenic bacteria, which could transfer electrons directly to an electrode; this marked a significant milestone for the field [7].

14.1.1.5 Expanding Applications

During the 1990s, an idea emerged about using MFCs for wastewater treatment. This was a groundbreaking approach as it combined the dual benefits of waste degradation and electricity generation. In this decade also, there was increased interest in the formation of the biofilm on electrodes by the electroactive bacteria and its impact on the MFC performance [6].

14.1.1.6 Early Twenty-First Century and the Rapid Advancements in the Applications of Microbial Fuel Cells

During the 2000s, the research and development of MFCs accelerated, translated into the researchers producing numerous studies about different types of configurations and applications in various fields. Some key developments included the miniaturized MFCs for applications as portable power sources and biosensors; the exploration of different substrates and the impact they had on the MFC's efficiency, substrates such as wastewater, organic waste, and synthetic compounds; the integration of MFCs as hybrid systems with other renewable energy technologies, such as solar cells, to enhance overall energy production [7].

14.1.1.7 Microbial Fuel Cell's Scalability as a New Concept

In the year 2004, a group of researchers led by the scientist Bruce E. Logan at Penn State University made very significant contributions to the field by demonstrating the potential for scaling up MFCs for practical applications. Their work highlighted the feasibility of using MFCs for wastewater treatment and bioenergy production on a large scale [8].

14.1.1.8 More Recent Developments

During the 2010s, some technological innovations in MFCs were made, contributing to making them of commercial interest. Between these technological innovations, advances in materials science made possible the development of more efficient and durable electrodes and membranes for MFCs [6]. Afterwards, a powerful idea of combining MFC technology with desalination systems introduced the concept of microbial desalination cells (MDCs), to produce clean water and electricity simultaneously. Thanks to these technologies and developments, the commercial interest in MFCs grew, having numerous startups and companies exploring the technology for applications in remote sensing, environmental monitoring, and portable power generation [9].

14.1.1.9 Optimization and Sustainability of Microbial Fuel Cells as Main Interest

Since 2020, the research on MFCs has focused on optimizing the general performance of the cell while also reducing costs and improving the sustainability of this technology. Some optimization strategies include the development of novel materials, such as nanostructured electrodes and advanced catalysts. By doing so, the idea aims to enhance electron transfer efficiency and increase power output. Another strategy is to integrate the MFCs into smart grid systems and renewable energy infrastructures to enhance energy efficiency and sustainability [10, 11].

14.2 Wastewater

Wastewater can be defined as any water that has been adversely affected in quality by some kind of anthropogenic influence. It is referred to as the water that has been discharged after being used in homes, businesses, industries, and other sources. This kind of water contains a wide range of contaminants, including physical, chemical, and biological pollutants, and requires treatment before it can be safely released into the environment or reused in some activities [12]. Depending on the source, wastewater can be classified as follows.

14.2.1 Domestic Wastewater

Domestic wastewater or sewage is the water that is discharged from households and residential areas; this water contains human waste, like feces and urine, and food scraps, detergents, or other household chemicals. This type of wastewater is typically high in organic matter; it also has nutrients such as nitrogen or phosphorus and has a lot of pathogens [13].

14.2.2 Industrial Wastewater

Industrial wastewater is produced, as the name suggests, from industrial activities and manufacturing processes. This type of wastewater contains a variety of pollutants depending on the industry, such as heavy metals, toxic chemicals, organic compounds, and suspended solids. Industrial wastewater can have an extremely variable composition and may require specialized treatment methods to remove specific contaminants. Wastewater of industrial origin or industrial effluents also contain contaminants and byproducts such as chemicals, fats, oils, organic matter, pathogenic microorganisms, toxic chemicals, and heavy metals [14] that make the water unsuitable for direct release into the environment or for use, thus requiring specific treatment [15, 16].

14.2.3 Municipal Wastewater

Municipal wastewater is a mixture of domestic sewage and industrial wastewater that is collected by municipal sewer systems. This type of wastewater contains pollutants from both residential and industrial sources, which typically need treatment at wastewater treatment plants that are designed to handle a broad spectrum of contaminants, either chemical or biological ones [16, 17].

14.2.4 Agricultural Wastewater

The wastewater that comes from agricultural activities, including runoff from fields and livestock operations, is called agricultural wastewater, and it contains fertilizers, pesticides, animal waste, and soil sediments. This type of wastewater has a high amount of nutrients like nitrogen and phosphorus; it also contains organic matter, potentially harmful chemicals, and microorganisms that may have pathogenic characteristics [18].

14.2.5 Storm Wastewater

Stormwater can also be considered wastewater because it comes from the runoff from rainwater or snowmelt that flows over urban and rural areas. It becomes a type of wastewater because, as it passes through, it can pick up pollutants, such as oil, grease, heavy metals, and debris from streets, lawns, and other surfaces. This type of wastewater often contains suspended solids and various pollutants, depending on the land use and weather conditions [19].

14.2.6 Combined Wastewater

Combined wastewater is a mixture of domestic wastewater, industrial wastewater, and stormwater, which contains a combination of the pollutants found in these different sources. This type of wastewater is particularly difficult to treat as it requires complex treatment processes to handle the diverse range of pollutants [12].

Figure 14.1 represents every type of wastewater that was previously described above. Each type of wastewater possesses unique challenges for its treatment and disposal or reuse, needing tailored approaches to effectively manage and mitigate the environmental impact that they could generate. To effectively treat wastewater, a set of processes and specific equipment is used to purify and condition the water before releasing it again into the environment or before reusing it. Wastewater treatment systems are fundamental for industries as they ensure that the water meets quality and safety standards for use in manufacturing processes, cooling, boilers, and washing, among others [15].

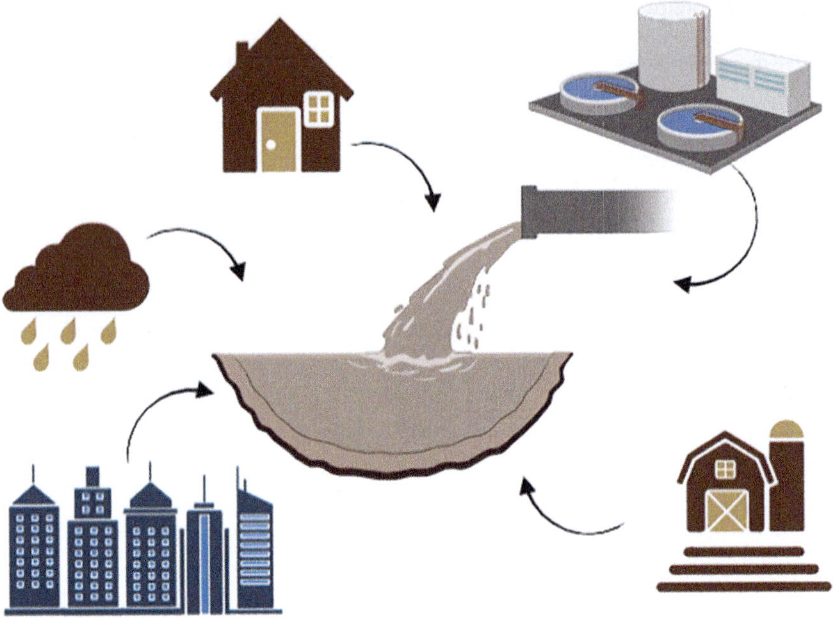

Figure 14.1 Types of wastewater. Starting from the top, clockwise: Industrial wastewater, agricultural wastewater, municipal wastewater, storm wastewater, and domestic wastewater; all of them may form combined wastewater. Source: Created by the author with BioRender.com.

There are different types of treatments for wastewater, and they depend totally on the characteristics of the incoming water as well as the requirements for specific industrial operations, so the choice of technologies and processes will vary according to the industry and treatment objectives [20]. For example, the disinfection treatment in wastewater eliminates pathogens within it through processes like ozonation, chlorination, or ultraviolet (UV) irradiation [14]; another example is using electrocoagulation treatments, which helps when the wastewater has metals, oils, dyes, and grease, by using electrical power to treat the water and separate the contaminants from it [15].

Diverse solutions for water pollution exist, and all of the actions carried out in each of them have the primary objective of removing both organic and inorganic analytes dissolved in the aqueous phase. Nevertheless, searching for new alternatives to solving water pollution is vital for this nonrenewable resource, which leads to the development of the integration of techniques in energy generation with those already known in pollutant reduction. Several efforts have been made, for example, to unify methods that generate electrical energy, vital for the operation of equipment, with advanced oxidation treatment methods or other decontamination strategies [21]. So here is where the MFCs enter the game, as they are an excellent proposal to remove dissolved contaminants in the aqueous phase.

14.3 Microorganisms

The smallest life forms that exist on our planet are called microorganisms; these extremely bast beings are extraordinarily small, their size can vary between 100 and 200 μm, they can be one-cell organisms or multi-cell microorganisms, and some of them even possess the ability to form colonies. An important feature of microorganisms is that they have a simple structure, which makes them adaptable to basically every environment; this means that they are present in all ecosystems and even wins them the category of the most abundant type of life in our entire planet [22].

Due to the variability and variety that all of them possess, microorganisms, also known as microbes, are distributed among several kingdoms at the same time: on the one hand, some microorganisms are part of the Monera Kingdom, and here is where all types of bacteria are found; then we have the Protista Kingdom, where we find protozoans; and microorganisms are even found in the Fungi Kingdom, where we can find them as fungus or mold in their microscopic form [23]. In Figure 14.2, a general example of the different types of microorganisms from each kingdom is schematized.

It is true that some microbes can be harmful to human beings. These types of microorganisms are named pathogens and may be extremely dangerous, but most

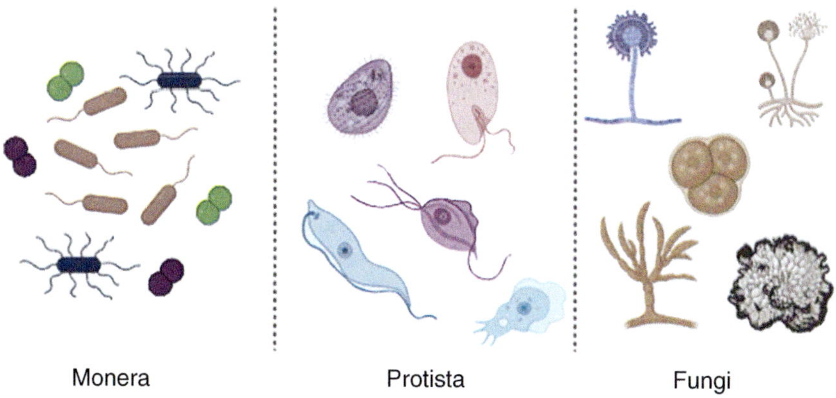

Figure 14.2 Microorganisms, a general view. In the Monera Kingdom, bacteria can be found; as part of the Protista Kingdom, protozoans are present; mold, fungus, and yeast are part of the Fungi Kingdom. Source: Created by the author with BioRender.com.

of them are also indispensable for life maintenance and can be widely used for industrial applications due to their unique enzymes or metabolic pathways that can be applied in different scenarios to obtain selective products or specific benefits [24]. For example, nowadays, half of the known enzymes used for commercial applications come from fungi and yeasts, while enzymes from bacteria are the second most used, followed by enzymes obtained from plants; the enzymes that come from any type of microorganisms are preferable to enzymes that come from animals because they have better stability and activeness, making them a better and more suitable option for industrial applications, thereby increasing the use of microorganisms and their enzymes in the production of high-value products; this is important as the microbe market continues to grow at a rapid pace [23].

While working with microbes, irrespective of the field, it is important to ensure a controlled environment for their optimal growth; this means having the proper temperature, pH, nutrients, and light, among other factors, and understanding that all microbes have specific characteristics and requirements, such as whether the absence of oxygen is needed for optimal growth or not; sometimes, this controlled environment may also need specific instruments to carry out the procedures and ensure that the microorganisms will work as we want them to [22].

Nowadays, microorganisms have been studied by researchers to harvest energy in a more sustainable and affordable way; some possible applications of microbes for energy are ethanol or methanol production, solar cells, energy storage, biodiesel, fuel cells, and electrosynthesis, among others. Using microbes for diverse purposes is always an eco-friendly alternative because they can consume organic wastes with a high concentration of multiple nutrients like sugars, lipids, or proteins, without needing purification methods, which is important because this way it is possible to use more cost-effective treatment methods [25, 26].

Many future applications of this microbe technology have been hypothesized, for example, the installation of microbial energy plants in all major cities worldwide to try to minimize the power demand of these places with a nonhazardous alternative that does not contribute to pollution or greenhouse gasses [27]. In order to make these types of applications a reality, it is necessary to acknowledge and solve some problems related to the use of microorganisms to generate energy. For example, while producing energy, these microorganisms also produce catabolic products that can end up affecting the efficiency of energy generation, making it necessary to remove these secondary products from the reactor when they reach a certain concentration [25].

There might be some concerns, but there is no doubt about the profound impact that using microorganisms in energy production may have in the near future.

This technology has a lot of potential because it is relatively cheaper, comes from renewable sources, and addresses waste management. Microbes can be part of fuel cells, they can produce value products like hydrogen, ethanol, and methane, among others, and they can certainly be a life-changing strategy as a resourceful opportunity for green alternatives of energy production [27].

14.4 Electrochemical Cells

To produce energy in a more sustainable way, several types of technology have emerged, for example, electrochemical cells. A direct-current electrochemical cell is formed by two electrical conductors called electrodes, and each of them is submerged in a suitable electrolyte solution. For current to flow in an electrochemical cell, it is necessary that the electrodes are externally connected by means of a metallic conductor and that the two electrolyte solutions are in contact to allow the movement of ions from one to the other; it is also important that an electron transfer reaction occurs at each of the electrodes [28].

In any electrochemical cell, both electrolyte solutions have to be connected by a salt bridge or a proton-exchange membrane (PEM), which isolates the contents of the two halves or the cell while maintaining an electrical contact between them. However, it is possible and sometimes even advantageous to construct cells in which the electrodes have a common electrolyte, thus eliminating the effect of junction potentials [29].

Electrochemical cells can be classified as Galvanic cells or as electrolytic cells. The main difference between them is that a Galvanic cell produces electrical energy, while an electrolytic cell consumes electrical energy. It is important to bear in mind that, in an electrochemical cell, the cathode is the electrode where reduction takes place, and the anode is the electrode where oxidation takes place [30].

In most electroanalytical methods, either the current in an electrochemical cell is measured at a fixed potential or the potential of a cell is measured while the current intensity is fixed at some constant value or zero. During an experimental process, it is only possible to control the potential of the cell at a desired value and measure the resulting current intensity, or vice versa, because by taking control of one variable, it becomes impossible to independently control the other variable [30]. It is important to say that the potential of an electrochemical cell is directly related to the activities of the reactants and products of the reaction in the cell, and indirectly to their molar concentrations; the performance of every electroanalytical method depends on the cell configuration, electrode material, configuration, and the technique that will be implemented [31].

14.5 Biofuel Cells

A biofuel cell (BFC) is a novel device that is able to use either microbes or microbial enzymes available in nature as catalysts to convert chemical energy into electrical energy [32]. As previously mentioned, the first time that a scientist was able to produce electric current from bacteria was in 1911 when a botanist named M. C. Potter used yeast and *Escherichia coli* in a rustic BFC, discovering that some microorganisms have electrogenic properties [4].

Some advantages of this type of fuel cell are that it is an environmentally friendly alternative because carbon dioxide and water are the main reaction products, and it produces zero-toxic pollutants. Another advantage of BFCs is that they are relatively easy to maintain as they only need to take care of the microbiome needs such as appropriate temperatures, atmospheric pressure, and a neutral pH medium to allow microbes to grow in optimal conditions; therefore, by attending to these factors the BFC can work properly. This alternative has the potential to be used inside the human body because it has good biocompatibility and can aid, in medical applications, for example, as an easy electricity supply for pacemakers or microsensors [33]; another advantage is that BFCs may work with several fuel types such as glucose, alcohols, and starch, and there are a vast amount of microorganisms and enzymes in nature that can be used to supply abundant catalytic materials for the BFC construction and performance [34].

If a BFC utilizes enzymes, it is renamed an enzyme biofuel cell (EBFC), but if it utilizes microorganisms, then it is called MFC. In EBFCs, the biofuel that is used can be a carbohydrate, an alcohol, or any other substance that exists in nature. When the oxidation reaction occurs under the catalysis of a certain enzyme, the electrons reach the cathode through the external circuit and the cathodic oxidant receives these electrons to produce current [32].

14.6 Yeast Fuel Cells

A yeast fuel cell (YFC) is a biological and electrochemical device that converts chemical energy from organic compounds into electrical energy using yeast, which is a type of fungi, as the biocatalyst. Yeast species are eukaryotic microorganisms that can be employed to break down organic substrates and produce electrons through their metabolic processes, and therefore they can be utilized in BFCs [35].

The process inside a YFC to produce electricity while breaking down the organic matter within the substrate begins when the yeast cells start metabolizing the organic substances in the anode chamber, releasing electrons; then these electrons travel through an external circuit from the anode to the cathode, generating an electric current that can be harnessed for power generation. The protons produced during the substrate oxidation move through the membrane between the cells to the cathode chamber and lastly, at the cathode chamber, these protons combine with the electrons and an electron acceptor, which is typically oxygen, to complete the reaction and produce energy, water, or other harmless byproducts [36].

Some advantages of this type of green technology are that these fuel cells need yeast, which is a widely available and inexpensive microorganism, making it a cost-effective biocatalyst species. This microorganism can also be easily cultured and maintained, requiring relatively simple growth conditions compared to other microorganisms and can metabolize a variety of organic substrates, including different types of sugars, alcohols, and organic acids. Nevertheless, it is important to keep in mind that the power output of an YFC is generally lower when compared to traditional fuel cells and other types of MFCs, so it can be challenging to maintain stable and efficient electron transference over long periods of time [37].

14.7 Microbial Fuel Cells

An MFC is a type of Galvanic cell that uses electroactive microorganisms, which are capable of generating electricity directly from organic compounds [38]. This technology is quite useful, and it is becoming of great interest nowadays because it aligns with some of the principles of green chemistry, aimed at reducing or even eliminating the use and generation of substances hazardous for the environment and human health [39] and also aligning with the UN Sustainable Development Goals (SDGs) that promote innovation and the adoption of practices that minimize the environmental impact, thus contributing to a more sustainable future [40].

The working principle of any MFC, represented in Figure 14.3, starts with oxidation at the anode, where the organic matter in this compartment is oxidized by the microorganisms present within. This process generates electrons, protons, and carbon dioxide. The electrons produced in the anodic chamber are transferred to the anode surface and travel through the external circuit to the cathode, generating an electric current. Then, reduction occurs at the cathode, where the electrons from the external circuit and the protons that have migrated through the PEM combine with an electron acceptor, which is usually oxygen, to form water [11].

In general, MFCs can treat wastewater by removing organic contaminants while simultaneously generating electricity. They also offer a sustainable method to generate electricity from organic waste materials and can be used to monitor the quality of water by detecting the presence of specific contaminants based on changes in the general electricity production [11, 38].

Figure 14.3 General structure of a double-chamber microbial fuel cell. The anodic chamber, in color green, mostly has anaerobic conditions, while the cationic chamber, in color blue, mostly has aerobic conditions. These chambers are connected by a proton-exchange membrane. Created by the author.

Some advantages of MFCs, compared to other green technologies, are that they can and have to use biodegradable organic matter to produce electricity, making them a renewable energy source. Also, they provide both wastewater treatment and energy generation and require a relatively simple setup when compared to other biological and electrochemical systems. However, there are some challenges that MFCs need to overcome: their current efficiencies are relatively low, and significant research is ongoing to improve power output. There is also the issue of scaling up these cells from laboratory-based systems to industrial applications because it is difficult, and some materials, especially for the cathode and membranes, can be expensive. It is necessary to mention that the electricity generation in these systems is directly influenced by factors such as the characteristics of the substrates present in the anodic chamber of the MFC [2].

Several classification strategies can be applied to MFCs, but it is possible to divide them easily into one of the following three types, according to their principle of electricity generation: hydrogen MFC, photoautotrophic MFC, and heterotrophic MFC [41].

In hydrogen MFCs, the cell can produce hydrogen as well as power, having microorganisms to produce the hydrogen directly from organic matter, and, at the same time, the hydrogen is oxidized by electrodes dispersed with chemical catalysts to generate electricity [41, 42]. In photoautotrophic MFCs, it is possible to convert optical energy into electricity by photosynthesis, and this occurs thanks to Cyanobacterium or other microorganisms with photosensitive qualities [43]. Finally, in heterotrophic MFCs, either anaerobic or facultative microorganisms are utilized to produce electricity by extracting electrons from organic dyes or other substrates, then these electrons are transferred to the electrodes, and finally the power is generated [44].

14.7.1 Exoelectrogenic Bacteria

Electrogenic or electroactive microorganisms are very small live forms with the capacity for generating electricity and can certainly be used in MFCs thanks to this particular property. These include a wide range of organisms like bacteria, archaea, and fungi. Bacteria found in MFCs are renowned as exoelectrogens due to their specific ability to transfer electrons outside the cell to the anode of the MFC. How capable a MFC is, in terms of performance and power capability, depends greatly on the kinetics of the electron transfer mechanism [2]. A generic view of the electron transport process in exoelectrogenic bacteria can be visualized in Figure 14.4.

Different electron transfer mechanisms exist for MFCs, and they can be divided into two main categories: direct electron transfer (DET) mechanism and mediated electron transfer (MET) mechanism. The DET mechanism takes place directly via membrane-bound cytochromes or electrically conductive nanowires [45]. On the other hand, in the MET mechanism, the electron transference happens by redox mediators or by the oxidation of reduced secondary metabolites, for example, flavins or phenazine [46].

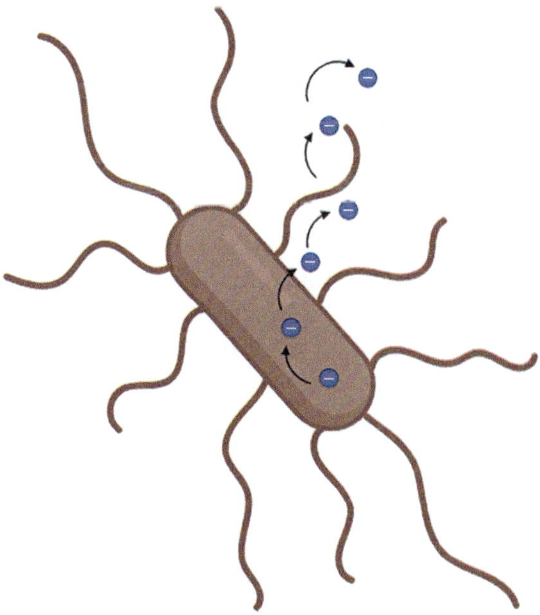

Figure 14.4 Generic view of the electron transport process in exoelectrogenic bacteria: electrons move from within the cell to the outer membrane and move through the pili to exit the bacteria, and then these electrons may arrive at a final acceptor. Source: Created by the author with BioRender.com.

The reduction of Fe(III) by electroactive microbes is a very important process in the anaerobic redox cycling of iron, which also includes the degradation of natural or contaminant organic matter present in diverse environments, such as freshwater, sediments, aquifers, or even in wastewater [38].

It is necessary to reiterate that exoelectrogens play a very important role in electron transfer mechanisms from the cell to the electrode, so the question is whether or not an exoelectrogenic community or exoelectrogenic isolates are better for the process [47]. Several Gram-negative bacterial species have been studied for their ability to couple anaerobic growth with the respiratory reduction of mineral oxides like Fe(III) and Mn(IV) [48]. In that regard, different pure isolates have been tested and reported as powerful exoelectrogens, including microorganisms like *Geobacter* sp. or *Shewanella* sp., with *Shewanella oneidensis* being an exoelectrogen standard nowadays [47].

The Gammaproteobacteria named *S. oneidensis MR-1* is a type of bacteria that has versatile metabolism pathways, which allows it to reduce insoluble metals, such as Fe(III) [49]. The phenotype for the reduction of Fe(III) present in most Shewanella species implicates decaheme c-type cytochrome interactions, which span the inner membrane, the periplasmic space, and the outer membrane of the cell [48, 50]. These species are also able to secrete soluble flavins that help facilitate electron transfer to metals and electrodes [50, 51].

There are other isolates reported as exoelectrogens, such as *Pseudomonas sp., Burkholderia sp., E. coli, Ochrobactrum anthropi, Rhodoferax ferrireducens, Citrobacter sp.,* and *Clostridium sticklandii.* However, further investigation is needed to fully understand the exoelectrogenic properties and the optimal growth conditions of these species so that they can be applied in power generation technologies, such as MFCs [38].

Instead of using isolates, some researchers rather focus on having not one but several exoelectrogenic species inside the anodic chamber of MFCs. By doing so, they believe that it would be better to have an exoelectrogenic microbial community to oxidize substrates and transfer the electrons generated to the anode electrode so that it can produce current in a more efficient way, thereby confirming that exoelectrogenic bacteria integrated with MFCs play a critical role in current generation. So, it is equally important to develop a microbial community with a sufficient and stable number of microbes in the anodic chamber [11, 52].

In order to optimize the MFCs, different methods have been studied, such as the application of a low anode potential, the variation of the external resistance, the use of selective enrichment media that may promote exoelectrogens to grow instead of other nonelectroactive species, or the iron enrichment methods that select microbes capable of reducing Fe(III) [47]. Having said that, it is necessary to fully understand that the selection pressures applied to the microorganisms within MFCs may not be strong enough to alter the microbial community so that the exoelectrogens are the predominant species, leading to a microbial composition mostly made of fermentative microbes which divert resources toward metabolite production instead of current generation and directly affect the MFC's conditions [52].

It has also been noted that the source of the inoculum for an MFC is important because its microbial community will be either partially or greatly responsible for the MFC's overall performance. Some sources where exoelectrogens can be found are natural, like seawater, and some can be created, like pure and mixed cultures isolated in laboratory conditions, but most MFCs use anaerobic sediments, activated sludge, or wastewater in the anodic chamber to function, which makes them an alternative source of clean energy while helping in the treatment of polluted water [47, 52].

14.7.2 Electron Transport Chain

Some bacteria do not always need oxygen to survive, which gives them the name of facultative anaerobes because, when oxygen is not available, they can undergo other metabolic processes that allow them to live under specific circumstances. For example, when oxygen is not available or there is not enough, the respiratory electron flow typically occurs at the inner membrane of microorganisms such as exoelectrogens, where the processes of oxidation and reduction can be linked to intracellular electron carriers, allowing them to generate a membrane potential [51, 53].

The electron transport chain (ETC) is part of cellular respiration and photosynthesis, which can be used as an alternative way by which electrons move to the outer membrane of the cell. This electron transfer process is essential for the generation of energy in the form of adenosine triphosphate (ATP), which is the energy currency of cells. This mechanism consists of a series of electron carriers and cytochromes that are responsible for making it work [54]. By making electrons available, this mechanism can be used for current production in MFCs.

Figure 14.5 depicts the working principle of ETC inside microorganisms, like exoelectrogens. During this transport mechanism, the cytochrome C reductase complex, also known as Complex III, releases the electron which is caught by a c-type cytochrome, reducing the Fe(III) located in the center of the cytochrome to Fe(II) momentarily; the c-type cytochrome then transports the electron and releases it to the cytochrome C oxidase complex, also named Complex IV, making the Fe(II) inside the cytochrome return to Fe(III) once again. The final step involves releasing the electron, again, from the fourth complex to the outside of the cell so that it can be received by the final electron acceptor, which can be the anode electrode in an MFC [49, 51].

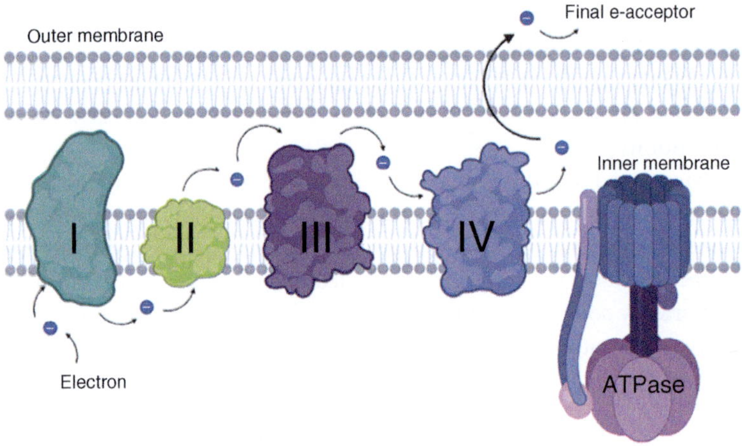

Figure 14.5 Electron transport chain: In exoelectrogenic bacteria, the electron goes from Complex I to Complex II, then passes from Complex II to Complex III (also known as Cyt C reductase); in order for the electron to pass to the next complex, a c-type cytochrome catches the electron and delivers it to Complex IV (also named Cyt C oxidase); finally, the electron goes from Complex IV to the outer membrane of the cell and arrives to the final electron acceptor. Source: Created by the author with BioRender.com.

14.7.3 Cytochromes

In the previous section, we have been referring a lot to cytochromes, so we might as well describe them to fully understand why they have an extremely important role in the ETC mechanism. Cytochromes are a class of heme proteins that play a crucial role in ETCs in several living organisms. They function as electron carriers, facilitating the transference of electrons between different molecules, in various biochemical pathways. This characteristic makes them essential components of the cellular machinery that drives the production of energy through electron transport processes because, by transferring electrons, cytochromes help generate a proton gradient across membranes, which is used to produce energy in the form of ATP, so the cell can survive [55]. In human beings, cytochromes are also involved in detoxification reactions in the liver [56].

Cytochromes contain at least one heme prosthetic group, which is a complex molecule with an iron atom at its center. The iron atom can alternate between different oxidation states (Fe^{2+} and Fe^{3+}, as previously described), allowing it to accept and donate electrons [55].

These molecules can be classified based on their heme group structure and the absorption spectra of their reduced forms. They come in various types, each with a specific role in different biological systems [57]. For example, a-type cytochromes can be found in living organisms that possess mitochondrial ETCs.

They function in the fourth complex of the ETC and contain "heme a" groups. Like a-type cytochromes, b-type cytochromes can also be found in the mitochondrial ETC, but they function in the second and the third complexes of the process. On the other hand, d-type cytochromes contain "heme d" groups and are found in some bacteria, while f-type cytochromes, with "heme f" groups, are found in the chloroplasts of plants and in Cyanobacteria, functioning in the ETC of photosynthesis [58].

Another type of cytochrome, found in either mitochondrial or bacterial ETCs, is the c-type cytochrome. This is a small, water-soluble globular heme protein that contains ten heme C groups and functions as a mobile electron carrier between the third and the fourth complexes in the ETC. This type of heme group has a small molecular weight of 12,284 Da and has an isoelectric point of 10, having a positive charge at a pH of 7. In its structure, it has four pyrrolic groups that give a nitrogen which coordinates to the Fe in a covalent bond [55]. Figure 14.6 presents the general molecular structure of a heme C group. Most exoelectrogenic bacteria have this type of cytochrome, making them extremely important for MFCs.

Two of the most studied c-type cytochromes worldwide are the cytochrome MtrC and the cytochrome OmcA, both of which are isolated from the exoelectrogenic microorganism *S. oneidensis*, previously described as a standard between exoelectrogenic species. These cytochromes are located on the extracellular face of the outer membrane of the living organism, where they can be directly involved in the electron transfer to the extracellular acceptors [18]. Both MtrC and OmcA cytochromes have 10 putative heme-binding sites and a lipoprotein consensus sequence, and they have also been previously demonstrated to play an important

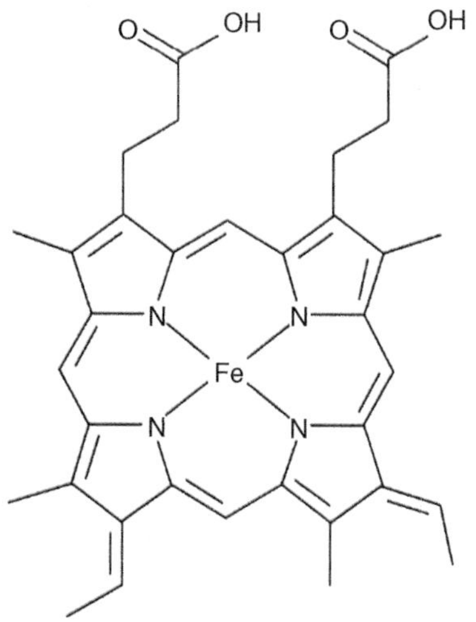

Figure 14.6 The molecular structure of a heme C group from a c-type cytochrome present in the bacterial electron transport chain. Created by the author.

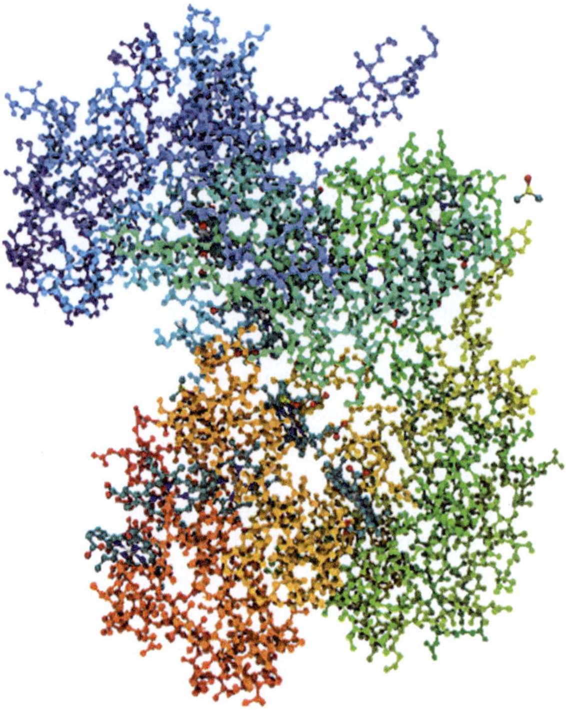

Figure 14.7 Crystal structure of the outer membrane decaheme cytochrome OmcA RCSB PDB-4LMH (Bank, RPD, n.d.). Source: Created by the author with BioRender.com.

role in the reduction of Fe(III) and Mn(III/IV) [59, 60]. The crystallized structure of the cytochrome C OmcA is shown in Figure 14.7, which was extracted from the Protein Data Bank under the code 4LMH.

14.8 Case Studies: Exoelectrogenic Bacteria in Microbial Fuel Cells

Since the beginning of MFC technology, the microorganisms within the substrate have been topics of interest due to their importance in the overall performance and optimization of any MFC. Nowadays, several research studies regarding exoelectrogens have been made and here we mention some examples:

14.8.1 Microbial Communities in Microbial Fuel Cells

Zhao et al. [61] conducted a study to evaluate the impact of electricity generation by the microbial communities present in an MFC. Their MFC was powered by macroalgal biomass in the anodic chamber, and part of their research included the sequencing of the 16S rRNA gene of the microorganisms within this chamber. This research group found out that the anodic biofilm in the MFC had a very complex composition, mainly of *Bacteroidetes*, *Firmicutes*, and *Proteobacteria*, followed by *Euryarchaeota*, *Deferribacteres*, *Spirochaetes*, *Chloroflexi*, and *Actinobacteria*.

14.8.2 Onion Residues and Microbial Fuel Cells

Another study conducted by Segundo et al. [62] showed that it was possible to generate a fair amount of electricity from onion residues in an MFC. With this strategy, they reported maximum peaks of 4.46 ± 0.06 mA of current and 0.99 ± 0.02 V of voltage. As an addition to their research, they also identified the microorganisms within the anodic chamber at a molecular level and found the presence of microbes like *Pseudomonas aeruginosa, Acinetobacter bereziniae, Stenotrophomonas maltophilia,* and *Yarrowia lipolytica* at the anode electrode biofilm.

14.8.3 Biofilm Analysis of Sewage Wastewater Microbial Fuel Cells

Bose et al. [63], in their evaluation of the generation of bioelectricity and the analysis of the biofilm on an MFC which was powered by sewage wastewater, identified six species of exoelectrogenic bacteria that were capable of degrading the organic matter present in the sewage wastewater. Within these species, the researchers found that the microorganism *Achromobacter xylosoxidans*, which is a form of Betaproteobacteria, could generate a good amount of bioelectricity while also decomposing very high concentrations of chromium and organic acids. This is extremely important because most wastewater types have large amounts of pollutants that may not be easy to degrade, especially heavy metals.

14.8.4 Electrical Power with Isolated Electrogenic Bacteria in Microbial Fuel Cells

A novel study by De la Cruz-Noriega et al. [2] showed how electrical power could be generated by double-chamber MFCs at a laboratory scale. These researchers used wastewater in the anodic chamber and, unlike other studies, they added isolated electrogenic bacteria to the anodic chamber; the isolates were of the species *S. maltophilia, A. bereziniae,* and *A. xylosoxidans.*

To evaluate the MFCs, they had three systems: the first one was the MFC-1, which included the isolated microorganism *S. maltophilia*; the second one was the MFC-2, with the isolated microorganism *A. bereziniae*; and the third one was the MFC-3, including the isolated microorganism *A. xylosoxidans*. In their results, they described their initial values of biochemical oxygen demand (BOD), being 640 mg/L for all systems. Then, after the MFCs worked for 30 days, the BOD values decreased to 5 mg/L in the MFC-1 system and to 3 mg/L in the MFC-2 and MFC-3 systems. Furthermore, this research team acknowledges MFC-3 as the biological electrochemical system that was able to generate the highest voltage, reporting a value of 1.01 ± 0.06 V. By evaluating the highest current, they discovered that it was developed by the MFC-1 system, with a value of 0.71 ± 0.02 mA. In terms of maximum conductivity, it was achieved by the MFC-2 system, with a value of 81 ± 2 mS/cm. The researchers also explained that the operating pH of all three systems was slightly alkaline.

Finally, they described how it was possible to achieve a maximum voltage of 8.58 V by connecting a series system of nine double-chamber MFCs. This system produced enough energy to even manage to light up, for more than 30 days, a 6-W light-emitting diode bulb.

14.9 Design of a Microbial Fuel Cell

The design of an MFC must bear in mind that it needs to prevent the reaction between the fuel at the anode and the oxidizing agent at the cathode because this affects the transfer of electrons in the circuit [64], and its performance will be directly related to the surface area of the exchange membrane and the surface area of the electrodes [65].

It is also necessary to consider that the type of material of the chambers must have low electrical conductivity levels so it can allow the passage of electrons exclusively through the selected electrodes. Using glass as the chamber material, for example, is a good idea because it is a resistive material, which makes it suitable [66]. Glass is not the only material that can be used for MFCs; if another material is considered, it must be adaptable to the operating conditions of the cell, knowing that it will be subjected to temperatures above 100°C during the sterilization process and it must allow heat conduction through the chamber. Likewise, it needs to be suitable for the optimal growth temperature of the microorganisms without damaging their chemical, physical, and mechanical stability [66]. Finally, the selected material of the chambers should not cause the adhesion of microorganisms to avoid biofilm formation [47].

14.9.1 Anode and Cathode of a Microbial Fuel Cell

On an MFC, the anodic oxidation process is generally carried out under anaerobic conditions due to the exoelectrogenic characteristics and based on the redox potential for the optimal growth of bacterial strains [10], which have better survival conditions precisely in an anaerobic system. Microorganisms colonize the anode and metabolize organic substrates like wastewater or organic waste, releasing electrons and protons as byproducts [10, 67].

On the other hand, the cathode is where the reduction reaction takes place. Electrons travel from the anode to the cathode through an external circuit, and protons move internally to the cathode. At the cathode, these electrons and protons combine with an oxidizing agent, typically oxygen, to form water. Hence, in the cathodic compartment of an MFC, the main character is oxygen because it is the most suitable electron acceptor for an electrochemical cell. This is due to its high standard reduction potential, availability, zero cost, and sustainability, and also because it does not generate secondary waste since its final product is water. Furthermore, it has been shown that oxygen reduction on carbon cathodes can be assisted by a bacterial film to increase the reaction kinetics [68]; therefore, aerobic conditions are advised. It is also necessary to understand that the choice of the cathode material greatly affects the performance of the MFC and it varies depending on the application and final purpose of the cell [69].

14.9.2 Single-Chamber Microbial Fuel Cell

A single-cell or single-chamber MFC is a type of biological and electrochemical system that uses microorganisms to convert chemical energy found in organic substrates directly into electrical energy. As the name may imply, it is made of one cell where both the anode and the cathode are present, but they are separated by a PEM that allows protons to pass through but prevents the anode and the cathode from directly touching each other. The general structure of a single-chamber MFC is depicted in Figure 14.8. The PEM also prevents the direct mixing of the substrates and the oxidants inside the anode and the cathode, which helps maintain a potential difference. The single cell's external circuit connects the anode and the cathode, making it possible for the electrons to flow from the first electrode to the second one, thus generating an electric current that can be captured and used [30].

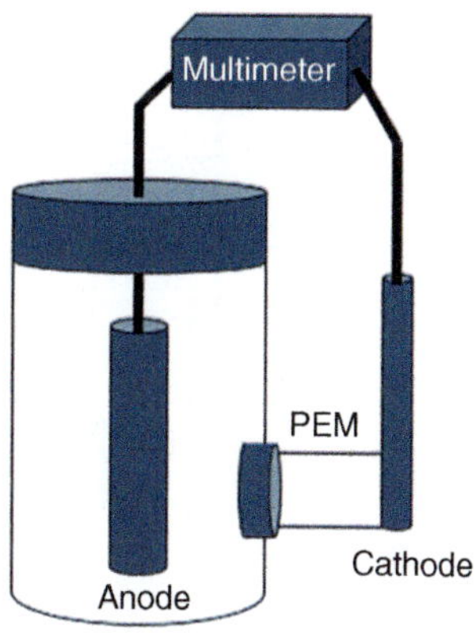

Figure 14.8 Single-chamber microbial fuel cell. This type of MFC has the anode inside the cell and the cathode is separated by a proton-exchange membrane. Created by the author.

A single-chamber MFC represents a promising technology in the field of renewable energy and environmental biotechnology as it combines the benefits of waste treatment and sustainable energy production [70].

14.9.3 Double-Chamber Microbial Fuel Cell

Another configuration of MFCs is the two-chamber MFC, also known as double-cell MFC or dual-chamber MFC, as depicted in Figure 14.9. This is a type of biological and electrochemical system that consists of two separate compartments: the first one for the anode and the second one for the cathode, which are separated by a PEM. The double-chamber MFC is reported to be the most appropriate configuration for the production of electrical energy because it allows the study of different factors that can affect the generation of electrical power, such as the influence of the electrode surface area, the contact area of the bridge or the proton exchange membrane with the chambers, and the compositions of different types of these ion-exchange membranes. It is important because, by having not only one but two chambers, it becomes easier to alter conditions on one of the compartments without altering the other one [66].

The external circuit of double-chamber MFCs connects the anode and the cathode, allowing electrons to flow from the first one to the second one and permitting them to generate an electric current, similar to single-chamber MFCs but with separate compartments. This arrangement typically includes a wire and a resistor or load to measure the current generated by the cell [30].

The double-cell MFC possesses some advantages compared to the single-cell MFC. For example, having separate anodic and cathodic chambers allows better control of the environmental conditions required in each compartment, which can

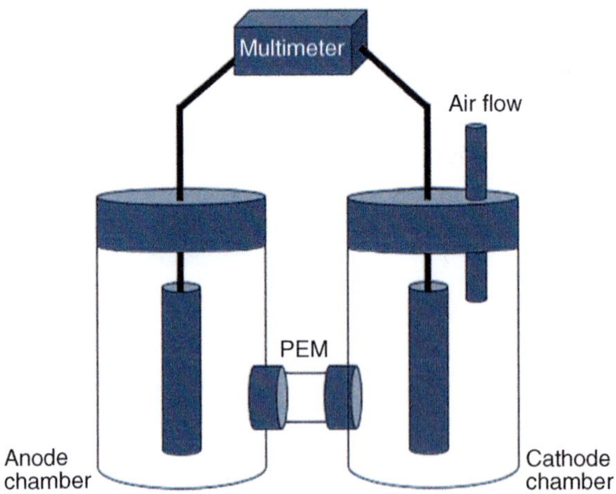

Figure 14.9 Double-chamber microbial fuel cell. This type of MFC has two cells separated by a proton-exchange membrane: the anodic chamber is always in anaerobic conditions and the cathodic chamber maintains aerobic conditions. Created by the author.

enhance the efficiency of the microbial processes and the overall performance of the MFC. Another advantage is that the separation of the chambers can reduce the mixing of the substrate and the electrolyte, present in the anodic and cathodic chambers, respectively, as well as the mixing of the electron acceptors, thereby increasing the efficiency of the electron transfer mechanism and power generation [71]. However, double-cell MFCs are more complex to construct and more difficult to operate when compared with single-cell MFCs, and they often have higher material costs as well as maintenance expenses. Scaling up double-cell MFCs from laboratory to industrial applications is possible but has significant challenges due to the need for large and durable PEMs and the complexity of managing two separate chambers. Another inconvenience is that the PEM can contribute to internal resistance, potentially limiting the overall efficiency of the MFC [72].

14.9.4 Proton-Exchange Membrane in Microbial Fuel Cells

Regardless of having a configuration of one-cell or a double-cell MFC, these systems require a membrane that allows only protons to move through both of the chambers (in a double-chamber MFC) or both sections of the chamber (in a single-cell MFC) and that prevents the direct mixing of substrates and oxidants (in both single- and double-chamber MFCs), which is essential to maintain a potential difference. The PEMs, which may be used in MFCs, can either be purchased or built. If purchased, common PEMs, such as Nafion, exist for this type of electrochemical cell. If built, there are many inexpensive ways to connect both the compartments of an MFC while also allowing the proton exchange to occur; for example, it is possible to use a glass tube bent into the shape of a U and to fill it with agar and salt. This is an economic ion-exchange membrane that might be useful in an MFC, but this type of built membrane may have the disadvantage of being a salt bridge that produces little energy due to its high internal resistance [66].

While designing the type of a PEM needed for the MFC, it is useful to understand that the greater the contact area of the exchange bridge or membrane with the chambers is, the better the MFC's performance will be. Regarding the length of the ion-exchange membrane or bridge, various studies have shown that the separation between the electrodes is not an influential parameter in the efficiency of any MFC [73].

14.9.5 Types of Substrates in Microbial Fuel Cells

In green technologies, the proper selection and management of substrates are critical factors for the efficiency of the system. In MFCs, optimizing factors such as the composition, the concentration, and the type of pretreatment of substrates, as well as considering the microbial compatibility, can lead to substantial improvements in electricity production and the overall efficiency of the MFC system [74].

Therefore, it is necessary to understand the different types of substrates that are available for MFCs and the impact they will have on the overall performance.

The first and most common substrate that can be implemented for MFCs is organic waste. This type of substrate includes food scraps, agricultural residues, and manure; it is rich in organic matter that can be decomposed by microorganisms and can be utilized by electroactive microorganisms to generate electricity. This waste is often abundant from diverse sources, and it is a very economical option, which makes it an attractive and cost-effective option for MFCs [74].

Using different types of wastewater, like domestic and industrial, is another viable alternative because all types of wastewater contain a variety of organic and inorganic compounds that can be metabolized by exoelectrogenic microorganisms in an MFC. In addition to generating electricity, treating wastewater by using it as a substrate for MFCs helps reduce the pollutant load in the water before releasing it into the environment [15].

Using synthetic organic compounds as substrates for MFCs can be another suitable option. The synthetic compounds that can be used are acetate, glucose, and lactate, as they are frequently reported in different literature due to their chemical composition, which is well-defined, and easy availability. These compounds allow us to have precise control over the experimental conditions in MFCs, making them a very good option to use in these bioelectricity systems [75, 76].

A novel strategy is implementing lignocellulosic materials as substrates in MFCs. The materials that can be used for this purpose are straw, wood residues, and bagasse because they contain either cellulose, hemicellulose, or lignin, and can be used as substrates after specific pretreatment methods that release the fermentable sugars within them [77].

Not only the selection of the substrate is important but also the preparation and processing of substrates so that they can be suitable for MFCs. Depending on the type of substrate, some may require pretreatment processes to increase their bioavailability and facilitate their degradation by microorganisms.

Some common pretreatment methods for substrates are divided into thermal, hydrothermal, and chemical treatments. These methods are used to break down complex structures or to use either enzymes or microorganisms to degrade certain components within the substrate. Another strategy is to ensure an optimal environment for the microorganisms by evaluating and controlling the substrate concentration and pH of the medium so that it can be adjusted when needed and can improve microbial activity as well as electron transference. It is even possible to dilute the substrate to prevent inhibition by high concentrations of organic compounds [41].

Another pretreatment method for substrates is solid removal and filtration, which is useful in the case where wastewater and solid organic waste are used as substrates because in some cases, the prefiltration or sedimentation is essential to remove large particles and suspended solids that may obstruct the MFC system [78].

The impact of substrates on MFC's performance depends directly on their composition. The chemical composition of the substrate directly influences MFC's efficiency because the substrates that contain high levels of easily biodegradable compounds, such as simple sugars and fatty acids, usually result in higher electricity production. On the other hand, when complex and recalcitrant substrates are used, they may require longer degradation times and may result in lower energy yields [41].

Not only the substrate concentration but also the composition is important. An optimal substrate concentration is crucial for the MFC's performance. If the substrate concentration is too low, then it may not provide enough organic matter to sustain microbial activity, but if the substrate concentration is too high, then it can cause inhibitory effects due to the accumulation of toxic byproducts [76].

The degradation kinetics are also important for MFCs because the rate of substrate degradation can affect the rate of electricity production. The substrates with the capacity for rapid degradation can provide a continuous flow of electrons, enhancing MFC's efficiency. However, if degradation is too rapid, there may be an accumulation of intermediate products that will inhibit microbial activity and ultimately affect the energy production of the cell [78].

Lastly, the microbial compatibility with the substrate is crucial; therefore, in the selection of the substrate, one should also consider how compatible it is with the microbial community present in the MFC. Some microorganisms may be more efficient at degrading certain substrates in comparison to others, so the appropriate selection of the substrate can significantly improve or diminish energy performance [76].

14.9.6 Electrodes

Electrodes in MFCs are essential for the energy production process and wastewater treatment in the cell, as the electrode is responsible for the correct electron transfer mechanism that will lead to electrical energy production, and it is also the site where the exoelectrogenic microorganisms will grow [79, 80].

In some studies, made at a laboratory scale, no direct relationship has been reported between the area or the volume dimensions of the electrodes and the generation of electrical energy, but it has been reported that having a larger electrode surface area makes a lower power density, optimizing MFC's general performance [64, 69].

Several types of materials have been studied to be used as electrodes in MFCs for some specific characteristics, such as electrical conductivity, corrosion resistance, chemical stability, large surface area, low cost, and bacterial colonization capacity [79, 80]. The type of material of the electrodes is very important for MFC's performance, so it must be considered that the roles of the anode and the cathode are different, which means that the electrodes may be of a different material in each chamber, if needed. For example, bacterial growth and colonization will occur at the anode so that electrons can be transported via an external circuit; then, the cathode receives such electrons and puts them in contact with oxygen and protons, creating an overpotential. Therefore, the anode and cathode electrode materials can be similar, can be the same, or can be different, but they are required to meet the characteristics of biocompatibility, corrosion resistance, high electrical conductivity, low cost, and good surface area so that they can be functional in any MFC, whether it is a single- or a double-cell type. Selecting the appropriate material is crucial because a high electrical conductivity of the electrode allows an effective electron transport mechanism and decreases the resistance of the cell [5].

Not only the material but also the type of surface area of the electrodes is important because the porosity and the pore size of the surface are both critical characteristics that can improve the interaction between the exoelectrogenic bacteria and the electrode inside the anodic chamber [5]. For this particular reason, carbon-based materials have been studied continuously, as their porous surfaces favor the process within MFCs. Some carbon-based materials that are widely used in this type of technology are carbon cloth, carbon brush, graphite plates, graphite paper, and granular carbon [68, 81]. Hence, carbon materials are commonly used as electrodes in MFCs and are highly effective for generating high power densities [80]. In Figure 14.10, some carbon materials used as anode electrodes for MFCs are shown.

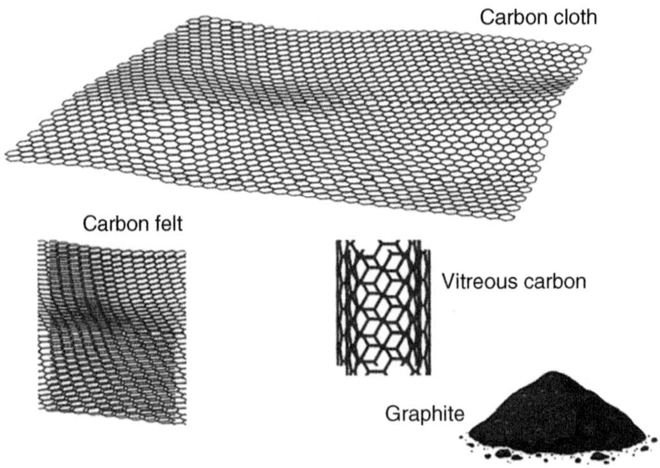

Figure 14.10 Commonly used carbon materials as anode electrodes for microbial fuel cells. Starting from the top of the figure: cloth, felt, vitreous carbon, and graphite. Source: Created by the author with BioRender.com.

As anode electrodes, the simplest materials are graphite plates, rods, and granules, as well as fibrous materials such as felt, cloth, paper, fibers, foam, and vitreous carbon [68]. The popularity of graphite electrodes relies on their low cost and ease of handling and that they have a high surface area, especially graphite felt is a great choice when the main goal is achieving electrodes with larger surface areas [64, 68]. Nowadays, several studies show that an increase in the current is directly proportional to the total surface area of carbon-based electrodes, with carbon felt being the best among all, followed by carbon foam and leaving pure graphite at the very end [68].

As an example, using reticulated vitreous carbon as electrodes can be a better option than using graphite felt because the surface areas that can be achieved are substantially higher due to the different pore sizes of this material. It is also possible to achieve higher surface areas through the use of packed layers of carbon granules or carbon beads instead of reticulated vitreous carbon [69].

Some metallic materials in electrodes can exhibit better conductive properties than some carbonaceous materials, so using this type of electrodes as anodes might also be a good option to have better performance in the MFC and a better effect on the electroactive bacterial growth [82, 83]. Choosing the appropriate type of metal is crucial. Metallic materials used in MFCs, such as stainless steel and titanium, are not a great option because they do not have good performance in energy production due to their smooth surface, which makes it difficult for the bacteria to colonize the material and it can even prevent any colonization at all. These materials can even prevent the electron transfer mechanisms. Having said that, metals do offer benefits such as improving the energy efficiency of the MFC, and hence they have been used in composite materials with graphite electrodes because the partial use of metals can contribute to better adhesion and can even reduce general costs in MFCs [80, 82].

An example of a metal that is commonly used in MFCs and that has been greatly studied is iron oxide (FeO). This metal stands out from the rest due to its promising electrical properties [80, 82], which are actually excellent electrical properties thanks to the electron transfer mechanism between Fe(II) and Fe(III), making it possible to achieve higher conductivity and to decrease the internal resistance of the biological and electrochemical system [80].

Therefore, electrical properties determine the electrode's ability to allow the passage of electrons, producing electrical current. The factors that determine how much the material opposes the passage of electrical energy are resistivity and conductivity. When comparing similar atomic carbon structures, such as amorphous carbon, graphite carbon, and diamond carbon, it has been shown that the lowest resistivity and the highest conductivity can be achieved by graphite carbon electrodes [21, 68]. However, using carbon nanotubes is a promising alternative because it can overcome the specific surface limitations of graphite, and their performance is 10 times higher than conventional carbon fibers [10].

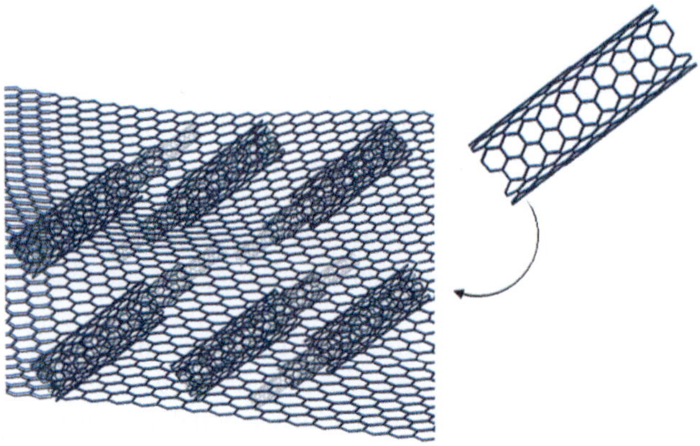

Figure 14.11 Carbon nanotube-textile (CNT-textile). This material is being studied as an electrode material due to its porous structure that allows the bacterial colonization of the anode as well as efficient electron transport. Source: Created by the author with BioRender.com.

Carbon nanotubes have been recently evaluated as electrode materials for MFCs, among other applications, due to their extraordinarily high conductivity, mechanical flexibility, and larger specific surface area [10]. For example, the research team of Xie et al. [84] described a highly conductive, biocompatible, dual-scale porous anode which is made from a carbon nanotube-textile (CNT-textile), a composite with the aim of obtaining a high-performance MFC. In this study, the authors reported that the macroscale porous structure of the interwoven textile fibers (general scheme visible in Figure 14.11) created enough space for efficient substrate transport and internal colonization by diverse microorganisms. The results of this CNT-textile model showed that it produced a maximum current density 157% higher, a maximum power density 68% higher, and energy recovery 141% higher than the traditional carbon cloth model, demonstrating how promising the material actually is.

14.9.7 Mediator Microbial Fuel Cells

MFCs use different mechanisms to transfer electrons from within the microorganisms to the anode electrode. When an MFC uses a synthetic compound to aid with the electron transfer process, it receives the name "mediator MFC" [85].

Mediator MFCs are useful when the bacteria in the anode are not able to perform the electron transfer mechanism efficiently. This mediator has the task of forming an extra redox pair when the electrochemical process occurs and it gets reduced first as it accepts an electron generated by the bacteria; then, the mediator delivers an electron at the surface of the anode, giving it to an electron acceptor and later oxidizing back to its original state so that the process may begin once again. At the same time, the electron acceptor also regenerates because it is oxidized as well by exposure to the oxygen that is present in the cathode because the cathodic chamber of the MFC must have aerobic conditions [65].

Some good mediators for MFCs include organometallic compounds as they can extract the electron from the bacteria and transfer it to the anode. The organometallic compounds commonly used as mediators in MFCs are ferrocene, derivatives of ferrocene, gallocyanin, and phenothiazines, among others [85, 86]. It is advantageous to use mediators in MFCs because they have good conductivity and low oxidation potential, causing less interference [65].

14.9.8 Un-Mediator or Mediator-Less Microbial Fuel Cells

Another mechanism that an MFC uses to transfer electrons from the microorganisms to the anode is the mediator-less mechanism, also known as un-mediator mechanism. In this type of MFC, the microorganisms can transfer the electron directly to the anode or can transfer it through the precursor that is self-synthesized [87].

When choosing mediator-less or un-mediator MFCs, it is advised to functionalize the anode's surface to enhance the efficiency and to have proper adhesion of the microorganisms while also increasing the surface area to improve the overall electron transfer process [5]. There are several methods and materials that can be used to functionalize the anode of the MFC, and the choice will depend greatly on the characteristics and finality of the MFC itself [88].

14.10 Microbial Fuel Cell's Areas of Improvement

MFCs are a promising technology that can be helpful in many circumstances in the near future; however, it is necessary to address the areas of improvement that this technology has to overcome.

One of the main issues of any MFC is that it may have a slow electron transfer mechanism, which impacts its efficiency due to the demining of the oxygen reduction reaction, which also enhances the overpotential of the process. Mediators such as phenazine or other chemicals can be used to improve the electron transfer rate and also the microbes themselves can naturally provide these compounds; in either way, these chemicals shuttle the electrons to the surface of the anode and improve the performance of the MFC [65]. Another possibility is to improve the catalytic properties of the cathode materials by doping them with nitrogen so that they can counter the slow step of oxygen reduction reaction at the cathode [89].

Some materials in MFCs are expensive, like the ones used for the cathode and anode electrodes, and even the PEM may be expensive when purchased instead of being manufactured. Hence, developing cheaper methods is crucial for the application of MFCs in daily life and daily industrial scenarios [2].

The power generation in these systems is influenced, among other factors, by the substrates present in the anodic section of the MFC [2]. This means that the type of waste that needs to be treated in the MFC will be important and will affect the system's performance directly. Whether it is wastewater, seawater, algae biomass, activated sludge, or other substrate source, the components and microorganisms present in this waste will have a major impact on the energy generation and the removal efficiency of pollutants. Therefore, it is necessary to evaluate the characteristics of the waste that the MFC will try to treat and to understand whether the substrate, which is organic matter, will be useful for the electroactive microorganisms [11].

In MFCs, current efficiencies may normally be low, making it necessary to improve power output. This is particularly troublesome because it makes it very difficult to scale up the MFCs from systems at a laboratory scale to industrial environments [2].

By acknowledging these areas of improvement and addressing each one of them, MFCs can soon be an optimal solution for energy production and water treatment when compared to other technologies.

14.11 Case Studies: Applications of Microbial Fuel Cells for Water Treatment

Several studies regarding water treatment and MFCs have been conducted in different fields. Some applications of these MFC technologies are as follows.

14.11.1 Microbial Fuel Cells for Drug Removal

As part of a study by Zhou et al. [90], they described the importance of treating wastewater that contains drugs used by humans and wrongfully discarded, such as antibiotics and their derivatives. This pharmaceutical waste can alter aquatic ecosystems if the wastewater is released into the environment without proper treatment. Therefore, this research team studied how MFCs could help with the biodegradation of sulfadimidine, which is an antibiotic, while also producing electrical energy. Their results showed that the MFC could have a degradation efficiency of 99% for this antibiotic.

14.11.2 Microbial Fuel Cells for Copper Removal

Wastewater from metallurgical industries is especially dangerous as it carries heavy metals and other pollutants that can damage ecosystems and threaten human health. In this regard, Djordjievski et al. explained how this type of wastewater is released into the Bor River, in Serbia, directly from copper mines without previous treatment, affecting the entire environment and the health of the people living near the area or people who use this river for different purposes. To attend to this problematic situation, this research team fabricated model solutions to mimic the copper concentration of this wastewater and treated it with an MFC. The results were shocking because they ended up achieving an efficiency of 99.42% [15].

14.11.3 Microbial Fuel Cells for Dye Removal

Textile industries produce wastewater that contains different types of dyes, which are very hard to separate and can be life-threatening for aquatic ecosystems. Therefore, it is necessary to treat this wastewater before releasing it into bodies of water. Patel et al. [20] described how MFCs can help remove these dyes, treating the water by microbial communities in a dynamic way while also producing electricity.

14.11.4 Microbial Fuel Cells for Domestic Wastewater Treatment

Rossi et al. [13] developed and studied an MFC with an amazing capacity of 850 L and air cathodes to evaluate the production of electricity and the treatment of domestic wastewater at a pilot scale. The wastewater was processed during a hydraulic retention time of 12 hours through a sequence of 17 brush anode modules and 16 cathode modules, and then the effluent had another treatment in a biofilter. After 6 months, the energy produced by this MFC was 0.46 ± 0.35 W, on average, and the current was 1.54 ± 0.90 A, with a coulombic efficiency of 9% With this MFC, up to 91% ± 6% of the chemical oxygen demand (COD) value and 91% of the BOD5 as well as some pathogenic bacteria like *E. coli* and fecal coliforms were eliminated.

14.12 Closing Argument and Perspectives of Microbial Fuel Cells

In the pursuit of sustainable and eco-friendly solutions, MFCs stand out as a groundbreaking technology with immense potential. These innovative systems leverage the metabolic processes of microorganisms to convert organic matter directly into electrical energy and, beyond energy production, MFCs offer a dual benefit by concurrently treating wastewater, thereby addressing two critical environmental issues with a single, plausible solution.

MFCs are a perfect example of green technology. By utilizing naturally present bacteria and other microorganisms to generate electricity, MFCs bypass the need for fossil fuels and their associated greenhouse gas emissions. This process not only mitigates climate change but also capitalizes on renewable resources, making it inherently sustainable. Moreover, MFCs operate at ambient temperatures and pressures, which reduces the energy footprint compared to conventional energy production methods.

The importance of MFCs extends beyond energy generation. Wastewater treatment is a pressing global challenge, with vast quantities of water needing to be purified before they can be safely released into the environment or reused for other purposes. Traditional wastewater treatment plants require a huge amount of energy and often rely on chemical processes that can produce secondary pollutants. In contrast, MFCs utilize organic waste present in the water as a fuel source, breaking down pollutants while simultaneously generating electricity. This not only reduces the load on traditional wastewater treatment facilities but also provides a sustainable method to manage organic waste.

MFCs rely greatly on the microorganisms present in the anodic chamber and the substrate within the wastewater that must be treated. The microorganisms responsible for the water treatment in MFCs are called exoelectrogens because they have the ability to transport electrons from within the cell to their outer membrane and to the exterior of the cell, where a final electron acceptor, such as the anode electrode, may receive it and produce energy. These exoelectrogenic microorganisms have the said ability thanks to the ETC mechanism which allows them, when oxygen is not abundant, to undergo this route and survive as anaerobic facultative organisms. The ETC in exoelectrogens uses C-type cytochromes to catch and release the electron from the third complex to the fourth complex within the cell, which makes these cytochromes a fundamental part of the process.

Despite the promising prospects, MFCs face several challenges that must be addressed to realize their full potential. Challenges such as the issue of scalability, the efficiency of electron transfer mechanisms, and the longevity of the microbial communities are areas that require further investigation and optimization. Additionally, the materials used for electrodes and membranes in MFCs can be very expensive and may degrade over time. Finding affordable, durable materials that maintain high performance is crucial for the widespread adoption of this technology.

To optimize MFC's performance, some carbon-based materials have been studied, such as carbon cloth, carbon felt, vitreous carbon, and even carbon nanotubes or CNT-textiles. These materials are promising and need further studies so that they can be widely used as effective electrodes in MFCs.

There is a need for more robust systems that are able to handle the variability in wastewater composition and flow rates, ensuring consistent and reliable performance of MFCs. Public awareness and regulatory frameworks need to evolve in order to support the integration of MFCs into existing energy and wastewater infrastructures. The benefits and potential of MFCs must be shared to foster investment and development in this promising field.

MFCs represent a transformative technology with the potential to revolutionize both energy production and wastewater treatment, and realizing their full potential will require overcoming significant technical and economic challenges. With continuous research, investment, and collaboration between academic fields and industries, this type of technology can play a pivotal role in our transition to a greener and more sustainable future.

Questions

1 What is the definition of wastewater?

2 What is a microbial fuel cell (MFC)?

3 Which types of microorganisms have the ability to transfer electrons?

4 How does the electron transport chain work?

5 What is the difference between a single-chamber and a double-chamber MFC?

6 How can we decide the type of electrodes to be used while designing an MFC?

7 Can MFCs be used at an industrial level?

References

1 Van Ruijven, B.J., De Cian, E., and Wing, I.S. (2019). Amplification of future energy demand growth due to climate change. *Nat. Commun.* 10 (1): https://doi.org/10.1038/s41467-019-10399-3.

2 De La Cruz-Noriega, M., Benites, S.M., Rojas-Flores, S. et al. (2023). Use of wastewater and electrogenic bacteria to generate eco-friendly electricity through microbial fuel cells. *Sustainability* 15 (13): 10640. https://doi.org/10.3390/su151310640.

3 Boretti, A. and Rosa, L. (2019). Reassessing the projections of the World Water Development report. *NPJ Clean Water* 2 (1): https://doi.org/10.1038/s41545-019-0039-9.

4 Potter, C.M. (1911). Electrical effects accompanying the decomposition of organic compounds. *Proc. R. Soc. Lond.* 84 (571): 260–276.

5 Santoro, C., Arbizzani, C., Erable, B., and Ieropoulos, I. (2017). Microbial fuel cells: from fundamentals to applications. A review. *J. Power Sources* 356: 225–244. https://doi.org/10.1016/j.jpowsour.2017.03.109.

6 Naseer, M.N., Zaidi, A.A., Khan, H. et al. (2021). Mapping the field of microbial fuel cell: a quantitative literature review (1970–2020). *Energy Rep.* 7: 4126–4138. https://doi.org/10.1016/j.egyr.2021.06.082.

7 Slate, A.J., Whitehead, K.A., Brownson, D.A., and Banks, C.E. (2019). Microbial fuel cells: an overview of current technology. *Renew. Sustain. Energy Rev.* 101: 60–81. https://doi.org/10.1016/j.rser.2018.09.044.

8 Logan, B.E. (2008). *Microbial Fuel Cells*. John Wiley & Sons.

9 Kim, Y. and Logan, B.E. (2013). Microbial desalination cells for energy production and desalination. *Desalination* 308: 122–130. https://doi.org/10.1016/j.desal.2012.07.022.

10 Iftimie, S. and Dumitru, A. (2019). Enhancing the performance of microbial fuel cells (MFCs) with nitrophenyl modified carbon nanotubes-based anodes. *Appl. Surf. Sci.* 492: 661–668. https://doi.org/10.1016/j.apsusc.2019.06.241.

11 Kurniawan, T.A., Liang, X., Muhammad, A. et al. (2022). Microbial fuel cells (MFC): a potential game-changer in renewable energy development. *Sustainability* 14 (24): 16847. https://doi.org/10.3390/su142416847.

12 Vallero, D.A. (2019). Wastewater. In: *Waste* (ed. T.M. Letcher and D.A. Vallero), 259–290. Elsevier. ISBN: 9780128150603.

13 Rossi, R., Hur, A.Y., Page, M.A. et al. (2022). Pilot scale microbial fuel cells using air cathodes for producing electricity while treating wastewater. *Water Res.* 215: 118208. https://doi.org/10.1016/j.watres.2022.118208.

14 Ghangrekar, M.M. and Chatterjee, P. (2018). Water pollutants classification and its effects on environment. In: *Carbon Nanotubes for Clean Water*, 11–26. Springer. https://doi.org/10.1007/978-3-319-95603-9_2.

15 Djordjievski, S., Yemendzhiev, H., Koleva, R. et al. (2022). Application of microbial fuel cell for simultaneous treatment of metallurgical and municipal wastewater - a laboratory study. *J. Serb. Chem. Soc.* 87 (6): 775–784. https://doi.org/10.2298/jsc211008009d.

16 Calvo, M.S. (2012). *Manual de las Aguas Residuales Industriales*. McGraw Hill.

17 Mason, S.A., Garneau, D., Suttton, R. et al. (2016). Microplastic pollution is widely detected in US municipal wastewater treatment plant effluent. *Environ. Pollut.* 218: 1045–1054. https://doi.org/10.1016/j.envpol.2016.08.056.

18 Lowrey, J., Brooks, M.S., and McGinn, P.J. (2014). Heterotrophic and mixotrophic cultivation of microalgae for biodiesel production in agricultural wastewaters and associated challenges—a critical review. *J. Appl. Phycol.* 27 (4): 1485–1498. https://doi.org/10.1007/s10811-014-0459-3.

19 Tran, N.H., Reinhard, M., Khan, E. et al. (2019). Emerging contaminants in wastewater, stormwater runoff, and surface water: application as chemical markers for diffuse sources. *Sci. Total Environ.* 676: 252–267. https://doi.org/10.1016/j.scitotenv.2019.04.160.

20 Patel, D., Bapodra, S.L., Madamwar, D., and Desai, C. (2021). Electroactive bacterial community augmentation enhances the performance of a pilot scale constructed wetland microbial fuel cell for treatment of textile dye wastewater. *Bioresour. Technol.* 332: 125088. https://doi.org/10.1016/j.biortech.2021.125088.

21 Fan, X., Gao, J., Pan, K. et al. (2018). Functional genera, potential pathogens and predicted antibiotic resistance genes in 16 full-scale wastewater treatment plants treating different types of wastewater. *Bioresour. Technol.* 268: 97–106. https://doi.org/10.1016/j.biortech.2018.07.118.

22 Ahamed, M.I. and Prasad, R. (2022). *Application of Microbes in Environmental and Microbial Biotechnology*. Springer.

23 Liu, X. and Kokare, C. (2017). Chapter 11—Microbial enzymes of use in industry. In: *Biotechnology of Microbial Enzymes: Production, Biocatalysis and Industrial Applications* (ed. G. Brahmachari, A.L. Demain, and J.L. Adrio), 267–298. Academic Press.

24 Gurung, N., Ray, S., Bose, S., and Rai, V. (2013). A broader view: microbial enzymes and their relevance in industries, medicine, and beyond. *Biomed Res. Int.* 2013: 329121. https://doi.org/10.1155/2013/329121.

25 Deval, A.S., Parikh, H.A., Kadier, A. et al. (2017). Sequential microbial activities mediated bioelectricity production from distillery wastewater using bio-electrochemical system with simultaneous waste remediation. *Int. J. Hydrogen Energy* 42 (2): 1130–1141. https://doi.org/10.1016/j.ijhydene.2016.11.114.

26 Gupta, A. and Verma, J.P. (2015). Sustainable bio-ethanol production from agro-residues: a review. *Renew. Sustain. Energy Rev.* 41: 550–567. https://doi.org/10.1016/j.rser.2014.08.032.

27 Puvanakrishnan, R., Sivasubramanian, S., and Hemalatha, T. (2019). *Microbial Technology: Concepts and Applications*. MJP Publisher.

28 Duan, C., Kee, R., Zhu, H. et al. (2019). Highly efficient reversible protonic ceramic electrochemical cells for power generation and fuel production. *Nat. Energy* 4 (3): 230–240. https://doi.org/10.1038/s41560-019-0333-2.

29 Tiwari, T., Kumar, M., Yadav, M. et al. (2019). Study of arrowroot starch-based polymer electrolytes and its application in MFC. *Stärke/Starch* 71 (7–8): https://doi.org/10.1002/star.201800313.

30 Skoog, D.A., Holler, F.J., and Crouch, S.R. (2017). *Principles of Instrumental Analysis*. Cengage Learning. ISBN: 9781337468039.

31 Cardoso, R.M., Mendonça, D.M., Silva, W.P. et al. (2018). 3D Printing for electroanalysis: from multiuse electrochemical cells to sensors. *Anal. Chim. Acta* 1033: 49–57. https://doi.org/10.1016/j.aca.2018.06.021.

32 Babadi, A.A., Wan, W.A., Chang, J.S. et al. (2019). High-performance enzymatic biofuel cell based on three-dimensional graphene. *Int. J. Hydrogen Energy* 44 (57): 30367–30374.

33 Jeon, W.Y., Lee, J.H., Dashnyam, K. et al. (2019). Performance of a glucose-reactive enzyme-based biofuel cell system for biomedical applications. *Sci. Rep.* 9: https://doi.org/10.1038/s41598-019-XXXXX-X.

34 Mano, N. and de Poulpiquet, A. (2018). O_2 reduction in enzymatic biofuel cells. *Chem. Rev.* 118 (5): 2392–2468.

35 Christwardana, M., Frattini, D., Duarte, K.D.Z. et al. (2019). Carbon felt molecular modification and biofilm augmentation via quorum sensing approach in yeast-based microbial fuel cells. *Appl. Energy* 238: 239–248. https://doi.org/10.1016/j.apenergy.2019.01.078.

36 Mardiana, U., Iccocent, C., Jarrer, H. et al. (2015). Electropolymerized neutral red as redox mediator for yeast fuel cell. *Int. J. Electrochem. Sci.* 10 (11): 8886–8898. https://doi.org/10.1016/s1452-3981(23)11145-x.

37 Christwardana, M., Frattini, D., Accardo, G. et al. (2018). Early-stage performance evaluation of flowing microbial fuel cells using chemically treated carbon felt and yeast biocatalyst. *Appl. Energy* 222: 369–382. https://doi.org/10.1016/j.apenergy.2018.03.193.

38 Chen, F., Li, J., Qin, D. et al. (2014). Characterization of exoelectrogenic bacteria enterobacter strains isolated from a microbial fuel cell exposed to copper shock load. *PLoS One* 9 (11): e113379. https://doi.org/10.1371/journal.pone.0113379.

39 Sheldon, R.A. (2017). Metrics of green chemistry and sustainability: past, present, and future. *ACS Sustain. Chem. Eng.* 6 (1): 32–48. https://doi.org/10.1021/acssuschemeng.7b03505.

40 United Nations, (2023). Global sustainable development report. https://sdgs.un.org/gsdr/gsdr2023 (accessed 25 May 2024)

41 Kadier, A., Jain, P., Lai, B. et al. (2020). Biorefinery perspectives of microbial electrolysis cells (MECs) for hydrogen and valuable chemicals production through wastewater treatment. *Biofuel Res. J.* 7 (1): 1128–1142.

42 Wan, L.L., Li, X.J., Zang, G.L. et al. (2015). A solar assisted microbial electrolysis cell for hydrogen production driven by a microbial fuel cell. *RSC Adv.* 5 (100): 82276–82281.

43 Kakarla, R. and Min, B. (2014). Photoautotrophic microalgae Scenedesmus obliquus attached on a cathode as oxygen producers for microbial fuel cell (MFC) operation. *Int. J. Hydrogen Energy* 39 (19): 10275–10283.

44 Jin, X., Guo, F., Ma, W. et al. (2019). Heterotrophic anodic denitrification improves carbon removal and electricity recovery efficiency in microbial fuel cells. *Chem. Eng. J.* 370: 527–535.

45 McGlynn, S.E., Chadwick, G.L., Kempes, C.P. et al. (2015). Single cell activity reveals direct electron transfer in methanotrophic consortia. *Nature* 526 (7574): 531–535. https://doi.org/10.1038/nature15512.

46 Duan, P., Pan, J., Du, W. et al. (2021). Activation of peroxymonosulfate via mediated electron transfer mechanism on single-atom Fe catalyst for effective organic pollutants removal. *Appl. Catal., B* 299: 120714. https://doi.org/10.1016/j.apcatb.2021.120714.

47 Nazeer, Z. and Fernando, E. (2022). A novel growth and isolation medium for exoelectrogenic bacteria. *Enzyme Microb. Technol.* 155: 109995. https://doi.org/10.1016/j.enzmictec.2022.109995.

48 Hartshorne, R.S., Reardon, C.L., Ross, D. et al. (2009). Characterization of an electron conduit between bacteria and the extracellular environment. *Proc. NatI. Acad. Sci. U. S. A.* 106 (52): 22169–22174. https://doi.org/10.1073/pnas.0900086106.

49 Pirbadian, S., Barchinger, S.E., Leung, K.M. et al. (2014). Shewanella oneidensis MR-1 nanowires are outer membrane and periplasmic extensions of the extracellular electron transport components. *Proc. Natl. Acad. Sci. U. S. A.* 111 (35): 12883–12888. https://doi.org/10.1073/pnas.1410551111.

50 Flynn, J.M., Ross, D.E., Hunt, K.A. et al. (2010). Enabling unbalanced fermentations by using engineered electrode-interfaced bacteria. *MBio* 1 (5): https://doi.org/10.1128/mbio.00190-10.

51 Baron, D., LaBelle, E., Coursolle, D. et al. (2009). Electrochemical measurement of electron transfer kinetics by Shewanella oneidensis MR-1. *J. Biol. Chem.* 284 (42): 28865–28873. https://doi.org/10.1074/jbc.m109.043455.

52 Aiyer, K.S. and Vijayakumar, B. (2020). A selection strategy for enhancing exoelectrogenic consortium towards improved power generation in microbial fuel cells. *Int. J. Energy Res.* 45 (4): 5318–5324. https://doi.org/10.1002/er.6154.

53 Wang, S., Qiu, L., Liu, X. et al. (2018). Electron transport chains in organohalide-respiring bacteria and bioremediation implications. *Biotechnol. Adv.* 36 (4): 1194–1206. https://doi.org/10.1016/j.biotechadv.2018.03.018.

54 Hein, S. and Simon, J. (2019). Bacterial nitrous oxide respiration: electron transport chains and copper transfer reactions. *Adv. Microb. Physiol.* 137–175. https://doi.org/10.1016/bs.ampbs.2019.07.001.

55 Tikhonova, T.V. and Popov, V.O. (2014). Structural and functional studies of multiheme cytochromes c involved in extracellular electron transport in bacterial dissimilatory metal reduction. *Biochemistry* 79 (13): 1584–1601. https://doi.org/10.1134/s0006297914130094.

56 Nebert, D.W., Wikvall, K., and Miller, W.L. (2013). Human cytochromes P450 in health and disease. *Philos. Trans. R. Soc. Biol. B Sci.* 368 (1612): 20120431. https://doi.org/10.1098/rstb.2012.0431.

57 Liu, F., Rotaru, A.E., Shrestha, P.M. et al. (2014). Magnetite compensates for the lack of a pilin-associated c-type cytochrome in extracellular electron exchange. *Environ. Microbiol.* 17 (3): 648–655. https://doi.org/10.1111/1462-2920.12485.

58 Cramer, W.A. and Kallas, T. (2016). *Cytochrome Complexes: Evolution, Structures, Energy Transduction, and Signaling*. Springer.

59 Shi, L., Chen, B., Wang, Z. et al. (2006). Isolation of a high-affinity functional protein complex between OmcA and MtrC: two outer membrane decaheme c-type cytochromes of Shewanella oneidensis MR-1. *J. Bacteriol.* 188 (13): 4705–4714. https://doi.org/10.1128/jb.01966-05.

60 Bank, R. P. D. (n.d.) RCSB PDB - 4LMH: Crystal structure of the outer membrane decaheme cytochrome OmcA. https://www.rcsb.org/structure/4lmh (accessed 28 May 2024).

61 Zhao, N., Jiang, Y., Alvarado-Morales, M. et al. (2018). Electricity generation and microbial communities in microbial fuel cell powered by macroalgal biomass. *Bioelectrochemistry* 123: 145–149. https://doi.org/10.1016/j.bioelechem.2018.05.002.

62 Segundo, R.F., De La Cruz-Noriega, M., Milly Otiniano, N. et al. (2022). Use of onion waste as fuel for the generation of bioelectricity. *Mol. Ther.* 27 (3): 625. https://doi.org/10.3390/molecules27030625.

63 Bose, D., Gopinath, M., Vijay, P. et al. (2019). Bioelectricity generation and biofilm analysis from sewage sources using microbial fuel cell. *Fuel* 255: 115815. https://doi.org/10.1016/j.fuel.2019.115815.

64 Isosaari, P. and Sillanpää, M. (2016). Use of sulfate-reducing and bioelectrochemical reactors for metal recovery from mine water. *Sep. Purif. Rev.* 46 (1): 1–20. https://doi.org/10.1080/15422119.2016.1156548.

65 Li, T., Song, H.L., Tang, X.L. et al. (2021). Biological detoxification and decolorization enhancement of azo dye by introducing natural electron mediators in MFCs. *J. Hazard. Mater.* 416: 125864. https://doi.org/10.1016/j.jhazmat.2021.125864.

66 Montaño, M. J. N., and Aragón, E. D. B., (2018). Implementación de una celda de combustible microbiana a escala laboratorio para generación de energía eléctrica. http://repository.uamerica.edu.co/bitstream/20.500.11839/6695/1/6131003-2018-1-IQ.pdf (accessed 25 May 2024).

67 Liu, Z., Liu, J., Zhang, S. et al. (2011). Microbial fuel cell based biosensor for in situ monitoring of anaerobic digestion process. *Bioresour. Technol.* 102 (22): 10221–10229. https://doi.org/10.1016/j.biortech.2011.08.053.

68 Dubey, R. and Guruviah, V. (2019). Review of carbon-based electrode materials for supercapacitor energy storage. *Ionics* 25 (4): 1419–1445. https://doi.org/10.1007/s11581-019-02874-0.

69 Walsh, F.C., Arenas, L.F., De León, C.P. et al. (2016). The continued development of reticulated vitreous carbon as a versatile electrode material: structure, properties and applications. *Electrochim. Acta* 215: 566–591. https://doi.org/10.1016/j.electacta.2016.08.103.

70 Santoro, C., Lei, Y., Li, B., and Cristiani, P. (2012). Power generation from wastewater using single chamber microbial fuel cells (MFCs) with platinum-free cathodes and pre-colonized anodes. *Biochem. Eng. J.* 62: 8–16. https://doi.org/10.1016/j.bej.2011.12.006.

71 Tamakloe, R.Y., Opoku-Donkor, T., Donkor, M.K. et al. (2015). Comparative study of double-chamber microbial fuel cells (DC-MFCs) using Mfensi clay as ion-exchange-partition: effect of electrodes. *Afr. J. Sci. Innovation Dev.* 7 (3): 207–210. https://doi.org/10.1080/20421338.2015.1043706.

72 Hidayat, A.R.P., Widyanto, A.R., Asranudin, A. et al. (2022). Recent development of double chamber microbial fuel cell for hexavalent chromium waste removal. *J. Environ. Chem. Eng.* 10 (3): 107505. https://doi.org/10.1016/j.jece.2022.107505.

73 Li, J., Chen, M., Cullen, D.A. et al. (2018). Atomically dispersed manganese catalysts for oxygen reduction in proton-exchange membrane fuel cells. *Nat. Catal.* 1 (12): 935–945. https://doi.org/10.1038/s41929-018-0164-8.

74 Pandey, P., Shinde, V.M., Deopurkar, R.L. et al. (2016). Recent advances in the use of different substrates in microbial fuel cells toward wastewater treatment and simultaneous energy recovery. *Appl. Energy* 168: 706–723. https://doi.org/10.1016/j.apenergy.2016.01.056.

75 Rahman, W., Yusup, S., and Mohammad, S.N.A.A. (2021). Screening of fruit waste as substrate for microbial fuel cell (MFC). *AIP Conf. Proc.* https://doi.org/10.1063/5.0043556.

76 Sotres, A., Tey, L., Bonmatí, A., and Viñas, M. (2016). Microbial community dynamics in continuous microbial fuel cells fed with synthetic wastewater and pig slurry. *Bioelectrochemistry* 111: 70–82. https://doi.org/10.1016/j.bioelechem.2016.04.007.

77 Nagesh, M., Senthilkumar, P., Jenifer Selvarani, A. et al. (2019). Electricity generation using carboxymethyl cellulose and kitchen waste as substrate by Exiguobacterium sp SU-5 in mediatorless microbial fuel cell. *J. Pure Appl. Microbiol.* 13 (4): 2151–2158. https://doi.org/10.22207/jpam.13.4.27.

78 Zhang, J., Zheng, P., Zhang, M. et al. (2013). Kinetics of substrate degradation and electricity generation in anodic denitrification microbial fuel cell (AD-MFC). *Bioresour. Technol.* 149: 44–50. https://doi.org/10.1016/j.biortech.2013.09.043.

79 Zhou, M., Chi, M., Luo, J. et al. (2011). An overview of electrode materials in microbial fuel cells. *J. Power Sources* 196 (10): 4427–4435. https://doi.org/10.1016/j.jpowsour.2011.01.012.

80 Muthukumar, H., Mohamed, S.N., Matheswaran, M., and Pugazhendhi, A. (2019). Effect of iron doped zinc oxide nanoparticles coating in the anode on current generation in microbial electrochemical cells. *Int. J. Hydrogen Energy* 44 (4): 2407–2416. https://doi.org/10.1016/j.ijhydene.2018.06.046.

81 Hussein, E.A., Mansour, A.B., and Bahgat, M. (2013). Metal electrodes and organic enrichment in doubled and single chambered microbial fuel cell (MFC) for electricity generation. *J. Biochem. Technol.* 4 (1): 554–560. https://jbiochemtech.com/article/metal-electrodes-and-organic-enrichment-in-doubled-and-single-chambered-microbial-fuel-cell-mfc-for-electricity-generation.

82 Baudler, A., Schmidt, I., Langner, M. et al. (2015). Does it have to be carbon? Metal anodes in microbial fuel cells and related bioelectrochemical systems. *Energy Environ. Sci.* 8 (7): 2048–2055. https://doi.org/10.1039/c5ee00866b.

83 Harshiny, M., Samsudeen, N., Kameswara, R., and Matheswaran, M. (2017). Biosynthesized FeO nanoparticles coated carbon anode for improving the performance of microbial fuel cell. *Int. J. Hydrogen Energy* 42 (42): 26488–26495. https://doi.org/10.1016/j.ijhydene.2017.07.084.

84 Xie, X., Hu, L., Pasta, M. et al. (2010). Three-dimensional carbon nanotube–textile anode for high-performance microbial fuel cells. *Nano Lett.* 11 (1): 291–296. https://doi.org/10.1021/nl103905t.

85 Huang, W., Chen, J., Hu, Y. et al. (2017). Enhanced simultaneous decolorization of azo dye and electricity generation in microbial fuel cell (MFC) with redox mediator modified anode. *Int. J. Hydrogen Energy* 42 (4): 2349–2359. https://doi.org/10.1016/j.ijhydene.201.

86 Pandit, S., Chandrasekhar, K., Kakarla, R. et al. (2017). Basic principles of microbial fuel cell: technical challenges and economic feasibility. In: *Microbial Applications*, vol. 1, 165–188. Springer.

87 Kang, C.S., Eaktasang, N., Kwon, D.Y. et al. (2014). Enhanced current production by Desulfovibrio desulfuricans biofilm in a mediator-less microbial fuel cell. *Bioresour. Technol.* 165: 27–30. https://doi.org/10.1016/j.biortech.2014.03.148.

88 You, S., Ma, M., Wang, W. et al. (2016). 3D Macroporous nitrogen-enriched graphitic carbon scaffold for efficient bioelectricity generation in microbial fuel cells. *Adv. Energy Mater.* 7 (4): https://doi.org/10.1002/aenm.201601364.

89 Yang, W., Peng, Y., Zhang, Y. et al. (2019). Air cathode catalysts of microbial fuel cell by nitrogen-doped carbon aerogels. *ACS Sustain. Chem. Eng.* 7 (4): 3917–3924. https://doi.org/10.1021/acssuschemeng.8b05000.

90 Zhou, Y., Zhu, N., Guo, W. et al. (2018). Simultaneous electricity production and antibiotics removal by microbial fuel cells. *J. Environ. Manage.* 217: 565–572. https://doi.org/10.1016/j.jenvman.2018.04.013.

15

Economic Feasibility of Membrane Technology

Asmaa Ahmed[1,2], Mussad M. Alzahrani[3], Muhammad Mubashir[4], Ahmed Refaat[5], and Mohamed Elsakka[1,2]

[1]*Mechanical Power Engineering Department, Faculty of Engineering, Port Said University, Port Said, Egypt*
[2]*Faculty of Engineering, East Port Said National University, Salam Misr City, Port Said, Egypt*
[3]*Mechanical and Energy Engineering Department, Imam Abdulrahman Bin Faisal University, Dammam, Saudi Arabia*
[4]*Water Technologies Innovation Institute and Research Advancement, Jubail, Saudi Arabia*
[5]*Electrical Engineering Department, Faculty of Engineering, Port Said University, Port Said, Egypt*

15.1 Introduction

Membrane technology is considered a cornerstone of present water treatment and seawater desalination technologies [1, 2]. This is attributable to its flexibility and high efficiency in separation processes [3]. This technology utilizes porous membranes to selectively filter out contaminants from water [4, 5]. Therefore, pore size, solution/solvent charge, and other physical characteristics are assumed to be essential factors that affect the efficiency of separation processes. Key membrane technologies include reverse osmosis (RO), forward osmosis (FO), membrane distillation (MD), nanofiltration (NF), ultrafiltration (UF), and microfiltration (MF) [5, 6]. Each process has different operating conditions, making it suitable for specific applications.

RO is considered one of the most outstanding membrane separation technologies as it is extensively employed in desalination and water purification processes [7]. This technology depends on applying a pressure gradient across a semi-permeable membrane. The pressure typically varies between 40 and 80 bars, depending on the salinity of the feedwater and whether it's seawater or brackish water [8, 9]. The applied pressure should be higher than the osmotic pressure of the feed solution. This forces water molecules to traverse the membrane while effectively retaining other dissolved solids [10]. These solids could be dissolved solids (salts or heavy metals), organic contaminants (neutral organic matter

Water Treatment and Desalination: Membrane Separation using Renewable Energy Resources, First Edition. Edited by Mohamed Bassyouni.

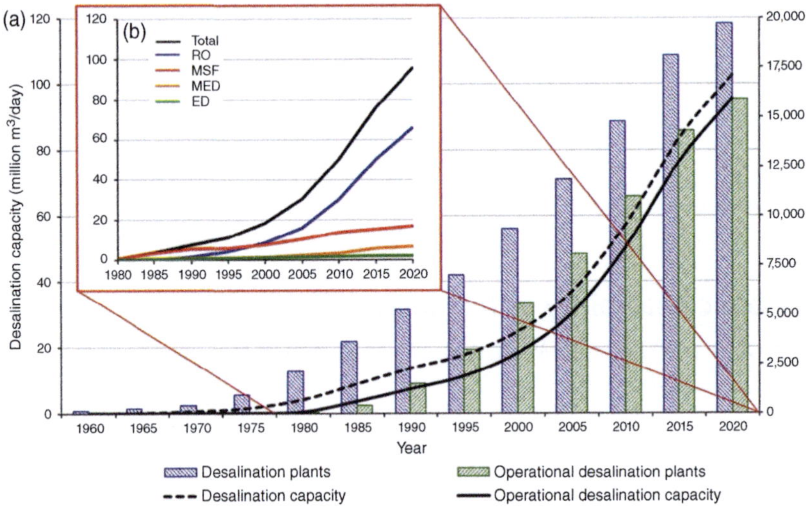

Figure 15.1 Global desalination over time. (a) The number and capacity of all desalination plants, including those currently in operation, and (b) the operational capacity segmented by different desalination technologies. Source: Jones et al. [15]/with permission of Elsevier.

[NOM], pesticides, herbicides, and/or pharmaceutical and personal care products [PPCPs]), and other impurities (fluoride and nitrate/nitrite) [11, 12]. The pore size of the RO membrane usually varies between 0.1 and 1 nm, enabling the process to generate highly purified water [13]. In addition to the desalination of seawater and brackish water, RO technology plays a fundamental role in treating and reusing advanced wastewater used in the pharmaceutical industry and industrial water, municipal water reclamation, and environmental and water resource management [14]. However, this method is energy-intensive, which requires significant power input due to the high-pressure conditions [15]. It is also prone to membrane fouling, which impacts the membrane's lifetime and separation efficiency. This indeed necessitates rigorous feedwater pretreatment and periodic maintenance to mitigate fouling and sustain system performance. Figure 15.1 shows trends in the types of installed desalination plants from 1960 to 2020, with RO plants holding the most significant share, according to Jones et al. [15]. The data reveal an exponential increase in the number of installed RO plants, surpassing other desalination technologies over the same period due to their advantages.

Given Saudi Arabia's arid climate, water desalination is regarded as one of the most effective solutions. Membrane technology in Saudi Arabia, particularly for desalination, began in the 1970s [5]. The country developed large-scale RO plants which are handled by the Saudi Water Authority (SWA). By the 1990s,

Saudi Arabia had become one of the world's leaders in RO desalination, with the technology being applied across multiple plants on the Red Sea and Arabian Gulf coasts [16]. The Shoaiba desalination plant, located on the Red Sea coast, is one of the largest in the world and a key example of the country's reliance on the RO membrane technology. It provides freshwater to millions of people in the Makkah and Jeddah regions. Over the past few decades, Saudi Arabia has made strides not only in adopting the RO membrane technology but also in advancing its development through research and innovation. King Abdullah University of Science and Technology (KAUST) has become a hub for membrane research, with a focus on improving desalination membranes to enhance efficiency and reduce energy consumption. In partnership with global companies (LG, Tory, Alfa Laval) and research institutions, KAUST and SWA (formally, saline water conversion corporation) are working on the next generation of RO membranes and hybrid systems that combine RO with other conventional technologies. These innovations aim to reduce the overall challenges of RO and desalination while making the process more cost-effective with a higher production rate. In addition, the effort to transition from thermal to RO desalination powered by renewable resources is becoming a mandate, particularly in Saudi Arabia, due to RO's energy efficiency and cost-effectiveness [17]. Globally, this transition reduces the needed energy from 10–15 kWh to 3–4 kWh/m^3 and the cost as low as \$0.40/m^3. The Institute of Water Technology Innovation and Advanced Research (WTIIRA), a highly reputed research institute in Saudi Arabia, mastered RO technology's lowest energy consumption record at 2.27 kWh/m^3 [18].

On the contrary, FO utilizes natural osmotic pressure gradients offering lower energy compared to RO. Using a semi-permeable membrane, water molecules move from a lower-concentration feed solution toward a higher-concentration draw solution [19]. This process is driven solely by the osmotic pressure differential, eliminating the need for external hydraulic pressure [20]. The pore size of the FO membrane is in the range of 0.4–1 nm [21], which enables the membrane to effectively block solutes such as salts and small organics. Several applications can utilize FO technology such as desalination, wastewater treatment, and concentration processes in the food and beverage industry [22]. In addition to the lower energy consumption of this process, it offers a low fouling potential and a higher tolerance for feedwaters with elevated salinity or foulants [23]. On the other hand, recovering and reusing the draw solution remain a critical factor in optimizing process performance [23]. Besides, FO provides a low permeation rate during the production process [24]. Therefore, ongoing advancements in draw solute chemistry and membrane design will likely enhance FO technology's scalability and efficiency. Instead of depending on hydraulic pressure or drawing a solution, MD is based on thermal gradients that exploit the vapor pressure difference across a hydrophobic membrane [25, 26]. The temperature difference on the two sides of the membrane vaporizes the feedwater, allowing the vapor to pass through the membrane and condense on the cooler (permeate) side. MD is used in the desalination of high-salinity water and wastewater treatment with high-temperature waste streams [27]. It also potentially integrates with low-grade waste heat or renewable thermal energy. However, one of its drawbacks is the significant heating required for the water as well as the possibility of membrane wetting, which can negatively impact the performance of the process [16].

In terms of filtration processes, NF operates under pressure conditions lower than RO, typically within the range of 4–35 bars [24]. Positioned between RO and UF, NF strikes a balance between water quality and energy consumption [28]. The pore sizes of NF membranes vary and are usually >10 nm. These small pores allow the passage of water and small ions while excluding larger particles, including viruses, bacteria, multivalent ions such as calcium and magnesium, and organic molecules [29]. This technology is particularly well suited for applications such as hard water softening by removing divalent ions, wastewater treatment to eliminate organic contaminants, and pretreatment for RO systems to reduce fouling. One advantage of this technology is its lower operating pressures, which reduce energy use [30]. It also removes larger contaminants while retaining essential monovalent ions (minerals) in the water. However, NF is less effective for desalination than RO because it doesn't remove all dissolved salts. Like other membrane systems, NF can also suffer from fouling, which reduces its performance over time. In SWA, NF plays a significant role in the separation of divalent ions, while it allows monovalent ions to pass through it and, thus, cannot be separated from water.

UF is another type of filtration process that operates at lower pressures typically between 1 and 7 bars [31, 32]. While NF targets smaller contaminants, UF focuses primarily on removing larger particles, colloids, and suspended solids. With pore sizes ranging from 10 to 100 nm, UF effectively filters out bacteria, viruses, and other macromolecules. UF is widely used as a pretreatment step for RO to mitigate fouling as well as for drinking water purification and industrial wastewater treatment. Its major advantages include the ability to eliminate a wide range of contaminants without chemical additives and its flexibility as a stand-alone system or as part of more advanced membrane technologies like RO. However, like other filtration methods, UF also has issues with fouling, especially with feedwater high in suspended solids. It is also ineffective at removing dissolved salts or very small particles.

Following UF, MF represents the coarsest membrane filtration technique, designed to target even larger contaminants than those removed by UF. The membrane pore size ranges from 100 to 1000 nm, effectively removing larger particles, including bacteria, algae, and suspended solids. MF operates at very low pressures, usually between 0.7 and 7 bars [33]. It is often used as a pretreatment step before more advanced filtration systems, such as UF, NF, or RO [34]. MF is highly efficient in microbial removal and significantly reduces water turbidity, which makes it suitable for a range of applications. This includes pretreatment for advanced membrane processes, drinking water purification, and clarification in the food and beverage industry [35]. Its main advantages include minimal energy consumption, owing to the low operational pressures and its ability to effectively eliminate larger particulates and microbes [36]. Although fouling is typically easier to manage in MF than in RO or NF systems, it remains a concern. Table 15.1 provides a comparison of the different membrane technologies discussed. Expanding on the previous discussion, this chapter explores the economic feasibility of membrane technologies by examining strategies to improve energy efficiency and system design and the integration of renewable energy sources (RESs). In addition, it provides a detailed analysis of different economic models. It also investigates how renewable energy such as solar and wind can reduce the energy footprint and improve economic sustainability. Following this, an analysis of the challenges impacting the economic viability of these systems along with suggested solutions for sustainable implementation is presented.

15.2 Integration of Renewable Energy Resources into Membrane Systems

RESs are increasingly incorporated into membrane-based desalination and water treatment systems to address global water scarcity and environmental concerns. Utilizing renewable energy for desalination processes not only reduces the carbon footprint of water treatment but also enhances the long-term cost-efficiency of these systems [37]. For example, desalination in arid and semi-arid regions is characterized by high energy consumption. This increases the dependence on fossil fuels, exacerbating operational costs and environmental impacts. Therefore, utilizing the RESs will offer a sustainable approach to disconnecting water treatment operations from the instability and environmental damages of conventional energy markets [38]. Also, in coastal and remote locations such as isolated islands where access to the grid is either limited or economically unfeasible, RESs present a decentralized solution. Figure 15.2 illustrates that the most commonly utilized renewable energy desalination methods are solar- and wind-based RO, representing 32% and 19% of the market, respectively [39].

15.2.1 Common Renewable Energy Sources for Membrane Technologies

15.2.1.1 Solar Energy

Photovoltaic (PV) technology and solar thermal energy can be utilized in membrane desalination processes. PV systems convert solar energy into electrical power, which can be employed to drive pressure-driven membrane processes,

Table 15.1 Comparison between different membrane technologies.

Technology	Pore size	Pressure	Removes	Applications	Strengths	Limitations
RO	<1 nm	40–80 bars	Dissolved salts and ions	Desalination and high-purity water	• Produces high-quality water • Effective for a wide range of solids • Industry standards for desalination	• High energy consumption • Membrane fouling issues • Requires extensive pretreatment
FO	0.4–1 nm	Low or no external pressure; relies on osmotic pressure gradients	Solutes (salts and small organics)	Industrial water treatment and desalination applications	• Lower energy consumption • Flexible draw solution	• Draw solution recovery • Lower flux rates
NF	> 10 nm	4–35 bars	Organic matter, multivalent ions	Water softening and organic removal	• Lower pressure than RO • Removes larger contaminants • Allows beneficial minerals to pass	• Less effective for desalination • Prone to fouling
UF	10–100 nm	1–7 bars	Viruses and colloids	Water purification and pretreatment	• Removes a wide range of contaminants • No chemicals required • Suitable for standalone or pretreatment	• Fouling concerns with high solids • Not effective for dissolved salts
MF	100–1000 nm	0.7–7 bars	Bacteria and large particles	Pretreatment and clarification	• Low energy requirement • Effective for large particles • Easy to manage fouling	• Ineffective for dissolved solids • Limited range of contaminant removal
MD	0.1–0.4 μm	Temperature driven	Nonvolatile contaminants	High- salinity water treatment and integration with thermal energy	• High level of purification • Suitable for high-salinity water • Can use low-grade waste heat • Potential for renewable energy integration	• Energy-intensive process • Membrane wetting issues • Scaling and fouling challenges

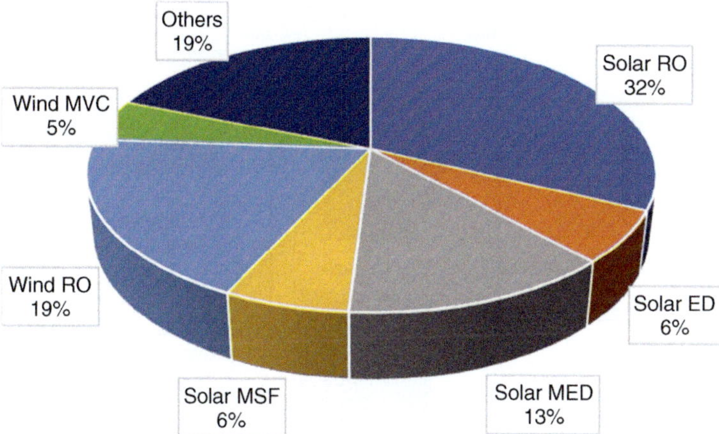

Figure 15.2 The most common renewable energy resources used in seawater desalination processes as reported by Alawad et al. [37]. Source: Alawad et al. [37]/with permission of Elsevier.

such as RO and NF [40–42]. For example, recently, SWA has integrated a solar PV power system into Jubail 3B RO plant to reduce reliance on the grid and to provide a cleaner alternative for the RO process. PVs can drive the pumps responsible for increasing the feed pressure before entering the unit. These efforts have contributed significantly to enhancing the economic feasibility of membrane desalination in the Kingdom of Saudi Arabia. On the other hand, solar thermal energy is particularly effective for powering thermal desalination methods, like MD. In these methods, thermal energy evaporates the water, which allows it to pass through a hydrophobic membrane and then condense into purified water. In addition, the integration of PV systems with solar thermal collectors in hybrid solar setups significantly enhances overall efficiency by addressing both electrical and thermal energy requirements. These systems have proven their economic and environmental feasibility [25, 43–45]. Mohamed and Papadakis [46] reported a PV-RO system integrated with battery energy storage, designed to produce 12 m^3 of freshwater per day, at a cost of 7.22 $/$m^3$. Al-Karaghouli and Kazmerski [47] featured a solar village desalination plant at Al-Khafji, Saudi Arabia, which is based on a RO process driven by solar PV systems with a production capacity of 60,000 m^3/d and an approximate cost of $0.50–$0.60/m^3. Rezk et al. [48] developed a hybrid RO water desalination plant with a production capacity of 150 m^3/d, powered by a combination of PV and fuel cells (FCs), for irrigation in remote areas, as presented in Figure 15.3. The study indicated that the CO_2 emissions avoided by using this system, compared to a diesel-fueled alternative, amounted to 70,974 kg annually.

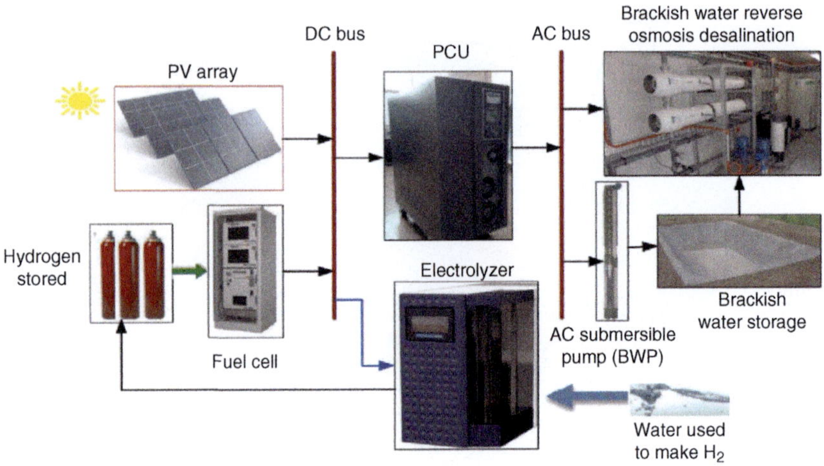

Figure 15.3 A hybrid desalination system powered by solar energy as proposed by Rezk et al. [48]. Source: Rezk et al. [48]/with permission of Elsevier.

15.2.1.2 Wind Energy

For coastal areas with significant wind conditions, wind power serves as a valuable resource for desalination processes. Wind turbines (WTs) convert wind energy into electrical power, which can then be harnessed to operate the RO and NF systems. Despite the challenges posed by the intermittent nature of wind energy, advances in energy storage and smart grid technologies have addressed these issues. As a result, wind energy has become an effective and reliable source of water treatment in suitable locations. Several researchers have considered this type of RESs due to its availability throughout the day [49–55]. In their recent study, Elsakka et al. [56] analyzed an RO system driven by wind energy for freshwater production in coastal cities. The system was based on energy storage by pneumatic approach, as illustrated in Figure 15.4, which shows its economic feasibility with the lowest desalination costs reaching 0.65 $/m^3$.

15.2.1.3 Geothermal Energy

For areas with significant geothermal activity, geothermal energy offers a highly efficient and reliable renewable resource for membrane desalination, especially for thermally driven processes, such as MD [38]. One of the advantages of this renewable energy resource is that it is not subject to the intermittency associated with solar or wind energy. Geothermal resources are dependent on the earth's mantle, sun heat, and lithology to categorize the geographical locations as low (<100°C: domestic use), medium (100–150°C: binary cycle and heating systems), and high

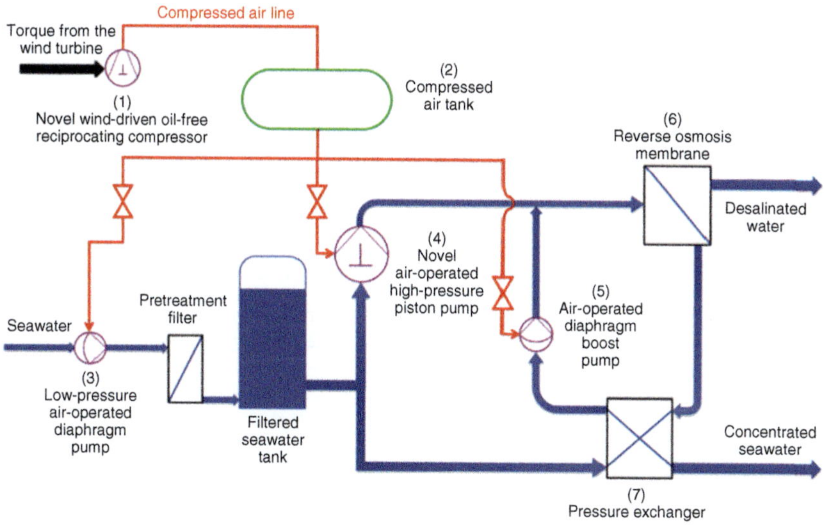

Figure 15.4 A wind-powered RO system with compressed air storage proposed by Elsakka et al. [56]. Source: Elsakka et al. [56]/with permission of Elsevier.

enthalpy (>150°C: steam turbines), where geothermal power is both readily accessible and economically viable [57]. In some cases, geothermally driven desalination systems are more cost-effective compared to solar-driven processes [58]. For example, SWA has developed a pilot scale project to induce the geothermal water energy for the MD process during water desalination and brine concentration. This plant is being operated using geothermal energy and is producing freshwater and brine. Figure 15.5 shows the pilot-scale MD project using geothermal energy developed by SWA.

Figure 15.6 illustrates a hybrid renewable energy system (solar and geothermal) designed for freshwater production. A high-pressure geothermal source enables the direct utilization of shaft power for mechanically driven desalination systems such as RO [47], while high-temperature geothermal energy can be used in MD processes [59].

15.2.2 Technological Advancements Enhancing Renewable Integration into Membrane Systems

Several technological advancements have considerably accelerated the integration of RESs into membrane desalination and water treatment systems [60]. This integration focuses on enhancing system performance and economic feasibility. In SWA's plants, RO remains the dominant technology which is coupled with renewable sources. The Shoaiba and Al Khobar plants, for instance, utilize RO

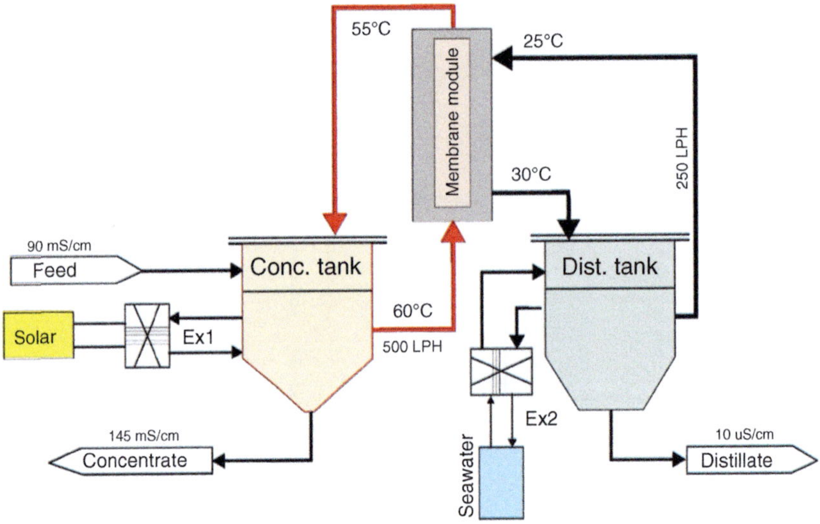

Figure 15.5 A pilot-scale MD using geothermal energy developed by SWA.

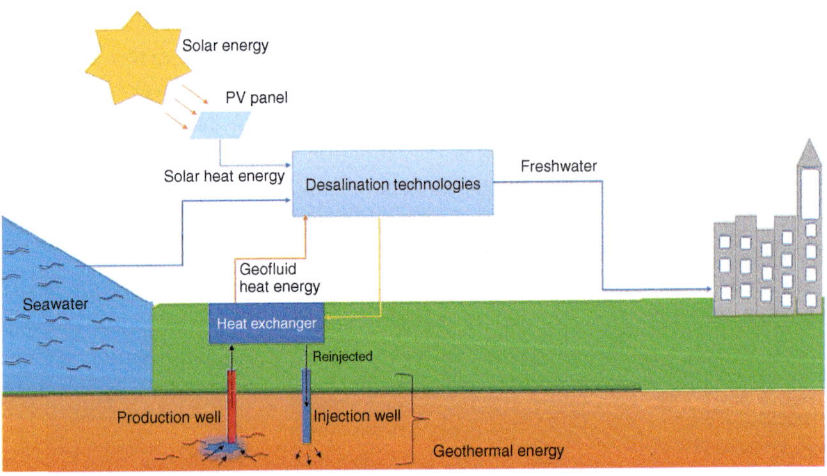

Figure 15.6 A hybrid solar-geothermal energy desalination system adapted from the study by Prajapati et al. [58]. Source: Prajapati et al. [58]/with permission of Elsevier.

systems powered by conventional energy sources, but recent expansions incorporate renewable sources like solar power. The blending of RESs helps mitigate energy costs and allows for more sustainable operations. A detailed description is provided in the following section.

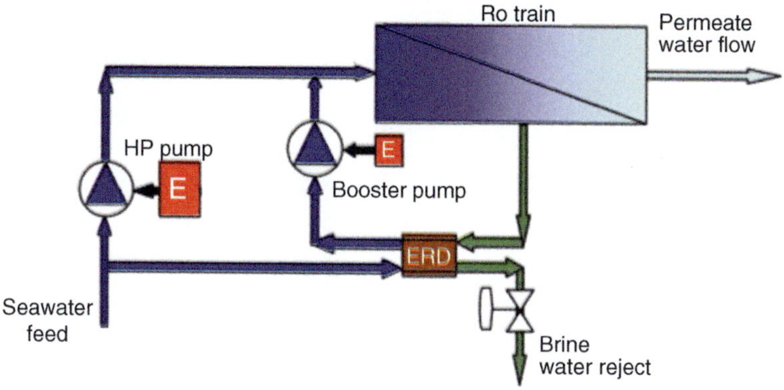

Figure 15.7 Integration of ERD into RO system as proposed by Peñate et al. [62]. Source: Peñate et al. [62]/with permission of Elsevier.

15.2.2.1 Energy Recovery Devices

Especially in RO systems which are considered pressure-driven membrane processes, energy recovery devices (ERDs) have greatly enhanced the system performance as proven in the literature [61]. The working idea of the ERDs, such as pressure exchangers (isobaric technology with higher efficiency) (Figure 15.7), is based on harnessing mechanical energy from the high-pressure brine reject stream and exchanging it with the inlet stream to the RO system [63]. This can save energy requirements by up to 60%, making it ideal for systems that depend on RESs which produce lower electricity outputs in comparison with conventional fossil-fuel plants [64]. Therefore, such devices are crucial for improving the system's economic feasibility. Although ERDs, such as Pelton wheel/Francis turbines and hydraulic turbochargers (HTCs), were implemented in the early 1980s and 1990s, the pressure exchanger has the highest energy recovery efficiency of around 95% and is still effectively incorporated in RO desalination systems [65].

15.2.2.2 Hybrid Energy Systems

For enhancing the functionality of desalination systems, integrating more than one renewable energy resource can result in reliable, cost-effective, and environmentally sustainable energy generation solutions, particularly for arid regions that rely massively on desalination systems to overcome water scarcity [66, 67]. For example, the integration of PVs with wind, as presented in Figure 15.8, or geothermal energy has emerged as an effective approach to overcoming the intermittency issues associated with individual RESs. In other words, hybrid systems improve energy supply by combining daytime solar generation with wind energy in the evening. They can also use geothermal or hydropower as a stable base load. In

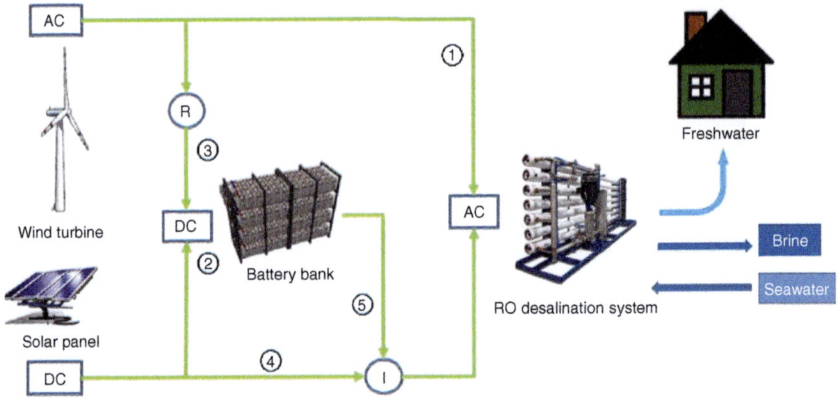

Figure 15.8 An RO desalination system driven by hybrid renewable energy configuration. Source: Esfahani et al. [68]/with permission of Elsevier.

addition, hybrid systems are particularly suitable for large-scale desalination facilities that require continuous operation to meet the high water demand.

15.2.2.3 Advanced Membrane Materials

Recent innovations in membrane materials, especially the use of nanomaterials and advanced surface modifications, have greatly improved membrane system performance. These advancements help reduce fouling, increase permeability, and enhance chemical resistance [69]. The top five advanced membranes are graphene oxide (GO), aquaporin-embedded membranes, metal–organic frameworks (MOFs), thin-film nanocomposites (TFNs), and carbon nanotube (CNT)-based membranes. CNT and aquaporin membranes allow the highest water permeability, succeeded by GO, TFN, and MOF. Generally, salt rejection exceeds 90%, with CNT membranes achieving up to 99%, TFN membranes ranging from 95% to 99%, and aquaporin membranes exceeding 95%. These membranes demonstrate the highest rejection rates. However, cost is a challenging factor in both aquaporin membranes and TFN, where GO and MOF face major hurdles related to long-term stability. Regarding high-efficiency applications, GO and TFN offer an excellent trade-off between cost-effectiveness and performance, which demonstrate considerable potential for lowering energy consumption due to their high flux rates and reduced liability to fouling [70–74]. Therefore, SWA has initiated the project on the scale-up of GO-based membranes for water desalination to enhance the water production in which an 8-inch spiral wound membrane is being produced and tested at a pilot scale at Jubail pilot plant, Saudi Arabia. This membrane rejects 99% of salts with a higher flux of 17 LMH/bar. In addition to GO material-based membranes, the development of anti-biofouling and anti-scaling coatings has also prolonged the membrane's lifespan, minimizing operational disruptions and maintenance and replacement costs [75, 76]. Therefore, SWA has developed a pilot-scale anti-fouling membrane project in collaboration with MOLYMEM (UK). The project focuses on applying a 2D MoS_2 coating to membrane support. This coating enhances anti-fouling properties and achieves 99.9% rejection of dissolved salts.

15.3 Cost Dynamics and Economic Feasibility of Membrane Systems

15.3.1 Investment Requirements

According to Nassrullah et al. [38], desalination is the most energy-intensive water treatment process with significant energy costs and an annual consumption of 75.2 TWh. Depending on the energy source, the investment costs for these systems vary considerably between conventional and renewable sources. Figure 15.9 shows that energy accounts for the largest portion of the operational and maintenance expenses (OPEX) in desalination processes, making up 46.6%. The energy costs are primarily electrical, constituting 84.8%, with a smaller portion being thermal energy at 15.2% [47]. Systems that rely on conventional energy sources such as grid electricity or fossil fuels usually have lower initial energy-related costs. This is due to the well-established infrastructure for power generation and distribution. On the other hand, RES-based membrane systems often involve higher initial costs. This is mainly due to the need for specialized infrastructure, such as PV panels, WTs, or geothermal plants. In addition, these systems require the integration of other components. This includes energy storage technologies (such as batteries or thermal energy storage), inverters, and advanced control systems to manage the intermittent nature of RESs [77]. These components increase the initial capital cost requirements. However, these higher upfront costs can be balanced by long-term savings in energy costs and environmental benefits.

15.3.1.1 Equipment, Infrastructure, and Installation Costs

The costs related to equipment, infrastructure, and installation represent the key components of the overall capital investment in desalination plants. In conventional systems, this includes the installation of pressure vessels, pumps, membranes, and piping. However, in RES-based membrane systems, additional equipment for energy generation and storage must be considered [78]. For example, solar energy systems require PV modules, mounting structures, and wiring to integrate with the desalination plant's power system, while wind-powered systems impose turbines, substations, and extensive cabling. All of

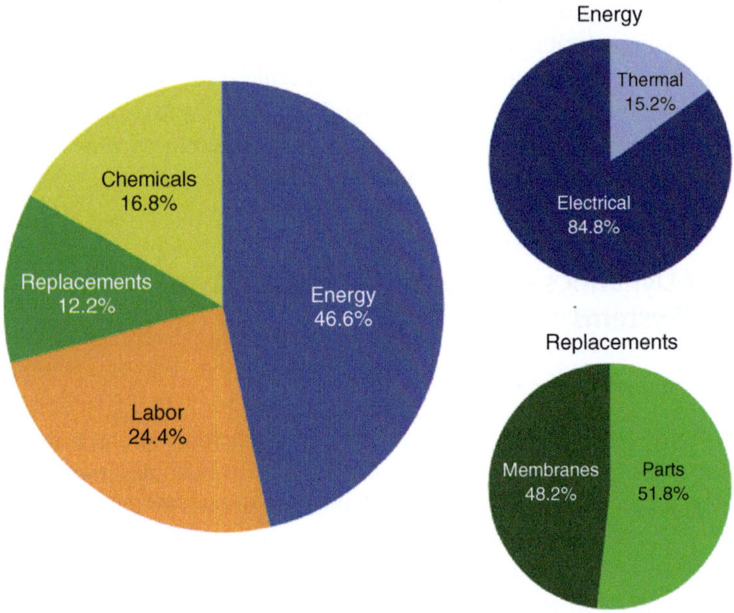

Figure 15.9 Common breakdown of operational costs in desalination processes. Source: Nassrullah et al. [38]/with permission of Elsevier.

these components increase capital expenditures. Meanwhile, the cost of 8-inch-RO membrane (spiral wound) is $470/module, while the cost of an 8-inch hollow fiber membrane is $2,000/module. Furthermore, installation costs for renewable energy infrastructure are generally higher due to the specialized equipment and technical expertise required [79]. Moreover, RESs often require significant land acquisition, especially for solar arrays or wind farms, which can markedly impact the total investment depending on the project's scale and location.

15.3.1.2 Economic Considerations in Plant Design and Sizing

Plant size considerably affects both initial capital investment and ongoing operational expenses. Larger facilities usually achieve cost-efficiencies through equipment procurement, installation, and maintenance. For example, according to SWA, the total project cost of the Shoaiba power and desalination plant in Saudi Arabia is $5 billion. The total project cost of the Jubail 3B RO desalination plant in Saudi Arabia is $692 million. It was designed to produce 570,000 m^3 of water per day in which renewable energy and RO desalination are coupled. However, combining renewable energy introduces complexity, as both the membrane system and renewable energy infrastructure must be carefully sized to achieve a balance between the energy supply and system demand. Given the energy variability (intermittent nature) of renewable energy, the capacity of the energy generation system should align with the desalination plant's load requirements as well as account for energy storage capabilities to ensure continuous desalination [78, 80]. However, oversizing the renewable energy system could lead to increased capital costs, while undersizing may cause operational inefficiencies and increased downtime. In hybrid systems, where multiple renewable sources are integrated, the design should consider the different combined energy resources to ensure continuous and efficient plant operation along with electrical synchronization and a grid-connection approach. To optimize this balance, advanced modeling and simulation techniques must be utilized at the early design stages to identify the most cost-effective configuration for RES-based membrane desalination systems.

15.3.1.3 Operational and Maintenance Costs

The preventive and predictive maintenance requirements for membrane systems powered by RESs generally surpass those for conventional systems due to the need for constant repair of the renewable infrastructure [81]. For example, PV panels require regular cleaning to eliminate dust and debris that can hinder sunlight and decrease efficiency, especially in areas with high soiling rates. Likewise, WTs entail frequent inspections and maintenance of mechanical parts, including blades, bearings, and gearboxes. This is to ensure their optimal functionality and avoid operational failures. Furthermore, membrane systems, regardless of their energy source, require routine maintenance to address operational-related issues, such as fouling, scaling, and membrane degradation. Therefore, using advanced membrane technologies, such as those with low-fouling or anti-scaling properties, can help minimize maintenance needs and reduce overall operational costs.

15.3.1.4 Downtime and Lifespan of Renewable Energy Components

The operational reliability of desalination systems powered by renewable energy is directly related to the durability and downtime of key renewable energy components. For instance, PV panels normally have a lifespan of 20–25 years. In addition, their efficiency gradually declines over time. WTs that typically last between 15 and 20 years face more rapid wear and tear on their mechanical parts [81]. These lifespans extensively impact the long-term financial agenda of desalination facilities. Also, in systems that lack sufficient energy storage, interruptions in RES generation can lead to reduced or inconsistent water output. Another concern is the downtime due to maintenance or system malfunctions, which could diminish the availability of renewable energy for desalination, adversely affecting water production. This is considered an issue, especially in areas with high water demands where even brief disruptions in desalination operations can lead to considerable social and economic consequences. Concisely, PV panels and WTs decay over relatively longer time periods than renewable energy transmission infrastructure, which is essential to electrify membrane systems.

15.3.2 Levelized Cost of Water Models

To compare the costs of conventional and renewable energy-based membrane desalination systems, a levelized cost of water (LCOW) model can be followed. Capital expenditures (CAPEX), capital recovery factor (CRF), and OPEX are used to calculate the cost of desalinated water.

15.3.2.1 Conventional Energy-Based Membrane Systems

For conventional systems, the annualized total cost C_{conv} is the sum of $CAPEX_{conv}$ multiplied by CRF, along with $OPEX_{conv}$. It can be expressed as follows [82]:

$$C_{conv} = CAPEX_{conv} \times CRF + OPEX_{conv} \tag{15.1}$$

In this context, the capital investment represents the initial costs of purchasing the membrane system and connecting it to the conventional electricity grid. Operational expenses include the cost of energy consumption (usually grid electricity or fossil fuels), labor costs, and other recurring costs. Additionally, maintenance costs cover membrane system replacement and servicing. Each of these components can be detailed as follows:

$$CAPEX_{conv} = C_{mem} + C_{inst} + C_{grid} \tag{15.2}$$

where

C_{mem} is the cost of the membrane system, C_{inst} is the cost of installation, and C_{grid} is the cost of the grid connection and power supply infrastructure. Additionally, CRF is a financial term used to calculate the amount of a fixed annual payment required to recover the initial investment ($CAPEX_{conv}$) over a specified period (n), with interest rate (r) taken into account:

$$CRF = \frac{r(1+r)^n}{((1+r)^n - 1)} \tag{15.3}$$

On the other hand, the operational and maintenance costs can be calculated using the following equation:

$$OPEX_{conv} = P_{mem} \times t \times C_{elec} + C_{main} + C_{labor} + C_{membrane,rep} \tag{15.4}$$

where

P_{mem}, t, and C_{elec} are the power consumption of the membrane system (kW), operating time (hr), and cost of electricity ($\$/kWh$), respectively. C_{labor} is the labor cost, while $C_{membrane,rep}$ represents the cost of membrane replacement. In addition,

C_{main} characterizes the system maintenance costs, which can be calculated using the following equation:

$$C_{main} = C_m \times t \tag{15.5}$$

where

C_m is the maintenance cost per unit time (\$/year) and t denotes the operational period considered for maintenance. By using the above terms, the LCOW can be calculated using the following equation:

$$LCOW = \frac{C_{conv}}{Q_{p,ann}} \tag{15.6}$$

where

$Q_{p,\,ann}$ is the annual freshwater produced/treated from the system.

15.3.2.2 Renewable Energy-Based Membrane Systems

For RE-based systems, the total cost C_{ren} can be expressed similarly but with additional factors to account for renewable energy infrastructure and storage [56]:

$$C_{ren} = CAPEX_{ren} \times CRF + OPEX_{ren} \tag{15.7}$$

The $CAPEX_{ren}$ includes the initial cost of both the membrane system and renewable energy generation and storage infrastructure. In addition, $OPEX_{ren}$ represents the operational and maintenance costs, which are mainly related to the storage, membrane, and renewable energy components. The $CAPEX_{ren}$ can be represented using the following equation:

$$CAPEX_{ren} = C_{mem} + C_{inst} + C_{renewables} + C_{storage} \tag{15.8}$$

where

$C_{renewables}$ represents the cost of renewable energy infrastructure (for example, solar panels or WTs). In addition, $C_{storage}$ denotes the cost of energy storage systems (such as batteries or thermal energy storage). On the other hand, the OPEX can be estimated using the following equation:

$$OPEX_{ren} = P_{storage} \times t \times C_{storage/m} + C_{labor} + C_{main} + C_{membrane,rep} \tag{15.9}$$

where

$P_{storage}$ and $C_{storage/m}$ are the energy used for storage (kW) due to the system's dependence on renewable energy and the cost related to energy storage maintenance or management (\$/kW/year). The total maintenance cost, C_{main}, is expressed using the following equation:

$$C_{main} = (C_{mem} + C_{m,ren}) \times t \tag{15.10}$$

In the above-mentioned equation, C_{mem} denotes the maintenance cost of the membrane system, while $C_{m,ren}$ refers to the maintenance cost of the renewable energy infrastructure (for instance, solar panels or WTs). Then, by using the above terms, the LCOW using RESs can be calculated using the following equation:

$$LCOW = \frac{C_{ren}}{Q_{p,ann}} \quad (15.11)$$

15.3.2.3 Levelized Water Cost Model for Reverse Osmosis System

In addition to what has been mentioned earlier, there are several models dedicated to calculating the cost of the produced water from the RO system. These models consider specific terms, such as the permeate flux, recovery ratio, membrane characteristics, and electricity cost. These models have been proposed by some studies [83–86], where the capital investment, $C_{I\&P}$(\$), of the intake and pretreatment stage is expressed as follows:

$$C_{I\&P} = 997 \times Q_{I\&P}^{0.8} \quad (15.12)$$

where

$Q_{I\&P}$(m^3/day) is the inlet flow rate to the intake stage. Hence, the operating cost, $OC_{I\&P}$(\$/day), of this stage can be calculated using the following relation:

$$OC_{I\&P} = \frac{0.028 \times P_{I\&P} \times Q_{I\&P} \times C_{elect}}{\eta_{I\&P}} \times lf \quad (15.13)$$

where

$P_{I\&P}$, C_{elect}, and lf are the intake pressure (bar), stage electricity cost (\$/kWh), and load factor, respectively. Hence, $\eta_{I\&P}$ is the efficiency of the intake pump. Other costs including the costs incurred on high-pressure pumps, energy recovery devices, and booster pumps can be estimated using the following relations [83–86]. The capital cost of the high-pressure pump, C_{HP}(\$) is:

$$C_{HP} = \frac{Q_f \times (393,000 + 10,710 \times P_f)}{30,000} \quad (15.14)$$

where

Q_f(m^3/day) is the flow rate of the feed and P_f(bar) is the feed pressure to the RO system. On the other hand, the operating cost of the high-pressure pump, OC_{HP}(\$/day), is estimated using the following relation:

$$OC_{HP} = \frac{0.028 \times P_f \times Q_f \times C_{elect}}{\eta_{HP}} \times lf \quad (15.15)$$

Similarly, the capital cost of the booster pump is as follows:

$$C_{BP} = \frac{(Q_f - Q_p) \times (393,000 + 10,710 \times P_f)}{60,000} \quad (15.16)$$

Also, the operating cost of the booster pump is as follows:

$$OC_{BP} = \frac{0.028 \times \Delta P_f \times \eta_{ERD} \times (Q_f - Q_p) \times C_{elec}}{\eta_{BP}} \times lf \quad (15.17)$$

where

ΔP_f is the feed pressure difference through the pressure vessel (bar), η_{ERD} is the energy recovery device efficiency, and η_{BP} is the booster pump efficiency. In addition, the capital cost of the energy recovery device is calculated using the following relation:

$$C_{ERD} = \frac{(Q_f - Q_p) \times (393,000 + 10,710 \times P_f)}{40,000} \tag{15.18}$$

Considering that there is a linear relationship between the membrane cost and its area, the capital cost of the RO membrane is expressed using the following relation [87]:

$$C_{mem} = C_{1,mem} \times A_{mem} \tag{15.19}$$

where

$C_{1,mem}$ is the cost of one membrane ($/m^2) and A_{mem} is the total membrane area (m^2). Therefore, the total cost of instruments, $C_{instruments}$, can be expressed as follows:

$$C_{instruments} = C_{I\&P} + C_{HP} + C_{BP} + C_{ERD} + C_{mem} \tag{15.20}$$

The site cost can represent 20% of the total instrumentation cost, while the indirect capital costs are set at 30% of the total direct costs [85]:

$$C_{site} = 20\% \times C_{instruments} \tag{15.21}$$

$$InC = 30\% \times (C_{instruments} + C_{site}) \tag{15.22}$$

Consequently, the total capital cost can be expressed as follows:

$$CAPEX = C_{instruments} + C_{site} + InC \tag{15.23}$$

The annual operating cost of power (OC_{power}, $/year) and membranes ($OC_{mem}$, $/year) and other costs ($OC_{other}$, $/year) can be expressed using the following relations [83–85]:

$$OC_{power} = (OC_{I\&P} + OC_{HP} + OC_{BP}) \times 365 \tag{15.24}$$

$$OC_{mem} = 20\% \times C_{mem} \tag{15.25}$$

$$OC_{other} = 30\% \times OPEX \tag{15.26}$$

$$OPEX = OC_{power} + OC_{mem} + OC_{other} \tag{15.27}$$

Then, the LCOW of the RO system can be calculated using the following relation:

$$LCOW = \frac{CAPEX \times CRF + OPEX}{Q_{p,ann} \times lf} \tag{15.28}$$

15.3.3 Sensitivity Analysis of Energy Prices and Other Key Factors

Several factors can impact the cost profile of desalination plants. Therefore, sensitivity analysis plays an important role in predicting the effects of these parameters on water costs across different scenarios [78, 81]. The energy price, especially when the desalination plant relies on conventional resources, is one of these factors. Hence, the fluctuations in fuel prices can cause major variations in OPEX, which impacts the economic feasibility of the plant. On the other hand, for plants dependent on RESs, the sensitivity analysis may consider the variation in the available natural resources (for example, variations in solar irradiance, clearness index, and wind velocity), which in turn impact the produced power that drives the desalination processes. In addition, advances in the cost of membrane materials, hydraulic equipment, or chemicals can shift the cost dynamics. For example, a 10% reduction in the cost of PV panels can significantly reduce the payback period (PBP) and improve overall financial returns. However, a 10% increase in renewable energy systems can increase the PBP, making other systems more competitive. Therefore, conducting a sensitivity analysis for these types of projects or studies is fundamental. This enables decision makers to assess potential risks and make informed adjustments to optimize performance under varying conditions.

15.3.4 Economic and Environmental Benefits of Renewable Energy Integration

15.3.4.1 Impact on Greenhouse Gas Emissions

Conventional desalination plants are responsible for significant emissions of CO_2 due to their energy-intensive operations. These plants can emit up to 6.7 kg of CO_2 per cubic meter of desalinated water [88]. In contrast, RE-based desalination systems generate minimal greenhouse gas (GHG) emissions. For instance, solar-powered plants utilize clean, abundant sunlight to drive desalination processes. This approach eliminates the carbon footprint associated with fossil fuel combustion. Furthermore, wind-powered desalination plants similarly leverage wind energy, further contributing to a reduction in CO_2 emissions. The cumulative effect of reduced emissions is particularly critical in regions facing strict environmental regulations or international climate commitments. In such areas, renewable energy integration supports both national and global efforts to combat climate change. The reduction in GHG emissions, $CO_2Savings$, by using renewable energy instead of fossil fuels can be calculated as [89]:

$$CO_2Savings = Q_p \times (CO_{2,conv} - CO_{2,ren}) \tag{15.29}$$

where

$CO_{2,conv}$ and $CO_{2,ren}$ are the CO_2 emissions of conventional desalination systems and RE-based systems, respectively.

15.3.4.2 Long-Term Financial Benefits

After installation, RESs like solar or wind are almost cost-free, resulting in very low ongoing energy expenses. In contrast, conventional plants face high and variable energy costs in their operating expenses because they depend on fossil fuels. For example, solar-powered desalination systems in regions with high solar irradiance can reduce energy costs by up to 80% in comparison with conventional fossil fuel-powered plants. In SWA's case, plants driven by renewable resources have witnessed a 20–30% reduction in energy costs compared to those relying solely on conventional energy. The lifespan of membranes has also improved through better pretreatment processes, reducing the frequency of replacement and driving down long-term operational costs. As a result, these systems become attractive from an economic perspective, especially in the face of rising global energy prices and energy supply instability [80]. Furthermore, RE-based desalination plants benefit from long-term price stability as they are insulated from the unpredictability of fossil fuel markets. This ensures more predictable and manageable operating costs, which contribute to improved financial planning and resource allocation for plant operators. The PBP is an important metric for evaluating the economic feasibility of such systems. The PBP for RE-based systems is generally between 8 and 15 years. This depends on the specific technology employed, the geographic location, and the availability of renewable resources. After this period, the plant continues to operate at reduced OPEX, resulting in considerable cost savings over the entire lifetime. The PBP of a system can be calculated using the following equation:

$$PP = \frac{CAPEX_{ren}}{OPEX_{conv} - OPEX_{ren}} \tag{15.30}$$

15.4 Challenges and Barriers to Economic Feasibility: An Analysis

The integration of RESs into membrane desalination systems presents numerous potential benefits. However, several technological, financial, and market barriers hinder their widespread adoption.

15.4.1 Technological Barriers

RESs like solar and wind can be unreliable, which makes it difficult to provide a steady energy supply for membrane desalination. RO and other membrane systems require continuous operation to maintain efficiency and avoid frequent membrane fouling. Energy storage systems such as batteries and fuel cells could mitigate these fluctuations. However, they remain costly and have a limited lifetime, which increases the capital expenses of the system. Also, although the use of ERDs is important in enhancing energy efficiency in desalination, they must be compatible with variable energy inputs from RESs to achieve the maximum economic and technical efficiency.

15.4.2 Financial Barriers

Energy storage units, PV panels, and WTs are all examples of renewable energy infrastructure that require significant upfront capital investment. Although the OPEX for RE-based systems are lower due to reduced energy costs, the CAPEX of such installations remain a restriction. This is especially true for developing regions where desalination is critically needed. Furthermore, access to financing for RE-based systems is limited due to perceived risks in the financial sector. In addition, the limited long-term data on RES integration with water treatment systems slow investment and increase financing costs.

15.4.3 Market and Infrastructure Limitations

One of the limitations that hinders the widespread adoption of RESs is the lack of established supply chains for renewable energy components and skilled labor. In addition, local infrastructure may not be well developed and ready enough to support the deployment of large-scale renewable installations. Also, the absence of regulations or incentives to promote the use of renewables in desalination delays adoption further.

15.5 Suggested Strategies to Overcome Challenges

Several approaches can be implemented to overcome these obstructions:

- Developing advanced energy storage systems with lower costs and longer lifetime could significantly reduce the intermittency problem in RESs. Furthermore, integrating smart grid technologies to better manage the fluctuating energy supply could enhance the stability and reliability of these desalination systems.
- Governments and international organizations could offer subsidies, low interest rates, or tax incentives. These measures could encourage investments in zero-emission desalination systems.
- Training programs focused on RESs and membrane technologies would help develop the required workforce. Capacity-building initiatives could further strengthen local expertise in these fields. Creating regional markets for renewable energy components would also help lower expenses. Additionally, establishing supportive infrastructure could reduce costs associated with transportation and installation.

15.6 Conclusion

This chapter has conducted a comprehensive exploration of the integration of diverse membrane technologies with renewable energy systems. A broad spectrum of common RESs including solar, wind, and geothermal are examined for their suitability in desalination applications. Particular attention is given to the technological advancements that have significantly enhanced the reliability, efficiency, and cost-effectiveness of renewable energy-based desalination systems. Pilot-scale examples related to renewable energy used by the SWA in Saudi Arabia are also analyzed. Financial considerations including investment requirements, operational costs, and the lifespan of such systems are analyzed. In SWA's case, the integration of RESs such as solar power has reduced these costs further. For example, plants driven by renewable resources have witnessed a 20–30% reduction in energy costs compared to those relying solely on conventional energy resources. LCOW models are presented for both conventional and renewable energy-based desalination plants. The economic and environmental implications of these systems are examined, considering factors such as energy consumption and GHG emissions. Finally, the chapter delves into the challenges and barriers that may hinder the widespread adoption of renewable energy-based desalination. This includes intermittent energy supply, grid integration issues, and regulatory barriers. Potential solutions to overcome these obstacles were proposed, such as the development of hybrid energy systems, improved grid infrastructure, and supportive policy frameworks. By addressing these challenges and leveraging the advancements in membrane technologies and RESs, it is anticipated that these systems will play a key role in meeting the growing global demand for freshwater. At the same time, they will promote sustainable, cost-effective, and environmentally responsible water management practices.

Questions

1 What factors influence the efficiency of membrane separation processes in desalination?

2 Explain why reverse osmosis (RO) has become the predominant technology globally, especially in arid regions.

3 How do energy recovery devices (ERDs) improve the economic feasibility of RO desalination systems?

4 What advantages do hybrid energy systems offer for large-scale desalination facilities in arid regions?

5 How do investment costs differ between conventional and renewable energy sources (RESs) for desalination systems?

6 Why is the sizing of renewable energy systems critical in designing RES-based desalination plants?

7 What are some maintenance challenges associated with desalination systems powered by RESs?

8 How does sensitivity analysis help in assessing the economic feasibility of desalination plants?

Acknowledgments

The authors extend their gratitude to the Science, Technology and Innovation Funding Authority (STDF) of Egypt for their generous support of this chapter under Grant IDs: 46931 and 47373.

References

1 Werber, J.R., Deshmukh, A., Elimelech, M. et al. (2016). The critical need for increased selectivity, not increased water permeability, for desalination membranes. *Environ. Sci. Technol. Lett.* 3: https://doi.org/10.1021/acs.estlett.6b00050.

2 Cay-Durgun, P. and Lind, M.L. (2018). Nanoporous materials in polymeric membranes for desalination. *Curr. Opin. Chem. Eng.* 20: https://doi.org/10.1016/J.COCHE.2018.01.001.

3 Qasim, M., Darwish, N.N., Mhiyo, S. et al. (2018). The use of ultrasound to mitigate membrane fouling in desalination and water treatment. *Desalination* 443: https://doi.org/10.1016/J.DESAL.2018.04.007.

4 Shalaby, S.M., Zayed, M.E., Hammad, F.A. et al. (2024). Recent advances in membrane distillation hybrids for energy-efficient process configurations: technology categorization, operational parameters identification, and energy recovery strategies. *Process Saf. Environ. Prot.* 190: https://doi.org/10.1016/J.PSEP.2024.07.098.

5 Amy, G., Ghaffour, N., Li, Z. et al. (2017). Membrane-based seawater desalination: present and future prospects. *Desalination* 401: https://doi.org/10.1016/J.DESAL.2016.10.002.

6 Mansi, A.E., El-Marsafy, S.M., Elhenawy, Y. et al. (2023). Assessing the potential and limitations of membrane-based technologies for the treatment of

oilfield produced water. *Alex. Eng. J.* 68: https://doi.org/10.1016/J.AEJ.2022.12.013.

7 Maddah, H.A., Alzhrani, A.S., Bassyouni, M. et al. (2018). Evaluation of various membrane filtration modules for the treatment of seawater. *Appl. Water Sci.* 8: https://doi.org/10.1007/S13201-018-0793-8/FIGURES/16.

8 Soliman, M.N., Guen, F.Z., Ahmed, S.A. et al. (2021). Energy consumption and environmental impact assessment of desalination plants and brine disposal strategies. *Process Saf. Environ. Prot.* 147: https://doi.org/10.1016/J.PSEP.2020.12.038.

9 Okamoto, Y. and Lienhard, J.H. (2019). How RO membrane permeability and other performance factors affect process cost and energy use: a review. *Desalination* 470: https://doi.org/10.1016/J.DESAL.2019.07.004.

10 Cherif, H. and Belhadj, J. (2018). Environmental life cycle analysis of water desalination processes. In: *Sustainable Desalination Handbook* (ed. V.G. Gude), 527–559. Elsevier.

11 Han, S.-H. and Chang, I.-S. (2013). Comparison of nitrate and fluoride removals between reverse-osmosis, nano-flitration, electro-adsorption, elecero-coagulation in small water treatment plants. *J. Korea Academia-Ind. coop. Soc.* 14: https://doi.org/10.5762/KAIS.2013.14.4.2027.

12 Khoo, Y.S., Goh, P.S., Lau, W.J. et al. (2022). Removal of emerging organic micropollutants via modified-reverse osmosis/nanofiltration membranes: a review. *Chemosphere* 305: https://doi.org/10.1016/J.CHEMOSPHERE.2022.135151.

13 Eloffy, M.G., El-Sherif, D.M., Abouzid, M. et al. (2021). Proposed approaches for coronaviruses elimination from wastewater: membrane techniques and nanotechnology solutions. *Nanotechnol. Rev.* 11: https://doi.org/10.1515/ntrev-2022-0001.

14 Harby, K., Emad, M., Benghanem, M. et al. (2024). Reverse osmosis hybridization with other desalination techniques: an overview and opportunities. *Desalination* 581: https://doi.org/10.1016/J.DESAL.2024.117600.

15 Jones, E., Qadir, M., van Vliet, M.T.H. et al. (2019). The state of desalination and brine production: a global outlook. *Sci. Total Environ.* 657: https://doi.org/10.1016/J.SCITOTENV.2018.12.076.

16 Liu, H., Jacob, K.I., Wang, Y. et al. (2024). Scaling in membrane distillation (MD): current state of art insight on mechanisms and membrane design. *Desalination* 581: https://doi.org/10.1016/J.DESAL.2024.117539.

17 Ministry of Environment, Water and Agriculture. (2018). Saudi Arabia's National Water Strategy 2030.

18 Ministry of Environment, Water and Agriculture. (2024). Water Innovation in Saudi Arabia.

19 Wilson, A.D. and Stewart, F.F. (2013). Deriving osmotic pressures of draw solutes used in osmotically driven membrane processes. *J. Membr. Sci.* 431: https://doi.org/10.1016/J.MEMSCI.2012.12.042.

20 Zhu, T., Zhang, Y., Li, Y. et al. (2023). Assessment of organic micropollutants rejection by forward osmosis system using interpretable machine learning-assisted approach: a new perspective on optimization of multifactorial forward osmosis process. *J. Environ. Chem. Eng.* 11: https://doi.org/10.1016/J.JECE.2023.110847.

21 Eyvaz, M., Arslan, S., İmer, D. et al. (2018). Forward osmosis membranes—a review: part I. In: *Osmotically Driven Membrane Processes—Approach, Development and Current Status* (ed. H. Du, A. Thompson, and X. Wang). IntechOpen. https://doi.org/10.5772/INTECHOPEN.72287.

22 Alaswad, S.O. and Almatrafi, E. (2024). Artificial intelligence applications in forward osmosis for water treatment: recent developments and research directions. *Desalin. Water Treat.* 319: https://doi.org/10.1016/j.dwt.2024.100551.

23 Mahto, A., Aruchamy, K., Meena, R. et al. (2021). Forward osmosis for industrial effluents treatment—sustainability considerations. *Sep. Purif. Technol.* 254: https://doi.org/10.1016/J.SEPPUR.2020.117568.

24 Yuchen, D.U., Pramanik, B.K., Zhang, Y. et al. (2022). Recent advances in the theory and application of nanofiltration: a review. *Curr. Pollut. Rep.* 8: https://doi.org/10.1007/s40726-021-00208-1.

25 Jawed, A.S., Nassar, L., Hegab, H.M. et al. (2024). Recent developments in solar-powered membrane distillation for sustainable desalination. *Heliyon* 10: https://doi.org/10.1016/J.HELIYON.2024.E31656.

26 Elhenawy, Y., Fouad, K., Bassyouni, M. et al. (2023). Design and performance a novel hybrid membrane distillation/humidification–dehumidification system. *Energy Convers. Manag.* 286: https://doi.org/10.1016/J.ENCONMAN.2023.117039.

27 Ahmad, N.A., Goh, P.S., Yogarathinam, L.T. et al. (2020). Current advances in membrane technologies for produced water desalination. *Desalination* 493: https://doi.org/10.1016/J.DESAL.2020.114643.

28 Guest, J.S., Skerlos, S.J., Barnard, J.L. et al. (2009). A new planning and design paradigm to achieve sustainable resource recovery from wastewater. *Environ. Sci. Technol.* 43: https://doi.org/10.1021/ES9010515/SUPPL_FILE/ES9010515_SI_001.PDF.

29 Tavangar, T., Karimi, M., Rezakazemi, M. et al. (2020). Textile waste, dyes/inorganic salts separation of cerium oxide-loaded loose nanofiltration polyethersulfone membranes. *Chem. Eng. J.* 385: https://doi.org/10.1016/J.CEJ.2019.123787.

30 Liu, Y., Ran, J., Guo, Q. et al. (2024). Advancements in nanofiltration fouling phenomenon: from water treatment to salt lakes environments. *Desalination* 583: https://doi.org/10.1016/J.DESAL.2024.117649.

31 Russo, F., Galiano, F., Pedace, F. et al. (2020). Dimethyl isosorbide as a green solvent for sustainable ultrafiltration and microfiltration membrane preparation. *ACS Sustain. Chem. Eng.* 8: https://doi.org/10.1021/ACSSUSCHEMENG.9B06496/SUPPL_FILE/SC9B06496_SI_001.PDF.

32 Dai, R., Han, H., Wang, T. et al. (2021). Cleaning-healing-interfacial polymerization strategy for upcycling real end-of-life polyvinylidene fluoride microfiltration membranes. *ACS Sustain. Chem. Eng.* 9: https://doi.org/10.1021/ACSSUSCHEMENG.1C03481/SUPPL_FILE/SC1C03481_SI_001.PDF.

33 Dahanayaka, M. and Chew, J.W. (2022). Organic solvent permeation through negatively charged graphene oxide membranes. *ACS Sustain. Chem. Eng.* 10: https://doi.org/10.1021/ACSSUSCHEMENG.1C06824/SUPPL_FILE/SC1C06824_SI_001.PDF.

34 Sánchez-Arévalo, C.M., Vincent-Vela, M.C., Luján-Facundo, M.-J. et al. (2023). Ultrafiltration with organic solvents: a review on achieved results, membrane materials and challenges to face. *Process Saf. Environ. Prot.* 177: https://doi.org/10.1016/J.PSEP.2023.06.073.

35 Shi, W., Yang, C., Qiu, M. et al. (2022). A new method for preparing α-alumina ultrafiltration membrane at low sintering temperature. *J. Membr. Sci.* 642: https://doi.org/10.1016/J.MEMSCI.2021.119992.

36 Castro-Muñoz, R., Boczkaj, G., Gontarek, E. et al. (2020). Membrane technologies assisting plant-based and agro-food by-products processing: a comprehensive review. *Trends Food Sci. Technol.* 95: https://doi.org/10.1016/J.TIFS.2019.12.003.

37 Alawad, S.M., Mansour, R.B., Al-Sulaiman, F.A. et al. (2023). Renewable energy systems for water desalination applications: a comprehensive review. *Energy Convers. Manag.* 286: https://doi.org/10.1016/J.ENCONMAN.2023.117035.

38 Nassrullah, H., Anis, S.F., Hashaikeh, R. et al. (2020). Energy for desalination: a state-of-the-art review. *Desalination* 491: https://doi.org/10.1016/J.DESAL.2020.114569.

39 Ghaffour, N., Bundschuh, J., Mahmoudi, H. et al. (2015). Renewable energy-driven desalination technologies: a comprehensive review on challenges and potential applications of integrated systems. *Desalination* 356: https://doi.org/10.1016/J.DESAL.2014.10.024.

40 Ahmed, A., Elsakka, M., Elminshawy, N. et al. (2023). Recent advances in floating photovoltaic systems. *Chem. Rec.* 23: https://doi.org/10.1002/tcr.202300229.

41 Ahmed, A., Alzahrani, M., Shanks, K. et al. (2022). Reliability and temperature limits of the focal spot of a primary optical component for an ultra-high concentrated photovoltaic system. *AIP Conf. Proc.* 2550: https://doi.org/10.1063/5.0099091.

42 Elminshawy, N.A.S., Ahmed, A., Osama, A. et al. (2024). The potential of optimized floating photovoltaic system for energy production in the Northern Lakes of Egypt. *Eng. Anal. Bound. Elem.* 161: https://doi.org/10.1016/J.ENGANABOUND.2024.01.022.

43 Mito, M.T., Ma, X., Albuflasa, H. et al. (2019). Reverse osmosis (RO) membrane desalination driven by wind and solar photovoltaic (PV) energy: state of the art and challenges for large-scale implementation. *Renew. Sust. Energ. Rev.* 112: https://doi.org/10.1016/j.rser.2019.06.008.

44 Elmarghany, M.R., Radwan, A., Abdelhay, Y. et al. (2023). Experimental study of a standalone membrane water desalination unit fully powered by solar energy. *Desalination* 553: https://doi.org/10.1016/J.DESAL.2023.116476.

45 Tlili, I., Sajadi, S.M., Baleanu, D. et al. (2022). Flat sheet direct contact membrane distillation study to decrease the energy demand for solar desalination purposes. *Sustain. Energy Technol. Assess.* 52: https://doi.org/10.1016/J.SETA.2022.102100.

46 Mohamed, E.S. and Papadakis, G. (2004). Design, simulation and economic analysis of a stand-alone reverse osmosis desalination unit powered by wind turbines and photovoltaics. *Desalination* 164: https://doi.org/10.1016/S0011-9164(04)00159-6.

47 Al-Karaghouli, A. and Kazmerski, L.L. (2013). Energy consumption and water production cost of conventional and renewable-energy-powered desalination processes. *Renew. Sust. Energ. Rev.* 24: https://doi.org/10.1016/J.RSER.2012.12.064.

48 Rezk, H., Sayed, E.T., Al-Dhaifallah, M. et al. (2019). Fuel cell as an effective energy storage in reverse osmosis desalination plant powered by photovoltaic system. *Energy* 175: https://doi.org/10.1016/J.ENERGY.2019.02.167.

49 Abdullah, A.S., Panchal, H., Alawee, W.H. et al. (2023). Methods used to improve solar still performance with generated turbulence for water desalination-detailed review. *Results in Eng.* 19: https://doi.org/10.1016/j.rineng.2023.101251.

50 Khanjanpour, M.H. and Javadi, A.A. (2021). Optimization of a horizontal axis tidal (HAT) turbine for powering a reverse osmosis (RO) desalination system using computational fluid dynamics (CFD) and Taguchi method. *Energy Convers. Manag.* 231: https://doi.org/10.1016/j.enconman.2021.113833.

51 Delgado-Torres, A.M., García-Rodríguez, L., del Moral, M.J. et al. (2020). Preliminary assessment of innovative seawater reverse osmosis (SWRO) desalination powered by a hybrid solar photovoltaic (PV)—tidal range energy system. *Desalination* 477: https://doi.org/10.1016/j.desal.2019.114247.

52 Rashidi, M.M., Mahariq, I., Murshid, N. et al. (2022). Applying wind energy as a clean source for reverse osmosis desalination: a comprehensive review. *Alex. Eng. J.* 61: https://doi.org/10.1016/J.AEJ.2022.06.056.

53 Ma, Q. and Lu, H. (2011). Wind energy technologies integrated with desalination systems: review and state-of-the-art. *Desalination* 277: https://doi.org/10.1016/J.DESAL.2011.04.041.

54 Manesh, M.H.K., Davadgaran, S., Rabeti, S.A.M. et al. (2024). Optimal 4E evaluation of an innovative solar-wind cogeneration system for sustainable power and fresh water production based on integration of microbial desalination cell, humidification-dehumidification, and reverse osmosis desalination. *Energy* 297: https://doi.org/10.1016/j.energy.2024.131256.

55 Raninga, M., Mudgal, A., Patel, V. et al. (2024). Energy, exergy and economic (3E) analysis of solar thermal energy assisted cascade Rankine cycle for reverse osmosis. *Energy* 293: https://doi.org/10.1016/j.energy.2024.130571.

56 Elsakka, M.M., Refaat, A., Alzahrani, K.M. et al. (2024). Techno-economic assessment of a novel wind-powered RO system with a compressed air energy storage for water desalination. *Energy* 311: https://doi.org/10.1016/J.ENERGY.2024.133296.

57 Xia, L. and Zhang, Y. (2019). An overview of world geothermal power generation and a case study on China—the resource and market perspective. *Renew. Sust. Energ. Rev.* 112: https://doi.org/10.1016/J.RSER.2019.05.058.

58 Prajapati, M., Shah, M., Soni, B. et al. (2021). Geothermal-solar integrated groundwater desalination system: current status and future perspective. *Groundw. Sustain. Dev.* 12: https://doi.org/10.1016/J.GSD.2020.100506.

59 Orfi, J., Najib, A., Ali, E. et al. (2017). Membrane distillation and reverse osmosis based desalination driven by geothermal energy sources. *Desalin. Water Treat.* 76: https://doi.org/10.5004/DWT.2016.11378.

60 Bamasag, A., Almatrafi, E., Alqahtani, T. et al. (2023). Recent advances and future prospects in direct solar desalination systems using membrane distillation technology. *J. Clean. Prod.* 385: https://doi.org/10.1016/J.JCLEPRO.2022.135737.

61 Schunke, A.J., Hernandez Herrera, G.A., Padhye, L. et al. (2020). Energy recovery in SWRO desalination: current status and new possibilities. *Front. Sustain. Cities.* 2: https://doi.org/10.3389/FRSC.2020.00009/BIBTEX.

62 Peñate, B. and García-Rodríguez Lourdes, L. (2011). Energy optimisation of existing SWRO (seawater reverse osmosis) plants with ERT (energy recovery turbines): technical and thermoeconomic assessment. *Energy* 36: https://doi.org/10.1016/J.ENERGY.2010.09.056.

63 Urrea, S.A., Reyes, F.D., Suárez, B.P. et al. (2019). Technical review, evaluation and efficiency of energy recovery devices installed in the Canary Islands desalination plants. *Desalination* 450: https://doi.org/10.1016/J.DESAL.2018.07.013.

64 Nashed, A. and Van Dang, B. (2022). Solar and water: high-technology readiness technologies. In: *Solar-Driven Water Treatment: Re-engineering and*

Accelerating Nature's Water Cycle (ed. O. Mahian, J. Wei, R.A. Taylor, and S. Wongwises). Academic Press. https://doi.org/10.1016/B978-0-323-90991-4.00007-4.

65 Takabatake, H., Taniguchi, M., Kurihara, M. et al. (2021). Advanced technologies for stabilization and high performance of seawater RO membrane desalination plants. *Membranes* 11: 138. https://doi.org/10.3390/MEMBRANES11020138.

66 Esmaeilion, F. (2020). Hybrid renewable energy systems for desalination. *Appl. Water Sci.* 10: https://doi.org/10.1007/S13201-020-1168-5/TABLES/11.

67 Saboori, H. (2023). Hybrid renewable energy powered reverse osmosis desalination—minimization and comprehensive analysis of levelized cost of water. *Sustain. Energy Technol. Assess.* 56: https://doi.org/10.1016/J.SETA.2023.103065.

68 Esfahani, I.J., Lee, S.C., Yoo, C.K. et al. (2015). Extended-power pinch analysis (EPoPA) for integration of renewable energy systems with battery/hydrogen storages. *Renew. Energy* 80: https://doi.org/10.1016/J.RENENE.2015.01.040.

69 Raza, S., Hayat, A., Bashir, T. et al. (2024). Recent progress in green thin film membrane based materials for desalination: design, properties and applications. *Desalination* 59: https://doi.org/10.1016/J.DESAL.2024.117973.

70 Yu, F., Shi, H., Shi, J. et al. (2020). High-performance forward osmosis membrane with ultra-fast water transport channel and ultra-thin polyamide layer. *J. Membr. Sci.* 616: https://doi.org/10.1016/J.MEMSCI.2020.118611.

71 Xu, Z., Li, P., Li, N. et al. (2021). Constructing dense and hydrophilic forward osmosis membrane by cross-linking reaction of graphene quantum dots with monomers for enhanced selectivity and stability. *J. Colloid Interface Sci.* 589: https://doi.org/10.1016/J.JCIS.2021.01.004.

72 Shan, M., Kang, H., Xu, Z. et al. (2019). Decreased cross-linking in interfacial polymerization and heteromorphic support between nanoparticles: towards high-water and low-solute flux of hybrid forward osmosis membrane. *J. Colloid Interface Sci.* 548: https://doi.org/10.1016/J.JCIS.2019.04.014.

73 Liu, G., Jin, W., Xu, N. et al. (2015). Graphene-based membranes. *Chem. Soc. Rev.* 44: https://doi.org/10.1039/c4cs00423j.

74 Perreault, F., De Faria, A.F., Elimelech, M. et al. (2015). Environmental applications of graphene-based nanomaterials. *Chem. Soc. Rev.* 46: https://doi.org/10.1039/C5CS00021A.

75 Liu, Q., Xu, G.R., Das, R. et al. (2019). Inorganic scaling in reverse osmosis (RO) desalination: mechanisms, monitoring, and inhibition strategies. *Desalination* 468: https://doi.org/10.1016/J.DESAL.2019.07.005.

76 Subramani, A. and Jacangelo, J.G. (2014). Treatment technologies for reverse osmosis concentrate volume minimization: a review. *Sep. Purif. Technol.* 122: https://doi.org/10.1016/J.SEPPUR.2013.12.004.

77 Amer, A.E., Ahmed, A., Nabil, T. et al. (2024). Performance evaluation of latent heat storage unit using novel hook-shaped fins: experimental investigation. *J. Energy Storage* 97: https://doi.org/10.1016/J.EST.2024.112747.

78 Eke, J., Yusuf, A., Giwa, A. et al. (2020). The global status of desalination: an assessment of current desalination technologies, plants and capacity. *Desalination* 495: https://doi.org/10.1016/J.DESAL.2020.114633.

79 Shokri, A. and Sanavi Fard, M. (2023). Techno-economic assessment of water desalination: future outlooks and challenges. *Process Saf. Environ. Prot.* 169: https://doi.org/10.1016/J.PSEP.2022.11.007.

80 Abdelkareem, M.A., Assad, M.E.H., Sayed, E.T. et al. (2018). Recent progress in the use of renewable energy sources to power water desalination plants. *Desalination* 435: https://doi.org/10.1016/J.DESAL.2017.11.018.

81 Tran, T.T.D. and Smith, A.D. (2017). Evaluation of renewable energy technologies and their potential for technical integration and cost-effective use within the U.S. energy sector. *Renew. Sust. Energ. Rev.* 80: https://doi.org/10.1016/J.RSER.2017.05.228.

82 Romero-Ternero, V., García-Rodríguez, L., Gómez-Camacho, C. et al. (2005). Thermoeconomic analysis of a seawater reverse osmosis plant. *Desalination* 181: https://doi.org/10.1016/J.DESAL.2005.02.012.

83 Vince, F., Marechal, F., Aoustin, E. et al. (2008). Multi-objective optimization of RO desalination plants. *Desalination* 222: https://doi.org/10.1016/J.DESAL.2007.02.064.

84 Marcovecchio, M.G., Aguirre, P.A., Scenna, N.J. et al. (2005). Global optimal design of reverse osmosis networks for seawater desalination: modeling and algorithm. *Desalination* 184: https://doi.org/10.1016/J.DESAL.2005.03.056.

85 Malek, A., Hawlader, M.N.A., Ho, J.C. et al. (1996). Design and economics of RO seawater desalination. *Desalination* 105: https://doi.org/10.1016/0011-9164(96)00081-1.

86 Choi, Y.J., Lee, S., Koo, J. et al. (2016). Evaluation of economic feasibility of reverse osmosis and membrane distillation hybrid system for desalination. *Desalin. Water Treat.* 57: https://doi.org/10.1080/19443994.2016.1152648.

87 Maskan, F., Wiley, D.E., Johnston, L.P.M. et al. (2000). Optimal design of reverse osmosis module networks. *AICHE J.* 46: 946–954.

88 Cornejo, P.K., Santana, M.V.E., Hokanson, D.R. et al. (2014). Carbon footprint of water reuse and desalination: a review of greenhouse gas emissions and estimation tools. *J. Water Reuse Desalin.* 4: https://doi.org/10.2166/WRD.2014.058.

89 Suraparaju, S.K. and Natarajan, S.K. (2021). Productivity enhancement of single-slope solar still with novel bottom finned absorber basin inserted in phase change material (PCM): techno-economic and enviro-economic analysis. *Environ. Sci. Pollut. Res.* 28: https://doi.org/10.1007/s11356-021-13495-4.

16

Practical Insights, Commercial Installations, and Future Projections

Asmaa Ahmed[1,2], *Mohamed Elsakka*[1,2], *Yasser Elhenawy*[1], *Asif Taher*[3], *Tapas Mallick*[3], *and Ahmed Amer*[1]

[1] *Mechanical Power Engineering Department, Faculty of Engineering, Port Said University, Port Said, Egypt*
[2] *Faculty of Engineering, East Port Said National University, Salam Misr City, Port Said, Egypt*
[3] *Environment and Sustainability Institute, University of Exeter, Penryn Campus, Cornwall, UK*

16.1 Introduction

Membrane-based water desalination is considered an important technology in addressing global water scarcity challenges, especially when the demand for freshwater increases and conventional water resources become more constrained. This chapter provides an overview of the technical, material, and operational aspects of membrane-based desalination systems. It also offers insights into real-world applications and future forecasts. The chapter begins by exploring the technical aspects of these systems with a focus on important factors that impact system performance, such as membrane permeability, fouling issues, and energy efficiency. The chapter also covers membrane fouling and scaling, which are the main operational challenges that affect system lifetime and efficiency. This includes advanced cleaning protocols and fouling mitigation strategies. In addition, cost management in operations and maintenance (O&M) emphasizes the need for efficient monitoring and optimization techniques to reduce operational expenses. Hence, innovative membrane materials are of importance. This includes advanced materials, such as graphene membranes, carbon nanotubes (CNTs), and bio-inspired materials. This can offer several enhancements to desalination processes. These improvements involve membrane selectivity, permeability, and durability. This section also highlights the potential of nanomaterials and functional composites to reduce energy consumption, the primary goal of which is improving fouling resistance and expanding the use of membrane desalination (MDS) in harsh environments. The chapter also gives information on MDS systems applied

Water Treatment and Desalination: Membrane Separation using Renewable Energy Resources, First Edition. Edited by Mohamed Bassyouni.

in various regions, showing the practical feasibility of integrating renewable energy sources (RESs). The chapter further emphasizes the future directions of MDS. This includes the development of hybrid energy systems with membrane technologies like reverse osmosis (RO) and membrane distillation (MD). These advances enable continuous off-grid operation in remote areas. Moreover, this chapter discusses the environmental and socioeconomic impacts of desalination. This is more relevant to developing countries where access to freshwater can accelerate economic growth and improve public health. Finally, the chapter focuses on political and regulatory overlooks that will form the future of MDS.

16.2 Technical Aspects of Membrane Systems

MDS technologies are highly efficient techniques. However, their performance, effectiveness, and durability are affected by numerous technical aspects. Thus, obtaining a complete understanding of these elements is necessary to improve the design and operation of such systems.

16.2.1 Membrane Performance

The effectiveness of the membrane is evaluated by two important parameters [1]. The first one is water flux which is defined as the amount of water passing through the membrane per unit area [2]. Salt rejection is another aspect that reveals how much salt is filtered out from the feedwater [3]. The material and structural details of the membrane are necessary in determining these factors. Polymeric membranes are usually used because they are affordable and adaptable. However, they tend to foul over time [4]. Instead, more advanced options like thin-film composite (TFC) and ceramic membranes deliver better performance, especially under challenging working conditions [5].

16.2.2 Fouling and Scaling Challenges

Fouling and scaling are the most important issues in the operation of MDS systems [6–8]. Fouling happens when particles, organic compounds, or microorganisms build up on the membrane's surface. This leads to a decrease in water flow, higher transmembrane pressure, and greater energy consumption [9]. Fouling is classified into organic, biological, colloidal, and inorganic [10]. Each of these types requires specific prevention methods. Inorganic scaling occurs when dissolved salts crystallize and form scale on the membrane, which reduces its efficiency and lifespan [11]. Therefore, this type is challenging in high-salinity environments. To tackle fouling and scaling, extensive pretreatment of feedwater

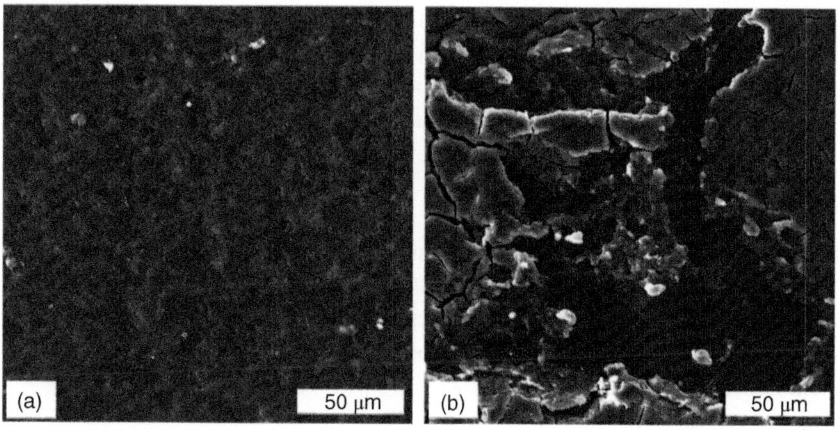

Figure 16.1 Fouling of NF270 membrane by (a) organic sodium alginate and (b) colloidal polystyrene carboxylate latex particles. Source: Mahlangu et al. [13]/with permission of Elsevier.

is of importance, often including filtration, chemical treatments, and anti-scaling use [12]. Figure 16.1 shows the impact of organic sodium alginate and colloidal polystyrene carboxylate latex particles on the membrane surface as organic and colloidal fouling examples, respectively. Therefore, innovations in membrane technology such as hydrophilic coatings and anti-fouling materials are being developed to prevent these ongoing issues.

16.2.3 Energy Consumption and Efficiency

Energy consumption is a key challenge in MDS, especially for RO systems. They typically require 3–6 kWh/m^3 for seawater desalination, with energy costs making up as much as 40% of total operational expenses [14]. Recent advancements have significantly boosted energy efficiency, including energy recovery devices (ERDs) that capture and reuse energy from high-pressure brine [15, 16]. Hybrid systems that integrate RESs with MDS induce additional opportunities for lowering energy demands and improving sustainability [17]. Furthermore, the development of high permeable membranes and refined system designs also helps reduce energy consumption even further.

16.2.4 System Design and Configuration

The design and setup of an MDS system play a critical role in determining its operational efficiency. Key aspects like the arrangement of membrane modules, the flow pattern (whether crossflow or dead-end filtration), and the number of stages in the desalination process all influence the system's overall performance [18]. A well-designed system ensures that feedwater is evenly distributed, reduces concentration polarization (which lowers water flux near the membrane), and enables efficient recovery of permeate water [19].

16.2.5 Operation and Maintenance

Proper O&M of MDS systems are essential for maintaining performance. This helps in avoiding equipment breakdowns and maintaining low operational costs. Therefore, successful O&M practices help the system run at optimal capacity while prolonging the lifespan of important components, especially the membranes [20, 21].

16.2.6 Cleaning Protocols

Regular cleaning of membranes is a vital aspect of ongoing O&M, especially in environments prone to heavy fouling or scaling. The cleaning schedule and methods used are determined by the type and extent of the fouling or scaling present [22]. It can be said that the effectiveness of cleaning procedures is determined by how well they restore water flow and lower transmembrane pressure while avoiding any lasting damage to the membrane's overall performance. Standard cleaning procedures include the following.

16.2.6.1 Chemical Cleaning

One typical method of cleaning involves applying either acidic or alkaline solution to break down fouling substances. Acidic solutions are typically utilized to eliminate inorganic scaling, whereas alkaline solutions are better suited for removing organic fouling [23]. It's important to choose cleaning agents that are safe for the membrane material to avoid any potential damage [24–26]. Figure 16.2 compares membranes under three different conditions: new, fouled, and chemically cleaned. The membranes that underwent chemical cleaning effectively restored the performance of both first-pass RO and second-pass nanofiltration (NF) membranes. However, it was less successful in removing deposited solids from the first-pass membranes [27].

16.2.6.2 Backwashing

This method uses a reverse flow of water through the membrane to release and remove accumulated foulants. Backwashing is a useful strategy for preserving permeate flow in ultrafiltration (UF) and NF systems. However, it is less frequently applied in RO systems because of the higher pressures required [28, 29]. Figure 16.3 illustrates the impact of backwashing on the surface of a

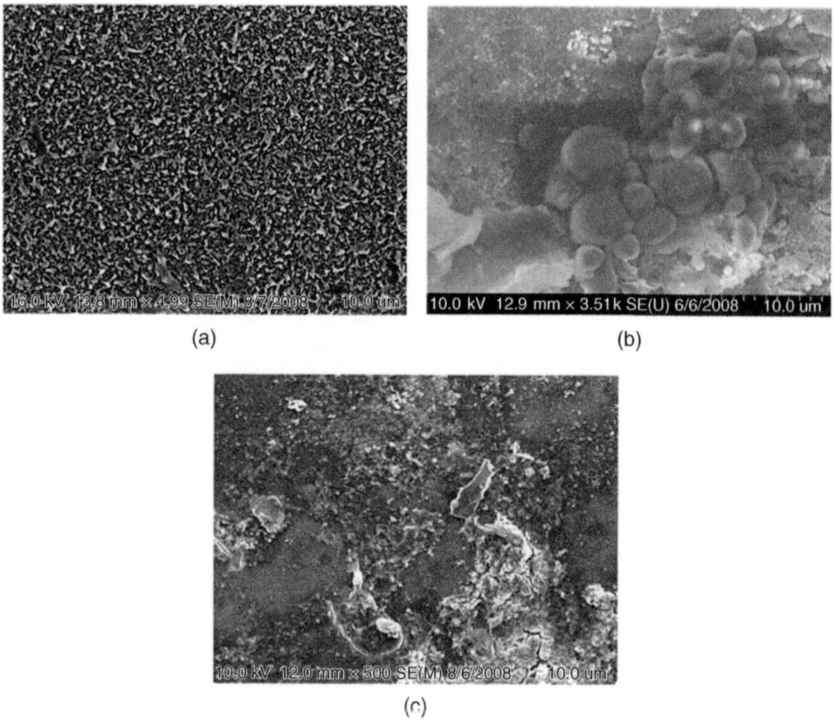

Figure 16.2 Comparison of scanning electron microscope (SEM) images of membranes from a desalination plant under three conditions: (a) new, (b) fouled, and (c) chemically cleaned. Source: Tanuwidjaja et al. [27]/with permission of Elsevier.

membrane used in filtration systems. The photographs show that fouling has been significantly reduced, resulting in a much cleaner and smoother surface [30].

16.2.6.3 Air Scouring

Air scouring involves introducing air bubbles to mechanically clean the membrane surface, Figure 16.4, which helps inhibit the accumulation of fouling. This method is frequently used alongside chemical cleaning to improve the removal of contaminants [32, 33].

16.2.7 Membrane Longevity and Replacement

A membrane's lifetime operation is affected by different factors: operating conditions, the quality of feedwater, the extent of fouling, and the frequency of cleaning. Usually, membranes have a lifespan of 5 years [34]. However, this differs based on the application and maintenance strategies employed. To extend

Figure 16.3 Impact of backwashing on the membrane surface: (a) before the process and (b) after the process. Source: Hwang et al. [30]/with permission of Elsevier.

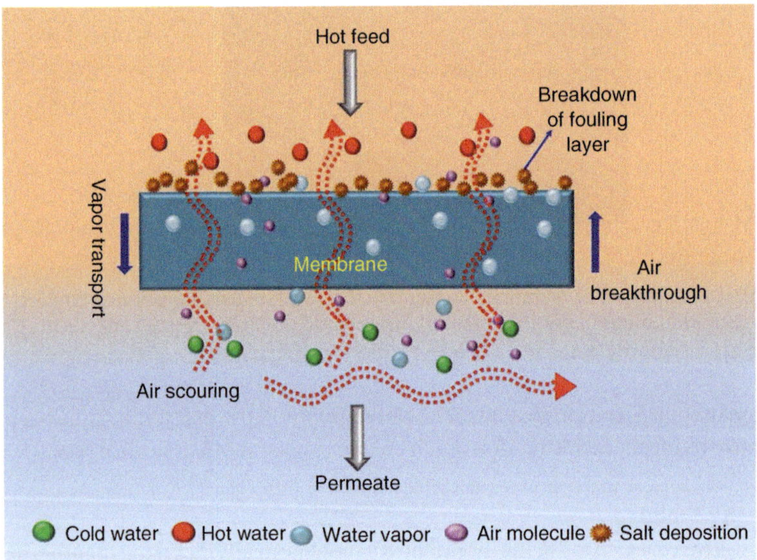

Figure 16.4 Schematic representation of the working principle of the air scouring process. Source: Bhoumick et al. [31]/with permission of Elsevier.

the life of a membrane, it is important to appropriately pretreat feedwater. This includes filtration to eliminate particulates using chemicals to mitigate fouling and scaling. Using these techniques ensures optimum operating conditions and prevents overwhelming the membrane [35]. In addition, regular performance monitoring is key for detecting the early signs of membrane wear. This includes checking pressure differentials and permeate quality. Besides, when membranes reach the end of their lifetime, they need to be replaced. This replacement can represent main capital expenses, especially in large desalination plants. Current research aims to enhance the durability of membranes by developing more durable materials. These materials include ceramic- and graphene-based options [36]. This also aims to reduce the costs associated with replacements.

16.2.8 System Monitoring and Optimization

Regular monitoring and analysis of system data are important for MDS systems to work efficiently. It's important to track key performance indicators (KPIs), such as permeate flow rate, salt rejection, pressure differential, and energy usage [37]. This can be achieved by using advanced control systems that employ artificial intelligence (AI) and machine learning. This is also implemented to predict fouling occurrences, optimize cleaning routines, and adjust system settings in real time for enhanced efficiency. In addition, maintenance approaches should be applied [38]. This includes routine checks for pumps, valves, and pressure vessels, besides regular calibration of sensors and instruments. Scheduling downtime for maintenance reduces the possibility of unexpected failures as well. It also allows for complete cleaning and inspection of components [39, 40].

16.2.9 Cost Management in Operation and Maintenance

The O&M costs account for a considerable portion of the overall lifecycle expenses of a desalination plant [41]. Hence, it is imperative to effectively manage energy usage, membrane replacement, cleaning using chemicals, and labor costs. This helps maintain the economic sustainability of the desalination process. This can also optimize energy efficiency through ERDs using predictive maintenance tools. This assists in decreasing downtime and employing automation to reduce labor needs [42]. The MDS systems can achieve permanent sustainability and operational effectiveness through these measures.

16.3 Novel Membrane Materials

Developing new membrane materials is needed to address persistent issues such as fouling and energy use for long-term operational reliability [43]. This section discusses recent advancements in membrane materials. It highlights their distinctive characteristics and capacity to improve the efficiency and sustainability of the desalination methods.

16.3.1 Graphene-Based Membranes

Graphene oxide (GO) and reduced graphene oxide (rGO) have attracted attention as promising materials for RO membranes. This interest is due to their unique

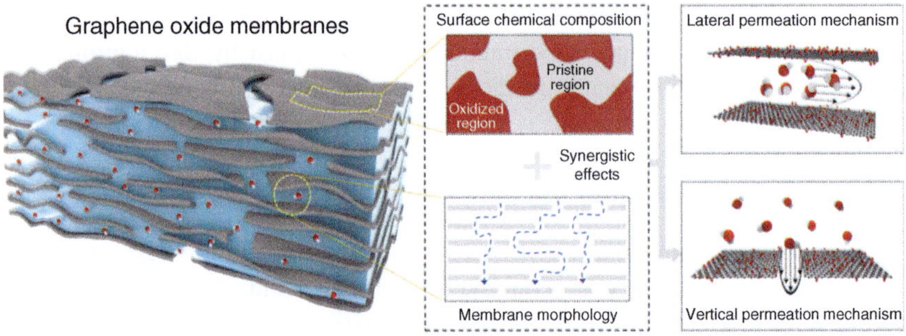

Figure 16.5 The fundamental structure of GO membranes and their mass transport pathways. Source: Lee et al. [44]/with permission of Elsevier.

two-dimensional structure and strong mechanical properties. In addition, GO membranes provide high water permeability while still effectively rejecting salts, as illustrated in Figure 16.5. Moreover, the functionality of GO with hydrophilic groups enhances its anti-fouling capabilities by reducing the accumulation of organic matter and microbial growth [45, 46]. However, they are still prone to inorganic scaling, especially in high-salinity water. Therefore, pretreatment processes such as anti-scaling treatment and careful feedwater management are necessary to lessen scaling issues. On the other hand, the ultra-thin nature of these membranes helps lower energy consumption. Less pressure is required to achieve the same water flux compared to traditional TFC membranes [47]. However, maintaining these membranes in high-pressure RO systems remains challenging as mechanical stability is a concern over prolonged periods. In addition, scalability and long-term stability remain other challenges that need to be addressed before these membranes can be widely adopted. In order to overcome these barriers, advances in cost-effective production methods such as roll-to-roll manufacturing should be implemented [48].

16.3.2 Carbon Nanotube Membranes

CNTs have attracted interest because of their excellent mechanical strength, large surface area, and the ability to transport water immediately through their hollow pores. In RO applications, CNT-based membranes, Figure 16.6, offer high water flux and strong salt rejection. They also have the potential to lower energy use [50–54]. Adding CNTs to TFC membranes also improves resistance to fouling, especially from organic materials. However, more research is necessary to improve the alignment of CNTs in membranes and guarantee stable performance over long periods of implementation. One of the scalability barriers of these membranes is that large-scale manufacturing processes capable of producing uniform CNT membranes remain underdeveloped [55]. In addition, current

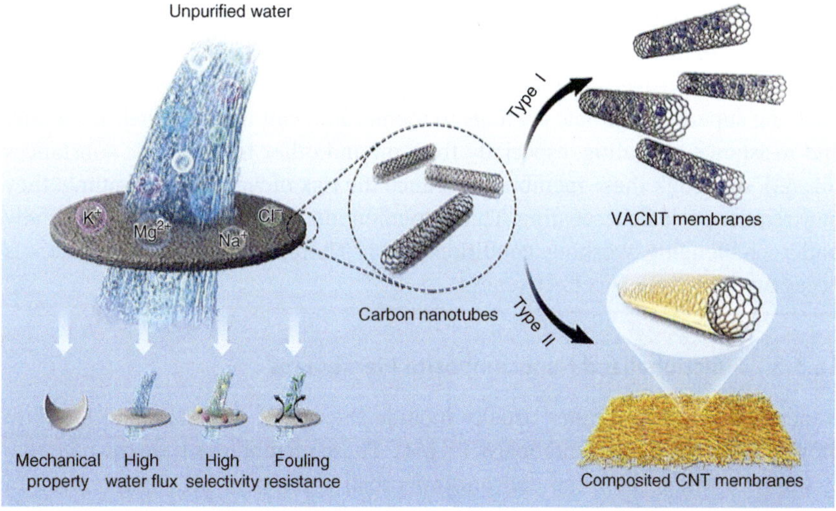

Figure 16.6 CNT membranes enhance water purification and membrane performance. Source: Wang et al. [49]/MDPI/CC BY 4.0.

fabrication techniques are labor-intensive and expensive, which limit their application in large-scale desalination plants.

16.3.3 Aquaporin-Based Biomimetic Membranes

Aquaporins are natural proteins that help water molecules cross through membranes with unparalleled selectivity and efficiency [56]. Biomimetic membranes, which use aquaporins, have been created to replicate these biological processes for RO desalination. These membranes offer much higher water flux, sometimes up to three times more than traditional RO membranes, while still rejecting salts effectively [57]. Moreover, the bio-inspired design of aquaporin membranes helps lower energy use. However, correct pretreatment of feedwater and regular maintenance are essential to prevent biofouling and scaling. In addition, due to its bio-nature, these membranes may require more rigorous operational protocols compared to conventional membranes. Taking these conditions into account, producing aquaporin-based membranes on a large scale is still a challenge, and ensuring their durability in tough conditions remains an additional concern [58].

16.3.4 Superhydrophobic Membranes

As mentioned previously, a key issue in MD is membrane wetting where the feedwater crosses the pores and interrupts vapor transport. Materials such as polytetrafluoroethylene (PTFE) and fluorinated silica can provide better resistance to wetting due to their high water contact angles [59, 60]. These membranes exhibit high performance in high-salinity environments, minimizing the chance of pore wetting and maintaining strong water vapor flow. Additionally, applying superhydrophobic coatings to membranes can increase their durability and resistance to fouling, especially from oil and other hydrophobic substances [61, 62]. Although these membranes reduce the risk of wetting and fouling, they may require periodic re-coating. This is to maintain their performance, especially under challenging working conditions [63]. Therefore, regular inspection is necessary for long-term effectiveness.

16.3.5 Functionalized Nanocomposite Membranes

Nanocomposite membranes usually include the use of materials like titanium dioxide (TiO_2) and silver nanoparticles [64]. These membranes have the potential to improve the anti-fouling and anti-microbial abilities of MD membranes. In terms of TiO_2, it has photocatalytic properties that break down organic fouling and microbial growth on the membrane. This helps maintain long-term stability [65]. Adding these nanomaterials also boosts the membrane's hydrophobicity and lowers the risk of pore wetting which is a major challenge in MD systems. The inclusion of nanoparticles into polymer matrices is a well-established process, which makes these membranes more applicable [66].

16.3.6 Zwitterionic Membranes

Zwitterionic polymers are membranes that contain both positive and negative charges. These polymers have been introduced to improve the anti-fouling characteristics of the UF membranes [67]. One of their advantages is the strong hydration layers that prevent the accumulation of foulants, mainly proteins and other organic materials [68]. In comparison with traditional UF membranes, these membranes show better water flow and are more resistant to fouling. This could reduce the frequency of cleanings and extend their overall life cycle. However, regular monitoring is still necessary to ensure that fouling does not occur over time, mainly in hard feedwater conditions [69]. In terms of large-scale production, the cost of these materials may still be a dominating consideration.

16.3.7 Nanofiber-Based Membranes

Electrospun nanofiber membranes presented in Figure 16.7 provide a distinct method for UF. They offer high porosity, a large surface area, and adjustable pore sizes. These membranes are made from materials like polyacrylonitrile (PAN)

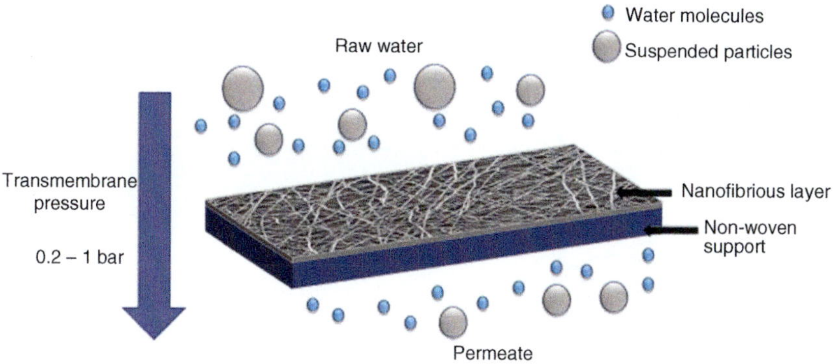

Figure 16.7 Electrospun nanofiber utilization in microfiltration process. Source: Faisal et al. [70]/with permission of Elsevier.

or polyvinylidene fluoride (PVDF) [71]. They show high water flow and lower fouling due to their open structure. These membranes can also include functional materials, such as anti-microbial agents or photocatalysts. These materials improve their performance and make them more stable during usage [72].

16.3.8 Metal–Organic Framework Membranes

Metal–organic frameworks (MOFs) (Figure 16.8) are porous crystalline materials with adjustable pore sizes and large surface areas. These membranes provide excellent selectivity for separating ions and removing organic compounds. This allows them to be well-suited for NF applications. These applications include water softening, heavy metal removal, and pharmaceutical filtration. Moreover, adding MOFs to NF membranes has been proven to improve water flow and increase the rejection of solutes, mainly heavy metals and organic contaminants [74]. They also help reduce fouling due to their smooth surfaces, which decreases the buildup of organic substances. The production of MOF membranes on a large scale is still being tested as making large defect-free membranes is still a challenge [75]. One of the advantages of these membranes is the resistance to organic fouling because of their smooth surfaces and high selectivity. However, it is important to pretreat feedwater and conduct regular maintenance to prevent scaling issues [76].

16.3.9 Mixed-Matrix Membranes

Mixed-matrix membranes (MMMs) integrate polymer materials with inorganic fillers like zeolites or silica nanoparticles. These types of membranes have been extensively studied for NF applications. Therefore, it can be said that these

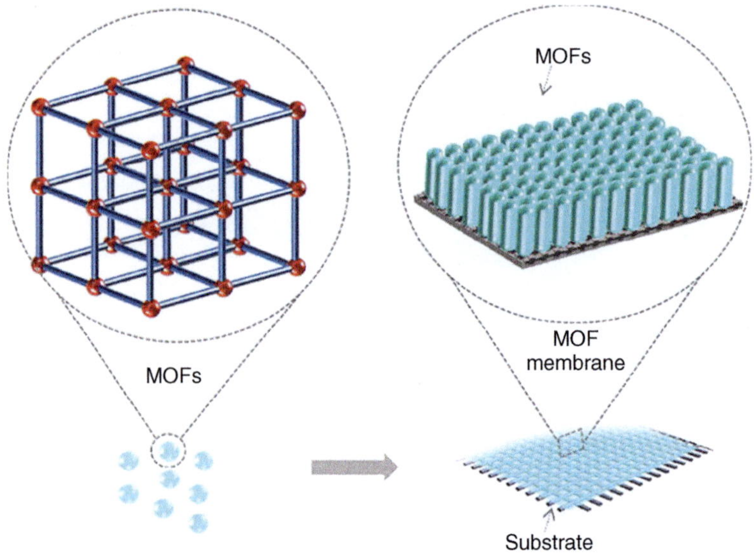

Figure 16.8 Construction of MOF membranes. Source: Deng et al. [73]/with permission of Elsevier.

membranes combine the benefits of both polymer and inorganic components, which results in better permeability and selectivity. In addition, they are particularly efficient at removing organic pollutants and divalent ions. They also boost resistance to fouling. Current research is focused on improving how fillers are distributed and interact within the polymer structure. This aims to optimize the performance of these membranes further [77].

16.4 Real-World Installations

As discussed previously, RO is considered the most commonly employed MDS method due to its energy efficiency, scalability, and capacity to remove various salts and contaminants from both seawater and brackish water. This section reviews some of the notable examples of commercial RO desalination plants. It highlights the importance of this technology in meeting worldwide water needs.

16.4.1 Al Montazah Solar-Powered Desalination Plant, Egypt

Sharm El-Sheikh's desalination plant operates off-grid by using solar energy to generate 18,000 m^3 of freshwater per day mainly for the city's tourism industry. The system employs RO technology and is powered by a 2.8-MW solar photovoltaic

(PV) array. To maintain consistent water production during times of limited sunlight, the plant is equipped with an energy storage system. This eco-friendly setup meets essential water needs while reducing the plant's environmental footprint. It is estimated to offset 54.11 t of CO_2 over its operational life [78–80].

16.4.2 Al Khafji Solar-Powered Desalination Plant, Saudi Arabia

Another remarkable example of a renewable energy-based system is the Al Khafji desalination plant. It is one of the world's largest solar-powered desalination facilities. This plant supplies 60,000 m^3 of potable water to 150,000 persons daily in Saudi Arabia's arid regions. The plant is powered by a 15-MW PV system assisted by an energy storage solution to ensure consistent operation [81]. This sustainable approach emphasizes Saudi Arabia's role in addressing water scarcity in an environmentally friendly manner.

16.4.3 Perth Desalination Plant, Australia

Perth's RO desalination plant in Australia is a facility powered by renewable energy, requiring a total electricity capacity of 24 MW. Furthermore, the plant has a production requirement ranging from 4.0 kWh to 6.0 kWh/kL. To meet this demand, the Emu Downs Wind Farm generating 80 MW is utilized for energy production. This wind farm comprises 48 wind turbines located 30 km east of Cervantes. In terms of financial considerations, the total cost of the project amounted to AUS$387 million, while the annual operating costs remained under $20 million. In addition, the estimated cost of water is approximately $1.17/kL. This plant is located in Kwinana approximately 25 km south of Perth. It has an initial capacity of 140,000 m^3/day with the potential to expand to 250,000. This capacity makes it the largest desalination plant in the southern hemisphere and the largest renewable energy-powered facility. The plant is expected to supply 17% of Perth's water needs. It will provide an annual total of 45 GL for a population of 1.5 million people [82].

16.4.4 Masdar City Solar Desalination Plant, UAE

Masdar City is an urban development in the UAE that concentrates on sustainability by incorporating RESs into its infrastructure. The solar desalination plant in Ghantoot uses a combination of PV and RO technologies to meet the city's water requirements. The plant has been in operation for 18 months, producing between 1,500 and 2,500 m^3 of water per day. The installed capacity includes a 5-MW solar PV system and solar thermal systems [83].

16.4.5 Small-Scale Solar-Powered Desalination, Cape Verde

Cape Verde is located off the coast of West Africa and faces persistent challenges related to water scarcity. To address these issues, a small-scale solar-powered desalination plant was installed to provide potable water to local communities. This project aims at offering a decentralized solution by harnessing the region's abundant solar energy. In Porto Novo, the system employs RO technology and is powered by a 55-kW solar PV installation system [84]. The solar plant helps in reducing operational costs at the Porto Novo Power Station. This leads to a 5% decrease in the cost of producing drinking water [85]. As a result of this, around 10,000 residents now have improved and more affordable access to clean water. This positively impacts their health and overall well-being. Correspondingly, in Praia on Santiago Island, an RO system operates with a capacity of 1,200 m^3 of water per day. This contributes to Cape Verde's water security efforts further [85].

16.4.6 Jebel Ali Desalination Plant, UAE

The Jebel Ali desalination plant in Dubai ranks among the largest RO plants globally with the ability to produce around 240,000 m^3 of water daily. It is part of the broader Jebel Ali Power and Desalination Complex. The plant combines RO and multi-stage flash (MSF) desalination technologies to meet the region's water and energy demands. This plant is essential for ensuring Dubai's water security, supplying freshwater for residential, industrial, and agricultural purposes. ERDs are used to lower energy usage, which enhances the overall cost-efficiency of the desalination process. Furthermore, advanced pretreatment systems such as UF membranes are employed to prevent membrane fouling and scaling. This is to enhance the durability of the RO membranes [86].

16.4.7 Carlsbad Desalination Plant, USA

The Carlsbad desalination plant located in California stands as the largest desalination installation in the USA. It has a daily output of 190,000 m^3 of water. It plays a key mission in supporting San Diego County's water resources, especially during droughts. The plant provides almost 10% of the area's drinking water, serving more than 300,000 residents. The plant uses an advanced RO process with a multi-stage pretreatment system. This involves microfiltration and chemical treatment to reduce membrane fouling. Moreover, pressure exchanger energy recovery systems are employed to reduce energy consumption [87].

16.4.8 Shuwaikh Desalination Plant, Kuwait

The Shuwaikh desalination plant in Kuwait uses RO technology to treat seawater. This installation produces 136,000 m^3 of water daily, supporting Kuwait's water security strategy. It serves an arid region with scarce natural freshwater resources. The plant employs advanced X-flow RO membranes to serve approximately 450,000 people. It also uses pressure exchangers to reduce energy usage and improve efficiency. In addition, it includes UF in its pretreatment process to protect RO membranes from fouling and extend their lifespan [88].

16.4.9 Salalah Desalination Plant, Oman

The Salalah desalination plant which is located in the city of Salalah produces 113,650 m^3 per day to serve both residential and industrial sectors. The plant meets the city's water demands by using high-efficiency RO membranes. The main aim is to maintain consistent water quality even in challenging seawater conditions. The plant also employs a complete multi-stage pretreatment process to ensure long-term operational reliability and minimize membrane fouling. This comprises dissolved air flotation and sand filtration to reduce the occurrence of membrane cleaning [89].

16.4.10 Gold Coast Desalination Plant, Queensland

The Gold Coast desalination plant has a daily capacity of 133 ML, serving the domestic water needs of around 665,000 people. Each RO unit is equipped with dual-work exchanger ERDs, capable of recovering up to 97% of the unused energy from the brine. The plant offsets its carbon emissions by utilizing renewable energy certificates. Seawater undergoes an initial screening to eliminate particles exceeding 3 mm, followed by pretreatment using a coagulant to filter out suspended solids through tanks filled with coal and sand. After the two-stage seawater reverse osmosis (SWRO) process removes salts and other fine particles, the water is re-mineralized with lime and carbon dioxide and disinfected with chlorine [90].

16.5 Future Projections for Membrane Desalination

As the global need for freshwater increases, MDS technologies are expected to improve significantly. These advancements will make the systems more efficient, sustainable, and easier to access. Therefore, based on the previous discussion, this section looks at the expected future trends in MDS.

16.5.1 Technological Advancements

One of the most promising areas in desalination technology is the development of new materials for membranes. Current research is focused on creating membranes that offer better permeability, selectivity, fouling resistance, and durability. Innovations in nanomaterials, GO, and biomimetic membranes are leading these efforts [45, 48]. High-permeability membranes developed through advances in nanotechnology improve selectivity. They also reduce energy consumption and boost system efficiency. For example, graphene-based membranes are strong and highly permeable. This quality allows RO systems to operate at lower pressures. By combining these advanced membranes with RESs, both capital and operational costs can be reduced significantly over time. Another emerging approach is hybrid membrane systems which integrate technologies like forward osmosis (FO) and RO [91]. When powered by RESs, these systems can provide potential energy savings [92]. They also extend the lifespan of membranes and enhance economic viability further.

In addition, AI and machine learning are expected to play an important role in improving the efficiency of MDS systems. AI can monitor system performance, predict fouling in membranes, and optimize energy use in real time. This will result in more efficient operations [37, 38]. In terms of predictive maintenance, AI can forecast when membranes are likely to foul or degrade. This will enable proactive maintenance that extends membrane life and reduces downtime. This will also increase the overall productivity. Furthermore, machine learning models can help optimize energy use by continuously adjusting operational settings to minimize energy consumption while maintaining water quality.

16.5.2 Renewable Energy Integration

The integration of renewable energy, especially solar and wind, with MDS systems is expected to be a major focus for future development [93]. These hybrid solar–wind-powered desalination systems offer the potential to reduce dependence on fossil fuels. They also lower operational costs and decrease the carbon footprint of water production. Hybrid systems that combine PV, concentrated solar power (CSP), and wind energy with RO or MD are becoming more feasible as renewable technology progresses [17, 94]. These systems utilize solar power during the day and wind energy at night. This allows them to provide a consistent freshwater supply in regions that are off-grid or have limited access to conventional energy sources. Figure 16.9 illustrates a hybrid system that combines both sources to power an RO unit, as proposed by Khan et al. [95]. In addition, new developments in ocean-based systems like tidal and wave energy could improve the economic feasibility of renewable energy-powered desalination. This can offer potential for coastal desalination plants by harnessing the natural movement of water with minimal environmental impact. However, the economic and technical viability of these systems needs further investigation.

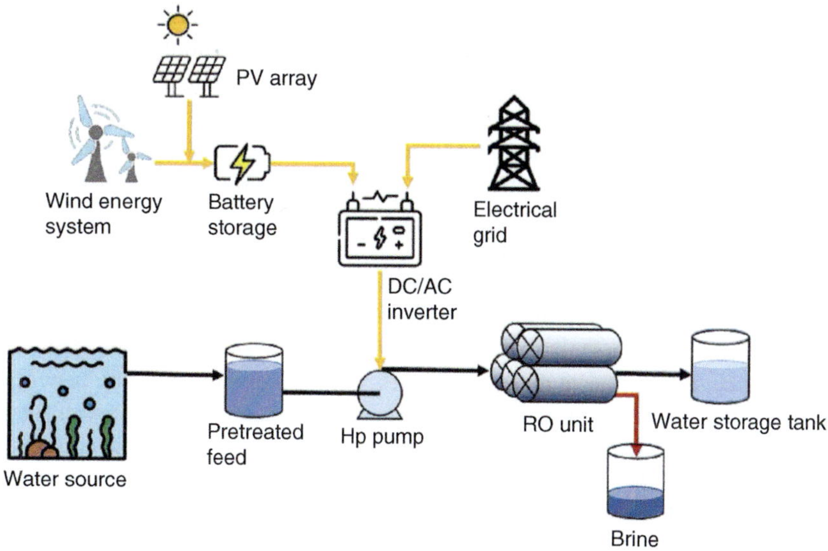

Figure 16.9 Schematic of a hybrid RO system powered by solar and wind energies. Source: Khan et al. [95]/with permission of Elsevier.

16.5.3 Environmental and Social Impacts

The change to renewable energy-powered desalination has the potential to significantly reduce the environmental impact of water production. Current membrane-based desalination methods especially RO are energy-intensive and often depend on fossil fuels. By incorporating solar/wind and other RESs, future desalination systems could achieve nearly zero carbon emissions [96]. MDS can also offer significant socioeconomic advantages, especially for regions that face water scarcity. A consistent supply of freshwater can support economic growth, enhance public health, and promote agricultural stability in areas with chronic water shortages. For agriculture and industry, desalination can provide a reliable water source which is valuable in regions affected by drought or with limited access to freshwater. This stability helps reinforce food security and economic resilience. Moreover, in areas where clean drinking water is scarce, desalination can reduce the incidence of waterborne diseases. This improves public health and reduces healthcare costs.

16.5.4 Policy and Regulatory Trends

Governments and international bodies are acknowledging the renewable energy role in desalination. New policies are developed to support the use of renewables in water production. These initiatives are driven by the need to reduce emissions and improve water security. For example, the European Union's Green Deal emphasizes reducing carbon emissions in sectors such as desalination. This offers financial incentives for projects that combine renewable energies to lower their environmental impact [97]. Also, Middle Eastern countries like Saudi Arabia and the UAE have set aspiring targets, in which they aim to power their desalination plants using renewable energy, aiming at reducing dependence on fossil fuels while addressing water shortages [98].

Investing in research and development (R&D) is also important to improve renewable desalination technologies. Efforts should focus on improving ERDs, developing advanced membranes, and creating hybrid systems for more efficient performance. Pilot projects and case studies will help demonstrate the benefits of renewable desalination technologies. Successful examples can provide key insights into cost savings, return on investment, and environmental advantages. These insights set the stage for broader market adoption.

Collaboration between the public and private sectors can drive progress further by reducing financial risks and encouraging investment. Public–private partnerships can combine government funding with private innovation. This collaboration helps expand renewable desalination projects. Governments may also introduce stricter rules on energy consumption and brine disposal to encourage the use of eco-friendly technologies. Managing brine effectively, for example, through zero liquid discharge (ZLD) systems, will be critical. This approach will ensure that desalination is both practical and environmentally sustainable [96]. Moreover, regulations will likely promote the acceptance of energy-efficient and renewable technologies in desalination plants. This could be accomplished through tax credits, subsidies, or mandatory renewable energy quotas for new projects.

16.6 Conclusion

This chapter highlighted the importance of MDS, especially RO, in mitigating global water scarcity. It also discussed several challenges like fouling and scaling. These problems remain significant operational issues, which affect system longevity and efficiency and increase maintenance costs. To prolong system life and maintain efficiency, it is necessary to use advanced fouling mitigation strategies. These include improved cleaning methods and the development of membranes that resist fouling better. Technological advancements in membrane materials are paving the way for improved performance. Innovations such as graphene, CNTs, and nanocomposites can enhance permeability and durability and reduce energy consumption. The chapter also explored hybrid systems that merge multiple renewable sources with advanced membrane technologies.

This could offer sustainable off-grid water production in remote regions. The use of AI and machine learning is to optimize system operations and predict fouling. These technologies also enhance energy efficiency and strengthen the future of MDS. Moreover, the socioeconomic and environmental impacts of renewable energy-powered desalination especially in water-scarce regions cannot be neglected. Desalination can support agriculture, industry, and public health. This could be done by providing a reliable freshwater supply while reducing waterborne diseases and improving food security. As policy and regulatory trends evolve, there is increased focus on energy recovery and brine management. Incentives for renewable integration will also drive the future growth and sustainability of MDS technologies.

Questions

1. What are the two main performance metrics for membrane desalination (MDS) systems and why are they important?

2. Why are polymeric membranes commonly used in MDS systems, and what are their limitations?

3. What are fouling and scaling, and how do they affect MDS systems?

4. How does chemical cleaning restore membrane performance in MDS systems?

5. What is the role of air scouring in MDS maintenance?

6. What is the purpose of a backwashing process in filtration systems?

7. What are the primary advantages of graphene oxide (GO) membranes in desalination applications? What is the key challenge associated with their use?

8. How do carbon nanotube (CNT) membranes benefit desalination systems? What is the significant obstacle to their large-scale application?

Acknowledgments

The authors extend their gratitude to the Science, Technology and Innovation Funding Authority (STDF) of Egypt for their generous support of this chapter under Grant IDs: 46931 and 47373.

References

1 Qasim, M., Badrelzaman, M., Darwish, N.N. et al. (2019). Reverse osmosis desalination: a state-of-the-art review. *Desalination* 459: https://doi.org/10.1016/J.DESAL.2019.02.008.

2 Maddah, H.A., Alzhrani, A.S., Bassyouni, M. et al. (2018). Evaluation of various membrane filtration modules for the treatment of seawater. *Appl. Water Sci.* 8: https://doi.org/10.1007/S13201-018-0793-8/FIGURES/16.

3 Kucera, J. (2023). *Reverse Osmosis*, 3e. Wiley.

4 Pinnau, I. and Freeman, B.D. (1999). Formation and modification of polymeric membranes: overview. *ACS Symp. Ser.* 744: https://doi.org/10.1021/BK-2000-0744.CH001.

5 Li, D., Yan, Y., Wang, H. et al. (2016). Recent advances in polymer and polymer composite membranes for reverse and forward osmosis processes. *Prog. Polym. Sci.* 61: https://doi.org/10.1016/J.PROGPOLYMSCI.2016.03.003.

6 Philibert, M., Villacorte, L.O., Ekowati, Y. et al. (2024). Fouling and scaling in reverse osmosis desalination plants: a critical review of membrane autopsies, feedwater quality guidelines and assessment methods. *Desalination* 592: https://doi.org/10.1016/J.DESAL.2024.118188.

7 Voutchkov, N. (2010). Considerations for selection of seawater filtration pretreatment system. *Desalination* 261: https://doi.org/10.1016/J.DESAL.2010.07.002.

8 Salinas-Rodríguez, S.G., Schippers, J.C., Amy, G.L. et al. (2021). *Seawater Reverse Osmosis Desalination: Assessment and Pre-treatment of Fouling and Scaling*. IWA Publishing. https://doi.org/10.2166/9781780409863.

9 Mansi, A.E., El-Marsafy, S.M., Elhenawy, Y. et al. (2023). Assessing the potential and limitations of membrane-based technologies for the treatment of oilfield produced water. *Alexandria Eng. J.* 68: https://doi.org/10.1016/J.AEJ.2022.12.013.

10 Penña, N., Gallego, S., Del Vigo, F. et al. (2013). Evaluating impact of fouling on reverse osmosis membranes performance. *Desalin. Water Treat.* 51: https://doi.org/10.1080/19443994.2012.699509.

11 Dhakal, N., Abushaban, A., Mangal, N. et al. (2020). Membrane fouling and scaling in reverse osmosis. In: *Membrane Desalination* (ed. A. Sapalidis), 325–344. CRC Press.

12 Schippers, J.C., Salinas-Rodríguez, S.G., Kennedy, M.D. et al. (2021). Fouling and pre-treatment. In: *Seawater Reverse Osmosis Desalination: Assessment and Pre-treatment of Fouling and Scaling*, 59–83. IWA Publishing. https://doi.org/10.2166/9781780409863_0059.

13 Mahlangu, O.T., Nthunya, L.N., Motsa, M.M. et al. (2023). Fouling of high pressure-driven NF and RO membranes in desalination processes: mechanisms

and implications on salt rejection. *Chem. Eng. Res. Des.* 199: https://doi.org/10.1016/J.CHERD.2023.09.037.

14 Feo-García, J., Pulido-Alonso, A., Florido-Betancor, A. et al. (2024). Cost studies of reverse osmosis desalination plants in the range of 23,000–33,000 m^3/day. *Water* 16: 910. https://doi.org/10.3390/W16060910.

15 Feo, J., Sadhwani, J.J., Alvarez, L. et al. (2013). More efficient production line with Desalination plants using reverse osmosis. *Desalin. Water Treat.* 51: https://doi.org/10.1080/19443994.2012.713565.

16 Ghaffour, N. (2009). The challenge of capacity-building strategies and perspectives for desalination for sustainable water use in MENA. *Desalin. Water Treat.* 5: https://doi.org/10.5004/DWT.2009.564.

17 Nurjanah, I., Chang, T.T., You, S.J. et al. (2024). Reverse osmosis integrated with renewable energy as sustainable technology: a review. *Desalination* 581: https://doi.org/10.1016/J.DESAL.2024.117590.

18 Peñate, B. and García-Rodríguez, L. (2012). Current trends and future prospects in the design of seawater reverse osmosis desalination technology. *Desalination* 284: https://doi.org/10.1016/J.DESAL.2011.09.010.

19 Ng, W.H., Liang, Y.Y., Weihs, G.A.F. et al. (2023). Quantifying the potential of pressure retarded osmosis advanced spacers for reducing specific energy consumption in hybrid desalination. *J. Water Process Eng.* 55: https://doi.org/10.1016/J.JWPE.2023.104197.

20 Al-Obaidi, M.A., Rasn, K.H., Aladwani, S.II. et al. (2022). Flexible design and operation of multi-stage reverse osmosis desalination process for producing different grades of water with maintenance and cleaning opportunity. *Chem. Eng. Res. Des.* 182: 525–543.

21 Mappas, V., Vassiliadis, V.S., Dorneanu, B. et al. (2022). Maintenance scheduling optimisation of reverse osmosis networks (RONs) via a multistage Optimal Control reformulation. *Desalination* 543: https://doi.org/10.1016/J.DESAL.2022.116105.

22 Madaeni, S.S., Mohamamdi, T., Moghadam, M.K. et al. (2001). Chemical cleaning of reverse osmosis membranes. *Desalination* 134: https://doi.org/10.1016/S0011-9164(01)00117-5.

23 Hoek, E.M.V., Weigand, T.M., Edalat, A. et al. (2022). Reverse osmosis membrane biofouling: causes, consequences and countermeasures. *NPJ Clean Water* 5 (1): doi: 10.1038/s41545-022-00183-0.

24 Trägårdh, G. (1989). Membrane cleaning. *Desalination* 71: https://doi.org/10.1016/0011-9164(89)85033-7.

25 Ang, W.S., Tiraferri, A., Chen, K.L. et al. (2011). Fouling and cleaning of RO membranes fouled by mixtures of organic foulants simulating wastewater effluent. *J. Membr. Sci.* 376: https://doi.org/10.1016/j.memsci.2011.04.020.

26 Mohammadi, T., Madaeni, S.S., Moghadam, M.K. et al. (2003). Investigation of membrane fouling. *Desalination* 153: https://doi.org/10.1016/S0011-9164(02)01118-9.

27 Tanuwidjaja, D., Jin, X., Huang, X. et al. (2020). Comparison of membrane fouling and cleaning in one-pass reverse osmosis and two-pass nanofiltration approaches to seawater desalination. *Desalin. Water Treat.* 193: https://doi.org/10.5004/DWT.2020.25832.

28 Chen, J., Kim, S.L., Ting, Y.P. et al. (2003). Optimization of membrane physical and chemical cleaning by a statistically designed approach. *J. Membr. Sci.* 219: https://doi.org/10.1016/S0376-7388(03)00174-1.

29 Chang, H., Liang, H., Qu, F. et al. (2017). Hydraulic backwashing for low-pressure membranes in drinking water treatment: a review. *J. Membr. Sci.* 540: https://doi.org/10.1016/J.MEMSCI.2017.06.077.

30 Hwang, K.J., Chan, C.S., Tung, K.L. et al. (2009). Effect of backwash on the performance of submerged membrane filtration. *J. Membr. Sci.* 330: https://doi.org/10.1016/J.MEMSCI.2009.01.012.

31 Bhoumick, M.C., Roy, S., Mitra, S. et al. (2021). Synergistic effect of air sparging in direct contact membrane distillation to control membrane fouling and enhancing flux. *Sep. Purif. Technol.* 272: https://doi.org/10.1016/J.SEPPUR.2021.118681.

32 Psoch, C. and Schiewer, S. (2006). Direct filtration of natural and simulated river water with air sparging and sponge ball application for fouling control. *Desalination* 197: https://doi.org/10.1016/J.DESAL.2005.11.027.

33 Gao, W., Liang, H., Ma, J. et al. (2011). Membrane fouling control in ultrafiltration technology for drinking water production: a review. *Desalination* 272: https://doi.org/10.1016/J.DESAL.2011.01.051.

34 Scholes, C.A. (2015). Durability of membrane (lifetime of membrane). Enrico Drioli and Lidietta Giorno *Encyclopedia of Membranes.* 1–2 Springer https://doi.org/10.1007/978-3-642-40872-4_820-4.

35 Li, N.N., Fane, A.G., Ho, W.S.W. et al. (2008). *Advanced Membrane Technology and Applications.* Wiley.

36 Yang, Z., Sun, P.F., Li, X. et al. (2020). A critical review on thin-film nanocomposite membranes with interlayered structure: mechanisms, recent developments, and environmental applications. *Environ. Sci. Technol.* 54: https://doi.org/10.1021/ACS.EST.0C05377/SUPPL_FILE/ES0C05377_SI_001.PDF.

37 Alaswad, S.O. and Almatrafi, E. (2024). Artificial intelligence applications in forward osmosis for water treatment: recent developments and research directions. *Desalin. Water Treat.* 319: https://doi.org/10.1016/j.dwt.2024.100551.

38 Gaudio, M.T., Coppola, G., Zangari, L. et al. (2021). Artificial intelligence-based optimization of industrial membrane processes. *Earth Syst. Environ.* 5: https://doi.org/10.1007/S41748-021-00220-X/FIGURES/13.

39 Li, L., Rong, S., Wang, R. et al. (2021). Recent advances in artificial intelligence and machine learning for nonlinear relationship analysis and process control in drinking water treatment: a review. *Chem. Eng. J.* 405: https://doi.org/10.1016/J.CEJ.2020.126673.

40 Fan, M., Hu, J., Cao, R. et al. (2018). A review on experimental design for pollutants removal in water treatment with the aid of artificial intelligence. *Chemosphere* 200: https://doi.org/10.1016/J.CHEMOSPHERE.2018.02.111.

41 Shokri, A. and Sanavi Fard, M. (2023). Techno-economic assessment of water desalination: future outlooks and challenges. *Process Saf. Environ. Prot.* 169: https://doi.org/10.1016/J.PSEP.2022.11.007.

42 Arenas Urrea, S., Reyes, F.D., Suárez, B.P. et al. (2019). Technical review, evaluation and efficiency of energy recovery devices installed in the Canary Islands desalination plants. *Desalination* 450: https://doi.org/10.1016/J.DESAL.2018.07.013.

43 Kochkodan, V. and Hilal, N. (2015). A comprehensive review on surface modified polymer membranes for biofouling mitigation. *Desalination* 356: https://doi.org/10.1016/J.DESAL.2014.09.015.

44 Lee, H.G., Jin, D., Shim, W. et al. (2020). Synergistic effects of crystal structure and surface chemistry of stacked graphene-oxide membranes on the water-permeation mechanism. *Desalination* 492: https://doi.org/10.1016/J.DESAL.2020.114603.

45 Dai, Y., Liu, M., Li, J. et al. (2022). Graphene based membranes for water desalination: a literature review and content analysis. *Polymers* 14: https://doi.org/10.3390/POLYM14194246/S1.

46 Ashfaq, M.Y., Al-Ghouti, M.A., Zouari, N. et al. (2020). Functionalization of reverse osmosis membrane with graphene oxide to reduce both membrane scaling and biofouling. *Carbon* 166: https://doi.org/10.1016/J.CARBON.2020.05.017.

47 Kong, P., Sun, Z., Gui, H. et al. (2024). Advances in the application of graphene oxide composite loose nanofiltration membranes for dye and salt separation. *J. Environ. Chem. Eng.* 12: https://doi.org/10.1016/J.JECE.2024.114278.

48 Liu, Y., Wang, J., Zhou, R. et al. (2024). Innovations and challenges in large-area graphene oxide membranes for seawater desalination: preparation techniques, bottlenecks, and module developments. *Desalination* 592: https://doi.org/10.1016/J.DESAL.2024.118177.

49 Wang, R., Chen, D., Wang, Q. et al. (2020). Recent advances in applications of carbon nanotubes for desalination: a review. *Nanomaterials* 10: 1203. https://doi.org/10.3390/NANO10061203.

50 Tankus, K.A., Issman, L., Stolov, M. et al. (2018). Electrotreated carbon nanotube membranes for facile oil-water separations. *ACS Appl. Nano Mater.* 1:

https://doi.org/10.1021/ACSANM.8B00442/SUPPL_FILE/AN8B00442_SI_001.PDF.

51 Rizzuto, C., Pugliese, G., Bahattab, M.A. et al. (2018). Multiwalled carbon nanotube membranes for water purification. *Sep. Purif. Technol.* 193: https://doi.org/10.1016/J.SEPPUR.2017.10.025.

52 Holt, J.K., Park, H.G., Wang, Y. et al. (2006). Fast mass transport through sub-2-nanometer carbon nanotubes. *Science* 312: https://doi.org/10.1126/SCIENCE.1126298/SUPPL_FILE/HOLT.SOM.PDF.

53 Majumder, M., Chopra, N., Andrews, R. et al. (2005). Enhanced flow in carbon nanotubes. *Nature* 438 (7064): https://doi.org/10.1038/438044a.

54 Li, S., Liao, G., Liu, Z. et al. (2014). Enhanced water flux in vertically aligned carbon nanotube arrays and polyethersulfone composite membranes. *J. Mater. Chem. A* 2: https://doi.org/10.1039/C4TA02119C.

55 Ihsanullah, A.A. (2019). Carbon nanotube membranes for water purification: developments, challenges, and prospects for the future. *Sep. Purif. Technol.* 209: https://doi.org/10.1016/J.SEPPUR.2018.07.043.

56 Tang, C.Y., Zhao, Y., Wang, R. et al. (2013). Desalination by biomimetic aquaporin membranes: Review of status and prospects. *Desalination* 308: https://doi.org/10.1016/J.DESAL.2012.07.007.

57 Reddy, A.S., Kalla, S., Murthy, Z.V.P. et al. (2022). Biomimetic membranes: advancements and applications—a minireview. *Bioresour. Technol. Rep.* 18: https://doi.org/10.1016/J.BITEB.2022.101047.

58 Fuwad, A., Ryu, H., Malmstadt, N. et al. (2019). Biomimetic membranes as potential tools for water purification: preceding and future avenues. *Desalination* 458: https://doi.org/10.1016/J.DESAL.2019.02.003.

59 Zhu, Z., Liu, Y., Hou, H. et al. (2018). Dual-bioinspired design for constructing membranes with superhydrophobicity for direct contact membrane distillation. *Environ. Sci. Technol.* 52: https://doi.org/10.1021/ACS.EST.7B06227/SUPPL_FILE/ES7B06227_SI_001.PDF.

60 Xu, S., Wang, Q., Wang, N. et al. (2021). Chemical fabrication strategies for achieving bioinspired superhydrophobic surfaces with micro and nanostructures: a review. *Adv. Eng. Mater.* 23: https://doi.org/10.1002/ADEM.202001083.

61 Nguyen-Tri, P., Tran, H.N., Plamondon, C.O. et al. (2019). Recent progress in the preparation, properties and applications of superhydrophobic nano-based coatings and surfaces: a review. *Prog. Org. Coat.* 132: https://doi.org/10.1016/J.PORGCOAT.2019.03.042.

62 Czyzyk, S., Dotan, A., Dodiuk, H. et al. (2020). Processing effects on the kinetics morphology and properties of hybrid sol-gel superhydrophobic coatings. *Prog. Org. Coat.* 140: https://doi.org/10.1016/j.porgcoat.2019.105501.

63 Liu, L., Xiao, Z., Liu, Y. et al. (2021). Understanding the fouling/scaling resistance of superhydrophobic/omniphobic membranes in membrane distillation. *Desalination* 499: https://doi.org/10.1016/J.DESAL.2020.114864.

64 Kumar, A., Gautam, Y.K., Singh, N. et al. (2024). State-of-the-art developments in surface functionalized carbon-based bio/nanocomposites for theranostic antibacterials, advanced bioimaging, and molecular bioelectronics inspired biosensing platforms. *J. Ind. Eng. Chem.* 140: https://doi.org/10.1016/J.JIEC.2024.05.065.

65 Maqbool, A., Hui, W., Xin, X. et al. (2020). Removal of heavy metals from urban soil using functionalized carbon-coated composite. *Int. J. Environ. Sci. Technol.* 17: https://doi.org/10.1007/S13762-020-02790-1/FIGURES/8.

66 Tofighy, M.A. and Mohammadi, T. (2020). Carbon nanotubes-polymer nanocomposite membranes for pervaporation. In: *Polymer Nanocomposite Membranes for Pervaporation*. (ed. S. Thomas, S.C. George, and T. Jose), 105–133. Elsevier. https://doi.org/10.1016/B978-0-12-816785-4.00005-7.

67 Gu, G., Yang, X., Li, Y. et al. (2025). Advanced zwitterionic polymeric membranes for diverse applications beyond antifouling. *Sep. Purif. Technol.* 356: https://doi.org/10.1016/J.SEPPUR.2024.129848.

68 Han, L., Tan, Y.Z., Xu, C. et al. (2019). Zwitterionic grafting of sulfobetaine methacrylate (SBMA) on hydrophobic PVDF membranes for enhanced anti-fouling and anti-wetting in the membrane distillation of oil emulsions. *J. Membr. Sci.* 588: https://doi.org/10.1016/J.MEMSCI.2019.117196.

69 Chiang, Y.C., Chang, Y., Higuchi, A. et al. (2009). Sulfobetaine-grafted poly(vinylidene fluoride) ultrafiltration membranes exhibit excellent antifouling property. *J. Membr. Sci.* 339: https://doi.org/10.1016/J.MEMSCI.2009.04.044.

70 Faisal, A.N., Chowdhury, Z.Z., Thomas, P. et al. (2023). Functionalized electrospun biobased polymeric materials in filtration. In: *Advanced Applications of Biobased Materials: Food, Biomedical, and Environmental Applications* (ed. S. Ahmed and Annu), 625–651. Elsevier. https://doi.org/10.1016/B978-0-323-91677-6.00009-X.

71 Janjhi, F.A., Chandio, I., Janwery, D. et al. (2025). A review on hydrophobic electrospun nanofibers-based materials and membranes for water treatment: challenges, outlook, and stability. *Sep. Purif. Technol.* 353: https://doi.org/10.1016/J.SEPPUR.2024.128370.

72 Zhao, K., Kang, S.X., Yang, Y.Y. et al. (2021). Electrospun functional nanofiber membrane for antibiotic removal in water: review. *Polymers* 13: 226. https://doi.org/10.3390/POLYM13020226.

73 Deng, Y., Wu, Y., Chen, G. et al. (2021). Metal-organic framework membranes: recent development in the synthesis strategies and their application in oil-water separation. *Chem. Eng. J.* 405: https://doi.org/10.1016/J.CEJ.2020.127004.

74 Tripathy, D.B. (2024). Recent advances on 2D metal organic framework (MOF) membrane for waste water treatment and desalination. *Desalination* 592: https://doi.org/10.1016/J.DESAL.2024.118183.

75 Kumar, J., Duraisamy, K., Rosaiah, P. et al. (2023). MXene nanocomposites for current trend applications: synthesis, properties, and future directions. *Journal of Molecular.* 394: https://doi.org/10.1016/j.molliq.2023.123787.

76 Shahid, M.U., Najam, T., Islam, M. et al. (2024). Engineering of metal organic framework (MOF) membrane for waste water treatment: synthesis, applications and future challenges. *J. Water Process Eng.* 57: https://doi.org/10.1016/J.JWPE.2023.104676.

77 Rezakazemi, M., Amooghin, A.E., Matsuura, T. et al. (2014). State-of-the-art membrane based CO_2 separation using mixed matrix membranes (MMMs): an overview on current status and future directions. *Prog. Polym. Sci.* 39: https://doi.org/10.1016/J.PROGPOLYMSCI.2014.01.003.

78 TAM Environmental Services (2023). References and Experiences. https://www.tam-es.com/wsImages/brochure%201.pdf (accessed 25 January 2025).

79 Environics (2023). Al Montazah Desalination Plant. https://environics.org/projects/eia-for-solarizegypt-pv-plant-al-montazah-co-for-tourism-investment-sharm-el-sheikh/ (accessed 25 January 2025).

80 Solarizegypt (2022). El Montazah Water Desalination & Treatment Plant. https://solarizegypt.com/portfolio/el-montazah-water-desalination-plant/ (accessed 25 January 2025).

81 SAVENER (2020). Solar Saline Water Reverse Osmosis AL-KHAFJI. https://www.savener.es/en/proyectos/solar-saline-water-reverse-osmosis-al-khafji/ (accessed 25 January 2025).

82 Power Technology (2020). Perth Wave Energy Project. https://www.power-technology.com/projects/perth-wave-energy-project/ (accessed 25 January 2025).

83 Masdar (2018). Renewable Energy Water Desalination Program. https://masdar.ae/en/renewables/our-projects/ghantoot-desalination-pilot-plant (accessed 25 January 2025).

84 Idro (2021). Praia, Santiago Island SWRO Project Data Sheet. https://www.idro.net/upload/blocchi/EC-WTP-Cape-Verde-Desalination-Project-EN-266.pdf (accessed 25 January 2025)

85 Aler-renovaveis (2021). Renewable Energy in Porto Novo's Desalination Plant, Cape Verde. https://www.aler-renovaveis.org/en/activities/publications/case-studies/renewable-energy-in-porto-novos-desalination-plant-cape-verde/ (accessed 25 January 2025).

86 Emirates News Agency (2022). SWRO Plant Jebel Ali. https://www.wam.ae/en/article/hszrczxj-swro-water-desalination-plant-jebel-ali-924 (accessed 25 January 2025).

87 San Diego County Water Authority (2025). Carlsbad Plant. https://www.sdcwa.org/your-water/local-water-supplies/seawater-desalination/.

88 PENTAIR (2024). Shuwaikh seawater RO plant. https://xflow.pentair.com/en/case-studies/shuwaikh.html.

89 SAVENER. https://www.savener.es/en/proyectos/sea-water-reverse-osmosis-salalah/ (accessed 25 January 2025).

90 Seqwater (2021). Desalination in South East Queensland- Fact Sheet. https://www.seqwater.com.au/sites/default/files/2021-11/2021%20-%20Fact%20Sheet%20-%20Gold%20Coast%20Desalination%20Plant.pdf.

91 Reddy, S., Dey, K., Rufuss, D.D.W. et al. (2024). Forward osmosis desalination: a critical review focussing on recent advancements in draw solution recovery techniques for enhanced efficiency and regeneration. *J. Environ. Chem. Eng.* 12: https://doi.org/10.1016/J.JECE.2024.113968.

92 Elhenawy, Y., Moustafa, G.H., Attia, A.M. et al. (2022). Performance enhancement of a hybrid multi effect evaporation/membrane distillation system driven by solar energy for desalination. *J. Environ. Chem. Eng.* 10: https://doi.org/10.1016/J.JECE.2022.108855.

93 Ahmed, A., Alzahrani, M., Shanks, K. et al. (2022). Reliability and temperature limits of the focal spot of a primary optical component for an ultra-high concentrated photovoltaic system. *AIP Conf. Proc.* 2550: https://doi.org/10.1063/5.0099091.

94 Elsakka, M.M., Refaat, A., Alzahrani, K.M. et al. (2024). Techno-economic assessment of a novel wind-powered RO system with a compressed air energy storage for water desalination. *Energy* 311: https://doi.org/10.1016/J.ENERGY.2024.133296.

95 Khan, M.A.M., Rehman, S., Al-Sulaiman, F.A. et al. (2018). A hybrid renewable energy system as a potential energy source for water desalination using reverse osmosis: a review. *Renew. Sustain. Energy Rev.* 97: https://doi.org/10.1016/J.RSER.2018.08.049.

96 Soliman, M.N., Guen, F.Z., Ahmed, S.A. et al. (2021). Energy consumption and environmental impact assessment of desalination plants and brine disposal strategies. *Process Saf. Environ. Prot.* 147: https://doi.org/10.1016/J.PSEP.2020.12.038.

97 European Parliament (2022). Green Deal: key to a climate-neutral and sustainable EU. https://www.europarl.europa.eu/topics/en/article/20200618STO81513/green-deal-key-to-a-climate-neutral-and-sustainable-eu (accessed 25 January 2025).

98 Ministry of Environment, Water and Agriculture. (2018). Saudi Arabia's National Water Strategy 2030. https://www.mewa.gov.sa/en/Ministry/Agencies/TheWaterAgency/Topics/Pages/Strategy.aspx

17

Advancements in Intensified Membrane Technology for Enhanced Desalination Process

Febio Dalanta[1], Osi Arutanti[1], and Tutuk Djoko Kusworo[2]

[1]Research Center for Chemistry, National Research and Innovation Agency, South Tangerang, Indonesia
[2]Department of Chemical Engineering, Faculty of Engineering, Diponegoro University, Semarang, Indonesia

17.1 Introduction

In the current global context, characterized by internationalization, sustainability, partnership, innovation, research, and development, meeting the increasing demand for progress in improvements, products, energy, and raw materials within the constraints of sustainable development has become a complex challenge. Process Intensification (PI), a strategy that aims to enhance industrial processes, making them more efficient, faster, and environmentally sustainable, offers strategic answers to this challenge [1]. The primary aim of PI is to significantly improve manufacturing activities, reduce unit size and energy requirements, increase overall efficiency, lower capital investments, enhance safety, minimize waste production, and reduce negative environmental impact [2, 3]. Achieving competitiveness and sustainability in industrial processes is crucial, and swift implementation of PI will help foster a robust economy. The substantial benefits of PI for industrial operations include improved energy efficiency, reduced gas emissions, and increased cost-effectiveness, making it a powerful influence across all industry sectors.

For instance, desalination has become a promising solution to combat freshwater scarcity in the twenty-first century. Reverse osmosis (RO), a standout method among various desalination techniques, offers a beacon of hope for producing drinkable water from seawater due to its efficiency and reliability [4]. RO functions by utilizing a pressure gradient to separate salts from water molecules using a semipermeable membrane. It is currently the most widely used technology for new large-scale desalination plants with capacities exceeding 1,000 cubic meters per day, and its integration into seawater treatment processes is continuously growing, paving the way for a more sustainable future [5]. PI plays

Water Treatment and Desalination: Membrane Separation using Renewable Energy Resources, First Edition. Edited by Mohamed Bassyouni.

a vital role in improving RO performance by significantly reducing equipment size and energy consumption, enhancing plant efficiency, reducing capital costs, minimizing waste generation, improving safety, and reducing environmental impacts. In many applications, membrane-based processes such as RO offer a high level of PI. This emphasis on the role of PI in improving RO performance will reassure you about the efficiency of desalination processes.

For almost three decades, PI has been acknowledged as an innovative engineering approach that has significantly contributed to industrial advancements, particularly in the area of membrane technology. Traditionally used for separating, concentrating, and purifying fluid both in liquid and gas systems [6, 7], membranes have become increasingly essential for modern applications. This technology is well aligned with the goals of PI due to its potential to surpass traditional methods, enable more efficient transfer and selection of specific compounds, and improve the effectiveness of separation-reactive processes [8]. In practice, several industrial operations have already replaced traditional separation units with membrane-based alternatives, leading to significant reductions in costs, energy consumption, and environmental impact [9]. This stress on the potential of membrane technology to surpass traditional methods will inspire you with the advancements in the field.

This chapter aims to highlight the strategic alignment of membrane technology with PI principles. It outlines the steps required for implementing PI strategies in intensified membrane applications, encompassing all relevant scales—from the molecular level to individual operating equipment and ultimately to the scale of an entire plant. Through this comprehensive examination, the chapter underscores the inspiring potential of membrane technology in driving PI goals across various temporal and spatial dimensions, thereby contributing significantly to the fields of industrial engineering, chemical engineering, and environmental management. This exploration of membrane technology's potential will leave you feeling optimistic about the future of desalination.

17.2 Process Intensification Applications in Membrane Technologies

In the early stages of PI, the primary goal was to significantly reduce the size of industrial plants, aiming to improve process performance while minimizing capital investment [10]. However, in the last decade, the focus of PI has shifted toward achieving sustainability in processes and products, with a strong emphasis on efficiency. This shift in focus, a significant trend in industrial engineering, chemical engineering, and environmental management aspects, is crucial for the future of industrial processes and products [10, 11]. Adopting PI principles entails using technologies that minimize energy consumption, raw material usage, and waste production. Although PI was initially applied in only a few sectors, it has more recently gained traction in numerous industries, such as petroleum, petrochemicals, agriculture, food, pharmaceuticals, electronics, and environmental management [12].

Membrane technology has popped up as a compelling solution to many modern industrial challenges, embodying PI principles by providing an alternative to conventional methods, performing more effective and selective separation ability [13, 14]. Membrane modification and transformation play a critical role in various applications, including potable water production, medicals, energy generation, packaging, pharmaceuticals, and diverse separation processes necessary for the manufacturing of many compounds, electronic products, and other consumable goods. At the heart of every membrane process lies a boundary layer formed by a specialized thin interface. This layer controls the transport between two phases, ensuring a balance between external pressures, fluid dynamics, and the intrinsic characteristics of the membrane material [4, 11].

Over the past decade, membrane applications have gained significant attention and are now competing with other separation technologies in terms of capacity, energy efficiency, selectivity, permeability, and cost. The versatility of membrane technologies in chemical processes is noteworthy as they can be used for molecular and charge separations, such as microfiltration (MF), ultrafiltration (UF), nanofiltration (NF), and reverse osmosis (RO), as well as for mass and energy transport between different phases, as seen in membrane contactors, pervaporation, crystallizers, and scrubbers. Moreover, membranes are employed in chemical reactions, including catalytic membranes, membrane reactors, and bioreactor-based membranes. This versatility underscores the wide range of applications of membrane technologies in various industrial scenarios.

The presence of thin membrane layers between different phases for mass and energy transport presents a significant opportunity to optimize a wide range of competitive chemical processes. This chapter will delve deeper into the recent advancements in membrane applications and their role in intensifying chemical processes, highlighting several successful cases of commercially adopted technologies. The potential of membrane technology in optimizing competitive chemical processes is a testament to its significant role in PI.

17.3 Applications of Intensified Membranes in Seawater Desalination

Membrane technology, particularly RO desalination, has proved to be one of the most successful applications of membrane processes and is clearly illustrative of PI [15–17]. At present, RO desalination plants are well accepted in the industry and compose about 56% of all the desalination units today, while only 35% of desalination units are thermal units [18]. The evaluation of RO technology is due to the fact it has lower energy needs, the recovery rates are higher, and the capital required to implement the technology is cheaper compared to thermal desalination. The RO technology does a better job of desalinating water and minimizing the energy consumption per gallon while producing less CO_2 emissions and concentrated brine by-products [19].

Nevertheless, the issues that are already seen in seawater desalination, including the necessity for cost-effective onshore desalination of brackish or saline water, define the drawbacks of the contemporary RO units [16, 20]. PI comes into focus at this juncture. It aims to design new units and methods that would dramatically reduce the size of equipment, energy demands, and waste production while enhancing the throughput capacity, making the technologies more cost-effective and sustainable. Over the last decade, membrane units have shown better prospects compared to conventional thermal desalination processes.

Moving forward, there is great potential to improve the existing seawater desalination process by making changes and adding some membrane operations. Multimembrane-integrated systems, due to their high flexibility, simplicity, and compatibility, are seen as a very permissive approach in the attempts to address the challenges and limitations of a single-membrane system and enhance the system's efficiency [19, 21]. For example, in the case of RO desalination, strict feedwater pretreatment is necessary to eliminate the particles that will impact the membrane. The previously used pretreatment techniques can also be expensive and cumbersome, but MF/UF (microfiltration/ultrafiltration) units, when used for pretreatment, are much more effective and operate within allowance budgets, making the membrane technologies favorable for future developments in the field [22, 23].

As membrane-based pretreatment processes offer several benefits, these technologies are becoming popular; more than 90 seawater reverse osmosis (SWRO) plants are already using one or other pretreatment technologies. More specifically, the use of membrane bioreactors (MBRs) as pretreatment of SWRO has attracted significant attention recently. MBRs, which are employed in wastewater treatment, utilize low-pressure membranes (MF/UF) to filter water from the activated sludge. This makes it possible for the treated water to be helpful in recycling and reusing in industries again. This tendency to search for more efficient and environmentally benign solutions puts a spotlight on the membrane technologies applied to seawater desalination. The use of MBRs in the pretreatment of seawater desalination, particularly in modifying and improving its outcomes without necessarily favorable conditions, as will be seen later, makes their integration into the operation extremely beneficial. Table 17.1 provides a detailed comparison between the membrane and conventional methods.

Table 17.1 General comparison of the desalination technology between the membrane and conventional methods [17, 20, 22, 23].

Parameter	MF/UF unit	Conventional method
Unit size	Significantly smaller compared to conventional units	Larger unit size compared to membrane-based separation units
Capital cost for RO	Lower, due to higher permeate flux, which reduces capital costs	Higher, as RO units require lower flux rates, leading to larger investment
Impact on Operational Costs for RO	Lower, due to reduced fouling risk, which decreases operating expenses	Higher, due to increased fouling risk, resulting in more cleaning, chemical use, and pressure requirements
Impact on Operating Energy Consumption for RO	Lower as reduced fouling leads to lower pressure requirements	Higher due to the tremendous pressure needed to counteract fouling formation
Pretreatment Energy Consumption	Higher, due to the energy required to generate pressure for filtration	Lower as gravity can be used for filtration (e.g., sand filters)
RO Membrane Replacement Frequency	Lower frequency as the feed water contains fewer foulants	Higher frequency as membranes are more easily damaged by severe fouling
Chemical Additive Usage	Requires fewer chemical additives	Requires a large amount of coagulants, flocculants, and disinfectants
Water Quality After Pretreatment	Higher quality as it effectively removes fine particles and contaminants	Lower quality may still contain impurities that can cause fouling in RO units
Maintenance Requirements	Lower, with less frequent need for chemical cleaning and backwashing	Higher due to frequent cleaning and possible mechanical repairs
Environmental Impact	Lower due to reduced chemical use and less waste generation	Higher because of more chemical waste and by-products
Installation Flexibility	More flexible and more accessible to integrate into existing systems	Less flexible, requires more space and specific configurations

17.3.1 Two-Dimensional (2D) Nanostructured Hybrid Membranes for Improvement of Desalination Processes

In the last few decades, membrane technologies have appeared as potential substitutes for conventional water treatment techniques because of their high adsorption characteristics and low energy demands. The membranes are most often designed from polymers, but the specificity of the majority of polymers is based on the sensitivity/porousness trade-off [24]. New materials have been synthesized to increase the separation performance in membrane operations. Especially two-dimensional (2D) nanomaterials, including graphene, MoS_2, and metal–organic frameworks, have been considered to be potential candidates for the fabrication of membranes for molecular and ionic selective separations [25]. By virtue of their atomic thickness that defines their plane and by their nanometer lateral dimensions, 2D materials present important advantages over bulk three-dimensional (3D) materials in terms of transport resistance and permeation flux. Nonetheless, the application of 2D nanomaterial-based membranes for highly efficient water desalination is an area of concern, mainly because of the swelling behavior of interlayer distances within the nanomaterial, which are prone to increase in water environments [26]. Such balances have mainly been achieved with much effort toward achieving great control of the interlayer spacing below lower nanometers and attempting to reduce the swelling behavior with a view to enhancing the ion rejection. One strategy for improving 2D-material membrane technology includes developing novel 2D-material membranes that are more water-stable and demonstrate superior separation efficiency.

17.3.1.1 MXene-Based Composite Membranes

Gogotsi and colleagues first reported on the first $Ti_3C_2T_x$ MXene in 2011, and since then, MXene has attracted a great deal of interest. Since then, there has been increasing focus on MXenes as single-layered 2D materials because of an array of tunable physiochemical characteristics [27]. $M_{n+1}X_nT_x$ (where $n = 1, 2, 3$) is a class of carbide, carbonitride, and nitrides, where M denotes an early transition metal, X is C or N, and T is functional groups such as –OH, –O, or –F atop the first four carbons, respectively. These materials are usually obtained from their precursor MAX phases by a process called chemical etching, where A stands for elements belonging to IIIA or IVA groups, as shown in Figure 17.1. When MXenes are produced by wet etching, the outer layers of these materials are chemically terminated with oxygen-containing functional groups. To date, about 30 sorts of MXenes have been fabricated with possibly more kinds because of the atomic structure of MAX phases [28]. Such versatility provides for extensive changes in the properties of MXenes, which makes them highly tunable.

MXenes exhibit excellent adsorption capacity when used as adsorbents for removing specific pollutants, owing to their functional groups and negatively

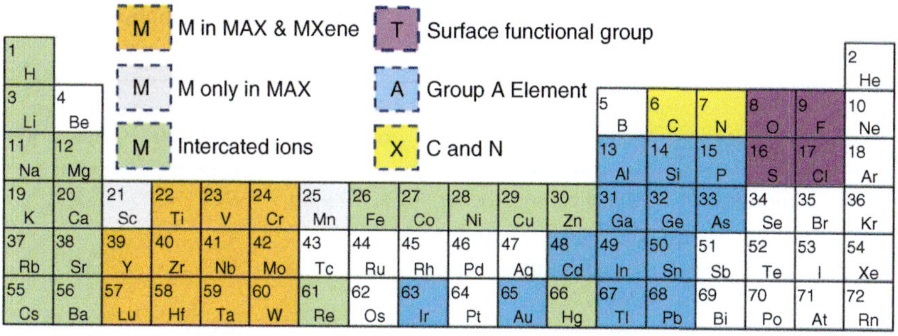

Figure 17.1 Developed elements of the MAX phase in the periodic table. Source: Magnuson and Mattesini [28]/with permission of Elsevier.

charged surfaces. Their unique properties have also made them highly suitable for creating environmentally friendly membranes with superior removal performance. Additionally, MXenes' exceptional electrical conductivity enables controlled ion permeability [29]. However, their separation performance tends to deteriorate when exposed to air and humidity, which affects their stability. The chemical stability of pristine MXenes can be improved by incorporating them with more stable materials, especially polymers, through methods such as in situ polymerization or ex situ incorporation, forming MXene–polymer composites.

Due to their favorable characteristics, ease of synthesis, and low manufacturing cost, MXenes are highly appealing for a range of applications [27]. MXene-based nanocomposites are increasingly recognized as ideal membrane materials for molecular extraction, particularly in water purification, due to their excellent hydrophilicity and resistance to fouling. However, only a limited number of studies have explored MXene-containing membranes for water desalination, with most research focusing on dye rejection and solvent dehydration [30]. Thanks to their outstanding physical, chemical, and structural properties, MXenes demonstrate superior stability and recyclability compared to other two-dimensional nanomaterials. Recently, attention has started to grow around MXenes and their composites for desalination, although the field is still underexplored. Table 17.2 compares the desalination capabilities of these nanomaterials.

Polymer and nanomaterial-based composite membranes show the favorable characteristics of both phases. Such membranes are made by dispersing nanoparticles in polymer matrices (the continuous phase, thus forming a dispersed phase structure). This approach takes advantage of both organic and inorganic materials, provides high permeance compared with other membranes, and shows relatively strong separation performance. For defect-free mixed matrix membranes (MMMs), some factors play a critical role, which includes the high compatibility of the fillers in the polymer solution as well as the stability of the

Table 17.2 Summary of desalination performance using different 2D-nanohybrid membranes.

2D nanocomposite	Polymer	Salt concentration	Salt rejection	References
Carbon nanotube (CNT)	Mixed cellulose ester (MCE)	1,167 ppm	90%	[31]
Ultrafine CNT (UCNT)	Mixed cellulose ester (MCE)	1,167 ppm	98%	[31]
Layered double hydroxide (LDH)	Polysulfone (PSf)	2,000 ppm	99.03%	[32]
Graphene	Polytetra-fluoroethylene/Teflon	40,000 ppm	94%	[33]
Laponite nanoclay	Polyamide	2,000 ppm	98.18%	[34]
CoFe-LDH/$Ti_3C_2T_x$	$Ti_3C_2T_x$ (MXene) film	80 mM	149.2 mg/g	[30]

dope solution. Polymer compounds like titanium silicalite (TS) and poly-ether ether ketone (PEEK) can have efficient miscibility with MXenes owing to their polarity, hydrophilic nature, and suitable surface morphological features, as a result of which strong polymer–MXene interaction occurs [24]. Furthermore, $Ti_3C_2T_x$ nanomaterials have good solubility in other aprotic solvents, including *N*-methyl pyrrolidone (NMP), dimethyl sulfoxide (DMSO), dimethylformamide (DMF), and water and ethanol. The presence of surface hydroxyl (–OH) groups in $Ti_3C_2T_x$ MXenes allows more chemical modifications to improve the interfacial compatibility between MXenes and polymers [26]. For example, works on mechanical strength enhancement and water permeability of the MMMs, particularly $Ti_3C_2T_x$/polvinyl alcohol (PVA), seem noteworthy for further application in the field.

The polyamide (PA) thin-film composite (TFC) membrane consists of a selective layer and a support layer. A very thin PA layer forms through interfacial polymerization (IP) between two monomeric units on a porous support layer [35]. Initially, the support is immersed in an aqueous diamine solution, typically *m*-phenylenediamine (MPD), to develop the PA layer. The coated support is then dipped into a monomer solution, such as trimethyl chloride (TMC) to enhance the selective layer's material transport capabilities. Nanoparticles, including MXenes, metal–organic frameworks (MOFs), and covalent organic frameworks (COFs), can be incorporated into the aqueous phase solution to improve performance further.

For example, incorporating $Ti_3C_2T_x$ MXenes led to increased water molecule diffusion, boosting water permeation from 1.73 to 2.53 L/m^2/h/bar and achieving a remarkable salt rejection rate of up to 98.5%. In another instance, a study used liquid exfoliation to synthesize MoS_2–MXene nanocomposite membranes, which were integrated into the selective layer of fine-film composite RO membranes [25]. The enhanced surface properties and reduced film thickness contributed to the membranes' superior performance. TFC membranes have gained widespread interest for applications in NF, RO, and forward osmosis (FO) processes. For example, using the IP method, a novel TFC RO membrane was developed by embedding $Ti_3C_2T_x$ MXenes into the selective layer. MPD solutions with varying $Ti_3C_2T_x$ MXene loadings (0.005–0.02 wt%) were prepared, and polysulfone supports were submerged in the MXene-enriched MPD solution. The MXene-loaded support was then immersed in a 1.0 wt% TMC solution to produce the final thin-film nanocomposite (TFN) membranes, as shown in Figure 17.2 [35]. Extensive surface modification improved the membrane's hydrophilicity and reduced surface roughness, contributing to enhanced performance.

An MXene-based $Fe_3O_4@Ti_3C_2T_x$ nanocomposite has developed through integrating Fe_3O_4 nanoparticles with $Ti_3C_2T_x$ MXenes for efficient salt rejection from water. The interior surface of the $Ti_3C_2T_x$ MXene nanocomposite acts as an electrode surface, facilitating enhanced electron transfer while providing numerous active sites for the uniform growth of Fe_3O_4 particles [36]. Coating

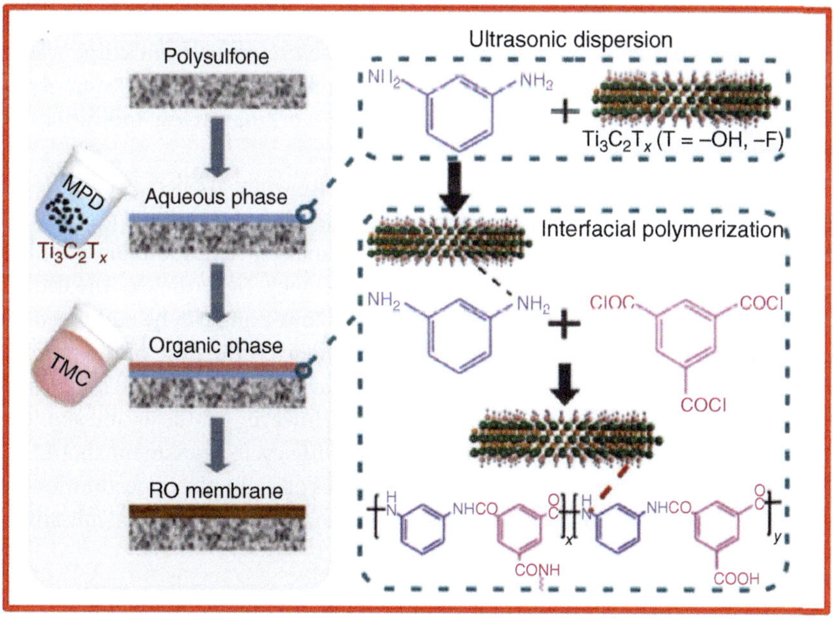

Figure 17.2 Preparation of the $Ti_3C_2T_x$ MXene membrane with detailed surface polymerization for the desalination process. Source: Wang et al. [35]/with permission of Elsevier.

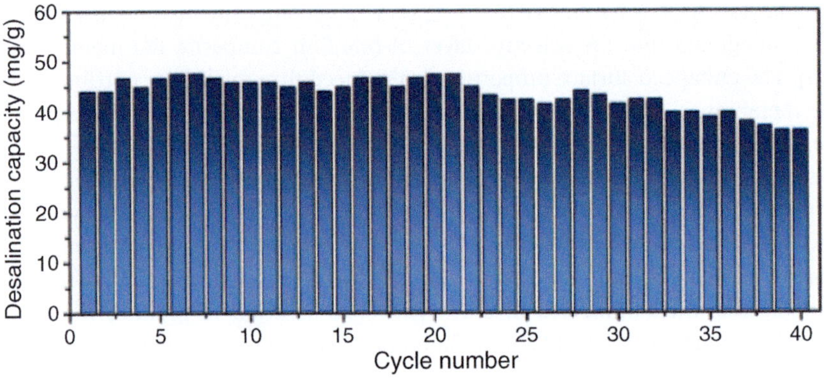

Figure 17.3 Long-term cycle test for the desalination capacity of $Fe_3O_4@Ti_3C_2T_x$ MXene composite. Source: Wang et al. [36]/with permission of Elsevier.

the $Ti_3C_2T_x$ MXene on the Fe_3O_4 nanoparticles helps protect the MXene from oxidation. The resulting MXene-based composite demonstrated exceptional desalination adsorption, reaching up to 44 mg/g, and exhibited excellent stability, maintaining its desalination performance over 40 working cycles with minimal reduction in capacity [25]. These impressive features highlight the high potential of this hybrid capacitive deionization system for salt rejection from saline water. Furthermore, the desalination capacity of the synthesized MXene-based material showed only a 2% decline after 40 continuous working cycles, indicating its durability and efficiency, as depicted in Figure 17.3.

MXenes demonstrated a high adsorption capacity when applied as adsorbents for selective removal of pollutants owing to well-developed functional groups and negatively charged surfaces. These properties also make MXenes suitable for green fabrication of membranes with high removal efficiency. Nevertheless, the intrinsic chemical stability of pristine MXene phases is lower and can be addressed by incorporating another chemically stable material. Given the time of the current research, further enhancements of MXenes and their composites could drastically transform the future of the desalination industry. However, synthesis and application issues are still daunting, and therefore, their life cycle assessment should be encouraged by more studies. Furthermore, MXenes offer unique opportunities for the author's vision to contribute to seawater desalination and ecological initiatives of the upcoming materials.

17.3.1.2 Carbon-Based Composite Membranes

Carbon-based membranes, including graphene oxide (GO) and carbon nanotube (CNT) membranes, have attracted extensive attention in desalination due to

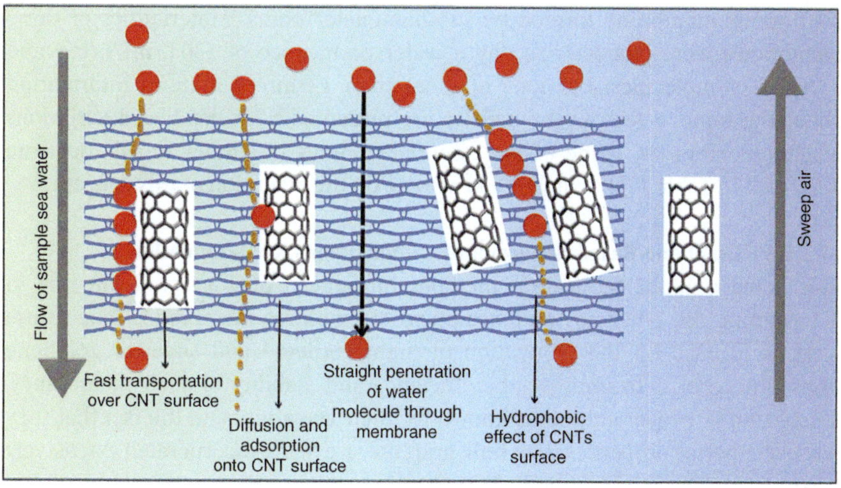

Figure 17.4 Specific process for the CNT-based membrane for desalination. Source: Chowdhury et al. [37]/with permission of Elsevier.

their high permeability, specific mechanical strength, and high molecular discriminative ability. These qualities make them efficient candidates for efficient desalination processes. Chowdhury et al. also provided a typical desalination process using CNTs, illustrating that the CNTs adsorb solutes due to their hydrophobic character and help leverage the vapor flow by not allowing water molecules to enter the membrane pores, as schematically shown in Figure 17.4 [37]. Kitano et al. have also done a comparative study of the usefulness of hypersonic nanoparticle deposition, proposing a novel process involving carbon nanomembranes in which they have achieved a far better efficiency with an excellent account of the stability of C-nanocomposites (CNMs) [38]. Yao et al. outlined the growth of molecular ultrathin films with sub-nanometer pores displaying exceptional water permeability and selectivity [39]. To fabricate CNMs, Rana et al. applied terphenyl thiol monolayers, which have a thickness of only 1.2 nm. These CNMs had highly ordered sub-nanometer pores that excluded nearly all the gases and liquids but let water in at a fast rate [40]. Remarkably, water had a comparably high permeance, while helium had a flux of only 2,500, thus indicating that water molecules easily traverse the channels in large groups. This makes CNMs particularly suitable for removing contaminants in water.

Building on this research, Yao et al. synthesized carbon nanomembranes from amorphous molecular layers for filterable molecular sieving and desalination. They used a scalable approach to synthesize CNMs from poly(4-vinyl biphenyl) (PVBP) onto SiO_2/Si wafers [39]. Thus, subjecting the amorphous PVBP layers to synthesis through electron irradiation led the researchers to form a 150 Å continuous membrane formed by sub-nanometer pores. Materialities of these membranes were characterized by a water permeance of 530 L/m^2/h/bar and a cutoff of molecules and ions of more than 1.0 nm. For more information about the topic, a few review articles are presented with detailed descriptions of nanomembranes. Some of the essential findings of studies in this field are described in Table 17.3, which is a brief note on the evolution of this concept.

17.3.1.3 Graphene-Based Nanocomposite Membranes for Desalination

It is noted that the fabrication methods introduced play a significant role in determining the efficiency and steadiness of graphene membranes. The future work is focused on the fabrication of highly ordered and superior graphene films—the goal is to improve the efficiency and stability of such membranes. Mono- and few-layer graphene, known for their versatility and the fact that they possess superior properties than bulk graphite, are being incorporated extensively in an extensive range of applications in photovoltaic (PV), optoelectronic, and electronic systems [33, 52]. They have high electron mobility, more significant than 200,000 cm^2/Vs, optical transparency close to 97.7%, high mechanical stiffness with Young's modulus comparable to 1,000 GPa, and high and relatively high thermal conductance varied within the range of 2,000–5,000 W/mK. The solid covalent links put in carbon atoms of graphene, alongside its large surface, in addition to the availability of edges and open channels, help in its flexibility and sturdiness in multiple uses [54, 55]. Possible applications incorporated in the biological aspect include disease biomarker identification and DNA hybridization in medical diagnosis, lithium-ion batteries, PV cells, and supercapacitors in energy conversion, desalination of seawater, and detection of pollutants such as heavy metals and hazardous gases [52, 56, 57].

Research has also looked at how laser features affect the formation of laser-induced graphene (LIG) filters as well as desalination and water filtration performances [52]. LIG filter materials have been characterized and modeled for specific uses in water purification duties such as dye decoloration and disinfection, along with interfacial evaporation-based desalination processes in both singular and multilayer forms. The discoveries pointed out that there are desirable characteristics for the polyether sulfone (PES)-LIG filters based on laser power, image density, platform, and membrane support. These parameters could be well-tuned to synthesize the PES-based LIG porous surface materials for enhancement in performance along with stability for different energy, environmental, and catalytic applications. Furthermore, research has been done on the relationship between the applied pressure, chemical functionalization, and pore size with the performance of the membrane. It has been shown that graphene membranes exclusively reject all the salt ions and allow water molecules to pass through at very high rates of (10–100 L/cm^2/day/MPa) [50, 52].

Table 17.3 Various types of nanocomposite membranes for the desalination process.

Modified membranes	Primary materials	Key features	Desalination ability (%)	References
Hybrid nanocomposite membranes	Polymeric membranes incorporated mostly with inorganic (sometimes organic) nanofillers	Surface modification, improved selectivity, increased permeability, chemical resistance, thermal resilience, and enhanced mechanical durability.	85–99	[41]
Thin film-based nanocomposite membranes	A polymer matrix, such as PA or PSf, is combined with nanoscale materials such as nanotubes and nanoparticles.	Improved permeability and physical durability, adjustable pore dimensions, compatibility with living tissue, surface customization, and increased surface area with minuscule pores.	>99	[42, 43]
Nanofiber membranes	Natural polymers, synthetic polymers, ceramics, and composites.	Increased porosity and strength, adjustable pore size, compatibility with living tissue, customizable surface, and expanded surface area with tiny pores.	80–99.99	[44–46]
Nanoporous membranes	A wide range of inorganic, organic, and composite particles.	Porosity, selective permeability, large surface area, adjustable pore size, resistance to harsh chemicals and temperature, flexibility, controllable surface chemistry, and expandability.	90–99.9	[47–49]
Carbon-based nanocomposite membranes	A variety of carbon allotropes, such as graphene, graphene oxide, carbon nanotubes, and activated carbon, exhibit unparalleled mechanical, thermal, electrical, optical, and chemical properties with unique structural characteristics. Unrivaled properties with distinctive structural dimensions.	Unrivaled mechanical, thermal, electrical, optical, and chemical properties with distinctive structural dimensions.	90–99.9	[50–53]

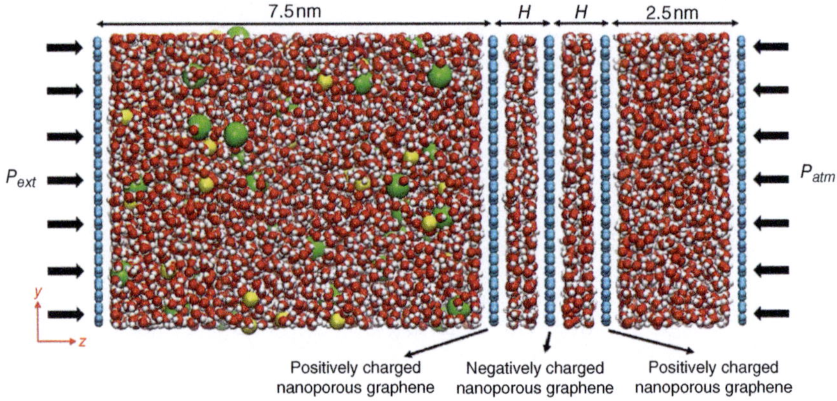

Figure 17.5 Schematic illustration of the seawater desalination process using charged multilayer graphene-based membranes. Source: Mortazavi et al. [58]/with permission of Elsevier.

These calculate more water fluxes in line with the atomic thickness and pore size of the membrane and, therefore, the adjustable selectivity of synthetic nanopores in graphene membranes for desalination approaches. Some of the trade-offs related to the creation and use of graphene-based nanomembranes are implicated by the question of energy and environmental efficiency: Sustainability is the key to energy saving, emission reduction, and environmentally friendly practice as Figure 17.5 represents the seawater desalination process using charged multilayer graphene-based membranes.

Mortazavi and colleagues have enhanced the performance of the graphene membranes by assigning an electric charge to them [58]. This work focuses on the impact of the charge density of the adsorbent and geometry features such as pore size and interlayer distance. Using molecular dynamic simulations, the researchers examined water desalination of multilayered graphene with applied charges. The researchers established that graphene with charged layers, pore size of 16.35 Å, has a water flow rate of 59.4% compared to uncharged multiple-layer graphene with a pore size of 12.6 Å. Also, the salt rejection efficiency increased from 67.27% to 94.54%. This enhancement was a result of the uniform distribution and accurate positioning of electric charges on the carbon atoms at the pore edges. The study also showed that if the two membrane layers were spaced apart further, any gain in electric charging meant for improving water flow ended up being reduced.

A further advancement in graphene-based materials is graphene oxide (GO). GO membranes exhibit inherent nanoscale pores of 0.3 nm, while the pores increase to 0.9 nm channel size in aqueous conditions. This enabled the passage of smaller particles, that is, those with a size smaller than 0.45 nm while rejecting larger ones.

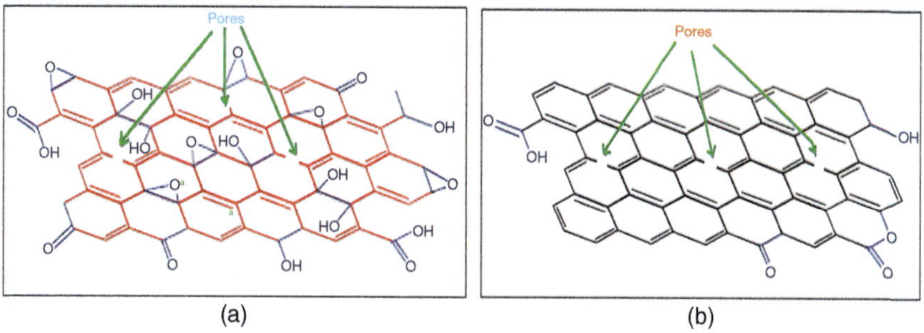

Figure 17.6 Chemical structure and pore characteristics of (a) GO and (b) rGO. Source: Sun et al. [55]/with permission of Elsevier.

Many methods exist for the deposition of GO membranes, such as layer-by-layer (LbL), vacuum filtration, spinning coating, and drop casting. Still, all these methods are pretty inexpensive and, hence, necessitate the use of comparatively cheap equipment. These 3D, highly porous, and functionalized GO microcombs contain either hydrophobic (CH_3) or hydrophilic (OH^-) functional groups [55, 59]. The literature review established that when the pore contains OH-functionalized groups, water flow is higher because the pores' paths are wider [55, 59].

When a process is employed to completely or partially eliminate functional oxygen groups in graphene, a modified material called reduced graphene oxide (rGO) is produced [60]. This process results in GO formation, which is known to possess epoxy, hydroxyl, and carboxyl facets on the graphene layers. These functional groups increase the hydrophilic and water-soluble properties of GO, which, in some processes, is more effective for use than pure graphene. As established from the interlayer spacing of about 0.34–0.37 nm, rGO has excellent potential in membrane application since it can filter out particular salt ions even at low sodium concentrations. Because of these differences, GO and rGO nanomembranes possess different mechanical, chemical, and electrical characteristics [60]. The first one is electrical conductivity: While rGO shows high ohmic values, GO can be considered as an insulator or semiconductor. It could be observed that compared to GO, rGO materials have a larger specific surface area after the elimination of oxygen-containing molecules. Also, rGO has a hydrophobic nature, while GO shows a hydrophilic nature of the material. In addition, rGO has better mechanical properties than GO, even though the reports in this area are somewhat conflicting. The schematic illustration for chemical and pore representation of GO and rGO is depicted in Figure 17.6.

Furthermore, new mixed membranes based on graphene oxide and polysulfone (GO-PSF) were synthesized through the method of wet-phase inversion, using GO content in the range from 0.25 to 2 wt%. These membranes were subjected to a solution containing sodium chloride at a concentration of 25,000 mg/L in the direct contact membrane distillation (DCMD) process. Nevertheless, the preformed membrane containing 2 wt% GO had the highest permeation flux, but the 1 wt% GO membrane had more significant porosity and higher salt rejection. Using such concentrations of sodium chloride in tests with RO concentrates, the treatments provided a membrane flux of 20.8 and 31.9 $L/m^2/h$, respectively [60]. This suggests that a specific level of GO is capable of removing excess ions without straining the membrane and the quality of the water produced. Moreover, hydroxyl groups are the other factor providing higher efficiency of sodium ion rejection; however, the influence of the hydroxyl groups on the chloride ion transport does not differ much. Hydrophilic groups formed lead to a decrease in the crystalline spacing of a GO layer. Thus, it is required to decrease the GO membrane to gain rGO. These elimination steps can eliminate functional groups, including carboxyl, epoxide, carbonyl, and hydroxyl groups. Generally, rGO is synthesized from high temperature or chemical treatment as a reduction process. A method for using temperature for the process of annealing is usually followed by the removal of oxygen from GO and using higher temperatures to remove oxygen or oxygen-containing bodies like –OH [56]. Nevertheless, this methodology needs much energy and it is difficult to set the desired level of oxidation deliberately. Chemical reduction requires a predetermined temperature and reducing agents, which include metal hydrides, hydroiodic acids, and hydrazine. It can be challenging to accurately target those functional groups, which is precisely what needs to be done. Some of the authors have pointed out that the fabrication of the nano-filter membranes using rGO has benefits when compared with those fabricated from GO. To sum it up, Table 17.4 summarizes the recent developments of GO and rGO-based membranes for desalination purposes.

17.3.2 Membrane Distillation for Desalination Process

Membrane distillation (MD) becomes one of many examples of an intensified membrane process in desalination. It can handle various types of water, including mine water, wastewater, radioactive wastewater, saline water, saltwater, and RO brine, using the MD method [67, 68]. MD appears to be among the most prospective technologies for seawater desalination, as schematically represented in Figure 17.7. Some of the benefits that make it more worthy to use include lower hydraulic pressure than that used in the RO process, high rejection of elements which are nonvolatile, lower operational temperatures than the distillation process, and, most importantly, its environmental friendliness in comparison to distillation processes [69].

In the case of the MD process, a hydrophobic membrane is located between the feed and the permeate. This membrane, because of its hydrophobicity, permits passage only of vapor molecules [70]. Therefore, the MD process can transform the wastewater and saline water to a better water quality where the undesirable

Table 17.4 Recent breakthroughs of GO- and rGO-based membranes for desalination purposes.

Membranes	Synthesis method	Permeability	Removal efficiency	References
Polyamide/GO	Covalently bonded	41.40 L/m^2/h	97.80% (NaCl)	[61]
PES/GO	Self-assembly	28 L/m^2/h	98% (NaCl)	[59, 62]
Polysulfone/GO	Layer-by-layer assembly	27.6 L/m^2/h/bar	46% (Na_2SO_4)	[63]
GO/PDA/Al_2O_3	Vacuum filtration	48.4 L/m^2/h	99.7% (NaCl)	[64]
Cationic porphyrin/GO	Vacuum filtration	9.3 L/m^2/h	87.7% (NaCl)	[65]
Ceramic GO	Vacuum filtration	14.40–58.60 L/m^2/h	31.4% (NaCl)	[57]
PANI/rGO	Phase inversion	27.12 L/m^2/h	82% (NaCl)	[60]
rGO/HNTs-PSS	Evaporation-induced assembly	8.8 L/m^2/h	4.7% ($MgCl_2$), 4.7% ($MgSO_4$), 14.3% (Na_2SO_4), and 6.8 (NaCl)	[54]
PDA/rGO	Chemical reduction	36.6 L/m^2/h	92% (NaCl)	[56]
MOF-801@ GO-PDMS membrane	Vacuum filtration	29.24 kg/m^2/h	99.99% (NaCl)	[66]

liquid may become a more concentrated solid solution in which all the component equivalents of the original liquid are compressed. The key to this separation is a vapor pressure difference resulting from the temperature difference between the high boiling point of the feed and the low boiling point of the permeate fraction. Furthermore, concentration polarization and temperature polarization yield different absolute concentrations and temperatures at the membrane surface than the bulk feeding solution and permeate. This can lead to a decline in the driving force and mass transfer and, hence, heat and mass transfer at the boundary layer of the membrane [68]. Consequently, it is crucial to analyze these phenomena more closely in order to better control the permeate side in the MD as well as in the membrane distillation crystallization (MDC) systems. The various types of MD processes include DCMD, vacuum membrane distillation (VMD), air-gap membrane distillation (AGMD), and sweep gas membrane distillation (SGMD). Two kinds of configurations are unique and have various benefits and drawbacks, as summarized in Table 17.5.

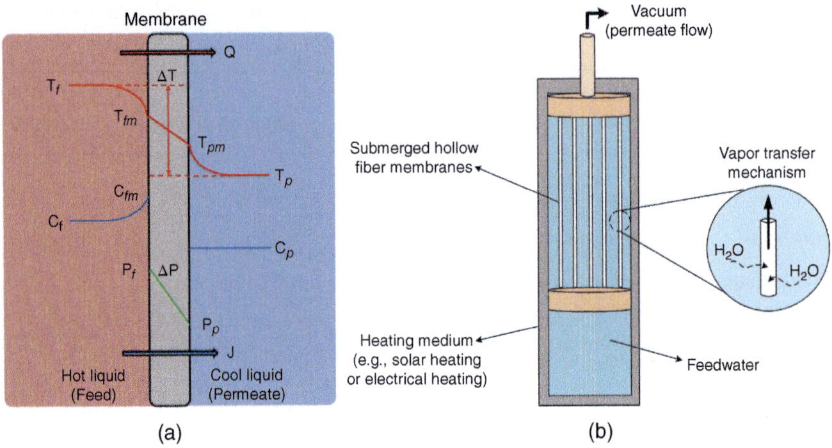

Figure 17.7 (a) Thermodynamic properties' profile of the MD process, and (b) the schematic illustration for the submerged MD process system. Source: Dalanta et al. [4]/with permission of Elsevier.

Table 17.5 Various applied MD processes with different materials and preparation methods.

MD material	Preparation method	Contact angle (degree)	Salt removal (%)	Permeability (L/m²/h)	References
PVDF	Phase inversion	148	99	87.40	[71]
PVDF	Perfluorinated polymers	135	99.9	83.40	[72]
PP	UV graft	138	97	3.32	[73]
PES	Sol–gel	119	99.3	44.70	[74]
PVDF	Electrospinning	154	99	5.80	[75]
PS	Electrospinning	113	99.9	31.00	[76]
PVC	Radiation-induced	114	99.98	50.50	[77]
PVDF	Electrospinning	139	99.9	10.70	[78]
SBS	Electrospinning	132	99.9	10.50	[79]
PES	Radiation-induced	114	99.98	50.50	[80]
PVDF	Electrospinning	130	99.9	49.00	[69]
PES	Plasma treatment	120	99.9	42.00	[81]
PVDF	Phase inversion	92	99.9	2.90	[82]

17.3.3 Photothermal and Solar-Powered Membrane Distillation for Desalination

Desalination and wastewater treatment are the best and most effective ways for water scarcity solutions. The desalination method called Solar Powered Membrane Distillation (SPMD) based on membranes made of photothermal materials has proved to be the most effective method of desalination. It has prospects for effectively utilizing low-quality as well as renewable energy sources. MD is a heat-operated process in which water at the feed side of the membrane evaporates and diffuses through pores of the hydrophobic membrane to the cold side, where it condenses into freshwater. Thus, MD is particularly effective for long-duration separation processes because of the advantages associated with this type of process [67, 83].

Some of the unique advantages of MD technology include the ability to remove highly saline water; these areas of application include the food industry, the pharmaceutical industry, and environmental conservation [67]. It can be practiced singly or in combination with other separation processes as the last process in separation operations. However, CMD has some limitations such as its promotion by large centralized pumping systems, heat loss during feed transfer, and high energy consumption in heating the bulk feed water. Conversely, SPMD systems require much less power and simplified equipment compared to the centralized instruction processing (CIP) systems, which have tremendous potential for laptops and integrated, portable computers designed for use in remote, off-grid environments. This approach not only minimizes heat loss by directly delivering the heat to the membranes' photothermal layers, but it also enhances thermal energy efficiency to approximately 60%. In contrast, the CMD achieves an efficiency of less than 10% [70, 83, 84].

One of the most significant concerns with the large-scale application of CMD is the fate of the thermal efficiency in the process due to the influence of temperature polarization. This is attributed to conduction and the characteristics of latent heat concerning the hydrophobic microporous membrane implemented during the operation. Considering the thermal conductivity of the membrane and the water vaporization process, a lower temperature at the surface feed interface T_{fm} is observed compared to the temperature of the bulk feed water T_f. Likewise, the temperature of the bulk distillate is less than at the temperature of the distillate interface, T_{dm}. This temperature polarization decreases the temperature gradient at the interface significantly, thus reducing the theoretical driving force across the two bulk phases of the membrane and water [85].

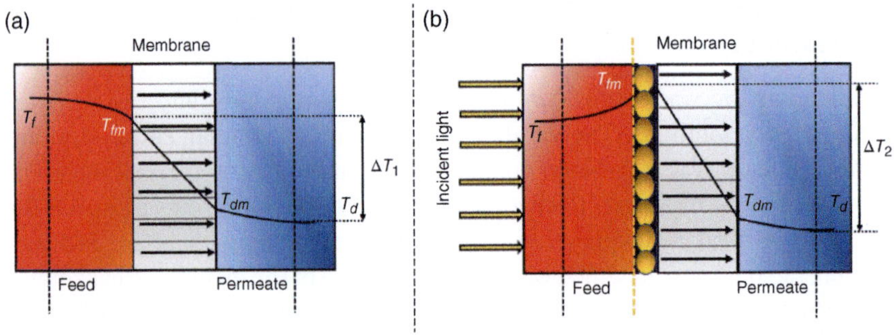

Figure 17.8 Profiles of temperature in (a) conventional MD and (b) solar-powered MD. Source: Jawed et al. [85]/with permission of Elsevier.

To overcome such problems and improve the performance of MD, various approaches have been studied to include the incorporation of photothermal membranes. These membranes contain 0.1–10 optical black nanoparticles or coatings on either side, which, when illuminated with light, readily transform the light energy to heat, giving a localized heat at the membrane interface [86, 87]. Photothermal membranes can enhance the rate of evaporation, minimize temperature difference, enhance mass transfer of fluids and vapor working with localized heating using light-absorbing materials, and allow selectivity of heating. This also increases the overall efficiency of the MD, where T_{fm} becomes higher than T_f in the same way T_d becomes higher than T_{dm}. The detailed temperature profiles of CMD and SPMD are shown in Figure 17.8.

The MD process is significantly energy-demanding on conventional energy forms such as fossil fuel energy. However, the employment of socially abundant, cheap, and clean energy sources remains a major strength of MD. Thus, the energy consumption of the MD process can be minimized by employing low-grade heat sources. Some of the available low-temperature heat sources that are under consideration for MD application include industrial and power plant waste heat, low-temperature solar thermal collectors, and geothermal potential. Due to unavoidable thermal losses in thermal desalination systems, district heating should be primarily used for commercial and residential buildings before being used in any other form of heat energy application. Industrial low-grade heat sources are regarded as more realistic, although there is scarce information about available waste heat quantity and quality [88, 89]. The utilization of the MD process can also be extended to minimize discharge temperature, safeguard water infrastructure, and achieve compliance. The main focus of the research has been on the design, installation, and testing of SPMD systems over the last two decades [86, 90]. For instance, Guillén et al. investigated a small-scale AGMD desalination system using a flat sheet polytetrafluoroethylene (PTFE) membrane with solar energy. A feed temperature of 85°C was used with a membrane surface area of 2.8–8.4 m^2 per module. With the latter, a daily permeate flux ranging from 156 to 168 kg/m^2 and a specific heat input of 810–2,220 kWh/m^3 were obtained. The research established that the enhancement of thermal efficiency requires a multistage campaign. In the same manner, Jantaporn et al. studied a compact, hybrid solar-powered vacuum multi-effect MD with a membrane surface area of 5.12m^2, obtained from Saudi Arabia, producing freshwater at a rate of 0.27–0.38 m^3/day [91].

Alobaidani et al. mentioned that the MD process is compatible with solar power systems well. This is primarily due to fluctuating and sporadic environmental conditions regarding MD operating conditions and its use of low-grade thermal energy [68]. MD, complemented with solar power, has thus become expedient and is being adopted worldwide. This configuration is best represented through systems like SPMD and serves as a model attributable to its off-grid electricity integration. In such a setup, the pressures are produced and generated separately by the system, and thermal as well as electrical power is delivered to the users, as illustrated in Figure 17.9.

The essential subsystems of the solar-powered DCMD system are solar thermal collectors or PV modules, feed and distillate tanks, the MD module, and circulation pumps. The standalone solar MD desalination system is similar in construction to the above, which is primarily that solar PV collectors provide the necessary electrical energy required for the MD process with DC batteries and inverters incorporated, rather than using a diesel generator [92]. Two coupling arrangements of solar energy for MD systems are shown here, where the thermal collectors apply heat to the feed water of the solar-coupled MD desalination units [93]. In these systems, all the necessary pumps and other equipment can be connected to an electrical grid or generator, and heat is supplied to the MD module either directly or through the heat exchanger.

The most efficient methods for creating a single layer for SPMD and CMD processes are non-solvent-induced phase separation and electrospinning [33, 76]. These methods are also used to synthesize an intermediate structure known as the MMMs, whereby the solid phase (nanomaterials) is incorporated in the matrix (polymers). A major strength of these fabrication techniques is that they allow the fabrication of MD membranes from different polymers. In contrast, commercial MD membrane synthesis is readily available only from polymers such as PVDF, PP, and PTFE. For instance, researchers recently utilized electrospinning to produce an S-coupled MD nanofibrous membrane from poly(styrene-butadiene-styrene), resulting in a membrane that is more hydrophobic compared to PTFE commercial membranes [33]. This advancement has opened up new possibilities for assessing newly synthesized hydrophobic polymers and developing a novel polymer mixture for MD applications. The solubility of different fluorinated copolymers in the

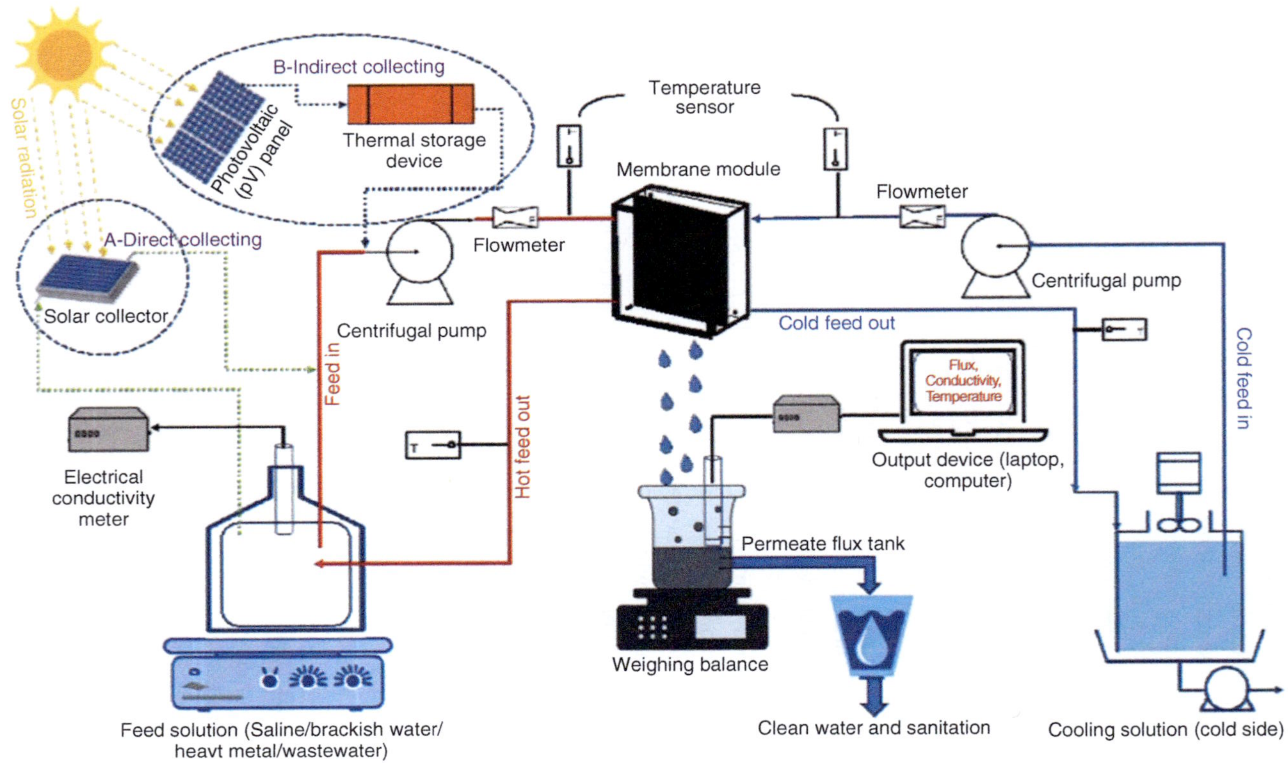

Figure 17.9 Schematic configuration of the direct and indirect collection techniques in SPMD. Source: Alobaidani et al. [68]/with permission of Elsevier.

solvents utilized for membrane fabrication is crucial for improving both mechanical strength and hydrophilicity. As a result, these copolymers can be employed in the production of microporous hydrophobic membranes for desalination purposes. Studies have sought to enhance performance by blending insoluble PTFE with soluble PVDF. Furthermore, studies show novel thin-film dual-layer PVDF/Ultem1 membranes for desalination application in VMD systems [82, 94]. Another fluoropolymer that shows excellent potential is Hyflon1 AD, which has been produced. Despite this progress more comprehensive studies are required to enhance the fabrication techniques of the membranes and to achieve improved performance of these distinctive polymers for MD. Table 17.6 summarizes the features and performance of current innovations in high-flux membranes for application in MD.

17.3.4 Electrodialysis Membranes for Desalination Process

Electrodialysis (ED) is an electro-membrane technique that is applied for the separation of ion-containing solutions through ion-exchange membranes [105, 106]. This operates through cross-flowing an electric field across a cation-exchange membrane followed by an anion-exchange membrane to form sections of anions and cations within a feed solution. In this process, positive ions are attracted to the cathode, and negative ions move toward the anode. Thus, effectively, the charged ions are separated out, as shown in Figure 17.10 and Figure 17.11. ED is applied more often for water desalination, including seawater and brackish water treatment, and for the concentration and purification in food, beverages, and pharmaceutical industries, ion-exchange processes, etc. [107–109]. However, seawater desalination poses several issues to ED systems because of the high ionic load and ion variety and high concentration of sodium, magnesium, and chloride ions in particular [110] Due to the high salinity, concentration polarization and scaling become potential problems, and thus, strong antifouling is needed. Also, seawater desalination brine belongs to the most recognizable focused waste streams produced by ED, which require further treatment to minimize environmental consequences.

Therefore, this shows that ED's efficiency in removing ions makes this technique appropriate for water purification and resource recovery. As has been observed, there are variations in the feed water used such that the salinity affects the ion transport rate and membrane selectivity of the ED process. Low salinity water is mainly less than seawater salinity, low in ion content, is easier to desalinate and involves low energy consumption because the membranes are less prone to fouling since scaling ions such as calcium and magnesium are present at lower concentrations. Nevertheless, sulfate concentrations above these levels still entail a scaling

Table 17.6 Summary of the features and performance of recent innovations in the high-flux membrane technology for membrane distillation (MD).

Membrane	Synthesis method	Permeability (kg/m^2h)	Membrane characteristic	Removal (%)	References
Polyvinylidene fluoride (PVDF) membrane	N-TIPS	40.9	δ: 105.0 μm ε: 86.0% LEP: 3.5 bar σ: 0.93 MPa	>99.99	[95]
Hydrophobic/hydrophilic PVDF-HFP/Nylon dual-layer membrane	Electrospinning	~36.0	r: 0.23 μm δt: 105 μm δb: 155 μm εt: 90.3% εb: 71.1% WCAt: 144.3° WCAb: <40.0°	>99.9	[96]
Hydrophobic/hydrophilic fluorinated PDA-Ag/PES membrane	Dip-coating	39.4 (4.5 wt%)	WCAt: 115.2° WCAb: 66.6°	99.99	[97]
Activated carbon enhanced hydrophobic/hydrophilic dual-layer PVDF-HFP/Nylon 6,6 ENM	Electrospinning	45.6	r: 0.79 μm δt: 40.0 μm δb: 160.0 μm ε: 90.5% WCAt: 142.7° WCAb: <40.0° LEP: 1.4 bar σ: 17.8 MPa	99.99	[98]

PTFE/PI bi-component ENM	Electrospinning	42.0	r: 1.0 μm δ: ~100 μm ε: 78.0% WCA: 150.0° LEP: 2.0 bar σ: 0.33 MPa	>99.9	[99]
Fluorosilane-coated TiO_2/PVDF-HFP ENM	Electrospinning	37.8 (7.0 wt%)	r: 0.75 μm δ: 100.0 μm ε: 91.6% WCA: 149.0° LEP: 1.0 bar σ: 7.0 MPa	98.0	[100]
rGO/PTFE membrane	Spray coating	~65.0	r: 0.15 μm WCA: 126.3° LEP: 0.93 bar	>99.99	[101]
CNT-immobilized on PTFE membrane	Spray coating	ND	ε: 72.0% WCA: 95.0°	>99.9	[102]
Octadecyltrimethoxysilane-modified SiO_2/PVDF ENM	Electrospinning	~35.0 (3.0 wt%)	r: 1.41 μm δ: 132.0 μm ε: 78.5% WCA: 162.6° LEP: 0.8 bar	99.9	[103]
Interweaving hierarchical PVDF-HFP/PET composite ENM	Electrospinning	36.9 (55 °C)	r: 1.0 μm δ: 80 μm ε: ND WCA: 135.3° LEP: 0.18 bar σ: 8.68 MPa	>99.99	[104]

Source: Dalanta et al. [4]/with permission of Elsevier.

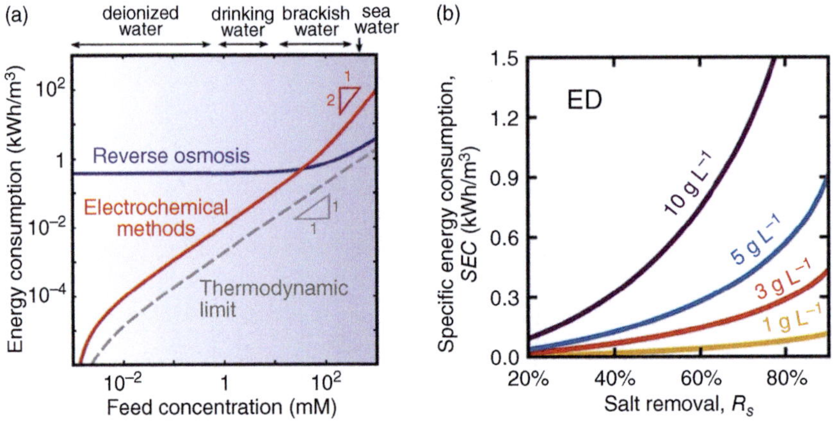

Figure 17.10 (a) Feed concentration vs. energy consumption profile, and (b) the specific energy consumption vs. salt removal correlation. Source: Alkhadra et al. [105]/with permission of American Chemical Society.

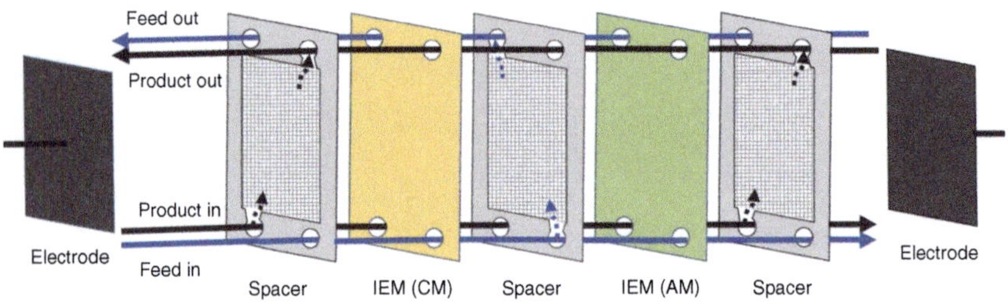

Figure 17.11 Schematic representation of an electrodialysis membrane unit. Note: IEM–ion-exchange membrane, CM–cationic membrane, and AM–anionic membrane. Source: Meng et al. [106]/Elsevier/CC BY 4.0.

threat, further making it mandatory to establish the fouling prevention technique [107, 111, 112].

ED has a range of uses for rare earth elements (REEs), which are actually a group of 17 chemical elements that possess interesting physical–chemical properties and are used across a multitude of industrial sectors [105–107, 112]. Using ED as a method, one of the most prominent areas of application is the separation of REEs from wastewater. Scientists have found that through the combination of the mentioned ED process and VMD process, the removal efficiency of REEs from ceric ammonium nitrate wastewater is possible. Furthermore, ED may regenerate REEs through selective ED and using heat from the brines. Besides, this process reduces the environmental effects resulting from wastewater discharge while at the same time proving economical from the perspective that valuable resources can be reused.

Reusing acids and bases is another essential application of ED with waste effluents as input materials. Plenty of industrial operations produce acidic and alkaline effluents that may contain soluble product species like inorganic salts, metals, organic acids, or inorganic salts. Research proves that ED membranes can potentially overcome these concerns by performing bipolar membranes which only allow specific ions to pass through. For example, studies have reported that ED can regenerate HCl from electroplating wastewater when high proton conductivity anion-exchange membranes are used. Furthermore, ED has been found to retrieve H_2SO_4 from the residual of flue gas of the desulfurization process by employing an ED membrane for separating water into hydroxide ions and protons.

The third application of ED involves concentrating or separating foods and pharmaceutical drugs that contain or require ionic compounds. ED membranes can efficiently isolate or separate biochemicals including amino acid, propionic acid, lactic acid, enzymes, antibiotics, minerals, and vitamins. Furthermore, ED has been employed to purify lactic acid from fermentation broth through the use of cation-exchange membranes that prevent the passage of lactate ions. Additionally, ED can separate amino acids from mixtures by utilizing monovalent-ion-permselective membranes, which are capable of distinguishing between amino acids based on their charge [106]. Table 17.7 provides the recent advancements in ED membrane applications based on PI principles.

Conclusively, ED is a structurally flexible and potentially effective membrane separation process amenable to a wide range of industries. However, the membrane-based FO still possesses several issues and limitations, such as membrane fouling, concentration polarization, water splitting, low selectivity for specific ions, and relatively high capital cost. Managing these problems has given rise to considerations for possible enhancements. Using spacers can help alleviate membrane fouling, while concentration polarization can be minimized by operating ED on a cyclic basis or by increasing the flow rates during operation. To avoid water splitting, one can incorporate selective membranes or operate at lower currents. Improving ion selectivity can also involve forming complexes or using modified membranes.

17.3.5 Future Challenges and Opportunities of Intensified Membrane Technology for Desalination

The potential for enhancing processes in membrane-based technology, especially in desalination, appears very promising, mainly when supported by renewable energy. With the increasing need for freshwater all over the world, desalination has become one of the most viable options available. However, many conventional

Table 17.7 Recent advancements in ED membrane applications based on process intensification principles.

The PI strategy of the ED method	Operating conditions	Findings & advantages of the ED process	Drawbacks/challenges	References
PFED for freshwater production	Conductivity (72 μS/cm); flow rate ratio of concentrate to dilute (1:1); flow rates (40–700 mL/min); power consumption (0.35 W/m^2); resistance (5 Ohm)	(1) Feasible from a technical perspective; (2) economically attractive; (3) removes the requirement for batteries; (4) the concentrated form of electrochemical desalination (ED) can be used as a feedstock for reverse electrodialysis (RED) to generate energy; (5) the voltage generated by RED stays constant.	(1) Concentration polarization was observed. (2) The membrane's permselectivity decreased at high concentrations. (3) The ED resistance is slightly higher than the internal resistance of the RED	[113]
Magnetic CEM incorporated with cobalt ferrite nanoparticles for chromium ions removal via ED.	—	$CoFe_2O_4$ nanoparticles. $CoFe^2O_4$ nanoparticles enhanced the membrane's hydrophilicity and conductivity, leading to improved permselectivity, transport number, ion-exchange capacity, and water content.	—	[114]

EDR for industrial RO brine treatment	Three sets of cells were used, with a current of 0.89 A. The active area of the membrane was 180 cm^2, and the conductivity was 10.5 ms/cm. The current was 0.75 A, and the voltage ranged from 0 to 20 V. The operation was conducted in batches over 8 h, resulting in a removal efficiency of 7.1% and a water recovery rate of 85%. The reduction of RO brine was 6.5 times, and there was a 50% reduction in conductivity.	(1) Able to produce highly concentrated brine. (2) There were no indications of organic fouling, even after running continuously for 6 d.	(1) The EDR resistance experienced an increase over 8 h of operation. (2) An inflection point was observed as a result of ion depletion. (3) Polarization was observed when operating at a current exceeding the limiting current. (4) Precipitation of CaCO3 occurred.	[115]
Penicillin G and sulfate ions separation using ED ultrafiltration configuration	The membrane's effective area was 214.7 cm^2, and the flow rate was 250 mL/min. The current intensities ranged from 0.2 to 0.4 A. The demineralization rate ranged from 69% to 84%. The laboratory scale process costs $10.7 for recovering 1 kg of penicillin G.	(1) No fouling present; (2) Recovering 203 tons of penicillin G saves 203,000 cubic meters of freshwater; (3) 3. Recovering 1,000 tons of penicillin G annually generates a yearly profit of $6.85 million.	(1) Resistance to stacking; (2) The pH level is altered by water splitting on the membrane surface	[116]
ED for inorganic acid recovery	The membrane's effective area is 0.785 cm^2, and there are two compartments, each with a volume of around 50 mL.	(1) Protons were transported against their concentration and potential gradients, which was undesirable. (2) The simulation results were in good agreement with the empirical findings.	(1) Acid leakage	[117]

(Continued)

Table 17.7 (Continued)

The PI strategy of the ED method	Operating conditions	Findings & advantages of the ED process	Drawbacks/challenges	References
Alpha-ketoglutaric (AKG) acid production using EDBM	In a two-chamber EDBM stack, the configuration consists of an anion-exchange membrane-bipolar membrane (AM-BM). The current efficiency is 71.8%, while the energy consumption is 3.72 kW/h/kg. The process takes 180 min to complete, with a flow rate of 5.6 L/h. The production of AKG is 4.83 g/L.	Under optimal conditions, there was a rise in AKG levels, an improvement in current efficiency, and a decrease in energy consumption.	More research is necessary to advance EDBM toward industrialization.	[118]
An electrodialysis process that uses both monovalent and divalent membranes for desalinating brackish groundwater.	Pilot-scale ED: Conductivity (4,626 μS/cm); 186 cell pairs of CEMs and AEMs; Effective membrane area 3,200 cm^2; Voltage (28–37 V); Energy consumption (1.3 kWh/kgal); Salt removal (0.42 and 0.74 kg salt/m^2/kWh at current density of 3.3 or 5.2 mA/cm^2, respectively); Water recovery 55%	(1) Elevated flow rates with selective membranes lead to greater salt removal due to increased current density. (2) The ion selectivity of selective membranes remains consistent across a broad spectrum of operating conditions. (3) Utilizing a pulsed electric field allows for the regulation of concentration polarization at the membrane interface. (4) Increased removal of inorganic elements is achieved with a higher applied voltage.	(1) Salt removal efficiency decreased with the flow rate increasing; (2) The monovalent selective membranes achieved the same desalting efficiency as the normal grade membranes during bench and pilot scale ED; (3) Stack resistance and membrane deposition [210]	[119]

Recovery of chemicals from a copper plating bath without cyanide using ED.	Flow rate (3.5 L/h); Produced a solution 5–6 times more concentrated than the initial one; Demineralization (90%); Extraction (>80%)	(1) The membrane characteristics were restored through cleaning processes. (2) Suitable for recovering water and materials (such as chemicals) and reducing waste generation.	(1) Backward diffusion; (2) Interactions between organic acid and membrane fixed groups	[120]
Pilot-scale reverse electrodialysis (RED) system fed with seawater and municipal effluent	1,000 cell pairs; total membrane area (250 m^2); Velocity (1.5 cm/s); Energy efficiency (11.4–13.3%); Power production (95.8 W)	(1) A source of sustainable and renewable energy.	(1) Spacers clogging; (2) Fouling due to inorganic precipitation	[121]
Electrodialysis for exclusive separation of sodium formate	The electrodialysis process includes four compartments and operates at a current density of 30 mA/cm^2, with a limiting current density range of 21.6–75 mA/cm^2. The flow rate ranges from 240 to 840 mL/h, achieving a current efficiency of 70%. The purity of the sodium formate is 87%, with a formate recovery of 69%. The energy consumption is measured at 96 kWh/m^3.	(1) The pH stayed constant without any significant changes. (2) The diffusion of formic acid and formate leakages did not impact the current efficiency.	(1) OH^- leakage through CEMs decreases the current efficiency	[122]

desalination techniques have been proven to be costly and energy-consuming and are based on fossil fuel energy sources, hence unsustainable to the environment. In this case, the integration of renewable energy into the desalination process can be regarded as a way of improving the sustainability and feasibility of this technology. By adopting clean energy technology like solar, wind, and hydroelectricity technologies, the energy used in desalinating this water can be significantly lowered and almost denotes a green energy value altogether. For example, PV systems or solar thermal technologies can be used to provide power for desalination plants to reduce their effect on the environment. It also has the added benefit of reducing operational expenditure and is in vice with efforts by the world to change gears toward green sources of energy.

In this regard, membrane technologies have the most potential in relation to the enhancement of the efficiency of desalination systems. It was found that using RO or FO membranes can allow the intensification of water recovery compared to energy consumption. When these technologies are integrated with renewables, they achieve improved overall performance with minimized environmental costs. Also, combined with desalination processes, the use of renewable energy sources will enable the design and implementation of efficient and synergistic schemes. For instance, integrating solar energy into the production of water through desalination can be a sustainable solution to the water problems in the world, especially in dry areas. Such an approach is practical not only in providing the necessary fresh-water within the shortest time possible but also in boosting energy independence and stability in the water infrastructure supply. Therefore, desalination, backed by renewable energy, is quite promising in the future of achieving sustainable industrial growth. Through membrane technologies and their combination with renewable energy systems, one can reinvent the desalination mechanisms so that they are efficient, economical, and sustainable. They have identified main trends for innovations within industries that include water usage and extraction, and they indicate that as industries integrate more of such innovations, they are likely to make strides to achieve the critical goal of creating a sustainable and secure water future.

17.4 Conclusion

Most of the fields, including applied chemistry, physics, electronics, computer engineering, and medical sciences, have undergone a drastic change since the Industrial Revolution, consequently leading to soaring market demands of different industrial usages, including seawater desalination. PI is critical for the development of new processes and their integration in order to capture enhanced intensification for achieving acceptable quality of desalinated water with reasonable effects on the environment. Emerging trends in two-dimensional (2D) membrane materials, MD, and membrane ED are considered disruptive to the desalination industry. 2D materials, including graphene and other nanosheets, are gaining industry attention for their uniquely high permeability and selectivity that improve the efficiency of desalination processes. It is possible to design and synthesize these materials to generate membranes that are capable of selectively filtering ions and even be resistant to fouling, aspects crucial to the efficiency of desalination systems. Another innovative technique that can be of great use and influence the application of these advanced membranes is MD. In this technique, the movement of water vapor across the membrane is done by taking advantage of a temperature gradient in the process of desalinating the water at minimal energy costs. The advancement of 2D materials into the MD systems also improves their efficiency and can be implemented at commercial levels of effectiveness. Also, membrane ED as a method of water desalination is seen as a prospect in the future, especially in view of the integration with renewable power sources. This process of carrying out the movement of ions through selective membranes under the influence of an electric field results in the separation of salt from water. The integration of novel 2D membranes in ED has the potential to increase the selectivity of ions to be transported and decrease the energy consumption of the process. Overall, the future development of seawater desalination is closely associated with the application of sustainable advanced 2D membrane materials, effective MD, and membrane ED technology. They also improve the performance and economic benefits of desalination processes. They are also in line with global strategies seeking to provide seawater desalination for human consumption and minimize the effects of such activities on the environment. As more industries adopt these innovative technologies, their collective impact will substantially improve water management strategies, thus shifting toward a more sustainable approach to fulfill global water needs.

Questions

1 What is Process Intensification (PI) and what are its primary objectives in industrial processes?

2 In the context of Process Intensification (PI) in membrane technology, how can PI principles be applied to enhance the performance of a membrane bioreactor (MBR) used in wastewater treatment?

3 How does the application of Process Intensification (PI) in RO desalination address the challenges in seawater desalination and what potential improvements are suggested?

4 What potential advancements in water desalination technology could be achieved if the stability issues of MXenes when exposed to air and humidity were fully addressed?

5 How might the integration of photothermal membranes in Solar Powered Membrane Distillation (SPMD) systems impact the future of desalination and wastewater treatment, particularly in remote, off-grid environments?

6 How does the integration of renewable energy sources and the use of membrane technologies contribute to the sustainability and efficiency of desalination processes?

Conflict of Interests

The authors declare that there is no conflict of interest.

Acknowledgments

The authors would like to express heartfelt gratitude to the National Research and Innovation Agency (BRIN), notably the Research Center for Chemistry, for their unwavering support and for providing the necessary facilities. Their commitment to advancing scientific inquiry has been instrumental in the successful completion of this chapter. The authors appreciate the fostering of an environment that encourages innovation and excellence in research.

References

1 Achouri, I.E. (2024). Interdisciplinary approaches in process intensification: building the next generation of engineers. *Chem. Eng. Process. Process Intensif.* 110007. https://doi.org/10.1016/j.cep.2024.110007.
2 Erfani, N., Baharudin, L., and Watson, M. (2024). Recent advances and intensifications in Haber-Bosch ammonia synthesis process. *Chem. Eng. Process. Process Intensif.* 204: 109962. https://doi.org/10.1016/j.cep.2024.109962.
3 Muscetta, M. (2024). Process intensification in metal recovery from solid waste: challenges, opportunities and recent advances. *Chem. Eng. Process. Process Intensif.* 204: 109937. https://doi.org/10.1016/j.cep.2024.109937.
4 Dalanta, F., Handoko, D.T., Hadiyanto, H., and Kusworo, T.D. (2024). Recent implementations of process intensification strategy in membrane-based

technology: a review. *Chem. Eng. Res. Des.* 202: 74–91. https://doi.org/10.1016/j.cherd.2023.12.014.

5 Bizualem, Y.D. and Nurie, A.G. (2024). A review on recent biodiesel intensification process through cavitation and microwave reactors: yield, energy, and economic analysis. *Heliyon* 10 (2): e24643. https://doi.org/10.1016/j.heliyon.2024.e24643.

6 Dal-Cin, M.M., McLellan, F., Striez, C.N. et al. (1996). Membrane performance with a pulp mill effluent: relative contributions of fouling mechanisms. *J. Membr. Sci.* 120 (2): 273–285. https://doi.org/10.1016/0376-7388(96)00151-2.

7 Kant Bhatia, S., Bhatia, R.K., Jeon, J.M. et al. (2021). An overview on advancements in biobased transesterification methods for biodiesel production: Oil resources, extraction, biocatalysts, and process intensification technologies. *Fuel* 285: 119117. https://doi.org/10.1016/j.fuel.2020.119117.

8 Tolley, T.E. and Fronk, B.M. (2020). Investigation of the benefits of diabatic microreactors for process intensification of the water-gas shift reaction within the steam reforming process. *Int. J. Hydrog. Energy* 45: https://doi.org/10.1016/j.ijhydene.2020.08.282.

9 Safdarnejad, S.M., Tuttle, J.F., and Powell, K.M. (2019). Development of a roadmap for dynamic process intensification by using a dynamic, data-driven optimization approach. *Chem. Eng. Process. Process Intensif.* 140: 100–113. https://doi.org/10.1016/j.cep.2019.04.002.

10 Cesari, L., Castel, C., and Favre, E. (2021). Membrane contactors for intensified gas-liquid absorption processes with physical solvents: a critical parametric study. *J. Membr. Sci.* 635: 119377. https://doi.org/10.1016/j.memsci.2021.119377.

11 Wasewar, K.L. (2021). Process intensification in wastewater treatments: basics of process intensification and inorganic pollutants. In: *Contamination of Water* (ed. A. Ahmad, S.I. Siddiqui, and P. Singh), 313–337. Elsevier. https://doi.org/10.1016/B978-0-12-824058-8.00011-6.

12 Davoodbeygi, Y., Askari, M., Salehi, E., and Kheirieh, S. (2023). A review on hybrid membrane-adsorption systems for intensified water and wastewater treatment: process configurations, separation targets, and materials applied. *J. Environ. Manag.* 335: 117577. https://doi.org/10.1016/j.jenvman.2023.117577.

13 Poonguzhali, E., Kapoor, A., and Prabhakar, S. (2023). Membrane assisted process intensification and optimization for removal and recovery of phenol from industrial effluents. *Sep. Purif. Technol.* 319: 124026. https://doi.org/10.1016/j.seppur.2023.124026.

14 Nadimi, M., Shahrooz, M., Wang, R. et al. (2024). Process intensification with reactive membrane distillation: a review of hybrid and integrated processes. *Desalination* 573: 117182. https://doi.org/10.1016/j.desal.2023.117182.

15 Lim, Y.J., Nadzri, N., Lai, G.S., and Wang, R. (2024). Demystifying the compaction effects of TFC polyamide membranes in the desalination of hypersaline brine via high-pressure RO. *J. Membr. Sci.* 707: 122950. https://doi.org/10.1016/j.memsci.2024.122950.

16 Shalaby, M.S., Abdallah, H., Wilken, R. et al. (2024). Economic assessment for TFC-RO membranes production for water desalination. *Chem. Eng. Sci.* 288: 119805. https://doi.org/10.1016/j.ces.2024.119805.

17 Chen, C., Yang, Y., Graham, N.J.D. et al. (2024). A comprehensive evaluation of the temporal and spatial fouling characteristics of RO membranes in a full-scale seawater desalination plant. *Water Res.* 249: 120914. https://doi.org/10.1016/j.watres.2023.120914.

18 Fajardo-Diaz, J.L., Morelos-Gomez, A., Cruz-Silva, R. et al. (2022). Antifouling performance of spiral wound type module made of carbon nanotubes/polyamide composite RO membrane for seawater desalination. *Desalination* 523: 115445. https://doi.org/10.1016/j.desal.2021.115445.

19 Abdelgaied, M., Abdullah, A.S., Kabeel, A.E., and Abosheiasha, H.F. (2022). Assessment of an innovative hybrid system of PVT-driven RO desalination unit integrated with solar dish concentrator as preheating unit. *Energy Convers. Manag.* 258: 115558. https://doi.org/10.1016/j.enconman.2022.115558.

20 Pazouki, P., Sidhu, J.P.S., Ipe, D.S. et al. (2022). Seawater dilution desalination with hybrid FO-RO and UF-RO: characterisation and assessment of pathogen removal efficacy. *Desalination* 525: 115509. https://doi.org/10.1016/j.desal.2021.115509.

21 van Rooij, F., Scarf, P., and Do, P. (2021). Planning the restoration of membranes in RO desalination using a digital twin. *Desalination* 519: 115214. https://doi.org/10.1016/j.desal.2021.115214.

22 Bodzek, M. and Konieczny, K. (2021). Nanomaterials in membrane water desalination. *Desalin. Water Treat* 214: 155–180. https://doi.org/10.5004/dwt.2021.26657.

23 Gharbia, A.S., Gharbia, S.S., Tamous, H. et al. (2021). Optimization modeling for the use of antiscalants in the brackish RO desalination plants. *Desalination Water Treat* 224: 12–20. https://doi.org/10.5004/dwt.2021.27176.

24 Fu, B., Sun, J., Wang, C. et al. (2021). MXenes: synthesis, optical properties, and applications in ultrafast photonics. *Small* 17 (11): https://doi.org/10.1002/smll.202006054.

25 Gao, L., Li, C., Huang, W. et al. (2020). MXene/polymer membranes: synthesis, properties, and emerging applications. *Chem. Mater.* 32 (5): 1703–1747. https://doi.org/10.1021/acs.chemmater.9b04408.

26 Liu, G., Shen, J., Liu, Q. et al. (2018). Ultrathin two-dimensional MXene membrane for pervaporation desalination. *J. Membr. Sci.* 548: 548–558. https://doi.org/10.1016/j.memsci.2017.11.065.

27 Solangi, N.H., Mubarak, N.M., Karri, R.R. et al. (2023). Applications of advanced MXene-based composite membranes for sustainable water desalination. *Chemosphere* 314: 137643. https://doi.org/10.1016/j.chemosphere.2022.137643.

28 Magnuson, M. and Mattesini, M. (2017). Chemical bonding and electronic-structure in MAX phases as viewed by X-ray spectroscopy and density functional theory. *Thin Solid Films* 621: 108–130. https://doi.org/10.1016/j.tsf.2016.11.005.

29 Mozafari, M., Arabi Shamsabadi, A., Rahimpour, A., and Soroush, M. (2021). Ion-selective MXene-based membranes: current status and prospects. *Adv. Mater. Technol.* 6 (10): https://doi.org/10.1002/admt.202001189.

30 Lei, J., Xiong, Y., Yu, F., and Ma, J. (2022). Flexible self-supporting CoFe-LDH/MXene film as a chloride ions storage electrode in capacitive deionization. *Chem. Eng. J.* 437: 135381. https://doi.org/10.1016/j.cej.2022.135381.

31 Yang, H.Y., Han, Z.J., Yu, S.F. et al. (2013). Carbon nanotube membranes with ultrahigh specific adsorption capacity for water desalination and purification. *Nat. Commun.* 4 (1): 2220. https://doi.org/10.1038/ncomms3220.

32 Mutharasi, Y., Zhang, Y., Weber, M. et al. (2021). Novel reverse osmosis membranes incorporated with co-Al layered double hydroxide (LDH) with enhanced performance for brackish water desalination. *Desalination* 498: 114740. https://doi.org/10.1016/j.desal.2020.114740.

33 Huang, L., Pei, J., Jiang, H., and Hu, X. (2018). Water desalination under one sun using graphene-based material modified PTFE membrane. *Desalination* 442: 1–7. https://doi.org/10.1016/j.desal.2018.05.006.

34 Zhao, Q., Zhao, D.L., and Chung, T. (2020). Nanoclays-incorporated thin-film nanocomposite membranes for reverse osmosis desalination. *Adv. Mater. Interfaces* 7 (8): https://doi.org/10.1002/admi.201902108.

35 Wang, X., Li, Q., Zhang, J. et al. (2020). Novel thin-film reverse osmosis membrane with MXene Ti3C2T embedded in polyamide to enhance the water flux, anti-fouling and chlorine resistance for water desalination. *J. Membr. Sci.* 603: 118036. https://doi.org/10.1016/j.memsci.2020.118036.

36 Wang, K., Chen, L., Zhu, G. et al. (2022). Ferroferric oxide@titanium carbide MXene heterostructure with enhanced sodium storage ability for efficient hybrid capacitive deionization. *Desalination* 522: 115420. https://doi.org/10.1016/j.desal.2021.115420.

37 Chowdhury, Z.Z., Pal, K., Sagadevan, S. et al. (2018). Electrochemically active carbon nanotube (CNT) membrane filter for desalination and water purification. In: *Emerging Technologies for Sustainable Desalination Handbook* (ed. V.G. Gude), 333–363. Elsevier. https://doi.org/10.1016/B978-0-12-815818-0.00010-2.

38 Kitano, H., Takeuchi, K., Ortiz-Medina, J. et al. (2020). Enhanced desalination performance in compacted carbon-based reverse osmosis membranes. *Nanoscale Adv.* 2 (8): 3444–3451. https://doi.org/10.1039/D0NA00263A.

39 Yao, Z., Meyerbröker, N., Qi, Y. et al. (2023). Scalable synthesis of carbon nanomembranes from amorphous molecular layers. *ACS Appl. Mater. Interfaces* 15 (34): 41101–41108. https://doi.org/10.1021/acsami.3c07369.

40 Rana, A.K., Gupta, V.K., Saini, A.K. et al. (2021). Water desalination using nanocelluloses/cellulose derivatives based membranes for sustainable future. *Desalination* 520: 115359. https://doi.org/10.1016/j.desal.2021.115359.

41 Cheng, Y., Xia, C., Garalleh, H.A.L. et al. (2023). A review on optimistic development of polymeric nanocomposite membrane on environmental remediation. *Chemosphere* 315: 137706. https://doi.org/10.1016/j.chemosphere.2022.137706.

42 Shukla, A.K., Alam, J., Alhoshan, M.S. et al. (2021). Thin-film nanocomposite membrane incorporated with porous Zn-based metal–organic frameworks: toward enhancement of desalination performance and chlorine resistance. *ACS Appl. Mater. Interfaces* 13 (24): 28818–28831. https://doi.org/10.1021/acsami.1c05469.

43 Zhang, N., Song, X., Jiang, H., and Tang, C.Y. (2021). Advanced thin-film nanocomposite membranes embedded with organic-based nanomaterials for water and organic solvent purification: a review. *Sep. Purif. Technol.* 269: 118719. https://doi.org/10.1016/j.seppur.2021.118719.

44 Cui, J., Li, F., Wang, Y. et al. (2020). Electrospun nanofiber membranes for wastewater treatment applications. *Sep. Purif. Technol.* 250: 117116. https://doi.org/10.1016/j.seppur.2020.117116.

45 Feng, C., Khulbe, K.C., Matsuura, T. et al. (2008). Production of drinking water from saline water by air-gap membrane distillation using polyvinylidene fluoride nanofiber membrane. *J. Membr. Sci.* 311 (1–2): 1–6. https://doi.org/10.1016/j.memsci.2007.12.026.

46 Swanckaert, B., Geltmeyer, J., Rabaey, K. et al. (2022). A review on ion-exchange nanofiber membranes: properties, structure and application in electrochemical (waste)water treatment. *Sep. Purif. Technol.* 287: 120529. https://doi.org/10.1016/j.seppur.2022.120529.

47 Adiga, S.P., Jin, C., Curtiss, L.A. et al. (2009). Nanoporous membranes for medical and biological applications. *WIREs Nanomed. and Nanobiotechnol.* 1 (5): 568–581. https://doi.org/10.1002/wnan.50.

48 Mijangos, C., Hernández, R., and Martín, J. (2016). A review on the progress of polymer nanostructures with modulated morphologies and properties, using nanoporous AAO templates. *Prog. Polym. Sci.* 54–55: 148–182. https://doi.org/10.1016/j.progpolymsci.2015.10.003.

49 Xu, J., Meng, K., Niu, Y. et al. (2023). Deciphering the electronic-level mechanism of Na+ transport in a graphdiyne desalination membrane with periodic nanopores. *Desalination* 546: 116183. https://doi.org/10.1016/j.desal.2022.116183.

50 Chadha, U., Selvaraj, S.K., Thanu, S.V. et al. (2022). A review of the function of using carbon nanomaterials in membrane filtration for contaminant removal from wastewater. *Mater. Res. Express* https://doi.org/10.1088/2053-1591/ac48b8.

51 Patel, K.D., Singh, R.K., and Kim, H.-W. (2019). Carbon-based nanomaterials as an emerging platform for theranostics. *Mater. Horiz.* 6 (3): 434–469. https://doi.org/10.1039/C8MH00966J.

52 Kumari, P., Tripathi, K.M., Jangir, L.K. et al. (2021). Recent advances in application of the graphene-based membrane for water purification. *Mater. Today Chem.* 22: 100597. https://doi.org/10.1016/j.mtchem.2021.100597.

53 Kusworo, T.D., Dalanta, F., Aryanti, N., and Othman, N.H. (2021). Intensifying separation and antifouling performance of PSf membrane incorporated by GO and ZnO nanoparticles for petroleum refinery wastewater treatment. *J. Water Proc. Eng.* 41: 102030. https://doi.org/10.1016/J.JWPE.2021.102030.

54 Zhu, L., Wang, H., Bai, J. et al. (2017). A porous graphene composite membrane intercalated by halloysite nanotubes for efficient dye desalination. *Desalination* 420: 145–157. https://doi.org/10.1016/j.desal.2017.07.008.

55 Sun, L., Yu, D., Yang, L. et al. (2022). Improvements in multifunctional graphene oxide-based separation membranes: mechanism, modification and properties. *Mater. Today Commun.* 33: 104274. https://doi.org/10.1016/j.mtcomm.2022.104274.

56 Yang, E., Kim, C.-M., Song, J. et al. (2017). Enhanced desalination performance of forward osmosis membranes based on reduced graphene oxide laminates coated with hydrophilic polydopamine. *Carbon* 117: 293–300. https://doi.org/10.1016/j.carbon.2017.03.005.

57 Chu, K.H., Fathizadeh, M., Yu, M. et al. (2017). Evaluation of removal mechanisms in a graphene oxide-coated ceramic ultrafiltration membrane for retention of natural organic matter, pharmaceuticals, and inorganic salts. *ACS Appl. Mater. Interfaces* 9 (46): 40369–40377. https://doi.org/10.1021/acsami.7b14217.

58 Mortazavi, V., Moosavi, A., and Nouri-Borujerdi, A. (2022). Water desalination by charged multilayer graphene membrane: a molecular dynamics simulation. *J. Mol. Liq.* 355: 118953. https://doi.org/10.1016/j.molliq.2022.118953.

59 Hegab, H.M. and Zou, L. (2015). Graphene oxide-assisted membranes: fabrication and potential applications in desalination and water purification. *J. Membr. Sci.* 484: 95–106. https://doi.org/10.1016/j.memsci.2015.03.011.

60 Akin, I., Zor, E., Bingol, H., and Ersoz, M. (2014). Green synthesis of reduced graphene oxide/polyaniline composite and its application for salt rejection by polysulfone-based composite membranes. *J. Phys. Chem. B* 118 (21): 5707–5716. https://doi.org/10.1021/jp5025894.

61 Ramanathan, T., Abdala, A.A., Stankovich, S. et al. (2008). Functionalized graphene sheets for polymer nanocomposites. *Nat. Nanotechnol.* 3 (6): 327–331. https://doi.org/10.1038/nnano.2008.96.

62 Zhao, C., Xu, X., Chen, J., and Yang, F. (2014). Optimization of preparation conditions of poly(vinylidene fluoride)/graphene oxide microfiltration membranes by the Taguchi experimental design. *Desalination* 334 (1): 17–22. https://doi.org/10.1016/j.desal.2013.07.011.

63 Ganesh, B.M., Isloor, A.M., and Ismail, A.F. (2013). Enhanced hydrophilicity and salt rejection study of graphene oxide-polysulfone mixed matrix membrane. *Desalination* 313: 199–207. https://doi.org/10.1016/j.desal.2012.11.037.

64 Xu, Y., Gao, X., Wang, Q. et al. (2016). Highly stable MIL-101(Cr) doped water permeable thin film nanocomposite membranes for water treatment. *RSC Adv.* 6 (86): 82669–82675. https://doi.org/10.1039/C6RA16896E.

65 Xia, S. and Ni, M. (2015). Preparation of poly(vinylidene fluoride) membranes with graphene oxide addition for natural organic matter removal. *J. Membr. Sci.* 473: 54–62. https://doi.org/10.1016/j.memsci.2014.09.018.

66 Mao, Y., Xu, J., Chen, H. et al. (2023). Hydrophobic metal-organic framework@graphene oxide membrane with enhanced water transport for desalination. *J. Membr. Sci.* 669: 121324. https://doi.org/10.1016/j.memsci.2022.121324.

67 Bundschuh, J., Kaczmarczyk, M., Ghaffour, N., and Tomaszewska, B. (2021). State-of-the-art of renewable energy sources used in water desalination: present and future prospects. *Desalination* 508: 115035. https://doi.org/10.1016/j.desal.2021.115035.

68 Alobaidani, S., Curcio, E., Macedonio, F. et al. (2008). Potential of membrane distillation in seawater desalination: thermal efficiency, sensitivity study and cost estimation. *J. Membr. Sci.* 323 (1): 85–98. https://doi.org/10.1016/j.memsci.2008.06.006.

69 Khayet, M., García-Payo, M.C., García-Fernández, L., and Contreras-Martínez, J. (2018). Dual-layered electrospun nanofibrous membranes for membrane distillation. *Desalination* 426: 174–184. https://doi.org/10.1016/j.desal.2017.10.036.

70 Ghaffour, N., Bundschuh, J., Mahmoudi, H., and Goosen, M.F.A. (2015). Renewable energy-driven desalination technologies: a comprehensive review on challenges and potential applications of integrated systems. *Desalination* 356: 94–114. https://doi.org/10.1016/j.desal.2014.10.024.

71 Kuo, C.-Y., Lin, H.-N., Tsai, H.-A. et al. (2008). Fabrication of a high hydrophobic PVDF membrane via nonsolvent induced phase separation. *Desalination* 233 (1–3): 40–47. https://doi.org/10.1016/j.desal.2007.09.025.

72 Kong, J. and Li, K. (2001). Preparation of PVDF hollow-fiber membranes via immersion precipitation. *J. Appl. Polym. Sci.* 81 (7): 1643–1653. https://doi.org/10.1002/app.1595.

73 Majidi Salehi, S., Di Profio, G., Fontananova, E. et al. (2016). Membrane distillation by novel hydrogel composite membranes. *J. Membr. Sci.* 504: 220–229. https://doi.org/10.1016/j.memsci.2015.12.062.

74 Khayet, M. (2011). Membranes and theoretical modeling of membrane distillation: a review. *Adv. Colloid Interf. Sci.* 164 (1–2): 56–88. https://doi.org/10.1016/j.cis.2010.09.005.

75 Ingole, P.G., Choi, W., Kim, K.H. et al. (2014). Synthesis, characterization and surface modification of PES hollow fiber membrane support with polydopamine and thin film composite for energy generation. *Chem. Eng. J.* 243: 137–146. https://doi.org/10.1016/j.cej.2013.12.094.

76 Ke, H., Feldman, E., Guzman, P. et al. (2016). Electrospun polystyrene nanofibrous membranes for direct contact membrane distillation. *J. Membr. Sci.* 515: 86–97. https://doi.org/10.1016/j.memsci.2016.05.052.

77 Bin Abid, M., Wahab, R.A., Salam, M.A. et al. (2023). Desalination technologies, membrane distillation, and electrospinning, an overview. *Heliyon* 9 (2): e12810. https://doi.org/10.1016/j.heliyon.2023.e12810.

78 Kebria, M.R.S., Rahimpour, A., Salestan, S.K. et al. (2020). Hyper-branched dendritic structure modified PVDF electrospun membranes for air gap membrane distillation. *Desalination* 479: 114307. https://doi.org/10.1016/j.desal.2019.114307.

79 Duong, H.C., Chuai, D., Woo, Y.C. et al. (2018). A novel electrospun, hydrophobic, and elastomeric styrene-butadiene-styrene membrane for membrane distillation applications. *J. Membr. Sci.* 549: 420–427. https://doi.org/10.1016/j.memsci.2017.12.024.

80 Nurwaha, D., Han, W., and Wang, X. (2013). Effects of processing parameters on electrospun fiber morphology. *J. Text. Inst.* 104 (4): 419–425. https://doi.org/10.1080/00405000.2012.738791.

81 Geise, G.M., Falcon, L.P., Freeman, B.D., and Paul, D.R. (2012). Sodium chloride sorption in sulfonated polymers for membrane applications. *J. Membr. Sci.* 423–424: 195–208. https://doi.org/10.1016/j.memsci.2012.08.014.

82 Efome, J.E., Baghbanzadeh, M., Rana, D. et al. (2015). Effects of superhydrophobic SiO_2 nanoparticles on the performance of PVDF flat sheet membranes for vacuum membrane distillation. *Desalination* 373: 47–57. https://doi.org/10.1016/j.desal.2015.07.002.

83 Son, H.S., Shahzad, M.W., Ghaffour, N., and Ng, K.C. (2020). Pilot studies on synergetic impacts of energy utilization in hybrid desalination system: multi-effect distillation and adsorption cycle (MED-AD). *Desalination* 477: 114266. https://doi.org/10.1016/j.desal.2019.114266.

84 Gopi, G., Arthanareeswaran, G., and Ismail, A.F. (2019). Perspective of renewable desalination by using membrane distillation. *Chem. Eng. Res. Des.* 144: 520–537. https://doi.org/10.1016/j.cherd.2019.02.036.

85 Jawed, A.S., Nassar, L., Hegab, H.M. et al. (2024). Recent developments in solar-powered membrane distillation for sustainable desalination. *Heliyon* 10 (11): e31656. https://doi.org/10.1016/j.heliyon.2024.e31656.

86 Warsinger, D.M., Nejati, S., and Juybari, H.F. (2021). Energy efficiency metrics in membrane distillation. *Adv. Water Desalin.* 263–288. https://doi.org/10.1142/9789811226984_0008.

87 Francis, L., Ghaffour, N., Alsaadi, A.A., and Amy, G.L. (2013). Material gap membrane distillation: a new design for water vapor flux enhancement. *J. Membr. Sci.* 448: 240–247. https://doi.org/10.1016/j.memsci.2013.08.013.

88 Politano, A., Argurio, P., Di Profio, G. et al. (2017). Photothermal membrane distillation for seawater desalination. *Adv. Mater.* 29 (2): https://doi.org/10.1002/adma.201603504.

89 Wang, W., Shi, Y., Zhang, C. et al. (2021). Solar seawater distillation by flexible and fully passive multistage membrane distillation. *Nano Lett.* 21 (12): 5068–5074. https://doi.org/10.1021/acs.nanolett.1c00910.

90 Chauhan, V.K., Shukla, S.K., Tirkey, J.V., and Singh Rathore, P.K. (2021). A comprehensive review of direct solar desalination techniques and its advancements. *J. Clean. Prod.* 284: 124719. https://doi.org/10.1016/j.jclepro.2020.124719.

91 Jantaporn, W., Ali, A., and Aimar, P. (2017). Specific energy requirement of direct contact membrane distillation. *Chem. Eng. Res. Des.* 128: 15–26. https://doi.org/10.1016/j.cherd.2017.09.031.

92 Shafieian, A. and Khiadani, M. (2019). A novel solar-driven direct contact membrane-based water desalination system. *Energy Convers. Manag.* 199: 112055. https://doi.org/10.1016/j.enconman.2019.112055.

93 Antonetto, G., Morciano, M., Alberghini, M. et al. (2021). Synergistic freshwater and electricity production using passive membrane distillation and waste heat recovered from camouflaged photovoltaic modules. *J. Clean. Prod.* 318: 128464. https://doi.org/10.1016/j.jclepro.2021.128464.

94 Ma, Q., Ahmadi, A., and Cabassud, C. (2022). Comparative study of small-scale flat-plate direct contact membrane distillation and vacuum membrane distillation modules with integrated direct solar heating. *Desalination* 529: 115633. https://doi.org/10.1016/j.desal.2022.115633.

95 Xiao, T., Wang, P., Yang, X. et al. (2015). Fabrication and characterization of novel asymmetric polyvinylidene fluoride (PVDF) membranes by the non-solvent thermally induced phase separation (NTIPS) method for membrane distillation applications. *J. Membr. Sci.* 489: 160–174. https://doi.org/10.1016/j.memsci.2015.03.081.

96 Zhao, L., Wu, C., Lu, X. et al. (2020). Theoretical guidance for fabricating higher flux hydrophobic/hydrophilic dual-layer membranes for direct contact membrane distillation. *J. Membr. Sci.* 596: 117608. https://doi.org/10.1016/j.memsci.2019.117608.

97 Ghorabi, S., Ashtiani, F.Z., Karimi, M. et al. (2021). Development of a novel dual-bioinspired method for synthesis of a hydrophobic/hydrophilic polyethersulfone coated membrane for membrane distillation. *Desalination* 517: 115242. https://doi.org/10.1016/j.desal.2021.115242.

98 Zhao, L. and Zhao, L. (2018). Activated carbon enhanced hydrophobic/hydrophilic dual-layer nanofiber composite membranes for high-performance direct contact membrane distillation. *Desalination* 446: 59–69. https://doi.org/10.1016/j.desal.2018.09.002.

99 Ju, J., Li, Z., Lv, Y. et al. (2020). Electrospun PTFE/PI bi-component membranes with robust 3D superhydrophobicity and high water permeability for membrane distillation. *J. Membr. Sci.* 611: 118420. https://doi.org/10.1016/j.memsci.2020.118420.

100 Lee, E.-J., An, A.K., He, T. et al. (2016). Electrospun nanofiber membranes incorporating fluorosilane-coated TiO_2 nanocomposite for direct contact membrane distillation. *J. Membr. Sci.* 520: 145–154. https://doi.org/10.1016/j.memsci.2016.07.019.

101 Jia, Y., Xu, G., An, X., and Hu, Y. (2022). Robust reduced graphene oxide composite membranes for enhanced anti-wetting property in membrane distillation. *Desalination* 526: 115549. https://doi.org/10.1016/j.desal.2022.115549.

102 Bhadra, M., Roy, S., and Mitra, S. (2016). Flux enhancement in direct contact membrane distillation by implementing carbon nanotube immobilized PTFE membrane. *Sep. Purif. Technol.* 161: 136–143. https://doi.org/10.1016/j.seppur.2016.01.046.

103 Nthunya, L.N., Gutierrez, L., Verliefde, A.R., and Mhlanga, S.D. (2019). Enhanced flux in direct contact membrane distillation using superhydrophobic PVDF nanofibre membranes embedded with organically modified SiO_2 nanoparticles. *J. Chem. Technol. Biotechnol.* 94 (9): 2826–2837. https://doi.org/10.1002/jctb.6104.

104 An, X. (2020). Fabrication of interweaving hierarchical fibrous composite (iHFC) membranes for high-flux and robust direct contact membrane distillation. *Desalination* 477: 114264. https://doi.org/10.1016/j.desal.2019.114264.

105 Alkhadra, M.A., Su, X., Suss, M.E. et al. (2022). Electrochemical methods for water purification, ion separations, and energy conversion. *Chem. Rev.* 122 (16): 13547–13635. https://doi.org/10.1021/acs.chemrev.1c00396.

106 Meng, J. (2024). Application of electrodialysis technology in nutrient recovery from wastewater: a review. *J. Water Proc. Engineering* 65: 105855. https://doi.org/10.1016/j.jwpe.2024.105855.

107 Deboli, F., Van der Bruggen, B., and Donten, M.L. (2022). A versatile chemistry platform for the fabrication of cost-effective hierarchical cation and anion exchange membranes. *Desalination* 535: 115794. https://doi.org/10.1016/j.desal.2022.115794.

108 Kovalenko, A.V. (2023). Mathematical modeling of electrodialysis of a dilute solution with accounting for water dissociation-recombination reactions. *Desalination* 550: 116398. https://doi.org/10.1016/j.desal.2023.116398.

109 Liu, R., Wang, Y., Wu, G. et al. (2017). Development of a selective electrodialysis for nutrient recovery and desalination during secondary effluent treatment. *Chem. Eng. J.* 322: 224–233. https://doi.org/10.1016/j.cej.2017.03.149.

110 Doornbusch, G., van der Wal, M., Tedesco, M. et al. (2021). Multistage electrodialysis for desalination of natural seawater. *Desalination* 505: 114973. https://doi.org/10.1016/j.desal.2021.114973.

111 Shi, J., Gong, L., Zhang, T., and Sun, S. (2022). Study of the seawater desalination performance by electrodialysis. *Membranes (Basel)* 12 (8): 767. https://doi.org/10.3390/membranes12080767.

112 Chen, Q.-B., Xu, Y., Li, P.F. et al. (2022). An emerging pilot-scale electrodialysis system for desalination of SWNF permeate: evaluating the role of typical factors. *Desalination* 542: 116064. https://doi.org/10.1016/j.desal.2022.116064.

113 Luo, F., Wang, Y., Jiang, C. et al. (2017). A power free electrodialysis (PFED) for desalination. *Desalination* 404: 138–146. https://doi.org/10.1016/j.desal.2016.11.011.

114 Hosseini, S.M., Sohrabnejad, S., Nabiyouni, G. et al. (2019). Magnetic cation exchange membrane incorporated with cobalt ferrite nanoparticles for chromium ions removal via electrodialysis. *J. Membr. Sci.* 583: 292–300. https://doi.org/10.1016/j.memsci.2019.04.069.

115 Zhao, D., Lee, L.Y., Ong, S.L. et al. (2019). Electrodialysis reversal for industrial reverse osmosis brine treatment. *Sep. Purif. Technol.* 213: 339–347. https://doi.org/10.1016/j.seppur.2018.12.056.

116 Lu, H., Zou, W., Chai, P. et al. (2016). Feasibility of antibiotic and sulfate ions separation from wastewater using electrodialysis with ultrafiltration membrane. *J. Clean. Prod.* 112: 3097–3105. https://doi.org/10.1016/j.jclepro.2015.09.091.

117 Jia, Y., Li, F., Chen, X., and Wang, M. (2018). Model analysis on electrodialysis for inorganic acid recovery and its experimental validation. *Sep. Purif. Technol.* 190: 261–267. https://doi.org/10.1016/j.seppur.2017.08.067.

118 Szczygiełda, M. and Prochaska, K. (2017). Alpha-ketoglutaric acid production using electrodialysis with bipolar membrane. *J. Membr. Sci.* 536: 37–43. https://doi.org/10.1016/j.memsci.2017.04.059.

119 Xu, X., He, Q., Ma, G. et al. (2018). Selective separation of mono- and di-valent cations in electrodialysis during brackish water desalination: bench and pilot-scale studies. *Desalination* 428: 146–160. https://doi.org/10.1016/j.desal.2017.11.015.

120 Scarazzato, T., Panossian, Z., Tenório, J.A.S. et al. (2018). Water reclamation and chemicals recovery from a novel cyanide-free copper plating bath using electrodialysis membrane process. *Desalination* 436: 114–124. https://doi.org/10.1016/j.desal.2018.01.005.

121 Nam, J.-Y., Hwang, K.S., Kim, H.C. et al. (2019). Assessing the behavior of the feed-water constituents of a pilot-scale 1000-cell-pair reverse electrodialysis with seawater and municipal wastewater effluent. *Water Res.* 148: 261–271. https://doi.org/10.1016/j.watres.2018.10.054.

122 Selvaraj, H., Aravind, P., and Sundaram, M. (2018). Four compartment mono selective electrodialysis for separation of sodium formate from industry wastewater. *Chem. Eng. J.* 333: 162–169. https://doi.org/10.1016/j.cej.2017.09.150.

18

Nuclear Power for Water Desalination

Mauricio Chocrón[1,2,3], María Cecilia Conti[2,4], and Natalia Quici[5,6]

[1] *Departamento Química y Procesos en Instalaciones Nucleares, Gerencia de Coordinación de Proyectos Tecnológicos Nucleares, Gerencia de Energía Nuclear, Comisión Nacional de Energía Atómica, Buenos Aires, Argentina*
[2] *Instituto Balseiro, Universidad de Cuyo, Comisión Nacional de Energía Atómica, Rio Negro, Argentina*
[3] *Facultad de Ingeniería, Universidad de Buenos Aires, Buenos Aires, Argentina*
[4] *Gerencia de Coordinación de Proyectos Tecnológicos Nucleares, Gerencia de Energía Nuclear, Comisión Nacional de Energía Atómica, Buenos Aires, Argentina*
[5] *División Química de la Remediación Ambiental, Gerencia Química, Centro Atómico Constituyentes, Comisión Nacional de Energía Atómica, CONICET, Buenos Aires, Argentina*
[6] *Centro de Tecnologías Químicas, Departamento de Ingeniería Química, FRBA-UTN, Buenos Aires, Argentina*

18.1 Introduction

The crucial nature of water access is undisputed. The United Nations in its *World Water Development Report* 2024 [1] highlighted the crucial role of water access in assuring basic human needs, supporting health and economic development, supporting food and energy security, and defending environmental integrity. Unfortunately, nowadays, the world is facing a water crisis as half of the global population suffers from water shortages. It is important to think about water shortage and water stress differences. Water shortage refers to the impact of low water availability per person compromising every small marginal demand, whereas water stress refers to the impact of high water use relative to water availability [2].

Groundwater accounts for 99% of the liquid freshwater on Earth and currently supplies half of the total water extracted globally for domestic purpose, including drinking water for most rural populations who lack access to public or private water delivery systems. Additionally, it accounts for approximately 25% of the water used for irrigation [3].

A recent estimate indicates that the global volume of fresh groundwater is 15.9 million km^3 [4]. When adding 27 million km^3 from ice sheets in Antarctica, 3 million km^3 from Greenland ice sheets, and 158 thousand km^3 from glaciers,

Water Treatment and Desalination: Membrane Separation using Renewable Energy Resources, First Edition. Edited by Mohamed Bassyouni.

the total available freshwater reaches 46 million km^3, which accounts for only 3% of the Earth's total water.

A 2016 study analyzing data from the twentieth century [2] showed that global water consumption rose from 358 km^3 per year in the 1900s to 1,500 km^3 per year in the 2000s, alongside a population increase to 6.5 billion (nearly four times higher than a century earlier). This increment in water consumption was particularly pronounced in the Australia–Pacific region, Eastern Europe, Central Asia, and Western Europe, while it remained stable in Southern Africa and South America and declined in the Middle East. Projections for water demand by 2050 suggest an increase of over 40%, 1.8 billion people are expected to reside in regions facing water scarcity [5], representing 35–51% of the global urban population [6].

Water scarcity can be seen as an engineering challenge, and sustainability undoubtedly requires the advancement of technologies to achieve the UN's Sustainable Development Goal 6, "Ensure availability and sustainable management of water and sanitation for all" [7]. In this context, water desalination is an alternative to safe water provision for communities where other options do not fulfill the water demands. Seawater desalination serves as a virtually limitless and climate-independent water source, capable of supplementing water supplies beyond the natural hydrological cycle [8]. This is particularly beneficial in coastal regions where traditional water sources are inadequate or overexploited [9, 10].

18.2 Seawater Desalination and Related Technologies

According to the European Commission [11], it is estimated that, on the global scale, there were more than 21,000 seawater desalination plants in 2022, resulting in a daily global production of 99 million m^3/day of desalinated water.

Commercial technologies for desalination can be mainly categorized as filtration (membrane) and thermal processes depending on the primary process used to separate water from salt.

Reverse osmosis (RO) is a membrane desalination process that uses pressure-driven separation employing a semipermeable membrane to separate brine. RO requires power to exceed the osmotic pressure to reverse the natural direction of water flow.

On the other hand, one of the thermal desalination processes is the multieffect distillation (MED). MED takes place in a series of evaporation and condensation stages at progressively lower temperatures that minimize fouling and scaling issues [12]. This technology utilizes low-temperature heat for distillation, enabling the use of low-grade heat sources ($<100^\circ C$) for operation. In MED, the separation is achieved through the evaporation of water in a multistage system. The process initiates with the seawater being heated by an external heat source before being introduced to the first in a series of evaporators (called effects), where it is sprayed onto the surface of evaporator tubes reaching its boiling point. Externally supplied steam flows through these tubes. In each effect, the pressure is reduced so that the water boils at lower temperatures, allowing the steam produced in the first effect to heat the next effect. The steam generated in each effect condenses into liquid water, which is collected as fresh water. This condensation process releases latent heat, which is used to preheat the incoming seawater in the subsequent effect. In the final effect, the concentrate brine may either be further processed or discharged, depending on the system design. MED systems are characterized as energy-efficient, capable of providing high quality freshwater, suitable for drinking and industrial use, and scalable, making them adaptable to variable needs.

The combination of a RO with MED is one of the most common hybrid systems [13] used to reduce cost and achieve water requirements. In this case, brine rejected by the RO process is fed to a MED [14].

18.3 Nuclear Desalination

Water desalination is an energy-intensive process, and, today, it mostly uses fossil fuel sources. Besides the cost, this is translated into significant greenhouse gases emission that, if not reduced, hinders the technology sustainability. In 2015, the United Nations Climate Change Conference COP21, held in Paris, launched the Global Clean Water Desalination Alliance, dedicated to take actions to reduce CO_2 emission in the desalination industry [15]. In the concept paper they published, the analysis of different scenarios showed that when a compounded growth rate of water desalination of 6.1% per year was assumed, the estimated CO_2 emissions for global water desalination plants for 2040 was 200 Mtonne CO_2 per year against 75 Mtonne CO_2 per year calculated for 2015. Under this climate initiative, the application of clean energy sources for water desalination is strongly supported. In this context, nuclear-powered desalination offers significant advantages in terms of cost and carbon footprint compared to fossil fuel alternatives. As an example, the production of 1 m^3 of freshwater by RO has an energy demand of 3–6 kWh/m^3. The carbon emissions related to this energy demand when RO is associated with fossil fuel (natural gas, diesel, or coal) are in the range of 1.23–6.30 kg CO_2, whereas if RO is associated with Nuclear Power Plants (NPPs), the carbon emissions are reduced to the range from 0.009–0.24 kg CO_2 [16].

Transferring thermal heat using the steam generated in a NPP is useful to different industrial applications, depending on the available temperature. For example, pressurized water, boiling water, pressurized heavy water, and light water graphite-moderated reactors require low temperature (300°C or higher), whereas high temperature reactors need temperatures as high as 1,000°C [17].

These uses of steam are not only attractive since they allow us to meet different needs of society in addition to electricity generation, but they also allow greater use of the heat generated by the plant, improving thermodynamic performance by reducing the heat that is released into the environment.

Also, besides the Global Clean Water Desalination Alliance, the International Atomic Energy Agency (IAEA) promotes the use of nuclear energy in nonelectrical applications and maintains various international forums and congresses aimed at strengthening it. In recent years, various international agencies have shown interest in expanding their development programs on issues related to nonelectrical applications coupled with low-power reactors. The use of nuclear energy (heat or electricity) to produce other value-added products employing facilities located near the NPPs, called nuclear cogeneration, has generated satisfactory results in several countries around the world. Nuclear desalination (ND) is one of the most important of these technologies.

Since 2020, the IAEA [18] has developed and made available different tools for desalination economic evaluation and thermodynamic optimization, including what they called the ND toolkit which gives access to all ND IAEA activities.

There is extensive experience in desalination plants coupled with nuclear reactors, and important studies on the subject, as several countries such as Japan, the USA, India, and Pakistan have achieved an operational experience over 250 reactor years [16, 19].

The development of this type of dual-purpose plants has been promoted in different countries with the objective of using the electrical and thermal energy generated by the plant to obtain drinking water, such is the case of the plants in Japan (1,000–2,000 m^3/day), Kazakhstan (80,000 m^3/day), and the Diablo Canyon plant in the United States (4,500 m^3/day) [17].

There are several nuclear reactor types than can be coupled with desalination plants. Table 18.1 provides information on the different types of nuclear power reactors used for this purpose, whereas Table 18.2 lists several NPPs associated with desalination systems to produce drinking water [25].

18.3.1 Uses and Consumption of Water in a Nuclear Power Plant (NPP)

NPPs require large quantities of water to supply the primary, secondary, auxiliary, and cooling process systems, among others. This makes the water consumption an essential issue when the construction of a NPP is projected.

In an NPP, water is both consumed and withdrawn. When consumed, the water is either evaporated or contaminated, rendering it unsuitable for direct return to its original source. During withdrawal, water is sourced in a manner that may either result in consumption without return or allow for its return to the original source compliant with applicable environmental regulations [26]. The type of

Table 18.1 Types of nuclear power reactors than can be coupled to desalination plants.

Type of reactor	Acronym	Main characteristics
Liquid Metal Fast Reactor	LMFR	This reactor uses liquid metal, such as sodium or lead, as a coolant, and there is no need to use a moderator. It is possible to achieve higher temperatures in the reactor vessel, making this type of reactor very fuel-use efficient [20].
Pressurized Water Reactor	PWR	This type of reactor is cooled by hot high-pressure water, which also acts as a moderator. It uses slightly enriched uranium or MOX as a fuel. The water in the reactor core is kept under high pressure to prevent it from boiling, allowing it to reach high temperatures. The heat is then transferred to a secondary water loop, which generates steam to drive a turbine [21].
Boiling Water Reactor	BWR	In this type of reactor, water, used as coolant, is brought to boiling within the reactor core, producing steam that drives the turbine [21].
Pressurized Heavy-Water Reactor	PHWR	These reactors use natural uranium as fuel and heavy water, which contains deuterium, as a more efficient moderator than light water [22].
Nuclear Heating Reactor	NHR	NHRs are integral pressurized water reactors (i.e., they have integral primary circuits and light water as a primary coolant). They typically use a simplified design with a focus on inherent safety features and can operate at lower temperatures and pressures [23].
High-Temperature Gas-cooled Reactor	HTGR	This type of reactor uses helium as a coolant and graphite as a moderator. It operates at high temperatures, which allows for efficient electricity generation and the potential for applications like hydrogen production. HTGRs are known for their inherent safety due to their high-temperature tolerance [24].

cooling system, the thermal efficiency of the NPP, the need for service water, etc., are factors that determine the water requirement for the NPP. Even when from an environmental point of view, NPP has the important advantage of having practically zero CO_2 emissions, they require greater water withdrawal than other types of thermal power stations.

Water demand in a NPP can be classified into three primary categories [26]:

- **Cooling Water:** This water is used to remove residual heat released into the environment after the generation of useful work in the turbine. Cooling water demands account for most of the total water consumption at the plant.

Table 18.2 Nuclear power plants coupled with desalination technologies.

Reactor type	Location	Desalination technology	Capacity (m³/día)
LMFR	Kazakhstan (Aktau)	MED/MSF	80,000
PWRs	Japan (OhiTakahama, Ikata, Genkai)	MED/MSF/RO	1,000–2,000
	South Korea	MED	40,000
	Argentina	MED/RO	12,000
	Russia	MED/RO	NA
BWR	Japan (Kashiwazaki)	MSF	NA
PHWR	India (Kalpakkam)	MSF/RO	6,300
	Pakistan (KANUPP)	MEE	4,800
NHR	China	MEE	160,000
HTGR	South Africa, France, and the Netherlands	MEE/RO	not available

NA: not available.

For a 100 MWe plant, the cooling water requirement would be approximately 46 m^3/s.

- **Demineralized Water:** Required for inventory and consumption in process systems for the reactor's primary, secondary, and auxiliary circuits. A 100 MWe plant would typically require between 50 and 150 m^3/day.
- **Potable Water:** Used for general plant services, sanitation, fire systems, etc. The demand for potable water in a 100 MWe plant would range from 200 to 600 m^3/day.

The precise values for each type of water will depend on the temperature of the cooling water, the temperature difference with the steam in the condenser, and the condenser efficiency.

Given the need to increase the availability of freshwater and industrial-quality water for both the consumption of populations and industries in general, as well as for NPP generating electrical energy, water desalination has grown substantially some years ago [27] and remains relevant today [28], especially when a country faces the need of increasing freshwater resources due to water stress in the region and, simultaneously, has the objective to decarbonize their energy systems [16]. Cogeneration can provide nearby populations with both electricity and potable water for domestic and industrial use. Additionally, it enables the NPP to self-supply demineralized water, eliminating the need for traditional ion exchange resins commonly used in power plants. This shifts not only reduces environmental impacts but also contribute to a decrease in carbon dioxide emissions, supporting a more sustainable operation.

18.3.2 Hybrid Systems Coupled with a Nuclear Power Plant

The coupling of hybrid desalination systems, combining thermal and membrane desalination processes, with power generation systems has been studied for at least 20 years [29–31]. This approach involves utilizing the steam produced by the NPP for the desalination process. Hybrid systems have the potential to increase water recovery, reduce overall specific energy consumption, reduce cost, and meet various water qualities that could be required [12, 32].

18.3.3 Case Study: Technical and Economic Aspects of a Desalination Plant Integrated to a Small Modular Reactor for Nuclear Cogeneration

To integrate nuclear electricity generation with simultaneous water production via desalination processes, a hybrid configuration of RO + MED technologies as an integrated Desalination Plant was proposed. The cost of water production was estimated through a site selection and evaluation, considering the local economy and population of a coastal city in Argentina.

Finally, a pilot demonstration project is presented, highlighting the design and operational functions, as well as the primary objectives of the experimental tests.

18.3.3.1 Connection of an Integrated Desalination Plant to the Secondary Circuit of the NPP in a Hybrid Configuration

Previous work [25, 33] was mainly focused on the production of drinkable water using a RO system connected to a NR, relying only on electricity. More recently, similar production was demonstrated through a MED plant, with a Gain Output Ratio (GOR) of 10:1.

In this setup, a turbine extraction line supplies the condenser/evaporator (IC1/IC2) of an intermediate circuit (IC), which subsequently provides the necessary heat for the MED process. Vapor from the turbine extraction is diverted to IC1 and returns as liquid to the Secondary Circuit (SC) main condenser. The objective of IC2 is the generation of steam to feed the MED device.

18.3.3.2 Main Characteristics of the Small Size Reactor Plant

The plant taken as reference is divided into three circuits: the primary circuit, composed of the reactor and the vapor generators; the secondary circuit, where the electricity is generated; and the tertiary circuit, which acts as a refrigeration circuit. In our analysis, the SC is the most relevant part as it has the potential to couple with the desalination plant.

18.3.3.3 Main Aspects of the NPP Secondary Circuit

The SC of the Small Size Reactor (SSR) plant is responsible for extracting the heat generated in the reactor core, which is then used to produce electricity while also releasing excess heat to the environment. The SC is constituted by three processing subsystems: the main steam system, the condensate system, and the main feedwater system.

The turbogenerator includes a single turbine and an electric generator for electric power production. The turbine features two steam extraction points: E1, the low-pressure extraction, and E2, the high-pressure extraction. The exhaust steam from the turbine is directed into the condenser through the dome, where it flows over the tubes and condenses as it cools. Cooling water from the tertiary system circulates inside the tubes, facilitating the condensation process. Table 18.3 outlines the main characteristics of the steam system.

The thermo-hydraulic circuit operates at a pressure of 5 bar, with a high temperature of 130°C and a low temperature of 90°C [34]. In IC1, steam extracted from the turbine condenses, transferring heat to pressurized water, which functions as a preheater. The resulting condensate is returned to the SC as subcooled liquid. The IC2 exchanger acts as a steam generator, producing saturated steam at 80°C for the MED plant. This steam releases its condensation heat to the desalination process water and exits as a saturated liquid. A 10°C temperature differential is maintained between the pressurized water in the coupling circuit and the steam supplied to the MED plant.

The production of water was investigated in a cogeneration mode aimed at producing both potable water through RO and reactor-quality water using a MED device. Four configurations for coupling the RO system with the NPP and the MED process were explored (Figure 18.1). The objective was to recover heat from one stream to raise the inlet temperature of the RO feedwater as a higher temperature improves the performance of the membranes [35]. Consequently, two of the coupling configurations (A and B) use a dedicated heat exchanger (IC3) to preheat the inlet stream with water from the IC. The third configuration (C) extracts part of the water from the mechanical compression outlet and directs it to the RO plant.

Table 18.3 Energy and mass balance of the SC turbine.

	Turbine inlet	High-pressure extraction (E2)	Low-pressure extraction (E1)	Discharge to the main condenser
Pressure (bar)	47	9.47	3.18	0.085
Temperature (°C)	314.4	177.5	135.5	42.7
Mass flow (kg/s)	43.69	3.49	5.33	34.87

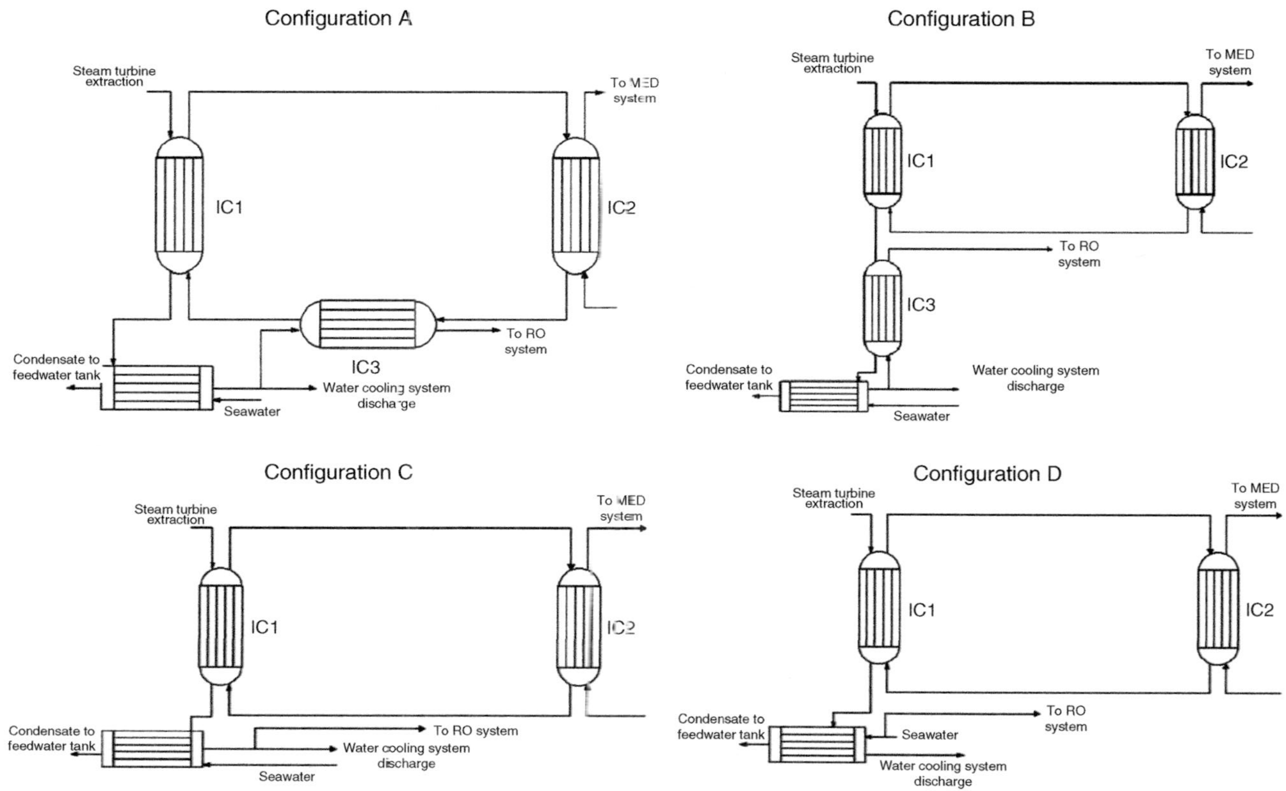

Figure 18.1 Configurations evaluated for integrating a hybrid system (RO + MED) with an SSR.

The fourth configuration (D) serves as the reference case. The studied schemes are outlined as follows:

- Configuration A: Feedwater to the RO plant comes from the discharge of the main condenser and is preheated in the coupling circuit by means of a third exchanger IC3 in series with IC1 and IC2.
- Configuration B: The feedwater to the RO plant comes from the discharge of the main condenser and is preheated by a third exchanger IC3 in series with IC1 and the main condenser.
- Configuration C: The incorporation of a third heat exchanger for preheating the feedwater to the RO plant is ruled out. Only the waste heat released in the main condenser is used.

To provide a comparison with the results obtained in schemes A, B, and C, the evaluation of scheme D is proposed:

- Configuration D: The RO plant is fed with seawater without preheating.

The effects of variables such as pressure, permeate flow, inlet temperature, and process configuration on the energy efficiency of RO (Figure 18.2) have been analyzed using ROSA Software© [36].

The findings from this analysis were considered when selecting the RO system for the experimental rig. Figure 18.3 displays the specific energy consumption (SEC) for each of the evaluated configurations (A, B, C, and D).

The key finding is that configuration C (Figure18.3) is the simplest and most efficient option as the other configurations do not offer a substantial improvement in overall efficiency. Furthermore, the alternative configurations would require additional equipment and pumps, leading to increased fixed and operating costs.

The considerations supporting the hybrid system are:

- **Production of Nuclear Reactor-Quality Water:** The hybrid system enables the production of NR-quality water, potentially eliminating the need for a makeup water plant. This change could also reduce reliance on ion exchange resins, chemicals, and associated pollution. A cost comparison will be presented to assess these benefits.
- **Enhanced Flexibility by Blending:** The MED system only produces demineralized water. By blending this water with RO water, the system gains greater flexibility. This approach can lower the pressure required for the RO process, allowing for higher salinity outputs and reducing the frequency of membrane replacement.
- **Consolidation of Makeup and Discharge Location** for both plants, simplifying operations and logistics.
- **Residual Heat Recovery:** The system recovers residual heat from the main condenser to preheat the RO feedwater, improving energy efficiency.

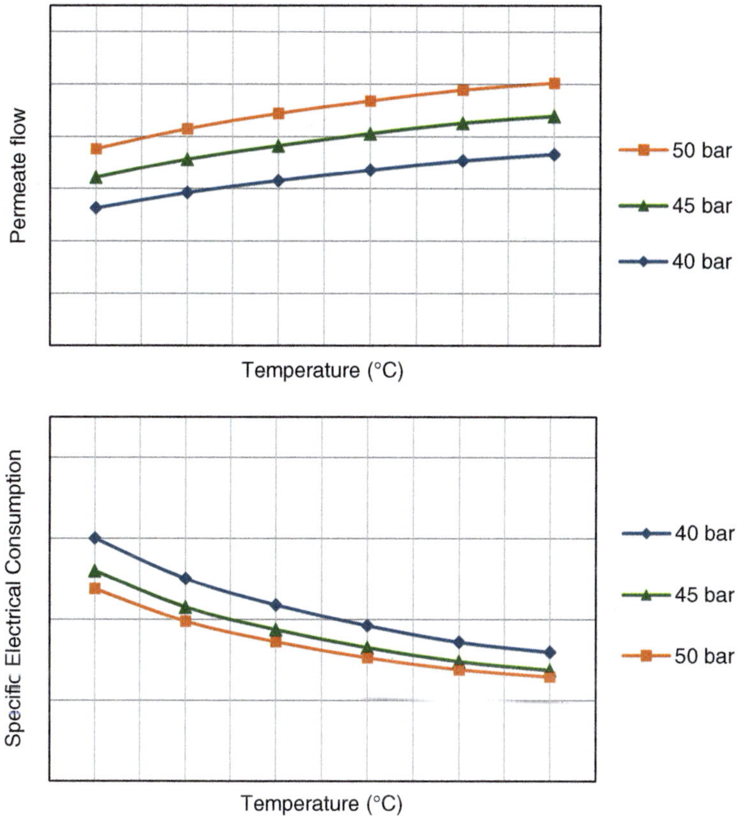

Figure 18.2 Effects of the inlet temperature on permeate flow and specific electric consumption analyzed using ROSA Software©.

Considering a SSR with a thermal output of 100 MWt and an electrical output of 27 MWe, diverting 12 MWt, and with a given GOR of 15:1, the NR-MED plant can supply sufficient resources for a target population of 80,000. This setup provides electricity and produces 6,000 m^3/day of demineralized water (with a salinity of less than 10 ppm). In operation, the NR only consumes 29 m^3/day of demineralized water. The demineralized water is then blended with RO water, which has salinity of approximately 500 ppm, to produce drinkable water.

The intermediate coupling circuit primarily functions to transfer energy from the balance of the plant (BoP) to the MED plant [37]. The feed for the RO plant is sourced from the discharge of the main condenser (Figure 18.4).

18.3.3.4 Description of the Intermediate Circuit

The IC consists of two independent branches, each handling 50% of the load and supplied with steam from the SSR-BoP. Each branch is equipped with a pump,

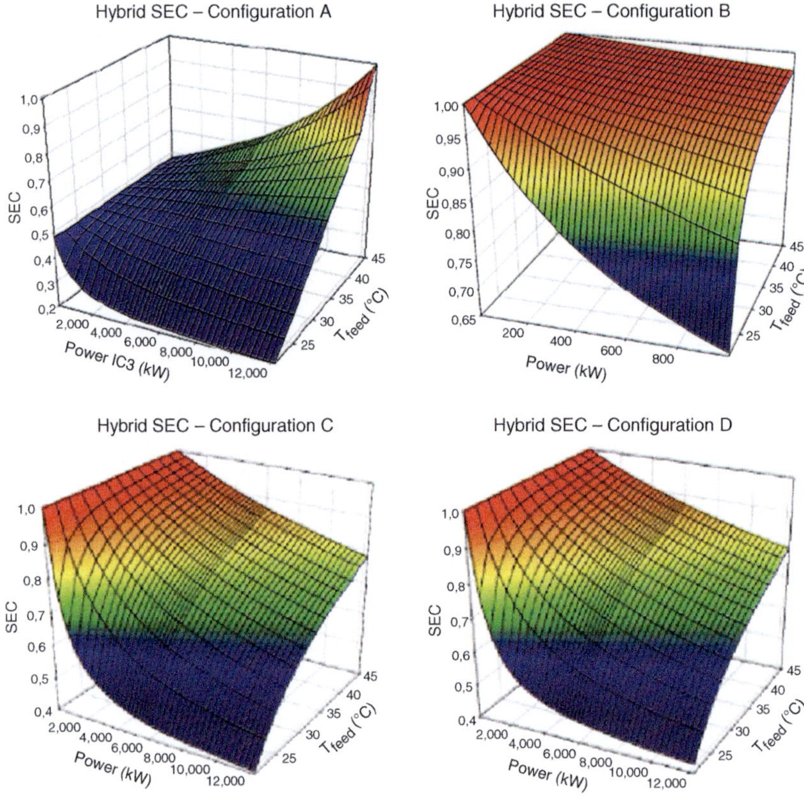

Figure 18.3 Hybrid specific electrical consumption of the different coupling configurations analyzed.

while a third pump (B-3) is kept in reserve, contributing to the system availability. In case of failure, the pump can direct flow to either of the parallel branches through interconnected piping and valves.

Each branch carries subcooled water transferring thermal energy from the condenser-subcooler IC1A1/A2 to the evaporator IC2. Additionally, there are derivations, alongside the pumps and heat exchangers, to operate in bypass mode either manually or remotely using motorized valves. Each pump is equipped with a minimum flow rate recycling line. In turn, each pump can direct flow to the other circuit. The following components are available for pressurization:

- Centrifugal pumps B-6/7 that fill the circuit.
- Pressurizers and surge tanks E-33/E-5 with nitrogen. Their primary function is to accommodate volume (and level) variations caused by changes in the temperature of the intermediate fluid.
- Drainage lines are located at specific points.

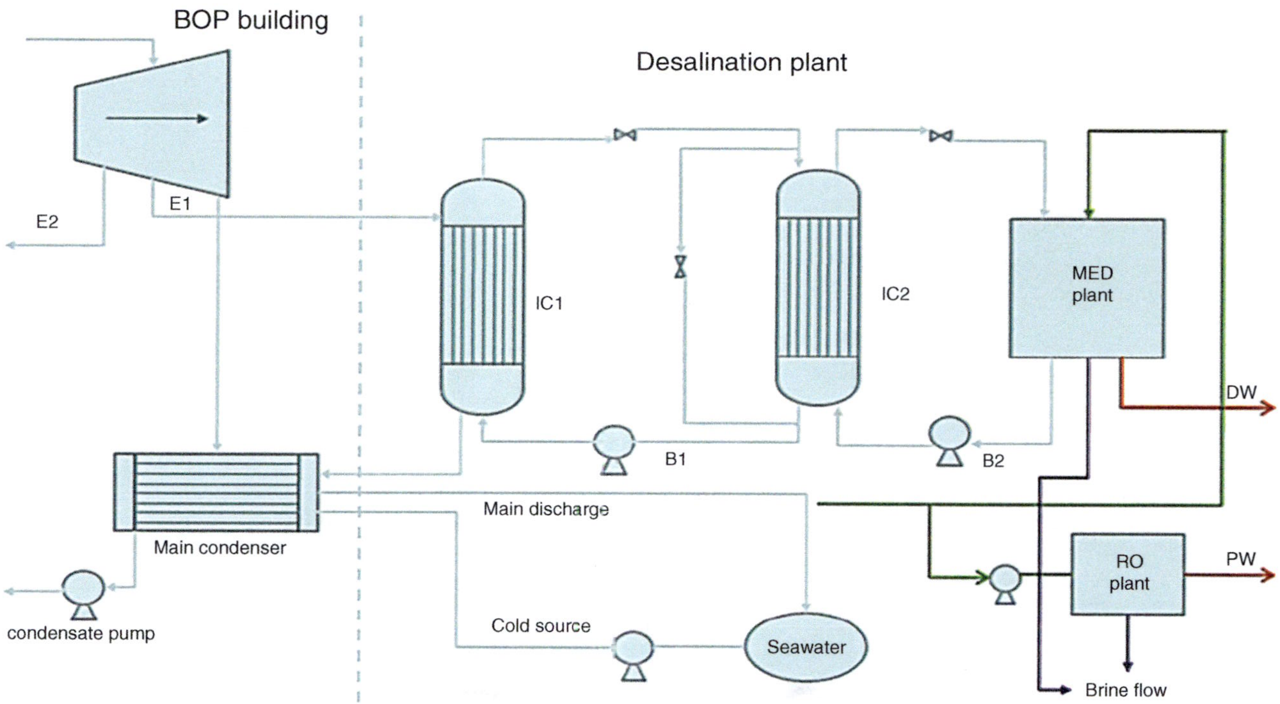

Figure 18.4 Secondary circuit showing the integration of the RO system with the MP-MEE and the IC for isolation.

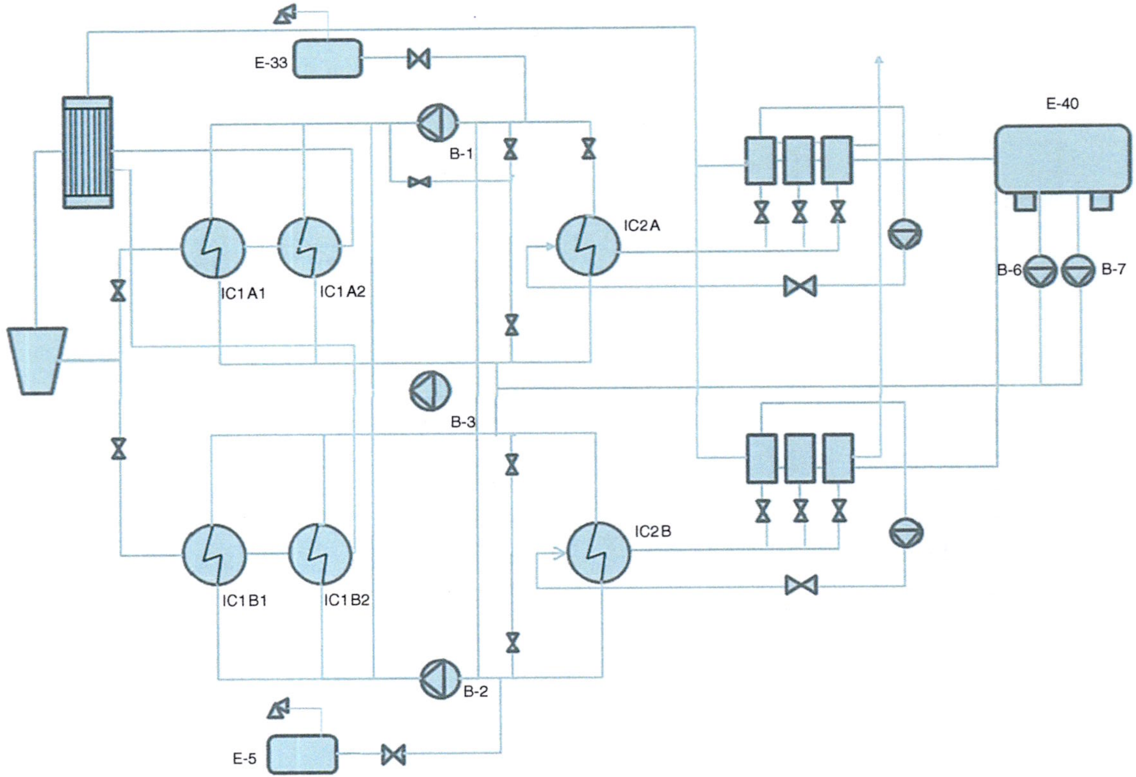

Figure 18.5 Intermediate circuit.

The circuit is equipped with dosing pumps for additives (ammonia/hydrazine) and a sampling cabinet. The evaporators IC2A/B produce low temperature steam (~80°C) to supply heat to the first stage of the MED devices. The condensate of each branch is recycled using the B-4/5pumps, respectively. The concentrate is directed to the discharge of the main condenser. The water produced by the MED is stored in tank E-40, from where it can be supplied to the process systems of the SSR and BoP (Figure 18.5).

18.3.3.5 Main Components and Dimensions

Tables 18.4–18.6 outline the details of the main components of the system.

Table 18.4 Condenser, subcooler, and evaporator details.

Equipment	Condenser IC1A1/B1	Subcooler IC1A2/B2	Evaporator IC2A/B
Type	Horizontal one step tubes and one step shell (condensation) CEM	Horizontal one step tubes and one step shell (hot fluid) CEM	Vertical evaporator in shell (BEM)
Material of the shell	Carbon steel	Carbon steel	Carbon steel
Fluid of the shell	Demineralized water	Demineralized water	Demineralized water
Material of the tubes	Carbon steel	Carbon steel	Carbon steel
Fluids of the tubes	Demineralized water	Demineralized water	Demineralized water
Inlet temperature (Tin, °C)	135	135	117
Outlet temperature (Tout)	135	95	85
Inlet temperature (Tin, °C)	85	85	80
Outlet temperature (Tout)	124	95	80
W (kg/s)	2.65	2.65	40.5
w (kg/s)	30	10	2.4
Power (MW)	5.1	0.44	5.44
Uglobal (W/m^2K)	2,000	800	1,050
ΔTlogor ΔT (°C)	25.7	21.6	16
Area (m^2)	100	25	323

Table 18.5 Pressurizer-surge and demineralized water tank details.

Tank	Pressurizer-surge	Demineralized water (product)
Fluid	Demineralized water	Demineralized water
Volume (m^3)	4	500
Pop (bar)	5	5
Material	Carbon steel	Carbon steel (with liner)

Table 18.6 Main piping details.

Line	Intermediate circuit (hot/cold branch)	Steam to MEE
DN#	200	254
Sch	40	20
Material	Carbon steel	Duplex or ferritic steel

18.3.3.6 Pretreatment

One of the challenges of RO membrane desalination is its reduced efficiency processing effluents with low concentrations of suspended solids, which can lead to membrane fouling. This fouling not only compromises membrane performance but also increases energy consumption due to higher osmotic pressure. As a result, effective pretreatment is crucial to enhance the feasibility and efficiency of RO systems. Both conventional methods such as coagulation, flocculation, and scale inhibition and nonconventional methods, including ultrafiltration (UF), microfiltration, and nanofiltration, have demonstrated improvements in operational performance. These pretreatment approaches help to minimize fouling and enhance the removal of suspended solids and dissolved organic matter from seawater [38].

Figure 18.6 illustrates the concept of the pretreatment trains used for RO and modular plate (MP)-MED raw water, sourced from the main condenser discharge, which is several degrees Celsius warmer than the intake temperature. Depending on the process and requirements, the water undergoes multiple pretreatment stages. Additionally, these flowsheets, like those of the IC system, also serve as a reference for evaluating capital costs.

18.3.3.7 Siting and Cost of Water Production

To estimate the water production costs for the SSR + ND plant, we considered the current population's electricity and freshwater demand, as well as the needs of

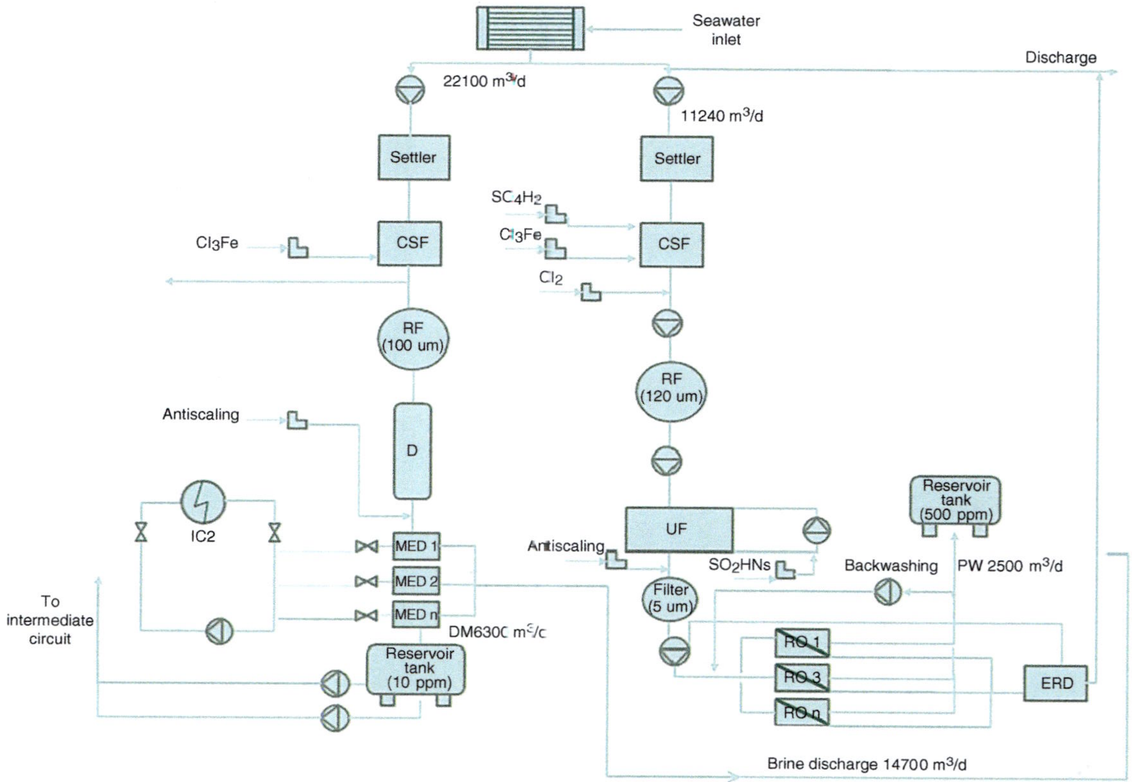

Figure 18.6 Diagram of the pretreatment system for the RO and MP-MEE plants. CSF: coagulation–sedimentation–flocculation, D: deaerator, ERD: energy recovery device, RF: rotative filter, UF: ultrafiltration.

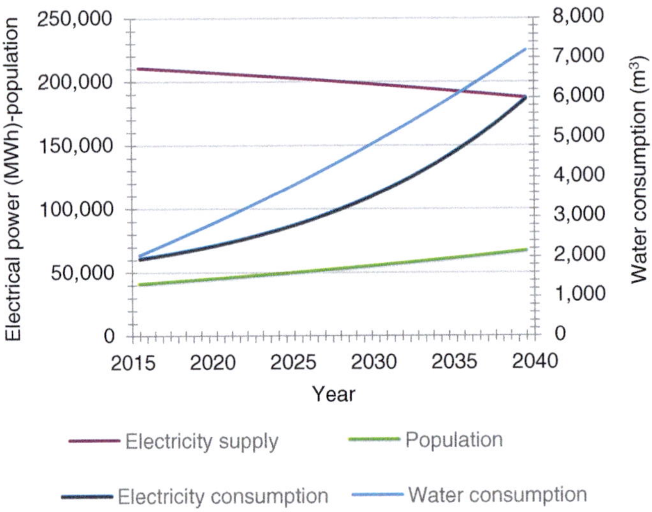

Figure 18.7 Evolution of water and electricity consumption in a specific site of Argentina considering a population growth of 2%, projected to 2040.

local industries. An extrapolation to the year 2040 was conducted to project future capacity, assess capital costs, and mitigate investment risk (Figure 18.7).

The site selection process involved analyzing the extent, salinity, and temperature of water bodies along the Atlantic coast, as well as continental water resources such as precipitations, surface water, and basins. Additionally, factors such as population density and the layout of the Argentine National Electrical System (comprising high-voltage transmission lines, distribution networks, and transformer stations) were taken into account.

The cost of water per cubic meter is determined by both operational and capital costs. To provide an initial estimate of the capital cost, the main contributors have been identified, including the water intake, intermediate circuit, RO plant, MED plant, concentrate discharge, and storage facilities for potable and demineralized water, along with associated construction costs. A detailed breakdown of the capital costs is presented in Table 18.7, while the operational costs are detailed in Table 18.8.

18.3.3.8 Pilot Demonstration Project

An experimental rig has been designed to conduct thermal and mechanical tests (using membrane devices) in both single and hybrid configurations. This rig also serves as a demonstration installation for several applications, including supplying makeup water for a 100 MWt nuclear reactor with once-through steam generators that have a limited blowdown rate, and providing reactor-quality

Table 18.7 Capital costs in USD.

Plant	Element	Design lifetime (years)	Cost (USD)
Intermediate circuit (IC)	Equipment	30	620,000
	Pumps	30	125,000
	I&C	15	50,000
	Piping and valves	30	200,000
	Pressurizers	30	80,000
	Total IC		1,075,000
Water intake (WI)	Pumping system	10	280,000
	Piping	25	10,000
	Pretreatment	10	14,700
	Filters	30	130,000
	Total WI		434,700
Reverse osmosis (RO)	UF equipment (membranes)	5	1,200,000
	UF backwashing	10	10,000
	Filters	10	560,000
	High pressure pumps with ERD	15	1,700,000
	RO membranes	7	280,000
	I&C	15	18,600
	Total RO		3,768,600
Multieffect distillation (MED)	Deaerator	30	60,000
	Equipment	30	1,700,000
	Piping and valves	30	100,000
	I&C	15	150,000
	Total MED		2,010,000
Discharge brine	Pumping system	15	170,000
	Piping	30	220,000
	Total discharge brine		390,000
Storage demi water (SDW)	Pumping system	30	92,000
	Piping	30	11,000
	Storage tank	30	31,600
	Total SDW		134,600
Storage potable water (SPW)	Pumping system	30	92,000
	Piping	30	11,000
	Intermediate storage tank	30	18,000
	Total SPW		121,000
Construction	Detailed engineering	30	770,000
	Assembling and installation	30	1,600,000
	Civil engineering	30	725,000
	Total construction		3,095,000
	Total capital investment		11,028,900
	Depreciation (USD/yr)		723,630

Table 18.8 Operating cost (*based on market price).

Item	Cost
Total electrical electricity	0.6 USD/m^3
Chemicals	0.48 USD/m^3
Maintenance	0.08 USD/m^3
Others	0.06 USD/m^3
Staff	1,295 USD/d
Produced flowrate	9,867 m^3/d
Depreciation	2,010 USD/d
Total operating cost	1.6 USD/m^3

water for primary and auxiliary systems during normal operation. Additionally, it will supply makeup water for a 10 MWt nuclear reactor with vertical recirculating steam generators, which require full blowdown flow rates and nuclear reactor makeup quality water for the primary and auxiliary circuits. Furthermore, the rig will demonstrate the production of potable water for staff and/or the surrounding population and will be part of a permanent laboratory that provides high-quality water to the existing facilities.

The experimental rig comprises a heat source that simulates a turbine steam extraction, along with connections to the MP-MED and RO systems. The heat source can supply up to 30 KW of steam (150°C, 6 bar, 40 kg/h) or hot water (4 m^3/h). It features both a condenser and a heat exchanger, allowing operation either with steam in natural convection or with recirculating hot water in forced convection.

Figure 18.8 presents a scheme of the rig, illustrating different components such as the boiler, condenser (located at the top), counter-current tube heat exchanger, and instrumentation.

Figures 18.9–18.12 show the different operating modes in the full flow sheets, along with the connections to the MED and RO systems.

The UF + RO circuit serves as an experimental platform for personnel training, equipping technical staff with essential skills for the start-up, normal operation and maintenance of UF, microfiltration (MF), and RO membranes, including the necessary chemical pretreatment process. This training is seamlessly integrated with water production in a hybrid configuration linked to the NPP, ensuring that personnel gain practical experience in real-world operations.

A series of operational tasks are planned to enhance staff experience and response capabilities addressing various plant events. The UF + RO system is

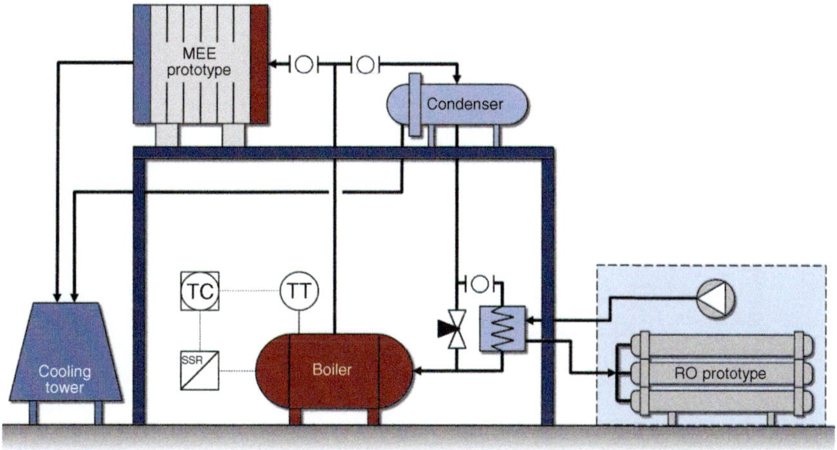

Figure 18.8 Scheme of the experimental RIG coupled to MED and RO systems.

designed to treat a flow rate of 0.9 m^3/h, accommodating various feedwater qualities, including seawater, brackish water, and freshwater sourced from potential sites within the country, with a maximum operating temperature of 40°C. The production capacity of the system ranges from 0.15 to 0.20 m^3/h, and the system comprises two modules: (1) UF and (2) RO.

The UF module plays a critical role in the plant by providing essential pretreatment for the water supplied to the RO system, thereby safeguarding its integrity and long-term operational efficiency. This module is equipped with two PDVF (polyvinylidene fluoride) membranes operating in parallel (Toray Model–HFU 1,010 N), with a total surface area of 7 m^2. Inline differential pressure meters are installed to monitor the pressure drop across the membranes. Furthermore, a chemical dosing and testing system is incorporated for water clarification, with biocide dosing anticipated upstream of the RO module to prevent biological fouling. The RO module is also equipped with inline instrumentation for monitoring differential pressure and flow rates. The water produced by the UF module is temporarily stored in a tank (TK-001) with a capacity of 1,000 L, facilitating a smooth transition to the subsequent treatment process.

The feedwater line to the RO system draws water from the TK-001 tank and is filtered using a microfiltration unit (CF-100) with disposable cartridges (4.5 in × 20 in AMTEC, 5 μm). To ensure optimal performance, differential pressure across the prefilter is also monitored continuously. Additionally, a secondary chemical dosing system has been installed to evaluate various anti-scalants and assess their effectiveness in controlling membrane fouling.

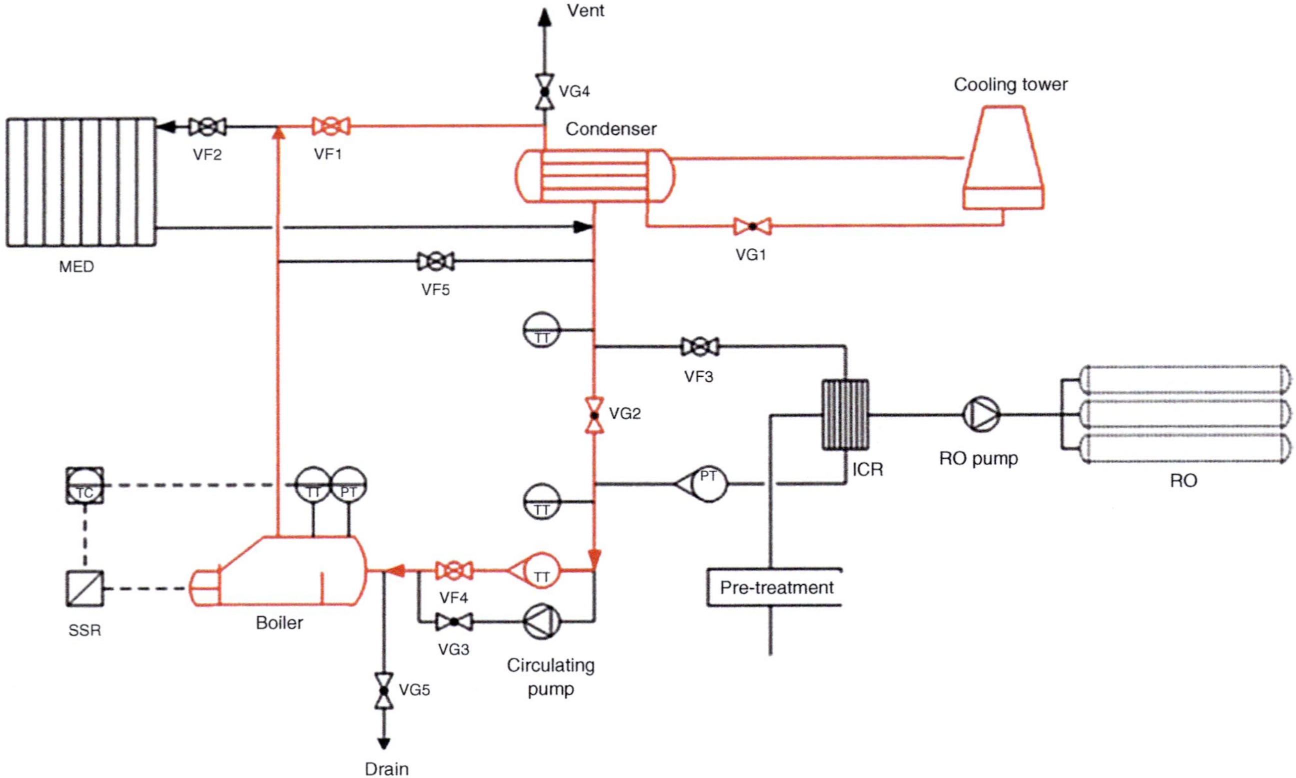

Figure 18.9 Experimental rig startup operative.

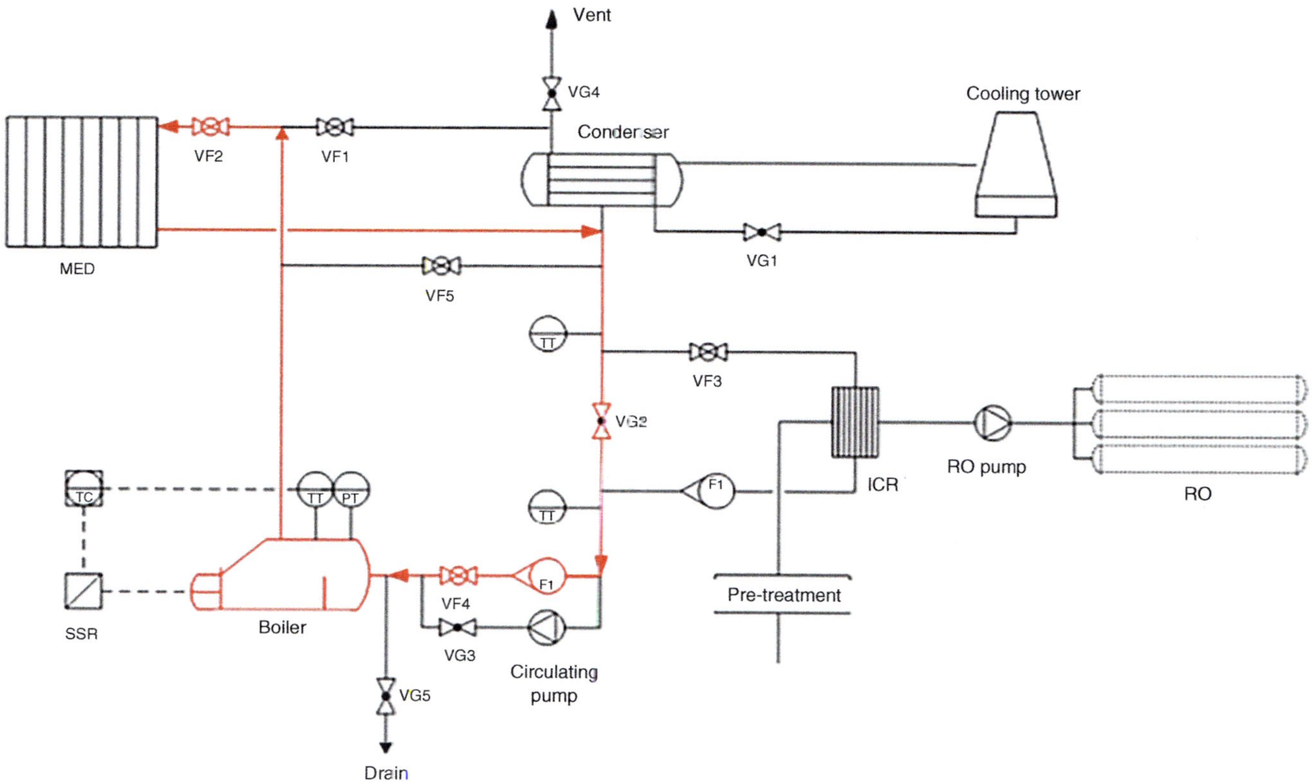

Figure 18.10 Experimental rig MED operative mode.

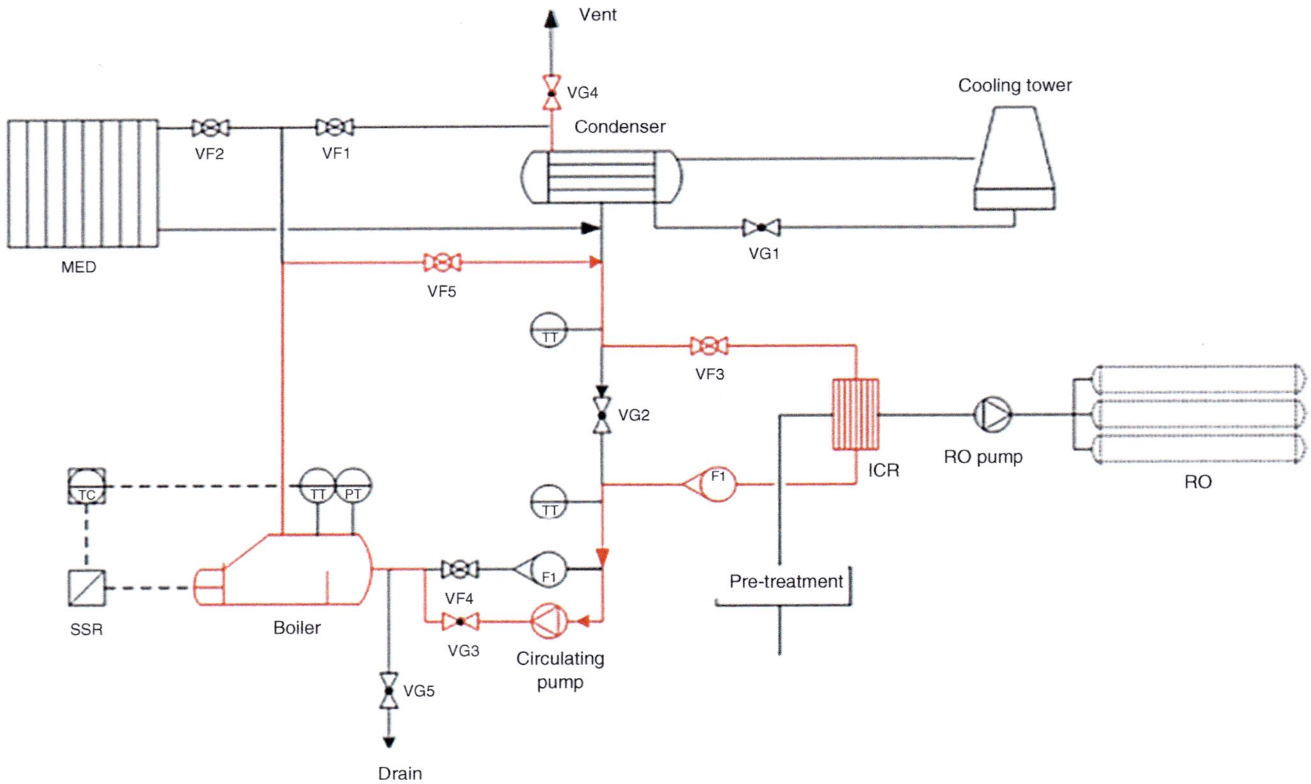

Figure 18.11 Experimental rig RO operative mode. *TC: temperature controller, TT: temperature transmitter, PT: pressure transmitter, F: flow meter, VF1/2/3/4/5: valve code, VG1/2/3/4/5: valve code, SSR: solid state relay, MED: Multi Effect Distillation, ICR: Heat Exchanger, RO: Reverse Osmosis.*

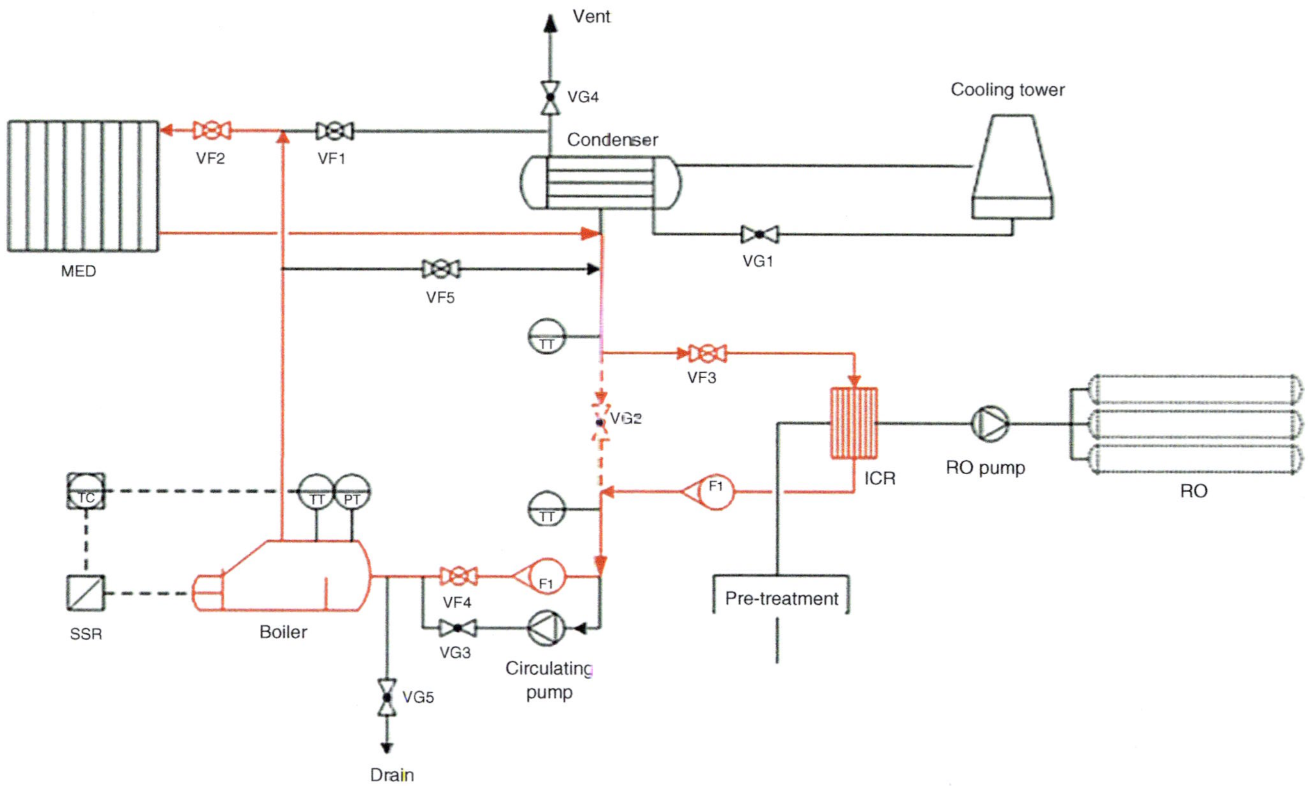

Figure 18.12 Experimental rig hybrid operative mode. *TC: temperature controller, TT: temperature transmitter, PT: pressure transmitter, F: flow meter, VF1/2/3/4/5: valve code, VG1/2/3/4/5: valve code, SSR: solid state relay, MED: Multi Effect Distillation, ICR: Heat Exchanger, RO: Reverse Osmosis.*

The skid is equipped with two multistage centrifugal pumps in series (models GRUNDFOS CMG 1-7 and CRNE 1-23 HS) to ensure adequate water inlet pressure for the low and high salinity modules. Four pressure vessels, interconnected through piping and valves, facilitate both series and parallel configurations of the RO membranes. The circuit is operated with two types of RO membranes (4 in × 40 in TORAY TMG10 and TORAY TM810L) for testing different water qualities. The permeate line is equipped with instrumentation to measure pressure, flow rate, and conductivity, while the concentrate line features similar monitoring capabilities.

The UF + RO model is connected to the heating rig via the heat exchanger (ICR) located downstream of the pump (P^{-100}). The ultra-filtered water is preheated by means of the boiler/heater with variable power, reaching temperatures of up to 40°C. This configuration aims to analyze the impact of temperature on the process variables. The pilot plant was designed and constructed entirely by a local company in Argentina.

18.4 Conclusion

In this chapter, a technical and economic feasibility study of a hybrid seawater desalination plant, coupled to a nuclear reactor, is presented for supplying water and electricity to a coastal city in the Patagonian Region of Argentina. The proposed hybrid installation combines a RO plant and a MED plant to produce various water qualities.

The results show that the production of 6,367 m^3/day of water by the distillation plant simultaneously with the production of 3,500 m^3/day by the RO system is technically feasible. This results in a total water production of 9,867 m^3/day, with an associated SEC of 11.3 kWh/m^3 and a final product quality of 150 ppm

Among the main advantages of the integrated desalination plant described, it is noteworthy that the SEC is significantly reduced compared to configurations relying solely on MED. This reduction leads to improved operating costs. Additionally, the coupling takes advantage of residual thermal energy from the NPP, resulting in lower environmental impact. The blending of produced water decreases the demand on membranes, resulting in a higher total dissolved solids' concentration in the RO devices.

The hybrid configuration combines the benefits of both thermal and membrane methods, significantly reducing operating costs, and the production of different water qualities, making it suitable for both industrial and domestic consumption.

An estimate of the water production cost for the MED + RO configuration coupled to SSR was approximately 1.6 USD/m^3. The cost structure is mainly associated with the following distribution: 38% electricity, 30% annual investment equivalent fee, 11% chemical pretreatments, 12% staff, 4% maintenance, and 5% miscellaneous expenses.

The finding demonstrates that ND coupled to a SSR represents a viable configuration for producing both freshwater and electricity for small cities in Argentina, fulfilling both technical and economic feasibility requirements.

Acknowledgment

The authors wish to express their gratitude to the Comisión Nacional de Energía Atómica (National Commission of Atomic Energy – CNEA) and the IAEA International Atomic Energy Agency for the support provided during the development of this work. The technical collaboration and knowledge exchange that made this study possible are also gratefully acknowledged.

Questions

1 What are the main benefits of coupling a nuclear reactor with a hybrid desalination plant in terms of environmental impact and operational efficiency?

2 How does the hybrid RO + MED configuration achieve lower energy consumption compared to a single desalination method?

3 Why is pretreatment necessary for the reverse osmosis process, and what methods are commonly used?

4 What could be the projected economic benefits of hybrid desalination for small coastal cities, based on the case study?

5 Describe the main cost components in the water production using a hybrid MED + RO plant coupled with a nuclear reactor.

References

1 United Nations (2024). *Water for Prosperity and Peace.* United Nations Educational, Scientific and Cultural Organization.
2 Kummu, M., Guillaume, J.H.A., de Moel, H. et al. (2016). The world's road to water scarcity: shortage and stress in the 20th century and pathways towards sustainability. *Sci. Rep.* 6: https://doi.org/10.1038/srep38495.
3 United Nations (2022). *Groundwater: Making the Invisible Visible.* United Nations Educational, Scientific and Cultural Organization.
4 Ferguson, G., McIntosh, J.C., Warr, O. et al. (2021). Crustal groundwater volumes greater than previously thought. *Geophys. Res. Lett.* 48 (16): https://doi.org/10.1029/2021GL093549.

5 Eliasson, J. (2015). The rising pressure of global water shortages. *Nature* 517 (7532): 6–6. https://doi.org/10.1038/517006a.

6 He, C., Liu, Z., Wu, J. et al. (2021). Future global urban water scarcity and potential solutions. *Nat. Commun.* 12 (1): https://doi.org/10.1038/s41467-021-25026-3.

7 United Nations (2022). *The Sustainable Development Goals Report 2022.*

8 Lin, S., Zhao, H., Zhu, L. et al. (2021). Seawater desalination technology and engineering in China: a review. *Desalination* 498: 114728. https://doi.org/10.1016/j.desal.2020.114728.

9 McDonald, R.I., Green, P., Balk, D. et al. (2011). Urban growth, climate change, and freshwater availability. *Proc. Natl. Acad. Sci. USA* 108 (15): 6312–6317. https://doi.org/10.1073/pnas.1011615108.

10 McDonald, R.I., Weber, K., Padowski, J. et al. (2014). Water on an urban planet: urbanization and the reach of urban water infrastructure. *Global Environ. Change* 27 (1): 96–105. https://doi.org/10.1016/j.gloenvcha.2014.04.022.

11 European Commission Blue Economy Observatory (2024). Desalination. https://blue-economy-observatory.ec.europa.eu/eu-blue-economy-sectors/desalination_en.

12 Shatilla, Y. (2020). Nuclear desalination. In: *Nuclear Reactor Technology Development and Utilization*, 247–270. Elsevier. https://doi.org/10.1016/B978-0-12-818483-7.00007-X.

13 International Atomic Energy Agency (2000). *Introduction of Nuclear Desalination A Guidebook*. Vienna: IAEA.

14 World Nuclear Association (2024). Information Library. https://world-nuclear.org/information-library/non-power-nuclear-applications/industry/nuclear-desalination.

15 Global Clean Water Desalination Alliance (2015). Global clean water desalination alliance 'H_2O minus CO_2' CONCEPT PAPER (accessed 1 December 2015).

16 El-Emam, R.S., Ozcan, H., Bhattacharyya, R., and Awerbuch, L. (2022). Nuclear desalination: a sustainable route to water security. *Desalination* 542: 116082. https://doi.org/10.1016/j.desal.2022.116082.

17 Cleveland, J.C. and McDonald, A. (2008). *Advanced Applications of Water Cooled Nuclear Power Plants.* International Atomic Energy Agency.

18 International Atomic Energy Agency (2024). Nuclear desalination. https://www.iaea.org/topics/non-electric-applications/nuclear-desalination (accessed October 2024).

19 Misra, B.M. (2007). Seawater desalination using nuclear heat/electricity—prospects and challenges. *Desalination* 205 (1–3): 269–278. https://doi.org/10.1016/j.desal.2006.03.555.

20 Rowinski, M.K., White, T.J., and Zhao, J. (2015). Small and medium sized reactors (SMR): a review of technology. *Renew. Sustainable Energy Rev.* 44: 643–656. https://doi.org/10.1016/j.rser.2015.01.006.

21 Belyakov, N. (2019). Modern nuclear power plant. In: *Sustainable Power Generation*, 291–325. Elsevier. https://doi.org/10.1016/B978-0-12-817012-0.00024-4.

22 Smith, J.C. (2022). Steam generators. In: *Pressurized Heavy Water Reactors*, 229–273. Elsevier. https://doi.org/10.1016/B978-0-12-822054-2.00019-2.

23 Dong, Z., Liu, M., Jiang, D. et al. (2018). Automatic generation control of nuclear heating reactor power plants. *Energies (Basel)* 11 (10): 2782. https://doi.org/10.3390/en11102782.

24 Iwatsuki, J., Kunitomi, K., Mineo, H. et al. (2021). Overview of high temperature gas-cooled reactor. In: *High Temperature Gas-Cooled Reactors*, 1–16. JSME Series on Thermal and Nuclear Power Generation. https://doi.org/10.1016/B978-0-12-821031-4.00001-4.

25 International Atomic Energy Agency (2007). *Status of Nuclear Desalination in IAEA Member States*. IAEA.

26 Khamis, I. (2012). *Efficient Water Management in Water Cooled Reactors*. International Atomic Energy Agency.

27 IAEA TECDOC Series (2015). *New Technologies for Seawater Desalination Using Nuclear Energy*. International Atomic Energy Agency.

28 Mukhtar, A., Saqib, S., Ullah, S. et al. (2021). Nuclear desalination. *Sustain. Mater. Syst. Water Desalin.* 121–135. https://doi.org/10.1007/978-3-030-72873-1_8.

29 Hamed, O.A. (2005). Overview of hybrid desalination systems—current status and future prospects. *Desalination* 186 (1–3): 207–214. https://doi.org/10.1016/j.desal.2005.03.095.

30 Khan, S.U.D., Khan, S.U.D., Haider, S. et al. (2017). Development and techno-economic analysis of small modular nuclear reactor and desalination system across Middle East and North Africa region. *Desalination* 406: 51–59. https://doi.org/10.1016/j.desal.2016.05.008.

31 Emamdoost, N., Jafarian, A., and Kouhikamali, R. (2020). Implementing multiple-effect distillation and reverse osmosis thermal coupling to improve desalination process performance in combined water and power plants. *Energy Convers. Manage.* 221: https://doi.org/10.1016/j.enconman.2020.113176.

32 Eke, J., Yusuf, A., Giwa, A., and Sodiq, A. (2020). The global status of desalination: an assessment of current desalination technologies, plants and capacity. *Desalination* 495: https://doi.org/10.1016/j.desal.2020.114633.

33 International Atomic Energy Agency (2007). *Economics of Nuclear Desalination: New Developments and Site Specific Studies*. IAEA.

34 Conti, M.C., Labollita, S., Chocrón, M., and Halpert, S. (2012). Estudio Técnico de Factibilidad del Acople de una Planta Térmica de Desalinización de Aguas al Reactor Nuclear. CAREM 25.

35 Ghazaie, S.H., Sadeghi, K., Fedorovich, E. et al. (2020). Assessment of the integrated nuclear plant for electricity production and seawater desalination in Iran. *Desalin. Water Treat.* 188: 20–30. https://doi.org/10.5004/dwt.2020.25338.

36 Dow Water & Process Solutions 2016. Rosa (c). Reverse Osmosis Analysis Design Software. Dow Chemical.

37 International Atomic Energy Agency (2001). *Safety Aspects of Nuclear Plants Coupled with Seawater Desalination Units, IAEA-TECDOC-1235*. Vienna: IAEA.

38 Jamaly, S., Darwish, N.N., Ahmed, I., and Hasan, S.W. (2014). A short review on reverse osmosis pretreatment technologies. *Desalination* https://doi.org/10.1016/j.desal.2014.09.017.

Key Answers for Questions

Chapter 1

1 Global water scarcity and the need for efficient, cost-effective, and sustainable methods of providing clean water are some of the reasons behind this importance. Some advantages that come with membrane technologies include lower energy consumption, higher efficiency, and target-specific contaminants.

2 Membrane distillation (MD), forward osmosis (FO), reverse osmosis (RO), osmotically assisted reverse osmosis (OARO), membrane capacitive deionization (MCDI), microfiltration (MF), ultrafiltration (UF), and nanofiltration (NF).

3 RO operates by applying pressure to force water through a semi-permeable membrane while leaving salts and impurities behind. FO relies on natural osmotic pressure that draws water from a less-concentrated solution to a more-concentrated one through a semi-permeable membrane. The difference lies mainly in the driving forces: external pressure in RO, whereas osmotic pressure in FO.

4 MD uses a hydrophobic membrane separating water vapor from a heated saline solution that then condenses to form distilled or pure water. In contrast, traditional distillation involves boiling and re-condensing of water, which typically requires more energy.

5 OARO combines RO with FO using osmotic pressure to aid the RO process, thereby reducing hydraulic pressures required as well as energy usage and hence improving efficiency in this way.

Water Treatment and Desalination: Membrane Separation using Renewable Energy Resources,
First Edition. Edited by Mohamed Bassyouni.

6 Membrane capacitive deionization (MCDI) is an innovative electrochemical process aimed at desalination which removes ions from water and stores them in capacitive porous electrodes. It is an energy-effective, low-pressure based method, and can selectively remove specific ions, which makes it more advantageous than RO among other techniques.

7 Considerations include contaminant types, required water quality, energy efficiency, operational costs, durability, as well as environmental impact.

8 Membrane technology can be used to improve energy efficiency, reduce operational costs, and treat different sources of water while reducing environmental impacts, making desalination processes more accessible and sustainable at the global level.

Chapter 2

1 Reverse osmosis, forward osmosis, electrodialysis, and nanofiltration. Of these, reverse osmosis is the process with the highest desalination efficiency of up to 99% rejection.

2 Organic polymers, for example, polyamide, are the most common desalination membrane materials, especially for reverse osmosis membranes. However, they are prone to fouling and have weak resistance to chemicals. Mixed-matrix membranes that incorporate functional nanomaterials like metal–organic frameworks or inorganic materials into the polymer matrix in the selective membrane layer are emerging materials being explored for membrane desalination.

3 Thin-film composite (TFC) membrane has a multilayer membrane design whereby a very thin active layer is coated on the top of a porous substrate that serves as mechanical support. This design allows the active/selective membrane layer to be made extremely thin, in the thickness range of a few hundred nanometers or even much less, such that the water flux through the active layer could be enlarged significantly. Essentially, it enables the same membrane area to achieve higher water flux, offering much higher process efficiency.

4 Scanning electron microscopy (SEM) and atomic force microscopy (AFM) are the key characterization instruments for obtaining the surface morphological details of a TFC membrane.

5 Water flux and salt rejection. Water flux is measured as the surface area normalized volumetric flow rate, and salt rejection is calculated as the fraction of salt ions remaining in the retentate without permeating through the membrane.

Chapter 3

1
- The first law of thermodynamics (FLT) states that energy can be transferred from one form to another but can neither be created nor be destroyed. It quantifies energy transferred in the system regardless of its quality using energy analysis.
- The second law of thermodynamics (SLT) identifies the energy flow direction from the higher energy content to the lower one, which clarifies the energy quality or exergy.

2 Exergy is a Greek word consisting of ex, which means from, and ergon, which means work.

3 There are two common types of exergies: chemical and thermomechanical exergies. Chemical exergy is defined as the maximum theoretical work that can be obtained from the system when the concentration (W) of each component in the system differs from its concentration (W_0) at environmental temperature and pressure. The thermomechanical (physical) exergy, on the other hand, can be defined as the maximum work that can be attained from the system when the temperature (T) and pressure (P) of the system differ from the temperature (T_0) and pressure (P_0) of the dead state.

4 Thermodynamic analysis is a known methodology for optimizing thermal systems' performance using energy or exergy methods.

5
- Energy analysis is based on FLT, which characterizes desalination systems in terms of energetic efficiency and compares the output energy relative to the input energy.
- Exergy analysis evaluates the quality of the usable energy by comparing system and environmental conditions.

6
- Conventional exergy analysis provides useful details about the system's thermodynamic performance and effectively determines each of the components' exergy destruction, i.e., exergy loss or energy quality loss. In other words, each component is treated as an independent entity.

- Advanced exergy analysis provides additional information compared to conventional exergy analysis. In advanced exergy analysis, each system component's rate of destruction is divided into avoidable, unavoidable, endogenous, and exogenous categories.

7
- Endogenous exergy destruction is the exergy destruction that occurs within the k-th component when operating under real conditions while other components operate under ideal conditions.
- Exogenous exergy destruction occurs due to the exergy destruction in other components when operating under real conditions while the k-th component operates under ideal conditions.
- Avoidable exergy destruction is the part of exergy destruction that reveals the largest potential for system improvement.
- Unavoidable exergy destruction is defined as the least exergy destruction that cannot be decreased.

8
- Total exergy is the maximum/minimum theoretical work obtainable/consumed when the system and the surroundings reach complete equilibrium (physical, chemical, and mechanical). It comprises physical, chemical, potential, and kinetic exergy.

$$Ex_t = Ex_{PH} + Ex_{CH} + Ex_{po} + Ex_{ke}$$

9 $P_o = 1\,\text{atm}$, $S_{wo} = 35\,\text{g/L}$, $T_o = 25°\text{C}$.

10 Exergy destruction is considered as the exergy loss due to the entropy generation in the system.

11 Fuel exergy can be defined as the exergy that is consumed in the system to produce a desired product.

12 Product exergy is the useful output exergy of the required product.

13 The second law efficiency of the desalination plant is defined as the ratio of the minimum work of separation required for the desalination process to the actual work consumed.

$$\eta_{II} = \frac{W_{min}}{W_{act}}$$

14 The minimum work of the desalination process is the least reversible work that is required for the separation process of a desalination system when the exergy destruction equals zero.

15 The endogenous avoidable exergy destruction is the exergy destruction that occurs due to the irreversibilities in the k-th component, which can be decreased by k-th component performance improvement.

16 The endogenous unavoidable exergy destruction is the exergy destruction that cannot be minimized by technical enhancement.

17 The exogenous avoidable exergy destruction is the exergy destruction that occurs due to irreversibilities in the system and other components excluding the k-th component, which can be avoided by system configuration change.

18 The exogenous unavoidable exergy destruction is the exergy destruction that occurs due to irreversibilities in the system and other components excluding the k-th component, which cannot be avoided.

19 An exergy point that the reverse osmosis process for water desalination suffers from is the exergy destruction across throttling valves. A study reported that replacing throttling valves with pressure exchangers that convey pressure energy from a high-pressure stream to a low-pressure stream can decrease the losses in exergy by 94% and enhance second law efficiency by increasing it to 4.9% instead of 4.3%.

20 The highest destruction in the FO system is in the pumping system.

21 The maximum entropy generation contribution was in the membranes by 80%, followed by concentrated and diluted channels by 16% and then membrane boundary layers by 4%.

Chapter 4

1 Solar water evaporation technology utilizes solar energy, a rich and ubiquitous resource, to address the water and energy crisis in an environmentally friendly manner. This technology employs solar-driven seawater desalination systems that not only produce clean freshwater but also generate electricity through various energy conversion methods, such as photovoltaics (PVs), thermoelectric, and mechanical energy conversion. By integrating water and energy technologies, this approach reduces the carbon footprint while enhancing the efficiency of water and power supply.

2 Solar seawater desalination uses sun's thermal energy to evaporate seawater, and then collects the freshwater through a condensation process. This method

includes solar stills and solar collectors. Solar stills directly heat water using sunlight, while solar collectors gather and concentrate solar energy to heat water, producing steam.

3 The combination of PV technology and seawater desalination works by capturing solar energy with PV cells and converting it into electrical energy. Simultaneously, the waste heat generated by the PV cells is used to drive a multistage membrane distillation (MD) system for seawater desalination. The advantages of this system include: (1) efficient utilization of the full solar spectrum; (2) simultaneous production of freshwater while effectively cooling the PV cells, enhancing their efficiency; (3) reduced carbon footprint and land use, achieving the dual production of water and electricity.

4 The thermoelectric effect refers to the phenomenon where a temperature gradient across thermoelectric materials generates a voltage. In solar water evaporation systems, thermoelectric generators are integrated into the solar evaporators. By utilizing the temperature gradient between the evaporator surface and the cooling water, the thermal energy is converted into electrical energy, enhancing the system's overall energy utilization efficiency. This approach reduces heat loss while increasing water evaporation rates and power output.

5 "Blue energy" refers to the energy harnessed from the salinity difference between seawater and freshwater. In solar water evaporation systems, continuous evaporation increases the salt concentration near the interface compared to the bulk water, creating a salinity gradient. This gradient drives ion diffusion through ion-selective membranes, generating electricity. This method leverages the high-salinity gradient produced during evaporation to enhance power output, achieving efficient energy conversion.

6 The integration of mechanical energy conversion into solar water evaporation systems uses the pressure generated by steam during evaporation to act on piezoelectric materials. This pressure causes mechanical deformation, which is then converted into electrical energy through the piezoelectric effect. Such systems include key components like solar interface evaporators, liquid mediums, piezoelectric materials, electrodes, and circuits. The generated steam exerts pressure on the piezoelectric materials, creating a potential difference collected as electrical energy, thus achieving efficient solar energy conversion.

7 Solar seawater desalination faces several challenges in practical applications, including high initial construction costs, solar energy utilization efficiency being affected by weather and geographical location, and the need for specialized technical support for the maintenance and operation of desalination equipment. Additionally, in some areas, the large-scale application of desalination systems may be limited by land resource availability and environmental impacts.

Chapter 5

1 Renewable energy, such as wind and wave power, is used to provide the electricity required for membrane desalination processes like reverse osmosis. It is considered sustainable because it reduces reliance on fossil fuels, lowering greenhouse gas emissions, and offers a renewable and environmentally friendly solution to address the global water scarcity crisis.

2 Fluctuations in renewable energy sources lead to inconsistent power supply, affecting the stability of the desalination process, especially in systems like reverse osmosis that require constant pressure. Strategies to mitigate these effects include using energy storage systems (batteries, pumped hydro), implementing backup power sources (like diesel generators), and designing modular desalination systems that can adjust capacity based on available power.

3 Key site considerations include wind speed and consistency, proximity to water sources, land availability, and environmental regulations. These factors directly impact plant performance, as higher and more consistent wind speeds improve energy generation and reduce the need for energy storage, while proximity to water sources and suitable land reduce infrastructure costs and environmental impact.

4 Direct-contact membrane distillation (DCMD) is a thermal-based desalination process where heated saline water passes through a membrane that allows water vapor to pass while retaining salts. Waste heat from wind turbines can be harnessed by using a heat exchanger to transfer the heat to the saline water, raising its temperature and facilitating the distillation process. This improves the overall efficiency of the system by utilizing energy that would otherwise be lost.

5
- **Challenges**: Intermittent wind energy requires reliable energy storage or backup power; desalination plants need to be optimized to match variable wind power output; ensuring energy efficiency and economic feasibility is also a concern.

- **Solutions**: Incorporating advanced energy storage systems like batteries or pumped hydro can mitigate power fluctuations. Developing control systems to adjust desalination operation based on energy availability and integrating grid power for backup can improve reliability.

6 Energy storage is crucial to smoothing intermittent wind energy and ensuring the continuous operation of desalination plants. Common storage technologies include batteries, which store excess electricity, and pumped hydro storage, where excess energy is used to pump water into elevated reservoirs, which can be released later to generate power.

7 Reverse osmosis is a membrane-based process that removes salts and impurities from seawater by forcing it through a semi-permeable membrane under high pressure. Wind energy can be integrated into RO systems either by generating electricity that powers the RO pumps or by directly driving the pumps via mechanical coupling. In the latter case, energy storage may be used to manage fluctuations in wind energy.

8 Direct-driven desalination systems mechanically couple wind turbines or wave energy converters directly to the pumps used in desalination, bypassing the need for electricity generation. This reduces energy losses associated with electrical conversion and improves system efficiency. Traditional setups rely on electricity generated by wind turbines to power the desalination process, often requiring energy storage systems to manage fluctuations in wind energy.

9 Hybrid renewable energy systems improve the reliability of desalination plants by combining multiple energy sources, ensuring a more consistent power supply. For example, solar energy can complement wind energy during periods of low wind, reducing the need for energy storage and increasing the overall system uptime. This leads to more efficient water production and potentially lower operational costs by optimizing the use of renewable energy sources throughout the day.

10 • **Advantages of wind-powered systems**: Wind turbines are a mature technology and can be economically deployed in coastal regions. They provide clean energy, incur low operational costs, and can produce water continuously by storing freshwater during periods of high wind energy.

- **Advantages of wave-powered systems**: Wave energy is more consistent than wind energy, especially in coastal regions. It provides a higher energy density and directly converts mechanical energy into hydraulic pressure for desalination, making the process more efficient.
- **Disadvantages**: Wind energy is intermittent, requiring energy storage solutions, which add to the cost and complexity. Wave-powered systems, though stable, are still an emerging technology and can be expensive to install and maintain in harsh marine environments.

11 Scaling up wave-powered desalination systems presents challenges, such as the need for robust structures that can withstand harsh marine environments, the complexity of managing variable wave energy, and the higher initial capital costs. Potential solutions include developing modular systems that can be expanded gradually, improving energy recovery devices to enhance efficiency, and advancing materials and designs to reduce maintenance and operational costs.

12 The DELBUOY system uses ocean wave motion to power a pump that pressurizes seawater, which is then processed through an RO membrane. Factors influencing its performance include wave height and frequency, the efficiency of the pump, the capacity of the buoy system to withstand rough seas, and the overall system design, such as the size and configuration of the membrane modules.

13 Energy recovery devices capture the energy from the high-pressure brine produced during desalination and use it to pressurize the incoming seawater, reducing the overall energy consumption of the system. Examples of ERDs include the pressure exchanger (PX), turbochargers, and Clark pumps. These devices significantly improve the energy efficiency of desalination processes by reducing the need for additional energy input.

14 Environmental considerations include potential impacts on marine ecosystems from brine discharge, noise pollution from turbines, visual impacts on the coastline, and land use requirements for onshore components of the system. Careful planning is necessary to mitigate these effects, including locating desalination plants away from sensitive areas and designing brine disposal systems that minimize ecological disruption.

Chapter 6

1 Geothermal energy has a superior capacity factor when compared to other renewable energy sources. The capacity factor for geothermal energy can reach up to 95% if operated in a closed-loop configuration. The capacity factor for the common renewable energy sources is less than one-third of the value for geothermal energy. For instance, the capacity factors for wind, tidal, and photovoltaic cells are 30%, 30%, and 20%, respectively. A high capacity factor motivates the scalability and investment of a technology. Therefore, in countries where geothermal sources are available near an abundant water source, geothermal energy can be considered a more effective solution over other renewable energy sources.

2 Geothermal water with a temperature range of 40–70°C is suitable for low-temperature desalination technologies, including membrane distillation, solar stills, humidification–dehumidification, and high vacuum multi-effect distillation. Medium-grade sources (>70°C) are suitable for multistage flash and multi-effect evaporation processes. The highest enthalpy sources (up to 200°C) can be used for cogeneration processes in large-scale desalination plants.

3
- Geothermal energy can be used to generate the electrical power necessary for operating the high-pressure pumps in RO-based desalination plants.
- The hot geothermal fluid can heat the feed used in thermal technologies, such as multi-effect and multistage flash evaporators.
- The cooled geothermal fluid can be further desalinated in a membrane distillation cell before being re-injected back into the reservoir.

4 Physical water scarcity means the depletion of water resources due to excess need and inadequate water management, whereas economic water scarcity is the lack of feasible technologies or resources to produce potable water, which is the case in least-developed countries with high-salinity water resources.

Geothermal energy can, therefore, be considered as a viable solution to eliminate economic water scarcity because it is a low-cost energy source. Interestingly, the regions striving against high water stress correspond with high-temperature geothermal sources.

5

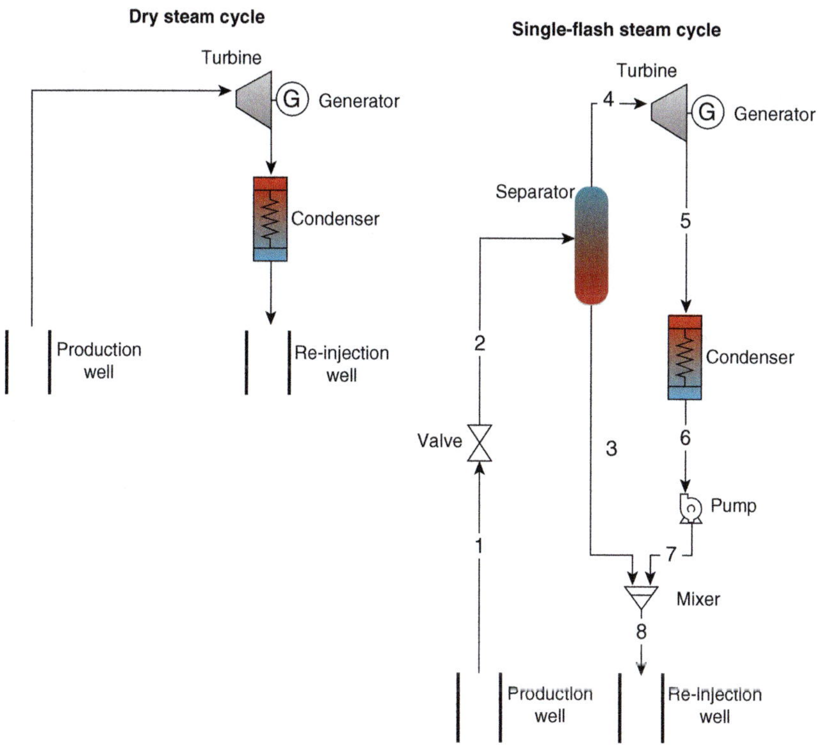

6 The primary criteria for choosing a quality geothermal resource include:

- for a cogeneration scheme, high temperature is necessary for improved power generation efficiency;
- a significant amount of stored thermal energy to extend the life of the reservoir;
- availability of potential energy difference between the production wells and the disposal wells to enable discharge under gravity;
- generates almost neutral pH fluids for low rates of corrosion in power plants and wells;
- high permeability to guarantee sufficient outputs from each well;
- a minimal chance of hydrothermal outbursts and volcanic activity.
- the distance from power transmission cables or loads.

7 For geothermal desalination facilities to be implemented successfully, crucial elements must be taken into consideration. Land use, geological dangers, atmospheric releases, waste heat, water footprint, and socioeconomic issues are the main concerns. Certain events and their impacts may have little chance of happening. Regardless, the environmental impacts of geothermal power and water facilities are subject to a variety of environmental restrictions.

Chapter 7

1 The global demand for freshwater is driven by population growth, changing consumer behavior, and economic expansion. These factors have increased water usage eightfold in the last century. Desalination is considered vital because only 0.3% of Earth's freshwater is easily accessible, and over half of this is already in use. Desalination provides a reliable method to converting saline water into freshwater, especially for regions with limited natural water resources.

2 Conventional desalination is energy-intensive, relying heavily on fossil fuels, which are costly and environmentally harmful. Bioenergy, especially biogas, offers a renewable alternative by providing the necessary heat and electricity for desalination. Biogas reduces carbon emissions, minimizes operational costs, and supports waste management. It is particularly useful for making desalination more sustainable and cost-efficient.

3 The main sources of bioenergy include:
- **Agricultural residues:** Crops like wheat straw and corn stover can be burned or biologically treated to produce energy.
- **Municipal solid waste (MSW):** Organic MSW serves as a feedstock for biogas, reducing landfill waste.
- **Forestry products:** Cellulose-rich lignocellulosic waste is pretreated for bioenergy production.
- **Animal waste:** Anaerobic digestion of manure reduces environmental pollution and generates biogas.
- **Algae:** Rapidly growing algae provide a high yield of bioenergy and are efficient in anaerobic digestion.
  These sources help replace fossil fuels, lowering greenhouse gas emissions, and enhance the circular economy.

4 Biogas is produced through anaerobic digestion, where organic matter decomposes under oxygen-free conditions. The efficiency of biogas production depends on:

- **Digester design:** Determines mixing and fermentation efficiency.
- **Temperature:** Optimal ranges (35–55°C) maximize microbial activity.
- **Hydraulic retention time (HRT):** Time spent by the feedstock in the digester affects gas yield.
- **C/N ratio:** A balanced carbon-to-nitrogen ratio (20–30.5) enhances digestion.
- **pH levels:** Methanogenic bacteria require a pH range of 6.8–8.5.

  Optimizing these factors ensures higher biogas yields and improved system efficiency.

5 The main desalination methods are:
- **Thermal methods (e.g., multistage flash distillation)**: They use heat to evaporate and condense water. These are simple but energy-intensive.
- **Membrane methods (e.g., reverse osmosis)**: They use a semi-permeable membrane to filter water. They are efficient but require high electricity.
- **Electrodialysis:** This method uses an electric field to move ions through membranes. It is useful for brackish water but limited to specific applications.

  These methods face challenges, such as high energy consumption, environmental impacts from brine disposal, and substantial operational costs.

6 The integration of biogas with desalination systems provides:
- **Economic advantages**: Lower operational costs due to the availability of low-cost biogas feedstock. Biogas-powered systems reduce energy costs by up to 30% compared to fossil fuels.
- **Environmental advantages**: Significant reduction in greenhouse gas emissions (up to 70%) and better waste management through anaerobic digestion. It supports a circular economy by converting organic waste into renewable energy.

7 Challenges include:
- **Infrastructure adaptation:** Existing systems are designed for fossil fuels, requiring modifications for biogas integration.
- **Feedstock variability:** Inconsistent availability and the quality of feedstocks affect biogas yields.
- **Biogas purification:** Removal of impurities like CO_2 and H_2S is critical for efficient energy use.

Future research should focus on:
- optimizing anaerobic digestion parameters (e.g., pH, temperature);
- developing advanced biogas purification technologies; and

- conducting life-cycle assessments to evaluate environmental impacts.
- enhancing public awareness and governmental support to promote adoption.

Chapter 8

1 The primary focus is to provide a comprehensive overview and detailed exploration of various emerging desalination techniques that utilize waste heat and highlight the potential of these innovative methods to revolutionize the field of water desalination, making it more sustainable and efficient.

2 MD utilizes a hydrophobic membrane to separate water vapor from saline solutions. Waste heat is employed to facilitate the evaporation process, allowing water vapor to pass through the membrane while retaining the salt and other impurities. This method effectively uses waste heat to produce freshwater.

3 Multi-effect distillation (MED) distinguishes itself by using successive stages of evaporation and condensation to desalinate seawater. Each stage operates at a lower pressure and temperature, utilizing waste heat sources to power the process. This method enhances energy efficiency and increases freshwater production.

4 AD relies on a thermally driven process where water vapor adheres to solid adsorbent materials. Waste heat is used to regenerate the adsorbent, releasing the adsorbed water vapor, which is then condensed to produce freshwater. This method leverages waste heat to drive the adsorption–desorption cycle efficiently.

5 HDSs integrate multiple desalination technologies, such as reverse osmosis (RO) and thermal processes, to optimize both energy efficiency and water production. By combining different methods, HDS can utilize waste heat more effectively and achieve higher overall performance compared to single-technique systems.

6 Low-temperature thermal desalination (LTTD) leverages low-grade waste heat sources, such as solar energy or industrial waste heat, to evaporate seawater at relatively mild temperatures. This method takes advantage of the abundant low-temperature waste heat to produce freshwater with minimal energy input.

7 Waste heat-powered desalination techniques offer practical solutions to mitigate water scarcity by providing a sustainable and energy-efficient means of producing freshwater. By utilizing waste heat, these methods reduce the dependency on traditional energy sources, lower operational costs, and enhance the availability of clean water in regions with abundant waste heat resources.

8 Continuous research and development are crucial for refining the efficiency, scalability, and affordability of emerging desalination technologies. By advancing these aspects, the technologies can become more viable and accessible, bringing us closer to a sustainable future where clean water is readily available for all.

9 Emerging desalination techniques address the challenge of energy consumption by utilizing waste heat, an often-overlooked resource, to power desalination processes. This approach reduces the reliance on conventional energy sources, thereby lowering overall energy usage and making desalination more sustainable and cost-effective.

10 Waste heat-powered desalination techniques are particularly relevant in regions endowed with abundant waste heat resources, such as industrial zones and areas with significant solar energy potential. These regions can harness the available waste heat to produce freshwater, addressing local water scarcity issues effectively and sustainably.

11 Emerging desalination techniques powered by waste heat contribute to achieving the Sustainable Development Goals (SDGs) by promoting sustainable water management and access to clean water (SDG 6). By utilizing waste heat, these methods reduce energy consumption and operational costs, aligning with the goals for affordable and clean energy (SDG 7) and industry innovation (SDG 9). Furthermore, these techniques support climate action (SDG 13) by minimizing the carbon footprint associated with traditional desalination processes, thereby contributing to a more sustainable and resilient future.

Chapter 9

1 Seawater contains valuable minerals like potassium, magnesium, calcium, lithium, strontium, and bromine, which are economically recoverable using membrane crystallization due to controlled crystallization under optimal conditions.

2 Membrane crystallization offers advantages over traditional methods by selectively crystallizing minerals at lower energy costs, higher efficiency, and reduced environmental impact, making large-scale mineral recovery economically viable.

3 Membrane crystallization can concentrate lithium ions effectively by controlling the superstation and crystallization processes, which is crucial given lithium's low concentration in seawater.

4 Supersaturation is essential in membrane crystallization, as it drives the crystallization process by inducing nucleation, which is required for forming solid crystals from dissolved minerals in seawater.

5 Hydrophobic membranes create a selective barrier that prevents liquid penetration while allowing vapor to pass, aiding in the controlled crystallization of dissolved minerals.

6 Membrane crystallization can economically benefit from the high purity and controlled crystallization processes, reducing posttreatment costs, offering sustainable mineral recovery, and supporting a circular economy.

7 Membrane crystallization minimizes land disruption and pollution associated with traditional mining, offering an eco-friendly alternative for mineral recovery from seawater, especially in resource-limited regions.

8 Materials like PVDF, PSF, and PP have unique thermal and chemical resistances, influencing membrane crystallization efficiency by affecting fouling, permeability, and durability.

9 Challenges include membrane fouling, scaling, pore wetting, and energy costs, which require membrane modification and operational optimization for efficient mineral recovery.

10 Antisolvent crystallization introduces an antisolvent to enhance crystal formation, improving particle size control and morphology in pharmaceutical and coating applications.

11 Membrane distillation crystallization (MDCr) combines membrane distillation with crystallization, allowing for controlled supersaturation, enhanced particle quality, and process separation efficiency while using lower temperatures.

12 Sodium, calcium, magnesium, and potassium are commercially recovered due to their high concentrations, while lithium and other minerals are promising due to advancements in crystallization technologies.

13 By recovering water and minerals from wastewater, membrane crystallization promotes zero liquid discharge (ZLD), minimizing waste and environmental pollution through comprehensive resource utilization.

14 Higher feed temperatures increase crystallization flux by raising the vapor pressure across the membrane, facilitating faster mineral recovery and improving process efficiency.

15 Polymeric membranes are chosen for their chemical stability, resistance to fouling, and thermal durability, which are critical for sustainable and effective crystallization processes.

16 An optimal flow rate reduces temperature polarization, enhances heat transfer, and supports crystal size control, which is crucial for quality crystal production.

17 Recovered minerals have applications across agriculture, pharmaceuticals, cosmetics, aerospace, and electronics, contributing to various industries sustainably.

18 Membrane crystallization allows for controlled crystallization processes that enhance purity, particle size, and morphology, benefiting pharmaceutical ingredient production.

19 Membrane crystallization aids in efficiently managing desalination brines by extracting valuable minerals, reducing environmental risks, and supporting circular economy principles.

20 A reduced air gap improves flux by increasing the temperature gradient, enhancing mineral recovery rates and supporting high-efficiency crystallization.

Chapter 10

1 The RO membrane system is a pressure-driven process that depends on applying high pressure on the membrane's feeding side to overcome the osmotic pressure and produce high-purity water. ED and EDR are voltage-driven processes. They need power supply to produce electric potential, which is the influencing factor in separating salts from water based on the positive and negative charges of ions.

2 There are two methods by which heat is transported via the membrane matrix in MD: conduction heat which represents 20–50% of the heat loss because there is no induced flux through the membrane and latent heat of vaporization according to mass transfer through the membrane.

3 The reverse osmosis (RO) technique requires pressure to separate salts from water through the membrane, while the forward osmosis (FO) technique is a technology based on differences in salinity gradients. The FO system consists of a selective membrane and two side liquids: feed which is considered freshwater and the other side is a draw solute, which is mostly hypersaline (brine) water. According to the osmotic pressure, the freshwater migrates to the draw solute to dilute it. The pressure-retarded osmosis (PRO) technique facilitates diluted solution migration through a membrane via osmosis from the feeding side to the brine stream under pressure.

4 The brine pressure is used to pressurize the feedwater, where the low-pressure feedwater fills the rotor chamber and the high-pressure brine pressurizes and displaces the feedwater.

5 For example, the system of RO/DCMD/RED, where the RO single-stage electrical energy consumption was 2.3 kWh/m^3 without using PX and after using PX, it was 1.54 kWh/m^3. However, using the RO brine concentration as a feed for DCMD provided no sign of electrical energy consumption, but the thermal energy consumption was observed to be 4.2 kWh/m^3 at a feeding temperature of 60°C. The integration with RED using high-concentration brine from DCMD produced the lowest energetic point at a temperature of 40°C, which was 2.9 kWh/m^3, with a reduction of 23% compared with the energetic point of 4.23 kWh/m^3 without integration with RED.

Chapter 11

1 According to osmotic energy, water moves from a low-salinity side (feed solution) toward a high-salinity side (draw solution) through a semi-permeable membrane. The accumulated water in the draw side can drive a turbine to produce mechanical energy, which is further converted into electrical energy.

2 PRO utilizes the natural salinity gradient between water sources and does not emit greenhouse gases or pollutants, making it a renewable, sustainable, and environmentally friendly source of energy. Moreover, it can be hybridized with desalination processes, improving energy recovery and enhancing the sustainability of water management practices, especially in regions with water scarcity.

3 The performance of PRO depends on the higher water flux rate across the membrane, affected by the membrane's characteristics such as thickness, porosity, tortuosity, and surface charge; therefore, the membrane has a crucial role in the PRO process. Selectivity of membranes is important to maintain the osmotic process by allowing water passage while rejecting solutes. Furthermore, the membrane material impacts its stability, mechanical strength under higher hydraulic pressure, and its compatibility with other chemicals. Therefore, optimizing the properties, design, and material of the membrane in a cost-effective way can directly affect the efficiency of PRO.

4 Fouling is a substantial challenge in the PRO process, caused by different types of foulants accumulating on the membrane's surface. This buildup decreases the water flux through escalating internal concentration polarization, thereby directly impacting the permeability of the membrane and overall efficiency. As fouling progresses, it further increases energy costs and maintenance requirements. Researchers have investigated several strategies to improve fouling resistance, such as modifying the orientation of the membrane, optimizing operational conditions, pretreating feedwater to remove particulates, and integrating anti-fouling coatings or nanomaterials on membranes. Employing these strategies aims to mitigate fouling and enhance PRO's performance.

5 For economic viability, PRO should generate a competitive amount of net energy output, and the expenses related to the advanced membrane, fouling mitigation, and maintenance should be reduced to make it a more attractive source of energy for large-scale applications.

6 There are several challenges for implementing PRO on a large scale including the performance of the membrane such as its durability, stability, scalability, water permeability, salt rejection, fouling resistance, and concentration polarization, which must be optimized. Design of the system like optimizing flow rate, pressure and maximizing harvesting energy while minimizing the costs, is another challenge that needs to be mitigated along with the pretreatments and its operational costs. Addressing these challenges is crucial to enhancing the feasibility of the PRO system on a large scale.

Chapter 12

1 TRNSYS software was used for modeling thermal desalination systems.

2 The study focused on air-gap membrane distillation (AGMD).

3 Hot climates, like in Adrar, produce the most distillate water (18.08 L/h), while cold climates, like in Whitehorse, produce the least (4.12 L/h).

4 Ethylene glycol improved heat transfer and increased distillate production by 27.7%.

5 Increasing heat exchanger efficiency from 20% to 90% raised distillate production from 7.63 to 40.63 L/h.

6 The AGMD pilot plant used a membrane surface area of 14.4 m^2.

7 Solar collectors were analyzed in Adrar (Algeria), Port Said (Egypt), and Whitehorse (Canada).

8 The maximum distillate production in Adrar was 18.08 L/h.

9 The water–ethylene glycol mixture outperformed water alone due to its higher heat capacity, improving distillate production.

10 The study concludes that fluid choice and climatic conditions significantly impact the performance of thermal desalination systems.

Chapter 13

1 The costs include capital costs (piping and valves, instruments and controls, tanks and frames, miscellaneous, pumps, membrane, and pressure vessel) and operating costs (membrane replacement, energy, chemicals, concentrate disposal, maintenance, and pretreatment).

2 Net present value (NPV), internal rate of return (IRR), and payback period (PBP).

3 They include the capital costs (mainly equipment) and operating costs (energy, chemicals, waste disposal, replacements, and labor), while the benefits are mainly in the form of revenue from water sales.

4 The factors could be based on conventional processes which may cause air pollution, land appropriation, and/or water improvement, as well as from resource recovery-based processes, such as thermal energy recovery, incineration, reclaimed water, phosphorus, and ash reuse.

5 Membrane technologies can be used to reclaim wastewater and capture CO_2, though selectivity and permeance are challenging to attain. Current membrane processes are not only cleaner than traditional methods but also use toxic chemicals as well as nondegradable materials. Sustainable membrane materials are critical for development. Salinity gradient power (SGP) is a promising clean energy alternative, and nanopore-based membranes have the capability of a high level of energy harvesting.

6 Renewable energy is being integrated into water treatment systems, focusing in particular on desalination. A study showed that hybrid systems, with solar and wind power mixed together, are more efficient than single-source systems. A study optimizing a brackish water RO system discovered that hybrid systems produce higher water volumes and are overall more efficient than a conventional RO system. Moreover, another study was conducted to examine the production of green hydrogen by desalination and propose four configurations utilizing water electrolysis based on PEM and renewable energy-powered desalination. It was found that wind-powered seawater reverse osmosis (SWRO) systems were the most cost-effective. The third study looked at advanced renewable energy-driven water desalination technologies using solar energy. Through the assessment of the efficiency of different hybrid desalination systems, they found that hybrid systems can have extremely high energy savings. A study also compared biogas steam reforming's (BGSMR's) environmental footprint with that of all other biohydrogen-generating methods. BGSMR has a slightly more carbon footprint than the MS but has the potential for waste valorization and less water consumption, making it a more sustainable option.

Chapter 14

1 We call wastewater any water that is affected greatly, in terms of quality, by anthropogenic influence. This water is discharged from houses, businesses, and industries, among other sources, and contains several as well as variable

contaminants, such as physical, chemical, and biological pollutants. Wastewater requires special treatment before it can be safely released into the environment or before it can be reused.

2 A microbial fuel cell, also named MFC, is a type of electrochemical or Galvanic cell that uses electroactive microorganisms to generate electricity from organic compounds. Microbial fuel cells can treat any type of wastewater because they remove organic contaminants while generating electricity at the same time. These devices are a sustainable option with multiple applications.

3 The microorganisms with the capacity of generating electricity are called exoelectrogens; other names for these living forms include electrogenic or electroactive microorganisms. These beings have the specific ability to transfer electrons to their outside so that they can be received by a final acceptor; in microbial fuel cells, these electrons are received by the anode of the cell. Some renowned and powerful exoelectrogens are *Geobacter* sp. and *Shewanella oneidensis*.

4 The electron transport chain is a mechanism that can be found as part of cellular respiration and photosynthesis. It is used as an alternative way by which electrons move to the outer membrane of the cell, essential for the generation of energy in the form of ATP. This process consists of a series of electron carriers and cytochromes to make it work.

5 As the name suggests, a single-chamber microbial fuel cell is made of one cell which consists of an anode and a cathode separated by a proton-exchange membrane (PEM) that prevents the direct mixing of substrates and oxidants, helping maintain a potential difference. The external circuit connects the anode and the cathode, allowing electrons to flow from the first to the second, generating an electric current.
On the other hand, a double-chamber microbial fuel cell has two separate compartments, one for the anode and one for the cathode, which are separated by a proton-exchange membrane (PEM). These types of cells allow the study of different factors that can affect the generation of electrical energy because they make it easier to alter conditions on one of the cells without altering the other one. The external circuit connects the anode and the cathode, generating an electric current.

6 The electrodes in microbial fuel cells are responsible for the correct electron transfer mechanism that will lead to electrical energy production in the system; at the same time, they are the site where the exoelectrogenic

microorganisms will grow. Therefore, it is necessary to consider some characteristics while designing a microbial fuel cell. First, in terms of size, it has been reported that having a larger electrode surface area makes a lower power density. Second, in terms of materials, it is important to consider their electrical conductivity, corrosion resistance, chemical stability, large surface area, low cost, and bacterial colonization capacity. Finally, it should be kept in mind that the roles of the anode and the cathode are different, which means that the electrodes may be of different material in each chamber, if needed.

7 Yes, it is possible to escalate microbial fuel cells from laboratory to industrial applications, but it is necessary to consider that doing so will take a large amount of effort and resources, it can be expensive, and it needs to address multiple inconveniences before having a working industrial system. Therefore, it is necessary to promote research that focuses on addressing these disadvantages.

Chapter 15

1 Several factors affect the separation processes by membranes, such as pore size, membrane material, solution/solvent charge, feedwater quality, and operating conditions. These conditions include temperature and pressure.

2 In arid regions, the RO process provides a reliable desalination method due to its high efficiency, relatively lower energy consumption compared to thermal methods, and ability to produce highly purified water. In addition, RO has a smaller environmental footprint compared to thermal desalination methods because it requires lower energy consumption per cubic meter than other methods.

3 ERDs such as pressure exchangers harness mechanical energy from the high brine reject stream and transfer it to the inlet stream. This can reduce the energy requirements by 60%, which in turn reduces the desalinated water cost. This energy-saving capability is valuable for systems utilizing renewable energy sources which often produce lower electricity outputs compared to conventional fossil-fuel plants.

4 Hybrid energy systems address the intermittency of individual energy sources. This integration could provide a more stable and reliable energy supply. This approach is suitable for large-scale desalination facilities that need continuous operation to meet the high-water demand.

5 Conventional energy systems incur lower initial costs due to established power infrastructure. On the other hand, renewable energy systems require higher capital investment for specified components like photovoltaic (PV) panels or wind turbines. However, by achieving lower energy expenses and environmental benefits, renewable systems can accomplish long-term savings.

6 The sizing of renewable energy systems must balance the energy supply with the desalination plant's demand. In addition, energy storage needs have to be considered to ensure continuous operation. Moreover, oversizing can lead to increased capital costs, while undersizing may cause inefficiencies and operational downtime.

7 Renewable energy infrastructure such as PV panels and wind turbines requires regular maintenance. This includes cleaning, inspections, and mechanical repairs to ensure optimal performance. Membrane systems also need ongoing maintenance to prevent fouling, scaling, and degradation, irrespective of the energy source used.

8 Sensitivity analysis evaluates how changes in key economic factors affect the overall costs of water production. This helps decision makers predict risks and adjust strategies to optimize plant performance under different scenarios.

Chapter 16

1 The two primary performance metrics are water flux and salt rejection. Water flux measures the volume of water passing through the membrane. On the other hand, salt rejection assesses the membrane's ability to filter out salts from feedwater.

2 Polymeric membranes are popular due to their affordability and adaptability. However, they are prone to fouling over time. This decreases membrane performance, which requires regular maintenance. In addition, they have low operating temperatures (below 100°C).

3 Fouling occurs when particles, organic compounds, or microorganisms accumulate on the membrane. This decreases the water flow and increases energy use. On the other hand, scaling happens when salts crystallize on the membrane, especially in high-salinity environments. This impacts the membrane's efficiency and lifespan adversely.

4 Chemical cleaning uses acidic or alkaline solutions to break down scaling and fouling effectively. This allows for restoring the water flow and lowering

transmembrane pressure. This process must be carefully managed to avoid damage to the membrane.

5 Air scouring introduces air bubbles to dislodge foulants from the membrane surface. This process is usually combined with chemical cleaning. The main aim is to enhance the removal of contaminants and prolong membrane performance.

6 Backwashing reverses the water flow to remove foulants accumulated on the membrane surface. This helps maintain a steady permeate flow. It is commonly used in UF and NF systems although less in RO systems due to higher pressure requirements.

7 Graphene oxide (GO) membranes provide high water permeability and effective salt rejection due to their unique structure. Their hydrophilic properties also help reduce fouling. However, they are prone to inorganic scaling in high-salinity water, necessitating careful pretreatment.

8 Carbon nanotube (CNT) membranes offer high water flux and strong salt rejection along with improved fouling resistance. However, the high cost and complexity of large-scale CNT membrane production are major barriers to widespread adoption.

Chapter 17

1 Process intensification (PI), the philosophy of process design, aims at the intensification of chemical processes through both efficiency and sustainability. The primary objectives of PI in industrial processes are:
- **Efficiency improvement:** Reducing the size of the equipment used reduces the capital and operational costs, making it possible for PI to increase the efficiency of an industrial process.
- **Energy reduction:** Through PI, it is possible to have a very significant impact in reducing energy use in industrial processes and, therefore, simultaneously reducing carbon emissions and their related problems.
- **Waste minimization:** To encourage a more sustainable and environmentally friendly industry, PIN strategies often involve process redesign consisting of ways to eliminate waste production.
- **Safety enhancement:** In addition, PI can also help make industrial processes safer by reducing the size of the equipment and the amount of hazardous materials utilized.

In a nutshell, PI is a strategy aimed at industry process optimization covering economic, environmental, and safety aspects simultaneously. In the modern world, it is a key tool for industries wanting to effectively and efficiently meet the growing demands for sustainability.

2 The performance of a membrane bioreactor (MBR) used for wastewater treatment can be significantly enhanced by means of process intensification (PI). This can be achieved through several strategies:

- **Membrane material and design:** Novel membrane materials, as well as their novel design, can enhance the selectivity and the permeability of the MBR further toward
- **More efficient separation of contaminants** from the wastewater. For example, MBR performance can be enhanced by the use of nanocomposite or mixed-matrix membranes.
- **Module configuration:** Reduction of membrane fouling, a major downside of MBR operation, is possible with the use of innovative module configurations, such as rotating or vibrating membranes. It can increase the lifespan of the membrane as well as reduce maintenance costs.
- **Process integration:** Such an integration with other unit operations like advanced oxidation and anaerobic digestion provides the possibility of enhanced performance of the overall wastewater treatment process. It can increase water quality and decrease energy costs.
- **Operational optimization:** Improving the performance of the MBR can be achieved by optimizing operational parameters, such as hydraulic residence times or sludge residence times. It can be accomplished through the application of advanced control strategies and real-time monitoring systems.

3 Process intensification (PI) in reverse osmosis (RO) desalination addresses the challenges in seawater desalination by aiming to design new units and methods that dramatically reduce the size of equipment, energy demands, and waste production while enhancing the throughput capacity. This makes the technologies more cost-effective and sustainable. Potential improvements suggested include the use of multimembrane-integrated systems due to their high flexibility, simplicity, and compatibility. These systems can address the challenges and limitations of a single-membrane system and enhance the system's efficiency. Additionally, the use of membrane bioreactors (MBRs) as pretreatment methods has attracted significant attention. MBRs, which are employed in wastewater treatment, utilize low-pressure membranes to filter water from the activated sludge, making the treated water useful for recycling and reusing in industries. This approach makes membrane technologies favorable for future developments in the field.

4 If the stability issues of MXenes when exposed to air and humidity were fully addressed, it could potentially revolutionize water desalination technology. MXenes, with their excellent adsorption capacity, unique properties, and exceptional electrical conductivity, could be used to create highly efficient and environmentally friendly desalination membranes. These membranes could offer superior removal performance and controlled ion permeability, leading to more effective and efficient water desalination processes. Furthermore, the improved stability could enhance the longevity and recyclability of these membranes, making them a more sustainable option for water desalination.

5 The integration of photothermal membranes into solar-powered membrane distillation (SPMD) systems could significantly enhance the future of desalination and wastewater treatment, especially in remote, off-grid environments. Photothermal membranes, which transform light energy to heat, can enhance the rate of evaporation, minimize temperature difference, enhance mass transfer of fluids and vapor, and allow selectivity of heating. This not only improves the overall efficiency of the membrane distillation process but also reduces energy consumption by utilizing low-grade heat sources. Furthermore, SPMD systems require much less power and simplified equipment compared to other systems, making them ideal for use in remote, off-grid environments. With the continuous research and development in this field, it is expected that these systems will become more efficient and widely adopted, providing a sustainable solution to water scarcity.

6 The integration of renewable energy sources into the desalination process significantly enhances its sustainability and feasibility. By using clean energy technologies such as solar, wind, and hydroelectricity, the energy required for desalination can be substantially reduced, thereby promoting a green energy value. This not only lessens the environmental impact but also decreases operational expenditure, aligning with global efforts to transition toward green energy sources. Membrane technologies, specifically reverse osmosis or forward osmosis, further enhance the efficiency of desalination systems. They allow for the intensification of water recovery relative to energy consumption, leading to improved overall performance and minimized environmental costs when integrated with renewable energy sources.
Moreover, the combination of renewable energy sources and desalination processes enables the design and implementation of efficient and synergistic schemes. For instance, the integration of solar energy into desalination can provide a sustainable solution to water scarcity, particularly in arid regions. This approach is not only practical in supplying freshwater in the shortest possible time but also boosts energy independence and stability in the water

infrastructure supply. In conclusion, desalination backed by renewable energy, particularly when combined with membrane technologies, holds great promise for the future of sustainable industrial growth. As industries integrate more of such innovations, they are likely to make strides toward the critical goal of creating a sustainable and secure water future.

Chapter 18

1 Coupling a nuclear reactor with a hybrid desalination plant, such as the combined RO and multi-effect distillation (MED) system, leverages residual thermal energy from the nuclear power plant. This reduces the need for additional energy sources, thereby decreasing greenhouse gas emissions and improving operational efficiency.

2 The hybrid RO + MED configuration optimizes energy use by utilizing RO for brackish water and MED for higher salinity seawater, taking advantage of their respective strengths. RO is more energy-efficient for lower salinities, while MED can use the thermal energy from the nuclear reactor effectively. The integration of both systems reduces the specific energy consumption (SEC) needed for desalination, achieving a combined SEC of around 11.3 kWh/m^3, which is lower than using MED alone.

3 Pretreatment is essential in RO desalination to prevent membrane fouling, which can increase energy consumption and maintenance costs. Typical pretreatment methods include coagulation, flocculation, and the use of scale inhibitors, as well as advanced filtration techniques, such as microfiltration and ultrafiltration.

4 The study estimates a production cost of 1.6 USD/m^3 for water produced by the MED + RO configuration, primarily due to the efficient use of nuclear reactor waste heat and optimized water production processes. Also, the proposed hybrid system can also fulfill the water demand of the NPP.

5 The cost of water production in the MED + RO hybrid plant is distributed across several main components: electricity accounts for 38%, annual investment fees for 30%, chemical pretreatments for 11%, staff costs for 12%, maintenance for 4%, and miscellaneous expenses for 5%. This breakdown highlights the importance of electricity and capital investment in the overall cost structure, underlining the need for efficient energy use and well-planned investment to maintain cost-effectiveness.

Index

a

acetogenesis 179
acidogenesis 179
acrylonitrile butadiene styrene (ABS) 92
actual energy production, PRO 337–338
adsorption-cooling and desalination (ACD) 230
adsorption desalination (AD)
  capacitive deionization technology 229
  composites and innovations 229
  cooling materials 203–204
  cooling systems 230
  energy and heat management 226–227
  heat transfer enhancement 229
  mechanics of 226
  metal–organic networks, ZIF family 228–229
  natural zeolites 228
  research on 227
  system overview 227
  using carbon nanotubes 227–228
  vermiculite 229–230
  water vapor and capillary adsorption 230
advanced exergy analysis 41, 43–45
AGMD *see* air gap membrane distillation (AGMD)
agricultural wastewater 429
agriculture residues 174
air gap membrane distillation (AGMD) 306, 307, 410
  membrane distillation 4–5
  three-dimensional parameters 385–386
  TRNSYS 377–378
algae 175
Al Khafji solar-powered desalination plant, Saudi Arabia 511
Al Montazah solar-powered desalination plant, Egypt 510–511
anaerobic baffled reactor (ABR) 177
anaerobic digestion (AD) 175, 176
  acetogenesis 179
  acidogenesis 179
  carbon to nitrogen 181
  hydraulic retention time 180
  hydrolysis 178–179
  methanogenesis 179
  pH value 181
  temperature 180
anaerobic sequencing batch reactor (ASBR) 177
animal waste material 175

Water Treatment and Desalination: Membrane Separation using Renewable Energy Resources, First Edition. Edited by Mohamed Bassyouni.

anion exchange membranes (AEMs) 13, 49–50, 132, 305
antisolvent crystallization 271–272
aquaporin-based biomimetic membranes 507
atmospheric emissions 164–165
atomic force microscopy (AFM) 32
Australia, geothermal desalination 160–161
autonomous wind-powered desalination
  with backup 126–127
    challenges and trade-offs 128
    diesel generators 127–128
    energy storage 128
    modular desalination capacity 129
  without backup 129–130

b

batch digesters 176
bioenergy
  anaerobic digestion process
    acetogenesis 179
    acidogenesis 179
    carbon to nitrogen 181
    hydraulic retention time 180
    hydrolysis 178–179
    methanogenesis 179
    pH value 181
    temperature 180
  bioenergy sources 174–175
  biogas
    combined heat and power system 183
    electricity generation 182
    heat production 182–183
    production 175–178
  environmental and economic aspects
    economic analysis 189–190
    sustainability 189
  integration
    biogas utilization in 186–188
    conventional desalination methods 184–185
    economic considerations 186
    energy intensity 185
    environment impacts 185–186
  overview of 173–174
  recommendations 192
  research gaps and needs 190–191
  technical challenges and limitations 190
  types of 174
biofuel cell 434
biogas
  applications of 182
  combined heat and power systems 183
  electricity generation 182
  heat production 182–183
  production 175–176
    advanced digesters 177–178
    batch digesters 176
    continuous digesters 177
    plug flow digester 177
  utilization
    energy generation (heat and electricity) 186–187
    hybrid systems 187–188
biohydrogen generation 418
biomass 173
biomethane 181
brackish water reverse osmosis and forward osmosis (BWRO-FO) 53

c

capacitive deionization (CDI) 13–14, 229, 408
capillary adsorption 230
capital costs 400–401
carbon-based composite membranes 536–538
carbon-based materials 79–81

carbon footprint
bioenergy 189
greenhouse gas emissions 353
carbon nanotubes (CNTs)
adsorption desalination 227–228
membranes 506–507
carbon quantum dots (CQDs) 350
carbon to nitrogen (C/N) ratio 181
Carlsbad desalination plant, USA 512
cation exchange membranes (CEMs) 14, 49–50, 132, 305
chemical exergy 40, 42
China's mega-desalination plant 323–325
closed-loop pressure-retarded osmosis (CLPRO) 334
C-nanocomposites (CNMs) 537
combined heat and power (CHP) systems 183
computational fluid dynamics (CFD)-based exergy study 49
concentrating solar power (CSP) systems 69
concentration polarization (CP) 338, 343–344
concentrative crystallization 260, 270–271
constant current (CC) 14
constant voltage (CV) 14
continuous digesters 177
continuous stirred-tank reactors (CSTRs) 177
conventional energy-based membrane systems 482–483
conventional exergy analysis 41, 42–43
conventional life cycle costing (CLCC) 404
converter unit models 384
cooling water 577–578
cost–benefit ratio (CBR) 403
cost dynamic economic feasibility
energy price, sensitivity analysis 486
investment requirements
equipment, infrastructure, and installation costs 479–480
operational and maintenance costs 481
plant design and size 480–481
renewable energy components 481–482
levelized water cost model
conventional energy-based membrane systems 482–483
renewable energy-based membrane systems 483–484
reverse osmosis system 484–485
renewable energy integration
greenhouse gas emissions 486
long-term financial benefits 487
covalent organic frameworks (COFs) 26, 27
crystallization
concentrative 270–271
process of 259
cytochromes 440–442

d

demineralized water 578
desalination mechanisms 33 34
direct contact membrane distillation (DCMD)
capital costs 400
hybrid desalination systems 315
membrane distillation 4–5, 306, 307
renewable energy integration 415
wind turbines 134
direct-driven desalination 130–132
direct electron transfer (DET) mechanism 437
distillation, bioenergy 184
domestic wastewater 428, 455

double-chamber microbial fuel cell 436, 446–447
downward vapor collection 87–89

e

eco-friendliness 111
economic aspects
  capital costs 400–401
  economic viability 402–403
  life cycle cost assessments 404–406
  operational costs 401–402
economic feasibility
  cost dynamics
    investment requirements 479–482
    levelized cost of water models 482–485
    renewable energy integration 486–487
    sensitivity analysis of energy prices 486
  financial barriers 488
  forward osmosis 469
  global desalination over time 468
  market and infrastructure limitations 488
  membrane technology 472
  nanofiltration 470
  obstructions 488
  renewable energy resources
    geothermal energy 474–475
    solar energy 471–474
    wind energy 474
  renewable integration 475–476
    advanced membrane materials 478–479
    energy recovery devices 477
    hybrid energy systems 477–478
  technological barriers 487–488
economic viability 402–403
electricity 94
  biogas 182, 186–187
  hydrovaltic effect on 105–107
  mechanical energy conversion for 104–105
  salinity gradients for 107–109
electrochemical cells 433
electrodes, MFC 449–452
electrodialysis (ED) 49–51, 305
  bioenergy 184–185
  intensification process 549, 552–553, 554–557
  wind-powered membrane desalination 132–134
electrodialysis reversal (EDR) 305
electron transport chain (ETC) 439–440
electrospun nanofiber membranes (NFMs) 29–31, 508
energy assessment 351–352
energy consumption 407–409
  desalination processes 303–304
  membrane system technical aspects 501
energy intensity 185
energy recovery device (ERD) 477
  high-efficiency pumps 312
  hydraulically driven pumping in series 309–310
  parallel pumping with hydraulic drive 310–312
  pumping from hydraulic to mechanical assistance 309
energy recovery, wave-powered desalination 140–141
energy storage, direct-driven desalination 131–132
environmental aspects 406
  energy consumption 407–409
  environmental impact assessment 411–413
  waste treatment 409–410
  water treatment 407

environmental impact assessment (EIA) 411–413
environmental life cycle costing (ELCC) 404, 405
environmental sustainability, PRO 353–354
enzyme biofuel cell (EBFC) 434
ethylene glycol mixture 381–382
evaporation-induced generator (EIG) 105
excess thermal energy 93
exergy analysis 40–42
  advanced 41, 43–45
  chemical exergy 40
  conventional 41, 42–43
  desalination process 39–40
  electrodialysis 49–51
  forward osmosis 48
  hybrid systems 51–53
  membrane distillation 48–49
  reverse osmosis 46–48
  terminology 44–45
  thermomechanical (physical) exergy 40
exoelectrogenic bacteria, MFC 437–439
  electrical power with isolated electrogenic bacteria 443–444
  microbial communities 442–443
  onion residues 443
external concentration polarization (ECP) 335, 344

f
fabrication of membrane desalination 24
  electrospun nanofiber membranes 29–31
  thin-film composite via interface polymerization 28–29
first law of thermodynamics (FLT) 40
flat-plate collectors (FPCs) 375, 379
flat-sheet membranes 347
flat sheet modules 217–218
forestry products 175
forward osmosis (FO) 6–8
  advantages and challenges of 8–9
  exergy analysis 48
forward osmosis and multi-effect distillation (FO-MED) 51
Fourier transform infrared (FTIR) 32
free energy of mixing 336
freeze desalination (FD) 322
functionalized nanocomposite membranes 508

g
geological hazards 164
geothermal desalination
  environmental impacts 161–163
    atmospheric emissions 164–165
    geological hazards 164
    land use 164
    noise and social impacts 165
    waste heat releases 164
    water footprint 165
  forecast 157–158
    in Australia 160–161
    in Middle East 159–160
    in USA 158–159
  global energy status of 150–151
  important parameters in 152
  for membrane processes 156–157
  thermophysical systems 154–156
  utilization schemes 153–154
geothermal energy 474–475
  dry steam 153
  energy production systems 154
  flash steam 153
  global status of 150–151
geothermal PRO hybrid system 361
Gibbs free energy 260, 332, 336

Global Clean Water Desalination Alliance 575, 576
global water shortage 149
Gold Coast desalination plant, Queensland 513
graphene-based membranes 505–506
graphene-based nanocomposite membranes 538–542
graphene oxide (GO) 478, 541–543
greenhouse gas (GHG) emissions 173, 353–354, 486
green pressure-retarded osmosis 364

h

harnessing waste heat 134
heat exchanger efficiency 392–393
heat exchanger fluid
  comparative analysis 382
  ethylene glycol 381
  water as 380–381
  water–ethylene glycol mixture 381–382
heat exchanger models 384
heat production 182–183
heat transfer enhancement 229
highest occupied molecular orbital (HOMO) 79
high-flux membrane technology 550–551
high-pressure recovery
  high-efficiency pumps 312
  hydraulically driven pumping in series 309–310
  parallel pumping with hydraulic drive 310–312
  pumping from hydraulic to mechanical assistance 309
hollow-fiber membranes 347–348
hybrid desalination systems (HDS) 230
  advantages of 231–232
  benefits of 231
  capacitive deionization with pressure RO 318–320
  challenges and considerations 231
  energy sources used 232
  innovations in 233–235
  integrated membrane system between RO and PRO 316–318
  renewable energy 232–233
  RO/MD/RED 315–316
  salt water and potable water 204
  zero liquid discharge 320–323
hybrid energy sources
  energy recovery device 309–312
  integrated renewable energy 312–314
hybrid energy systems 477–478
hybridized materials 81–83
hybrid systems
  China's mega-desalination plant 323–325
  energy consumption for 303–304
  exergy analysis 51–53
  hybrid energy sources *see* hybrid energy sources
  membrane technology
    electrodialysis and electrodialysis reversal 305
    membrane distillation 305–308
    processes 308–309
    reverse osmosis 304–305
  nuclear energy 314–315
  pressure-retarded osmosis
    geothermal 361
    liquid-phase ion-stripping 359
    membrane bioreactor 359
    membrane distillation 358–359
    multigeneration 361–363
    reverse osmosis 356–357
    solar ponds 360–361
    thermosiphon systems 359–360
hydraulic retention time (HRT) 180

hydrogen bonds, water molecules 67
hydrolysis 178–179
hydrophilic membranes 2
hydrophobic microporous membranes 278
hydropower co-generation devices 101

i

Inductively Coupled Plasma Mass Spectrometry (ICP-MS) 86
industrial wastewater 428
integrated evaporative hydrovaltic generator (IEHVG) 107
integrated hybrid membrane system with enhanced water treatment (IHMS-EWT) 407
integrated renewable energy 312–314
intensification membrane technology
  conventional methods 531
  electrodialysis membranes 549, 552–553, 554–557
  future challenges and opportunity 553, 558
  membrane distillation for 542–544
  photothermal and SPMD 545–549
  process intensification 528–529
  reverse osmosis 527–528
  seawater desalination 529–531
  2D nanostructured hybrid membranes
    carbon-based composite membranes 536–538
    graphene-based nanocomposite membranes 538–542
    MXenes-based composite membranes 532–536
interdisciplinary 110
interface polymerization 28–29
interfacial photothermal heating 70–72
interference scattering microscopy (iSCAT) 26
internal concentration polarization (ICP) 335
internal heat recovery 307
ion exchange membranes (IEMs) 132

j

Janus-interface solar steam generator (J-SSG) 86–87
Janus membranes 219–220
Jebel Ali desalination plant, UAE 512

l

large-scale implementation PRO system 365–367
laser-induced graphene (LIG) 538
levelized water cost model
  conventional energy-based membrane systems 482–483
  renewable energy-based membrane systems 483–484
  reverse osmosis system 484–485
life cycle assessment 354–355
life cycle cost analysis 354–355
life cycle cost assessments (LCCAs) 404–406
lignocellulosic wastes 175
linear Fresnel reflectors (LFRs) 69
linear low-density polyethylene (LLDPE) 92
liquid–liquid phase separation 26
liquid-phase ion-stripping, PRO system 359
localized surface plasmon resonance (LSPR) 74–75
lowest unoccupied molecular orbital (LUMO) 79
low salt rejection reverse osmosis (LSRO) 11
low-temperature thermal desalination (LTTD) 235–236
  advantages and future trends 244

low-temperature thermal desalination (LTTD) (*contd.*)
case studies and prototypes 244
challenges and innovations 236–238
design of 237
enhancements and innovations 239–240
environmental and ecological considerations 240
evaporation rate models 241
heat exchanger design and optimization 241
heat transfer in condensers 241
innovations and efficiency enhancements 238–239
reducing energy consumption 240
solar energy integration 242
sustainable water 205
waste heat from industrial processes 242, 243
water quality in 240

m

Marangoni effect 73
Masdar City solar desalination plant, UAE 511
maximum power point tracking (MPPT) 366
mediated electron transfer (MET) mechanism 437
mediator microbial fuel cells 452–453
membrane 184
membrane-based desalination 210
membrane bioreactors (MBRs) 359, 530
membrane brine concentrator (MBC) 320
membrane capacitive deionization (MCDI) 13–14
membrane crystallization (MCr)
antisolvent crystallization 271–272
application of 282–284
challenges and opportunities for 280–282
concentrative crystallization 270–271
economic viability and sustainability 284–286
flowchart 266
membrane distillation crystallization 266–269
operating parameters and process optimization 276–277
feed flow, recirculation rates and air gap 277–278
fouling 278–280
membrane characteristics 278
solution supersaturation 277
temperature 277
osmotic membrane crystallization 269–270
processes 259–262
membrane desalination 95
desalination mechanisms 33–34
environmental and social impacts 515
fabrication of 24
electrospun nanofiber membranes 29–31
thin-film composite via interface polymerization 28–29
forward osmosis 6–8
advantages and challenges of 8–9
fundamentals of 2–4
integrated processes 315–316
material types 24–26
membrane capacitive deionization 13–14
membrane distillation
advantages and challenges 5–6
configurations 4–5
microfiltration 15

morphological, packing, and chemical structures 31–32
nanofiltration 16
performance evaluation 32–33
policy and regulatory trends 515–516
renewable energy integration 514–515
reverse osmosis 9–10
advantages and challenges of 10–11
osmotically assisted reverse osmosis 11–13
state-of-the-art membrane processes 23–24, 25
surface properties 32
technological advancements 513–514
thermal and mechanical properties 32
ultrafiltration 15
membrane distillation (MD) 4, 200–201
advantages and challenges 5–6
cleaning membrane 222–223
configurations 4–5, 216–217
exergy analysis 48–49
flat sheet modules 217–218
fouling membrane 221–222
hybrid desalination systems 305–308
intensification membrane 542–544
membrane classification 219–221
membrane crystallization 267–269
pretreatment for feed 223–225
principle and operating conditions 213–215
PRO system 358–359
spiral wound modules 218–219
tubular modules 218
membrane distillation crystallization (MDCr) 266–269, 278–280
membrane fabrication techniques 220
membrane fouling 278–280, 344
membrane materials
polyacrylonitrile 273
polyamide 275–276
polypropylene 275
polysulfone 272–273, 274
polyvinylidene fluoride 273, 275
membrane system technical aspects
air scouring 503
backwashing 502–503
chemical cleaning 502, 503
cost management in operation and maintenance 505
energy consumption and efficiency 501
fouling and scaling challenges 500–501
membrane longevity and replacement 503–505
membrane performance 500
operation and maintenance 502
system design and configuration 501–502
system monitoring and optimization 505
membrane technology
pressure-retarded osmosis
flat-sheet membranes 347
hollow-fiber membranes 347–348
membrane development 346–347
nanomaterial role 348–351
Memstill process 308
metal nanoparticles 74–77
metal-organic frameworks (MOFs) 25, 228–229, 509
methanogenesis 179
MFC *see* microbial fuel cell (MFC)
microbes 431
microbial fuel cell (MFC) 435–437
areas of improvement 453–454
biofuel cell 434

microbial fuel cell (MFC) (*contd.*)
closing argument and perspectives 455–457
cytochromes 440–442
design of
anode and cathode of 444–445
double-chamber 446–447
electrodes 449–452
mediator microbial fuel cells 452–453
proton-exchange membrane in 447
single-chamber 445–446
substrates types 447–449
un-mediator or mediator-less 453
electrochemical cells 433
electron transport chain 439–440
exoelectrogenic bacteria 437–439, 442–444
history and development of
developments 427
early twenty-first century 427
expanding applications 427
experiments 426
first observation 426
late twentieth century 426–427
mid-twentieth century 426
optimization and sustainability 428
scalability 427
microorganisms 431–433
wastewater 428–431
water treatment 454–455
yeast fuel cell 434–435
microfiltration (MF) 15
microorganisms 431–433
Middle East, geothermal desalination 159–160
minerals, seawater 257
concentration elements 262–265
extraction methods 258
importance of 258
seawater recovery 258
minimum liquid discharge (MLD) techniques 323
mixed matrix membranes (MMMs) 25, 509–510
moisture-resistant thermal bridging component (MRTBC) 98
multigeneration PRO system 361–363
multiple-effect distillation (MED) 51–52, 574
configurations 211
industrial residual energy 202–203
limitations and economics 211–213
principle and operating conditions 210–211
multistage condensation device 89–93
multistage flash distillation (MSF) 1
multistage membrane distillation (MSMD) system 99, 101
Municipal SolidWaste (MSW) 175
municipal wastewater 429
MXenes-based composite membranes 532–536

n

nanocomposite membranes 539
nanofiber-based membranes 508–509
nanofiltration (NF) 16
narrow-bandgap semiconductors 77
natural gas steam methane reforming (NGSMR) 418
natural zeolites 228
net present value (NPV) 355
NFMs *see* electrospun nanofiber membranes (NFMs)
non-phase change *see* membrane-based desalination
novel membrane materials
aquaporin-based biomimetic membranes 507

carbon nanotube membranes 506–507
functionalized nanocomposite membranes 508
graphene-based membranes 505–506
metal–organic framework membranes 509
mixed-matrix membranes 509–510
nanofiber-based membranes 508–509
superhydrophobic membranes 507–508
zwitterionic membranes 508
nuclear desalination (ND) 576
nuclear energy 314–315
Nuclear Power Plants (NPPs)
hybrid systems coupled 579
nuclear desalination 575–576
seawater desalination 574–575
small modular reactor
hybrid secondary circuit configuration 579
intermediate circuit 583–587
main components and dimensions 587–588
pilot demonstration project 590–598
pretreatment 588
secondary circuit main aspects 580–583
sitting and cost of water production 588–590
small size reactor plant 579
uses and consumption of water 576–579
nylon *see* polyamide (PA)

o

ocean environment 136–137
Ocean Thermal Energy Conversion (OTEC) 236
omniphobic membranes 219
onshore deployment strategy 137–140
open humidification–dehumidification (OHD) 234
open-loop pressure-retarded osmosis (OLPRO) 334
operating parameters and process optimization 276–277
feed flow, recirculation rates and air gap 277–278
fuling 278–280
membrane characteristics 278
solution supersaturation 277
temperature 277
operational costs (OpEx) 401–402
oscillating water column (OWC) 135
osmosis-based technology 340–342
osmotically assisted reverse osmosis (OARO) 11–13
osmotic membrane crystallization (OMC) 269–270
osmotic membrane distillation crystallization (OMDC) 270
osmotic pressure
fouling challenges 344–345
and water flux 339–340

p

parabolic trough collectors (PTCs) 69
permeate-gap membrane distillation (PGMD) systems 408
Perth desalination plant, Australia 511
pharmaceutically active compounds (PhACs) 409
phase change *see* thermal desalination
photon absorption process 64–65
photothermal conversion process 65
photothermal distillation (PD) technology 95
photothermal heating 69–70

photothermal interface evaporation 110–111
photothermal membrane distillation (PMD) 95
photothermal water evaporation 73–74
  carbon-based materials 79–81
  hybridized materials 81–83
  metal nanoparticles 74–77
  multifunctional integration 111
  semiconductor materials 77–79
  smart and automated 110
photovoltaic power generation principles 95–102
photovoltaics (PVs) 312, 354, 471, 538
pH value 181
piezoelectric nanogenerators (PENGs) 105
plug flow digester 177
polvinyl alcohol (PVA) 91
poly(vinylidene fluoride-*co*-hexafluoropropylene) (PVDF-HFP) 91
polyacrylonitrile (PAN) 273
polyamide (PA) 28, 275–276
polyaniline (PANI) 78
polydopamine (PDA) 73
polymeric membranes 278
polypropylene (PP) 275
polysulfone (PSF) membranes 272–273, 274
polyvinylidene fluoride (PVDF) 273, 275, 277
population explosion 414
potable water 578
power take-off (PTO) 135
pressure-retarded osmosis (PRO) 7, 51
  actual energy production of 337–338
  applications 363–364
  challenges and future directions
    large-scale implementation 365–367
    technical challenges and solutions 364–365
  configurations of 334–335
  cost contributors 352
  energy assessment 351–352
  environmental sustainability 353–354
  free energy of mixing 336
  fundamentals
    concentration polarization 343–344
    fouling challenges in osmosis 344–345
    osmosis-based technology 340–342
    osmotic pressure and water flux 339–340
    water transport and theory of rejection 342–343
  greenhouse gas emissions 353–354
  historical development and evolution 332–333
  hybrid system
    geothermal 361
    liquid-phase ion-stripping 359
    membrane bioreactor 359
    membrane distillation 358–359
    multigeneration 361–363
    reverse osmosis 356–357
    solar ponds 360–361
    thermosiphon systems 359–360
  integrated membrane system 316–318
  life cycle assessment and cost analysis 354–355
  maximum energy production 338–339
  membrane technology
    flat-sheet membranes 347
    hollow-fiber membranes 347–348
    membrane development 346–347

nanomaterial role 348–351
overview 331–332
power density factors 335
principles of operation 333
water and salt flux 336–337
PRO *see* pressure-retarded osmosis (PRO)
process intensification (PI) 528–529
proton-exchange membrane (PEM) 433, 447
pump models 384
PV-membrane distillation evaporation crystallizer (PMDEC) 93

r

reduced graphene oxide (rGO) 541–543
renewable economic feasibility 475–476
advanced membrane materials 478–479
energy recovery devices 477
hybrid energy systems 477–478
renewable energy
based membrane systems 483–484
costs 413
hybrid desalination systems 232–233
renewable energy integration
comparative study 416–418
economic benefits 414–416
greenhouse gas emissions 486
long-term financial benefits 487
MDS projections 514–515
renewable energy costs 413
technical challenges and solutions 414
renewable energy resources (RES)
economic feasibility
geothermal energy 474–475
solar energy 471–474
wind energy 474
renewable natural gas 181
reverse electrodialysis (RED) 108
integrated processes 315–316
membrane technology 305
reverse osmosis (RO) 9–10
advantages and challenges of 10–11
bioenergy 184
exergy analysis 46–48
hybrid desalination systems
capacitive deionization with pressure 318–320
integrated membrane system 316–318
integrated processes 315–316
membrane technology 304–305, 308
intensification process 527–528
levelized water cost model 484–485
osmotically assisted reverse osmosis 11–13
PRO system 356–357
wind-powered membrane desalination 124–126
autonomous with backup 126–129
autonomous without backup 129–130
direct-driven desalination 130–132
reverse osmosis with pressure-retarded osmosis (RO-PRO) 51
reverse salt flux (RSF) 339

s

Salalah desalination plant, Oman 513
salinity-gradient power (SGP) 414
salinity gradients 107–109
scanning electron microscopy (SEM) 31, 502, 503
seawater desalination 574–575
Seawater Desalination with an Autonomous Wind Energy System (SDAWES) project 133
seawater evaporation and collection process 66–67

seawater minerals
  concentration elements 262–265
  extraction methods 258
  importance of 258
  recovery 258
seawater reverse osmosis (SWRO) plants 530
second lawof thermodynamics (SLT) 40
semiconductor materials 77–79
semipermeable membrane 2, 9, 184
shale gas-produced water (SGPW) 285–286
Shewanella oneidensis 438
Shuwaikh desalination plant, Kuwait 512–513
side vapor condensation 86–87
single-chamber microbial fuel cell 445–446
single-stage condensation device
  downward vapor collection 87–89
  side vapor condensation 86–87
  upward vapor collection 83–86
small modular reactor (SMR)
  components and dimensions 587–588
  hybrid secondary circuit configuration 579
  intermediate circuit 583–587
  pilot demonstration project
    hybrid operative mode 597
    MED operative mode 595
    operating cost 592
    RIG coupled to MED and RO systems 593
    RO operative mode 596
    startup operative 594
  pretreatment 588
  secondary circuit main aspects 580–583
  sitting and cost of water production 588–590, 591
  small size reactor plant 579
small-scale solar-powered desalination, Cape Verde 512
small size reactor (SSR) 579, 581
social life cycle costing (SLCC) 404, 405
solar collectors 378–380
solar combined membrane distillation (SCMD) 376
solar domestic hot water (SDHW) systems 376
solar-driven seawater desalination systems 94
solar energy
  hybrid desalination system power 474
  integration 242
  PRO power 359–360
  PV systems 471, 473
solar evaporation-piezoelectric (SE-PZ) technology 104
solar heat
  direct 69
  global annual average 64
solar-induced interfacial steam generation (SISG) 83
solar membrane distillation (SMD) 375
solar pond PRO system 360–361
solar-powered desalination plant
  Al Khafji, Saudi Arabia 511
  Al Montazah, Egypt 510–511
  Carlsbad, USA 512
  Gold Coast, Queensland 513
  Jebel Ali, UAE 512
  Masdar City, UAE 511
  Perth, Australia 511
  Salalah, Oman 513
  Shuwaikh, Kuwait 512–513
  small-scale, Cape Verde 512
Solar Powered Membrane Distillation (SPMD) 545–549
solar power towers (SPTs) 69

solar-to-freshwater conversion 109
solar water desalination
  historical development 67–68
    direct solar heating 69
    interfacial photothermal heating 70–72
    photothermal heating 69–70
    3D interfacial photothermal heating 72–73
  solar water evaporation process
    definition of 63–64
    photon absorption process 64–65
    photothermal conversion process 65
    seawater evaporation and collection process 66–67
solar water evaporation
  definition of 63–64
  hydrovaltic effect on electricity generation 105–107
  with mechanical energy conversion for electricity generation 104–105
  photon absorption process 64–65
  photothermal conversion process 65
  photothermal water evaporation materials 73–74
    carbon-based materials 79–81
    hybridized materials 81–83
    metal nanoparticles 74–77
    semiconductor materials 77–79
  with photovoltaic power generation 95–102
  with salinity gradients for electricity generation 107–109
  seawater evaporation and collection process 66–67
  with thermoelectric power generation 102–104
  water condensation and collection devices
    multistage condensation device 89–93
    single-stage condensation device 83–89
spiral wound modules 218–219
stormwater 429
superhydrophobic membranes 219, 507–508
surge wave energy converter (SWEC) 138–139
sustainability 111
Sustainable Development Goals (SDGs) 435
sustainable energy conversion applications 94
  solar water evaporation
    hydrovaltic effect on electricity generation 105–107
    with mechanical energy conversion for electricity generation 104–105
    with photovoltaic power generation 95–102
    with salinity gradients for electricity generation 107–109
    with thermoelectric power generation 102–104
sweeping gas membrane distillation (SGMD) 4–5
sweeping gas membrane distillation-bubble column dehumidifier (SGMD-BCD) 52–53

t

techno-economic assessment (TEA) 400
temperature polarization 91
thermal collector systems 390–391
thermal desalination
  distillation 184

thermal desalination (*contd.*)
heat exchanger fluids 380
multi-effect distillation 209
thermal energy storage (TES) systems 242
thermal interconnection layer (TIL) 98
thermal management and evaporation system (TMES) technology 105
thermodynamic analysis 40
thermoelectric power generation 102–104
thermogalvanic cells (TGCs) 108
thermogravimetric analysis (TGA) 32
thermomechanical (physical) exergy 40
thermophysical geothermal systems 154–156
thermosiphon systems 359–360
thermostatic sweeping gas membrane distillation (TSGMD) 5
thin film composite (TFC)
membranes 346, 349–350
via interface polymerization 28–29
thin-film nanocomposite (TFN) 28, 348
3D AGMD parameters 385–386
3D interfacial photothermal heating 72–73
$Ti_3C_2T_x$ MXene membrane 534–535
Transient System Simulation Tool (TRNSYS) software
desalination system model 384–385
distillate water, inlet and outlet temperatures 388–390
heat capacity fluid influence 391–392
heat exchanger efficiency effect 392–393
heat exchanger fluid
comparative analysis 382
ethylene glycol 381
water as 380–381
water–ethylene glycol mixture 381–382
materials and methods
air-gap membrane distillation 377–378
solar collectors 378–380
thermal desalination systems 380
meteorological parameters 386–388
thermal collector systems 390–391
3D air-gap membrane distillation parameters 385–386
weather characteristics, locations 383
transmission electron microscopy (TEM) 32
triboelectric nanogenerators (TENGs) 105
tubular modules 218
2D-nanohybrid membranes 534
2D nanostructured hybrid membranes
carbon-based composite membranes 536–538
graphene-based nanocomposite membranes 538–542
MXenes-based composite membranes 532–536

u

ultrafiltration (UF) 15
ultrafiltration and reverse osmosis (UF-RO) 401
ultralong carbon nanotubes (UCNTs) 227
United Arab Emirates 160
United Nations' Sustainable Development Goals (SDGs) 206
up-flow anaerobic sludge blanket (UASB) reactor 177
upward vapor collection 83–86
USA, geothermal desalination 158–159

V

vacuum membrane distillation (VMD) 4–5, 402
Van't Hoff's formula 339
variable frequency drivers (VFDs) 133
vermiculite (VMT) 229–230
vortex induced vibration (VIV) 237

W

waste heat-powered desalination method
  adsorption desalination *see* adsorption desalination (AD)
  diagram 199–200
  hybrid desalination systems 230–235
  low-temperature thermal desalination 235–236
    advantages and future trends 244
    challenges and innovations 236–238
    experimental studies 239–240
    innovations and efficiency enhancements 238–239
    integrating renewable energy and industrial heat sources 242–244
    mathematical models and heat transfer optimization 241–242
  membrane distillation
    configurations 216–217
    fouling and cleaning of membranes 221–223
    membrane classification 219–221
    module types 217–219
    pretreatment for feed 223–225
    principle and operating conditions 213–215
  multiple-effect distillation
    configurations 211
    limitations and economics 211–213
    principle and operating conditions 210–211
waste heat releases 164
waste management 189
waste treatment 409–410
wastewater
  agricultural wastewater 429
  domestic wastewater 428
  industrial wastewater 428
  municipal wastewater 429
  stormwater 429
  types of 429–431
wastewater treatment (WWT) 404, 412, 416
water condensation and collection devices
  multistage condensation device 89–93
  single-stage condensation device 83–89
water–ethylene glycol mixture 381–382
water footprint 165
water molecules, hydrogen bonds 67
water-related sustainability 207
water scarcity 121, 375, 574
water treatment (WT) 407
  microbial fuel cell
    for copper removal 455
    for domestic wastewater treatment 455
    for drug removal 454
    for dye removal 455
water vapor adsorption 230
wave energy converter (WEC) 135, 138, 140
wave-powered membrane desalination 135–136
  energy recovery 140–141
  integration in ocean environment 136–137
  onshore deployment strategies 137–140

wave-powered RO (WPRO) desalination system 136, 137
weather models 384
wind energy 474
Wind Energy Marine Unit (WEMU) 132
wind energy systems 122, 124
wind-powered membrane desalination
 electrodialysis 132–134
 plant design challenges 123
 plant design considerations 123–124
 reverse osmosis 124–126
  autonomous with backup 126–129
  autonomous without backup 129–130
  direct-driven desalination 130–132
 via harnessing waste heat of wind turbines 134
wind power technology 122
wind turbines (WTs) 122, 134, 474

x
X-ray diffraction (XRD) spectroscopy 32, 282

y
yeast fuel cell (YFC) 434–435

z
zeolites (ZIFs) 228–229
zero liquid discharge (ZLD) 320–323
zwitterionic membranes 508

Printed and bound by CPI Group (UK) Ltd, Croydon, CR0 4YY
17/09/2026
14964233-0003